CIM Series in Mathematical Sciences

Volume 1

The CIM Series in Mathematical Sciences is published on behalf of and in collaboration with the Centro Internacional de Matemática (CIM) in Coimbra, Portugal. Proceedings, lecture course material from summer schools and research monographs will be included in the new series.

More information about this series at
http://www.springer.com/series/11745

Jean-Pierre Bourguignon • Rolf Jeltsch •
Alberto Adrego Pinto • Marcelo Viana
Editors

Dynamics, Games and Science

International Conference and Advanced School Planet Earth, DGS II, Portugal, August 28–September 6, 2013

Springer

Editors
Jean-Pierre Bourguignon
IHES Le Bois-Marie
Bures-sur-Yvette, France

Rolf Jeltsch
Department of Mathematics
ETH Zürich
Seminar für Angewandte Mathematik
Zürich, Switzerland

Alberto Adrego Pinto
Department of Mathematics
University of Porto
Faculty of Sciences
Porto, Portugal

Marcelo Viana
Instituto de Matemática Pura e Aplicada
IMPA
Rio de Janeiro, Brazil

ISSN 2364-950X ISSN 2364-9518 (electronic)
CIM Series in Mathematical Sciences
ISBN 978-3-319-16117-4 ISBN 978-3-319-16118-1 (eBook)
DOI 10.1007/978-3-319-16118-1

Library of Congress Control Number: 2015945812

Mathematics Subject Classification (2010): 37-02, 37-06, 91-01, 91-02

Springer Cham Heidelberg New York Dordrecht London

Printed on acid-free paper

Springer International Publishing AG Switzerland is part of Springer Science+Business Media (www.springer.com)

Foreword

I was very honored to be invited by Professor Alberto Adrego Pinto to the lecture at the Advanced School Planet Earth, Dynamics, Games and Science II (DGS II) from August 28 to August 30, 2013, and to speak at the associated international conference from September 2 to September 4, 2013. I have known Alberto since I was a graduate student at the CUNY Graduate Center in the 1980s. After both of us completed our Ph.D. degrees, we worked on a similar subject: smooth rigidity for one-dimensional dynamical systems and its generalization to Anosov dynamical systems of the two-torus, for many years. I was impressed by his work with his collaborators, using techniques and methods in smooth dynamical systems to develop many excellent results on smooth rigidity. Meanwhile, my collaborators and I tried to develop smooth rigidity into symmetric rigidity by applying techniques and methods in quasiconformal mappings theory and Teichmüller theory, and to build up a new Teichmüller theory for dynamical systems. So I knew that attending the DGS II would be stimulating and rewarding. Also, I knew that Alberto is an outstanding organizer and has the talent to lead a successful advanced school and conference, and indeed, his organizational skills and talents were proven again. I had a wonderful time in Lisbon, Portugal, and enjoyed many fruitful discussions with Alberto and his Portuguese colleagues. In particular, Alberto and his collaborators explained to me their work in game theory and some basic facts about its related Nash equilibrium, and we discussed some differences and similarities between the Nash equilibrium and the Gibbs equilibrium from a dynamical systems point of view. The Advanced School DGS II and the Conference DGS II were very successful. The Advanced School DGS II and the Conference DGS II in Portugal were parts of the international year of the Mathematics of Planet Earth 2013 (MPE 2013) held under the patronage of UNESCO. The activities at the Advanced School DGS II and the Conference DGS II were held in the beautiful city of Lisbon. The Advanced School DGS II was held in Escola Superior de Economia e Gestão, Universidade Técnica de Lisboa (ISEG-UTL), Lisbon, Portugal, and the Conference DGS II was held in the renowned Calouste Gulbenkian Foundation, Lisbon, Portugal. This was my third trip to the city of Lisbon. The previous two were only for one or two days, but this one was for a week. On this trip, I not only had many fruitful discussions with other lecturers and speakers and Portuguese mathematicians, but I also visited a museum in Lisbon where I learned more about Prince Henry the Navigator and his school of navigation, where some of the leading geographers, cartographers, astronomers, and mathematicians of the fifteenth century from various parts of Europe came to work; and participants were trained in navigation, map-making, and science, including mathematics. The school of navigation started the Portuguese as well as the European exploration of new lands. So, following the scientific tradition of Portugal, this volume contains the broad mathematical themes of this conference on dynamical systems and game theory. It contains samples of the numerous talks and presentations given at the Advanced School DGS II and the Conference DGS II. The reader will find many interesting topics in dynamical systems and game theory, including many interdisciplinary contributions from economics, population dynamics, ecology, healthcare, disease epidemics, cell biology, and physics. This volume will also encourage and help the reader to explore "new lands" in various scientific areas. Finally, I would like to express my thanks to Alberto Adrego Pinto, Jean-Pierre Bourguignon, Rolf Jelstch, and Marcelo Viana for their efforts in editing and putting together this important volume.

Yunping Jiang
Distinguished Professor of Mathematics
Department of Mathematics
Queens College of the City University of New York
65-30 Kissena Blvd, Queens, NY 11367-1597, USA
Department of Mathematics
The Graduate Center of the City University of New York
365 Fifth Avenue, New York, NY 10016, USA

Foreword

I was quite pleased, and honored, to be asked by Alberto Pinto to speak at the International Conference and Advanced School Planet Earth, Dynamics, Games and Science II (DGS II) and to lecture in the advanced school that accompanied the conference from 28 August to 6 September, 2013. I had met Alberto at several conferences over the previous years and was well aware of the high-quality work that he and many of his Portuguese colleagues were doing in many branches of mathematics and science. So I knew that attending DGS II would be stimulating and rewarding. I was, however, unaware of Alberto's extraordinary organizational and leadership talents, as were displayed by these events. The extent of the diverse activities organized under Alberto's leadership as president of the international Center of Mathematics (CIM), together with Irene Fonseca (president of the CIM's scientific council) and a large number of Portuguese mathematicians, universities, institutions and organizations, is quite remarkable. These activities constitute an outstanding contribution to the international year of the Mathematics of Planet Earth 2013 (MPE 2013), held under the patronage of UNESCO, during which mathematical organizations, universities, and research centers around the world hosted conferences, workshops, schools, and long-term programs intended to showcase the ways in which the mathematical sciences can be useful in addressing our planet's many problems.

A highlight of the MPE 2013 activities centered in Portugal was the DGS II conference and the associated advanced schools, held at the facilities of the renowned Calouste Gulbenkian Foundation and the Escola Superior de Economia e Gestão, Universidade Técnica de Lisboa, respectively. These locations in the beautiful city of Lisbon are wonderful venues for scientific meetings and their hospitality was greatly enjoyed by all. The broad mathematical themes of the conference were dynamics and game theory. The chapters in this volume constitute a sampling of the numerous talks and presentations held during this event. A casual glance at the table of contents will show the reader a list of contributions to the mathematical development of game theory and dynamical systems as well as inter-disciplinary contributions from numerous scientific fields , including economics, population dynamics, ecology, healthcare, disease epidemics, cell biology, and physics. Game theory's roots were in economics and the contributions in this volume show that its vibrant role in economics continues unabated. More recently, game theory and methodologies inspired by game theoretic ideas have made foundational contributions to other disciplines, the life sciences being a notable example. For example, extensions of game theoretic methods to dynamic settings have been and continue to be developed in order to model and understand evolutionary and adaptive processes in biology, with impacts ranging from the evolution of antibiotic resistance of pathogens to large-scale ecosystems.

This volume serves well to illustrate the many roles that mathematics can play in addressing a wide variety of scientific problems that relate to our planet earth. I am confident that the reader will be inspired by the contributions and will be stimulated to learn more about the goals of MPE 2013. I want to thank Alberto and his fellow editors, Jean-Pierre Bourguignon, Rolf Jelstch, and Marcelo Viana, for their efforts in putting this important volume together.

Jim Michael Cushing
Professor of Mathematics
Interdisciplinary Program in Applied Mathematics
Department of Mathematics
University of Arizona
Tucson, AZ 85721, USA

Preface

As the International Center for Mathematics (CIM) celebrated its 20th anniversary on the 3rd of December 2013, it is the perfect opportunity to look back on this past year, which has undoubtedly been one of the most ambitious and eventful ones in its history. With the support of our associates from 13 leading Portuguese universities, our partners at the University of Macau, and member institutions such as the Portuguese Mathematical Society, in 2013 the CIM showed yet again the importance of a forum such as this for bringing together leading Portuguese-speaking scientists and researchers from around the world.

The hallmark project of the year was the UNESCO-backed International Program Mathematics of Planet Earth (MPE) 2013, which the CIM participated in as a partner institution. This ambitious and global program was tasked with exploring the dynamic processes underpinning our planet's climate and man-made societies, and with laying the groundwork for the kind of mathematical and interdisciplinary collaborations that will be pivotal to addressing the myriad issues and challenges facing our planet now and in the future. The CIM heeded the MPE's call to action by organizing two headline conferences in March and September of 2013: the "Mathematics of Energy and Climate Change" conference in Lisbon in the spring, and the conference "Dynamics, Games, and Science II" in the fall. Both were held at the world-renowned Calouste Gulbenkian Foundation in Lisbon, one of more than 15 respected Portuguese foundations and organizations that enthusiastically supported the CIM conferences. As well as the conferences themselves, well attended "advanced schools" were held before and after each event: at the Universidade de Lisboa in the spring, and at the Universidade Técnica de Lisboa in the fall.

These conferences succeeded in bringing together some of the most accomplished mathematical and scientific minds from across the Portuguese-speaking world and beyond, while also serving as a launch pad for one of the CIM's most exciting endeavors in years: the new CIM Series in Mathematical Sciences, which will include lecture notes and research monographs and be published by Springer-Verlag. "The collaboration with Springer will bring mathematics developed in Portugal to a global audience," CIM President Alberto Adrego Pinto said at the time of the announcement, "and will help strengthen our contacts with the international mathematics community."

These first two volumes in the series, consisting of review articles selected from work presented at the "Mathematics of Energy and Climate Change" and "Dynamics, Games, and Science" conferences, reflect the CIM's international reach and standing. Firstly, they are characterized by an impressive roster of mathematicians and researchers from across the United States, Brazil, Portugal, and several other countries whose work will be included in the volumes.

The authors are complemented by the editorial board responsible for this first installment, a world-renowned "quartet" consisting of: president of the European Research Council Jean-Pierre Bourguignon from the École Polytechnique; former Société Mathématiques Suisse and European Mathematical Society president Rolf Jeltsch from the ETH Zurich; current Sociedade Brasileira de Matemática president Marcelo Viana from Brazil's Instituto Nacional de Matemática Pura e Aplicada; and CIM president Alberto Adrego Pinto from the Universidade do Porto.

While the MPE program was a major focus of the CIM's activities in 2013, the center also organized a number of further events aimed at fostering closer ties and collaboration between mathematicians and other scientists, mainly in Portugal and other Portuguese-speaking countries. In this context the CIM held the 92nd European Study Group with Industry meeting, part of a vital series held throughout Europe to encourage and strengthen the connections between mathematics and industry. As the MPE program made clear, humanity faces all manner of challenges, both man-made and natural, and though industry is attempting to overcome them, in many cases mathematics and science are far better suited to the task. Yet it is often industry that delivers the kinds of innovative ideas that will launch the next great scientific and technological revolutions, and which academia must adapt to. The potential for dialogue and cooperation between academia and industry is in fact so great that I have now made it one of the core initiatives in my presidency of the US-based Society for Industrial and Applied Mathematics (SIAM).

As we look back at the successful year the CIM had in 2013, we should also bear in mind the dramatic changes currently taking place in the world, changes that above all the mathematical sciences—including statistics, operational research, and computer science—will be called upon to address. Foremost among them is the rise of Big Data, especially as it relates to national security, finance, medicine, and the Internet (among other fields), which has come to dominate research in many scientific sectors and requires new analytical tools, which mathematics can provide. This new landscape will require an unparalleled level of partnership between science and industry, and is what prompted the European Commission to recently announce its Europe 2020 Growth Strategy, which calls for investment in groundbreaking research, innovation in industry, and the cultivation of a new generation of scientists. It is no coincidence that these three pillars are at the core of the CIM's own mission, and the CIM series in Mathematical Sciences will provide the ideal platform for communicating and broadening the impact of the CIM's activities with regard to these global challenges.

President of CIM Scientific Council Irene Fonseca

Acknowledgements

We thank all the authors of the chapters and we thank all the anonymous referees. We are grateful to Irene Fonseca for contributing the preface of this book. We thank the Executive Editor for Mathematics, Computational Science and Engineering at Springer-Verlag Martin Peters for invaluable suggestions and advice throughout this project. We thank João Paulo Almeida, Ruth Allewelt, Joana Becker, João Passos Coelho, Ricardo Cruz, Helena Ferreira, Isabel Figueiredo, Alan John Guimarães, Filipe Martins, José Martins, Bruno Oliveira, Telmo Parreira, Diogo Pinheiro, and Renato Soeiro for their invaluable help in assembling this volume and for their editorial assistance.

Alberto Pinto would like to acknowledge the financial support of Centro Internacional de Matemática (CIM), Ciência Viva (CV), Calouste Gulbenkian Foundation, Fundação para a Ciência e a Tecnologia, LIAAD-INESC TEC, USP-UP project, IJUP and Mathematics Department, Faculty of Sciences, University of Porto, Portugal.

Alberto Pinto gratefully acknowledges the financial support to the conclusion of this project provided by the FCT—Fundação para a Ciência e a Tecnologia (Portuguese Foundation for Science and Technology) within project UID/EEA/50014/2013 and ERDF—European Regional Development Fund through the COMPETE Program (operational program for competitiveness) and by National Funds through the FCT—Fundação para a Ciência e a Tecnologia (Portuguese Foundation for Science and Technology) within Project "Dynamics and Applications" with reference PTDC/MAT/121107/2010.

Bures-sur-Yvette, France	Jean-Pierre Bourguignon
Zürich, Switzerland	Rolf Jeltsch
Porto, Portugal	Alberto Adrego Pinto
Rio de Janeiro, Brazil	Marcelo Viana

Contents

Corruption, Inequality and Income Taxation

Elvio Accinelli and Edgar J. Sánchez Carrera

Abstract It is recognized that corrupt behavior determines the institutional types of an economic system where an institution is ruled out by economic agents (e.g. officials-public or private) abusing their role to procure gain for themselves (rent-seeking activities) or somebody else. In this vein, we study an evolutionary model of institutional corruption. We show that income inequality and income taxation are the main factors (explanatory variables) for fighting institutional corruption. We conclude with some feasible policies on institutions, beliefs and incentives to combat the corruption.

1 Introduction

A large number of papers on the causes and consequences of corruption have been published (for a survey, see [3, 5, 7, 8], among others). Bardhan [3] notes that corruption appears relevant in undeveloped economies where the organization of the State is inefficient, democratic control of the civil community over government actions is absent, and bureaucrats have wide discretionary power (see also [2]). The literature about the long-run economic consequences of corruption (see [4, 11]) focuses on rent seeking in the provision of public services. A government official controls the offer of a service against private demand, and then he/she has some discretionary power on the offer and can restrict it in several ways (e.g. denying permission or delaying its release). Bribes are the extra-price charged by bureaucrats to private customers, and arise like rents. The economic consequences of this phenomenon concern distortions in resources allocation mainly in terms of less private investment, and a reduced rate of human capital formation. For example, Ehrlich [4] states that corruption is an economic activity that requires some political capital. Effort devoted to the accumulation of this kind of knowledge has an alternative use in human capital production. Corruption reduces economic growth through a negative influence on investments in human capital.

E. Accinelli (✉) • E.J. Sánchez Carrera
Autonomous University of San Luis Potosi, San Luis, México
e-mail: elvio.accinelli@eco.uaslp.mx; edgar.carrera@uaslp.mx

J.-P. Bourguignon et al. (eds.), *Dynamics, Games and Science*, CIM Series in Mathematical Sciences 1, DOI 10.1007/978-3-319-16118-1_1

While a large proportion of corrupt practices are illegal, in this paper we do not consider a legal approach to the definition of corruption since not all corrupt practices are illegal and not all illegal activities are corrupt practices. In fact, Jain [7] identifies three categories of corruption grand involving political elite, bureaucratic involving corrupt practices by appointed bureaucrats, and legislative corruption involving how legislative votes are influenced by the private interest of the legislator. The three types of corruption differ only in terms of the decisions that are influenced by corrupt practices.

However, few are the articles studying strategic fundamentals that cause corrupt behavior in a society. Hence, the aim of this paper is to describe the evolution of corruption behavior in a society. Our approach is based on recognizing of what economists call incentives or psychologists reinforcement for choosing a certain behavior. When individuals need to choose an action or future behavior between several possible, they are pressed by different kind of incentives and penalties. We understand corruption as a possible behavior followed by several individuals in a given population (see [1]). Accinelli and Carrera [1] pointed out that individuals under the pressure of incentives and penalties need to choose one of two antagonistic possible behaviors, being corrupt or non-corrupt. When individuals choose driven by imitation, but they have not complete information, however they must choose and do this base upon its own beliefs.

In this paper we assume that individuals are no completely informed about the payoff of his/her choices, but they are rational in the sense that they choose with higher probability the behavior that they understand has in each moment, the highest expected value. In our model, we consider a distribution of income for the population, and strategic interaction between people who pay taxes and officials who control such tax compliance. The baseline approach of our model comes from [6] that examine the implications of corruptibility and the potential abuse of authority for the effects and optimal design of (potentially non-linear) tax collection schemes. Hindriks et al. find that the distributional effects of corruption and tax evasion are regressive, hence for the poor have little to gain from evading taxes and are at the same time vulnerable to over-reporting of their incomes; for the rich, the converse is true. The government can Levy progressive taxes without reducing its own payoff by creating countervailing incentives in the form of commissions: the parties are tempted to understate income to evade progressive taxes, and tempted to overstate income to raise the commission payments.

The central authority problem's is to choose a system of fines and capture the corrupt behavior of the auditors, in order to discourage this kind of behavior. We call evaders to citizens who do not pay taxes and corrupts to auditors accept bribes. However, as long as the central planner sets the optimal policy on the basis that all citizens pay taxes, every deviation of this situation imply a deviation of the optimal fiscal policy, with repercussion in the social welfare. Consequently if in the society there exist evaders, the established taxation is not optimal, and this fact becomes in its turn in a new incentive for citizens choosing to be evaders. Nonetheless, if this subpopulation with the time, tends to become smaller, then the perception of social welfare from each of the social groups increases as the share of the population that pays taxes increases. Consequently the action to pay taxes is perceived as not prejudicial like in other cases. In this way, any incentives to do not comply with tax obligations tend to disappear and any basis of corrupt auditors. In what follows, we analyze the impact on the decisions made by different social groups, of the possible policies defined by the central planner.

Our goal is to explain the structural evolution of corrupt behavior in a given society as the result of individuals' decisions influenced by the behavior of the others members of this society. Along this evolutionary process, at every time, individuals make their choices about their future behavior, the result of this process is the social evolution. In particular we analyze the interaction between the tax authority and citizens, to study the evolution of corrupt behavior as the result of individuals' beliefs and institutional policies.

The remainder of this paper is organized as follows. Section 2 develops a game-theory model related to tax evasion and corruption in the tax inspectorate. Section 3 is devoted to study the evolutionary dynamics of corrupt behavior and taxpayers. Finally, Sect. 4 contains some implications of the results and discusses their application to economy.

2 The Model

Consider an economy where institutions are ruled out by two populations, namely citizens and auditors.

Citizens are required to pay taxes, however only those following a non-corrupt behavior meet this requirement. We shall say that are evaders, or corrupts, those citizens who do not pay taxes. Consequently, the population of citizens, C, is composed by: tax evaders or corrupt, C_C, and tax payer or non-corrupt, C_N. There are tax audits done in each period. The task of auditors is to monitor tax compliance of citizens. The population of auditors, P, is composed in turn, by: corrupt, P_C, and non-corrupt, P_N. Non-corrupt auditors are those that make their job according to the national tax compliance laws. Corrupt auditors do not do their job according to the law, and they take bribes from citizens evaders. Moreover:

1. Citizens are distributed according to their levels of income, denoted by a set Y, and a probability that a citizen $x \in C$ has income lower or equal to $y \in Y$ is:

$$P(y(x) \leq y) = P(y).$$

   Note that this probability corresponds to the fraction of citizens with income lower than or equal to y. We assume that according to their income level, citizens are divided into n different groups, $I_1, I_2 \dots, I_n$, thus $y : C \to I$ where $I = \{I_1, \dots I_n\}$, so $x \in I_i$ if and only if $y(x) \in y(I_i) = (\underline{y}_i, \bar{y}_i)$ where $\underline{y}_i$ is de lower income of a citizen in class I_i and $\bar{y}_i$ is the higher income of a citizen in such class. By $n(I_i)$ we denote the share of citizens in the level I_i.

2. We consider that the central planner has implemented a proportional taxation policy, so all citizen should pay taxes proportional to their income, $\tau(y) \cdot y(x)$, where $0 < \tau(y) < 1$. By $y(x)$ we denote the income of the citizen x. That is, the central authority sets rates by income levels, i.e. $\tau(y) = \tau(I_i)$ for all $y \in I_i, i = 1, \ldots, n$. So that the total amount paid as tax by a x-citizen with income level $y \in I_i$ is equal to $\tau(I_i) \cdot y(x)$.
3. We consider that income distribution is constant over time, but the percentage of taxpayers is time-variant. Hence, in every period of time t we represent by:
   - $\alpha(t)$ the share of citizens taxpayers,
   - $\gamma(t)$ the share of non-corrupt auditors.
   - $\beta(t) = 1 - \alpha(t)$ is the share of tax evaders, and $\delta(t) = 1 - \gamma(t)$ represents the share of corrupt auditors.
4. In this vein, we state the following.

   Definition 1 Let us define the index, ι_c, as a measure of total illegal behavior in the economy, i.e.

$$\iota_c(t) = \beta(t) + \delta(t).$$

5. We denote by *underliney* and $\bar{y}$ the lowest and the highest income levels, respectively. Thus the distribution of taxes $P_\alpha(y)$ is supported in the interval $[\underline{y}, \bar{y}]$. Total income due to taxes collected in time t is:

$$T_t = \int_{\underline{y}}^{\bar{y}} \tau(y)y(x)dP_\alpha(y),$$

   The subscript α indicates that the total of citizens paying $\tau(y) \cdot y(x)$ depends on the total share of taxpayers, given by α.
6. We consider that the tax audit is performed, in each period, on citizens with certain probability $P_A \in [0, 1]$. Thus tax evasion is punishable and let us denote by $m > 0$ the fine imposed by a non-corrupt auditor on a citizen tax evader when s/he is audited.
7. The model takes into account the possibility that a briber may bargain with the auditor some money in exchange for not revealing the evasion. This bargain has been succeeded when a corrupt auditor meets an evader, and then the corrupt auditor gets a bribe equal to $\bar{B} = k\tau(y)y(x) > 0, \forall\ 0 < k < 1$.
8. However the central authority can detect to the illegal behavior and consequently punishing the corrupt auditor. The fine imposed to the corrupt auditor by the central authority is $M > 0$, and $p_M \in [0, 1]$ is the probability that the corrupt auditor is detected. Hence, we can state that the sum of the probabilities $\gamma(t) + p_M \geq 0$ measures the efficiency of the central authority as guarantors of legal behavior.

9. If a non-corrupt auditor meets a tax evader citizen, s/he is facing the monitoring cost, $c(\alpha, \gamma) > 0$, by punishing the evader. This cost is a decreasing function of α, and increasing with γ, and convex in both variables. Such a monitoring cost corresponds to the work associated with this process, and it increases as the number of evaders or the number of corrupt officers is increasing. This somehow shows that the incentives to behave legally changes according to the profile distribution of economic agents (see the above Item 3–4, Definition 1). To counteract this negative action about the behavior of public officials, can be doing, for instance, paying a premium to those officials who fulfill their duties. In many Latin American countries, there is a prize to presenting a right fiscal report, and it is paid to employees who are not cheating.
10. If corruption is punished, the total amount received by the payment of fines is transferred to improve the social welfare. The total money obtained by the central authority is the sum of the total money of taxes collected plus the total amount received from fines. The total amount of fines is a random variable, W, with expected value $\bar{W}$. So, the central authority has an expected total national revenue:

$$R_t = T_t + \bar{W}_t > 0.$$

Individuals, P and C, have some utility due to the tax system and national revenue. That is, utilities of auditors and citizens,

$$u_P(\alpha, R) > 0 \text{ and } u_x(\alpha, R) > 0,$$

depend on the total national revenue, R, and on the share, α, of taxpayers.

Therefore, under the above considerations, if the policy of the central planner is given, then individual (expected money-metric) utility functions (or expected payoffs) are given by:

$$
\begin{aligned}
&u_{C_{Nx}}(\alpha) = u_x(\alpha, R) + (1 - \tau(y))y(x), && (A)\\
&u_{C_{Cx}}(\alpha, \gamma) = u_x(\alpha, R) - P_A\big[\gamma\big(m + (1 - \tau(y))y(x)\big) + \big(1 - \gamma\big)\big((1 - k\tau(y))y(x)\big)\big], && (B)\\
&u_{P_C}(\alpha) = u_p(\alpha, R) + (1 - \alpha)\big[\textstyle\sum_{i=1}^{n} k\tau(I_i)y(I_i)n_i\big] - P_M M, && (C)\\
&u_{P_{NC}}(\alpha, \gamma) = u_p(\alpha, R) - (1 - \alpha)c(\gamma). && (D)
\end{aligned}
\tag{1}
$$

Note that these utilities can change over time if the share populations change, and α and γ are the only endogenous variables while the parameters R, $\tau(y)$, m are exogenously determined by the central authority. The parameter k is a constant fixed by the corrupt auditors. The first equation (A) is the utility function of a taxpayer with income $y(x)$. The second one (B) is the utility function corresponding to a citizen tax-evader, with income $y(x)$, $0 < k < 1$ correspond to the proportion of the tax that citizen tax-evader must pay to a corrupt auditor with probability $(1 - \gamma)$. With probability γ, an evader is revealed by a non-corrupt auditor and must pay a fine m plus the amount owned. The third one (C) is the utility function of the corrupt auditor. We assume that with probability $(1 - \alpha)$ the audited citizens are evaders, and in this case the auditor takes bribes. The last one (D) the utility of a non-corrupt auditor. we assume that an honest auditor must perform certain management when confronting an individual evader. This management has a cost, which we assume decreases when the number of honest auditors. This management should only be performed when confronting an individual evader, otherwise we assume it is zero, i.e., $\frac{\partial c(\alpha,\gamma)}{\partial \gamma} < 0$. Obviously, the probability to pay this cost, decrease when the number of honest citizen increase. Either because the probability of facing a citizen evader decreases or because the cost is shared between more auditors

Remark 1 A citizen chooses to be a non-corrupt, i.e. he/she is taxpayer, if $u_{C_{Nx}}(\alpha) > u_{C_{Cx}}(\alpha, \gamma)$ which holds when:

$$\tau(y) \leq \frac{y(x)(1 + P_A) + mP_A\gamma}{y(x)\big[1 - P_A(\gamma k - k - \gamma)\big]},$$

where $\tau(y)$ is a threshold value indicating a social limit, beyond which the utility of an honest citizen with income y surpasses the associated utility to the corrupt behavior. This threshold value makes reference to the highest income tax rate that the central authority should impose for not favoring the evader behavior. Note that $\tau'(y) < 0$ means that citizens with higher incomes are more likely to become evaders.

Remark 2 An auditor chooses to be a non-corrupt if $u_{P_{NC}}(\alpha, \gamma) > u_{P_C}(\alpha)$ which holds when:

$$p_M > \frac{(1-\alpha)[\sum_{i=1}^{n} k\tau(I_i)y(I_i)n_i] + (1-\alpha)c(\gamma)}{M},$$

and so the difference $u_{P_{NC}}(\alpha, \gamma) - u_{P_C}(\alpha)$ is positive and it is increasing either when $p_M$or M are large enough.

We assume that the level of social welfare increases with the total national revenue and with the share of taxpayers, i.e.

$$\frac{\partial u_j}{\partial R}(\alpha, R) > 0 \quad and \quad \frac{\partial u_j}{\partial \alpha}(\alpha, R) > 0 \;\; for\ all\ \ j \in \{C, P\},$$

and that the functions $u_j(\alpha, R)$ are concave with respect to R, i.e.

$$\frac{\partial^2 u_j(\alpha, R)}{\partial R^2} < 0,$$

where auditors and citizens do not value equally the welfare obtained by taxes, this assumptions is considered in the fact that $u_C(\alpha, R)$ is not necessarily equal to $u_P(\alpha, R)$.

Central authority should fix the optimal tax rate assuming that every citizen pay taxes. So this is not longer optimal in the presence of citizens evaders. Suppose the share of taxpayers in time t is $\alpha(t) = \alpha$. Consider in addition that $P_\alpha(I_i)$ correspond to the proportion of citizens in the level $I_i,\ i = 1, \ldots, n$ that in time t are paying taxes. The level of income of each group (or social class) is symbolized by $y(I_i)$. Then in terms of income, the expected amount of tax collected can be written as:

$$T_\alpha(t) = \sum_{i=1}^{n} \tau(I_i)[y(I_i) - y(I_{i-1})P_\alpha(I_i, t)],$$

where as we said $P_\alpha(I_i, t)$ represents the percentage of citizens with income $y(I_i)$ that are taxpayers, in time t. While total (potential) amount collected corresponds to:

$$T_{\alpha=1} = \sum_{1}^{n} \tau(I_i)[y(I_i) - y(I_{i-1})P_{\alpha=1}(I_i)]$$

where $P_{\alpha=1}(I_i)$ is the share of taxpayers citizen with income $y_{I_{i+1}}$ while $P(I_i)$ is the total share of individuals with such income in the population, so $P(I_i) \geq P_\alpha(I_i, t)$ for all t, with equality if and only if $\alpha = 1$.

From now on to facilitate the scripture, if not strictly necessary, we suppress the variable t although all values depend on the distribution of populations, which certainly change over time.

The utility of a citizen x,q who is a taxpayer, is given by the Eq. (1A), and it can be written as:

$$u_{C_{Nx}}(\alpha, \tau) = u_x\left(\alpha, \sum_{1}^{n} \tau(y(I_i)[y(I_i) - y(I_{i-1})]P_\alpha(I_i) + \bar{W}\right) + (1 - \tau(I_j))y(I_j).$$

To simplify the notation we denote by τ_j the optimal tax corresponding to the income level equal to $y_{I_j},\ j = 1, 2 \ldots, n$ i.e. $\tau_j = \tau(I_j)$. The next proposition offers an important result.

Proposition 1 *As the gap between social classes (measured by the different income levels I_i), is increasing (more income inequality), the greater the tax evasion (more corruption).*

Proof If the policy maker considers that every citizen pay taxes according with their income, i.e.: $\alpha = 1$, then the utility function depends only on taxes $\tau = (\tau_1, \ldots, \tau_n)$, and these are fixed by the central authority. More precisely, if $\alpha = 1$ then $R(\tau) = T_{\alpha=1}(\tau)$ and so, the optimal tax rate $\tau^*(y)$ (the optimal policy for the central planner) must verify the equations:

$$\frac{\partial u_{C_{Nx}}}{\partial \tau_j}(1, R(\tau^*)) = \frac{\partial u_{C_{Nx}}}{\partial R}(1, R(\tau^*))\frac{\partial R}{\partial \tau_j}(\tau^*)$$

$$= \frac{\partial u_{C_{Nx}}}{\partial R}\left(1, \sum_1^n \tau_i^*[y(I_i) - y(I_{i-1})]P_{\alpha=1}(I_i)\right)[y(I_j) - y(I_{j-1})]P_{\alpha=1}(I_j) - y(I_j) = 0 \tag{2}$$

or equivalently,

$$\frac{\partial u_{C_{Nx}}}{\partial R}\left(1, \sum_1^n \tau_i^*[y(I_i) - y(I_{i-1})]P_{\alpha=1}(I_i)\right)\frac{\Delta y(I_j)}{y(I_j)}P_{\alpha=1}(I_j) = 1, \tag{3}$$

for all $j = 1, \ldots, n$; where $\Delta y(I_j) = y(I_j) - y(I_{j-1})$ is the income gap between the social classes I_j and I_{j-1}. Note that if we assume $\alpha = 1$, then $P_{\alpha=1}(I_j)$ is equal to the total percentage of citizens with income $y \leq I_j$. Given that the utility function is strictly concave in R it follows that $\frac{\partial^2 u_{C_{Nx}}}{\partial \tau_j^2} < 0$ so, τ_j^* is a maximum. In conclusion, by Eq. (3) it follows that the number of citizens that are willing to be taxpayers is a decreasing function of the gap between social classes I_j and I_{j-1}. Hence, as lower is the gap between social classes, the lower is the tax evasion.

As we argue previously, auditors may have interest in to coexist with evaders (Eqs. (1C) and (1D)). It follows from these equations that the interest of auditors in this complicity tends to decrease when the possibility of being caught in their illegal actions is increasing. This argues in favor of auditing and administrative controls, because they are part of public activities aimed at ensuring the normal functioning of the institutions. Another problem is the cost of establishing a convenient mechanism to punish the illegal activity of evaders and corrupt auditors. As we will prove in the next section, it is possible to establishing an adequate system of monitoring, based on probabilities and fines, enabling to ensure that shares of taxpayers and no-corrupt auditors may evolve positively.

3 On the Evolutionary Dynamics of the Model

We consider in this section that citizens and auditors play an asymmetric contest evolutionary game. The possible strategies for each individual in each subpopulation are to implement a corrupt behavior or a legal (honest) behavior. Since players are rational they will choose between these two pure strategies, according with the perception of the rewards and possible punishment that, each election imply. The strategic election is influenced on one hand, by the behavior of their peers (imitative behavior), and on the other hand this strategic election is strongly influenced by the behavior of the counterpart, i.e. the corrupt behavior of the citizens is encouraged by the corrupt behavior of the auditors, and reciprocally.

To analyze the evolution of legal behavior by citizens and auditors, we admit that the tax rate imposed by the policy maker is optimal, and then we introduce a dynamical system of imitation based on the well known model of [9], where the parameters of this dynamical system are strongly related with the degree of efficiency of the monitoring system (see Remarks 1–3). Therefore, consider that:

1. Citizens imitate the behavior of their leader neighbors or successful people, and they perceive the possibilities to be punished or not. This fact is captured by the parameters b and f in the dynamical system (4), see below. According with their beliefs, they will choose the most profitable behavior. So, this beliefs are strongly related with the perception of the citizens of the governmental efficiency to capture the illegal actions.
2. It is natural also to assume that the growth rate of corrupt auditors, $\frac{\dot{\delta}}{\delta}$, increases with their relative weight in the auditors' population and decreases with the number of citizens who are not willing to give bribes. Recall that $\beta = 1 - \alpha$ and $\gamma = 1 - \delta$, and all these variables must be non-negative in every time.
3. If in some time t_f, $\alpha(t_f) = 1$, then for all $t > t_f$, implies that $\dot{\delta}(t) < 0$ and $\dot{\alpha}(t) = 0$. Reciprocally, if in a given time all auditor is corrupt then every citizen is an evader, and then $\dot{\alpha}(t) < 0$ and $\dot{\delta}(t) = 0$.

To obtain the evolution for the share of corrupt behavior in a given time $t = t_0$, assuming that in $t = t_0$, $0 < \alpha(t_0) < 1$ and $0 < \delta(t_0) < 1$. This situation can be medelled using the following system of differential equations:

$$\begin{aligned} \dot{\alpha} &= \alpha(a + b\alpha - c\delta), \\ \dot{\beta} &= -\dot{\alpha}, \\ \dot{\delta} &= \delta(d - e\alpha + f\delta), \\ \dot{\gamma} &= -\dot{\delta}. \end{aligned} \tag{4}$$

where a, b, c, d, e, f are positive constants, and the magnitude of these parameters are in direct relation with the policy implemented by the central authority, in particular with the amount of fines and the probability that the corrupt behavior can be caught and punished (see Remarks 1–2).

The study of many evolutionary models, social or biological, is based on the determination and analysis of parameters or combinations of parameters that determine the change in the qualitative behavior of the solutions of a system of differential equations. In our case this study corresponds to the central authority, who must find those combinations that achieve a better social growth. To ensure that the appropriate values of the parameters, prevail in society, she must design a suitable social, or economic policy. Certainly, this is not a simple task, however the identification of these parameters and the role they play in social evolution, can help in the process of fulfilling this goal.

For instance in the first equation, the parameter c represents the negative weight that the corrupt auditors play in the evolution of the society. It follows that the social influence of this group, measured by c, decreases if m or $p(m)$ increase. The parameter b represents the importance of the imitation inside the subpopulation of the taxpayers, this value increases with the difference: $u_{CI_x}(\alpha, I, t) - u_{CNI_x}(\alpha, I)$. We assume that the citizens or auditors, can change their behavior followed up to the present, if and only if there exists in society, a different behavior that may be imitated. This leads us to conclude that if in time $t = t_f$ for instance $\alpha(t_f) = 1$ then $\alpha(t) = 1$ for all $t \geq t_f$. Analogously, for the other cases, i.e., if $\beta(t_f) = 1$ then $\beta(t) = 1$ for all $t \geq t_f$ and the same for the auditors. So, the dynamical system (4) can be reformulated as:

$$\dot{\alpha} = \begin{cases} \alpha(a + b\alpha - c\delta), \; if \; 0 < \alpha(t_0) < 1 \\ 0 \qquad\qquad\qquad \forall\, t \geq t_f : \; \alpha(t_f) = 1 \; or \; \alpha(t_f) = 0. \end{cases}$$
$$\dot{\delta} = \begin{cases} \delta(d - e\alpha + f\delta), \\ 0 \qquad\qquad\qquad \forall\, t \geq t_f : \; \delta(t_f) = 1 \; or \; \delta(t_f) = 0. \end{cases} \tag{5}$$

The parameter b, measures the effect of the imitation in the behavior of the citizens. The growth rate of the legal behavior, is higher when greater the influence of imitation in social behavior, measured by b.

$$\frac{\partial(\frac{\dot{\alpha}}{\alpha})}{\partial \alpha} = b > 0.$$

The intensity of this parameters, depends strongly on the difference between the utilities of the tax payers citizens and evaders. This shows that it is possible, for the central authority, to design a policy to ensure legal behavior on the part of citizens, (see Remark 1) which as we will see impact favorably also in the behavior of auditors (see below Eq. (6)).

The pernicious effect on the society, of the corrupt behavior of auditors is strongly related with the parameter c. Note that

$$\frac{\partial(\frac{\dot{\alpha}}{\alpha})}{\partial \delta} = -c < 0.$$

The parameter e, measure the rate at which decreases, by the effect of an increased legal behavior of citizens, the corrupt behavior of auditors, and the parameter f measure the rate at which increases by the effect of an increased illegal activity of the auditors, the corrupt behavior of auditors. This parameter is strongly related with the role that the imitation plays in the society.

$$\frac{\partial(\frac{\dot{\delta}}{\delta})}{\partial\alpha} = -e < 0 \ \ and \ \ \frac{\partial(\frac{\dot{\delta}}{\delta})}{\partial\delta} = f > 0. \tag{6}$$

Note that the incentives of the auditors to follow a corrupt behavior decrease, as increase the percentage of citizens following the legal behavior.

Therefore:

Remark 3 If $\delta > \frac{a+b}{\alpha}$ then, $\dot{\alpha} < 0$ the growth rate of taxpayers will be negative. So, only in the case where δ is large enough it is possible to observe an increased illegal activity of the citizens. This means that in the absence of the corrupt auditors, tax evasion tends to disappear.

Remark 4 If we get that in given period of time $t = t_0$, the number of taxpayers is large enough, $\alpha(t_0) > \frac{c\delta-a}{b}$, then the citizens prefer to pay taxes. In the case where $\alpha(t_0) > \frac{c-a}{b}$ this preference dos not depend on the number of corrupt auditors. This possibility is given in the case where $\delta = 1$

This show, once again, that the main characteristics of the differential system are strongly related with the policy of incentives chosen by the central planner.

The system (4) represents the structural dynamics of the behaviors of corrupt auditors and taxpayer citizens. According with this evolutionary system the index of corruption in the society (see Definition 1) evolves according with de differential equation:

$$\dot{\iota}_c(t) = (1 - \dot{\alpha}) + \dot{\delta}.$$

From the dynamical system (5) we obtain de following proposition

Proposition 2 *Coexist both corrupt and non-corrupt officials and citizens in the economy. The relative share of every group depend on the policy followed by the central authority.*

To see this, the system (4) admits the following nullclines:

$$\begin{aligned} \mu &: \ a + b\alpha - c\delta = 0, \\ \nu &: \ d - e\alpha + f\delta = 0. \end{aligned} \tag{7}$$

defined in a closed bounded region, and suppose a positive half path which lies entirely within that region.

Proof Note that the region $[0, 1] \times [0, 1]$ is invariant for this differential equation system (4). Which means that, the solution of the system will remain permanently in this region. Firstly, consider that the phase plane portrait of the nullclines (see Fig. 1) is: Then, one of the different situations below is true:

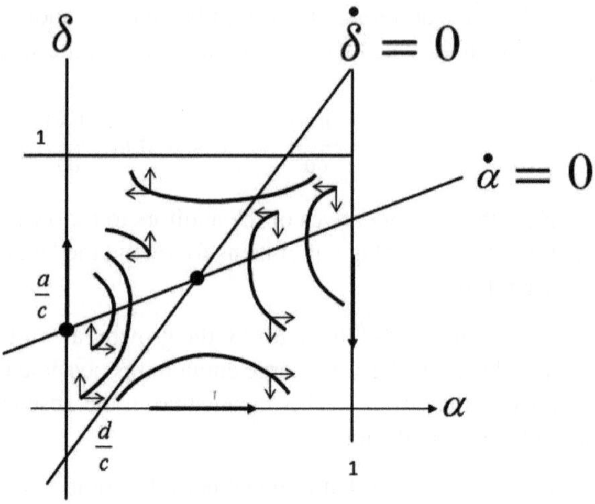

Fig. 1 Nullclines do intersect in $[0, 1] \times [0, 1]$

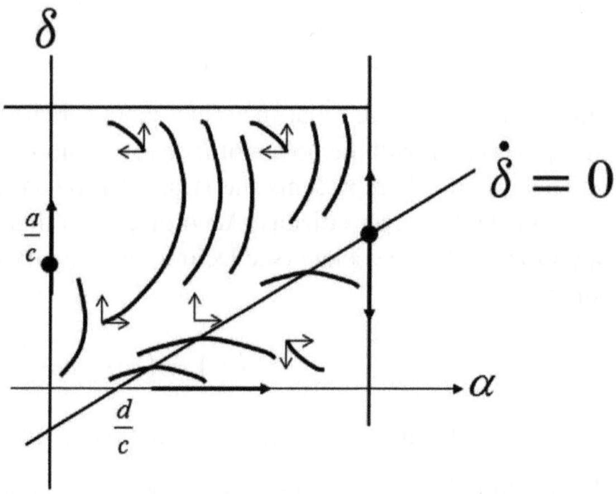

Fig. 2 Nullclines do not intersect in $[0, 1] \times [0, 1]$ with $0 < \frac{d}{c} < 1$

1. The nullclines do intersect in $[0, 1] \times [0, 1]$, see Fig. 1, and this is the case if $\frac{e}{f} < 1$ and $\frac{b}{c} < \frac{e}{f}$.
2. The nullclines do not intersect in $[0, 1] \times [0, 1]$, these cases are represented in Fig. 2 and Fig. 3, then

   a. μ is below ν, this is the case if $\frac{a}{c} < 1$ and $\frac{b}{c} < \frac{e}{f}$, or
   b. $\frac{a}{c} > 1$ and $\frac{e}{f} < 1$, or
   c. $\frac{a}{c} < 1$ and $\frac{e}{f} > 1$.

The more relevant case is 2(a). Note that in this case, the initial relation between the percentages of honest citizens and corrupt auditors is the clue to that allows to

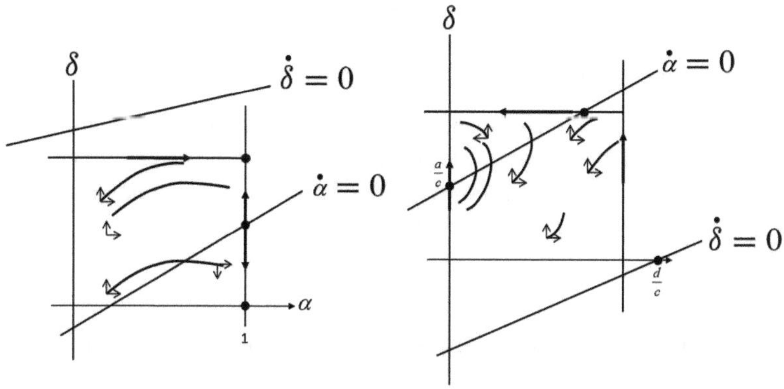

Fig. 3 Nullclines do not intersect in [0, 1] × [0, 1]

understanding the future evolution of the population. So, if the initial values verify the relationships

$$0 < \alpha(t_0) < -\frac{f\delta(t_0) + d}{e}$$

and

$$1 > \delta(t_0) > \frac{a + b\alpha(t_0)}{c}$$

then the population evolves in such way that all citizen does not like to be a taxpayer and all auditor is corrupt. But if the initial number of honest citizens is large enough

$$\alpha(t_0) > -\frac{f\delta(t_0) + d}{e}$$

and

$$\delta(t_0) < \frac{a + b\alpha(t_0)}{c}$$

the society evolves to an idyllic world without corruption. However, more realistic are the situations in which:

$$0 < \alpha(t_0) < -\frac{f\delta(t_0) + d}{e} \quad and \quad 0 < \delta(t_0) < \frac{a + b\alpha(t_0)}{c}$$

or

$$\alpha(t_0) > -\frac{f\delta(t_0) + d}{e} \quad and \quad \delta(t_0) > \frac{a + b\alpha(t_0)}{c},$$

because, in this case, the society evolves to a steady state in which there exists a positive percentage of corrupt auditors and evaders together with a positive percentage of honest auditors and taxpayers see (Fig. 1).

The final distribution to which society arrives, in the case given by Proposition 2, depends strongly, on the ability of the government to develop successful institutional policies. In terms of our model, choosing a good policy means to implement the right values for the main parameters, see Remarks 1–4.

4 Concluding Remarks: Institutions, Beliefs and Incentives

In this paper we show that there is a positive relationship between income inequality and corruption. We also show that the evolutionary dynamics yields results such that corruption prevails and/or coexists with non-corrupt behavior. But we can also get the result of the eradication of corruption and tax evasion.

The evolution of corruption is the result of a free choice made by individuals in a society. This choice is based on beliefs originated in the perception of the actual world. These beliefs may be wrong or not, but define the future behavior of individuals and thus the evolution of society. As [10] explained, all rational model includes explicitly the beliefs, and the decision making as a process in two steps:

1. The step one entails the creation of a model on how the world is and how it evolves in the future. This model establishes relationships between actions an consequences.
2. In the second step, given such a model individuals choose the behavior or the action that they prefer. This choice is an individual fact, and in principle individuals do not consider the social implications of these choices. They do it only taken account their individual beliefs and preferences.

   So, what is the role of institutions in this process?

- Institutions must ensure that rational choice is freely made by individuals depending on their personal interests and does not become counterproductive or pernicious in the long period.
- Individuals not fully informed about the future performance of their current behavior can choose rationally according with the information available and their belief, but these can be wrong or incorrect when evaluated in light of the complete information. For this purpose, institutions should design a policy of incentives, rewards and penalties that citizens do choose correctly in full use of their talents and abilities, compensating in this way, for the lack of information available to citizens.

In terms of the dynamical system the responsibility of the central authority is to put the society in the basin of attraction of a desirable state. To do this she must consider the possibility to change the parameters, in such way that the initial conditions remain in the interior of such basin of attraction. If the central authority is not able to obtain this result, then nothing will change and we are caught in a poverty trap characterized by a system of institutional corruption.

Acknowledgements The author "E. Accinelli" wishes to acknowledge the support of CONACYT through the project CB-167004.

The author "E.J. Sánchez Carrera" wishes to acknowledge the support Secretary of Research and Graduate UASLP through project FAI 2015.

References

1. Accinelli, E., Sanchez, C.E.: Corruption driven by imitative behavior. Econ. Lett. **117**(1), 84–87 (2012)
2. Azariadis, C., Lahiri, A.: Do Rich Countries Choose Better Governments? Working paper, UCLA (1997)
3. Bardhan, P.: Corruption and development: a review of issues. J. Econ. Lit. **XXXV**, 1320–46 (1997)
4. Ehrlich, I.: Bureaucratic corruption and endogenous economic growth. J. Polit. Econ. **107**(S6), S270–S293 (1999)
5. Gupta, S., Davoodi, H., Alonso-Terme, R.: Does corruption affect income inequality and poverty? IMF Working Paper No. WP/98/76 (1998)
6. Hindriks, J., Keen, M., Muthoo, A.: Corruption, extortion and evasion. J. Public Econ. **74**(3), 395–430 (1999)
7. Jain, A.K.: Corruption: a review. J. Econ. Surv. **15**(1), 71–121 (2001)
8. Li, H., Xu, L.C., Zou, H.: Corruption, income distribution and growth. Econ. Polit. 12(2): 155–185 (2000)
9. Lotka, A.J.: Elements of Physical Biology. Publisher Williams and Wilkins Company (1925). Reprinted by Dover in 1956 as Elements of Mathematical Biology
10. Savage, L.: The Foundations of the Statics. Dover, New York (1972)
11. Shleifer, A., Vishny, R.W.: Corruption. Q. J. Econ. 108, 599–617 (1993)

Discrete Symmetric Planar Dynamics

B. Alarcón, S.B.S.D. Castro, and I.S. Labouriau

Abstract We review previous results providing sufficient conditions to determine the global dynamics for equivariant maps of the plane with a unique fixed point which is also hyperbolic.

1 Introduction

The Discrete Markus-Yamabe Question is a problem concerning discrete dynamics, formulated in dimension n by Cima et al. [9] as follows:
[DMYQ(n)] *Let* $f : \mathbf{R}^n \longrightarrow \mathbf{R}^n$ *be a* C^1 *map such that* $f(0) = 0$ *and for any* $x \in \mathbf{R}^n$, $Jf(x)$ *has all its eigenvalues with modulus less than one. Is it true that* 0 *is a global attractor for the discrete dynamical system generated by* f*?*

It is known that the answer is affirmative in dimension 1 and there are counter-examples for dimensions higher than 2, see [8, 14].

In dimension 2, Cima et al. [9] prove that an affirmative answer is obtained when f is a polynomial map, and provide a counter example which is a rational map. After this, research on planar maps focused on the quest for minimal sufficient conditions under which the DMYQ has an affirmative answer. Alarcón et al. [6] use the existence of an invariant embedded curve joining the origin to infinity to show the global stability of the origin. Symmetry is a natural context for the existence of

B. Alarcón (✉)
Departamento de Matemática Aplicada, Instituto de Matemática e Estatística, Universidade Federal Fluminense, Rua Mário Santos Braga, S/N, Campus do Valonguinho, CEP 24020 140 Niterói, RJ, Brasil
e-mail: balarcon@id.uff.br

S.B.S.D. Castro
Centro de Matemática da Universidade do Porto and Faculdade de Economia do Porto, Rua Dr. Roberto Frias, 4200-464 Porto, Portugal
e-mail: sdcastro@fep.up.pt

I.S. Labouriau
Centro de Matemática da Universidade do Porto, Rua do Campo Alegre 687, 4169-007 Porto, Portugal
e-mail: islabour@fc.up.pt

J.-P. Bourguignon et al. (eds.), *Dynamics, Games and Science*, CIM Series in Mathematical Sciences 1, DOI 10.1007/978-3-319-16118-1_2

such a curve, and this led us to a symmetric approach to this problem and to the results in [1–5] that we review in this article.

The present article studies maps f of the plane which preserve symmetries described by the action of a compact Lie group. In this setting we characterise the possible local dynamics near the unique fixed point of f, that we assume hyperbolic. We establish for which symmetry groups local dynamics extends globally. For the remaining groups we present illustrative examples.

2 Preliminaries

This section consists of definitions and known results about topological dynamics and equivariant theory. These are grouped in two separate subsections, which are elementary for readers in each field, containing material from the corresponding sections of [1–5] and is included here for ease of reference.

2.1 *Topological Dynamics*

We consider planar topological embeddings, that is, continuous and injective maps defined in $\mathbf{R}^2$. The set of topological embeddings of the plane is denoted by $\mathrm{Emb}(\mathbf{R}^2)$.

Recall that for $f \in \mathrm{Emb}(\mathbf{R}^2)$ the equality $f(\mathbf{R}^2) = \mathbf{R}^2$ may not hold. Since every map $f \in \mathrm{Emb}(\mathbf{R}^2)$ is open (see [12]), we will say that f is a homeomorphism if f is a topological embedding defined onto $\mathbf{R}^2$. The set of homeomorphisms of the plane will be denoted by $\mathrm{Hom}(\mathbf{R}^2)$. When $\mathscr{H}$ is one of these sets we denote by $\mathscr{H}^+$ (and $\mathscr{H}^-$) the subset of orientation preserving (reversing) elements of $\mathscr{H}$.

We denote by $\mathrm{Fix}(f)$ the set of fixed points of a continuous map $f : \mathbf{R}^2 \to \mathbf{R}^2$.

Let $\omega(p)$ be the set of points q for which there is a sequence $n_j \to +\infty$ such that $f^{n_j}(p) \to q$. If $f \in Hom(\mathbf{R}^2)$ then $\alpha(p)$ denotes the set $\omega(p)$ under f^{-1}.

Let $f \in \mathrm{Emb}(\mathbf{R}^2)$ and $p \in \mathbf{R}^2$. We say that $\omega(p) = \infty$ if $\|f^n(p)\| \to \infty$ as n goes to ∞, where $\|\cdot\|$ denotes the usual Euclidean norm. Analogously, if $f \in \mathrm{Hom}(\mathbf{R}^2)$, we say that $\alpha(p) = \infty$ if $\|f^{-n}(p)\| \to \infty$ as n goes to ∞.

A map $f \in \mathrm{Emb}(\mathbf{R}^2)$ is *dissipative* if there exists a compact set $W \subset \mathbf{R}^2$ that is positively invariant and attracts uniformly all compact sets. This means that $f(W) \subset W$ and for each $x \in \mathbf{R}^2$,

$$dist(f^n(x), W) \to 0, \qquad \text{as } n \to \infty$$

uniformly on balls $\|x\| \leq r$, $r > 0$. Observe that in the case $f \in \mathrm{Hom}(\mathbf{R}^2)$ the dissipativity of f means that ∞ is asymptotically stable for f^{-1}.

We say that $0 \in \mathrm{Fix}(f)$ is a *local attractor* if its basin of attraction $\mathscr{U} = \{p \in \mathbf{R}^2 : \omega(p) = \{0\}\}$ contains an open neighbourhood of 0 in $\mathbf{R}^2$ and that 0 is a *global attractor* if $\mathcal{U} = \mathbf{R}^2$. The origin is a *stable fixed point* if for every neighborhood U of 0 there exists another neighborhood V of 0 such that $f(V) \subset V$ and $f(V) \subset U$. Therefore, the origin is an *asymptotically local (global) attractor* or a *(globally) asymptotically stable fixed point* if it is a stable local (global) attractor. See [7] for examples.

We say that $0 \in \text{Fix}(f)$ is a *local repellor* if there exists a neighbourhood V of 0 such that $\omega(p) \notin V$ for all $0 \neq p \in \mathbf{R}^2$ and a *global repellor* if this holds for $V = \mathbf{R}^2$.

The origin is an *asymptotically global repellor* if it is a global repellor and, moreover, if for any neighbourhood U of 0 there exists another neighbourhood V of 0, such that, $V \subset f(V)$ and $V \subset f(U)$.

When the origin is a fixed point of a C^1-map of the plane, the origin is a *local saddle* if the two eigenvalues of Df_0, α, β, are both real and verify $0 < |\alpha| < 1 < |\beta|$. In case the two eigenvalues are strictly positive the origin is called a direct saddle. We define the origin to be a *global (topological) saddle* for a C^1–homeomorphism if additionally its stable and unstable manifolds $W^s(0,f)$, $W^u(0,f)$ are unbounded sets that do not accumulate on each other, except at 0 and ∞, and such that

$$\mathbf{R}^2 \setminus (W^s \cup W^u \cup \{\infty\}) = U_1 \cup U_2 \cup U_3 \cup U_4,$$

where for all $i = 1, \ldots, 4$ $\;U_i \subset \mathbf{R}^2$ is open connected and homeomorphic to $\mathbf{R}^2$ verifying:

(i) either $f(U_i) = U_i$ or there exists an involution $\varphi : \mathbf{R}^2 \to \mathbf{R}^2$ such that $(f \circ \varphi)(U_i) = U_i$
(ii) for all $p \in U_i$ both $\|f^n(p)\| \to \infty$ and $\|f^{-n}(p)\| \to \infty$ as n goes to ∞.

We say that $f \in \text{Emb}(\mathbf{R}^2)$ has *trivial dynamics* if $\omega(p) \subset \text{Fix}(f)$, for all $p \in \mathbf{R}^2$. Moreover, we say that a planar homeomorphism has trivial dynamics if both $\omega(p), \alpha(p) \subset \text{Fix}(f)$, for all $p \in \mathbf{R}^2$.

Let $f : \mathbf{R}^N \to \mathbf{R}^N$ be a continuous map. Let $\gamma : [0, \infty) \to \mathbf{R}^2$ be a topological embedding of $[0, \infty)$. As usual, we identify γ with $\gamma([0, \infty))$. We will say that γ is an *f-invariant ray* if $\gamma(0) = (0, 0)$, $f(\gamma) \subset \gamma$, and $\lim_{t \to \infty} \|\gamma(t)\| = \infty$.

Proposition 1 (Alarcón et al. [6]) *Let $f \in Emb^+(\mathbf{R}^2)$ be such that $Fix(f) = \{0\}$. If there exists an f-invariant ray γ, then f has trivial dynamics.*

Corollary 1 *Let $f \in Hom^+(\mathbf{R}^2)$ be such that $Fix(f) = \{0\}$. If there exists an f-invariant ray γ, then for each $p \in \mathbf{R}^2$, as n goes to $\pm\infty$, either $f^n(p)$ goes to 0 or $\|f^n(p)\| \to \infty$.*

In order to explain the construction of examples in Sect. 5 we need to introduce the concept of prime end.

We say that $f : \mathbf{R}^2 \to \mathbf{R}^2$ is an *admissible homeomorphism* if f is orientation preserving, dissipative and has an asymptotically stable fixed point with proper and unbounded basin of attraction $U \subset \mathbf{R}^2$. Note that U is non empty, so the proper condition follows when the fixed point is not a global attractor. Since $f(U) = U$, we can obtain automatically the unboundedness condition if we suppose that f is area contracting.

Let $f : \mathbf{R}^2 \to \mathbf{R}^2$ be an admissible homeomorphism and consider the compactification of f to the Riemann sphere $f : \mathbf{S}^2 \to \mathbf{S}^2$. Hence $U \subset \mathbf{S}^2 = \mathbf{R}^2 \cup \{\infty\}$. A *crosscut* C of U is an arc homeomorphic to the segment $[0, 1]$ such that $a, b \notin U$ and $\dot{C} = C \setminus \{a, b\} \subset U$, where a and b are the extremes of C. Every crosscut divides U into two connected components homeomorphic to the open disk $\mathrm{d} = \{z \in \mathbb{C} : |z| < 1\}$.

Let x^* be a point in U. For convenience we will consider only the crosscut such that $x^* \notin C$. We denote by $D(C)$ the component of $U \setminus C$ that does not contain x^*. A *null-chain* is a sequence of pairwise disjoint crosscuts $\{C_n\}_{n \in \mathbf{N}}$ such that

$$\lim_{n \to \infty} \operatorname{diam}(C_n) = 0 \text{ and } D(C_{n+1}) \subset D(C_n),$$

where $\operatorname{diam}(C_n)$ is the diameter of C_n on the Riemann sphere.

Two *null-chains* are *equivalent* $\{C_n\}_{n \in \mathbf{N}} \sim \{C_n^*\}_{n \in \mathbf{N}}$ if given $m \in \mathbf{N}$

$$D(C_n) \subset D(C_m^*) \text{ and } D(C_n^*) \subset D(C_m),$$

for n large enough. A *prime end* is defined as a class of equivalence of a null-chain and the space of prime ends is

$$\mathbf{P} = \mathbf{P}(U) = \mathscr{C} / \sim,$$

where $\mathscr{C}$ is the set of all null-chains of U.

The disjoint union $U^* = U \cup \mathbf{P}$ is a topological space homeomorphic to the closed disk $\bar{\mathrm{d}} = \{z \in \mathbb{C} : |z| \leq 1\}$ such that its boundary is precisely $\mathbf{P}$.

It is well studied in [13] that the Theory of Prime Ends implies that an admissible homeomorphism f induces an orientation preserving homeomorphism $f^* : \mathbf{P} \to \mathbf{P}$ in the space of prime ends. This topological space is homeomorphic to the circle, that is $\mathbf{P} \simeq \mathbf{T} = \mathbb{R}/\mathbb{Z}$, and hence the rotation number of f^* is well defined, say $\bar{\rho} \in \mathbf{T}$. The *rotation number* for an admissible homeomorphism is defined by $\rho(f) = \bar{\rho}$.

2.2 Equivariant Theory

Let Γ be a compact Lie group acting on $\mathbf{R}^2$, that is, a group which has the structure of a compact C^∞-differentiable manifold such that the map $\Gamma \times \Gamma \to \Gamma$, $(x, y) \mapsto xy^{-1}$ is of class C^∞. The following definitions and results are taken from Golubitsky et al. [10], especially Chapter XII, to which we refer the reader interested in further detail.

We think of a group mostly through its action or representation on $\mathbf{R}^2$. A *linear action* of Γ on $\mathbf{R}^2$ is a continuous mapping

$$\Gamma \times \mathbf{R}^2 \to \mathbf{R}^2$$

$$(\gamma, p) \mapsto \gamma p$$

such that, for each $\gamma \in \Gamma$ the mapping ρ_γ that takes p to γp is linear and, given $\gamma_1, \gamma_2 \in \Gamma$, we have $\gamma_1(\gamma_2 p) = (\gamma_1\gamma_2)p$. Furthermore the identity in Γ fixes every point. The mapping $\gamma \mapsto \rho_\gamma$ is called the *representation* of Γ and describes how each element of Γ transforms the plane.

We consider only standard group actions and representations. A representation of a group Γ on a vector space V is *absolutely irreducible* if the only linear mappings on V that commute with Γ are scalar multiples of the identity.

Given a map $f : \mathbf{R}^2 \to \mathbf{R}^2$, we say that $\gamma \in \Gamma$ is a *symmetry* of f if $f(\gamma x) = \gamma f(x)$. We define the *symmetry group* of f as the biggest closed subset of $GL(2)$ containing all the symmetries of f. It will be denoted by Γ_f.

We say that $f : \mathbf{R}^2 \to \mathbf{R}^2$ is Γ*-equivariant* or that f *commutes* with Γ if

$$f(\gamma x) = \gamma f(x) \quad \text{for all} \quad \gamma \in \Gamma.$$

It follows that every map $f : \mathbf{R}^2 \to \mathbf{R}^2$ is equivariant under the action of its symmetry group, that is, f is Γ_f-equivariant.

Let Σ be a subgroup of Γ. The *fixed-point subspace* of Σ is

$$\mathrm{Fix}(\Sigma) = \{p \in \mathbf{R}^2 : \sigma p = p \text{ for all } \sigma \in \Sigma\}.$$

If Σ is generated by a single element $\sigma \in \Gamma$, we write $Fix\langle\sigma\rangle$.

We note that, for each subgroup Σ of Γ, $\mathrm{Fix}(\Sigma)$ is invariant by the dynamics of a Γ-equivariant map ([10], XIII, Lemma 2.1).

When f is Γ-equivariant, we can use the symmetry to generalize information obtained for a particular point. This is achieved through the *group orbit* Γx of a point x, which is defined to be

$$\Gamma x = \{\gamma x : \ \gamma \in \Gamma\}.$$

Lemma 1 *Let $f : \mathbf{R}^2 \to \mathbf{R}^2$ be Γ-equivariant and let p be a fixed point of f. Then all points in the group orbit of p are fixed points of f.*

Proof If $f(p) = p$ it follows that $f(\gamma p) = \gamma f(p) = \gamma p$, showing that γp is a fixed point of f for all $\gamma \in \Gamma$.

The relation between the group action and the Jacobian matrix of an equivariant map f is obtained through the following

Lemma 2 *Let $f : V \to V$ be a Γ-equivariant map differentiable at the origin. Then $Df(0)$, the Jacobian of f at the origin, commutes with Γ.*

Proof Since f is Γ-equivariant we have $f(\gamma.v) = \gamma f(v)$ for all $\gamma \in \Gamma$ and $v \in V$. Differentiating both sides of the equality with respect to v, we obtain $Df(\gamma.v)\gamma = \gamma Df(v)$ and, evaluating at the origin gives $Df(0)\gamma = \gamma Df(0)$.

3 Symmetries in the Plane

In this section, we describe the consequences for the local dynamics arising from the fact that a map is equivariant under the action of a compact Lie group Γ. These are patent in the structure of the Jacobian matrix at the origin, obtained using Lemma 2.

Since every compact Lie group in $GL(2)$ can be identified with a subgroup of the orthogonal group $O(2)$, we need only be concerned with the groups we list below.

Compact Subgroups of $O(2)$

- $O(2)$, acting on $\mathbf{R}^2 \simeq \mathbb{C}$ as the group generated by θ and κ given by

$$\theta.z = e^{i\theta}z, \quad \theta \in S^1 \qquad \text{and} \qquad \kappa.z = \bar{z}.$$

- $SO(2)$, acting on $\mathbf{R}^2 \simeq \mathbb{C}$ as the group generated by θ given by

$$\theta.z = e^{i\theta}z, \quad \theta \in S^1.$$

- D_n, $n \geq 2$, acting on $\mathbf{R}^2 \simeq \mathbb{C}$ as the finite group generated by ζ and κ given by

$$\zeta.z = e^{\frac{2\pi i}{n}}z \qquad \text{and} \qquad \kappa.z = \bar{z}.$$

- $\mathbf{Z}_n$, $n \geq 2$, acting on $\mathbf{R}^2 \simeq \mathbb{C}$ as the finite group generated by ζ given by

$$\zeta.z = e^{\frac{2\pi i}{n}}z.$$

- $\mathbf{Z}_2(\langle\kappa\rangle)$, acting on $\mathbf{R}^2$ as

$$\kappa.(x, y) = (x, -y).$$

Since most of our results depend on the existence of a unique fixed point for f, it is worthwhile noting that the group actions we are concerned with are such that $\mathrm{Fix}(\Gamma) = \{0\}$. Therefore, if f is Γ-equivariant then $f(0) = 0$.

If the representation is absolutely irreducible, we know that $Df(0)$ is a multiple of the identity and thus it has one real eigenvalue of geometric multiplicity two. Therefore, the origin is locally either an attractor or a repellor. We have the following

Lemma 3 *The standard representation on* $\mathbf{R}^2$ *is absolutely irreducible for* $O(2)$ *and* D_n *with* $n \geq 3$ *and for no other subgroup of* $O(2)$.

Proof The proof follows by direct computation.

- $O(2)$: the generators of this group are θ and κ and it suffices to find the linear matrices that commute with both. A real matrix

$$\begin{pmatrix} a & b \\ c & d \end{pmatrix}$$

commutes with κ if and only if $b = c = 0$. In order for such a matrix to commute with any rotation it must be

$$\begin{pmatrix} a & 0 \\ 0 & d \end{pmatrix} \begin{pmatrix} \cos\theta & -\sin\theta \\ \sin\theta & \cos\theta \end{pmatrix} = \begin{pmatrix} \cos\theta & -\sin\theta \\ \sin\theta & \cos\theta \end{pmatrix} \begin{pmatrix} a & 0 \\ 0 & d \end{pmatrix}$$

which holds when $a = d$ or $\sin\theta = 0$ for all $\theta \in S^1$. Therefore, the action of $O(2)$ is absolutely irreducible.
- $SO(2)$: the elements of $SO(2)$ are rotation matrices which commute with any other rotation matrix, also non-diagonal ones.
- D_n, $n \geq 3$: see the proof for $O(2)$. In the last step, we must have $a = d$ or $\sin 2\pi/n = 0$ which is never satisfied for $n \geq 3$. Hence, the action is absolutely irreducible.
- $\mathbf{Z}_n$, $n \geq 3$: as for $SO(2)$, any rotation matrix commutes with the rotation of $2\pi/n$, including non-diagonal ones.
- $\mathbf{Z}_2(\langle\kappa\rangle)$: see the proof for $\kappa \in O(2)$ to conclude that linear commuting matrices are diagonal but not necessarily linear multiples of the identity.
- $\mathbf{Z}_2$: all linear maps commute with $-Id$.
- $D_2 = \mathbf{Z}_2 \oplus \mathbf{Z}_2(\langle\kappa\rangle)$: as above, $\mathbf{Z}_2$ introduces no restrictions and for commuting with κ it suffices that the map is diagonal.

The following result is then a straightforward consequence of the previous proof.

Lemma 4 *The linear maps that commute with the standard representations of the subgroups of* $O(2)$ *are rotations and homotheties (and their compositions) for* $SO(2)$ *and* $\mathbf{Z}_n$, $n \geq 3$, *linear multiples of the identity for* $O(2)$ *and* D_n, $n \geq 3$, *any linear map for* $\mathbf{Z}_2$ *and maps represented by diagonal matrices for the remaining groups.*

Proof The only linear maps that were not already explicitly calculated in the previous proof are those that commute with rotations. We have

$$\begin{pmatrix} a & b \\ c & d \end{pmatrix} \begin{pmatrix} \cos\theta & -\sin\theta \\ \sin\theta & \cos\theta \end{pmatrix} = \begin{pmatrix} \cos\theta & -\sin\theta \\ \sin\theta & \cos\theta \end{pmatrix} \begin{pmatrix} a & b \\ c & d \end{pmatrix}$$

if and only if either $\sin\theta = 0$ for all $\theta \in S^1$ or $a = d$ and $b = -c$. Hence, the only maps commuting with either $SO(2)$ or $\mathbf{Z}_n$, $n \geq 3$, are rotations and homotheties and their compositions.

With the results obtained so far, we are able to describe the Jacobian matrix at the origin for maps equivariant under each of the groups above.

Proposition 2 (Proposition 2.3 in [4]) *Let f be a planar map differentiable at the origin. The admissible forms for the Jacobian matrix of f at the origin are those given in Table 1 depending on the symmetry group of f.*

Furthermore, the symmetry constrains the normal form as described in [3, Theorem 2.1] and in the next result, and its consequences for the linear part of f appear in Table 1.

Proposition 3 (Proposition 3.1 in [3]) *Let Γ be a compact Lie group acting on $\mathbf{R}^2$. Assume Γ is the symmetry group of a polynomial map f.*

(i) If $\kappa \in \Gamma$, then f does not answer the DMYQ(2) in the affirmative unless f is of the form:

$$f(x, y) = \begin{pmatrix} d_1 & 0 \\ 0 & d_2 \end{pmatrix} \begin{pmatrix} x \\ y \end{pmatrix} + y^2 p(y^2) \begin{pmatrix} 1 \\ 0 \end{pmatrix} .$$

Table 1 Compact subgroups of $O(2)$ and the admissible forms of the Jacobian at the origin of maps equivariant under the standard action of each group. If in addition the Jacobian at the origin is hyperbolic, then this determines the local stability

Symmetry group	$Df(0)$	Hyperbolic local dynamics
$O(2)$	$\begin{pmatrix} \alpha & 0 \\ 0 & \alpha \end{pmatrix}$ $\alpha \in \mathbf{R}$	Attractor/repellor
$SO(2)$	$\begin{pmatrix} \alpha & -\beta \\ \beta & \alpha \end{pmatrix}$ $\alpha, \beta \in \mathbf{R}$	Attractor/repellor
D_n, $n \geq 3$	$\begin{pmatrix} \alpha & 0 \\ 0 & \alpha \end{pmatrix}$ $\alpha \in \mathbf{R}$	Attractor/repellor
$\mathbf{Z}_n$, $n \geq 3$	$\begin{pmatrix} \alpha & -\beta \\ \beta & \alpha \end{pmatrix}$ $\alpha, \beta \in \mathbf{R}$	Attractor/repellor
$\mathbf{Z}_2$	Any matrix	Saddle/attractor/repellor
$\mathbf{Z}_2(\langle\kappa\rangle)$	$\begin{pmatrix} \alpha & 0 \\ 0 & \beta \end{pmatrix}$ $\alpha, \beta \in \mathbf{R}$	Saddle/attractor/repellor
$D_2 = \mathbf{Z}_2 \oplus \mathbf{Z}_2(\langle\kappa\rangle)$	$\begin{pmatrix} \alpha & 0 \\ 0 & \beta \end{pmatrix}$ $\alpha, \beta \in \mathbf{R}$	Saddle/attractor/repellor

(ii) If there is an element $\zeta \in \Gamma$ of order $n \geq 3$, then f does not answer the DMYQ(2) in the affirmative unless f is linear. Moreover, the linear part of f is either a homothety or a rotation matrix.

4 Dynamics: Local to Global

Figure 1 illustrates the dynamics near the origin of equivariant maps for several symmetry groups. A common feature of Fig. 1a–d is the existence of at least one symmetry axis. This axis is the subspace fixed by a reflection and hence it is invariant under the dynamics. Such a fixed-point subspace naturally contains an invariant ray (see [4, Lemma 3.3]). This allows us to use Proposition 1 to obtain the following results:

Proposition 4 (Proposition 3.4 in [4]) *Let $f \in Emb(\mathbf{R}^2)$ have symmetry group Γ with $\kappa \in \Gamma$, such that $Fix(f) = \{0\}$. Suppose one of the following holds:*

(a) $f \in Emb^+(\mathbf{R}^2)$ and f does not interchange connected components of $\mathbf{R}^2 \setminus Fix\langle\kappa\rangle$.
(b) $Fix(f^2) = \{0\}$.

Then for each $p \in \mathbf{R}^2$ either $\omega(p) = \{0\}$ or $\omega(p) = \infty$.

The next example shows that assumption (b) in Proposition 4 is necessary in the case where f interchanges connected components of $\mathbf{R}^2 \setminus \mathrm{Fix}\langle\kappa\rangle$.

Example Consider the map $f : \mathbf{R}^2 \to \mathbf{R}^2$ defined by

$$f(x,y) = \left(-ax^3 + (a-1)x, -\frac{y}{2}\right) \quad 0 < a < 1.$$

It is easily checked that f has symmetry group D_2 and verifies (see Fig. 2):

1. $f \in \mathrm{Emb}^+(\mathbf{R}^2)$ is an orientation-preserving diffeomorphism.
2. $\mathrm{Spec}(f) \cap [0, \infty) = \emptyset$.
3. 0 is a local hyperbolic attractor.
4. $\mathrm{Fix}(f^2) \neq \{0\}$.

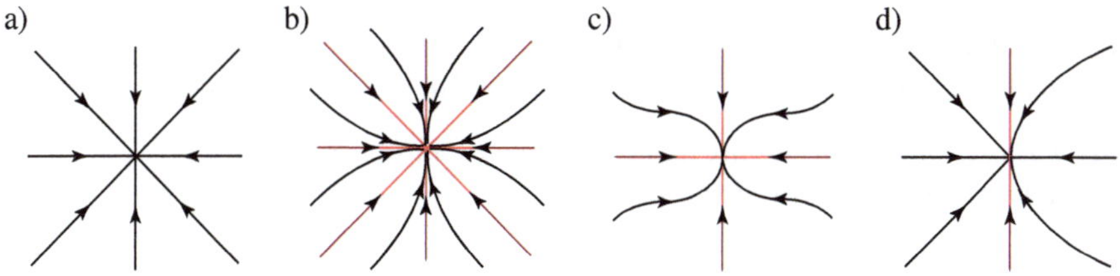

Fig. 1 Local/global attractor with symmetry: (**a**) $O(2)$; (**b**) D_4 (without symmetries $\mathbf{Z}_8$ or $SO(2)$); (**c**) D_2 (without symmetry D_4); (**d**) $\mathbf{Z}_2\langle\kappa\rangle$ (without symmetry D_4)

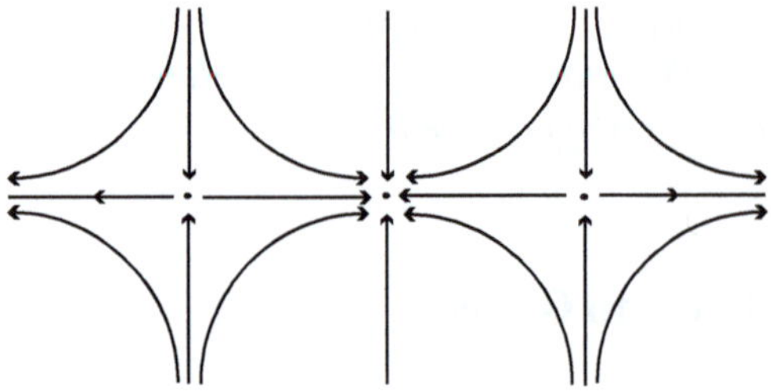

Fig. 2 A local attractor which is not a global attractor due to the existence of periodic orbits

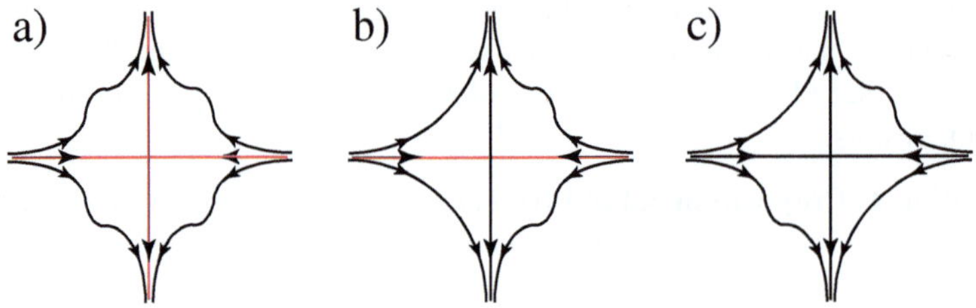

Fig. 3 Local/global saddle with symmetry: (**a**) D_2; (**b**) $\mathbf{Z}_2(\kappa)$; (**c**) $\mathbf{Z}_2$

Theorem 1 (Theorem 3.5 in [4]) *Let* $f \in Emb(\mathbf{R}^2)$ *be dissipative with symmetry group* Γ *with* $\kappa \in \Gamma$ *such that* $Fix(f) = \{0\}$. *Suppose in addition that one of the following holds:*

(a) $f \in Emb^+(\mathbf{R}^2)$ *and* f *does not interchange connected components of* $\mathbf{R}^2 \setminus Fix\langle\kappa\rangle$.
(b) *There exist no* 2*-periodic orbits.*

Then 0 *is a global attractor.*

Corollary 2 (Corollary 3.6 in [4]) *Suppose the assumptions of Theorem 1 are verified and* f *is differentiable at* 0. *If every eigenvalue of* $Df(0)$ *has norm strictly less than one, then* 0 *is a global asymptotic attractor.*

For analogous results concerning a repellor see [4].

For the groups $\mathbf{Z}_2$, $\mathbf{Z}_2(\langle\kappa\rangle)$ and $D_2 = \mathbf{Z}_2 \oplus \mathbf{Z}_2(\langle\kappa\rangle)$ the origin may also be a saddle as illustrated in Fig. 3. For D_2, we have:

Proposition 5 ([5]) *Let* $f \in Hom(\mathbf{R}^2)$ *be a* C^1*-homeomorphism with symmetry group* D_2 *such that* $Fix(f) = \{0\}$. *Suppose also that one of the following holds:*

(a) f *is orientation preserving and* 0 *is a direct saddle.*
(b) *There exist no* 2*-periodic orbits.*

Then if 0 *is a local saddle, then* 0 *is a global saddle.*

In order to obtain a global saddle for f with symmetry group either $\mathbf{Z}_2$ or $\mathbf{Z}_2(\langle\kappa\rangle)$, we need the additional assumption that f is a diffeomorphism, see [5].

5 Strictly Local Dynamics

Figure 4 shows the local dynamics for maps equivariant under the action of groups that do not contain a reflection. These are $SO(2)$ and $\mathbf{Z}_n$. For these groups, local dynamics of attractor/repellor type does not necessarily extend to global dynamics, as we proceed to indicate.

We use examples referring to a local attractor. Examples with a local repellor may be obtained considering f^{-1}.

The dynamics of an $SO(2)$-symmetric embedding is mostly determined by its radial component, as can be seen by writing f in polar coordinates as $f(\rho, \theta) = (R(\rho, \theta), T(\rho, \theta))$. It is easily shown that since f is $SO(2)$-equivariant, the radial component $R(\rho, \theta)$ only depends on ρ and $R \in Emb(\mathbf{R}^+)$. The fixed points of the radial component are invariant circles for f hence knowledge about local dynamics does not contribute to the description of global dynamics unless $\text{Fix}(R) = \{0\}$.

The next two theorems show how a local attractor may be prevented from being a global attractor in a $\mathbf{Z}_n$-equivariant problem. Thus the examples in Fig. 4a, b may or may not extend to the whole plane.

Theorem 2 (Theorem 3.1 in [2]) *For each $n \geqslant 2$ there exists $f : \mathbf{R}^2 \to \mathbf{R}^2$ such that:*

(a) f is a differentiable homeomorphism;
(b) f has symmetry group $\mathbf{Z}_n$;
(c) $Fix(f) = \{0\}$;
(d) The origin is a local attractor;
(e) There exists a periodic orbit of minimal period n.

The idea of the proof is to start with a $\mathbf{Z}_4$-equivariant example due to Szlenk (see [9]), sharing the same properties. Each quadrant of the plane is invariant under the map f_4 of this example. We deform the first quadrant into a sector of the plane, of angle $2\pi/n$ and then use the $\mathbf{Z}_n$ symmetry to cover the rest of the plane, as illustrated in Fig. 5. The main difficulty is to prove that the result is a differentiable homeomorphism.

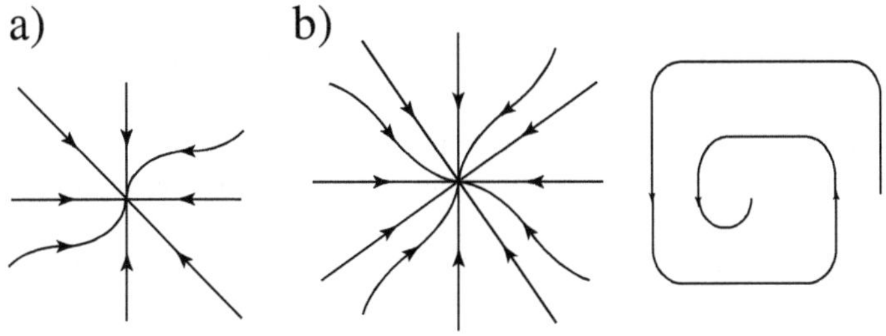

Fig. 4 Local attractor with symmetry: (**a**) $\mathbf{Z}_2$; (**b**) $\mathbf{Z}_4$

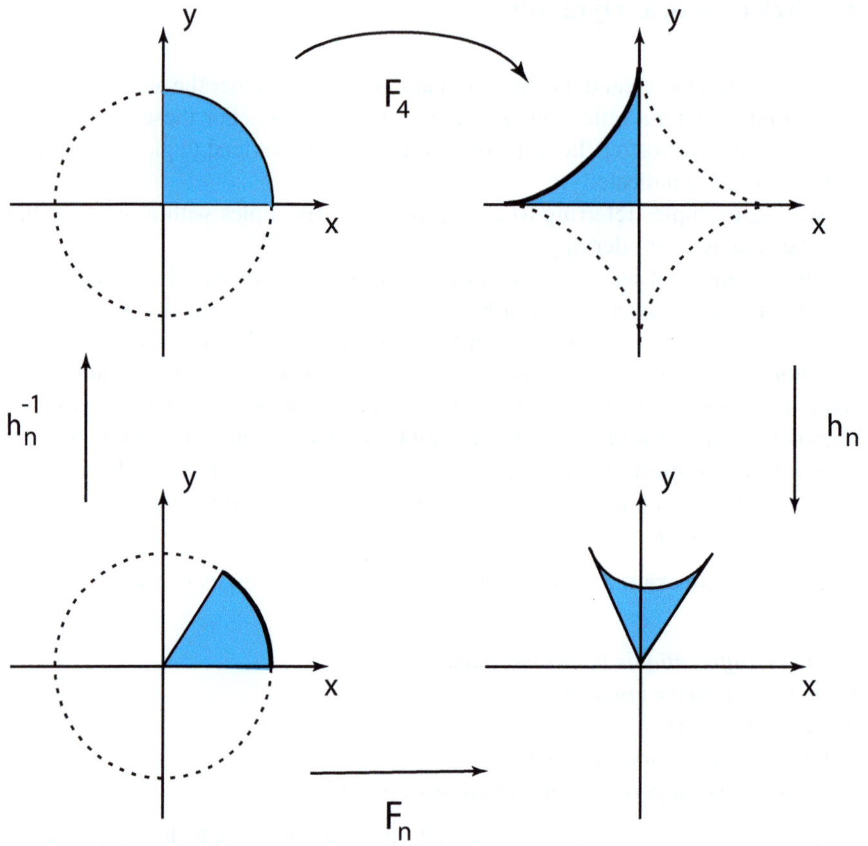

Fig. 5 Construction of the $\mathbf{Z}_n$-equivariant example F_n in a fundamental domain of the $\mathbf{Z}_n$-action, shown here for $n = 6$

The $\mathbf{Z}_m$-equivariant homeomorphisms constructed in Theorem 2 have rotation number $1/m$. So we might be led to think that the presence of the $\mathbf{Z}_m$-symmetry implies that the rotation number of the homeomorphism should be rational. One consequence would be that the asymptotically stable fixed point is a global attractor if and only if there are no periodic points different from the fixed point.

The next result shows that this is false. We prove in [1] the existence of $\mathbf{Z}_m$-equivariant and dissipative homeomorphisms in the plane with an asymptotically stable fixed point such that the induced map in the space of prime ends is conjugated to a Denjoy map, which is also $\mathbf{Z}_m$-equivariant. The idea is to construct $\mathbf{Z}_m$-equivariant Denjoy maps in the circle and then, in the context of symmetry, to reproduce the construction used to prove the following:

Proposition 6 (Proposition 2 in [11]) *Given $w \in (0, 1) \setminus \mathbf{Q}$ and a Denjoy map ϕ, there exists an admissible map f with rotation number $\bar{w}$ and such that f^* is topologically conjugated to ϕ.*

Observe that two admissible homeomorphisms f_1, f_2 with the same basin of attraction U verify that

$$(f_1 \circ f_2)^* = f_1^* \circ f_2^*.$$

Let f be an admissible homeomorphisms with basin of attraction U. Suppose f is Z_m-equivariant and U is also invariant by $R_{\frac{1}{m}}$. Hence, the following holds:

$$f^* \circ R^*_{\frac{1}{m}} = R^*_{\frac{1}{m}} \circ f^*.$$

Since $R^*_{\frac{1}{m}}$ is a periodic homeomorphism of $\mathbf{T}$ with rotation number $1/m$, then $R^*_{\frac{1}{m}}$ is conjugated to the linear rotation $R_{\frac{1}{m}}$ and f^* is said to be $\mathbf{Z}_m$-*equivariant in the space of prime ends.*

Theorem 3 (Theorem 4.2 in [1]) *Given an irrational number $\tau \notin \mathbf{Q}$, there exists a $\mathbf{Z}_m$−equivariant and admissible homeomorphism in $\mathbf{R}^2$ with rotation number $\bar{\tau} \in \mathbf{T}$ that induces a Denjoy map in the circle of prime ends which is also $\mathbf{Z}_m$−equivariant.*

Hence, [1] shows that for $\mathbf{Z}_m$-equivariant homeomorphisms one cannot guarantee that the rotation number is rational and proves the existence of $\mathbf{Z}_m$-equivariant homeomorphisms with some complicated and interesting dynamical features.

Acknowledgements The research of all authors at Centro de Matemática da Universidade do Porto (CMUP) had financial support from the European Regional Development Fund through the programme COMPETE and from the Portuguese Government through the Fundação para a Ciência e a Tecnologia (FCT) under the project PEst-C/MAT/UI0144/2011. B. Alarcón was also supported by grant MICINN-12-MTM2011-22956 of the Ministerio de Ciencia e Innovación (Spain).

References

1. Alarcón, B.: Rotation numbers for planar attractors of equivariant homeomorphisms. Topological Methods Nonlinear Anal. **42**(2), 327–343 (2013)
2. Alarcón, B., Castro, S.B.S.D., Labouriau, I.S.: A local but not global attractor for a $\mathbf{Z}_n$-symmetric map. J. Singularities **6**, 1–14 (2012)
3. Alarcón, B., Castro, S.B.S.D., Labouriau, I.S.: The discrete Markus-Yamabe problem for symmetric planar polynomial maps. Indag. Math. **23**, 603–608 (2012)
4. Alarcón, B., Castro, S.B.S.D., Labouriau, I.S.: Global dynamics for symmetric planar maps. Discrete Contin. Dyn. Syst. A **37**, 2241–2251 (2013)
5. Alarcón, B., Castro, S.B.S.D., Labouriau, I.S.: Global saddle for planar diffeomorphisms (in preparation)
6. Alarcón, B., Guíñez, V., Gutierrez, C.: Planar embeddings with a globally attracting fixed point. Nonlinear Anal. **69**(1), 140–150 (2008)

7. Bhatia, N.P., Szegö, G.P.: Stability Theory of Dynamical Systems. Springer, New York (2002)
8. Cima, A., van den Essen, A., Gasull, A., Hubbers, E., Mañosas, F.: A polynomial couterexample to the Markus-Yamabe conjecture. Adv. Math. **131**, 453–457 (1997)
9. Cima, A., Gasull, A., Mañosas, F.: The discrete Markus-Yamabe problem. Nonlinear Anal. **35**, 343–354 (1999)
10. Golubitsky, M., Stewart, I., Schaeffer, D.G.: Singularities and Groups in Bifurcation Theory, Vol. 2. Applied Mathematical Sciences, vol. 69. Springer, New York (1985)
11. Hernández-Corbato, L., Ortega, R., Ruiz del Portal, F.R.: Attractors with irrational rotation number. Math. Proc. Camb. Philos. Soc. **153**, 59–77 (201)
12. Ortega, R.: Topology of the plane and periodic differential equations. http://www.ugr.es/~ecuadif/fuentenueva.htm (2008)
13. Pommerenke, Ch.: Boundary Behaviour of Conformal Maps. Springer, Heiderberg (1992)
14. van den Essen, A., Hubbers, E.: A new class of invertible polynomial maps. J. Algebra **187**, 214–226 (1997)

Decision Analysis in a Model of Sports Pricing Under Uncertain Demand

Alberto A. Álvarez-López and Inmaculada Rodríguez-Puerta

Abstract We consider a model, due to Andersen and Nielsen (Econ Lett 118(2):262–264, 2013), concerning the behavior of a risk-averse sports team under uncertainty in demand: the team chooses a value for the price of its ticket, but the ticket demand is stochastic at the moment of decision. For this model, we carry out a decision analysis by studying several comparative-static effects not considered by the authors in their paper. Specifically, we examine the effect of changes in the team's risk aversion, and also the effect of a variation in the risk of the random demand. Furthermore, we enhance the model by considering a proportional profit tax, and we study the effect of a variation in the tax rate. We derive some conditions under which the sports team finds optimal to reduce the ticket price as a consequence of a rise in the tax rate.

1 Introduction

In [2], Andersen and Nielsen present a relevant static model of decision for a risk-averse sports team under uncertain demand. The team chooses a price for its ticket (for a particular match), but the decision must be made facing a stochastic demand for tickets. They find that the team will price in the inelastic part of the demand curve, according to empirical evidence coming from the sports economics literature (see the references in [2]). The authors also consider a comparative-static effect: the influence on the optimal ticket price of a variation in fixed costs.

A.A. Álvarez-López (✉)
Department of Quantitative Applied Economics II, UNED, Paseo Senda del Rey, 11, Madrid 28040, Spain
e-mail: aalvarez@cee.uned.es

I. Rodríguez-Puerta
Department of Economics, Quantitative Methods and Economic History, Pablo de Olavide University, Carretera de Utrera Km. 1, Sevilla 41013, Spain
e-mail: irodpue@upo.es

J.-P. Bourguignon et al. (eds.), *Dynamics, Games and Science*, CIM Series in Mathematical Sciences 1, DOI 10.1007/978-3-319-16118-1_3

With this model as the starting point, we carry out a decision analysis for the sports team. To this end, we will examine the influence on the optimal ticket price of two kinds of variation: in the team's risk aversion, and in the risk of the random demand. In addition, we will enhance the model with the consideration of a proportional profit tax, and we will study the effect of changes in its rate.

In Sect. 2, we formulate the model and derive its main result. The presentation is the same as that given by Andersen and Nielsen in their paper, except for some relevant methodological details. Namely, in this section we start making use of a lemma (Lemma 2) that allows us to obtain useful bounds for the expectation of products of random variables. This lemma is uncommon in the literature, but applying it simplifies and enhances the usual Sandmo's methodology (see [5]).

With Sect. 3, we properly start the decision analysis for the sports team by studying the effect of a change in the team's risk aversion. As a straightforward application of it, we also study the effect of a variation in fixed costs.[1]

Section 4 focuses on the effect of a variation in the risk of the random demand, expressed as a variation in its standard deviation.

In Sect. 5, the proportional profit tax is included in the model, and the effect of a variation in its rate is examined.

Finally, Sect. 6 presents some concluding remarks.

Throughout this paper, we shall denote the Arrow–Pratt measures of risk aversion by r_u (absolute) and R_u (relative) for a firm with a Bernoulli utility function u. In addition, we will make use of the acronyms CARA, DARA, IARA, CRRA, etc., in the usual manner: to stand for *c*onstant, *d*ecreasing or *i*ncreasing, *a*bsolute or *r*elative, *r*isk *a*version.

2 The Model

We consider a sports team that faces an uncertain demand. The team must fix a ticket price. Given a ticket price $p \geqslant 0$, the *actual* demand is postulated in the following form:

$$x(p,\varepsilon) = f(p)\varepsilon + \alpha(\varepsilon - 1), \tag{1}$$

where $f(p) > 0$ is the theoretical non-random demand for the price p, $\varepsilon > 0$ is a non-degenerate random variable with $\mathsf{E}[\varepsilon] = 1$, and $\alpha > 0$ is a (non-random)

[1]The effect of fixed costs is examined in [2] only for the case of a team that exhibits decreasing absolute risk aversion.

number.[2] Ex post profits are given by

$$\pi(p, \varepsilon) = pf(p)\varepsilon + p\alpha(\varepsilon - 1) - B,$$

where $B > 0$ are fixed costs.[3]

The sports team's attitude towards risk is modeled by a Bernoulli utility function u, sufficiently regular (at least of class $\mathscr{C}^2$ on $\mathbb{R}$) and such that $u' > 0$ and $u'' < 0$. In particular, the team is risk averse. The expected utility of choosing the price p is given by $U(p) \equiv \mathsf{E}\big[u\big(\pi(p, \varepsilon)\big)\big]$. The team maximizes this function U on the interval $[0, +\infty)$:

$$\max_{p \in [0, +\infty)} U(p). \tag{2}$$

Writing $M(p) \equiv f(p) + pf'(p)$, the first- and second-order derivatives of U take the form:

$$U'(p) = \mathsf{E}\big[u'(\pi)\big(M(p)\varepsilon + \alpha(\varepsilon - 1)\big)\big],$$
$$U''(p) = \mathsf{E}\big[u''(\pi)\big(M(p)\varepsilon + \alpha(\varepsilon - 1)\big)^2 + u'(\pi)M'(p)\varepsilon\big].$$

We assume that $U''(p) < 0$ for each $p \in [0, +\infty)$,[4] and thus the function U is strictly concave on $[0, +\infty)$. Furthermore, we have that $U'(0) = u'(-B)f(0) > 0$, and thereby there is no corner solution. That is, if the maximization problem (2) admits a solution, this solution is positive and unique, and is characterized by the first-order condition $U'(p) = 0$.

Before studying the properties of the problem (2), let us consider the sports team's problem in absence of uncertainty, that is, as if the demand were certain and solely described by f. This problem is exactly that of maximizing expected profits, which takes the form

$$\max_{p \in [0, +\infty)} pf(p) - B. \tag{3}$$

The objective function of this maximization problem is simply the team's total income, and its first derivative is the marginal revenue: $M(p) = f(p) + pf'(p)$. From now on, we assume that the maximization problem (3) has a unique solution p_0. We have that $p_0 > 0$ (since $M(0) = f(0) > 0$), and $M(p_0) = 0$ (first-order condition).

[2] According to [2, p. 262], the number α "determines the strength of the additive shift relative to the multiplicative shift." On the other hand, it is assumed that f is a function of class $\mathscr{C}^2$ on $[0, +\infty)$.

[3] In [2], the fixed costs are denoted by F. The marginal costs are assumed to be null (this is usual in the literature on Sports Economics, as pointed out in [2]). In addition, capacity restrictions of the team's stadium are not considered.

[4] A sufficient condition for $U'' < 0$ is the following: $f' < 0$ and $f'' \leqslant 0$.

Notice that, at a price for which the marginal revenue is null (as it is for p_0), the elasticity of demand equals -1.

Coming back to the problem (2), the following lemma gives us the sign of one of the terms in the expression of $U'(p)$.

Lemma 1 *For all* $p \in (0, +\infty)$, *we have that* $\mathsf{E}\big[u'\big(\pi(p,\varepsilon)\big)\alpha(\varepsilon - 1)\big] < 0$.

Proof Fix $p > 0$. We apply Lemma 2 (see Appendix) with the random variable $X = \varepsilon - 1$, and the real functions $\psi \equiv 1$ and $\phi(s) = u'\big(pf(p)(s+1) + p\alpha s - B\big)$: since $u'' < 0$ and $pf(p) + p\alpha > 0$, the function ϕ is strictly decreasing, and we obtain:

$$\mathsf{E}\big[u'(\pi)(\varepsilon - 1)\big] < u'(-B) \cdot \mathsf{E}[\varepsilon - 1] = 0,$$

and hence the result is proven (recall that $\alpha > 0$). □

The following proposition now establishes that the maximization problem (2) has indeed a solution, and compares it to that of the problem (3).

Proposition 1 *There exists a price* $p^* > 0$ *such that* p^* *is the (unique) solution of the maximization problem* (2). *Furthermore:* $p^* < p_0$.

Proof We know that $M(p_0) = 0$, and hence $U'(p_0) = \mathsf{E}\big[u'\big(\pi(p_0,\varepsilon)\big)\alpha(\varepsilon - 1)\big]$. According to Lemma 1, we have that $U'(p_0) < 0$. As $U'(0) > 0$, there exists a point $p^* \in (0, p_0)$ such that $U'(p^*) = 0$. This point $p^* > 0$ is the unique solution of the problem (2), and $p^* < p_0$. □

That is: *on facing uncertain demand, the risk-averse sports team reduces the optimal price of its tickets.*

Now, as a result of the equality $U'(p^*) = 0$ and Lemma 1, we deduce:

$$\mathsf{E}\big[u'(\pi^*)M(p^*)\varepsilon\big] = -\,\mathsf{E}\big[u'(\pi^*)\alpha(\varepsilon - 1)\big] > 0,$$

where $\pi^* \equiv \pi(p^*, \varepsilon)$. Thus $M(p^*) > 0$. This means that *the sports team prices in the inelastic part of the demand curve.*

These two results correspond to those presented by Andersen and Nielsen in [2].

3 Effect of a Change in Risk Aversion

In this section, we wish to study the effect on the optimal ticket price of a change in the team's risk aversion. To this end, we consider a new Bernoulli utility function v in the same conditions as u, and define $V(p) \equiv \mathsf{E}\big[v\big(\pi(p,\varepsilon)\big)\big]$. We assume that the absolute risk aversion for v is strictly greater than that for u, in the sense that $r_v > r_u$ (that is: $r_v(s) > r_u(s)$ for all s). Also, we denote by p_u^* and p_v^* the corresponding optimal prices.

The following proposition holds:

Proposition 2 *Assume that $r_v > r_u$. We have that $p_v^* < p_u^*$.*

Proof Fix $p > 0$ such that $M(p) > 0$. We can write:

$$U'(p) = \mathsf{E}\big[u'(\pi)\big(M(p)\varepsilon + \alpha(\varepsilon - 1)\big)\big] = \mathsf{E}\Big[v'(\pi)\frac{u'(\pi)}{v'(\pi)}\big(M(p)\varepsilon + \alpha(\varepsilon - 1)\big)\Big].$$

Now, consider the real functions

$$\xi(s) = u'\Big(p\big(f(p) + \alpha\big)\frac{s+\alpha}{M(p)+\alpha} - p\alpha - B\Big),$$
$$\psi(s) = v'\Big(p\big(f(p) + \alpha\big)\frac{s+\alpha}{M(p)+\alpha} - p\alpha - B\Big),$$

and set $\phi = \xi/\psi$. Thus $\psi > 0$; also, as $(u'/v')' = (u'/v')(r_v - r_u) > 0$ (recall the hypothesis), the function ϕ results to be strictly increasing. By applying Lemma 2 (see Appendix) to these functions and the random variable $X = M(p)\varepsilon + \alpha(\varepsilon - 1)$, we obtain:

$$U'(p) = \mathsf{E}\Big[v'(\pi)\frac{u'(\pi)}{v'(\pi)}\big(M(p)\varepsilon + \alpha(\varepsilon - 1)\big)\Big] > \phi(0)\,\mathsf{E}\big[v'(\pi)\big(M(p)\varepsilon + \alpha(\varepsilon - 1)\big)\big].$$

Writing this inequality at $p = p_v^*$ yields:

$$U'(p_v^*) > \phi(0)\,\mathsf{E}\big[v'\big(\pi(p_v^*, \varepsilon)\big)\big(M(p_v^*)\varepsilon + \alpha(\varepsilon - 1)\big)\big] = 0,$$

according to the first-order condition for p_v^* with the utility function V. That is: $U'(p_v^*) > 0 = U'(p_u^*)$. Since the function U' is strictly decreasing, it follows that $p_v^* < p_u^*$. □

Thus, *if risk aversion increases, the optimal price for the ticket decreases.*

An Application: Effect of a Variation in Fixed Costs The effect on the optimal ticket price of an increase in fixed costs can be readily studied as a consequence of Proposition 2. To see this, assume that fixed costs rise from its original value B to a new value B', and set $\beta = B' - B > 0$. We can consider the utility $v(t) = u(t + \beta)$; depending on whether the team exhibits DARA, CARA, or IARA, we have: $r_u \geqslant r_v$, $r_u = r_v$, or $r_u \leqslant r_v$, respectively. With an easy generalization of Proposition 2, we obtain the following conclusion: *an increase in fixed costs reduces, leaves constant, or increases, the optimal ticket price depending on whether the sports team exhibits DARA, CARA, or IARA, respectively.*

A reasonably intuitive explanation of this behavior can be given. Assume the IARA case, for instance. The team gets poorer when fixed costs increase. As the team exhibits IARA, its risk aversion decreases. As a result, the team will price at a higher value.[5]

4 Variations in the Risk of the Random Demand

A variation in the risk of a random variable can be formulated under several approaches. For instance, in terms of stochastic dominance (usually, first- or second-order stochastic dominance). Here we will focus on the effect of a variation in the standard deviation. To this end, set: $\varepsilon = 1 + \sigma\delta$, where $\sigma > 0$ is a number and δ is a random variable with $\mathsf{E}[\delta] = 0$ and $\mathsf{Var}[\delta] = 1$, so that $\mathsf{Var}[\varepsilon] = \sigma^2$. We wish to examine the effect on the optimal ticket price p^* of a variation in the standard deviation σ.[6]

We denote by $\mathrm{d}p^*/\mathrm{d}\sigma$ the corresponding comparative-static effect of σ on p^*. The following proposition establishes a sufficient condition for this effect to be negative.

Proposition 3 *If the function u'' is decreasing, then* $\mathrm{d}p^*/\mathrm{d}\sigma < 0$.

Proof According to the Implicit Function Theorem applied to the first-order condition $U'(p^*) = 0$, we have that $\mathrm{d}p^*/\mathrm{d}\sigma$ equals:

$$-\frac{\mathsf{E}\big[u''(\pi^*)p\big(f(p)+\alpha\big)\delta\big(M(p)+M(p)\sigma\delta+\alpha\sigma\delta\big)+u'(\pi^*)\big(M(p)+\alpha\big)\delta\big]}{U''(p^*)}$$

(recall that $\pi^* = \pi(p^*, \varepsilon)$). As $U''(p^*) < 0$, the sign of $\mathrm{d}p^*/\mathrm{d}\sigma$ is the same as that of its numerator. The numerator can be written in the following form:

$$p\big(f(p)+\alpha\big)M(p)\,\mathsf{E}\big[u''(\pi^*)\delta\big]+p\big(f(p)+\alpha\big)\big(M(p)+\alpha\big)\sigma\,\mathsf{E}\big[u''(\pi^*)\delta^2\big] \\ +\big(M(p)+\alpha\big)\,\mathsf{E}\big[u'(\pi^*)\delta\big].$$

Notice that each factor written out of the expectation operator is positive. Now the second addend is negative, and also the third addend according to Lemma 1 (note that $\delta = (\varepsilon - 1)/\sigma$). About the first addend, it is negative or null provided that u'' is decreasing, as a result of applying Lemma 2 taking $\psi \equiv 1$, the decreasing

[5] As pointed out in footnote 1, Andersen and Nielsen also study the effect of fixed costs (for the DARA case) in their paper (see [2, p. 263]). They give a direct proof of the result.

[6] Notice that the standard deviation of the random variable $x(p, \varepsilon)$ (see (1)) is $\big(f(p)+\alpha\big)\sigma$. Thus a variation in σ is indeed equivalent to a variation in the standard deviation of the random demand x.

function $\phi(s) = u''\big(pf(p)(1+\sigma s) + p\alpha\sigma s - B\big)$, and the random variable $X = \delta$:

$$\mathsf{E}\big[u''(\pi^*)\delta\big] \leqslant \phi(0)\,\mathsf{E}[\delta] = 0.$$

Finally, indeed $\mathrm{d}p^*/\mathrm{d}\sigma < 0$. □

5 Consideration of a Proportional Profit Tax

In this section, we consider that there exists a proportional profit tax at a rate τ, with $0 < \tau < 1$. Therefore, after-tax profits are given by

$$\pi_\tau(p,\varepsilon) = (1-\tau)\big(pf(p)\varepsilon + p\alpha(\varepsilon-1) - B\big),$$

and the new utility to be maximized is $U_\tau(p) = \mathsf{E}\big[u\big(\pi_\tau(p,\varepsilon)\big)\big]$. We assume, as before with the no-tax model, that $U''_\tau < 0$, and eventually that the new maximization problem admits a solution p^*_τ that is positive and unique. We would like to study the effect on the optimal ticket price p^*_τ of a variation in the tax rate τ.

Write $\mathrm{d}p^*_\tau/\mathrm{d}\tau$ to stand for the corresponding comparative-static effect of τ on p^*_τ. The following proposition gives us its sign, which is closely related to the relative risk aversion of the sports team.

Proposition 4 *Depending on whether the sports team exhibits DRRA, CRRA, or IRRA, we have that* $\mathrm{d}p^*_\tau/\mathrm{d}\tau \leqslant 0$, $\mathrm{d}p^*_\tau/\mathrm{d}\tau = 0$, *or* $\mathrm{d}p^*_\tau/\mathrm{d}\tau \geqslant 0$, *respectively.*

Proof From the first-order condition $U'_\tau(p^*_\tau) = 0$, we obtain:

$$\frac{\mathrm{d}p^*_\tau}{\mathrm{d}\tau} = -\frac{\mathsf{E}\big[-u''(\pi^*_\tau)\cdot\pi^*_\tau\cdot\big(M(p^*_\tau)\varepsilon + \alpha(\varepsilon-1)\big)\big]}{U''_\tau(p^*_\tau)},$$

where $\pi^*_\tau \equiv \pi_\tau(p^*_\tau,\varepsilon)$. As $U''_\tau(p^*_\tau) < 0$, the sign of $\mathrm{d}p^*_\tau/\mathrm{d}\tau$ is the same as that of its numerator. We can write:

$$\mathsf{E}\big[-u''(\pi^*_\tau)\cdot\pi^*_\tau\cdot\big(M(p^*_\tau)\varepsilon+\alpha(\varepsilon-1)\big)\big] = \mathsf{E}\big[u'(\pi^*_\tau)\,R_u(\pi^*_\tau)\,\big(M(p^*_\tau)\varepsilon+\alpha(\varepsilon-1)\big)\big].$$

Now, consider the following real functions:

$$\psi(s) = u'\left((1-\tau)\left(p^*_\tau\big(f(p^*_\tau)+\alpha\big)\frac{s+\alpha}{M(p^*_\tau)+\alpha} - p^*_\tau\alpha - B\right)\right),$$

$$\phi(s) = R_u\left((1-\tau)\left(p^*_\tau\big(f(p^*_\tau)+\alpha\big)\frac{s+\alpha}{M(p^*_\tau)+\alpha} - p^*_\tau\alpha - B\right)\right).$$

We have that $\psi > 0$, and also that ϕ is decreasing, constant, or increasing, according as the team exhibits DRRA, CRRA, or IRRA, respectively.[7] Furthermore, consider the random variable $X = M(p_\tau^*)\varepsilon + \alpha(\varepsilon - 1)$. In the DRRA case, for instance, from Lemma 2 we can deduce:

$$\mathsf{E}\big[u'(\pi_\tau^*)\, R_u(\pi_\tau^*)\, \big(M(p_\tau^*)\varepsilon + \alpha(\varepsilon - 1)\big)\big] \leqslant \phi(0)\, \mathsf{E}\big[u'(\pi_\tau^*)\big(M(p_\tau^*)\varepsilon + \alpha(\varepsilon - 1)\big)\big] = 0$$

(the last factor is null according to the first-order condition $U'_\tau(p_\tau^*) = 0$); that is: $\mathrm{d}p_\tau^*/\mathrm{d}\tau \leqslant 0$. For the other cases, the final result is obtained, *mutatis mutandis*, with the same proof. □

Thus we have just proven this property: *on facing an increase in the rate of a proportional profit tax, the optimal ticket price decreases, remains constant or increases depending on whether the sports team exhibits DRRA, CRRA or IRRA, respectively.*

The case of a reduction of price motivated by an increase in the tax rate (the DRRA case) is somewhat surprising. We find an intuitively plausible explanation. A rise in the tax rate implies a proportional decrease in profits. Since the team exhibits DRRA, a proportional reduction of profits makes the team more risk averse. This leads to a reduction in the ticket price.

6 Concluding Remarks

We consider a model, due to Andersen and Nielsen (see [2]), for a sports team that has to price its ticket under uncertain demand. For this model, we study some comparative-static effects not examined by the authors in their paper.

Firstly, we present the model and derive the main result obtained by Andersen and Nielsen, namely: the sports team will choose a value for the ticket price lower than the value chosen in absence of uncertainty, and doing this way, the team will price in the inelastic range of the demand curve. Our method of proof is slightly different.

Secondly, we study the effect of a variation in the team's risk aversion. We find that an increase in risk aversion produces a decrease in the optimal value of the ticket price. The analysis in this section easily finds an application on examining the effect of a variation in fixed costs: a rise in fixed costs will reduce, leave constant, or increase, the optimal ticket price depending on whether the team exhibits DARA, CARA, or IARA, respectively.

[7]Notice that $M(p_\tau^*) > 0$, since Lemma 1 remains valid if we write π_τ instead of π (the proof would be, *mutatis mutandis*, the same).

Next, we consider a variation in the risk of the random demand, postulated as a variation in its standard deviation. We find that, provided that the function u'' is decreasing, the optimal ticket price decreases when the standard deviation increases.

Finally, we enhance the model by considering a proportional profit tax. We examine the effect of a variation in the tax rate. We obtain that the optimal price decreases, remains constant, or increases, according as the sports team exhibits DRRA, CRRA, or IRRA, respectively. It is a remarkable fact that reducing the ticket price can be optimal for the sports team facing a rise in a tax rate.

Acknowledgements We are grateful to two referees for their very helpful comments. Álvarez-López would also like to thank the financial support provided by the Spanish Interministerial Commission of Science and Technology (CICYT: Comisión Interministerial de Ciencia y Tecnología), under the Project with the reference number ECO2012-39553-C04-01.

Appendix

The following lemma slightly generalizes a result taken from [3]:

Lemma 2 *Let ψ and ϕ be two real functions defined on $\mathbb{R}$ such that $\psi > 0$ and ϕ is increasing. If $\xi = \psi \cdot \phi$, and X is a real random variable such that the expectation $E[X\,\psi(X)]$ is finite, and such that the probability of the set $\{X \neq 0\}$ is positive, then:*

$$E[X\,\xi(X)] \geqslant \phi(0)\,E[X\,\psi(X)]\,,$$

and the reverse inequality holds when ϕ is decreasing. In addition, if ϕ is strictly increasing or strictly decreasing, the corresponding inequalities also hold strictly.

Proof See [4]: in Lemma 1, write $F \equiv 1$ and $Z = X$. See also [1]. □

References

1. Álvarez-López, A.A., Rodríguez-Puerta, I.: Teoría de la empresa bajo incertidumbre con mercado de futuros: el papel de los costes fijos y de un impuesto sobre los beneficios. Rect@ **10**(1), 253–265 (2009)
2. Andersen, P., Nielsen, M.: Inelastic sports pricing and risk. Econ. Lett. **118**(2), 262–264 (2013)
3. Lippman, S.A., McCall, J.J.: The economics of uncertainty: selected topics and probabilistic methods. In: Arrow, K.J., Intriligator, M.J. (eds.) Handbook of Mathematical Economics, vol. 1, chap. 6. North-Holland, Amsterdam (1982)
4. Rodríguez-Puerta, I., Sebastiá-Costa, F., Álvarez-López, A.A., Buendía, M.: Una herramienta de análisis teórico en la teoría de la empresa bajo incertidumbre. Revista de Métodos Cuantitativos para la Economía y la Empresa **11**, 33–40 (2011)
5. Sandmo, A.: On the theory of the competitive firm under price uncertainty. Am. Econ. Rev. **61**(1), 65–73 (1971)

Growth Diagrams and Non-symmetric Cauchy Identities on NW (SE) Near Staircases

Olga Azenhas and Aram Emami

Abstract The Robinson-Schensted-Knuth (RSK) correspondence is an important combinatorial bijection between two line arrays of positive integers (or non-negative integer matrices) and pairs of semi-standard Young tableaux (SSYTs). One of its applications, in the theory of Schur polynomials, is a bijective proof of the well known Cauchy identity. An interesting analogue of this bijection was given by Mason, where SSYTs are replaced by semi-skyline augmented fillings (SSAFs), originated in the Haglund-Haiman-Loehr formula for non-symmetric Macdonald polynomials. The latter object SSAF has the advantage of detecting the key of a SSYT which is easily read off from the SSAF shape. Using this analogue, we have previously considered the restriction of RSK correspondence to multisets of cells in a (truncated) staircase. The image is described by a Bruhat order inequality between the keys of the recording and the insertion fillings. This has allowed to derive a (truncated) triangular version of the Cauchy identity, due to Lascoux, where Schur polynomials are replaced by key polynomials or Demazure characters. We now consider the restriction of RSK to a near staircase, in French convention, where the top leftmost and the bottom rightmost cells and also possibly some cells in the diagonal layer are deleted. The image is described by additional Bruhat order inequalities, specified by the cells in the diagonal layer. The bijection is then used to extend the triangular version to near staircases, also a version due to Lascoux, where Demazure characters are now under the action of Demazure operators specified by the cells in the diagonal layer. Our analysis is made in the framework of Fomin's growth diagrams where a formulation of the Mason's analogue is given. This is then used to show how to pass from a triangular shape to a near staircase, via the action crystal operators, and how this affects the keys in the image of the RSK.

1 Introduction

The Robinson-Schensted-Knuth (RSK) correspondence [12] is a bijection between biwords (an array of two words), over two finite totally ordered alphabets, and pairs of semi-standard Young tableaux (SSYTs), of the same shape, with entries

O. Azenhas (✉) • A. Emami
CMUC, Department of Mathematics, University of Coimbra, 3001-501 Coimbra, Portugal
e-mail: oazenhas@mat.uc.pt; aramee@mat.uc.pt

J.-P. Bourguignon et al. (eds.), *Dynamics, Games and Science*, CIM Series in Mathematical Sciences 1, DOI 10.1007/978-3-319-16118-1_4

respectively in those alphabets. Let $\mathbb{N}$ denote the set of nonnegative integers, and, as usual, if n is a positive integer, let $[n]$ be the set $\{1,\dots,n\}$. Given a positive integer n, let m and k be fixed positive integers such that $1 \leqslant m \leqslant n$, $1 \leqslant k \leqslant n$ and $m+k \geqslant n+1$. Let $x=(x_1,\dots,x_n)$ and $y=(y_1,\dots,y_n)$ be two sequences of indeterminates. The well-known Cauchy identity [24] expresses the Cauchy kernel $\prod_{i=1}^{k}\prod_{j=1}^{m}(1-x_iy_j)^{-1}$ as a sum of products of Schur polynomials s_λ in $(x_1,x_2,\dots,x_k)$ and $(y_1,y_2,\dots,y_m)$, respectively,

$$\prod_{i=1}^{k}\prod_{j=1}^{m}(1-x_iy_j)^{-1}=\sum_{\lambda}s_\lambda(x_1,\dots,x_k)s_\lambda(y_1,\dots,y_m), \tag{1}$$

over all partitions λ of length $\leqslant \min\{k,m\}$. Schur polynomials in a finite number of indeterminates $(x_1,x_2,\dots,x_k)$ are indexed by partitions λ of length $\leqslant k$. They are combinatorially described by SSYTs of shape λ, over the alphabet $[k]$, see [7, 30, 31],

$$s_\lambda(x_1,\dots,x_k)=\sum_{\substack{T\ SSYT\\ sh(T)=\lambda}}x^T,$$

where $sh(T)$ denotes the shape of the SSYT, T, and $x^T := x_1^{c_1}\cdots x_k^{c_k}$, with c_i the multiplicity of i in T. Thereby, the right hand side of (1) can be written as $\sum_{(P,Q)}x^Py^Q$, where the sum runs over all pairs (P,Q) of SSYTs of the same shape with length $\leqslant \min\{k,m\}$. On the other hand, expanding the product of formal power series, on the left hand side of (1), and identifying each monomial x_iy_j, $i\in[k]$, $j\in[m]$, with the biletter $\binom{j}{i}$, the RSK correspondence, over the finite alphabets $[k]$ and $[m]$, provides a bijective proof for identity (1). Key polynomials or Demazure characters κ_α, with $\alpha\in\mathbb{N}^n$ [3, 19, 28], and Demazure atoms $\hat{\kappa}_\alpha$, with $\alpha\in\mathbb{N}^n$ [19, 26], defined in Sect. 6, both of which form a $\mathbb{Z}$-linear basis for the ring of integer polynomials $\mathbb{Z}[x_1,\dots,x_n]$. When the vectors $\alpha\in\mathbb{N}^n$ are anti-dominant, key polynomials lift the basis of Schur polynomials for the subring of symmetric polynomials $\mathbb{Z}[x_1,\dots,x_n]^{\mathfrak{S}_n}$, where $\mathfrak{S}_n$ denotes the symmetric group of degree n. The Cauchy kernel $\prod_{i=1}^{k}\prod_{j=1}^{m}(1-x_iy_j)^{-1}$ in (1) can be written as $\prod_{(i,j)\in(m^k)}(1-x_iy_j)^{-1}$, an expansion over the rectangle shape (m^k) of height k and width m. Cauchy kernels over arbitrary Ferrers shapes are no more symmetric in the indeterminates. Their expansions are not on the basis of Schur polynomials but rather over the basis of key polynomials and the basis of Demazure atoms. Lascoux has studied Cauchy kernel expansions over staircases which he then generalized for arbitrary Ferrers shapes [16]. For staircase shapes the expansion is explicit in the SSYTs for which Lascoux has provided both an algebraic proof, with Fu and Lascoux [6], and a combinatorial proof. The latter based on the fact that, in type A—not known for other Weyl groups [21]—RSK can be formulated in the language of bicrystals [11, 14, 16, 22]. For other shapes, the expansion is not entirely explicit in the SSYTs and only an algebraic explanation for the expansion was provided in [16].

Mason [25, 26] has defined an analogue of RSK where the output is a pair of semi-skyline augmented fillings (SSAFs) whose shapes, vectors in $\mathbb{N}^n$, are a rearrangement of each other, see Sects. 2 and 3. The SSAFs, combinatorial objects coming from the Haglund-Haiman-Loehr formula for non-symmetric Macdonald polynomials [9], are in bijection with SSYTs in a way that the shape detects the right key of a SSYT [19], see Section 3. Key polynomials and Demazure atoms (or standard bases) were first described combinatorially by Lascoux and Schützenberger in [18, 19], for which they have introduced the key notion of right key of a SSYT. Thus, they are also combinatorially described by SSAFs [26]

$$\hat{\kappa}_\nu(x) = \sum_{\substack{F\ SSAF\\ sh(F)=\nu}} x^F, \quad \kappa_\nu(x) = \sum_{\substack{F\ SSAF\\ sh(F)\leqslant\nu}} x^F, \tag{2}$$

where the inequality regarding the shape of F, $sh(F)$, is in the Bruhat order. In [1, 2], we have proved that the analogue of RSK correspondence, restricted to multisets of cells in the staircase of size n, gives pairs (F, G) of SSAFs, with entries $\leqslant n$, whose shapes satisfy $sh(G) \leqslant \omega sh(F)$ in Bruhat order, with ω the longest permutation of $\mathfrak{S}_n$. As a consequence, using (2), we can write

$$\prod_{\substack{i+j\leq n+1\\ 1\leqslant i,j\leqslant n}} (1-x_iy_j)^{-1} = \sum_{\substack{(F,G)\\ sh(G)\leqslant\omega sh(F)}} x^F y^G = \sum_{\nu\in\mathbb{N}^n} \hat{\kappa}_\nu(x)\kappa_{\omega\nu}(y), \tag{3}$$

where the sum runs over pairs (F, G) of SSAFs with entries $\leqslant n$. More generally, the analogue of RSK can be restricted to multiset of cells in a truncated staircase with height k and width m [1, 2, 4]. This allows to extend the triangular expansion (3) to truncated triangles. A bijective proof for the following non-symmetric Cauchy kernel expansion, over a truncated staircase of size n, with height k and width $m \geqslant k$—a special case of the more general formula for Ferrers shapes, due to Lascoux, in [16]—has been provided in [1, 2, 4]

$$\prod_{(i,j)\in} (1-x_iy_j)^{-1} = \sum_{\mu\in\mathbb{N}^k} \hat{\kappa}_\mu(x)\pi_{\sigma(\lambda,SE)}\kappa_{\omega\mu}(y). \tag{4}$$

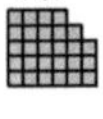

Here $\pi_{\sigma(\lambda,SE)}$ is the Demazure operator, see Sect. 6, indexed by $\sigma(\lambda, SE)$ a reduced expression of $\mathfrak{S}_n$, specified by the cells above the biggest staircase inside the truncated shape, as explained in [16]. Recall that Demazure operators π_i act on key polynomials κ_μ via elementary bubble sorting operators on the entries of the vector μ [28], see Sect. 6. It is then possible to determine explicitly $\pi_{\sigma(\lambda,SE)}\kappa_{\omega\mu}$, and write

$$(4) = \sum_{\mu\in\mathbb{N}^k} \hat{\kappa}_\mu(x)\kappa_{(0^{m-k},\alpha)}(y), \tag{5}$$

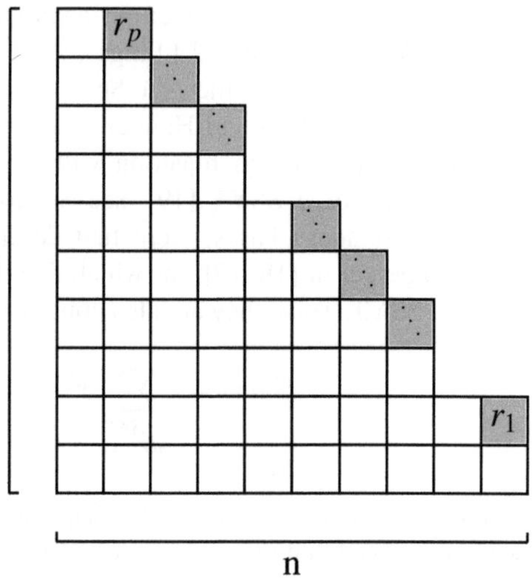

Fig. 1 A near staircase of size n with one layer of p *gray* cells, $1 \leqslant p < n$, sited on the stairs, at most one cell in each stair, avoiding the top and the basement. The label r_i indicates that the row index is $r_i + 1$, $1 \leqslant i \leqslant p$, counted from the *bottom* to the *top*

where α depends on μ in a certain way as explained in [2]. In particular, when $k = m = n$, the identity for staircase shapes (3) is recovered.

In this work, we give a combinatorial expansion for the non-symmetric Cauchy kernel $\prod_{(i,j)\in\lambda}(1-x_iy_j)^{-1}$, being the product over all cells (i,j) of the near staircase λ, in French convention, as shown in the Fig. 1. Theorem 5, in Sect. 7, extends the analogue of RSK on triangular shapes to multiset of cells in the near staircase. It produces pairs (F, G) of SSAFs with entries $\leqslant n$ such that the pair of shapes satisfy inequalities, in the Bruhat order, specified the p gray boxes sited on the stairs of staircase Fig. 1. This bijection allows the following combinatorial formula expansion in terms of pairs of SSAFs (SSYTs), see Theorem 6, Sect. 8,

$$\prod_{(i,j)\in\lambda}(1-x_iy_j)^{-1} = \sum_{(F,G)\in\mathscr{A}} x^F y^G + \sum_{1\leqslant z\leqslant p}\sum_{H_z}\sum_{(F,G)\in\mathscr{A}_z^{H_z}} x^F y^G$$

$$= \sum_{\nu\in\mathbb{N}^n}(\pi_{r_1}\dots\pi_{r_p}\hat{\kappa}_\nu(x))\kappa_{\omega\nu}(y).$$

For $z = 0, 1, \dots, p$, $H_z = \{i_1 < i_2 < \cdots < i_z\} \in \binom{[p]}{z}$, and

$$\mathscr{A}_z^{H_z} := \left\{ (F,G)\in SSAF_n^2 \colon \begin{matrix} sh(G)\not\leq\omega s_{r_{i_z}}\cdots\hat{s}_{r_{i_m}}\cdots s_{r_{i_1}}sh(F),\ m=1,2,\dots,z \\ sh(G)\leq\omega s_{r_{i_z}}\cdots s_{r_{i_1}}sh(F)\end{matrix} \right\},$$

where $SSAF_n$ denotes the set of all SSAFs with entries $\leqslant n$, and the inequalities are in the Bruhat order of $\mathfrak{S}_n$ ('^' means omission). In particular, $\mathscr{A} := \mathscr{A}_0^{\emptyset} := \{ (F,G)\in SSAF_n^2 \colon sh(G)\leq\omega sh(F) \}$. This provides a bijective proof for the identity

$$\prod_{(i,j)\in\lambda}(1-x_iy_j)^{-1} = \sum_{\nu\in\mathbb{N}^n}(\pi_{r_1}\dots\pi_{r_p}\hat{\kappa}_\nu(x))\kappa_{\omega\nu}(y) \tag{6}$$

an instance of the Lascoux's formula for an arbitrary Ferrers shape [16], in the case, λ is the near staircase in Fig. 1.

The paper is organised in eight sections as follows. In Sect. 2, we introduce the basic terminology of our combinatorial objects and the relationship between them. In Sect. 3, the RSK, reverse RSK and the RSK analogue, due to Mason, are defined, and it is shown that the key of a SSYT is easily read off from a SSAF. In Sect. 4, we recall the growth diagram version of reverse RSK and from that our growth diagram version for the RSK analogue is given. In Sect. 5, the definition of Bruhat order in $\mathfrak{S}_n$ and some basic properties are briefly recalled. In Sect. 6, Demazure operators, key polynomials or Demazure characters, and Demazure atoms as well as some of their properties are summarised. In Sect. 7, the RSK analogue under the action of crystal operators is analysed. The main results of this section, Theorems 4 and 3, detect how the key of a SSYT change in Bruhat order when cells are created in certain corners of a Ferrers shape. The restriction of the analogue of RSK to near stair shapes is described in Theorem 5. Finally, in the last section, this latter theorem is used to give a combinatorial proof of the Cauchy kernel expansion (6), due to Lascoux.

2 SSYT, Reverse SSYT and SSAF

Semi-skyline augmented fillings (SSAFs) have been introduced in [9, 10], to describe combinatorially (non-symmetric) Macdonald polynomials. In [26], Mason has defined a weight preserving bijection, ϱ, between reverse semi-standard Young tableaux (RSSYTs) and SSAFs. We shall use this map to define SSAFs. It allows later to translate the analogue of RSK [25] for growth diagrams via the usual reverse RSK.

2.1 SSYT and Reverse SSYT

A weak composition $\gamma = (\gamma_1 \dots, \gamma_n)$ is a vector in $\mathbb{N}^n$. A weak composition γ whose entries are in weakly decreasing order, that is, $\gamma_1 \geq \cdots \geq \gamma_n$, is said to be a partition. Every weak composition γ determines a unique partition obtained by arranging the entries in weakly decreasing order. More precisely, it is the unique partition in the orbit of γ regarding the usual action of symmetric group $\mathfrak{S}_n$ on $\mathbb{N}^n$. A partition $\lambda = (\lambda_1, \dots, \lambda_n)$ is identified with its Young diagram (or Ferrers shape) $dg(\lambda)$ in French convention, an array of left-justified cells (boxes) with λ_i cells in row i from the bottom, for $1 \leq i \leq n$. The cells are located in the diagram $dg(\lambda)$ by their row and column indices (i,j), where $1 \leq i \leq n$ and $1 \leq j \leq \lambda_i$.

A filling of shape λ (or a filling of $dg(\lambda)$), in the alphabet $[n]$, is a map $T : dg(\lambda) \rightarrow [n]$. A semi-standard Young tableau (SSYT) T of shape $sh(T) = \lambda$, in the alphabet $[n]$, is a filling of $dg(\lambda)$ weakly increasing in each row from left to right and strictly increasing up in each column. The column word of the SSYT T is the word

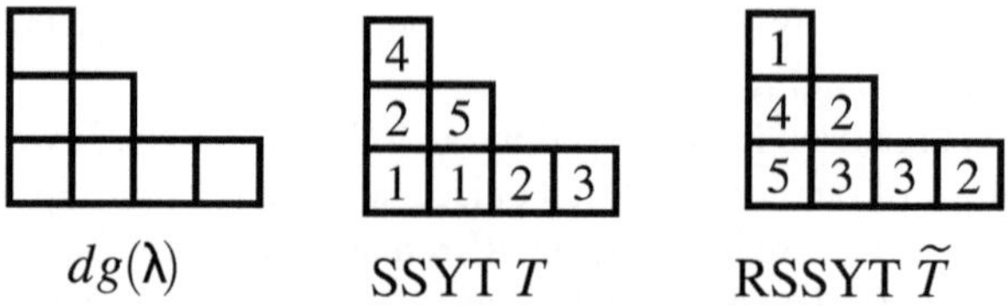

Fig. 2 The Ferrers diagram of $\lambda = (4, 2, 1)$, a SSYT and a RSSYT of shape λ, respectively, with contents $c(T) = (2, 2, 1, 1, 1)$ and $c(\widetilde{T}) = (1, 2, 2, 1, 1)$

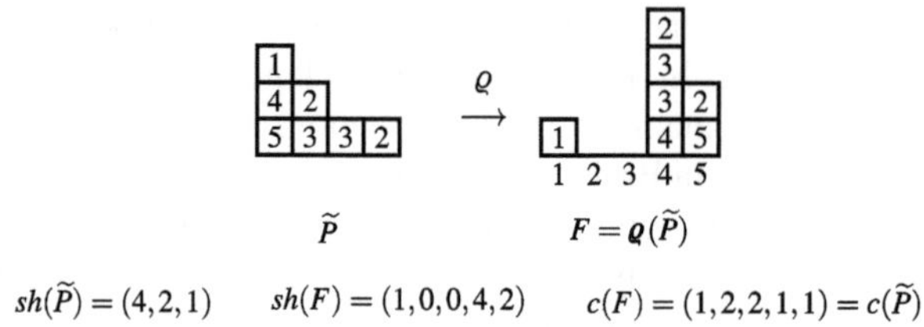

Fig. 3 The SSAF F corresponding to the RSSYT $\widetilde{P}$ defined by the weight preserving bijection ϱ

consisting of the entries of each column, read top to bottom and left to right. The content $c(T) = (\alpha_1, \ldots, \alpha_n)$ or weight of T is the content or weight of its column word, that is, α_i is the multiplicity of i in the column word of T, for all i. A key tableau is a SSYT such that the set of entries in the $(j+1)^{th}$ column is a subset of the set of entries in the jth column, for all j. There is a bijection [28] between weak compositions in $\mathbb{N}^n$ and keys in the alphabet $[n]$ given by $\gamma \rightarrow key(\gamma)$, where $key(\gamma)$ is the SSYT such that for all j, the first γ_j columns contain the letter j. Any key tableau is of the form $key(\gamma)$ where γ is the content and the shape is the unique partition in its $\mathfrak{S}_n$-orbit.

A reverse semi-standard Young tableau (RSSYT), $\tilde{T}$, of shape $sh(\tilde{T}) = \lambda$, in the alphabet $[n]$, is a filling of $dg(\lambda)$ such that the entries in each row are weakly decreasing from left to right, and strictly decreasing from bottom to top (Fig. 2).

2.2 SSAFs are in Bijection with RSSYTs

Fix $n \in \mathbb{N}$. A weak composition $\gamma = (\gamma_1, \ldots, \gamma_n)$ is visualised as a diagram consisting of n columns, with γ_j cells (boxes) in column j, for $1 \leqslant j \leqslant n$. Formally, the *column diagram* of γ is the set $dg'(\gamma) = \{(i,j) \in \mathbb{N}^2 : 1 \leq j \leq n, 1 \leq i \leq \gamma_j\}$ where the coordinates are in French convention, i indicates the vertical coordinate, indexing the rows, and j the horizontal coordinate, indexing the columns. (The prime reminds that the components of γ are the columns.) The number of cells in a column is called the height of that column, and a cell a in a column diagram is written $a = (i,j)$, where i is the row index and j the column index. The *augmented diagram* of γ, $\widehat{dg}(\gamma) = dg'(\gamma) \cup \{(0,j) : 1 \leq j \leq n\}$, is the column diagram with n extra cells adjoined in row 0. This adjoined row is called the *basement* and it always contains the numbers 1 to n in strictly increasing order. The shape of $\widehat{dg}(\gamma)$ is defined to be γ. The empty augmented diagram consists of the basement elements from 1 to n.

We now introduce the *semi-skyline augmented filling* (SSAF) object as the output of the injective map ϱ, in [26], acting on RSSYTs. Let $\tilde{P}$ be a RSSYT in the alphabet $[n]$. Define the empty semi-skyline augmented filling as the empty augmented diagram with basement elements from 1 to n. Pick the first column of $\tilde{P}$, say, P_1. Put all the elements of the first column P_1 to the top of the same basement elements in the empty semi-skyline augmented filling. The new diagram is called the semi-skyline augmented filling corresponding to the first column of $\tilde{P}$ and is denoted by SSAF. Assume that the first i columns of $\tilde{P}$, denoted $P_1, P_2, \dots, P_i$, have been mapped to a SSAF. Consider the largest element, a_1, in the $(i+1)$-th column P_{i+1}. There exists an element greater than or equal to a_1 in the i-th row of the SSAF. Place a_1 on top of the leftmost such element. Assume that the largest $k-1$ entries in P_{i+1} have been placed into the SSAF. The k-th largest element, a_k, of P_{i+1} is then placed into the SSAF. Place a_k on top of the leftmost entry b in row $k-1$ such that $b \geqslant a_k$ and the cell immediately above b is empty. Continue this procedure until all entries in P_{i+1} have been mapped into the $(i+1)$-th row and then repeat for the remaining columns of $\tilde{P}$ to obtain the semi-skyline augmented filling F.

It is clear that rotating F 90°, sliding down the boxes in each column, and reordering them, in decreasing order from bottom to top, we obtain $\tilde{P}$.

We can associate to each SSAF, F, a weak composition that records the length of the columns of F, and defines the shape of F, $sh(F)$. The content of the SSAF F is the vector $c(F) = c(\tilde{P}) \in \mathbb{N}^n$ whose i-th entry is the multiplicity of the letter i in the SSAF F (Fig. 3).

3 RSK, Reverse RSK, and the Analogue of RSK Detecting Keys

The Robinson-Schensted-Knuth (RSK) correspondence is a bijection between two line arrays of positive integers and pairs SSYTs of the same shape. Mason has introduced an interesting analogue of RSK where SSYTs are replaced by SSAFs [25]. This latter bijection has the advantage that the shapes of the pair of SSAFs exhibit the keys of the pair of SSYTs produced by RSK.

3.1 The Reverse RSK

The reverse Schensted insertion applied to the word $b_1 \dots b_m$, over the alphabet $[n]$, gives the reverse SSYT $\tilde{P}$. It consists of reversing the roles of $\leqslant$ and $\geqslant$ in defining the Schensted insertion of $b_1 \dots b_m$. Equivalently, if we apply the Schensted insertion to $-b_m, \ldots, -b_1$ to get the SSYT, $P(-b_m, \ldots, -b_1)$, and then change the sign in all entries of $P(-b_m, \ldots, -b_1)$, we obtain the reverse SSYT $\tilde{P}$ [31].

The two line array $w = \begin{pmatrix} j_1\, j_2 \cdots j_l \\ i_1\, i_2 \cdots i_l \end{pmatrix}$, such that $j_r < j_{r+1}$, and if $j_r = j_{r+1}$ then $i_r \leqslant i_{r+1}$, for all $1 \leqslant r \leqslant l-1$, where $i_r, j_r \in [n]$, is called a biword in lexicographic order (with respect to the first row) over the alphabet $[n]$. The reverse RSK (RRSK) algorithm is the obvious variant of the RSK algorithm [31]. We apply the RSK to the biword $\tilde{w} = \begin{pmatrix} -j_l \ldots -j_1 \\ -i_l \cdots -i_1 \end{pmatrix}$, instead of $w = \begin{pmatrix} j_1 \cdots j_l \\ i_1 \cdots i_l \end{pmatrix}$, to obtain a pair of semi-standard Young tableaux, and then change the sign in all entries of that pair of SSYTs. We will obtain a pair $(\tilde{P}, \tilde{Q})$ of reverse SSYTs.

3.2 *Analogue of Schensted Insertion and Reverse Schensted Insertion: SSAFs in Bijection with SSYTs by Assigning the Right Key*

The fundamental operation of the Robinson-Schensted-Knuth (RSK) algorithm [12] is Schensted insertion which is a procedure for inserting a positive integer k into a semi-standard Young tableau T. In [25] a similar procedure for inserting a positive integer k into a SSAF F is defined, which is used to describe an analogue of the RSK algorithm. Based on this analogue of Schensted insertion, a weight preserving and a shape rearranging bijection Ψ between SSYTs and SSAFs, over the alphabet $[n]$, is given. The bijection Ψ is defined to be the insertion, from right to left, of the column word of a SSYT, which consists of the entries of each column, read from top to bottom and left to right, into the empty SSAF with basement $1, \ldots, n$. The shape of $\Psi(T)$ provides the right key, $K_+(T)$, of T, a notion due to Lascoux and Schützenberger [19].

Theorem 1 [26] *Given an arbitrary SSYT T, let γ be the shape of $\Psi(T)$. Then $K_+(T) = key(\gamma)$.*

On the other hand, applying the reverse Schensted insertion to the column word of the SSYT, T, gives the RSSYT, $\tilde{T}$. Then $\varrho(\tilde{T})$ is a SSAF and $\varrho(\tilde{T}) = \Psi(T)$ [25]. We then have two equivalent weight preserving and shape rearranging bijections between SSYTs and SSAFs, see Fig. 4.

3.3 *RSK, Reverse RSK and Analogue of RSK for SSAFs*

Given the alphabet $[n]$, the RSK algorithm is a bijection between biwords in lexicographic order and pairs of SSYTs of the same shape over $[n]$. The analogue of Schensted insertion is applied in [25] to find an analogue Φ of the RSK to produce

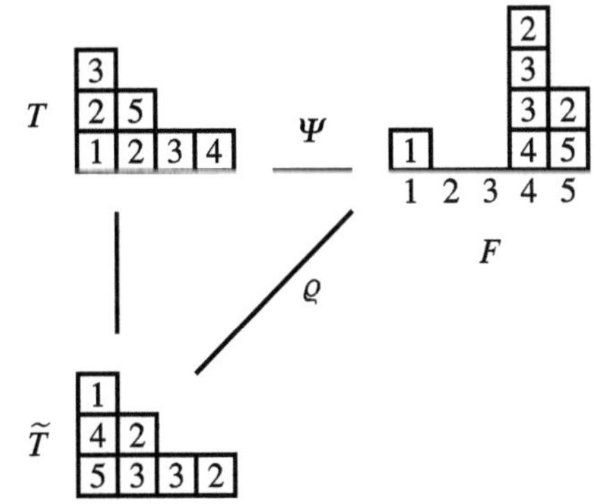

$$sh(F) = (1,0,0,4,2), \quad K_+(T) = key(1,0,0,4,2)$$

Fig. 4 $\widetilde{T}$ is the reverse Schensted insertion of T

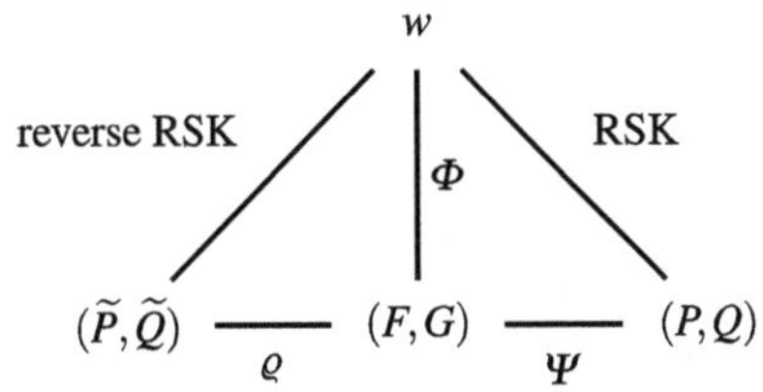

$$sh(F)^+ = sh(G)^+ = sh(P) = sh(Q) = sh(\widetilde{P}) = sh(\widetilde{Q})$$
$$key(sh(F)) = K_+(P),\ key(sh(G)) = K_+(Q)$$
$$c(F) = c(P) = c(\widetilde{P}),\ c(G) = c(Q) = c(\widetilde{Q})$$

Fig. 5 The analogue of RSK $\Phi = \varrho \circ RRSK = \Psi \circ RSK$ detects the keys of the pair (P, Q) of SSYTs produced by RSK

pairs of SSAFs. The map Φ defines a bijection between the set of all biwords w in lexicographic order in the alphabet $[n]$, and pairs of SSAFs whose shapes are rearrangements of a same partition in $\mathbb{N}^n$, and the contents are respectively those of the second and first rows of w. The bijection Φ applied to a biword w is the same as applying the reverse RSK to w and then applying ϱ to each reverse SSYT of the output pair $(\tilde{P}, \tilde{Q})$, that is, $\Phi(w) = (\varrho(\tilde{P}), \varrho(\tilde{Q}))$.

Corollary 1 [25, 26] *The RSK algorithm commutes with the above analogue Φ. That is, if (P, Q) is the pair of SSYTs produced by RSK algorithm applied to biword w, then $(\Psi(P), \Psi(Q)) = \Phi(w)$, and $K_+(P) = key(sh(\Psi(P)))$, $K_+(Q) = key(sh(\Psi(Q)))$.*

The relation between RSK, the reverse RSK and Φ, the analogue of RSK, is summarised in Fig. 5. In particular, it is clear that Φ the analogue of RSK also shares the symmetry of RSK.

4 Reverse RSK and Analogue of RSK in Terms of Fomin's Growth Diagrams

The formulation of RSK in terms of growth diagrams is due to Fomin [5], subsequently developed by Roby [29] and van Leewven [23], and applied to enumeration by Krattenthaler [13]. The bijection ϱ between SSAFs and RSSYTs allows a growth diagrammatic formulation of the analogue of RSK where SSYT are replaced with SSAFs [25] via reverse Schensted insertion. In this section we follow very closely [13, 31].

Let w be a biword in the lexicographic order over the alphabet $[n]$. We can represent a biword w in the $n \times n$ square diagram, by putting the number r in the cell (i,j) of the square grid, when the biletter $\begin{pmatrix} j \\ i \end{pmatrix}$ appears r times in the biword w. The rows are counted from bottom to top and the columns from left to right. For instance, if $w = \begin{pmatrix} 1\,1\,2\,3\,4\,4\,5\,7\,7 \\ 2\,7\,2\,4\,1\,3\,3\,1\,1 \end{pmatrix}$, with $n = 7$, then we obtain the 7×7 square diagram Fig. 6.

We would like to have a 01-filling of this diagram Fig. 6, that is, at most one 1 in each row and each column. To remedy this, the entries in the diagram are separated as we now explain. Construct a rectangle diagram with more rows and columns. The entries which are originally in the same column or in the same row are put in different columns and rows in the larger diagram. An entry m is replaced by m 1's in the new diagram, all of them placed in different rows and columns. The entries in a row are separated from bottom/left to top/right, and the 1's are represented by **X**'s. If there should be several entries in a column as well, separate entries in a column from bottom/left to top/right. In the cell with entry m, we replace m by a chain of m **X**'s arranged from bottom/left to top/right. The original n columns and n rows are indicated by thick lines, whereas the newly created columns and rows are indicated by thin lines. See Fig. 7.

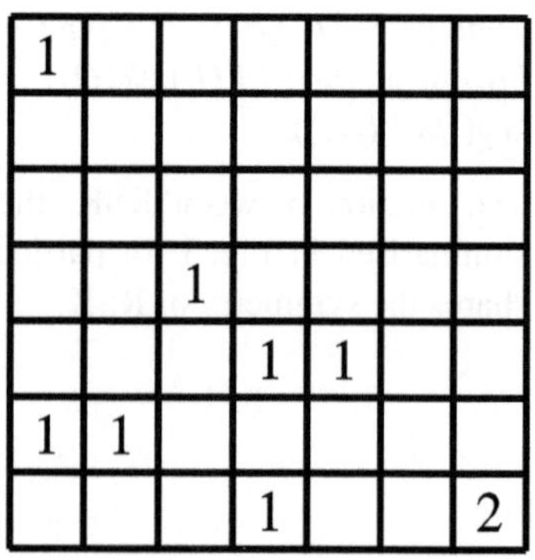

Fig. 6 Scanning the columns, from *left* to *right* and *bottom* to *top*, the biword w is recovered in lexicographic order

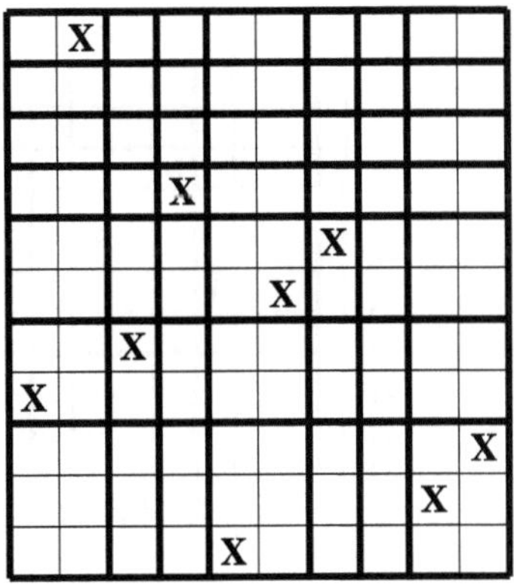

Fig. 7 The 01-filling representation of the biword w. The 1's are represented by *crosses*

To give an interpretation of reverse RSK in terms of growth diagrams, we start by assigning the empty partition $\emptyset$ to each point of a corner cell on the right column and on the top row of the 01-filling. Then assign partitions to the other corners inductively by applying the following local rules. Consider the cell below, labeled by the partitions ε, μ, ν, such that $\varepsilon \subseteq \mu$ and $\varepsilon \subseteq \nu$, where the containment means that the Ferrres shapes differ at most by one box. Then λ is determined as follows:

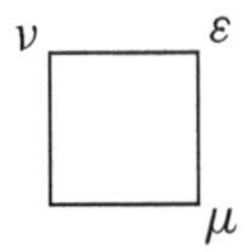

- If $\varepsilon = \mu = \nu$, and if there is no cross in the cell, then $\lambda = \varepsilon$.
- If $\varepsilon = \mu \neq \nu$, then $\lambda = \nu$.
- If $\varepsilon = \nu \neq \mu$, then $\lambda = \mu$.
- If ε, μ, ν are pairwise different, then $\lambda = \mu \cup \nu$, i.e., $\lambda_i = max\{\mu_i, \nu_i\}$.
- If $\varepsilon \neq \mu = \nu$, then λ is formed by adding a box to the $(k+1)$-st row of $\mu = \nu$, given that $\mu = \nu$ and ε differ in the k-th row.
- If $\varepsilon = \mu = \nu$, and if there is a cross in the cell, then λ is formed by adding a box to the first row of $\varepsilon = \mu = \nu$.

Applying the local rules leads to a pair of nested sequences of partitions on the left column and in the bottom row of the growth diagram. Let λ^i be the partition assigned to the i-th thick column on the bottom row of the growth diagram when we scan the thick columns from right to left, with the rightmost column being column 0. Then the bottom row labelling assigned to the thick columns of the growth diagram produces a sequence of partitions $\lambda^n \supseteq \cdots \supseteq \lambda^1 \supseteq \lambda^0 = \emptyset$, such that λ^i/λ^{i-1} is a horizontal strip. Let $\underline{\lambda}^i$ be the partition assigned to the i-th thick row on the left of the growth diagram when we scan the thick rows from top to bottom, with the top row being row 0. Then the left column labelling assigned to the thick rows of the growth diagram produces a sequence of partitions $\emptyset = \underline{\lambda}^0 \subseteq \lambda^1 \subseteq \cdots \subseteq$

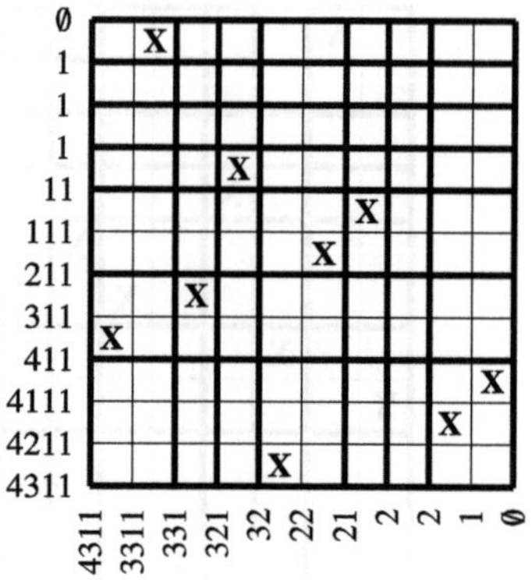

$$\widetilde{P} = \begin{array}{llll} 1 & & & \\ 3 & & & \\ 4 & 1 & 1 & \\ 7 & 3 & 2 & 2 \end{array} \qquad \widetilde{Q} = \begin{array}{llll} 1 & & & \\ 3 & & & \\ 5 & 4 & 2 & \\ 7 & 7 & 4 & 1 \end{array}$$

Fig. 8 The growth diagram of the reverse RSK of w, and the pair $(\tilde{P}, \tilde{Q})$ of RSSYTs, respectively, produced by the sequence of partitions on the *left column* and on the *bottom row*

$\underline{\lambda}^n$, such that $\underline{\lambda}^i/\underline{\lambda}^{i-1}$ is a horizontal strip. Filling in with $n+1-i$ respectively the cells of $\underline{\lambda}^i/\underline{\lambda}^{i-1}$, and λ^i/λ^{i-1}, for $i \geqslant 1$, produces the pair $(\tilde{P}, \tilde{Q})$ of RSSYTs of the same shape. Their contents are, respectively, those of the second and the first rows of w. See Fig. 8. This is the same as applying the reverse RSK to the biword w.

In addition there is a global description of the local rules as a consequence of a variant of Greene's theorem [8] and Theorem 2 in [13]. A SW-chain of a 01-filling is a sequence of 1's such that any 1 is below and to the left of the preceding 1 in the sequence. The length of a SW-chain is defined to be the number of 1's in the chain. Another way to find the nested sequences of partitions on the bottom and on the left of the growth diagram is just looking for the k SW-chains by using the following natural version of the Theorem 2 in [13].

Theorem 2 *Given a diagram with empty partitions labelling all the corners along the right side and the top side of a rectangle shape, which has been completed according to the reverse RSK local rules, the partition $\lambda = (\lambda_1, \lambda_2, \ldots, \lambda_l)$ labelling corner c satisfies the following property: For any k, the maximal cardinality of the union of k SW-chains situated in the rectangular region to the right and above c is equal to $\lambda_1 + \lambda_2 + \cdots + \lambda_k$. In particular, λ_1 is the length of the longest SW-chain in the rectangular region to the right and above c.*

The map ϱ, defined in Sect. 2.2, allows us to find the pair of SSAFs from the growth diagram corresponding to the reverse RSK. Recall that the shape of a SSAF is the weak composition that records the length of its columns. A partition on the left column or in the bottom row of the growth diagram is the shape of a RSSYT, a rearrangement of the shape of a SSAF. Consider the bottom row labelling $\lambda^n \supseteq \cdots \lambda^1 \supseteq \lambda^0 = \emptyset$ assigned to the thick columns of the growth diagram. For each $i = 1, \ldots, n$, let $\lambda^{i_{l_i-1}} \supseteq \cdots \supseteq \lambda^{i_1}$, with $\lambda^{i_{l_i}} := \lambda^i$, be the bottom sequence of partitions labelling the $l_i - 1$ thin columns, strictly in between the two thick columns $i-1$ and i. Start with the empty partition $\lambda^0 = \emptyset$, the rightmost partition of the bottom sequence in the growth diagram, and the empty SSAF with basement $[n]$. Proceed to the left along the bottom row. When we arrive to the partition λ^{i_j}, we put a cell, filled with $n-i+1$, in the leftmost possible place of the SSAF such that the shape of the new SSAF becomes a rearrangement of the partition λ^{i_j} and the decreasing property on the columns of the SSAF, from the bottom to the top, is preserved. At the end of the scanning of the bottom row, the SSAF G is obtained. Its shape is a rearrangement of the shape of $\tilde{Q}$.

Similarly, consider the left column labelling $\emptyset = \underline{\lambda}^0 \subseteq \underline{\lambda}^1 \subseteq \cdots \subseteq \underline{\lambda}^n$ assigned to the thick rows of the growth diagram. For each $i = 1, \ldots, n$, let $\underline{\lambda}^{i_1} \subseteq \cdots \subseteq \underline{\lambda}^{i_{l_i-1}}$, with $\underline{\lambda}^{i_{l_i}} := \underline{\lambda}^i$, be the left sequence of partitions labelling the $l_i - 1$ thin rows, strictly in between the two thick rows $i-1$ and i. At the end of the procedure, when the scanning of the left column is finished, the SSAF F is obtained. Its shape is a rearrangement of the shape of $\tilde{P}$. See Fig. 9.

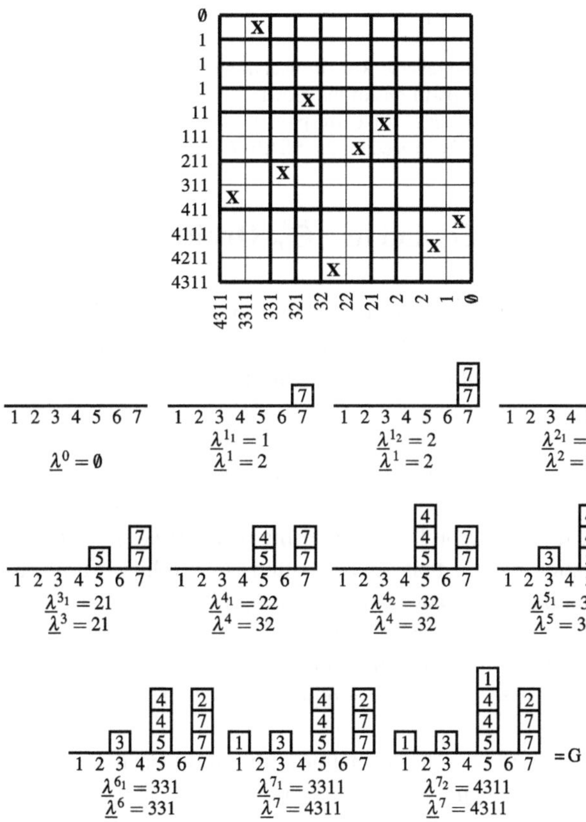

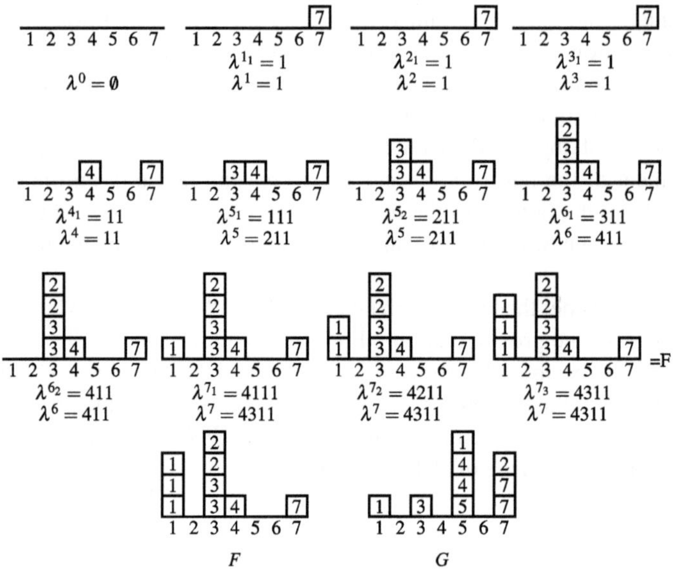

Fig. 9 The growth diagram of the reverse RSK of w and the growth of the corresponding pair of SSAFs. The *bottom row* chain of partitions of the growth diagram produces a sequence of SSAFs where *each shape* is a rearrangement of the corresponding partition. Similarly for the *left column* chain of partitions. At the end of the procedure the SSAFs G and F are respectively obtained

5 Bruhat Order in $\mathfrak{S}_n$ and in a $\mathfrak{S}_n$-Orbit

Let $\theta = \theta_1 \dots \theta_n \in \mathfrak{S}_n$, written in one line notation. A pair (i,j), with $i < j$, such that $\theta_i > \theta_j$, is said to be an inversion of θ, and $\ell(\theta)$ denotes the number of inversions of θ. The Bruhat order in $\mathfrak{S}_n$ is the partial order in $\mathfrak{S}_n$ defined by the transitive closure of the relations

$$\theta < t\theta, \text{ if } \ell(\theta) < \ell(t\theta), (t \text{ transposition}, \theta \in \mathfrak{S}_n).$$

We may write $\alpha < \beta$ in the Bruhat ordering of $\mathfrak{S}_n$ if $\ell(\alpha) < \ell(\beta)$ and $\beta = \tau\alpha$ for some permutation τ in $\mathfrak{S}_n$ that can be written as a product of transpositions each increasing the number of inversions when passing from α to β.

Let $\theta = s_{i_N} \cdots s_{i_1}$ be a decomposition of θ into simple transpositions $s_i = (i\ i+1), 1 \leqslant i < n$. When $N = \ell(\theta)$, the number N in a such decomposition is minimised, and we say that we have a reduced decomposition of θ. The longest permutation of $\mathfrak{S}_n$ is denoted by ω.

Let λ be a partition in $\mathbb{N}^n$. The Bruhat ordering of the orbit of λ, $\mathfrak{S}_n\lambda$, is defined by taking the transitive closure of the relations

$$\alpha < t\alpha, \text{ if } \alpha_i > \alpha_j, i < j, \text{ and } t \text{ the transposition } (i\,j), (\alpha \in \mathfrak{S}_n\lambda).$$

Given $\alpha \in \mathbb{N}^n$, a pair (i,j), with $i < j$, such that $\alpha_i < \alpha_j$, is called an inversion of α, and $\iota(\alpha)$ denotes the number of inversions of α. We may write $\alpha < \beta$ if $\iota(\alpha) < \iota(\beta)$ and $\beta = \tau\alpha$ for some permutation τ in $\mathfrak{S}_n$ that can be written as a product of transpositions each increasing the number of inversions when passing from α to β. The following lemma recalls some useful properties whose proof are a simple exercise.

Lemma 1 *Let α, $\beta \in \mathbb{N}^n$. Let $1 \leqslant p < n$ and $1 \leqslant r_1 < r_2 < \cdots < r_p < n$. Then*

(a) $\alpha_{r_1} < \alpha_{r_1+1} \Rightarrow s_{r_p} \dots s_{r_2}\alpha > s_{r_p} \dots s_{r_2}s_{r_1}\alpha$.
(b) $\alpha_{r_1} > \alpha_{r_1+1} \Rightarrow s_{r_p} \dots s_{r_2}s_{r_1}\alpha > s_{r_p} \dots s_{r_2}\alpha$.
(c) *if* $\alpha_{r_1} < \alpha_{r_1+1}$ *and* $(s_{r_{p-1}} \dots s_{r_2}\alpha)_{r_p} < (s_{r_{p-1}} \dots s_{r_2}\,\alpha)_{r_p+1}$ *it implies*

$$(s_{r_{p-1}} \dots s_{r_2}s_{r_1}\alpha)_{r_p} < (s_{r_{p-1}} \dots s_{r_2}s_{r_1}\alpha)_{r_p+1}.$$

(d) *if* $\beta \not\leq \omega s_{r_p} \cdots \hat{s}_{r_i} \cdots s_{r_2}s_{r_1}\alpha$, *for* $i = 1, 2, \dots, p$, *and* $\beta \leq \omega\, s_{r_p} \cdots s_{r_2}\, s_{r_1}\, \alpha$, *it implies ('^' means omission)*

$$(s_{r_t} \dots s_{r_2}s_{r_1}\alpha)_{r_{t+1}} < (s_{r_t} \dots s_{r_2}s_{r_1}\alpha)_{r_{t+1}+1}, \quad t = 0, 1, \dots, p-1.$$

6 Demazure Operators, Demazure Characters and Demazure Atoms

Isobaric divided difference operators [17], or Demazure operators [3], π_i and $\hat{\pi}_i$, $1 \leqslant i < n$, act on $\mathbb{Z}[x_1, \dots, x_n]$ by

$$\pi_i f = \frac{x_i f - s_i(x_i f)}{x_i - x_{i+1}}, \tag{7}$$

$$\hat{\pi}_i f = (\pi_i - 1)f = \pi_i f - f, \tag{8}$$

where the simple transposition s_i of $\mathfrak{S}_n$ acts on f swapping x_i with x_{i+1}, and 1 is the identity operator on $\mathbb{Z}[x_1, \dots, x_n]$. It follows from the definition that $\pi_i f = f$ and $\hat{\pi}_i f = 0$ if and only if $s_i f = f$. They both satisfy the commutation and the braid relations of $\mathfrak{S}_n$, $\pi_i\pi_j = \pi_j\pi_i$, $\hat{\pi}_i\hat{\pi}_j = \hat{\pi}_j\hat{\pi}_i$ for $|i-j| > 1$, and $\pi_i\pi_{i+1}\pi_i = \pi_{i+1}\pi_i\pi_{i+1}$, $\hat{\pi}_i\hat{\pi}_{i+1}\hat{\pi}_i = \hat{\pi}_{i+1}\hat{\pi}_i\hat{\pi}_{i+1}$. This guarantees that, for any permutation $\sigma \in \mathfrak{S}_n$, there exists a well defined isobaric divided difference $\pi_\sigma := \pi_{i_N} \cdots \pi_{i_2}\pi_{i_1}$ and $\hat{\pi}_\sigma := \hat{\pi}_{i_N} \cdots \hat{\pi}_{i_2}\hat{\pi}_{i_1}$, where $s_{i_N} \dots s_{i_2}s_{i_1}$ is any reduced expression of $\mathfrak{S}_n$. In addition, they satisfy the quadratic relations $\pi_i^2 = \pi_i$ and $\hat{\pi}_i^2 = -\hat{\pi}_i$.

The 0-Hecke algebra $H_n(0)$ of $\mathfrak{S}_n$, a deformation of the group algebra of $\mathfrak{S}_n$, is an associative $\mathbb{C}$-algebra generated by $T_1, \dots, T_{n-1}$ satisfying the commutation and the braid relations of the symmetric group $\mathfrak{S}_n$, and the quadratic relation $T_i^2 = T_i$ for $1 \leqslant i < n$. Setting $\hat{T}_i := T_i - 1$, for $1 \leqslant i < n$, another set of generators of the 0-Hecke algebra $H_n(0)$ is obtained. The sets $\{T_\sigma : \sigma \in \mathfrak{S}_n\}$ and $\{\hat{T}_\sigma : \sigma \in \mathfrak{S}_n\}$ are both linear bases for $H_n(0)$, where $T_\sigma = T_{i_N} \cdots T_{i_2} T_{i_1}$ and $\hat{T}_\sigma := \hat{T}_{i_N} \cdots \hat{T}_{i_2} \hat{T}_{i_1}$, for any reduced expression $s_{i_N} \cdots s_{i_2} s_{i_1}$ in $\mathfrak{S}_n$. Since Demazure operators (7) or bubble sort operators satisfy the same relations as T_i, and similarly for isobaric divided difference operators (8) and $\hat{T}_i$, the 0-Hecke algebra $H_n(0)$ of $\mathfrak{S}_n$ may be viewed as an algebra of operators realised either by any of the two isobaric divided differences, or by bubble sort operators, swapping entries i and $i+1$ in a weak composition α, if $\alpha_i > \alpha_{i+1}$, and doing nothing, otherwise.

Therefore, the two families $\{\pi_\sigma : \sigma \in \mathfrak{S}_n\}$ and $\{\hat{\pi}_\sigma : \sigma \in \mathfrak{S}_n\}$ are both linear bases for $H_n(0)$, and from the relation $\hat{\pi}_i = \pi_i - 1$, the change of basis from the first to the second is given by a sum over the Bruhat order in $\mathfrak{S}_n$, $\pi_\sigma = \sum_{\theta \leqslant \sigma} \hat{\pi}_\theta$ [15, 27]. Key polynomials and Demazure atoms can be defined through Demazure operators, $\kappa_\alpha = \pi_\sigma x^\lambda$ where $\alpha = \sigma\lambda$ and λ is a partition, and similarly $\hat{\kappa}_\alpha = \hat{\pi}_\sigma x^\lambda$ (assume σ a minimal coset representative modulo stabiliser of λ). Thereby, key polynomials or Demazure characters are decomposed into Demazure atoms [17, 19],

$$\kappa_\alpha = \sum_{\beta \leqslant \alpha} \hat{\kappa}_\beta. \tag{9}$$

We recall that Demazure operators π_i act on key polynomials κ_μ via elementary bubble sorting operators on the entries of the vector μ [28], that is,

$$\pi_i \kappa_\mu = \begin{cases} \kappa_{s_i\mu} & \text{if } \mu_i > \mu_{i+1} \\ \kappa_\mu & \text{if } \mu_i \leqslant \mu_{i+1} \end{cases}. \tag{10}$$

The action of Demazure operators on Demazure atoms $\hat{\kappa}_\mu$ is as follows

$$\pi_i \hat{\kappa}_\mu = \begin{cases} \hat{\kappa}_{s_i\mu} + \hat{\kappa}_\mu & \text{if } \mu_i > \mu_{i+1} \\ \hat{\kappa}_\mu & \text{if } \mu_i = \mu_{i+1} \\ 0 & \text{if } \mu_i < \mu_{i+1} \end{cases}. \tag{11}$$

7 Crystal Operators and Growth Diagrams

Biwords are multisets of cells in a Ferrers shape. We analyse the behaviour of the RSK analogue under the action of a crystal operator. The main results of this section, Theorems 4 and 3, detect how the key of a SSYT change in Bruhat order when cells are created in certain corners of a Ferrers shape.

7.1 *Crystal Operators on Biwords and the Analogue of RSK*

Crystal operators or coplactic operations e_r, f_r, $1 \leqslant r < n$, are defined on any word over the alphabet $[n]$ [20]. These operations can be extended to biwords in two ways. Either by considering w in lexicographic order, with respect to the first row, or to the second. Let $w = \begin{pmatrix} \mathbf{u} \\ \mathbf{v} \end{pmatrix}$ be in lexicographic order, with respect to the first row. Write $e_r w := \begin{pmatrix} \mathbf{u} \\ e_r\mathbf{v} \end{pmatrix}$ and, similarly, for $f_r w$. Let $\begin{pmatrix} \mathbf{k} \\ \mathbf{l} \end{pmatrix}$ be the biword w rearranged in lexicographic order, with respect to the second row. Write $e_r^* w := \begin{pmatrix} \mathbf{l} \\ e_r\mathbf{k} \end{pmatrix}$ and, similarly, for $f_r^* w$. The resulting biwords are still in lexicographic order with respect to the first row. (The $*$ recalls that the action is in the first row of w equipped with the appropriate order.) For details, see [14, 16, 20]. In this work we shall not consider the two actions simultaneously on w. We rather emphasise that if w^* denotes w with the rows $\mathbf{u}$ and $\mathbf{v}$ swapped, and rearranged in lexicographic order, with respect to the first row, that is, $w^* = \begin{pmatrix} \mathbf{l} \\ \mathbf{k} \end{pmatrix}$, then $e_r^* w = e_r w^*$. As our running example, consider the following biword in lexicographic order, with respect to the first row, over the alphabet [7],

$$w = \begin{pmatrix} 1 & 1 & 2 & 2 & 3 & 3 & 3 & 3 & 4 & 4 & 4 & 5 & 5 & 6 & 7 \\ 3 & 4 & 2 & 6 & 3 & 4 & 4 & 4 & 3 & 3 & 4 & 3 & 4 & 1 & 1 \end{pmatrix}. \tag{12}$$

The crystal operator e_3 acts on w through its action on the second row of w as follows. Ignore all entries different from 3 and 4, obtaining the subword 34344433434. Match, in the usual way, all 43 (in blue in the example below), remaining the subword $v' = 344$. Change to 3 the leftmost 4, giving 334. The image of the initial word $\mathbf{v}$ is obtained by replacing the subword v' in $\mathbf{v}$ with 334. For example, applying twice e_3 to w means to apply twice e_3 to the second row of w, and the subword 344 change to 333, obtaining

$$34344433434 \xrightarrow{e_3} 34334433434 \xrightarrow{e_3} 34334433433$$

The action of the crystal operator f_3 is defined similarly and $f_3^2 e_3^2 w = f_3 e_3 w = w$. Here, if, after the matching, the remaining word is empty then the operators e_r and f_r do nothing on w. Recall the representation of a biword in a rectangle diagram, defined in Sect. 4. We now represent a biword w in a Ferrers shape, see Fig. 10, by putting a cross "**X**" in the cell (i,j) of λ for each biletter $\begin{pmatrix} \mathrm{j} \\ \mathrm{i} \end{pmatrix}$ in w. A biletter is identified with the corresponding cell. The number of crosses is the multiplicity of the biletter in the biword. The biword w can be recovered, from this representation, in two ways. In lexicographic order with respect to the first row, scanning the

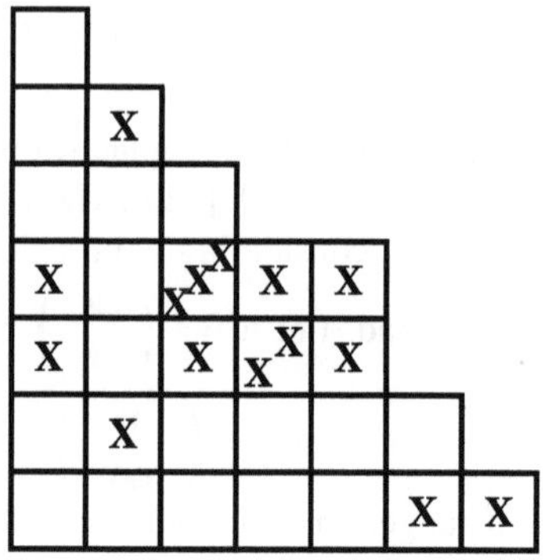

Fig. 10 Representation of the biword w (12) in a Ferrers shape. The cells marked with **X** are the biletters in w

columns of the Ferrers shape λ, from left to right, and bottom to top. In lexicographic order with respect to the second row, scanning the rows of the Ferrers shape λ, from bottom to top and left to right.

Let w be a biword in lexicographic order represented in the Ferrers shape λ. We introduce an operation Υ_r in the Ferrers shape λ, acting in the rows r and $r+1$ as follows. Consider the two row rectangle defined by the rows r and $r+1$ of λ. (If necessary add blank cells to the row $r+1$ such that rows r and $r+1$ have the same length. If there is more than one cross in the same cell then order the crosses, from left to right, and put each one in a different sub column made of thin lines. We say that a crossed thick cell in row $r+1$ and a crossed thick cell, to the SE, in row r, is a *factor* if in these two rows there is no crossed cell in the columns between them. In each factor match the rightmost cross in row $r+1$ with the leftmost cross in row r. Ignore the matched crosses. Repeat the procedure with the new factors until no factors are left. At this point slide down all the unmatched crosses from row $r+1$ to row r. See Figs. 11 and 12 where, for readers convenience, the crosses are represented with different colours to stress the matching.

This matching and sliding of crosses translate to the action of the operator e_r, as long as it is possible, on the second row of the biword w. The operator Υ_r is the analogue of applying m times the crystal operator e_r, to the second row of w, where m is the number of unmatched $r+1$ in the second row of w. Thereby, we also write $\Upsilon_r w = e_r^m w$ to mean the biword obtained by applying m times the crystal operator e_r, to the second row of w, where m is the number of unmatched $r+1$ in the second row of w. Similarly, we define the operator $\overline{\Upsilon}_r w := f_r^m w$, where m is the number of unmatched r in the second row of w.

Consider now the 01-filling representation of the biwords w and $\Upsilon_r w$ in the Ferrers shapes λ embedded in a rectangle shape. Apply the local rules, as defined in Sect. 4. Notice that in the 01-filling of w, we match a cross in row $r+1$ with a cross to the SE, in row r, such that in these two rows there is no unmatched cross in a column between them. These two growth diagrams have the same bottom sequences of partitions and the left sequences are different only in the partitions assigned to the

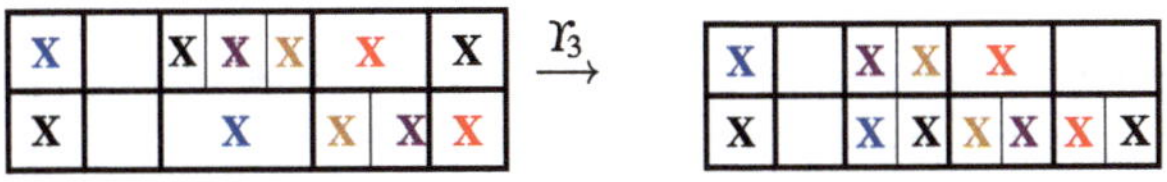

$$\begin{pmatrix} 1 & 1 & 3 & 3 & 3 & 3 & 4 & 4 & 4 & 5 & 5 \\ 3 & 4 & 3 & 4 & 4 & 4 & 3 & 3 & 4 & 3 & 4 \end{pmatrix} \underset{e_3^2}{\overset{f_3^2}{\substack{\longleftarrow \\ \longrightarrow}}} \begin{pmatrix} 1 & 1 & 3 & 3 & 3 & 3 & 4 & 4 & 4 & 5 & 5 \\ 3 & 4 & 3 & 3 & 4 & 4 & 3 & 3 & 4 & 3 & 3 \end{pmatrix}$$

Fig. 11 The double action of the crystal operators e_3 on the biword w (12) is visualized in the *rectangle* defined by the *two rows* 3 and 4 of the Ferrers shape λ. The cross matching is represented with *different colours*. The unmatched *black* **Xs** in the *top row* slide down. One has $\Upsilon_3 w = e_3^2 w$ and $f_3^2 \Upsilon_3 w = w$

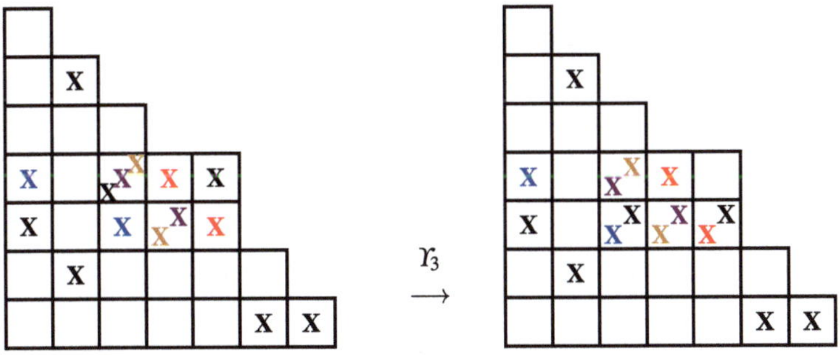

$$\begin{pmatrix} 1 & 1 & 2 & 2 & 3 & 3 & 3 & 3 & 4 & 4 & 4 & 5 & 5 & 6 & 7 \\ 3 & 4 & 2 & 6 & 3 & 4 & 4 & 4 & 3 & 3 & 4 & 3 & 4 & 1 & 1 \end{pmatrix} \overset{\Upsilon_3}{\longrightarrow} \begin{pmatrix} 1 & 1 & 2 & 2 & 3 & 3 & 3 & 3 & 4 & 4 & 4 & 5 & 5 & 6 & 7 \\ 3 & 4 & 2 & 6 & 3 & 3 & 4 & 4 & 3 & 3 & 4 & 3 & 3 & 1 & 1 \end{pmatrix}$$

Fig. 12 The cross matching in the *rows* 3 and 4 of λ is represented using *different colours*, where the *black* **Xs** are unmatched. On the *left*, scan λ along columns from *left* to *right* and *bottom* to *top* to obtain w. On the *right*, the new set of crossed cells of λ yields a new biword $\Upsilon_3 w$

rows r and $r+1$. In [20], it is proved that the bottom sequence is preserved by the operations e_r and f_r, when the entries of the first row of the biword w are distinct. In the 01-filling the biword read from the diagram has no repeated letters in both rows. (In fact the first row of the biword can be read as $[n]$, if the blank columns in the 01-filling are shrunk.) Thus the bottom sequence of the 01-filling of the growth diagram is preserved by those operations. Moreover, from the 01-filling and [20], we can assure the validity of left square diagram in Fig. 14 for any biword.

Let w_r and $\tilde{w}_r$ be the biwords that are obtained from w and $\Upsilon_r w$, after deleting all the biletters with bottom rows different from r and $r+1$. The translation of the movement of the cells in the Ferrers shape to the 01-filling is as follows. In the 01-filling of w_r move up, without changing of columns, the matched crosses of row $r+1$, say s crosses, to the top most s rows such that they form SW chain. Then slide down the remaining unmatched crosses, from row $r+1$ to row r, without changing of columns, such that these crosses and all the crosses of row r form a SW chain. The result is the 01-filling corresponding to $\tilde{w}_r$. See Fig. 13.

It is clear that the longest SW chain in the first k columns, from right to left, of the 01-filling of w_r and of $\tilde{w}_r$, has length equal to the total number of crosses in row

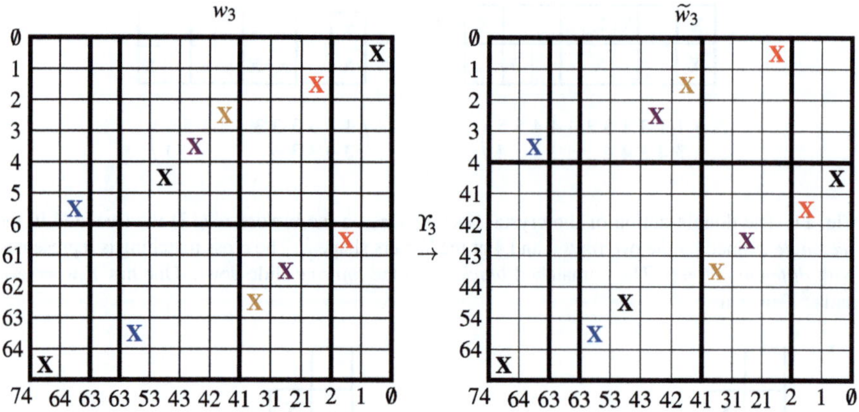

Fig. 13 The 01-fillings of w_3 and $\tilde{w}_3$ are respectively a blow-up of the *two row rectangles* defined by the *rows* 3 and 4 of the Ferrers shapes in Fig. 11. The *crosses* are represented using *different colors* according to the matching

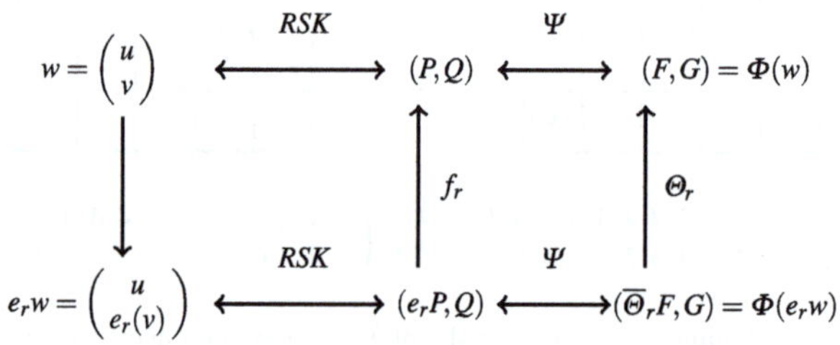

Fig. 14 $\Phi(w) = (F, G)$, $\Phi(e_r w) = (\overline{\Theta}_r F, G)$ and $\Phi(f_r w) = (\Theta_r F, G)$

r and row $r+1$, of those columns, minus the number of matched crosses in row $r+1$, of those columns. It means that the length of the longest SW chain in the first k columns, from right to left is preserved. Theorem 2 implies that the bottom sequences in growth diagrams corresponding to w_r and $\tilde{w}_r$ are the same. See Fig. 13.

Let Θ_r be the analogue operator of f_r, and $\overline{\Theta}_r$ the analogue of e_r for SSAFs, defined in [26]. Figure 14 shows the relation between the action of crystal operators e_r, f_r, their analogues $\overline{\Theta}_r$, Θ_r, the RSK and the analogue of RSK Φ. If F is SSAF, put $\Upsilon_r F = \overline{\Theta}_r^m F$ where m is the number of unmatched $r+1$ in the row reading (left to right and top to bottom) of the SSAF F. Similarly, put $\overline{\Upsilon}_r F = \Theta_r^m F$. Equivalently, if $F = \Psi(P)$ with P a SSYT, then $\Upsilon_r F = \Psi(e_r^m P)$ where m is the number of unmatched $r+1$ in P. See Fig. 14.

Next theorem is, therefore, a consequence of our discussion. Recall that $\Phi = \varrho \circ RRSK = \Psi \circ RSK$.

Theorem 3 *Let w be a biword in lexicographic order. If $\Phi(w) = (F, G)$ then $\Phi(\Upsilon_r w) = (\Upsilon_r F, G)$.*

Theorem 4 [4] *Let λ be a Ferrers shape where $\lambda_r = \lambda_{r+1} > \lambda_{r+2} \geqslant 0$, for some $r \geqslant 1$. Let w be a biword consisting of a multiset of cells of λ containing the cell $(r+1, \lambda_{r+1})$ with multiplicity at least one. Let $\Phi(w) = (F, G)$. If $sh(F) = \nu$ then $\nu_r < \nu_{r+1}$ and $sh(\Upsilon_r F) = s_r \nu$. Moreover, $\Upsilon_r w$ does not contain the biletter $\binom{\lambda_{r+1}}{r+1}$ and therefore it fits the Ferrers shape λ with the cell $(r+1, \lambda_{r+1})$ deleted.*

Figure 15 illustrates this theorem.

Transpose the Ferrers shape λ, and denoted it by $\overline{\lambda}$. This means that we swap the rows of w and then rearrange it in lexicographic order. This biword is w^* and is now represented in $\overline{\lambda}$. The move of crosses between the rows r and $r+1$ of λ is translated to a move of crosses on the columns r and $r+1$ of $\overline{\lambda}$. The translation of Υ_r to the columns r and $r+1$ of the Ferrers shape λ is $\Upsilon_r^* w := \Upsilon_r w^*$. The growth diagram of the 01-filling of w is transposed, through the secondary diagonal. The move of crosses on the rows is translated to a move of crosses on the columns. As a consequence of the symmetry of the growth diagram we have the following versions of Theorems 3 and 4.

Corollary 2 [4] *Let w be a biword in lexicographic order. If $\Phi(w) = (F, G)$ then $\Phi(\Upsilon_r^* w) = (F, \Upsilon_r G)$.*

Corollary 3 [4] *Let λ be a Ferrers shape and let $\overline{\lambda} = (\lambda'_1, \lambda'_2, \ldots, \lambda'_{\lambda_1})$ be the conjugate of λ where $\lambda'_r = \lambda'_{r+1} > \lambda'_{r+2}$. Let w be a biword consisting of a multiset of cells of λ containing the cell $(\lambda'_{r+1}, r+1)$ with multiplicity at least one. Let $\Phi(w) = (F, G)$. If $sh(G) = \nu$ then $\nu_r < \nu_{r+1}$ and $sh(\Upsilon_r G) = s_r \nu$. Moreover, $\Upsilon_r^* w$ does not contain the biletter $\binom{r+1}{\lambda'_{r+1}}$ and therefore it fits the Ferrers shape λ with the cell $(\lambda'_{r+1}, r+1)$ deleted.*

The following proposition says that under certain conditions the shape of a SSAF change.

Proposition 1 [4] *Let F be a SSAF with shape ν, and $\nu_r < \nu_{r+1}$, for some $r \geqslant 1$. Then $sh(\Upsilon_r F) = s_r \nu$.*

7.2 The Bijection

Next theorem characterizes the biwords whose biletters constitute a multiset of cells in a staircase possibly plus a layer of boxes sited on the stairs of the staircase, in French convention, leaving the top of the first column and the end of the first row free. See Fig. 1. This is the NW version because we use operations on the rows of a Ferrers shape. Rows are counted from SE to NW. A SE version also exists by performing operations on the columns. Columns are counted from NW to SE. Let $SSAF_n$ be the set of all SSAFs with basement $[n]$.

Theorem 5 (NW Inner Layer) *Let w be a biword in lexicographic order on the alphabet $[n]$, and let $\Phi(w) = (F, G) \in SSAF_n^2$, with $sh(F) = \nu$ and $sh(G) = \beta$.*

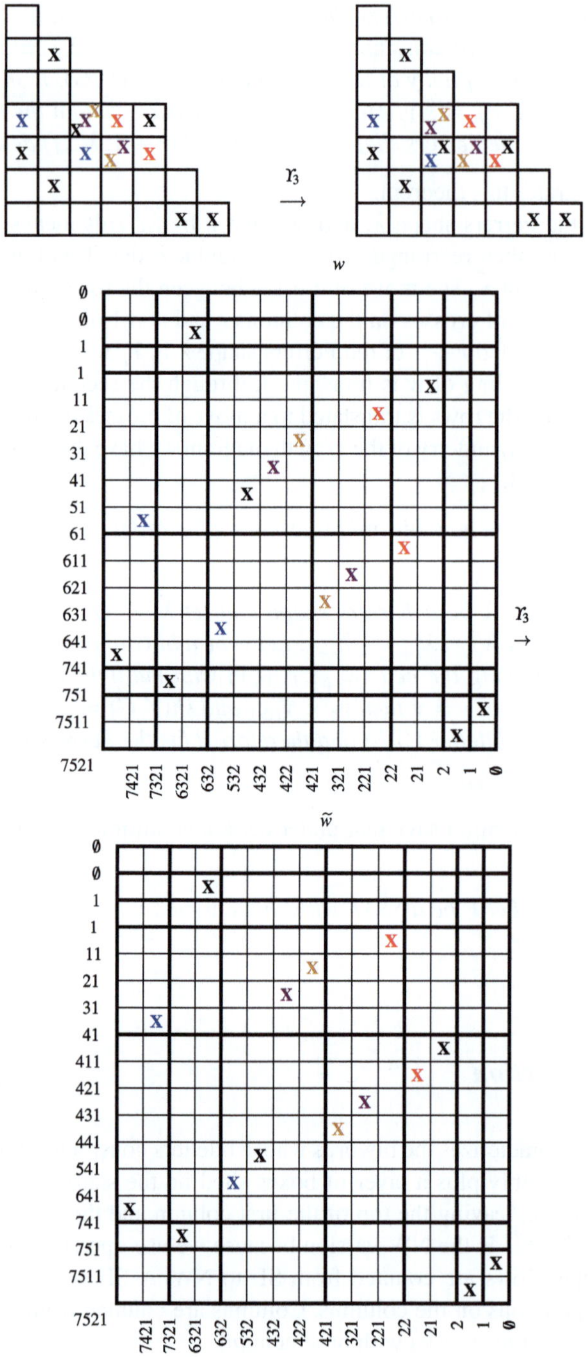

Υ3
w
∅ ∅ 1 1 11 21 31 41 51 61 611 621 631 641 741 751 7511 7521
7421 7321 6321 632 532 432 422 421 321 221 22 21 2 1 ∅
Υ3
w̃
∅ ∅ 1 1 11 21 31 41 411 421 431 441 541 641 741 751 7511 7521
7421 7321 6321 632 532 432 422 421 321 221 22 21 2 1 ∅

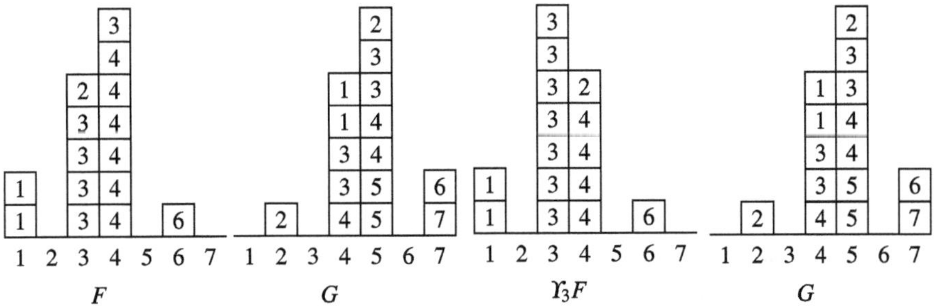

$sh(F) = (2,0,5,7,0,1,0)$ $\qquad$ $sh(\Upsilon_3 F) = (2,0,7,5,0,1,0) = s_3 sh(F)$

Fig. 15 The procedure of passing from a biword under the action of the operator Υ_r to a pair of SSAFs, where $n = 7$, $r = 3$. The biletters (i, j) satisfy $i + j \leqslant 8 + 1$ in w, and $i + j \leqslant 7 + 1$ in $\Upsilon_3 w$. In particular, the biletter $(r + 1, 5) = (4, 5)$ in w is transformed to $(r, 5) = (3, 5)$ in $\Upsilon_3 w$. The *crossed cells* are represented using *different colours* according to the matching

Let $0 \leqslant p < n$, $1 \leqslant r_1 < \cdots < r_p < n$. *Then* w *consists of a multiset of cells in the staircase of size* n *and the* p *cells* $\binom{n - r_1 + 1}{r_1 + 1}, \ldots, \binom{n - r_p + 1}{r_p + 1}$, *each with multiplicity at least one, if and only if*

(*a*) $\beta \leq \omega s_{r_p} \cdots s_{r_2} s_{r_1} \nu$,
(*b*) $\beta \not\leq \omega s_{r_p} \cdots \hat{s}_{r_i} \cdots s_{r_2} s_{r_1} \nu$, *for* $i = 1, 2, \ldots, p$, *where* $\widehat{\ }$ *means omission.*

Proof By induction on p. For $p = 0$, it is the main Theorem in [1, 2, 4]. For $p > 0$, we use Theorem 4 and Proposition 1. Let $p \geqslant 1$ and assume that the statement is true for $p - 1$.

Only if part. We have $\Phi(w) = (F, G)$, with $sh(F) = \nu$, and w has the p biletters $\binom{n - r_1 + 1}{r_1 + 1}, \binom{n - r_2 + 1}{r_2 + 1}, \ldots, \binom{n - r_p + 1}{r_p + 1}$, where $1 \leqslant r_1 < \cdots < r_p < n$. Applying Theorem 4 to the cell $(r_1 + 1, n - r_1 + 1)$, it follows that $\nu_{r_1} < \nu_{r_1+1}$, and if $\Phi(\Upsilon_{r_1} w) = (\Upsilon_{r_1} F, G)$, then $sh(\Upsilon_{r_1} F) = s_{r_1} \nu$ and $\Upsilon_{r_1} w$ does not contain the cell $(r_1 + 1, n - r_1 + 1)$. Therefore, $\Upsilon_{r_1} w$ fits the staircase and, in addition, it has the $p - 1$ biletters $\binom{n - r_2 + 1}{r_2 + 1}, \ldots, \binom{n - r_p + 1}{r_p + 1}$. From the *only if part* of the inductive hypothesis, if $\beta := sh(\Upsilon_{r_1} F) = s_{r_1} \nu$ then $sh(G) \leq \omega s_{r_k} \cdots s_{r_2} \beta$ and, for all $2 \leq i \leq k$, $sh(G) \not\leq \omega s_{r_k} \cdots \hat{s}_{r_i} \cdots s_{r_2} \beta$. Henceforth, $sh(G) \not\leq \omega s_{r_k} \cdots \hat{s}_{r_i} \cdots s_{r_2} s_{r_1} \nu$, for $i = 2, \ldots, k$, and $sh(G) \leq \omega s_{r_k} \cdots s_{r_2} s_{r_1} \nu$. It remains to prove that $sh(G) \not\leq \omega s_{r_k} \cdots s_{r_2} \nu$.

By contradiction suppose that $sh(G) \leq \omega s_{r_p} \cdots s_{r_2} \nu$. Since, for all $2 \leq i < p$, $sh(G) \not\leq \omega s_{r_p} \cdots \hat{s}_{r_i} \cdots s_{r_2} s_{r_1} \nu$ and, from Lemma 1, (*a*), $\omega s_{r_p} \cdots \hat{s}_{r_i} \cdots s_{r_2} \nu \leq \omega s_{r_p} \cdots \hat{s}_{r_i} \cdots s_{r_2} s_{r_1} \nu$, then for all $2 \leq i \leq p$, $sh(G) \not\leq \omega s_{r_p} \cdots \hat{s}_{r_i} \cdots s_{r_2} \nu$. Hence, we have, $\Phi(w) = (F, G)$, with $sh(F) = \nu$ such that, for all $2 \leq i \leq p$, $sh(G) \not\leq \omega s_{r_p} \cdots \hat{s}_{r_i} \cdots s_{r_2} \nu$, and $sh(G) \leq \omega s_{r_p} \cdots s_{r_2} \nu$. From the *if part* of the inductive hypothesis, w only has $p-1$ cells sited on the staircase of size n. This is false, and we must have $sh(G) \not\leq \omega s_{r_p} \cdots s_{r_2} \nu$.

If part. Let $1 \leqslant r_1 < \cdots < r_p < n$. Consider $\Phi(w) = (F, G) \in SSAF_n^2$ with $sh(F) = \nu$ such that $sh(G) \not\leq \omega s_{r_p} \cdots \hat{s}_{r_i} \cdots s_{r_2} s_{r_1} \nu$, for $i = 1, 2, \ldots, p$, and $sh(G) \leq \omega s_{r_p} \cdots s_{r_2} s_{r_1} \nu$. Then, by Lemma 1, (d), $(s_{r_t} \cdots s_{r_2} s_{r_1} \nu)_{r_{t+1}} < (s_{r_t} \cdots s_{r_2} s_{r_1} \nu)_{r_{t+1}+1}$, for $t = 0, \ldots, p-1$. Since $\nu_{r_1} < \nu_{r_1+1}$, by Proposition 1, one has $sh(\Upsilon_{r_1} F) = s_{r_1} \nu$. Let $s_{r_1} \nu =: \beta$, then $\Phi(\Upsilon_{r_1} w) = (\Upsilon_{r_1} F, G) \in SSAF_n^2$ with $sh(\Upsilon_{r_1} F) = \beta$ is such that $sh(G) \not\leq \omega s_{r_p} \cdots \hat{s}_{r_i} \cdots s_{r_2} \beta$, for $i = 2, \ldots, p$, and $sh(G) \leq \omega s_{r_p} \cdots s_{r_2} \beta$. From the *if part* of the inductive hypothesis, $\Upsilon_{r_1} w$ has the following $p-1$ cells sited on the staircase of size n, $\begin{pmatrix} n - r_2 + 1 \\ r_2 + 1 \end{pmatrix}, \ldots, \begin{pmatrix} n - r_p + 1 \\ r_p + 1 \end{pmatrix}$. But $\bar{\Upsilon}_r \Upsilon_r w = w$ and applying $\bar{\Upsilon}_r$ to $\Upsilon_r w$ will not change these $p-1$ biletters. Either it creates a cell, in row $r_1 + 1$, sited on the staircase of size n, thus the biletter $\begin{pmatrix} n - r_1 + 1 \\ r_1 + 1 \end{pmatrix}$, or does nothing above the staircase of size n. Suppose that we are in the last case, w has the same $p-1$ biletters, as $\Upsilon_r w$, sited on the staircase. Using the *only if part* of the inductive hypothesis, we have

(a) $sh(G) \not\leq \omega s_{r_p} \cdots \hat{s}_{r_i} \cdots s_{r_2} \nu$, for $i = 2, \ldots, p$, and
(b) $sh(G) \leq \omega s_{r_p} \cdots s_{r_2} \nu$.

The last condition (b) is in contradiction with our hypothesis because $sh(G) \not\leq \omega s_{r_p} \cdots s_{r_2} \nu$. Hence, w should have one more biletter sited on the staircase and since $\bar{\Upsilon}_{r_1}$ can only create the biletter $\begin{pmatrix} n - r_1 + 1 \\ r_1 + 1 \end{pmatrix}$, then w has the p biletters

$$\begin{pmatrix} n - r_1 + 1 \\ r_1 + 1 \end{pmatrix}, \begin{pmatrix} n - r_2 + 1 \\ r_2 + 1 \end{pmatrix}, \ldots, \begin{pmatrix} n - r_p + 1 \\ r_p + 1 \end{pmatrix}.$$

8 A Non-symmetric Cauchy Kernel Over Near Staircases

We give finally a bijective proof via the RSK analogue of the identity (6).

8.1 Some Notation and a Lemma

Given a finite set S and an integer $m \geqslant 0$, let $\binom{S}{m}$ denote the set of all m-element subsets of S.

Let $0 \leqslant p < n$ and $1 \leqslant r_1 < r_2 < \cdots < r_p < n$. For each $0 \leqslant z \leqslant p$, and each $H_z = \{i_1 < i_2 < \cdots < i_z\} \in \binom{[p]}{z}$, define

$$\mathscr{A}_z^{H_z} = \left\{ (F,G) \in SSAF_n^2 : \begin{array}{c} sh(G) \not\leq \omega s_{r_{i_z}} \cdots \hat{s}_{r_{i_m}} \cdots s_{r_{i_1}} sh(F),\ m=1,2,\ldots,z \\ sh(G) \leq \omega s_{r_{i_z}} \cdots s_{r_{i_1}} sh(F) \end{array} \right\}.$$

Put $\mathscr{A} := \mathscr{A}_0^{\emptyset} = \{ (F,G) \in SSAF_n^2 : sh(G) \leq \omega sh(F) \}$.

For each $z = 0, \ldots, p-1$, and $H_z = \{2 \leqslant i_1 < \cdots < i_z\} \in \binom{[2,p]}{z}$, where $[2,p] = [p] \setminus \{1\}$, let

$$\mathscr{B}_z^{H_z} := \left\{ (F,G) \in SSAF_n^2 : \begin{array}{c} sh(F)_{r_1} < sh(F)_{r_1+1} \\ sh(G) \not\leq \omega s_{r_{i_z}} \cdots \hat{s}_{r_{i_m}} \cdots s_{r_{i_1}} s_{r_1} sh(F),\ m=1,2,\ldots,z \\ sh(G) \leq \omega s_{r_{i_z}} \cdots s_{r_{i_1}} s_{r_1} sh(F) \end{array} \right\}.$$

Lemma 2 *Given $1 \leqslant p < n$, for each $z = 0, \ldots, p-1$, and $H_z = \{2 \leqslant i_1 < \cdots < i_z\} \in \binom{[2,p]}{z}$, let $H_{z+1}^1 := \{1\} \cup H_z$. Then*

$$\mathscr{B}_z^{H_z} = \{(F,G) \in \mathscr{A}_z^{H_z} : sh(F)_{r_1} < sh(F)_{r_1+1}\} \cup \mathscr{A}_{z+1}^{H_{z+1}^1}.$$

Let $x = (x_1, \ldots, x_n)$ and $y = (y_1, \ldots, y_n)$ be two sequences of indeterminates. From (11), with π_{r_1} the isobaric divided difference with respect to x, one has

$$\sum_{\nu \in \mathbb{N}^n} \pi_{r_1} \hat{\kappa}_\nu(x) = \sum_{\nu \in \mathbb{N}^n} \pi_{r_1} \sum_{\substack{F \in SSAF_n \\ sh(F)=\nu}} x^F = \sum_{\substack{\nu \in \mathbb{N}^n \\ \nu_{r_1} \geqslant \nu_{r_1+1}}} \pi_{r_1} \sum_{\substack{F \in SSAF_n \\ sh(F)=\nu}} x^F$$

$$= \sum_{\substack{\nu \in \mathbb{N}^n \\ \nu_{r_1} \geqslant \nu_{r_1+1}}} \sum_{\substack{F \in SSAF_n \\ sh(F)=\nu}} x^F + \sum_{\substack{\nu \in \mathbb{N}^n \\ \nu_{r_1} > \nu_{r_1+1}}} \sum_{\substack{F \in SSAF_n \\ sh(F)=s_{r_1}\nu}} x^F.$$

Thereby

$$\sum_{\nu \in \mathbb{N}^n} \left(\pi_{r_1} \sum_{\substack{(F,G) \in \mathscr{A}_z^{H_z} \\ sh(F)=\nu}} x^F y^G \right) = \sum_{\substack{(F,G) \in \mathscr{A}_z^{H_z} \\ sh(F)_{r_1} \geqslant sh(F)_{r_1+1}}} x^F y^G + \sum_{(F,G) \in \mathscr{B}_z^{H_z}} x^F y^G. \quad (13)$$

8.2 The Combinatorial Formula

In [16], Lascoux gives a Cauchy kernel expansion formula for any Ferrers shape which produces, in particular, the following Cauchy kernel expansion over the near staircase, on the NW part Fig. 1,

$$\prod_{(i,j)\in\lambda}(1-x_iy_j)^{-1}=\sum_{\nu\in\mathbb{N}^n}(\pi_{r_1}\dots\pi_{r_p}\hat{\kappa}_\nu(x))\kappa_{\omega\nu}(y).$$

Next theorem gives a bijective explanation.

Theorem 6 *Let* $0\leqslant p<n$ *and* $1\leqslant r_1<r_2<\cdots<r_p<n$. *Let* λ *be the near staircase Fig. 1. Then*

1.

$$\sum_{\nu\in\mathbb{N}^n}(\pi_{r_1}\dots\pi_{r_p}\hat{\kappa}_\nu(x))\kappa_{\omega\nu}(y)=\sum_{(F,G)\in\mathscr{A}}x^Fy^G+\sum_{1\leqslant z\leqslant p}\sum_{H_z}\sum_{(F,G)\in\mathscr{A}_z^{H_z}}x^Fy^G, \tag{14}$$

where $H_z\in\binom{[p]}{z}$.

2.

$$\prod_{(i,j)\in\lambda}(1-x_iy_j)^{-1}=\sum_{\nu\in\mathbb{N}^n}\left(\pi_{r_1}\dots\pi_{r_p}\hat{\kappa}_\nu(x)\right)\kappa_{\omega\nu}(y).$$

Proof
1. The proof is by induction on p. If $p=0$, we get,

$$\sum_{\nu\in\mathbb{N}^n}\hat{\kappa}_\nu(x)\kappa_{\omega\nu}(y)=\sum_{\nu\in\mathbb{N}^n}\sum_{\substack{(F,G)\in SSAF_n^2\\ sh(G)\leqslant\omega sh(F)\\ sh(F)=\nu}}x^Fy^G=\sum_{\nu\in\mathbb{N}^n}\sum_{\substack{(F,G)\in\mathscr{A}\\ sh(F)=\nu}}x^Fy^G=\sum_{(F,G)\in\mathscr{A}}x^Fy^G.$$

Let $p\geqslant 1$ and suppose that identity (14) is true for $p-1$ operators π_i. Then, since π_{r_1} is linear,

$$\sum_{\nu\in\mathbb{N}^n}(\pi_{r_1}\pi_{r_2}\dots\pi_{r_p}\hat{\kappa}_\nu(x))\kappa_{\omega\nu}(y)=\pi_{r_1}\left(\sum_{\nu\in\mathbb{N}^n}(\pi_{r_2}\dots\pi_{r_p}\hat{\kappa}_\nu(x))\kappa_{\omega\nu}(y)\right)$$

$$=\pi_{r_1}\left(\sum_{z=0}^{p-1}\sum_{H_z\in\binom{[2,p]}{z}}\sum_{(F,G)\in\mathscr{A}_z^{H_z}}x^Fy^G\right)=\sum_{z=0}^{p-1}\sum_{H_z\in\binom{[2,p]}{z}}\left(\sum_{\nu\in\mathbb{N}^n}\pi_{r_1}\sum_{\substack{(F,G)\in\mathscr{A}_z^{H_z}\\ sh(F)=\nu}}x^Fy^G\right)$$

$$=\sum_{z=0}^{p-1}\sum_{H_z\in\binom{[2,p]}{z}}\left(\sum_{\substack{(F,G)\in\mathscr{A}_z^{H_z}\\ sh(F)_{r_1}\geqslant sh(F)_{r_1+1}}} x^F y^G+\sum_{(F,G)\in\mathscr{B}_z^{H_z}} x^F y^G\right). \tag{15}$$

Using Lemma 2,

$$(15)=\sum_{z=0}^{p-1}\sum_{H_z\in\binom{[2,p]}{z}}\left(\sum_{\substack{(F,G)\in\mathscr{A}_z^{H_z}\\ sh(F)_{r_1}\geqslant sh(F)_{r_1+1}}} x^F y^G+\sum_{\substack{(F,G)\in\mathscr{A}_z^{H_z}\\ sh(F)_{r_1}<sh(F)_{r_1+1}}} x^F y^G+\sum_{(F,G)\in\mathscr{A}_{z+1}^{H_{z+1}^1}} x^F y^G\right)$$

$$=\sum_{z=0}^{p-1}\sum_{H_z\in\binom{[2,p]}{z}}\left(\sum_{(F,G)\in\mathscr{A}_z^{H_z}} x^F y^G+\sum_{(F,G)\in\mathscr{A}_{z+1}^{H_{z+1}^1}} x^F y^G\right)=\sum_{z=0}^{p}\sum_{H_z\in\binom{[p]}{z}}\sum_{(F,G)\in\mathscr{A}_z^{H_z}} x^F y^G.$$

2. Let λ_0 the biggest staircase inside of λ. Then, identifying x_iy_j with the biletter $\binom{j}{i}$, and using the bijection in Theorem 5, it follows that

$$\prod_{(i,j)\in\lambda}(1-x_iy_j)^{-1}=\prod_{(i,j)\in\lambda_0}(1-x_iy_j)^{-1}\prod_{i=1}^{p}(1-x_{r_i+1}y_{n-r_i+1})^{-1}$$

$$=\sum_{(F,G)\in\mathscr{A}} x^F y^G+\sum_{z=1}^{p}\sum_{H_z\in\binom{[p]}{z}}\sum_{(F,G)\in\mathscr{A}_z^{H_z}} x^F y^G.\square$$

The combinatorial expansion formula for the SE part, in the sense of [16], can be obtained by using the change of basis (9), or, in alternative, the SE version of Theorem 5 coming from Corollary 3.

Acknowledgements We thank to Robin Langer for asking us, during the Summer School on Algebraic and Enumerative Combinatorics in S. Miguel de Seide, whether we had a growth diagrammatic interpretation for the Mason's analogue of RSK to produce pairs of semi-skyline augmented fillings. The authors are grateful to the referees whose comments rather improved the final version of the manuscript.

This work was partially supported by the Centro de Matemática da Universidade de Coimbra (CMUC), funded by the European Regional Development Fund through the program COMPETE and by the Portuguese Government through the FCT—Fundação para a Ciência e a Tecnologia under the project PEst-C/MAT/UI0324/2011. The second author was also supported by Fundação para a Ciência e a Tecnologia (FCT) through the Grant SFRH/BD/33700/2009.

References

1. Azenhas, O., Emami, A.: Semi-skyline augmented fillings and non-symmetric Cauchy kernels for stair-type shapes. In: FPSAC'13, DMTCS proc. AS, pp. 981–992 (2013)
2. Azenhas, O., Emami, A.: An analogue of the Robinson-Schensted-Knuth correspondence and non-symmetric Cauchy kernels for truncated staircases. Eur. J. Combin. **46**, 16–44 (2015) [arXiv:1310.0341]
3. Demazure, M.: Désingularisation des variétés de Schubert généralisées. Ann. Sci. École Norm. Sup. **4**(7), 53–88 (1974)
4. Emami, A.: An analogue of the Robinson-Schensted-Knuth correspondence, growth-diagrams and non-symmetric Cauchy kernels. Ph.D. Thesis, Universidade de Coimbra, Coimbra (2014)
5. Fomin, S.: The generalised Robinson-Schensted-Knuth correspondence. J. Zap. Nauchn. Sem. Leningrad. Otdel. Mat. Inst. Steklov. (LOMI) **155**(195), 156–175 (1986)
6. Fu, A.M., Lascoux, A.: Non-symmetric Cauchy kernels for the classical groups. J. Comb. Theory, Ser. A **116**(4), 903–917 (2009)
7. Fulton, W.: Young Tableaux with Applications to Representation Theory and Geometry. London Mathematical Society Student Texts, vol. 35. Cambridge University Press, Cambridge (1997)
8. Greene, C.: An extension of Schensted's theorem. Adv. Math. **14**, 254–265 (1974)
9. Haglund, J., Haiman, M., Loehr, N.: A combinatorial formula for Macdonald polynomials. J. Am. Math. Soc. **18**, 735–761 (2005)
10. Haglund, J., Haiman, M., Loehr, N.: A combinatorial formula for non-symmetric Macdonald polynomials. Am. J. Math. **130**, 359–383 (2008)
11. Hong, J., Kang, S-J.: Introduction to Quantum Groups and Crystal Bases. Graduate Studies in Mathematics, vol. 42. American Mathematical Society, Providence (2002)
12. Knuth, D.E.: Permutations, matrices and generalized Young tableaux. Pac. J. Math. **34**, 709–727 (1970)
13. Krattenthaler, C.: Growth diagrams, and increasing and decreasing chains in fillings of ferrers shapes. Adv. Appl. Math. **37**, 404–431 (2006)
14. Kwon, J.H.: Crystal graphs and the combinatorics of Young tableaux. In: Hazewinkel, M. (ed.) Handbook of Algebra, vol. 6. North-Hollands, Amsterdam (2009)
15. Lascoux, A.: Anneau de Grothendieck de la variété de drapeaux. In: Cartier, P., Illusie, L., Katz, N.M., Laumon, G., Manin, Y.I., Ribet, K.A. (eds.) The Grothendieck Festschrift, vol. III, pp.1–34. Birkhäuser, Boston (1990)
16. Lascoux, A.: Double crystal graphs. Studies in memory of Issai Schur. In: Progress in Mathematics, vol. 210, pp. 95–114. Birkhäuser, Boston (2003)
17. Lascoux, A.: Polynômes (2012). http://phalanstere.univ-mlv.fr/~al/
18. Lascoux, A., Schützenberger, M.-P.: Tableaux and noncommutative Schubert polynomials. Funct. Anal. Appl. **23**, 63–64 (1989)
19. Lascoux, A., Schützenberger, M.-P.: Keys and standard bases, invariant theory and tableaux. IMA Vol. Math. Appl. **19**, 125–144 (1990)
20. Lascoux, A., Leclerc, B., Thibon, J.-Y.: The plactic monoid. In: Lothaire, M. (ed.) Algebraic Combinatorics on Words, chap. 6. Cambridge University Press, Cambridge (2002)
21. Lecouvey, C.: Combinatorics of crystal graphs for the root systems of types A_n, B_n, C_n, D_n and G_2. In: Kuniba, A., et al. (eds.) Combinatorical aspects of integrable systems. Based on the workshop, RIMS, Kyoto, Japan, July 26–30 (2004), vol. 17, pp. 11–41. Mathematical Society of Japan. MSJ Memoirs, Tokyo (2007)
22. Lecouvey, C.: Crystal bases and combinatorics of infinite rank quantum groups. Trans. Am. Math. Soc. **361**(1), 297–329 (2009)
23. Leeuwen, M.V.: Spin-preserving Knuth correspondences for ribbon tableaux. Electron. J. Combin. **12**(1), Article R10, 65 (2005)
24. Littlewood, D.E.: The theory of group characters. Clarendon Press, Oxford (1940)

25. Mason, S.: A decomposition of Schur functions and an analogue of the Robinson-Schensted-Knuth algorithm. Sém. Lothar. Combin. **57**, B57e, 24 (2006/2008)
26. Mason, S.: An explicit construction of type A Demazure atoms. J. Algebra Comb. **29**(3), 295–313 (2009)
27. Pons, V.: Interval structure of the Pieri formula for Grothendieck polynomials. Int. J. Autom. Comput. **23**(1), 123–146 (2013) [arXiv:1206.6204]
28. Reiner, V., Shimozono, M.: Key polynomials and flagged Littlewood-Richardson rule. J. Combin. Theory Ser. A **70**(1), 107–143 (1995)
29. Roby, T.W.: Applications and extensions of Fomin's generalization of the Robinson-Schensted correspondence to differential posets. Ph.D. Thesis, MIT, Cambridge (1991)
30. Sagan, B.: The Symmetric Group: Representations, Combinatorial Algorithms, and Symmetric Functions. Graduate Texts in Mathematics, vol. 203. Springer, New York (2001)
31. Stanley, R.P.: Enumerative Combinatorics, vol. 2. Cambridge Studies in Advanced Mathematics, vol. 62. Cambridge University Press, Cambridge (1998)

Clustering Techniques Applied on Cross-Sectional Unemployment Data

Carlos Balsa, Alcina Nunes, and Elisa Barros

Abstract Using a cross-section database that observes the Portuguese labour market in two different phases of the business cycle, the present paper aims to address the issue of the segmentation of the Portuguese labour market taking into account the heterogeneity resulting from different unemployment characteristics observed along the Portuguese geographical space and applying two optimization clustering methods: the k-means and the spectral methods. The k-means is a traditional optimisation clustering method applied to cluster data observations. Spectral clustering is an alternative method based on the computation of the dominant eigenvectors of a matrix related with the distance among data points. The results obtained by the two methods are not identical but are very close and show that, apart the economic phase of the cycle, Portugal presents two very different profiles of registered unemployment. One of them can be considered problematic because it presents a higher percentage of unemployed women, long duration unemployed and unemployed with low levels of formal education—these are the groups that present more difficulties in the labour market and for which is more difficult to find a job after losing one. The segmentation of the labour market is a reality and the labour market is not adjusting to the business cycle.

C. Balsa (✉)
Instituto Politécnico de Bragança (IPB), Bragança, Portugal

Centro de Estudos de Energia Eólica e Escoamentos Atmosféricos (CEsA) da Faculdade de Engenharia da Universidade do Porto, Porto, Portugal
e-mail: balsa@ipb.pt

A. Nunes
Instituto Politécnico de Bragança (IPB), Bragança, Portugal

Grupo de Estudos Monetários e Financeiros (GEMF) da Faculdade de Economia da Universidade de Coimbra, Coimbra, Portugal
e-mail: alcina@ipb.pt

E. Barros
Instituto Politécnico de Bragança (IPB), Bragança, Portugal
e-mail: ebarros@ipb.pt

J.-P. Bourguignon et al. (eds.), *Dynamics, Games and Science*, CIM Series in Mathematical Sciences 1, DOI 10.1007/978-3-319-16118-1_5

1 Introduction

Clustering aims the partition of a data set by bringing together similar elements in subsets, called clusters. The similarity depends on the distance between data points such that a reduced distance indicates that they are more similar among them than a larger one. Several distinct methods can be used to measure the distance among the elements of a data set [1]. Along this work we will consider the traditional Euclidian distance, i.e., the 2-norm of the differences between data points vectors.

The k-means [2] is an optimization method that partitions the data in exactly k clusters, previously determine. This is achieved in a sequence of steps which begins, for instance, with an initial partition randomly generated. In each step the cluster's centroid (arithmetic vector mean) is computed. The minimum distance between each data point and the clusters′ different centroids will decide the formation of new clusters. The formation of a new cluster implies assigning each observation to the cluster that presents the lowest distance. After that the centroids are (re)calculated and the former step is repeated until the moment each individual belongs to a stable cluster, i.e., when the sum of the squared distances to the centroid of all data points over all the clusters is minimized. The algorithm presents a rather fast convergence, but one cannot guarantee that the algorithm finds the global minimum [3].

Spectral methods, and in particular the spectral clustering algorithm, are useful when considering non-convex shaped subsets of points. Spectral clustering methods use the k dominant eigenvectors of a matrix, called affinity matrix, based on the distance between the observations. The idea is grouping data points in a lower-dimensional space described by those k eigenvectors [4]. The approach may not make a lot of sense, at first, since we could apply the k-means methodology directly without going through all the matrix calculations and manipulations. However, some analyses show that mapping the points to this k-dimensional space can produce tight clusters that can easily be found applying k-means [5].

The empirical application of the two cluster methods will be made using as observations 278 Portuguese mainland municipalities (*concelhos*) that will be classified regarding the type/characteristics of unemployment official registers. The set of observations, $x_1, \ldots, x_{278}$, that contains 278 vectors, whose 11 coordinates (variables that represent unemployment features) are the values for some of the indicators used to characterise Portuguese unemployment (gender, age classes, levels of formal education, situation relating unemployment and unemployment duration), is divided in k clusters. The idea is that the classification of observations resulting from the spectral method could be than compared to the classification given by the traditional k-means method.

The composition of the groups, in two different phases of the economic cycle, will be analysed. In the empirical analysis will be classify data for the year 2007, which corresponds to an expansion phase (2007 was a year of low unemployment) of the business cycle, and for the year 2012, which corresponds to a busyness cycle recession phase (2012 was a year with a high level of register unemployment). In 2012 data regarding unemployment in Portugal were disturbing—14% of the population which wanted to work, was looking for a job and, at the same time, was available to work was unemployed. According to Centeno and Novo [6], if until during the nineties the unemployment rate had a cyclical behaviour in this new century the Portuguese economy is not being able to change the structural nature of the registered unemployed, in particular in the positive phase of the business cycle as expected. The authors refer, for instance, the long term unemployment that remains a problem over time in the Portuguese economy due, among others, to low levels of formal education and generous unemployment benefits. They point out the labour market segmentation as the cause for the problem: the Portuguese labour market splits itself in a stronger and a weaker market in a vicious cycle of low productivity, qualification and remuneration. Starting from this point our aim is to verify the existence of this segmentation in two different economic periods confirming, or not, the idea of persistence of difficulties regarding the adjustment of the labour market to the business cycle. At the same time, the distribution of the registered unemployed features over space will also be analysed to verify if the Portuguese labour market segmentation persists just over time or if remains also over space—considering as units of space the Portuguese municipalities.

The results are analysed from both mathematical and economic points of view. The main goal is to find evidence regarding which method produces the best cluster partition and, accordingly, to understand if the resulting clusterisation makes sense either in terms of the spatial distribution of unemployment characteristics over a country's administrative territory and over time—particularly over the different phases of the business cycle—regarding the issue of segmentation of the Portuguese labour market refereed above. On one hand it is important to understand if the application of the cluster methodology could avoid *a priori* subjective grouping criteria as the one that just groups municipalities in administrative regions [7]. The problem of unemployment has traditionally been studied as a national phenomenon being the national unemployment rates considered as a consequence of national labour market characteristics. However the rates of unemployment at the regional level are very heterogeneous inside countries, particularly in Europe. According to Südekum [8], in Europe, regional labour market disparities within many countries are of about the same magnitude as differences between countries. Taking into account this findings is important to understand the regional dynamics of unemployment [9, 10]. On another hand is important to understand if the mathematical methodology could offer insights that help to stress the need to implement structural changes in the labour market since the normal stabilization mechanism of the business cycle seems not enough to overcome the segmentation problem. In sum, the idea is to understand if a particular cluster methodology for data mining analysis provides useful and suitable information that could be used to the development of national, regional or local unemployment policies.

The paper is divided as follows. The k-means method and the spectral clustering method are presented in Sects. 2 and 3, respectively. The methods description is followed by Sect. 4 where data and variables analysed are also presented and described. In Sect. 5 we move ahead toward the optimal number of clusters applying both selected methods. In Sect. 6 the results are presented and discussed. The groups obtained by the two methods, its composition and its evolution from 2007 to 2012 will be analysed. The concluding remarks can be found on Sect. 7.

2 The *k*-Means Method

We are concerned with m data observations $x_i \in \mathbb{R}^n$ that we want classify in k clusters, where k is predetermined. We organize the data as lines in a matrix $X \in \mathbb{R}^{m\times n}$. To describe the k-means method as proposed in [3] we denote a partition of vectors $x_1, \ldots, x_m$ in k clusters as $\prod = \{\pi_1, \ldots, \pi_k\}$ where

$$\pi_j = \{\ell : x_\ell \in \text{cluster} j\}$$

defines the set of vectors in cluster j. The centroid, or the arithmetic mean, of the cluster j is:

$$m_j = \frac{1}{n_j} \sum_{\ell \in \pi_j} x_\ell \tag{1}$$

where n_j is the number of elements in cluster j. The sum of the squared distance, in 2-norm, between the data points and the j cluster's centroid is known as the *coherence*:

$$q_j = \sum_{\ell \in \pi_j} \left\| x_\ell - m_j \right\|_2^2 \tag{2}$$

The closer the vectors are to the centroid, the smaller the value of q_j. The quality of a clustering process can be measured as the *overall coherence*:

$$Q\left(\prod\right) = \sum_{j=1}^{k} q_j \tag{3}$$

The k-means is considered an optimization method because it seeks a partition process that minimizes $Q(\prod)$ and, consequently, finds an optimal coherence. The problem of minimizing the *overall coherence* is NP-hard and, therefore, very difficult to achieve. The basic algorithm for k-means clustering is a two step heuristic procedure. Firstly, each vector is assigned to its closest group. After that, new centroids are computed using the assigned vectors. In the following version of k-means algorithm, proposed by Eldén [3], these steps are alternated until the changes in the *overall coherence* are lower than a certain tolerance previously defined.

The k-means algorithm

1. Start with an initial partitioning $\prod^{(0)}$ and compute the corresponding centroid vectors $m_j^{(0)}$ for $j = 1, \ldots, k$. Compute $Q(\prod^{(0)})$. Put $t = 1$.
2. For each vector x_i find the closest centroid. If the closest centroid is m_p^{t-1} assign i to $\pi_p^{(t)}$.
3. Compute the centroids $m_j^{(t)}$ for $j = 1, \ldots, k$ of the new partitioning $\prod^{(t)}$.
4. If $\left|Q(\prod^{(t)}) - Q(\prod^{(t-1)})\right| <$ tol, stop; Otherwise $t = t + 1$ and return to step 2.

Since it is an heuristic algorithm there is no guarantee that k-means will converge to the global minimum, and the result may depend on the initial partition $\prod^{(0)}$. To avoid this issue, it is common to run it multiple times, with different starting conditions choosing the solution with the smaller $Q\left(\prod\right)$.

3 Spectral Clustering Method

Let $x_1, \ldots, x_m$ be a m data observations set in a n-dimensional euclidian space. We want to group these m points in k clusters in order to have better within-cluster affinities and weaker affinities across clusters. The affinity between two observations x_i and x_j is defined by Ng et al. [11] as:

$$A_{ij} = e^{-\frac{\left\|x_i - x_j\right\|_2^2}{2\sigma^2}} \tag{4}$$

where σ is a scaling parameter that determines how fast the affinity decreases with the distance between x_i and x_j. The appropriate choice of this parameter is crucial [5]. In [11] we can find a description of a method able to choose the scaling parameter automatically.

The spectral clustering algorithm proposed by Ng et al. [11] is based on the extraction of dominant eigenvalues and their corresponding eigenvectors from the normalized affinity matrix $A \in \mathbb{R}^{m \times m}$. The components A_{ij} of A are given by Eq. (4), if $i \neq j$, and by $A_{ii} = 0$, if $i = j$. The sequence of steps is presented below in the Spectral Clustering Algorithm.

The spectral clustering algorithm

1. Form the affinity matrix A as indicated in Eq. (4).
2. Construct the normalized matrix $L = D^{-1/2}AD^{-1/2}$ with $D_{ii} = \sum_{j=1}^{m} A_{ij}$.
3. Construct the matrix $V = [v_1 v_2 \ldots v_k] \in \mathbb{R}^{m \times k}$ by stacking the eigenvectors associated with the k largest eigenvalues of L.
4. Form the matrix Y by normalizing each row in the $m \times k$ matrix V (i.e. $Y_{ij} = V_{ij} / \left(\sum_{j=1}^{k} V_{ij}^2\right)^{1/2}$).
5. Treat each row of Y as a point in $\mathbb{R}^k$ and group them in k clusters by using the k-means method.
6. Assign the original point x_i to cluster j if and only if row i of matrix Y was assigned to cluster j.

4 Data Description

The 278 data observations represents the Portuguese continental *concelhos*. Each data point have 11 coordinates representing characteristics of the unemployed register individuals. Indeed, the unemployed individuals registered in the Portuguese public employment services of the *Instituto de Emprego e Formação Profissional (IEFP)* present a given set of distinctive characteristics related with gender, age, formal education, unemployment spell (unemployment for less than a year or more than a year) and the situation related with the unemployment situation (unemployed individual looking for a first employment or for another employment). The observations and coordinates are observed in two points in time—2007 and 2012—which correspond to two different phases of the business cycle as mentioned above.

The referred characteristics are important determinants of unemployment and are important economic vectors regarding the development of public employment policies. National public policies benefit from being based on simple and objective rules however a blind application of these national policies across space (regions) and time could be ineffective if the addressed problem is not well explored and identified [12] at a regional and temporal level. For example, in many countries the labour market problems of large cities are quite different from those of rural areas—even when the unemployment rate is the same [13]. It is believed this is the case of the Portuguese economy. The same happens when the issue concerns different phases of the business cycle, i.e. when the problem is analysed over time. The impact of a crisis on the labour market varies across (and within) countries depending on the structure of the economy, institutions in place and policymakers response. In particular, the downturn in the business cycle has different implications for various segments of the population as defined by such characteristics as gender and age [14]. So, in terms of expected results and to avoid the waste of scarce resources, well targeted policies are more efficient. Consequently, the main strategies of the labour market policy, and in particular of the labour market measures, should be to deal with the situation at hand, across regions and time. For instance, it is easier to integrate an unemployed person into a job if the policy measure depends on the local labour market conditions [13] at a particular moment in time depending on the macroeconomic context. Even considering the hypothesis that the Portuguese market is a segmented market where the situation of groups more exposed to labour market problems persists over time independently of the business cycle phases.

A complete study of regional and temporal similarities (or dissimilarities) in a particular labour market, as the Portuguese, should not be limited by a descriptive analysis of the associated economic phenomena. It should also try to establish spacial and temporal comparison patterns among geographic areas and time periods in order to develop both national and regional public policies to fight the problem.

Indeed high unemployment indicators and regional inequalities are major concerns for European policy-makers since the creation of European Union. However, even if the problem is known the policies dealing with unemployment and regional inequalities have been few and weak [15]. In Portugal, in particular, there are some studies that try to define geographic, economic and social homogeneous groups [16]. Yet, to the best of our knowledge, there are no studies that offer an analysis of regional unemployment profiles. Other economies are starting to develop this kind of statistical analysis using as a policy tool the cluster analysis methodology [7, 17–19]. Regarding unemployment and business cycle there are several national and international studies that focus this subject however they not compare results obtained from the cluster methodology as this papers intends to do.

The data concerning the above mentioned characteristics are openly available in a monthly period base in the website of *IEFP* (http://www.iefp.pt/estatisticas/Paginas/Home.aspx). Additionally, the month of December gives information about the stock of registered unemployed individuals at the end of the respective year. In the case of this research work, data from unemployment registers in 2007 and 2012 have been used (2007 was a year of low register unemployment and 2012 was a year with a high level of register unemployment). The eleven variables available to characterise the individuals and that have been used here are divided in demographic variables and variables related with the labour market. These variables are dummy variables, measured in percentage of the total number of register individuals in a given *concelho*. We have a total of 278 observation vectors $x_1, \ldots, x_{278}$, which one with 11 unemployment characteristics presented in Table 1.

Women, individuals in a situation of long duration unemployment, younger or older unemployed individuals and the ones with lower formal education are the most fragile groups in the labour market and, consequently, are the most exposed

Table 1 Description of the register unemployed

Variable number	Unemployment variable
1	Female
2	Long duration unemployed
3	Unemployed looking for a new employment
4	Age lower than 25 years
5	Age between 25 and 35 years
6	Age between 35 and 54 years
7	Age equal or higher than 55 years
8	Less than 4 years of formal education
9	Between 4 and 6 years of formal education
10	Between 6 and 12 years of formal education
11	Higher education

to unemployment situations [14, 20]. They are also the most challenging groups regarding the development of public employment policies, namely the regional ones.

5 Determining the Clusters' Number

We begin by applying the two clustering methods to partition in k clusters the data points set $x_1, \ldots, x_m$, with $m = 278$ Portuguese mainland *concelhos* regarding the 11 chosen unemployment characteristics. As the optimal number of targeted groups is unknown *a priori*, we repeat the partition for $k = 2,\ 3,\ 4$ and 5 clusters.

To evaluate the quality of the results from the cluster methodology and to estimate the correct number of groups in our data set we resort the silhouette statistic framework. The silhouette statistic introduced by Kaufman and Rousseeuw [1] is a way to estimate the number of groups in a data set. Given observation x_i, the average dissimilarity to all other points in its own cluster is denoted as a_i. For any other cluster c, the average dissimilarity of x_i to all data points in cluster c is represented by $\bar{d}\,(x_i, c)$. Finally, b_i denote the minimum of these average dissimilarities $\bar{d}\,(x_i, c)$. The *silhouette width* for the observation x_i is:

$$s_i = \frac{(b_i - a_i)}{\max\{b_i,\ a_i\}}. \tag{5}$$

The *average silhouette width* is obtained by averaging the s_i over all observations:

$$\bar{s} = \frac{1}{m}\sum_{i=1}^{m} s_i. \tag{6}$$

If the *silhouette width* of an observation is large it tends to be well clustered. Observations with small *silhouette width* values tend to be those that are scattered between clusters. The *silhouette width* s_i in Eq. (5) ranges from -1 to 1. If an observation has a value close to 1, then it is closer to its own cluster than it is to a neighbouring one. If it has a *silhouette width* close to -1, then it is a sign that it is not very well clustered. A *silhouette width* close to zero indicates that the observation could just as well belong to its current cluster or one that is near to it.

The *average silhouette width* (Eq. (6)) can be used to estimate the number of clusters in the data set by using the partition with two or more clusters that yield the largest average silhouette width [1]. As a rule of thumb, it is considered that an *average silhouette width* greater than 0.5 indicates a reasonable partition of the data, and a value less than 0.2 would indicate that the data do not exhibit a cluster structure [5].

Figure 1 presents the average *silhouette width* value (Eq. (6)) corresponding to the case of seven different partitions of the data observations set, this is, $k =$

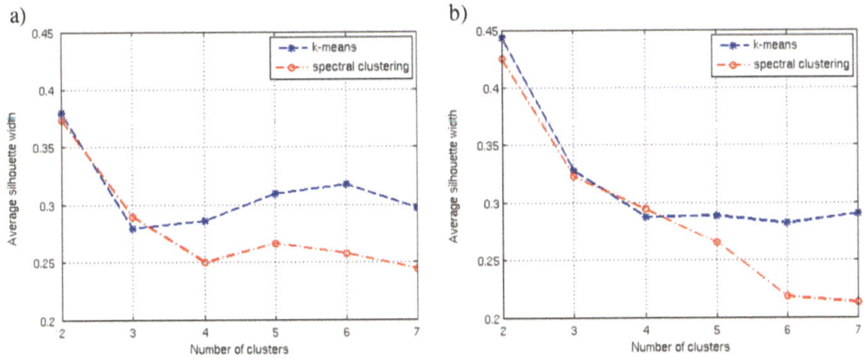

Fig. 1 Average *silhouette width* for $k = 2, 3, \ldots, 7$ clusters. (**a**) 2007, (**b**) 2012

2, 3, 4, 5, 6 and 7 clusters resulting from the application of the *k*-means and spectral method in 2007 and 2014.

In 2007 the best partition obtained with the application of the two methods occurs with $k = 2$. The two cluster solution presents a larger value of the average *silhouette width*, near 0.37. Nonetheless, the *average of the silhouette* is close but smaller than 0.5 which reveals that the data set does not seem to present a strong trend to be partitioned in two clusters. The computed value indicates that the distance between the two considered clusters is not very large.

In 2012 the best partition obtained with the application of the two methods occurs also with $k = 2$. The average *silhouette with* value is near 0.43, for the spectral method, and near 0.44 for the *k*-means method.

These values reveal that the data set presents a stronger trend to be partitioned in two clusters in 2012 than in 2007. Nevertheless, the results seem to stress the hypothesis defined in the beginning of this paper—the Portuguese labour market is segmented in two parts that persist over the business cycle being necessary more (in terms of labour policy) than just an adjustment of the economy. It could be concluded that more than cyclical the unemployment in Portugal seems to be structural.

6 Results' Analysis

In Sect. 5 we have seen that both spectral clustering method and *k*-means method indicate that the data are best partitioned into two clusters. Here we analyse the results of the classification of the 278 Portuguese mainland *concelhos* in two groups. We start by analysing the differences between clusters resulting from the application of different methods in Sect. 6.1, after that (Sect. 6.2) we compare the two clusters in order to identifies the two main unemployment profiles. Finally, in Sect. 6.3 we identify the differences between groups obtained for the years 2007 and 2012.

Table 2 Clusters' properties

Method	j	n_j	q_j	Q
(a) 2007				
k-Means	1	160	5.0954	8.9582
	2	118	3.8627	
Spectral	1	153	4.8862	8.9706
	2	125	4.0844	
(b) 2012				
k-Means	1	177	3.4161	5.3276
	2	101	1.9115	
Spectral	1	154	2.7511	5.3536
	2	124	2.6026	

6.1 Comparing Methods

The number of observations included in the two clusters corresponding to the years 2007 and 2012 partitions are presented in Table 2. Both methods produce a larger and a smaller cluster. However, the difference between the two clusters is higher in the case of k-means. Cluster 1 obtained by the k-means includes $n_1 = 160$ observations in 2007 and $n_1 = 177$ in 2012. The corresponding cluster obtained by the spectral method includes $n_1 = 153$ observations in 2007 and $n_1 = 154$ in 2012. The difference between cluster 1 and cluster 2 increases from 2007 to 2012 for both methods. In the case of k-means the difference increases from 42 to 76, while in the case of the spectral method the difference increases from 28 to 30.

We also note in Table 2 that the higher the number of observations n_j the greater the value of the local coherence q_j (Eq. (2)). For example, the difference of 23 observations for the first cluster in 2012 is reflected in the larger local coherence ($q_1 = 3.4161$) obtained with the k-means methods. The second cluster comprises $n_2 = 101$ observations and presents a local coherence of $q_2 = 1.9115$, for the k-means, and $n_2 = 124$ observations and a local coherence of $q_2 = 2.6026$ for the spectral method.

Although the differences between the computed coherence for each cluster, it is possible to observe that both methods achieve a very similar overall coherence (Eq. (3)). In 2007, $Q \approx 8.96$ for the k-means and $Q \approx 8.97$ for the spectral method. In 2012, $Q \approx 5.33$ for the k-means and $Q \approx 5.35$ for the spectral method. Both methods achieve approximately the same *overall coherence*.

As the two methods do not produce the same two clusters, we have identified the composition of each cluster to know which are the observations that remain in the same cluster independently of the method. Table 3 presents the number of repeated observations in the two clusters, both in 2007 and 2012. The cluster determined by one method includes, or is included in, the corresponding cluster determined by the other method. In 2007, for instance, the number of repeated observations in the cluster one is 153 which corresponds to the total number of observations

Table 3 Intersection of the two clusters

Cluster j	k-Means n_j	Spectral n_j	Repeated n_j
(a) 2007			
1	160	153	153
2	118	125	118
(b) 2012			
1	177	154	154
2	101	124	101

included in the same cluster ($n_1 = 153$) obtained by spectral method. In turn, the 118 observations assigned to second cluster by the k-means are also part of the same cluster determined by the spectral method ($n_2 = 118$).

The results presented in Table 3 show that clusters obtained by the two methods are very similar. In 2007, 153 observations are assigned to the first cluster and 118 assigned to the second by the two methods. There are only 7 observations whose allocation fluctuates with the method. This number represents about 2.5% of the total number of observations (278). The number of floating observations increases to 23 in 2012, which corresponds to 8.8% of the total number of observations.

6.2 *Comparing Clusters*

The comparison of observations' values in the two clusters over the years of 2007 and 2012 is presented in Fig. 2. The mean value obtained for the 11 variables, presented in Table 1, are plotted for each cluster. It is possible to observe that both methods retrieve clusters that present the same pattern. This reinforces the idea that the clusters produced by the two methods are very similar.

From Fig. 2 we can also observe that the big difference between cluster 1 and cluster 2 is mainly due to the value of the variables 2 (long duration unemployment), 9 (formal education lower than 4 year) and 10 (6–12 years of formal education). In the second cluster are gathered the Portuguese mainland *concelhos* that present a higher percentage of unemployed register individuals with long duration unemployed (variable 2) and unemployed with a low level of formal education (variable 9). In cluster 1 we have *concelhos* that presents unemployed register individuals with an higher level of formal education. This observation is made regardless of the phase of the economic cycle because the differences between the two clusters in 2007 are very similar to the differences in 2012, as stressed before.

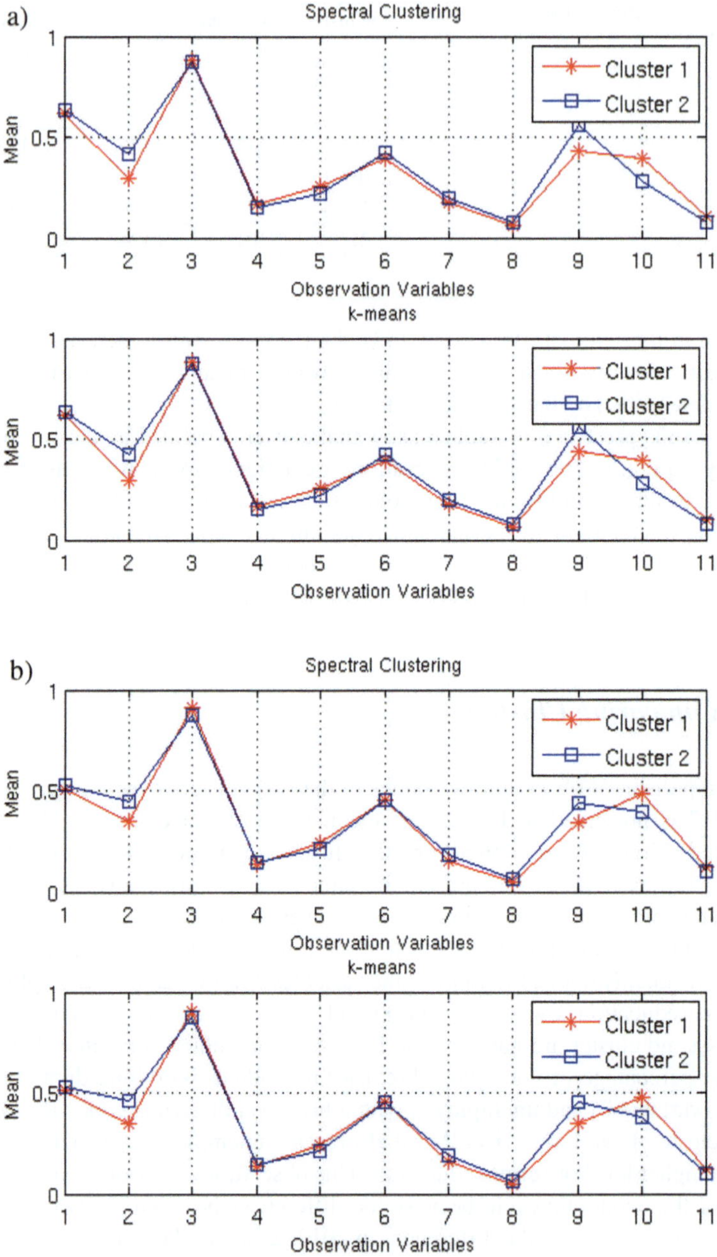

Fig. 2 Observations' mean by variable. (**a**) 2007, (**b**) 2012

Table 4 Comparison between 2007 and 2012

Cluster j	2007 n_j	2012 n_j	Repeated n_j
(a) k-Means			
1	160	177	133
2	118	101	74
(b) Spectral clustering			
1	153	154	118
2	125	124	89

6.3 Comparing Years

Table 4 allows to analyse the composition of the cluster considering the year in analysis. Cluster 1 increases, regarding the number of observations including on it, and cluster 2 decreases its number, from 2007 to 2012. This variation is more pronounced in the case of k-means application.

In Table 4 is also possible to observe that there are a large number of observations that are in the same cluster whatever the phase of the cycle in study. For example, in the case of the K-means application there are 133 observations that remain in the first cluster and 101 which remain in the second one. The high number of repeated observations indicates that an high number of *concelhos* follow the same unemployment profile pattern over the economic cycle apart the economic phase, stressing the division of the country in two distinct unemployment profiles that remain over time and economic crises.

The selection of only 2 years may not allow to capture the dynamics of the economic cycle but these results show that the Portuguese labour market seem not to adapt to changes in the economic and financial framework and may be subjected to an unemployment hysteresis phenomenon—an eminently microeconomics (firm level) phenomenon that spreads to a macroeconomic level [21]. The economic literature discusses that the link between growth and unemployment may be hysteretic [22]. The phenomenon arrives from, for instance, human capital depreciation, stigmatization by employers, loss of social networks and a strict employment protection legislation which, for example, makes Portugal a good case to study labour demand driven hysteresis [23].

The results presented in this research work seem to stress the existence of path dependent unemployment profiles (as happens for the employment [21]) determined by the history of previous economic cycle dynamics. As a result the labour market policies designed to fight unemployment should be not only address particular regional problems but should also present long-run effects to fight, for example, the human capital depreciation in more depressed regions and/or the stigmatization of particular labour market groups (as women and older people).

In Fig. 3 the mean value of the variables found in each cluster's repeated observations is compared. Figure 3a refers to the *concelhos* that remain in the

a)

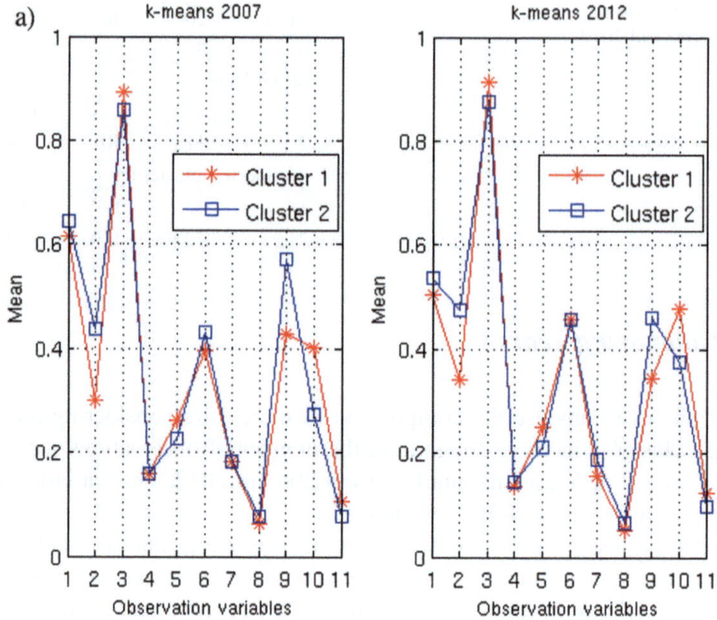

k-means

b)

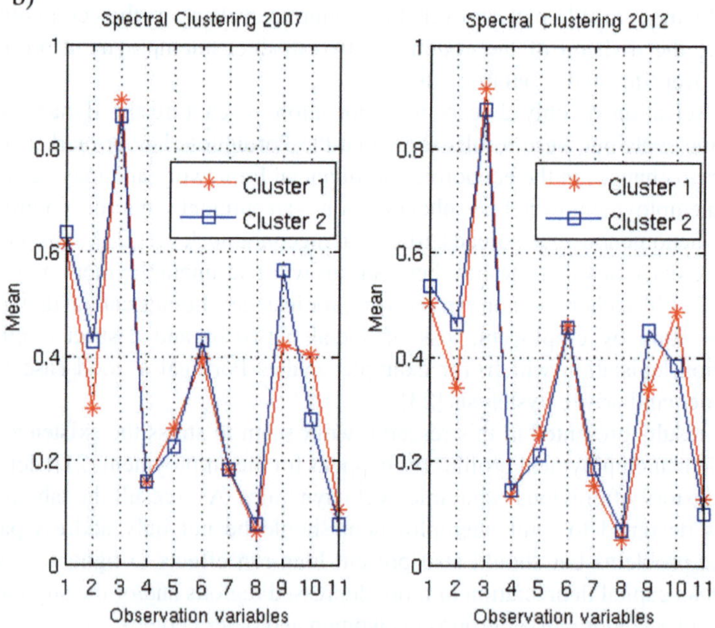

spectral

Fig. 3 Values of the repeated observations over time. (**a**) *k*-Means, (**b**) spectral same cluster, independently of the year where the k-means had been applied (see Table 4a), and Fig. 3b refers to those *concelhos* that are repeated when the spectral method had been applied (see Table 4b). It is clear that the main differences between the two cluster are due to the variables 1 (female unemployment), 2 (long duration unemployment), 9 (4–6 years of formal education) and 10 (6–12 years of formal education). In the second cluster (cluster 2) are gathered the Portuguese mainland *concelhos* that present a higher percentage of unemployed register individuals with more problematic characteristics—women, long duration unemployed individuals, individuals that are looking for a job for the first time (individuals with no connections with the labour market), individuals with more than 55 years and with lower number of years of formal education (for example, this cluster gathers the *concelhos* with a lower percentage of unemployed individuals with a higher education). From the labour market groups point of view, we can say that cluster 2 includes the groups of individual that are the most fragile.

The results illustrated in Fig. 3 confirm the existence of two distinct country's unemployment profiles. Despite the stage of the business cycle, that tends to align the unemployment registration rates regardless of the observed individual characteristics, it is possible to verify the existence of regional differences that should be studied and analysed carefully in order to make employment public policies more effective and efficient.

Comparing Fig. 3a with Fig. 3b we can conclude that k-means and spectral method denote the same overall data partition and, consequently, both can identify the two distinct unemployment profiles.

It has been concluded until now that there are two groups of *concelhos* with homogeneous profiles in terms of unemployment characteristics. In order to have an idea about the composition of each group, Table 5 presents the list of Portuguese mainland *concelhos* that remain in the same cluster independently of the applied cluster method and the analysed year. There are 118 *concelhos* always classified in cluster one and 74 always classified in cluster two. Such observation allows to conclude these are the most representative *concelhos* regarding each one of the two unemployment profiles. For a full characterisation of the *concelhos* within each cluster would be important to have information regarding other economic, social and demographic features however it is possible to find in cluster 2 well known depressed municipalities located in the more depressed Portuguese regions. In cluster 1 is possible to find municipalities located along the coast, the more developed area of the Portuguese economy. The remaining *concelhos*—the 86 ones not presented in the above table—move between the two unemployment profiles depending on the applied cluster method and the year in analysis. Such swing suggests that, over time, their profile of unemployment individuals is no as robust as the one found for the *concelhos* in the table. Nevertheless, they match one of the two identified profiles.

Table 5 *Concelhos* that remain in the same cluster independently of the applied cluster method and the analysed year

Cluster	*Concelhos*
1	Bragança, Caminha, Melgaço, Miranda do Douro, Monção, Ponte da Barca, Vila Nova de Cerveira, Vila Real, Almeida, Anadia, Aveiro, Batalha, Carregal do Sal, Castelo Branco, Celorico da Beira, Coimbra, Condeixa-a-Nova, Figueira da Foz, Figueira de Castelo Rodrigo, Guarda, Ílhavo, Leiria, Lousã, Mangualde, Mealhada, Meda, Miranda do Corvo, Montemor-o-Velho, Mortágua, Oliveira do Hospital, Penela, Pinhel, Pombal, Porto de Mós, Proença-a-Nova, Santa Comba Dão, São Pedro do Sul, Trancoso, Vila de Rei, Alcanena, Alcobaça, Alcochete, Alenquer, Almada, Almeirim, Amadora, Benavente, Bombarral, Cadaval, Caldas da Rainha, Cartaxo, Cascais, Constância, Ferreira do Zêzere, Golegã, Lisboa, Loures, Lourinhã, Mação, Mafra, Moita, Montijo, Nazaré, Odivelas, Oeiras, Ourém, Palmela, Peniche, Rio Maior, Santarém, Sardoal, Seixal, Sesimbra, Setúbal, Sintra, Sobral de Monte Agraço, Tomar, Torres Novas, Torres Vedras, Vila Franca de Xira, Vila Nova da Barquinha, Aljustrel, Alvito, Arraiolos, Beja, Borba, Campo Maior, Castelo de Vide, Cuba, Elvas, Eestremoz, Évora, Fronteira, Grândola, Montemor-o-Novo, Mora, Portalegre, Reguengos de Monsaraz, Santiago do Cacém, Sines, Sousel, Vendas Novas, Viana do Alentejo, Vila Viçosa, Albufeira, Aljezur, Castro Marim, Faro, Lagoa, Lagos, Loulé, Olhão, Portimão, São Brás de Alportel, Silves, Tavira, Vila do Bispo, Vila Real de Santo António
2	Alijó, Amarante, Armamar, Arouca, Baião, Barcelos, Boticas, Cabeceiras de Basto, Carrazeda de Ansiães, Castelo de Paiva, Celorico de Basto, Cinfães, Espinho, Fafe, Felgueiras, Freixo de Espada à Cinta, Gondomar, Guimarães, Lamego, Lousada, Marco de Canaveses, Mesão Frio, Moimenta da Beira, Mondim de Basto, Montalegre, Murça, Oliveira de Azemeis, Paços de Ferreira, Paredes, Penafiel, Peso da Régua, Ponte de Lima, Póvoa de Lanhoso, Póvoa de Varzim, Resende, Ribeira de Pena, Sabrosa, Santa Maria da Feira, Santa Marta de Penaguião, Santo Tirso, São João da Pesqueira, Tabuaço, Torre de Moncorvo, Trofa, Valongo, Valpaços, Vieira do Minho, Vila do Conde, Vila Nova de Famalicão, Vila Nova de Foz Côa, Vila Nova de Gaia, Vila Pouca de Aguiar, Vila Verde, Vimioso, Vinhais, Vizela, Albergaria-a-Velha, Arganil, Castanheira de Pera Castro Daire, Fornos de Algodres, Gouveia, Idanha-a-Nova, Penacova, Penamacor, Sátao, Seia, Avis, Castro Verde, Crato, Mourão, Ourique, Serpa, Monchique

7 Concluding Remarks

Results obtained with spectral clustering method are consistent with the k-means results. Both methods denote the same overall data partition over space (different geographic areas) and time (different phases of the business cycle). The cluster methodology being considered an exploratory data-analysis technique intended largely for generating rather than testing hypothesis [24] proves to be, in this empirical application, a powerful ally to start an analysis of a given labour market (in this particular case, the Portuguese).

From the economic point of view the clustering methodology helps to identifies the two main groups of Portuguese *concelhos* in terms of unemployment profiles. Both cluster methods seem to divide the total number of *concelhos* in two economic meaningful clusters that persists regardless of the business cycle phase denoting the segmentation of the Portuguese labour market and a possible unemployment hysteresis phenomenon. Indeed, these two clusters persist over time and space, in the Portuguese economy.

Apart the economic phase of the cycle, Portugal presents two very different profiles of registered unemployment. One of them can be considered problematic because it presents a higher percentage of unemployed women, long duration unemployed and unemployed with low levels of formal education—these are the groups that present more difficulties in the labour market and for which is more difficult to find a job after losing one. The two well defined groups could be object of distinct public policies—policies well targeted that can be more effective and efficient regarding the spatial and the economic context where they will be implemented and that be able to have not only short- but also long-run effects which could overcome the problems arising from path dependent unemployment profiles.

References

1. Kaufman, L., Rousseeuw, P.J.: Finding Groups in Data: An Introduction to Cluster Analysis. Wiley, New York (1990)
2. MacQueen, J.B.: Some methods for classification and analysis of multivariate observations. In: Proceedings of 5-th Berkeley Symposium on Mathematical Statistics and Probability, vol. 1, pp. 281–297. University of California Press (1967)
3. Eldén, L.: Matrix methods in data mining and pattern recognition. SIAM (2007)
4. Mouysset, S., Noailles, J., Ruiz, D.: Using a global parameter for Gaussian affinity matrices in spectral clustering. High Performance Computing for Computational Science—VECPAR 2008, pp. 378–390 (2008)
5. Martinez, W.L., Martinez, A.R., Solka, J.L.: Exploratory Data Analysis with MATLAB. CRC Press, Boca Raton (2010)
6. Centeno, M., Novo, Á.A.: Segmentação. Tema de Discussão, Boletim Económico de Primavera 2012 do Banco de Portugal **7**, 30 (2012)
7. Álvarez de Toledo, P., Núñez, F., Usabiaga, C.: Labour Market Segmentation, Clusters, Mobility and Unemployment Duration with Individual Microdata. MPRA Paper 46003, University Library of Munich, Germany (2013)
8. Südekum, J.: Increasing returns and spatial unemployment disparities. Pap. Reg. Sci. **84**, 159–181 (2005)
9. Garcilazo, J.E., Spiezia, V.: Regional nemployment clusters: neighborhood and state effects in Europe and North America. Rev. Reg. Stud. **37**(3), 282–302 (2007)
10. Altavilla, C., Caroleo, F.E.: Asymmetric Effects of National-based Active Labour Market Policies. Reg. Stud. **47**(9), 1482–1506 (2013)
11. Ng, A.Y., Jordan, M.I., Weiss, Y.: On spectral clustering: analysis and an algorithm. Adv. Neural Inf. Process. Syst. (NIPS) **14**, 849–856 (2002)
12. Campo, D., Monteiro, C.M.F., Soares, J.O.: The European regional policy and the socio-economic diversity of European regions: a multivariate analysis. Eur. J. Oper. Res. **187**, 600–612 (2008)
13. Blien, U., Hirschenauer, F., Van, P.T.H.: Classification of regional labour markets for purpose of labour market policies. Pap. Reg. Sci. **89**(4), 859–881 (2009)
14. Verick, S.: Who is hit hardest during a financial crisis? The Vulnerability of Young Men and Women to Unemployment in an Economic Downturn. IZA Discussion Papers, 4359 (2009)

15. Overman, H.G., Puga, D.: Unemployment Clusters across Europe's Regions and Countries. Econ. Policy **17**(34), 115–148 (2002)
16. Soares, J.O., Marques, M.M.L., Monteiro, C.M.F.: A multivariate methodology to uncover regional disparities: a contribution to improve European union and Governmental decisions. Eur. J. Oper. Res. **45**, 121–135 (2003)
17. Arandarenko, M., Juvicic, M.: Regional labour market differences in Serbia: assessment and policy recommendations. Eur. J. Comp. Econ. **4**(2), 299–317 (2007)
18. López-Bazo, E., Del Barrio, T., Artís, M.: Geographical distribution of unemployment in Spain. Reg. Stud. **39**(3), 305–318 (2005)
19. Nadiya, D.: Econometric and cluster analysis of potential and regional features of the labor market of Poland. Ekonomia **21**, 28–44 (2008)
20. Dean, A.: Tackling long-term unemployment amongst vulnerable groups. OECD Local Economic and Employment Development (LEED) Working Paper 2013/11, OECD Publishing (2013)
21. Mota, P.R., Vasconcelos, P.B.: Nonconvex Adjustments Costs, Hysteresis, and the Macrodynamics of Employment. J. Post Keynesian Econ. **36**(1), 93–112 (2012)
22. Lang, D., Peretti, C.: A strong hysteretic model of Okun's Law: theory and a preliminary investigation. Int. Rev. Appl. Econ. **23**(4), 445–462 (2009)
23. Mota, P.R., Varejão, J., Vasconcelos, P.B.: Hysteresis in the dynamics of employment. Metroeconomica **63**(4), 661–692 (2012)
24. Everitt, B.S.: Cluster Analysis. Wiley, London (1993)

A Note on the Dynamics of Linear Automorphisms of a Convolution Measure Algebra

A. Baraviera, E. Oliveira, and F.B. Rodrigues

Abstract Given a finite group G and $\nu \in \mathscr{P}(G)$, we study the dynamics of the linear automorphisms of a convolution measure algebra over G, $T_\nu(\mu) = \nu * \mu$. In order to understand and classify the asymptotic behavior of this dynamical system we provide an alternative to classical results, a very direct way to understand convergence of the sequence $\{\nu^n\}_{n\in\mathbb{N}}$, where $\nu^n = \underbrace{\nu * \ldots * \nu}_{n}$, through the subgroup generated by its support.

1 Introduction

The space of probabilities on a metric space G (or more generally, Radon Measures) has two natural classes of linear automorphisms. The first one is the push forward induced by some fixed map $f\colon G \to G$, (and it just takes in consideration the linear structure of the space of measures). It has been extensively studied by Sigmund (in [1]) and Komuro (in [9]); more recently it also appears in Kloeckner (in [8]) for example. The second one, when G is a topological group, is based on the convolution of two measures. In this case, the space of Radon measures is an infinite dimensional Banach algebra, with respect to the convolution operation, that is, a Measure Convolution Algebra (see [3, p. 73] and [11]). Hence the other natural linear automorphism is $T_\nu(\mu) = \nu * \mu$, for a fixed measure ν.

We propose to understand the topological dynamics of this map in this way. The iteration of T_ν led us to analyze the powers of convolutions of ν, since, from basic properties of the operation $*$, we obtain that iterating the map n times T_ν is the convolution $\nu^n * \mu$.

The problem of studying powers of convolution of probability measures has been explored in several papers in the last few years and has several applications in statistics and group theory (see [4, 7]). In a general setting, G is a compact topological group, $\mathscr{P}(G)$ is the set of all probability measures on G and $\nu \in \mathscr{P}(G)$.

A. Baraviera • E.R. Oliveira (✉) • F.B. Rodrigues
Instituto de Matemática-UFRGS, Avenida Bento Gonçalves, 9500 Porto Alegre, RS, Brazil
e-mail: baravi@mat.ufrgs.br; oliveira.elismar@gmail.com; fagnerbernardini@gmail.com

J.-P. Bourguignon et al. (eds.), *Dynamics, Games and Science*, CIM Series in Mathematical Sciences 1, DOI 10.1007/978-3-319-16118-1_6

The main goal of this paper is to establish direct conditions on the support of the measure ν, which is quite natural from the ergodic point of view, to ensure convergence of the sequence $\{\nu^n := \underbrace{\nu * \ldots * \nu}_{n}\}_{n\in\mathbb{N}}$.

We study the asymptotic behavior of the sequence $\{\nu_n\}_{n\in\mathbb{N}}$ on a finite group, with a complete description of the accumulation points of that sequence, that is, the limit sets of the dynamics T_ν. The main point in this note is that our presentation follows a dynamical point of view, and the main result is obtained with the use of the Perron-Frobenius Theorem (see [2]).

We would like to point out that our results on the convergence of power are not necessarily new, or that they replace the classical literature, but are just easier to compute and to apply. To the best of our knowledge, there is no direct way to extract this kind of characterization of the limit powers just from the necessary and sufficient conditions for convergence, that we find in previous works. Moreover, our characterization makes use of much more elementary results of analysis and algebra.

Many of the ideas developed here can be immediately applied to compact (or locally compact) topological groups (see Remark 6 for details), but the results will be more abstract and not computational since the probability spaces are not finite dimensional and the objects are given by existence theorems.

1.1 Main Result

In this text we present the following:

Theorem 1 *Let $G = \{g_0, \ldots, g_{n-1}\}$ be a finite group. If $\nu \in \mathscr{P}(G)$ is an acyclic probability and H is the subgroup generated by the support of ν, then*

$$\lim_{n\to\infty} \nu^n = \sum_{h\in H} \frac{1}{|H|}\delta_h.$$

We also get an interesting result when the probability measure ν is not acyclic, which we will then use in the last section in order to obtain a solution for the Choquet-Deny equation (see [5]).

2 Proof of Theorem 1

We will always denote by $(G = \{g_0, g_1, \ldots, g_{n-1}\}, \cdot)$ a finite group of order n where $g_0 = e$ is the neutral element of the operation "·".

Remember that the space of real continuous functions in G, $C(G,\mathbb{R})$ is identified with $\mathbb{R}^n$. We denote a function $f \in C(G,\mathbb{R})$ by the row vector

$$f(G) = (f(g_0), f(g_1), \ldots, f(g_{n-1})) \in \mathbb{R}^n.$$

As usual, the dual of $C(G,\mathbb{R})$ is identified with $(\mathbb{R}^n)^* \simeq \mathbb{R}^n$, is the space of signed measures over G, $C(G,\mathbb{R})' = \left\{\mu = \sum_{i=0}^{n-1} p_i \delta_{g_i},\ p = (p_0, p_1, \ldots, p_{n-1}) \in \mathbb{R}^n\right\}$.

In this work we denote $\int_G f d\mu = \sum_{i=0}^{n-1} p_i f(g_i) = \langle f(G), p\rangle$, where $\langle \cdot,\cdot\rangle$ is the usual product in $\mathbb{R}^n$. In this setting, if $\Delta_n = \{p \in \mathbb{R}^n \mid p_i \in [0,1]$, and $\sum_{i=0}^{n-1} p_i = 1\}$ then $\mathscr{P}(G) = \left\{\sum_{i=0}^{n-1} p_i \delta_{g_i} \in C(G,\mathbb{R})' \mid p \in \Delta_n\right\}$. If $\nu = \sum_{i=0}^{n-1} p_i \delta_{g_i}$ and $\mu = \sum_{i=0}^{n-1} q_i \delta_{g_i}$ we define their convolution as

$$(\nu * \mu)(f) = \int_G f d(\nu * \mu) = \int_G \int_G f(gh) d\nu(g) d\mu(h).$$

Defining $f(G^2)$ as

$$f(G^2) = \begin{bmatrix} f(g_0 g_0) & \cdots & f(g_0 g_{n-1}) \\ \vdots & \ddots & \vdots \\ f(g_{n-1} g_0) & \cdots & f(g_{n-1} g_{n-1}) \end{bmatrix}$$

we get a characterization of the convolution in coordinates.

Lemma 1 *If* $\nu = \sum_{i=0}^{n-1} p_i \delta_{g_i} \simeq p$ *and* $\mu = \sum_{i=0}^{n-1} q_i \delta_{g_i} \simeq q$ *then* $(\nu * \mu)(f) = \langle q, f(G^2) \cdot p\rangle$.

Proof Indeed,

$$(\nu * \mu)(f) = \int_G \sum_{i=0}^{n-1} p_i f(g_i h) d\mu(h) = \sum_{i=0}^{n-1} \sum_{j=0}^{n-1} q_i p_j f(g_i g_j) = \langle q, f(G^2) \cdot p\rangle.$$

Since $\mathscr{P}(G)$ is an affine space of codimension 1 in $C(G,\mathbb{R})'$ we know that $\nu(G \times G)$ is given by a bi-stochastic matrix. In order to get the next result, we define a new matrix obtained by a measure $\nu \simeq (p_0, \ldots, p_{n-1}) \in \mathscr{P}(G)$. Denoting

$$G^{-1} \times G = \begin{bmatrix} g_0^{-1} g_0 & \cdots & g_0^{-1} g_{n-1} \\ \vdots & \ddots & \vdots \\ g_0^{-1} g_{n-1} & \cdots & g_{n-1}^{-1} g_{n-1} \end{bmatrix},$$

then

$$\nu(G^{-1} \times G) = \begin{bmatrix} \nu(g_0^{-1}g_0) & \cdots & \nu(g_0^{-1}g_{n-1}) \\ \vdots & \ddots & \vdots \\ \nu(g_0^{-1}g_{n-1}) & \cdots & \nu(g_{n-1}^{-1}g_{n-1}) \end{bmatrix},$$

where $\nu(g_i^{-1} * g_j) = p_m$ if $g_i^{-1} * g_j = g_m$.

Lemma 2 *Given $\nu, \mu \in \mathscr{P}(G)$, then $\nu * \mu = \mu \cdot \nu(G^{-1} \times G)$.*

Proof If $\nu = \sum_{i=0}^{n-1} p_i \delta_{g_i}$ and $\mu = \sum_{i=0}^{n-1} q_i \delta_{g_i}$ we set $\nu * \mu = \sum_{k=0}^{n-1} a_k \delta_{g_k}$. From Lemma 1 we know that $a_k = \sum_{g_i g_j = g_k} p_i q_j = \sum_{i=0}^{n-1} \{q_i p_j \mid g_j = g_i^{-1} g_k\}$. Since the equation $g_i g_j = g_k$ has an unique solution, for a fixed k and for each i we have $j(i,k)$ well determined. It allows us to write, $a_k = q_0 \cdot p_{j(0,k)} + \ldots + q_{n-1} \cdot p_{j(n-1,k)}$. Using matrices we have

$$\begin{bmatrix} a_0 \cdots a_{n-1} \end{bmatrix} = \begin{bmatrix} q_0 \cdots q_{n-1} \end{bmatrix} \cdot \begin{bmatrix} p_{j(0,0)} & \cdots & p_{j(n-1,0)} \\ \vdots & \ddots & \vdots \\ p_{j(0,n-1)} & \cdots & p_{j(n-1,n-1)} \end{bmatrix}.$$

and we get the formula $\nu * \mu = \mu \cdot \nu(G^{-1} \times G)$.

Thus, in order to compute the powers of the convolution $\nu * \nu * \cdots$, we have $\nu^{m+1} := \underbrace{\nu * \ldots * \nu}_{m+1} = \nu \cdot \nu(G^{-1} \times G)^m$, so we can estimate the long time behavior of ν^n from the powers of the matrix $\nu(G^{-1} \times G)$.

Example 1 We consider $G = (\mathbb{Z}_3, +)$ and $\nu = (1/3, 1/4, 5/12)$. So

$$G^{-1} \times G = \begin{bmatrix} 0 & 1 & 2 \\ 2 & 0 & 1 \\ 1 & 2 & 0 \end{bmatrix} \text{ and } \nu(G^{-1} \times G) = \begin{bmatrix} 1/3 & 1/4 & 5/12 \\ 5/12 & 1/3 & 1/4 \\ 1/4 & 5/12 & 1/3 \end{bmatrix}.$$

Definition 1 A stochastic matrix $A = (a_{ij})$ is called primitive if there is $N \in \mathbb{N}$ such that all the entries of the matrix A^N are positive.

Definition 2 A matrix A with non-negative entries is called doubly-stochastic if its rows and columns sum 1.

The following will be very useful in what follows

Theorem 2 (Berman and Plemmons [2]) *If A is an $n \times n$ primitive and doubly stochastic matrix, then $\lim_{m\to\infty} A^m = \frac{1}{n} J$, where $J = (a_{ij})$, $a_{ij} = 1$ for all i, j.*

Definition 3 We say that G is generated by $g_1, \ldots, g_k \in G$ such that for all $g \in G$ we have that $g = g_1^{r_1} \cdots g_k^{r_k}$, with $r_j \in \{0, 1, \ldots, n\}$.

We recall the definition of the support of a given measure. Let G be a finite group and $\nu = (p_0, \ldots, p_{n-1}) \in \mathscr{P}(G)$; The support of ν is the set $\mathrm{supp}(\nu) = \{g_i \in G : \nu(g_i) = p_i > 0\}$. We denote by H the subgroup of G generated by supp(ν), i.e., $H = \langle \mathrm{supp}(\nu) \rangle$. In order to get the next result, Proposition 1, we need a new definition and some technical lemmas (Lemmas 3 and 4). We start with the definition:

Definition 4 (Acyclic) Given $\nu \in \mathscr{P}(G)$, we define the set $Z_+(\nu)^m$ by

$$Z_+(\nu)^m = \{g_{i_1} \ldots g_{i_m} : g_{i_k} \in \mathrm{supp}(\nu)\}.$$

Let H be the subgroup of G generated by supp(ν). We say that ν is an acyclic probability measure if there exists $N \in \mathbb{N}$ such that $Z_+(\nu)^N = H$. In particular, $Z_+(\nu)^1 = \mathrm{supp}(\nu)$.

We would like to observe that the acyclic property is similar to [4] for probabilities in matrices, but in that case the convergence is given by a rank theorem.

Example 2 Let $g \in G$ be an element of order 2 and $\nu = \delta_g$. In this case $H = \{e, g\}$ and

$$Z_+(\nu)^m = \begin{cases} e, & \text{if } m \text{ is even} \\ g, & \text{if } m \text{ is odd.} \end{cases}$$

From this property follows that ν is not an acyclic probability measure.

Example 3 Let H be a cyclic group generated by g and $\nu = \alpha\delta_e + (1 - \alpha)\delta_g$, $0 < \alpha < 1$. Then ν is acyclic. In fact, if $H = \{e, g, \ldots, g^{n-1}\}$, then

$$Z_+(\nu)^n - \{e^n, e^{n-1}g, e^{n-2}g^2, \ldots e^1 g^{n-1}\} = H.$$

Example 4 Let G be a finite abelian group of order n and $\nu \in \mathscr{P}(G)$. We can identify $\nu = \sum_i p_i \delta_{g_i} \simeq p = (p_0, \ldots, p_{n-1})$. If $Z_+(p) = \{g, h\}$ and $H = \langle g^{-1}h \rangle$, then ν is acyclic. In fact, to see this we only need to notice that $g^{n-k}h^k = (g^{-1}h)^k$.

Example 5 Let $H = \langle g_1, \ldots, g_k \rangle$ be a finitely generated abelian subgroup of G and $\nu \in \mathscr{P}(G)$. If $\nu \in \mathscr{P}(G)$ is such that $Z_+(\nu) = \{e, g_1, \ldots, g_k\}$ then ν is acyclic.

Remark 1 If we have that $|G| = n$ and the support of ν has more than $\frac{n}{2} + 1$ elements, then ν is acyclic; in particular $\nu(G^{-1} \times G)$ is primitive.

When the probability ν is acyclic, we have the following proposition:

Proposition 1 *Let $G = \{g_0, \ldots, g_{n-1}\}$ be a finite group, $\nu \in \mathscr{P}(G)$ acyclic and $H = \langle Z_+(\nu) \rangle$. It is possible to order the elements of G such that the matrix $\nu(G^{-1} \times G) = (\nu(H^{-1}g_i^{-1} \times g_j H))_{i,j}$ satisfies $\lim_{n\to\infty} \nu(G^{-1} \times G)^n = B$, where B is the matrix* *given by*

$$\begin{pmatrix} \frac{1}{|H|}J & 0 & \dots & 0 \\ 0 & \frac{1}{|H|}J & 0 & \dots \\ \vdots & \vdots & \vdots & \vdots \\ 0 & \dots & 0 & \frac{1}{|H|}J \end{pmatrix},$$

where 0 *is the null matrix of order* $|H|$ *and* J *is the matrix of order* $|H|$ *with all the coefficients equal to 1.*

Proof The proof of this result follows from the remark and lemmas below.

Remark 2 Let us consider an acyclic probability $\nu \in \mathscr{P}(G)$ and the subgroup H generated by $Z_+(\nu)$. We suppose that $|H| = n$. And we can take the equivalence classes determined by H in G, i.e., $gH = \{gh : h \in H\}$. And we know that G can be written as a disjoint union of the equivalence classes determined by H. Therefore we can write G as follows

$$\begin{aligned} G &= \{e, h_1, \dots, h_k, g_1h_1, \dots, g_1, g_1h_k, g_2h_1, \dots, g_2, g_2h_k, \dots, g_l, g_lh_1, \dots, g_lh_k\} \\ &= H \dot{\cup}\, g_1H \,\dot{\cup} \cdots \dot{\cup}\, g_lH, \end{aligned}$$

for certain group elements $g_1 = e, g_2, \dots, g_l \in G$ where $\dot{\cup}$ denotes the disjoint union and $g_iH \cap g_jH = \emptyset$ for $i \neq j$. Thus l is such that nl is the order of the group G. Then we have that the matrix $\nu(G^{-1} \times G)$ is given in blocks by

$$\begin{pmatrix} \nu(H^{-1} \times H) & \nu(H^{-1} \times g_1H) & \dots & \nu(H^{-1} \times g_lH) \\ \nu(H^{-1}g_1^{-1} \times H) & \nu(H^{-1}g_1^{-1} \times g_1H) & \dots & \nu(H^{-1}g_1^{-1} \times g_lH) \\ \vdots & \vdots & \vdots & \vdots \\ \nu(H^{-1}g_l^{-1} \times H) & \dots & \dots & \nu(H^{-1}g_l^{-1} \times g_lH) \end{pmatrix},$$

where the blocks in the diagonal are always the matrix $\nu(H^{-1} \times H)$.

Lemma 3 *The blocks* $\nu(H^{-1}g_i^{-1} \times g_jH)$ *are always the null matrix for* $i \neq j$.

Proof Take the block $\nu(H^{-1}g_i^{-1} \times g_2H)$ and notice that

$$\begin{aligned} \nu((h_r^{-1}g_i^{-1})(g_jh_s)) > 0 \Leftrightarrow (h_r^{-1}g_i^{-1})(g_jh_s) \in Z_+(\nu) \subset H \\ \Rightarrow g_i^{-1}g_j \in H \Rightarrow g_iH = g_jH, \end{aligned}$$

but it is a contradiction, since $g_iH \cap g_jH = \emptyset$.

By Lemma 3 we have that the powers of the matrix $\nu(G^{-1} \times G)$ are given by

$$\nu(G^{-1} \times G)^n = \begin{pmatrix} \nu(H^{-1} \times H)^n & 0 & \dots & 0 \\ 0 & \nu(H^{-1} \times H)^n & \dots & 0 \\ \vdots & \vdots & \vdots & \vdots \\ 0 & \dots & \dots & \nu(H^{-1} \times H)^n \end{pmatrix}.$$

Lemma 4 *The matrix $\nu(H^{-1} \times H)$ is primitive.*

Proof Let us consider the matrix $A = (a_{ij})_{i,j} := (\nu(h_i^{-1}h_j))_{i,j}$. Then we notice that

$$\begin{aligned} a_{ij} > 0 \quad & \Leftrightarrow h_i^{-1}h_j \in Z_+(\nu) \\ & \Leftrightarrow \exists \bar{h} \in Z_+(\nu) \text{ such that } h_i^{-1}h_j = \bar{h} \\ & \Leftrightarrow h_j = h_i\bar{h},\ \bar{h} \in Z_+(\nu). \end{aligned}$$

It implies that $a_{ij} > 0$ if and only if $h_j \in L_{h_i}(Z_+(\nu))$. As L_{h_i} is a bijection, in each line we have $|Z_+(\nu)|$ positive coefficients. Consider now A^2, which we denote by $A^2 = (a^2_{ij})_{i,j}$. Then we have that

$$\begin{aligned} a^2_{ij} > 0 \quad & \Leftrightarrow \sum_{k=0}^{n-1} \nu(h_i^{-1}h_k)\nu(h_k^{-1}h_j) \\ & \Leftrightarrow \exists k \in \{0, \dots, n-1\} \text{ such that } \nu(h_i^{-1}h_k)\nu(h_k^{-1}h_j) > 0 \\ & \Leftrightarrow \nu(h_i^{-1}h_k) > 0 \text{ and } \nu(h_k^{-1}h_j) > 0 \\ & \Leftrightarrow \exists h', h'' \in Z_+(\nu) \text{ such that } h_k = h_ih',\ h_j = h_kh'' \\ & \Leftrightarrow h_j = h_ih'h'' \\ & \Leftrightarrow h_j \in L_{h_i}(Z_+(\nu)^2). \end{aligned}$$

Again, we can see that A^2 has $|Z_+(\nu)^2|$ positive coefficients. Following by induction, if $A^n = (a^n_{ij})_{i,j}$, then $a^n_{ij} > 0 \Leftrightarrow h_j \in L_{h_i}(Z_+(\nu)^n)$. As ν is acyclic we have, from Definition 4, that there exists $N \in \mathbb{N}$ such that for $n > N$

$$a^n_{ij} > 0 \Leftrightarrow h_j \in L_{h_i}(Z_+(\nu)^n) = h_iH = H.$$

It implies that for $n > N$ the matrix $A^n = (a^n_{ij})_{i,j}$ has $|H|$ positive coefficients in each line. As the matrix A has order $|H|$ we see that A is primitive.

Lemma 5 *Let $\mu, \nu \in \mathscr{P}(G)$ and σ a permutation on G. Then we have that*

$$\mu \cdot \nu((\sigma(G))^{-1} \times \sigma(G)) = \mu \cdot \nu(G^{-1} \times G).$$

Proof We notice that the convolution does not depend on the way the group elements of G are ordered, then $\mu \cdot \nu((\sigma(G))^{-1} \times \sigma(G)) = \mu * \nu = \mu * \nu(G^{-1} \times G)$.

We would like to point out that $B = \lim_{n\to\infty} \nu(G^{-1} \times G)^n$ is also doubly stochastic and has always 1 as an eigenvalue.

Remark 3 The main fact used in the Lemma 5 was that the integral does not change under permutation of the group G, that is, the convolution depends of the operations between two elements and not of the position that we choose to identify a probability with a vector.

Using Lemma 5 to rearrange the group by the equivalence classes determined by H and making some permutation on the elements one can easily conclude that under the conditions of Proposition 1, the matrix $\nu(G^{-1} \times G) = (\nu(g_i^{-1}g_j))_{i,j}$ satisfies $\lim\limits_{n\to\infty} \nu(G^{-1} \times G)^n = B$, where B is the matrix given by

$$b_{ij} = \begin{cases} 0, & \text{if } g_i^{-1}g_j \notin H \\ \frac{1}{|H|}, & \text{if } g_i^{-1}g_j \in H \end{cases}$$

This completes the proof of Theorem 1.

Example 6 Take $G = \langle a\rangle \times \langle b\rangle$ isomorphic to $\mathbb{Z}_2 \times \mathbb{Z}_3$ and $\nu = p = \alpha\delta_e + (1-\alpha)\delta_b$, $0 < \alpha < 1$. So we have

$$\nu(G^{-1} \times G) = \begin{pmatrix} \alpha & 0 & 0 & 0 & (1-\alpha) & 0 \\ 0 & \alpha & 0 & (1-\alpha) & 0 & 0 \\ (1-\alpha) & 0 & \alpha & 0 & 0 & 0 \\ 0 & 0 & 0 & \alpha & 0 & (1-\alpha) \\ 0 & 0 & (1-\alpha) & 0 & \alpha & 0 \\ 0 & (1-\alpha) & 0 & 0 & 0 & \alpha \end{pmatrix}.$$

In that case, $H = Z_+(\nu) = \{e, b\}$ and $\langle Z_+(\nu)\rangle = \{e, b, b^2\}$, and by Proposition 1, reordering the elements of G as $\{e, a, b^2, ab, b, ab^2\}$ we have that

$$\lim_{n\to\infty} \nu(G^{-1} \times G)^n = \operatorname{diag}\left(\frac{1}{3}J, \frac{1}{3}J\right),$$

where J is a 3×3 matrix with all entries equal to 1. Then we have that $\lim\limits_{n\to\infty} \nu^n = \frac{1}{3}(\delta_e + \delta_b + \delta_{b^2})$.

Remark 4 If the probability $\nu \in \mathscr{P}(G)$ is not acyclic, then there exists a finite number of subsets of $\langle Z_+(\nu)\rangle$, let us say $K_1, \ldots, K_l$ such that for each $n \in \mathbb{N}$, $Z_+(\nu)^n = K_i$ for some $i \in \{1, \ldots, l\}$. Following the same computations used to get Theorem 1, it is possible to show that the sequence $\{\nu^n\}_{n\in\mathbb{N}}$ has l accumulation points and each of these accumulation points is a uniform probability measure supported on a set K_j.

Remark 5 Let G_1, G_2 be compact topological groups and $\phi : G_1 \to G_2$ a homomorphism of groups. It is easy to see that the push forward map $\phi_\sharp : \mathscr{P}(G_1) \to \mathscr{P}(G_2)$ given by $\phi_\sharp(\mu)(A) = \mu(\phi^{-1}(A))$ for all $A \subset G_2$, satisfies the following:

$$\phi_\sharp(\nu * \mu) = \phi_\sharp(\nu) * \phi_\sharp(\mu).$$

It implies that $\lim_{n\to\infty} \phi_\sharp(\nu^n) = \lim_{n\to\infty}(\phi_\sharp(\nu))^n$.

The next proposition guarantees the density of the set of acyclic probabilities. It shows how big this set is, in the sense of the topology of $\mathscr{P}(G)$.

Proposition 2 *Let $\nu_0 \in \mathscr{P}(G)$, where G is a finite group. Given $\varepsilon > 0$, there exists $\bar{\nu} \in \mathscr{P}(G)$ such that $\bar{\nu}$ is an acyclic probability and $d(\bar{\nu}, \nu_0) < \varepsilon$, i.e., the set of acyclic probabilities is dense in $\mathscr{P}(G)$.*

Proof Let $\varepsilon > 0$ and $\nu_0 = p = \sum_{i=0}^{k-1} p_i\delta_{h_i}$ with $\mathrm{supp}(\nu_0) = Z_+(p) = \{g \in G : \nu_0(g) > 0\}$ and $H = \langle Z_+(p)\rangle = \{h_0, \ldots, h_{k-1}\}$. Then we define $a = \min\{p_i : p_i > 0\}$ and $\bar{\varepsilon} = \frac{1}{2}\min\{\varepsilon, a\}$.We consider the measure $\bar{\nu} = \bar{p} = \sum_{i=0}^{k-1} \bar{p}_i\delta_{h_i}$, where

$$\bar{p}_i = \begin{cases} \frac{\bar{\varepsilon}}{k-|Z_+(p)|}, & \text{if } p_i = 0 \\ p_i - \frac{\bar{\varepsilon}}{|Z_+(p)|}, & \text{if } p_i > 0. \end{cases}$$

Obviously $\bar{\nu} \in \mathscr{P}(G)$ and as $d(\nu_0, \bar{\nu}) = \sum_i |p_i - \bar{p}_i| = \sum_{p_i=0} \frac{\bar{\varepsilon}}{k-|Z_+(p)|} + \sum_{p_i>0} \frac{\bar{\varepsilon}}{|Z_+(p)|} = 2\bar{\varepsilon} < \varepsilon$, so we get the result.

3 Application: Dynamics of T_ν

We start this section with the basic properties of the linear automorphism $T_\nu : \mathscr{P}(G) \to \mathscr{P}(G)$, given by

$$T_\nu(\mu) = \mu * \nu,$$

where $\nu \in \mathscr{P}(G)$ is a fixed probability.

Proposition 3 *The map T_ν is continuous in the weak topology, linear and its fixed points satisfy the Choquet-Deny equation, $\mu * \nu = \mu$.*

For a proof the reader can see for example [3, p. 73], [5] and [6]. What remains to be understood is the asymptotic behavior of T_ν.

Theorem 3 *Let $G = \{g_0, \ldots, g_{n-1}\}$ be a finite group. If $\nu \in \mathscr{P}(G)$ is an acyclic probability, $H = \langle Z_+(\nu)\rangle$ is the subgroup generated by the support of ν, then the*

$\omega-$limit set, here denoted by $L_\omega(\mu)$, that is the set of accumulation points of its orbit, is $L_\omega(\mu) = \sum_{h\in H} \frac{1}{|H|}\delta_h * \mu$, linear on μ. Moreover, μ is a recurrent point of the dynamics, that is, $\mu \in L_\omega(\mu)$, only if μ is a solution of the Choquet-Deny equation $\bar{\nu} * \mu = \mu$, where $\bar{\nu} = \lim_{n\to\infty} \nu^n$.

Proof Since $T^n(\mu) = \nu^n * \mu$ we have from Theorem 1 that $\{\bar{\nu} \mid \bar{\nu}$ is an accumulation points of $\nu^n\} = \{\sum_{h\in H} \frac{1}{|H|}\delta_h\}$ because ν is an acyclic. Thus

$$L_\omega(\mu) = \{\bar{\nu} * \mu \mid \bar{\nu} \text{ is an accumulation points of } \nu^n\} = \left\{\sum_{h\in H} \frac{1}{|H|}\delta_h * \mu\right\},$$

that is linear on μ. In particular $\mu \in L_\omega(\mu)$, only if μ is solution of the Choquet-Deny equation $\bar{\nu} * \mu = \mu$.

Example 7 We consider $G = (\mathbb{Z}_3, +)$ and $\nu = (1/3, 1/4, 5/12)$. So

$$G^{-1} \times G = \begin{bmatrix} 0 & 1 & 2 \\ 2 & 0 & 1 \\ 1 & 2 & 0 \end{bmatrix} \quad \text{and } \nu(G^{-1} \times G) = \begin{bmatrix} 1/3 & 1/4 & 5/12 \\ 5/12 & 1/3 & 1/4 \\ 1/4 & 5/12 & 1/3 \end{bmatrix}.$$

To find the fixed points for T_ν we need to solve the following equation:

$$\begin{bmatrix} q_0 & q_1 & q_2 \end{bmatrix} = \begin{bmatrix} q_0 & q_1 & q_2 \end{bmatrix} \cdot \begin{bmatrix} 1/3 & 1/4 & 5/12 \\ 5/12 & 1/3 & 1/4 \\ 1/4 & 5/12 & 1/3 \end{bmatrix}.$$

By linear algebra we have that there is only one solution for the above equation and it is given by $\mu_0 = \frac{1}{3}(\delta_0 + \delta_1 + \delta_2)$. So the unique fixed point is μ_0.

We also have that μ is recurrent only if $\lim_{n\to\infty} \nu^n * \mu = \mu$. But $\lim_{n\to\infty} \nu^n = \mu_0$, and it implies that μ is recurrent only if

$$\begin{bmatrix} q_0 & q_1 & q_2 \end{bmatrix} = \begin{bmatrix} q_0 & q_1 & q_2 \end{bmatrix} \cdot \begin{bmatrix} 1/3 & 1/3 & 1/3 \\ 1/3 & 1/3 & 1/3 \\ 1/3 & 1/3 & 1/3 \end{bmatrix},$$

and solving this equation the unique possibility is $\mu = \mu_0$.

Using Theorem 3, we will try to find conditions for two measures to have the same $\omega-$limit, where $\nu = \sum_i p_i\delta_{g_i}$ is an acyclic. First we observe that if $\mu = \sum_i q_i\delta_{g_i} \in \mathscr{P}(G)$, then $L_\omega(\mu) = \{\mu{\cdot}B\}$. If we identify μ with the vector $q = \sum_i q_i e_i$ in $\mathbb{R}^n$, where $\{e_i\}_{0\leq i\leq n-1}$ is the canonical basis of $\mathbb{R}^n$, we have that

$$q \cdot B = \Big(\sum_i q_i e_i\Big) \cdot B = \sum_i q_i\Big(e_i \cdot B\Big),$$

where $B = \bar{\nu}(G^{-1} \times G)$ for $\bar{\nu} = \lim_{n\to\infty} \nu^n$ as in Theorem 3. It implies that $L_\omega(\mu) = \sum_i q_i L_\omega(\delta_{g_i})$. So, to determine the $\omega-$limit of a measure it is enough to determine the $\omega-$limit of the measures δ_{g_i}, for all $g_i \in G$. Then we notice that if $H = \langle Z_+(p)\rangle$, $|H| = k$, $|G| = |H|l$, $\bar{\mu} = (q_0, \ldots, q_{n-1})$, $\mu = \delta_{g_0}$, and if we write $\alpha_0 = \sum_{i=0}^{k-1} q_i$, $\alpha_1 = \sum_{i=k}^{2k-1} q_i, \ldots, \alpha_l = \sum_{i=n-k-1}^{n-1} q_i$ then we have the equivalence $\mu \cdot B = \bar{\mu} \cdot B$ if and only if

$$\Big(\underbrace{\frac{1}{k}, \frac{1}{k}, \ldots, \frac{1}{k}}_{k}, 0, \ldots, 0\Big) = \Big(\underbrace{\frac{1}{k}\alpha_0, \ldots, \frac{1}{k}\alpha_0}_{k}, \underbrace{\frac{1}{k}\alpha_1, \ldots, \frac{1}{k}\alpha_1}_{k}, \ldots, \underbrace{\frac{1}{k}\alpha_l, \ldots, \frac{1}{k}\alpha_l}_{k}\Big)$$

$$\sum_{i=0}^{k-1} q_i = 1, \sum_{i=k}^{2k-1} q_i = 0, \ldots, \sum_{i=n-k-1}^{n-1} q_i = 0.$$

It implies that $L_\omega(\delta_{g_0}) = L_\omega(\mu)$ if and only if $\sum_{i=0}^{k-1} q_i = 1$, where $\mu = (1, 0, \ldots, 0)$. By the previous argument we can see that

$$L_\omega(\delta_{g_i}) = L_\omega(\delta_{g_0}) \text{ for } 0 \le i \le k-1,\ L_\omega(\delta_{g_i}) = L_\omega(\delta_{g_k}) \text{ for } k \le i \le 2k-1, \ldots,$$
$$L_\omega(\delta_{g_i}) = L_\omega(\delta_{g_{n-k-1}}) \text{ for } n-k-1 \le i \le n-1,$$

and from it, it follows that $L_\omega(\mu) = \sum_i q_i L_\omega(\delta_{g_i}) = \sum_{j=0}^{l} \alpha_j L_\omega(\delta_{g_{jk}})$; and if $\bar{\mu} = (q_0, \ldots, q_{n-1})$ and we take $\mu = \delta_{g_i}$ with $mk \le i \le (m+1)k - 1$,

$$L_\omega(\bar{\mu}) = L_\omega(\delta_{g_i}) \Leftrightarrow \alpha_m = \sum_{i=mk}^{(m+1)k-1} q_i = 1, \text{ and } \alpha_j = 0 \text{ for } j \neq m.$$

Finally, given $\mu = (q_0, \ldots, q_{n-1})$ and $\mu' = (q'_0, \ldots, q'_{n-1})$,

$$L_\omega(\mu) = L_\omega(\mu') \Leftrightarrow \sum_{i=0}^{k-1} q_i = \sum_{i=0}^{k-1} q'_i, \sum_{i=k}^{2k-1} q_i = \sum_{i=k}^{2k-1} q'_i, \ldots, \sum_{i=n-k-1}^{n-1} q_i = \sum_{i=n-k-1}^{n-1} q'_i.$$

Definition 5 Let $\nu \in \mathscr{P}(G)$ be an acyclic probability measure and $\eta \in \mathscr{P}(G)$. We call the basin of η the set $\{\mu \in \mathscr{P}(G) : \lim_{n\to\infty} T_\nu^n(\mu) = \eta\}$.

Example 8 Now we return to the Example 6 where $G = \{e, a, b^2, ab, b, ab^2\}$; in that particular situation, $\nu = p = \alpha\delta_e + (1-\alpha)\delta_b$, $0 < \alpha < 1$ is acyclic and we can rewrite G as $G = \{e, b, b^2, a, ab, ab^2\}$ according to the conjugation classes of $H = \{e, b, b^2\}$ as in Remark 2 in order to apply Lemma 3.

Then, given $\mu = (q_0, \ldots, q_5)$ and $\mu' = (q'_0, \ldots, q'_5)$, we have

$$L_\omega(\mu) = L_\omega(\mu') \Leftrightarrow \sum_{i=0}^{2} q_i = \sum_{i=0}^{2} q'_i, \text{ and } \sum_{i=3}^{5} q_i = \sum_{i=3}^{5} q'_i.$$

For instance, if $\mu' = (\frac{1}{4}, \frac{1}{2}, 0, \frac{1}{8}, 0, \frac{1}{8})$ we have

$$\begin{aligned}\lim_{n\to\infty} T_\nu^n(\mu') &= \tfrac{1}{3}\Big(q'_0 + q'_1 + q'_2, \ldots, q'_0 + q'_1 + q'_2, q'_3 + q'_4 + q'_5, \ldots, q'_3 + q'_4 + q'_5\Big) \\ &= \big(\tfrac{1}{4}, \tfrac{1}{4}, \tfrac{1}{4}, \tfrac{1}{12}, \tfrac{1}{12}, \tfrac{1}{12}\big) = \eta.\end{aligned}$$

So, the basin of attraction of $\eta = \big(\frac{1}{4}, \frac{1}{4}, \frac{1}{4}, \frac{1}{12}, \frac{1}{12}, \frac{1}{12}\big)$, that is, $\{\mu = (q_0, \ldots, q_5) \mid \lim_{n\to\infty} T_\nu^n(\mu) = \eta\}$ is given by

$$\begin{cases} q_0 + q_1 + q_2 = \frac{3}{4} \\ q_3 + q_4 + q_5 = \frac{1}{4} \\ q_0, \ldots, q_5 \quad \in [0, 1] \end{cases}$$

that is a convex region of an hyperplane in $\mathbb{R}^6$ of dimension 4; more precisely $q_0 = \frac{3}{4} - a - b, q_1 = a,\ q_2 = b, q_3 = \frac{1}{4} - c - d, q_4 = c, q_5 = d, a + b \le \frac{3}{4},\ c + d \le \frac{1}{4}$ and $a, b, c, d \in [0, 1]$ is the basin of attraction of $\eta = \big(\frac{1}{4}, \frac{1}{4}, \frac{1}{4}, \frac{1}{12}, \frac{1}{12}, \frac{1}{12}\big)$.

Actually, the next proposition shows that the basin of a measure $\eta \in \mathscr{P}(G)$ is always a convex subset of an hyperplane.

Proposition 4 *Let $\nu = p \in \mathscr{P}(G)$ be an acyclic probability measure and $H = \langle Z_+(p)\rangle$, with $|H| = k$ and $|G| = |H|l$. Given $\eta \in \mathscr{P}(G)$ with*

$$\eta = (\underbrace{q_0, \ldots, q_0}_{k}, \underbrace{q_1, \ldots, q_1}_{k}, \ldots, \underbrace{q_{l-1}, \ldots, q_{l-1}}_{k})$$

Then the basin of η is a convex subset of a hyperplane of dimension $\frac{n(k-1)}{k}$ in $\mathbb{R}^n$.

Proof To prove the convexity of the basin of a given η we only need to notice that if $\mu_1, \mu_2 \in \mathscr{P}(G)$ and $0 \le \alpha \le 1$, then

$$\begin{aligned} T_\nu(\alpha\mu_1 + (1-\alpha)\mu_2) &= (\alpha\mu_1 + (1-\alpha)\mu_2) \cdot \nu(G^{-1} \times G) \\ &= \alpha\mu_1 \cdot \nu(G^{-1} \times G) + (1-\alpha)\mu_2 \cdot \nu(G^{-1} \times G). \end{aligned}$$

Hence, if μ_1, μ_2 are in the basin of η, then

$$\lim_{n\to\infty} T_\nu^n(\alpha\mu_1 + (1-\alpha)\mu_2) = \alpha \lim_{n\to\infty} T_\nu^n(\mu_1) + (1-\alpha)\lim_{n\to\infty} T_\nu^n(\mu_2)$$
$$= \alpha\eta + (1-\alpha)\eta = \eta.$$

In order to prove the second part of the proposition, we notice that if $\mu = (q'_0, q'_1, \ldots, q'_{n-1})$ is in the basin of η, then $\lim_{n\to\infty} T_\nu^n(\mu) = \eta$ if and only if

$$\begin{cases} \dfrac{1}{k}\displaystyle\sum_{i=0}^{k-1} q'_i = q_0 \\ \vdots \\ \dfrac{1}{k}\displaystyle\sum_{i=n-k-1}^{n-1} q'_i = q_{l-1} \\ q_0, \ldots, q_{n-1} \in [0,1]. \end{cases}$$

If we forget the restriction $\sum_{i=0}^{n-1} q'_i = 1$ and $q_0, \ldots, q_{n-1} \in [0,1]$, we have a linear system which has n variables and l linearly independents equations. Then its space of solution is given by an hyperplane of dimension $n-l = n-\frac{n}{k} = \frac{n(k-1)}{k}$. Therefore the solution of the system above is a convex set given by the intersection of a hyperplane of dimension $\frac{n(k-1)}{k}$ with the simplex $\Delta_n = \{(x_0, \ldots, x_{n-1}) : \sum_i x_i = 1,\ x_i \in [0,1]\}$.

Thus we have a complete characterization of the limit set of the dynamics of a map T^ν, but the generical behaviour of this type of dynamical systems is given in Example 7. Indeed we can prove the following:

Theorem 4 *There is an open and dense set $\mathcal{O} \subset \mathcal{P}(G)$ such that for all $\nu \in \mathcal{O}$, $L_\omega(\mu) = \{\nu_0 * \mu\}, \forall \mu \in \mathcal{P}(G)$, where $\nu_0 = \frac{1}{|G|}\sum_{i=0}^{|G|-1} \delta_{g_i}$ is the unique fixed point of T_ν.*

Proof Consider the set of probabilities

$$\mathcal{O} = \left\{ \nu = \sum_{i=0}^{|G|-1} p_i \delta_{g_i} \in \mathcal{P}(G) \mid p_i > 0 \right\}.$$

Thus, for every $\nu \in \mathcal{O}$, ν is an acyclic probability because $H = G$, and Theorem 3 proves that the $\omega-$limit set by T_ν of $\mu \in \mathcal{P}(G)$, here denoted by $L_\omega(\mu)$, is $L_\omega(\mu) = \frac{1}{|G|}\sum_{i=0}^{|G|-1} \delta_{g_i} * \mu$. Moreover, μ is a recurrent point of the dynamics, that is, $\mu \in L_\omega(\mu)$, only if μ is solution of the Choquet-Deny equation $\nu_0 * \mu = \mu$, where $\nu_0 = \lim_{n\to\infty} \nu^n = \frac{1}{|G|}\sum_{i=0}^{|G|-1} \delta_{g_i}$. From the theory of doubly stochastic matrices, we know that the only solution is $\mu = \nu_0$, since every fixed point is recurrent, there is just one of them. Thus, we just need to prove that $\mathcal{O}$ is an open and dense set, which is trivial because its complementary set is

$$\mathcal{O}^c = \left\{ \nu = \sum_{i=0}^{|G|-1} p_i \delta_{g_i} \in \mathcal{P}(G) \mid \exists p_i = 0 \right\},$$

which is a finite union of algebraic sets in $\mathbb{R}^{|G|}$, so it is closed with empty interior, which concludes the proof.

Remark 6 For the general case, locally compact groups or semi-groups, we can ask the same questions about the dynamics of T_ν as before. However we can get existence results, or some characterization, but not explicit computations. For example, the set of fixed points of T_ν is the solution of the Choquet-Deny equation $\mu * \nu = \mu$, which can be explicitly solved for finite groups. In [10], Thm 2, they prove that μ satisfies $\mu * \nu = \mu$ if, and only if, $\mu * \delta_g = \mu$, for all $g \in S(\nu)$, where $S(\nu)$ is the closed semi-group generated by the support of ν. Also, the limit sets of T_ν are related to the accumulation points of ν^n for $n \in \mathbb{N}$. Again the results that we can find in the literature are of existence or characterization of the convergence, see for instance [7], Corollaries 2.1 and 2.2. Thus a generalization of Theorem 3 will be:

Theorem 5 *Let G be a compact topological group. If $\nu \in \mathcal{P}(G)$ then ω−limit set by T_ν of $\mu \in \mathcal{P}(G)$, here denoted by $L_\omega(\mu)$, is*

$$L_\omega(\mu) = \{\bar{\nu} * \mu \mid \bar{\nu} \textit{ is an accumulation points of } \nu^n\}.$$

Again, we can not describe this set in an unique way for all topological groups as we did for the finite ones.

Acknowledgements The first author is partially supported by CNPq, CAPES and FAPERGS.

References

1. Bauer, K., Sigmund, K.: Topological dymanics induced on the space of probability measures. Monatshefte fur Math. **79**, 81–92 (1975)
2. Berman, A., Plemmons, R.: Nonnegative Matrices in the Mathematical Sciences. Academic Press, New York (1979)
3. Bobrowski, A.: Functional Analysis for Probability and Stochastic Processes. An Introduction, pp. xii+393. Cambridge University Press, Cambridge (2005)
4. Chakraborty, S.: Cyclicity and weak convergence for convolution of measures on non-negative matrices. Indian J. Stat. **69**(2), 304–313 (2007)
5. Choquet, G., Deny, J.: Sur l'equation de convolution $\mu = \nu * \mu$. C. R. Acad. Sci. Paris **250**, 799–801 (1960)

6. Choquet, G., Deny, J.: Sur l'équation de convolution $\mu = \mu * \sigma$. C. R. Acad. Sci. Paris **250**, 799–801 (1960, in French)
7. Hognas, G., Mukherjea, A.: Probability Measures on Semigroups: Convolution Products, Random Walks, and Random Matrices. Plenum Press, New York (1995)
8. Kloeckner, B.: Optimal transport and dynamics of expanding circle maps acting on measures. Ergod. Theor. Dyn. Syst. **33**, 529–548 (2013)
9. Komuro, M., The pseudo orbit tracing properties on the space of probability measures. Tokyo J. Math. **7**(2), 461–468 (1984)
10. Székely, G.J., Zeng, W.-B.: The Choquet-Deny convolution equation $\mu = \mu * \sigma$ for probability measures on abelian semigroups. J. Theor. Probab. **3**(2), 361–365 (1990)
11. Wermer, J.: Banach algebras and analytic functions. Adv. Math. **1**(1), 51–102 (1961)

Periodic Homogenization of Deterministic Control Problems via Limit Occupational Measures

Martino Bardi and Gabriele Terrone

Abstract We consider optimal control problems where the dynamical system and the running cost are affected by fast periodic oscillations of the state variables. We show that, under suitable controllability and structure assumptions, it is possible to describe the limiting optimal control problem. The proofs make use of results in the theory of homogenization and singular perturbations of Hamilton-Jacobi equations.

1 Introduction

We consider an optimal control problem in $\mathbf{R}^N$ in which the dynamics

$$\begin{cases} \dot{x}(s) = f\left(x(s), \frac{x(s)}{\epsilon}, \alpha(s)\right) \\ x(0) = x \end{cases} \tag{1}$$

and the cost functional

$$J^{\epsilon}(t, x, \alpha) := \int_0^t l\left(x(s), \frac{x(s)}{\epsilon}, \alpha(s)\right) \mathrm{d}s + h(x(t)) \tag{2}$$

undergo fast periodic oscillations. The controls α are measurable functions taking values in a compact metric space A. The vector field $f : \mathbf{R}^N \times \mathbf{R}^N \times A \to \mathbf{R}^N$ is bounded, uniformly continuous, and Lipschitz-continuous in x uniformly with respect to α. The running cost $l : \mathbf{R}^N \times \mathbf{R}^N \times A \to \mathbf{R}$ and the terminal cost h :

M. Bardi
Dipartimento di Matematica, Università di Padova, via Trieste 63, 35121 Padova, Italy
e-mail: bardi@math.unipd.it

G. Terrone (✉)
Departamento de Matemática, Center for Mathematical Analysis, Geometry, and Dynamical Systems, Instituto Superior Técnico, Universidade Técnica de Lisboa, Av. Rovisco Pais, 1049-001 Lisboa, Portugal
e-mail: gterrone@math.ist.utl.pt

J.-P. Bourguignon et al. (eds.), *Dynamics, Games and Science*, CIM Series in Mathematical Sciences 1, DOI 10.1007/978-3-319-16118-1_7

$\mathbf{R}^N \to \mathbf{R}$ are given bounded uniformly continuous functions. Both $f(x,\cdot,\alpha)$ and $l(x,\cdot,\alpha)$ are $\mathbf{Z}^N$-periodic.

We are interested in understanding the behaviour of the solutions of this problem for very small $\epsilon > 0$. In particular we investigate the existence and the nature of a limit problem, independent of ϵ, which approximates in some sense the ϵ-problem. Similar issues in Calculus of Variations have a very wide literature and various notions of convergence were developed in that context, see, for instance, Braides and De Franceschi [10] and the references therein. Some particular control problems have been formulated as problems in Calculus of Variations with dynamical constraints and have been studied in that context [11, 12, 17], but the problem can be still considered largely open.

Here we study a notion of variational convergence based on the value function of the control problem. We recall that the value function is defined by the infimum of the cost functional among all trajectories, that is,

$$v^\epsilon\left(t,x\right) := \inf\left\{J^\epsilon\left(t,x,\alpha\right) \;\middle|\; x(\cdot) \text{ solves } (1)\right\}. \tag{3}$$

We say that the control problem with cost functional J^ϵ defined in (2) and dynamics (1) *converges as* $\epsilon \to 0$ *to the limit control problem* with cost functional

$$\overline{J}(t,x,\gamma) = \int_0^t \overline{l}\left(x(s),\gamma(s)\right) \mathrm{d}s + h(x(t)) \tag{4}$$

and dynamics

$$\dot{x}(s) = \overline{f}\left(x(s),\gamma(s)\right), \tag{5}$$

if the value function $v^\epsilon\left(t,x\right)$ converges locally uniformly to the value function of the limit control problem $v(t,x) := \inf\left\{\overline{J}(t,x,\gamma) \;\middle|\; x(\cdot) \text{ solves } (5)\right\}$.

In the sequel of the paper we look for $\overline{f}, \overline{l}$ and a constraint on the control functions $\gamma(\cdot)$ so that this kind of convergence occurs. We split the study in two parts. First we use that v^ϵ solves in viscosity sense the Cauchy problem

$$\begin{cases} \partial_t v^\epsilon + H\left(x, \frac{x}{\epsilon}, Dv^\epsilon\right) = 0 & \text{in } (0,+\infty)\times \mathbf{R}^N \\ v^\epsilon\left(0,x\right) = h(x) & \text{in } \mathbf{R}^N, \end{cases} \tag{6}$$

where the Hamiltonian H is given by

$$H(x,y,p) = \max_{\alpha\in A}\left\{-p\cdot f\left(x,y,\alpha\right) - l\left(x,y,\alpha\right)\right\},$$

see, e.g., [8]. This is a periodic homogenization problem for a Hamilton–Jacobi equation, which consists in finding an *effective Hamiltonian* $\bar{H}(x,p)$ and appropriate conditions such that $v^\epsilon(t,x)$ converges locally uniformly to a function $v(t,x)$, viscosity solution of a limiting Cauchy problem

$$\begin{cases} \partial_t v + \bar{H}(x, Dv) = 0 & \text{in } (0, +\infty) \times \mathbf{R}^N \\ v(0, x) = h(x) & \text{in } \mathbf{R}^N. \end{cases} \tag{7}$$

Results of this type go back to the seminal paper [21] and have been extensively studied in the last decades for many different problems within the theory of viscosity solutions; see [15, 16] and also [1, 2], and [3] where it was shown the connection with singular perturbation problems. This is a classical subject in ODEs and control, pioneered by Levinson and Tichonov, see [14, 19], and the references therein. In the context of singular perturbations Artstein and Gaitsgory introduced averaging techniques and the use of invariant measures and limit occupational measures; see [4–7, 18], and also [20] for connections between such method and the viscosity theory for Hamilton-Jacobi equations.

The second part of our strategy is a representation of the effective Hamiltonian $\bar{H}$ as a Bellman Hamiltonian for suitable dynamics and cost $\bar{f}, \bar{l}$ related to the data f, l of the original problem (1), (2). Then the uniqueness of viscosity solutions to (7) implies that the limit v of v^ϵ is in fact the value function of the problem (4), (5). Limit occupational measures play a crucial role in the construction of such limit system.

The paper is organized as follows. Section 2 discusses the simple case of uncontrolled dynamics to show the role of invariant measures of ergodic dynamical systems. In Sect. 3 we reformulate the homogenization problem as singular perturbation, introduce the limit occupational measures, and formulate the main representation result for the limit control problem, under suitable controllability conditions on the system (1). Section 4 deals with vector fields f in (1) that depend on $x(s)/\epsilon$ but not on $x(s)$, and show that the controllability assumptions of the preceding section can be weakened and in some cases the limit problem is very simple. Finally in Sect. 5 we briefly describe some generalizations to appear in [9] together with a more complete theory and detailed proofs.

2 Uncontrolled Problem and Invariant Measures

We assume in this section that the dynamics and running cost are not controlled, that is

$$f = f\left(x, \frac{x}{\epsilon}\right) \qquad \text{and} \qquad l = l\left(x, \frac{x}{\epsilon}\right). \tag{8}$$

Then the value function $v^\epsilon(t, x)$ coincides with the cost functional $J^\epsilon(t, x)$ and it solves the inhomogeneous transport equation

$$\begin{cases} \partial_t v^\epsilon - Dv^\epsilon \cdot f(x, \frac{x}{\epsilon}) = l(x, \frac{x}{\epsilon}) & \text{in } (0, +\infty) \times \mathbf{R}^N \\ v^\epsilon(0, x) = h(x) & \text{in } \mathbf{R}^N. \end{cases} \tag{9}$$

Classical results in ergodic theory (see [13, 22], [3, Sect. 3.1]) ensure that the dynamics

$$\dot{y}(t) = f(x, y(t)) \qquad y(0) = y \qquad x \in \mathbf{R}^N \text{ frozen.} \tag{10}$$

has an invariant probability measure μ_x. Here $y \in \mathbf{R}^N$, but since f and l are $\mathbf{Z}^N$ periodic, we define the averaged vector field and running cost by setting

$$\hat{f}(x) := \int_{\mathbf{T}^N} f(x, y) \mathrm{d}\mu_x(y), \qquad \hat{l}(x) := \int_{\mathbf{T}^N} l(x, y) \mathrm{d}\mu_x(y),$$

where $\mathbf{T}^N = \mathbf{R}^N / \mathbf{Z}^N$.

In the following Proposition we recover within the theory of homogenization of PDEs a result of the classical theory of averaging of ODE's.

Proposition 1 *Consider the problem (1)–(2) with f and l as in (8). Assume that for every x the dynamics (10) has a unique invariant measure μ, independent of x. Then, as $\epsilon \to 0$, the problem converges to the dynamics*

$$\dot{x}(s) = \hat{f}(x(s)) \tag{11}$$

with cost functional

$$\hat{J}(t, x) := \int_0^t \hat{l}(x(s)) ds + h(x(t)). \tag{12}$$

Proof We look for a solution of (9) of the form $v^\epsilon(t, x) = u^\epsilon(t, x, \frac{x}{\epsilon})$. Then $u^\epsilon(t, x, y)$ solves

$$\begin{cases} \partial_t u^\epsilon - \left(D_x u^\epsilon + \frac{1}{\epsilon} D_y u^\epsilon\right) \cdot f(x, y) = l(x, y) & \text{in } (0, +\infty) \times \mathbf{R}^{2N} \\ u^\epsilon(0, x, y) = h(x) & \text{in } \mathbf{R}^{2N}. \end{cases}$$

A direct computation shows that the function

$$w(t, y) = \int_0^t G(\bar{x}, y(s), \bar{p}, 0) \mathrm{d}s \qquad y(\cdot) \text{ solving (10) with } x = \bar{x}$$

is the unique viscosity solution of the evolutive problem

$$\partial_t w - D_y w \cdot f(\bar{x}, y) + G(\bar{x}, y, \bar{p}, 0) = 0 \quad \text{in } (0, +\infty) \times \mathbf{R}^N, \qquad w(0, y) = 0.$$

where $G(x, y, p, q) := -(p + q) \cdot f(x, y) - l(x, y)$. Since by assumption the invariant measure is unique, it is easy to check that the quotient $w(t, y)/t$ converges to a constant uniformly w.r.t. y as $t \to +\infty$. Then such constant is the appropriate value of the effective Hamiltonian at $(\bar{x}, \bar{p})$, see [3, Sect. 2.1]. By definition of invariance we also get

$$\bar{H}(\bar{x},\bar{p}) = \int_{\mathbf{T}^N} [-\bar{p}\cdot f(\bar{x},y) - l(\bar{x},y)]\,\mathrm{d}\mu(y) = -\bar{p}\cdot\hat{f}(x) - \hat{l}(x).$$

Moreover, by the theory of [2, 3], the upper and lower semilimits of u^ϵ are respectively a subsolution and a supersolution of

$$\begin{cases} \partial_t v - Dv\cdot\hat{f}(x) = \hat{l}(x) & \text{in } (0,+\infty)\times\mathbf{R}^N \\ v(0,x) = h(x) & \text{in } \mathbf{R}^N. \end{cases} \tag{13}$$

Observe that $\hat{f},\hat{l}$ are averages with respect to a measure independent of x. Then (13) satisfies the comparison principle between viscosity sub- and supersolutions, and its unique solution is the value function associated to problem (11)–(12)

$$\hat{v}(t,x) := \int_0^t \hat{l}(x(s))\mathrm{d}s + h(x(t)), \qquad x(\cdot) \text{ solving (11) with } x(0) = x.$$

Then the upper and lower semilimits of u^ϵ coincide and we conclude that the convergence of $v^\epsilon(t,x) = u^\epsilon(t,x,\frac{x}{\epsilon})$ to $\hat{v}(t,x)$ unique solution of (13), is locally uniform. □

Remark 1 In Proposition 1 we have assumed that the unique invariant measure of (10) is independent of x. This is verified when $f = f(y)$ and $\dot{y} = f(y)$ is a uniquely ergodic dynamical system. Another case in which this hypothesis is satisfied is when $f = f(x)$ and the following non-resonance condition holds:

$$f(x)\cdot k \neq 0 \quad \forall\, k\in\mathbf{Z}^N\backslash\{0\},\ x\in\mathbf{R}^N.$$

In this case, the unique invariant measure of (10) is the Lebesgue measure, $\hat{f} = f$ and

$$\hat{l}(x) = \int_{\mathbf{T}^N} l(x,y)\mathrm{d}y.$$

3 Controllable Dynamics and Limit Occupational Measures

By introducing the additional state variables $y = x/\epsilon$, we rewrite the dynamics (1) as the singularly perturbed control system

$$\begin{cases} \dot{x}(s) = f(x(s),y(s),\alpha(s)) & x(0) = x \\ \dot{y}(s) = \frac{1}{\epsilon}f(x(s),y(s),\alpha(s)) & y(0) = x/\epsilon \end{cases} \tag{14}$$

Rescaling time by $t = \epsilon s$, the dynamics for the fast variables y in (14) can be approximated by

$$\dot{y}(t) = f(x, y(t), \alpha(t)) \qquad y(0) = y, \tag{15}$$

where the slow variable x is frozen in its initial position. For any choice of control α and initial point $y \in \mathbf{R}^N$ there is a unique solution $y(\cdot)$ of the previous dynamics and we use it to define a measure over $\mathbf{R}^N \times A$ as

$$\mu_t := \frac{1}{t} \int_0^t \delta_{(y(s),\alpha(s))} \mathrm{d}s,$$

where δ is the Dirac's delta. These measures are called *occupational measure*, as they are probability measures giving the percentage of time interval $(0, t)$ spent by a trajectory of (15) in Borel subsets of $\mathbf{R}^N \times A$. We further define the set of *limit occupational measures* [18] or *limiting relaxed controls* [1] as the set of weak-star limits of occupational measures:

$$Z(x) := \left\{ \mu \;\middle|\; \mu = \lim_{n \to \infty} \mu_{t_n} \text{ weak-star, for some } t_n \to \infty,\ \alpha(\cdot),\ y \right\}.$$

If the dynamics (15) is *bounded-time controllable*—that is, any pair of points in $\mathbf{T}^N$ can be joined by a trajectory of (15) corresponding to a suitable choice of the control, in a uniformly bounded time—the set $Z(x)$ is nonempty, convex and compact with respect to the weak-star topology; see [18, 20]. We define the averaged vector field and running cost by integrating with respect to measures in $Z(x)$:

$$\bar{f}(x, \mu) := \int_{\mathbf{R}^N \times A} f(x, y, \alpha) \mathrm{d}\mu(y, \alpha),$$

$$\bar{l}(x, \mu) := \int_{\mathbf{R}^N \times A} l(x, y, \alpha) \mathrm{d}\mu(y, \alpha), \qquad \mu \in Z(x).$$

Proposition 2 *Assume that*

$$\begin{array}{c} \text{for any } x \text{ exists } \nu(x) > 0 \text{ s.t.} \\ B(0, \nu(x)) \subseteq \overline{\text{co}} f(x, y, A) \text{ for any } y. \end{array} \tag{16}$$

Then the optimal control problem (1)–(2) converges as $\epsilon \to 0$ to the problem with cost functional

$$\bar{J}(t, x, \mu) := \int_0^t \bar{l}(x(s), \mu(s)) ds + h(x(t)). \tag{17}$$

and dynamics given by the differential inclusion

$$\dot{x}(s) = \bar{f}(x(s), \mu(s)) \qquad \mu(s) \in Z(x(s)), \qquad x(0) = x. \tag{18}$$

Proof We look for a solution of (6) of the form $v^\epsilon(t,x) = u^\epsilon(t,x,\frac{x}{\epsilon})$. Then $u^\epsilon(t,x,y)$ solves in viscosity sense

$$\begin{cases} \partial_t u^\epsilon + G\left(x, y, D_x u^\epsilon, \frac{D_y u^\epsilon}{\epsilon}\right) = 0 & \text{in } (0,+\infty) \times \mathbf{R}^{2N} \\ u^\epsilon(0,x,y) = h(x) & \text{in } \mathbf{R}^{2N}, \end{cases} \tag{19}$$

where $G(x,y,p,q) := H(x,y,p+q)$. This PDE corresponds to a singularly perturbed control problem that was studied under the current assumptions in [20]. We recall here the main steps of the proof. The controllability condition (16) is equivalent to the following coercivity property with respect to q: for every $x, p \in \mathbf{R}^N$,

$$G(x,y,p,q) \geq \nu(x)|q| - C(1 + |p|) \quad \text{for any } y, q \tag{20}$$

for some $C > 0$ (see [3, Section 6.1]). This entails that G is ergodic, namely, $\bar{H}$ exists and the upper and lower semi-limits of u^ϵ are respectively a viscosity subsolution and supersolution of (7). Arguing as in [1, Theorem 7] it is possible to prove the following representation formula:

$$\bar{H}(x,p) = \max_{\mu \in Z(x)} \left\{-p \cdot \bar{f}(x,\mu) - \bar{l}(x,\mu)\right\}. \tag{21}$$

Since (20) also implies that $\bar{H}(x,p)$ is Lipschitz continuous (see [3, Proposition 6.4]), the comparison principle holds for problem (7) and u^ϵ converges locally uniformly as $\epsilon \to 0$ to $v(t,x)$, unique solution of (7); then v^ϵ also converges to the same function. To complete the proof it is necessary to check that the value function associated to (18)–(17), that is

$$\bar{v}(t,x) := \inf\left\{\bar{J}(t,x,\mu) \;\middle|\; x(\cdot) \text{ solves } (18)\right\},$$

is a viscosity solution of (7). To see this, one needs to take into account formula (21), to prove that $\bar{v}$ is continuous, that the multivalued map $f(x, Z(x))$ in (18) is Lipschitz continuous, and that the limiting dynamics admits trajectories defined for any positive time: we refer to [20] for the details. □

Remark 2 Although the controllability condition (16) does not hold in Sect. 2, the statement of Proposition 1 is consistent with that of Proposition 2 and provides an example in which the set of limiting occupational measures is explicit and it is a singleton.

4 Purely Oscillating Dynamics

In this section we study simpler expressions for the limiting dynamics when the vector field depends on $x(s)$ only via the oscillating terms $\frac{x(s)}{\epsilon}$, that is,

$$\begin{cases} \dot{x}(s) = f\left(\frac{x(s)}{\epsilon}, \alpha(s)\right) \\ x(0) = x. \end{cases} \tag{22}$$

4.1 Weakening the Controllability Conditions

For systems of the form (22) the dynamics of the oscillating variables in rescaled time is independent of x

$$\dot{y}(t) = f(y(t), \alpha(t)). \tag{23}$$

Consequently, the set of limiting relaxed controls is $Z(x) = Z$ independent of x. This permits to prove the convergence without the additional controllability assumption (16). As before, we set

$$\bar{f}(\mu) := \int_{\mathbf{R}^N \times A} f(y, \alpha) d\mu(y, \alpha), \qquad \mu \in Z.$$

Proposition 3 *Consider the optimal control problem (22)–(2) and assume that the system (23) is bounded-time controllable. Then, the problem converges as $\epsilon \to 0$ to the one with dynamics*

$$\dot{x}(s) = \bar{f}(\mu(s)), \qquad x(0) = x, \qquad \mu(s) \in Z \tag{24}$$

and cost functional $\bar{J}(t, x, \mu)$ as in (17).

Proof We proceed along the same lines and keep the same notations as in the proof of Proposition 2. The assumed bounded-time controllability of (23) implies that Z is a convex and compact subset of probability measures. Moreover, the upper and lower semi-limits of u^ϵ are, respectively, a subsolution and a supersolution of (7), with

$$\bar{H}(x, p) = \max_{\mu \in Z} \left\{ -p \cdot \bar{f}(\mu) - \bar{l}(x, \mu) \right\}.$$

Now, since f does not depend on x, $\bar{H}(x, p)$ satisfies regularity properties that guarantee the comparison principle for the effective Cauchy problem (7). Thus, the locally uniform convergence of u^ϵ—and then that of v^ϵ—to the value function

$$\bar{v}(t,x) := \inf\left\{\bar{J}(t,x,\mu) \;\middle|\; x(\cdot) \text{ solves } (24)\right\},$$

unique solution of (7) can be proved without requiring any extra controllability assumption. □

4.2 A Simple Degenerate Limit for Special Costs

Here we consider the special case that

$$l = l\left(\frac{x}{\epsilon}\right), \qquad h(x) = 0,$$

in (2), so that the cost functional is

$$J^\epsilon(t,x,\alpha) = \int_0^t l\left(\frac{x(s)}{\epsilon}\right) ds. \tag{25}$$

We also assume the following controllability condition, much weaker than condition (16),

$$\max_{\alpha \in A}\{-q \cdot f(y,\alpha)\} \geq 0 \quad \text{for any } y, q \in \mathbf{R}^N. \tag{26}$$

The next result says that in this case the limit control problem reduces to the static optimization problem of the running cost with respect to the state variables.

Proposition 4 *Consider the optimal control problem (22)–(25) and assume that the system (23) is bounded-time controllable and satisfies (26). Then, the value function* $v^\epsilon(t,x) := \inf J^\epsilon(t,x,\alpha)$ *converges locally uniformly as* $\epsilon \to 0$ *to*

$$\overline{v}(t) = t \min_{y \in \mathbf{T}^N} l(y).$$

Proof The value function v^ϵ is the unique solution of

$$\begin{cases} \partial_t v^\epsilon + \max\limits_{\alpha}\left\{-D_x v^\epsilon \cdot f\left(\frac{x}{\epsilon},\alpha\right)\right\} = l\left(\frac{x}{\epsilon}\right) & \text{in } (0,+\infty) \times \mathbf{R}^N \\ v^\epsilon(0,x) = 0 & \text{in } \mathbf{R}^N. \end{cases} \tag{27}$$

Set $H(y,q) := \max\limits_{\alpha}\{-q \cdot f(y,\alpha)\} - l(y)$. Condition (26) implies that $H(y,q) \geq H(y,0)$ for any y, q. Then, arguing by contradiction as in [3, Proposition 6.6], we get

$$\bar{H} = \max_{y \in \mathbf{R}^N} H(y,0) = \max_{y \in \mathbf{T}^N}\{-l(y)\} = -l(y_0),$$

for some $y_0 \in [0,1)^N$, since l is continuous and periodic. Then $v^\epsilon(t,x)$ converges locally uniformly to the unique solution of

$$\partial_t v + \bar{H} = 0 \qquad v(0) = 0,$$

which is $\bar{v}(t) = tl(y_0)$. □

5 Generalizations of the Results

We will show in the forthcoming paper [9] how the results described in this note can be generalized to prove variational convergence of optimal control problems with dynamics

$$\begin{cases} \dot{x}_1(s) = f_1\left(x(s), \frac{x_2(s)}{\epsilon}, \alpha_1(s), \alpha_2(s)\right) \\ \dot{x}_2(s) = f_2\left(x(s), \frac{x_2(s)}{\epsilon}, \alpha_2(s)\right) \\ x(0) = x, \end{cases} \tag{28}$$

and cost functional

$$\int_0^t l\left(x(s), \frac{x_2(s)}{\epsilon}, \alpha_1(s), \alpha_2(s)\right) \mathrm{d}s + h(x(t)). \tag{29}$$

The state variable x is divided here in two groups, x_1 and x_2. The dynamics for the oscillating variables, x_2, is controlled only by the α_2 component of the control variable (α_1, α_2). Problem (1) considered here corresponds to the particular choice $f_1 \equiv 0$ (then the dynamics for x_1 can be ignored) and $f_2 = f$.

The arguments sketched in the proof of Proposition 1 can be adapted to show convergence of (28)–(29) when the dynamics for x_2 is uncontrolled, i.e. $f_2 = f_2(x, x_2/\epsilon)$. A representation of the limiting optimal control problem can be provided in terms of invariant measures of the flow associated to the dynamics of fast oscillations.

The strategy described in Sect. 3 and the averaging result of Proposition 2 can be adapted to show convergence of optimal control problems like (28)–(29); assumption (16) must be satisfied with $f = f_2$ and $A = A_2$, the compact set of values for the α_2 components of the control. An analog of Proposition 3 holds, whenever f_2 depends on x_2/ϵ but not on x and the dynamics

$$\dot{y}(t) = f_2\left(y(t), \alpha_2(t)\right), \qquad y(0) = y$$

is bounded-time controllable.

The case treated in Sect. 4.2 can be generalized to (28)–(29) provided $f_i = f_i(x_1, \frac{x_2}{\epsilon}, \alpha_i)$ $(i = 1, 2)$, $l = l(x_1, \frac{x_2}{\epsilon}, \alpha_1)$, $h = h(x_1)$. We can show that an optimal control problem satisfying this partially decoupled structure admits a variational limit. As $\epsilon \to 0$, the oscillations x_2/ϵ play the role of a new control variable valued in the torus. More precisely, the limiting dynamics is governed by the drift $f_1 = f_1(x_1, y, \alpha_1)$ controlled by

$$(y, \alpha_1) : [0, \infty) \to [0, 1)^{N_2} \times A_1 \text{ measurable}$$

and the associated cost functional is

$$\int_0^t l(x_1(s), y(s), \alpha_1(s))\mathrm{d}s + h(x_1(t)).$$

Acknowledgements Martino Bardi was partially supported by the Fondazione CaRiPaRo Project "Nonlinear Partial Differential Equations: models, analysis, and control-theoretic problems" and the European Project Marie Curie ITN "SADCO—Sensitivity Analysis for Deterministic Controller Design". Gabriele Terrone was supported by the UTAustin-Portugal partnership through the FCT post-doctoral fellowship SFRH/BPD/40338/2007, CAMGSD-LARSys through FCT Program POCTI—FEDER and by grants PTDC/MAT/114397/2009, UTAustin/MAT/0057/2008, and UTA-CMU/MAT/0007/2009.

References

1. Alvarez, O., Bardi, M.: Viscosity solutions methods for singular perturbations in deterministic and stochastic control. SIAM J. Control Optim. **40**(4), 1159–1188 (2001/2002)
2. Alvarez, O., Bardi, M.: Singular perturbations of nonlinear degenerate parabolic PDEs: a general convergence result. Arch. Ration. Mech. Anal. **170**(1), 17–61 (2003)
3. Alvarez, O., Bardi, M.: Ergodicity, stabilization, and singular perturbations for Bellman-Isaacs equations. Mem. Am. Math. Soc. **204**(960), vi+77 (2010)
4. Artstein, Z.: Stability in the presence of singular perturbations. Nonlinear Anal. **34**(6), 817–827 (1998)
5. Artstein, Z.: Invariant measures of differential inclusions applied to singular perturbations. J. Differ. Equ. **152**(2), 289–307 (1999)
6. Artstein, Z., Gaitsgory, V.: Tracking fast trajectories along a slow dynamics: a singular perturbations approach. SIAM J. Control Optim. **35**(5), 1487–1507 (1997)
7. Artstein, Z., Gaitsgory, V.: The value function of singularly perturbed control systems. Appl. Math. Optim. **41**(3), 425–445 (2000)
8. Bardi, M., Dolcetta, I.C.: Optimal control and viscosity solutions of Hamilton-Jacobi-Bellman equations. Birkhäuser Boston Inc., Boston (1997)
9. Bardi, M., Terrone, G.: Homogenization of some optimal control problems (preprint, to appear)
10. Braides, A., De Franceschi, A.: Homogenization of Multiple Integrals. Oxford University Press, Oxford (1998)
11. Buttazzo, G., Maso, G.D.: Γ-Convergence and optimal control problems. J. Optim. Theory Appl. **38**(3), 385–407 (1982)
12. Buttazzo, G., Drakhlin, M.E., Freddi, L., Stepanov, E.: Homogenization of optimal control problems for functional-differential equations. J. Optim. Theory Appl. **93**(1), 103–119 (1997)
13. Cornfeld, I.P., Fomin, S.V., Sinaĭ, Y.G.: Ergodic Theory. Springer, New York (1982)
14. Dontchev, A.L., Zolezzi, T.: Well-Posed Optimization Problems. Lecture Notes in Mathematics, vol. 1543. Springer, Berlin (1993)

15. Evans, L.C.: The perturbed test function method for viscosity solutions of nonlinear PDE. Proc. R. Soc. Edinb. Sect. A **111**(3–4), 359–375 (1989)
16. Evans, L.C.: Periodic homogenisation of certain fully nonlinear partial differential equations. Proc. R. Soc. Edinb. Sect. A **120**(3–4), 245–265 (1992)
17. Freddi, L.: Γ-Convergence and chattering limits in optimal control theory. J. Convex Anal. **8**(1), 39–70 (2001)
18. Gaitsgory, V., Leizarowitz, A.: Limit occupational measures set for a control system and averaging of singularly perturbed control systems. J. Math. Anal. Appl. **233**(2), 461–475 (1999)
19. Kokotović, P., Khalil, H.K., O'Reilly, J.: Singular Perturbation Methods in Control. Classics in Applied Mathematics, vol. 25. Society for Industrial and Applied Mathematics (SIAM), Philadelphia (1999)
20. Terrone, G.: Limiting relaxed controls and averaging of singularly perturbed deterministic control systems. Dyn. Contin. Discrete Impuls. Syst. Ser. A **18**(5), 653–672 (2011)
21. Varadhan, S.R.S., Lions, P.-L., Papanicolaou, G.: Homogeneization of Hamilton-Jacobi equations (1986, Unpublished preprint)
22. Walters, P.: An Introduction to Ergodic Theory. Springer, New York (1982)

On Gradient Like Properties of Population Games, Learning Models and Self Reinforced Processes

Michel Benaim

Abstract We consider ordinary differential equations on the unit simplex of $\mathbb{R}^n$ that naturally occur in population games, models of learning and self reinforced random processes. Generalizing and relying on an idea introduced in Dupuis and Fisher (On the construction of Lyapunov functions for nonlinear Markov processes via relative entropy, 2011), we provide conditions ensuring that these dynamics are gradient like and satisfy a suitable "angle condition". This is used to prove that omega limit sets and chain transitive sets (under certain smoothness assumptions) consist of equilibria; and that, in the real analytic case, every trajectory converges toward an equilibrium. In the reversible case, the dynamics are shown to be C^1 close to a gradient vector field. Properties of equilibria -with a special emphasis on potential games—and structural stability questions are also considered.

1 Introduction

Let S be a finite set, say $S = \{1, \ldots, n\}$. A *rate matrix* over S is a $n \times n$ matrix L such that $L_{ij} \geq 0$ for $i \neq j$ and $\sum_j L_{ij} = 0$. We let $\mathbf{R}(S)$ denote the space of such matrices. For $x \in \mathbb{R}^n$ and $L \in \mathbf{R}(S)$ we let xL denote the vector defined by $(xL)_i = \sum_j x_j L_{ji}$.

Let

$$\Delta = \{x \in \mathbb{R}^n : x_i \geq 0, \sum_i x_i = 1\}$$

be the *unit simplex of probabilities* over S. In this paper we are interested in ordinary differential equations on Δ having the form

$$\frac{dx}{dt} = xL(x) := F(x) \tag{1}$$

M. Benaim (✉)
Institut de Mathématiques, University of Neuchâtel, Neuchâtel, Switzerland
e-mail: michel.benaim@unine.ch

J.-P. Bourguignon et al. (eds.), *Dynamics, Games and Science*, CIM Series in Mathematical Sciences 1, DOI 10.1007/978-3-319-16118-1_8

where $L : \Delta \mapsto \mathbf{R}(S)$ is a sufficiently smooth function. Such dynamics occur—through a natural averaging procedure- in models of games describing strategic interactions in a large population of players, as well as in certain models of learning and reinforcement. These models are usually derived from qualitative assumptions describing the "microscopic" behavior of anonymous agents, and it is usually believed or assumed that similar qualitative microscopic behaviors should lead to similar global dynamics. However there is no satisfactory general theory supporting this belief.

To be more precise, under the assumption that $L(x)$ is irreducible, there exists a unique "invariant probability" for $L(x)$, $\pi(x) \in \Delta$ characterized by

$$\pi(x)L(x) = 0. \tag{2}$$

Several models corresponding to different rate functions $x \mapsto L(x)$ have the same invariant probability function $x \mapsto \pi(x)$. For instance, to each population game (see Sect. 2.1) which average ODE is given by (1), there is a canonical way to define a learning process (see Sect. 2.2) which average ODE is given by

$$\frac{dx}{dt} = -x + \pi(x) := F_{\pi}(x), \tag{3}$$

but there is no evidence that the dynamics of (1) and (3) are related in general.

The purpose of this paper is to provide sufficient conditions on $\pi(x)$ ensuring that (1) has a gradient-like structure. This heavily relies on an idea introduced in [10] where it was shown that the relative entropy between x and $\pi(x)$ is a strict Lyapounov function for systems of Gibbs type. We extend this idea to other class of systems beyond systems of Gibbs type, including population games and reinforcement process with imitative dynamics, and investigate further dynamics properties.

To give the flavor of the results presented in this paper, let $\pi : \Delta \mapsto \dot{\Delta}$ be a smooth function mapping Δ into its relative interior. Let χ_{π} be the set of vector fields having the form given by (the right hand side of) (1), where for each x, $L(x)$ is irreducible and verifies (2). Note that χ_{π} is a convex set of vector fields on Δ and that $F_{\pi} \in \chi_{\pi}$.

Theorem A

For all $F \in \chi_{\pi}$.

(i) Equilibria (respectively non degenerate equilibria) of F coincide with equilibria (respectively non degenerate equilibria) of F_{π}.
(ii) In general, global dynamics of F and F_{π} are "unrelated." We construct an example for which F_{π} is globally asymptotically stable (every trajectory

converge toward a linearly stable equilibrium) while every non equilibrium trajectory for F converge to a limit cycle.

(iii) Assume that there exists a $C^k, k \geq 1$ strictly increasing function $s : \mathbb{R} \mapsto \mathbb{R}$ such that $x \in \dot{\Delta} \mapsto s((\frac{x_i}{\pi_i(x)}))_{i \in S}$ is the gradient (or quasi gradient) of some function $V : \dot{\Delta} \mapsto \mathbb{R}$. Then

   (a) V is a strict Lyapounov function for F (F is gradient-like) and verifies an *angle condition*,
   (b) Omega limit sets and chain-transitive sets of F are equilibria,
   (c) In the real analytic case, every solution to (1) converge toward an equilibrium,
   (d) In the reversible case, hyperbolic equilibria of F coincide with non degenerate critical points of V and, provided there are finitely many equilibria, F is C^1 close to a gradient vector field for a certain Riemannian metric,
   (e) The set χ_π is not (in general) structurally stable.

Section 2 describes a few examples that motivate this work. Section 3 contains some preliminary results and the main assumptions. Section 4 is devoted to Theorem $A, (ii)$; Sect. 5 to $(iii), (a), (b), (c)$; Sects. 6 and 7 to $(iii)(d)$ and Sect. 8 to $(iii)(e)$. Other results and examples are also discussed in these sections. For instance, in Sect. 6.1, the local dynamics (dynamics in the neighborhood of equilibria) of mean field systems associated to potential games is precisely described in term of Nash equilibria.

2 Motivating Examples

Throughout this section we see S as set of *pure strategies*. A Markov matrix over S is a $n \times n$ matrix K such that $K_{ij} \geq 0$ and $\sum_j K_{ij} = 1$. We let $\mathbf{M}(S)$ denote the sets of such matrices and we assume given a Lipschitz map

$$K : \Delta \mapsto \mathbf{M}(S).$$

For further reference we may call such a map a *revision protocol.* This terminology is borrowed from [23].

2.1 Population Games

Good references on the subject include [22] and the survey paper [23] from which some of the examples here are borrowed.

Consider a population of N *agents,* each of whom chooses a strategy in S at discrete times $k = 1, 2, \ldots$. Depending on the context, an agent can be a player, a set of players, a biological entity, a communication device, etc. The state of the system at time $k \in \mathbb{N}$ is the vector $X_k^N = (X_{k,1}^N, \ldots X_{k,n}^N) \in \Delta$ where $NX_{k,i}^N$ equals the number of agents having strategy i. The system evolves as follows. Assume that at time k the system is in state $X_k^N = x$. Then an agent is randomly chosen in the population. If the chosen agent is an i–strategist, he/she switches to strategy j with probability $K_{ij}(x)$. This makes $(X_k^N)_{k\geq 1}$ a discrete time Markov chain, which transition probabilities are

$$\mathbf{P}(X_{k+1}^N = x + \frac{1}{N}(e_j - e_i)|X_k^N = x) = x_i K_{ij}(x)$$

where $(e_1, \ldots, e_n)$ is the standard basis of $\mathbb{R}^n$. Let

$$L(x) = -Id + K(x). \tag{4}$$

By standard mean-field approximation (see [5, 15] for precise statements, and [22, 23] for discussions in the context of games), the process $\{(X_k^N) : kN \leq T\}$ can be approximated by the solution to (1) (with $L(x)$ given by (4)) with initial condition $x = X_0^N$, over the time interval $[0, T]$.

2.1.1 Revision Protocols

Assume, as it is often the case in economic or biological applications, that the population game is determined by a continuous *Payoff-function* $U : \Delta \mapsto \mathbb{R}^n$. The quantity $U_i(x)$ represents the payoff (utility, fitness) of an $i-$ strategist when the population state is x.

An *attachment-function* is a continuous map $w : \Delta \mapsto \mathbb{R}_+^{n^2}$. The weight $w_{ij}(x)$ can be seen as an a priori *attachment* of an i–strategist for strategy j. It can also encompasses certain constraints on the strategy sets. For instance $w_{ij}(x) = 0$ (respectively $w_{ij}(x) << 1$) if a move from i to j is forbidden (respectively costly). We call the attachment function *imitative* if

$$w_{ij}(x) = x_j \tilde{w}_{ij}(x) \tag{5}$$

Most, not to say all, revision protocols in population games fall into one of the two next categories:

(i) [Sampling]

$$K_{ij}(x) = \frac{w_{ij}(x) f(U_j(x))}{\sum_k w_{ik}(x) f(U_k(x))} \tag{6}$$

where f is a non negative increasing function.

(ii) [Comparison]

$$K_{ij}(x) = w_{ij}(x)g(U_i(x), U_j(x)) \text{ for } i \neq j \tag{7}$$
$$K_{ii}(x) = 1 - \sum_{j \neq i} K_{ij}(x)$$

where $g(u, v)$ is nonnegative, decreasing in u, increasing in v and such that $\sum_{j \neq i} K_{ij}(x) \leq 1$.

2.2 Processes with Reinforcement and Adaptive Learning

Suppose now there is only one single agent in the population. In the context of games, one can imagine that this agent consists of a finite set of players and that S is the cartesian product of the strategy sets of the players. let $X_k \in S$ denote the strategy of this agent at time k. Let $\mu_k \in \Delta$ denote the empirical occupation measure of (X_k) up to time k. That is

$$\mu_k = \frac{1}{k} \sum_{j=1}^{k} \delta_{X_j}$$

where $\delta : S \mapsto \Delta$ is defined by $\delta_i = e_i$. Suppose now that the agent revises her strategies as follows:

$$\mathbf{P}(X_{k+1} = j | X_0, \ldots, X_{k-1}, X_k = i) = K_{ij}(\mu_k).$$

The process (X_k) is no longer a Markov process but a *process with reinforcement* (see [20] for a survey of the literature on the subject). Using tools from stochastic approximation theory, it can be shown (see [1]) that, under certain irreducibility assumptions, the long term behavior of (μ_k) can be precisely related (see [1–3] and the brief discussion preceding Corollary 3) to the long term behavior of the differential equation on Δ

$$\frac{dx}{dt} = -x + \pi(x) \tag{8}$$

where $\pi(x) \in \Delta$ is the invariant probability of $K(x)$. Note that (8) can be rewritten as (1) with $L_{ij}(x) = \delta_{ij} - \pi_j(x)$.

3 Hypotheses, Notation, and Preliminaries

Let L be a rate matrix, as defined in the introduction. Then L is the infinitesimal generator of a continuous time Markov chains on S. A probability $\pi \in \Delta$ is called *invariant* for L if it is invariant for the associated Markov chain, or equivalently

$$\pi L = 0.$$

A sufficient condition ensuring that $\pi \in \Delta$ is invariant is that L is *reversible* with respect to π, meaning that

$$\pi_i L_{ij} = \pi_j L_{ji}.$$

The matrix is said *irreducible* if for all $i,j \in \{1,\ldots,n\}$ there exist some integer k and a sequence of indices $i = i_1, i_2, \ldots, i_{k-1}, i_k = j$ such that $L_{i_l,i_{l+1}} > 0$ for $l = 1, \ldots, k-1$.

An irreducible rate matrix admits a unique invariant probability which can be expressed as a rational function of the coefficients (L_{ij}) (see e.g. Chapter 6 of [11]).

The relative interior of Δ is the set

$$\dot{\Delta} = \{x \in \Delta : \ \forall i \in S, \, x_i > 0\}.$$

From now on we assume given a C^1 map[1] $L : \Delta \mapsto \mathbf{R}(S)$ satisfying the following assumption:

Hypothesis 1 (Standing Assumption) For all $x \in \dot{\Delta}, L(x)$ is irreducible.

We sometimes assume

Hypothesis 2 (Occasional Assumption) For all $x \in \Delta$, $L(x)$ is irreducible.

In view of the preceding discussion Hypotheses 1 and 2 imply the following

Lemma 1 *There exists a C^1 map $\pi : \dot{\Delta} \mapsto \dot{\Delta}$ such that for all $x \in \dot{\Delta}$, $y = \pi(x)$ is the unique solution to the equation*

$$yL(x) = 0, \ y \in \Delta.$$

If L is C^k, C^∞ or real analytic, the same is true for π. Under Hypothesis 2, π is defined on all Δ and maps Δ into $\dot{\Delta}$.

We let F_π denote the map defined as

$$F_\pi(x) = -x + \pi(x). \tag{9}$$

---

[1]By this we mean that L is the restriction to Δ of a C^1 map defined in a neighborhood of Δ in $\mathit{aff}(\Delta) = \{x \in \mathbb{R}^n : \sum_i x_1 = 1\}$.

Throughout, it is implicitly assumed that the domain of F_π is $\dot{\Delta}$ under Hypothesis 1 and Δ under Hypothesis 2.

We now consider the dynamics induced by (1). Without loss of generality, we may assume that (1) is defined on all $\mathbb{R}^n$ and induces a flow $\Phi = (\Phi_t)$ leaving Δ positively invariant. Indeed, by convexity of Δ, the retraction $r : \mathbb{R}^n \mapsto \Delta$ defined by $r(x) = \mathbf{argmin}_{y \in \Delta} \|x - y\|$ is Lipschitz so that the differential equation

$$\frac{dy}{dt} = yL(r(y)) \tag{10}$$

is Lipchitz and sub-linear on all $\mathbb{R}^n$. By standard results, it then induces a flow $\Phi : \mathbb{R} \times \mathbb{R}^n \mapsto \mathbb{R}^n$ where $t \mapsto \Phi(t, y) = \Phi_t(y)$ is the unique solution to (10) with initial condition y. For all $x \in \Delta$ and $t \geq 0$, $\Phi_t(x) \in \Delta$ and the map $t \in \mathbb{R}^+ \mapsto \Phi_t(x)$ is solution to (1).

In the following we sometime use the notation $\Phi_t(x) = x(t) = (x_1(t), \ldots, x_n(t))$. The *tangent space* of Δ is the space

$$T\Delta = \{u \in \mathbb{R}^n : \sum_{i=1}^{n} u_i = 0\}.$$

Lemma 2 *(i) There exists $\alpha \geq 0$ such that for all $x \in \Delta$ $x_i(t) \geq e^{-\alpha t} x_i(0)$. In particular, $\dot{\Delta}$ is positively invariant.*
(ii) If for all $x \in \partial\Delta$ $L(x)$ is irreducible, then $\Phi_t(\Delta) \subset \dot{\Delta}$ for all $t > 0$ and the dynamics (1) admits a global attractor

$$A = \bigcap_{t \geq 0} \Phi_t(\Delta) \subset \dot{\Delta}.$$

Proof (*i*) Let $\alpha - \sup_{x \in \Delta} -L_{ii}(x)$. For all $j \neq i$ and $x \in \Delta$

$$\dot{x}_i \geq -\alpha x_i + x_j L_{ji}(x).$$

Hence

$$x_i(t) \geq e^{-\alpha t}[x_i(0) + \int_0^t e^{\alpha s} x_j(s) L_{ji}(x(s)) ds] \geq e^{-\alpha t} x_i(0).$$

The second inequality is the first statement. From the first inequality and the continuity of $L(x(t))$ it follows that $x_i(t) > 0$ for all $t > 0$ whenever that $x_j L_{ji}(x) > 0$. Let now $x \in \partial\Delta$. Assume without loss of generality that $x_1 > 0$. By irreducibility there exists a sequence $1 = i_1, i_2, \ldots i_k = j$ such that $L_{i_l, i_{l+1}}(x) > 0$. Hence, by continuity, $L_{i_l, i_{l+1}}(x(t)) > 0$ for all t small enough. It then follows that $x_j(t) > 0$ for all $t > 0$. ■

Remark 1 Assumption 1 is not needed in Lemma 2.

Throughout we let

$$\mathbf{Eq}(F) = \{x \in \Delta : F(x) = 0\}$$

denote the *equilibria* set of F. Note that in view of the preceding Lemmas

$$\mathbf{Eq}(F) \cap \dot{\Delta} = \{x \in \dot{\Delta} : F_\pi(x) = 0\}$$

and, in case $L(x)$ is irreducible for all $x \in \Delta$, $\mathbf{Eq}(F) \subset \dot{\Delta}$.

An equilibrium p is called *non degenerate for F* provided the Jacobian matrix $DF(p) : T\Delta \mapsto T\Delta$ is invertible.

Lemma 3 *Let $p \in \mathbf{Eq}(F) \cap \dot{\Delta}$. Then p is non degenerate for F if and only if it is non degenerate for F_π.*

Proof Let $L^T(x) : T\Delta \mapsto T\Delta$ be defined by $L^T(x)h = hL(x)$. Then for all $x \in \dot{\Delta}$ $F(x) = xL(x) = (x - \pi(x))L(x) = L^T(x)(x - \pi(x))$. Hence at every equilibrium $p \in \dot{\Delta}$ $DF(p) = -L^T(p)(DF_\pi(p))$. By irreducibility, $L^T(p)$ is invertible (see Lemma 8 in the Appendix). Thus $DF(p)$ is invertible if and only if $DF_\pi(p)$ is invertible. ■

4 Dynamics of F and F_π are Generally Unrelated

While F and F_π have the same equilibria, they may have quite different dynamics as shown by the following example.

Suppose $n = 3$ so that Δ is the unit simplex in $\mathbb{R}^3$. Let G be a smooth vector field on Δ such that:

(i) G points inward $\dot{\Delta}$ on $\partial\Delta$,
(ii) Every forward trajectory of G converge to $p = (1/3, 1/3, 1/3)$,
(iii) $\jmath DG(p)\jmath^{-1} = \begin{pmatrix} -\eta & -1 \\ 1 & -\eta \end{pmatrix}$, $\eta > 0$,
where $\jmath : T\Delta \mapsto \mathbb{R}^2$ is defined by $\jmath(u_1, u_2, u_3) = (u_1, u_2)$.

It is easy to construct such a vector field.

Choose $\varepsilon > 0$ small enough so that $\varepsilon G(x) + x$ lies in $\dot{\Delta}$ for all $x \in \Delta$ and set $\pi(x) = \varepsilon G(x) + x$. Then, F_π and G have the same orbits.

Let W be a 3×3 symmetric irreducible matrix with positive off-diagonal entries. Set $L_{ij}(x) = W_{ij}\pi_j(x)$ for $i \neq j$ and $L_{ii}(x) = -\sum_{j \neq i} L_{ij}(x)$. The matrix $L(x)$ is an irreducible rate matrix, reversible with respect to $\pi(x)$. It follows from Lemmas 2 and 3 that $F(x) = xL(x)$ has a global attractor contained in $\dot{\Delta}$ and a unique equilibrium given by p.

Furthermore,

$$DF(p) = -L(p)^T DF_\pi(p) = -\varepsilon L(p)^T DG(p) = -\frac{\varepsilon}{3} WDG(p)$$

where the last equality follows from the definition of L and the fact that $\pi(p) = p$. To shorten notation, set $b = \frac{\varepsilon}{3} W_{12}, c = \frac{\varepsilon}{3} W_{13}$ and $d = \frac{\varepsilon}{3} W_{23}$. Then

$$JDF(p)J^{-1} = \begin{pmatrix} (b+2c) & c-b \\ d-b & (b+2d) \end{pmatrix} \begin{pmatrix} -\eta & -1 \\ 1 & -\eta \end{pmatrix}$$

The determinant of this matrix is positive and its trace equals

$$(c-d) - 2\eta(b+c+d).$$

If one now choose $c > d$ and η small enough, the trace is positive. This makes p linearly unstable. By Poincaré-Bendixson theorem, it follows that every forward trajectory distinct from p converges toward a periodic orbit.

Remark 2 It was pointed out to me by Sylvain Sorin and Josef Hofbauer that this example is reminiscent of the following phenomenon. Consider a population game which revision protocol takes the form

$$K_{ij}(x) = \frac{x_j}{R} \max(0, U_j(x) - U_i(x)) \text{ for } i \neq j$$

(here R is chosen so that $\sum_{j \neq i} K_{ij}(x) \leq 1$). This is a particular case of imitative pairwise comparison protocol (see Eq. (7)).

Then, the mean field ode is the classical replicator dynamics (see Example 3.2 in [23]):

$$\dot{x}_i = x_i(U_i(x) - \sum_{j \in S} x_j U_j(x)) \tag{11}$$

Here the rate matrix $L(x)$ is not irreducible and its set of invariant probabilities is easily seen to be the *Best Reply* set

$$BR(x) = conv(\{e_i : U_i(x) = \max_{j \in S} U_j(x)\}).$$

where *conv*(A) stands for the convex hull of A. The vector field (9) is not defined but can be replaced by the differential inclusion

$$\dot{x} \in -x + BR(x). \tag{12}$$

If one assume that $U_i(x) = \sum_j U_{ij}x_j$ with U the payoff matrix given by a Rock-Paper-Scissors game,

$$U = \begin{pmatrix} 0 & -1 & 1 \\ 1 & 0 & -1 \\ -1 & 1 & 0 \end{pmatrix};$$

Then $p = (1/3, 1/3, 1/3)$ is the unique equilibrium of (11) in $\dot{\Delta}$ (corresponding to the unique Nash equilibrium of the game) and every solution to (11) with initial condition in $\dot{\Delta}\setminus\{p\}$ is a periodic orbit. On the other hands, solutions to (12) converge to p. Phase portraits of these dynamics can be found in ([23], Section 5) and a detailed comparison of the replicator and the best reply dynamics is provided in [14].

5 Gradient Like Structure

For $u, v \in \mathbb{R}^n$ we let $\langle u, v\rangle = \sum_i u_i v_i$.

A map $h : \dot{\Delta} \mapsto \mathbb{R}^n$, is called a *gradient* if there exists a C^1 map $V : \dot{\Delta} \mapsto \mathbb{R}$ such that for all $x \in \dot{\Delta}$ and $u \in T\Delta$

$$\langle h(x), u\rangle = DV(x).u := \langle \nabla V(x), u\rangle.$$

It is called a *quasigradient* or a *α-quasigradient* if $x \mapsto \alpha(x)h(x)$ is a gradient for some continuous map $\alpha : \dot{\Delta} \mapsto \mathbb{R}^*_+$. That is

$$\alpha(x)\langle h(x), u\rangle = \langle \nabla V(x), u\rangle \tag{13}$$

for all $x \in \dot{\Delta}$ and $u \in T\Delta$.

Remark 3 If V is the restriction to $\dot{\Delta}$ of a C^1 map $W : \mathbb{R}^n \mapsto \mathbb{R}$, then $\nabla V(x)$ is the orthogonal projection of $\nabla W(x)$ onto $T\Delta$. That is

$$\nabla V_i(x) = \frac{\partial W}{\partial x_i}(x) - \frac{1}{n}\sum_{j=1}^{n} \frac{\partial W}{\partial x_j}(x), i = 1, \dots n.$$

Remark 4 A practical condition ensuring that h is a gradient is that

(a) h is the restriction to $\dot{\Delta}$ of a C^1 map $h : \mathbb{R}^n \mapsto \mathbb{R}^n$,
(b) For all $x \in \dot{\Delta}$ and $i, j, k \in \{1, \dots, n\}$

$$\frac{\partial h_i}{\partial x_j}(x) + \frac{\partial h_j}{\partial x_k}(x) + \frac{\partial h_k}{\partial x_i}(x) = \frac{\partial h_i}{\partial x_k}(x) + \frac{\partial h_k}{\partial x_j}(x) + \frac{\partial h_j}{\partial x_i}(x).$$

This follows from ([13], Theorem 19.5.5.)

Notation

We use the following convenient notation. If x, y are vectors in $\mathbb{R}^n$ and $s : \mathbb{R} \mapsto \mathbb{R}$ we let $x.y \in \mathbb{R}^n$ (respectively $\frac{x}{y}$ and $s(x)$) be the vector defined by $(xy)_i = x_i y_i$ (respectively $(\frac{x}{y})_i = \frac{x_i}{y_i}, s_i(x) = s(x_i)$).

5.1 Gradient Like Structure

A C^1 map $V : \dot{\Delta} \mapsto \mathbb{R}$ is called a *strict Lyapounov function* for F (or Φ) if for all $x \in \dot{\Delta}$

$$F(x) \neq 0 \Rightarrow \langle F(x), \nabla V(x) \rangle < 0.$$

Theorem 3 *Let s :$]0, \infty[\mapsto \mathbb{R}$ be a C^1 function with positive derivative and let $h^s : \dot{\Delta} \mapsto \mathbb{R}^n$ be the map defined by*

$$h^s(x) = s(\frac{x}{\pi(x)}).$$

Assume that h^s is a α-quasigradient. Then

(i) The map V (given by (13)) is a strict Lyapounov function for F on $\dot{\Delta}$;
(ii) The critical points of V coincide with $\mathbf{Eq}(F) \cap \dot{\Delta}$;
(iii) V satisfies the following angle condition*: For every compact set $K \subset \dot{\Delta}$ there exists $c > 0$ such that*

$$| \langle \nabla V(x), F(x) \rangle | \geq c \parallel \nabla V(x) \parallel \parallel F(x) \parallel$$

for all $x \in K$.

Remark 5 (Gibbs Systems) If $\pi(x)$ is a *Gibbs measure*,

$$\pi_{\beta,i}(x) = \frac{e^{-(U_i^0 + \beta \sum_j U_{ij} x_j)}}{Z(x)} \tag{14}$$

where $U = (U_{ij})$ is a symmetric matrix, $\beta \geq 0$, and

$$Z(x) = \sum_j e^{-(U_j^0 + \beta \sum_k U_{ik} x_k)},$$

parts (i) and (ii) of Theorem 3 have been proved in [10], Theorems 5.3 and 5.5. Here $s(t) = \log(t)$ and

$$V(x) = \sum_i x_i \log(x_i) + \sum_j U_j^0 x_j + \frac{\beta}{2} \sum_{ij} U_{ij} x_i x_j. \tag{15}$$

Proof of Theorem 3

Part (i) relies on the following Lemma.

Lemma 4 *Let L be an irreducible transition matrix with invariant probability π. Let $x \in \Delta, f_i = \frac{x_i}{\pi_i}$, $s(f)_i = s(f_i)$ and $c_f = \inf_i s'(f_i) > 0$. Then there exists $\lambda(L) > 0$ depending continuously on L such that*

$$\langle xL, s(f) \rangle \leq -c_f \lambda(L) Var_\pi(f)$$

where $Var_\pi(f) = \sum_i (f_i - 1)^2 \pi_i = \sum_i \frac{(x_i - \pi_i)^2}{\pi_i}$.

The proof of this lemma uses elementary convexity arguments and classical tools from Markov chain theory. It is proved in appendix. Applying this lemma with $L = L(x)$ and $\pi = \pi(x)$ gives

$$\langle F(x), \nabla V(x) \rangle < 0$$

unless $x = \pi(x)$.

(ii) The set $\mathbf{Eq}(F) \cap \dot{\Delta}$ coincides with fixed points of π in $\dot{\Delta}$. Let $x \in \dot{\Delta}$. $\nabla V(x) = 0 \Leftrightarrow h^s(x) \in \mathbb{R}1$ where 1 is the vector which components are all equal to 1. The function s being injective this is equivalent to $\frac{x_i}{\pi_i(x)} = \frac{x_j}{\pi_j(x)}$ for all i, j. That is $x = \pi(x)$.

(iii) Let $K \subset \dot{\Delta}$. By Lemma 4 (applied with $L = L(x)$ and $\pi = \pi(x)$) and continuity of the maps involved, there exists $c > 0$ depending on K such that

$$|\langle \nabla V(x), F(x) \rangle| \geq c \sum_i (x_i - \pi_i(x))^2 = c\|x - \pi(x)\|^2.$$

To prove the angle condition it then suffices to show that both $\|F(x)\|$ and $\|\nabla V(x)\|$ are bounded by some constant times $\|x - \pi(x)\|$. Now, $F(x) = xL(x) = xL(x) - \pi(x)L(x)$ so that

$$\|F(x)\| \leq c_1 \|x - \pi(x)\|$$

with $c_1 = \sup_{x \in \Delta} \|L(x)\|$.

By Lipschitz continuity of s and compactness, there exist $c_2, c_3 > 0$ depending on K such that

$$|s(\frac{x_i}{\pi_i(x)}) - s(1)| \leq c_2|\frac{x_i}{\pi_i(x)} - 1| \leq c_3|x_i - \pi_i(x)|.$$

Thus, for all $u \in T\Delta$ such that $\|u\| = 1$

$$\langle h^s(x), u\rangle = \langle h^s(x) - s(1)1, u\rangle \leq \|h^s(x) - s(1).1\| \leq c_3\|x - \pi(x)\|.$$

This implies that $\|\nabla V(x)\| \leq c_3\|x - \pi(x)\|$ and concludes the proof. ∎

The following result proves to be useful for certain dynamics leaving invariant the boundary of the simplex. Such dynamics occur in population games using imitative protocols (see Eq. (5)) as well as in certain models of vertex reinforcement (see Example 3 below).

For $x \in \Delta$ let $Supp(x) = \{x \in \Delta : x_i > 0\}$.

Proposition 1 *Assume that assumptions of Theorem 3 hold. Assume furthermore that*

(a) For all $x \in \Delta$

$$x_i = 0 \Rightarrow L_{ji}(x) = 0$$

and the reduced rate matrix $[L_{ij}(x)]_{i,j \in Supp(x)}$ is irreducible

(b) The maps $V : \dot{\Delta} \mapsto \mathbb{R}^n$ and $\alpha : \dot{\Delta} \mapsto \mathbb{R}^*_+$ (given by Eq. (13)) extend to C^1 (respectively continuous) maps $V : \Delta \mapsto \mathbb{R}^n$ and $\alpha : \Delta \mapsto \mathbb{R}^*_+$.

Then V is strict Lyapounov function for F on Δ.

Proof Let $T\Delta(x) = \{u \in T\Delta : u_i = 0 \text{ for } i \notin Supp(x)\}$. By assumption (a) the map $x \mapsto \pi(x)$ is defined for all $x \in \Delta$ continuous in x and $\pi_i(x) = 0 \Leftrightarrow x_i = 0$.

Therefore, using assumption (b), the equation

$$\forall x \in \dot{\Delta}, \forall u \in T\Delta\ \alpha(x)\langle h^s(x), u\rangle = \langle \nabla V(x), u\rangle$$

extends to

$$\forall x \in \Delta, \forall u \in T\Delta(x) \sum_{i \in Supp(x)} s(\frac{x_i}{\pi_i(x)})u_i = \langle \nabla V(x), u\rangle$$

Thus

$$\sum_{i \in Supp(x)} s(\frac{x_i}{\pi_i(x)})(xL(x))_i = \langle \nabla V(x), F(x)\rangle$$

for all $x \in \Delta$. By Lemma 4 the left hand side is nonpositive and zero if and only if $x_i = \pi_i(x)$ for all $i \in Supp(x)$. ■

Remark 6 Note that under the assumptions of Proposition 1, the angle inequality of Theorem 3 doesn't hold on the boundary of the simplex

Example 1 Let $W : \mathbb{R}^n \mapsto \mathbb{R}$ be a C^k map, $k \geq 1$. Suppose that for all $x \in \dot{\Delta}$

$$\pi_i(x) = \frac{f_i(x_i)\psi(\frac{\partial W}{\partial x_i}(x))}{\sum_{j=1}^n f_j(x_j)\psi(\frac{\partial W}{\partial x_j}(x))}$$

Then, Theorem 3 applies in the following cases:

Case 1 $\psi(u) = e^{-\beta u}$ with $\beta \geq 0$, and $f_i(t) > 0$ for all $t > 0$. It suffices to choose $s(t) = \log(t)$ and

$$V(x) = \sum_{i=1}^n x_i \log(x_i) - \sum_{i=1}^n \int_1^{x_i} \log(f_i(u))du + \beta W(x) \tag{16}$$

Then h^s is the gradient of V.
Case 2 $\psi(u) = u^\beta, \beta > 0, f_i(t) = t$ and $\frac{\partial W}{\partial x_i} > 0$ on $\{x \in \Delta : x_i > 0\}$. It suffices to choose $s(t) = -t^{-1/\beta}$ and

$$V(x) = -W(x).$$

Then h^s is the α-quasigradient of V with

$$\alpha(x) = [\sum_j x_j(\frac{\partial W}{\partial x_j})^\beta)]^{-1/\beta}.$$

Example 2 (Potential Games) Examples of applications of Example 1, case 1, are given by *Potential Games* (see [22] for an comprehensive presentation and motivating examples). We use the notation of Sect. 2. A *Potential Game* is a game for which the payoff function is such that for all $x \in \Delta$

$$U_i(x) = -\frac{\partial W}{\partial x_i}(x), i = 1 \ldots n$$

Consider a population game with a revision protocol given by (7). Suppose that the attachment matrix takes the form

$$w_{ij}(x) = f_j(x_j)\tilde{w}_{ij}(x)$$

with $\tilde{w}$ irreducible and symmetric. Let $\beta \geq 0$. Assume furthermore that $g(u, v)$ takes one of the following form:

Pairwise comparison

$$g(u, v) = \frac{e^{\beta(v-u)}}{1 + e^{\beta(v-u)}} \text{ or } g(u, v) = \min(1, e^{\beta(v-u)}),$$

Imitation driven by dissatisfaction

$$g(u, v) = e^{-\beta u},$$

Imitation of success

$$g(u, v) = e^{\beta v}.$$

In all these situations, $K(x)$, hence $L(x)$ is reversible with respect to $\pi_\beta(x)$ with

$$\pi_{\beta,i}(x) = \frac{f_i(x_i)e^{-\beta \frac{\partial W}{\partial x_i}(x)}}{\sum_j f_j(x_j)e^{-\beta \frac{\partial W}{\partial x_j}(x)}}.$$

Theorem 3 applies with V given by (16).

Remark 7 (Gibbs Systems, 2) A particular case of potential games is obtained with $W(x) = \frac{1}{2}\sum_{ij} U_{ij}x_ix_j$ with $U = (U_{ij})$ symmetric, and $f_i(x) = e^{-U_i^0}$. Here payoffs are linear in x :

$$U_i(x) = -\sum_j U_{ij}x_j$$

and we retrieve the situation considered in [10]. See Remark 5.

Example 3 (Vertex Reinforcement) Let K be the revision protocol defined by

$$K_{ij}(x) = \frac{A_{ij}x_j^\gamma}{\sum_k A_{ik}x_k^\gamma}$$

where A is a matrix with positive entries and $\gamma \geq 1$. For population games (see Sect. 2.1) this gives a simple model of imitation: an agent of type i, when chosen, switches to j with a probability proportional to the (*number of agents of type* j)$^\gamma$. For processes with reinforcement (as defined in Sect. 2.2) the probability to jump from i to j at time n is proportional to (*the time spent in* j *up to time* n)$^\gamma$. This later model called a *vertex reinforced random walks* was introduced by Diaconis and first analyzed in Pemantle [19] (see also [1] and [7] for more references on the subject).

When A is symmetric, $K(x)$ is reversible with respect to

$$\pi_i(x) = \frac{x_i^\gamma \sum_k A_{ik} x_k^\gamma}{\sum_{ij} A_{ij} x_i^\gamma x_j^\gamma} = \frac{x_i \frac{\partial W}{\partial x_i}}{\sum_j x_j \frac{\partial W}{\partial x_j}} \tag{17}$$

with

$$W(x) = \sum_{i,j} A_{ij} x_i^\gamma x_j^\gamma \tag{18}$$

We are then in the situation covered by Example 1, case 2, with $\psi(u) = u, f_i(t) = t, s(t) = -\frac{1}{t}$ and $V = -W$.

Both Theorem 3 and Proposition 1 apply.

Example 4 (Interacting Urn Processes) Closely related to vertex reinforced random walks are models of *interacting urns* (see [6, 8, 24]). For these models $\pi_i(x) = x_i \frac{\partial W}{\partial x_i}$ for some smooth function W. This is a particular case of Example 1, case 2.

5.2 Limit Sets and Chain Transitive Sets

Using Lasalle's invariance principle we deduce the following consequences from Theorem 3.

Corollary 1 *Assume that assumptions of Theorem 3 hold. Then every omega limit set of Φ contained in $\dot{\Delta}$ is a connected subset of* $\mathbf{Eq}(F) \cap \dot{\Delta}$.

Combining this results with Lemma 2 (ii) and Proposition 1 gives

Corollary 2 *Assume that one of the following condition hold:*

(a) Assumptions of Theorem 3 and Hypothesis 2 or;
(b) Assumptions of Proposition 1.

Then every omega limit set of Φ is a connected subset of $\mathbf{Eq}(F)$.

A set L is called *attractor free* or *internally chain transitive* provided it is compact, invariant and $\Phi|L$ has no proper attractor. For reinforced random processes like the ones defined in Sect. 2.2, limit sets of (μ_n) are, under suitable assumptions, attractor free sets of the associated mean field Eq. (8) (see [1]). More generally attractor free sets are limit sets of *asymptotic pseudo trajectories* (see [3]). It is then useful to characterize such sets. Note however that the existence of a strict Lyapounov function, doesn't ensure in general, that internally chain transitive sets consist of equilibria (see e.g. Remark 6.3 in [2]).

Corollary 3 *Assume that assumptions of Theorem 3 hold and that h^s is C^k for some $k \geq n - 2 = \dim(T\Delta) - 1$. Then every internally chain transitive set of Φ contained in $\dot{\Delta}$ is a connected subset of* $\mathbf{Eq}(F) \cap \dot{\Delta}$. *If we furthermore assume that $L(x)$ is irreducible for all $x \in \Delta$, then every internally chain transitive set of Φ is a connected subset of* $\mathbf{Eq}(F)$

Proof Let $C = \mathbf{Eq}(F) \cap \dot{\Delta}$ and $A \subset \dot{\Delta}$ an attractor free set. By Theorem 3, C coincide with critical points of V. By the assumption V is C^{k+1} so that by Sard's theorem (see [12]), $V(C)$ has empty interior. It follows (see e.g. Proposition 6.4 in [2]) that $A \subset C$. ■

5.3 Convergence Toward One Equilibrium

In case equilibria are isolated, Corollary 1 implies that every trajectory bounded away from the boundary converge to an equilibrium and that every trajectory converges in case $L(x)$ is irreducible for all x. However, when equilibria are degenerate, the gradient-like property is not sufficient to ensures convergence. There are known examples of smooth gradient systems which omega limit sets are a continuum of equilibria (see [18]). However, in the real analytic case, gradient like systems which verify an angle condition are known to converge.

Theorem 4 *Suppose that assumptions of Theorem 3 hold and that V is real analytic. Then every omega limit set meeting $\dot{\Delta}$ reduces to a single point.*

Proof Let p be an omega limit point. If V is real analytic, it satisfies a *Lojasiewicz inequality* at p in the sense that there exist $0 < \eta \leq 1/2, \beta > 0$ and a neighborhood $U(p)$ of p such that

$$|V(x) - V(p)|^{1-\eta} \leq \beta \|\nabla V(x)\|$$

for all x in a $U(p)$. Such an inequality called a "gradient inequality" was proved by Lojasiewicz [16] and used (by Lojasiewicz again) to show that bounded solutions of real analytic gradient vector fields have finite length, hence converge. When the dynamics is not a gradient, but only gradient like with V as a strict Lyapounov function, the same results holds provided that V satisfies an angle condition:

$$\langle \nabla V(x), F(x) \rangle \geq c\|F(x)\| \|\nabla V(x)\|$$

for all $x \in U(p)$. This is proved in [9] (see also [17], Theorem 7). ■

Example 5 (Gibbs Systems, 3) If π is given by (14) with U symmetric, V given by (15) is real analytic so that every solution to (1) converges toward an equilibrium.

6 Equilibria

Recall that point $p \in \mathbf{Eq}(F)$ is called *non degenerate* if the jacobian matrix $DF(p) : T\Delta \mapsto T\Delta$ is invertible. It is called *hyperbolic* if eigenvalues of $DF(p)$ have non zero real parts. If p is hyperbolic, $T\Delta$ admits a splitting

$$T\Delta = E_p^u \oplus E_p^s$$

invariant under $DF(p)$ such that the eigenvalues of $DF(p)|_{E_p^s}$ (respectively $DF(p)|_{E_p^u}$) have negative (respectively positive) real parts.

Point $p \in \mathbf{Crit}(V) = \{x \in \dot{\Delta} : \nabla(V)(x) = 0\}$ is called *non-degenerate* if $Hess(V)(p)$ the *Hessian* or V at p has full rank. In a suitable coordinate systems $Hess(V)(p)(u,u) = \sum_{i=1}^{n_+} u_i^2 - \sum_{j=1}^{n_-} u_j^2$ with $n_+ + n_- = dim(T\Delta) = n - 1$. The number n_- is called the *index* of p (with respect to V) and is written $\mathbf{Ind}(p, V)$.

Proposition 2 *Assume that assumptions of Theorem 3 hold. Let $p \in \mathbf{Eq}(F) \cap \dot{\Delta}$. Then*

(i) Point p is non degenerate if and only if it is a non degenerate critical point of V.
(ii) If furthermore L is C^2 and p is hyperbolic,

$$\dim(E_p^u) = \mathbf{Ind}(p, V).$$

Proof From Lemma 3, p is non degenerate if and only if $DF_\pi(p)$ is invertible and (see the proof of Lemma 3)

$$DF(p) = -L^T(p)DF_\pi(p) \tag{19}$$

Now, a direction computation (details are left to the reader) of the Hessian of V at x leads to

$$\langle Hess(V)(x)u, v\rangle = \alpha(x)\langle s'(\frac{x}{\pi(x)})(u - \frac{x}{\pi(x)}D\pi(x)u, v\rangle_{1/\pi(x)}$$

where $\langle u, v\rangle_{1/\pi}$ stands for $\sum_i u_i v_i \frac{1}{\pi_i}$. Since $p = \pi(p)$

$$\langle Hess(V)(p)u, v\rangle = \alpha(p)s'(1)\langle (I - D\pi(p))u, v\rangle_{1/p} \tag{20}$$

for all $u, v \in T\Delta$, This proves that $HessV(p)$ is non degenerate if and only if $(I - D\pi(p)) = -DF_\pi(p)$ is non degenerate and concludes the proof of the first part.

We now prove the second part. By the stable manifold theorem, there exists a (local) C^2 manifold W_p^s tangent to E_p^s at p positively invariant under Φ and such that for all $x \in W_p^s$ $\lim_{t\to\infty} \Phi_t(x) = p$. Clearly p is a global minimum of V restricted to W^s. For otherwise there would exists $x \in W_p^s$ such that

$$V(p) > V(x) > \lim_{t\to\infty} V(\Phi_t(x)) = V(p).$$

Since p is also a critical point $\nabla V(p) = 0$. Let $u \in E^s_p$ and let $\gamma :]-1, 1[\mapsto W^s_p$ be a C^2 path with $\gamma(0) = p, \dot{\gamma}(0) = u$. Set $h(t) = V(\gamma(t))$. Then $\dot{h}(0) = 0$ (because p is a critical point of V) and $h''(0) = \langle HessV(p)u, u\rangle$ is non negative because $h(t) \geq h(0)$.

On the other hand, by the spectral decomposition of $HessV(p)$ we can write $T\Delta = E^s_V \oplus E^u_V$ with $\langle HessV(p)u, u\rangle > 0$ (respectively < 0) for all $u \in E^s_V \setminus \{0\}$ (respectively $E^u_V \setminus \{0\}$). Thus, $E^s_p \cap E^u_V = \{0\}$ and, consequently, $\dim(E^s_p) + \dim(E^u_V) \leq \dim(T\Delta)$. Similarly $\dim(E^u_p) + \dim(E^s_V) \leq \dim(T\Delta)$. This proves that $\dim(E^u_p) = \dim(E^u_V) = \mathbf{Ind}(p, V)$. ■

Remark 8 This later proposition shows that in the neighborhood of an hyperbolic equilibrium p, $\dot{x} = F(x)$ and $\dot{x} = -\nabla V(x)$ are topologically conjugate. Indeed, part (ii) of the proposition implies that the linear flows $\{e^{tDF(p)}\}$ and $\{e^{tHess(V)(p)}\}$ are topologically conjugate (see e.g. Theorem 7.1 in [21]), and by Hartman-Grobman Theorem (see again [21]), nonlinear flows are locally conjugate to their linear parts in the neighborhood of hyperbolic equilibria. However, note that while eigenvalues of $Hess(V)(p)$ are reals there is no evidence that the same is true for $DF(p)$ in general. The next proposition proves that this is the case when $L(x)$ is reversible with respect to $\pi(x)$.

Proposition 3 *Let $p \in \mathbf{Eq}(F) \cap \dot{\Delta}$. Assume that assumptions of Theorem 3 hold and that $L(p)$ is reversible with respect to $\pi(p) = p$. Then there exists a positive definite bilinear form $g_0(p)$ on $T\Delta$ such that for all $u, v \in T\Delta$*

$$g_0(p)(DF(p)u, v) = -\langle Hess(V)(p)u, v\rangle$$

In particular

(i) $DF(p)$ has real eigenvalues,
(ii) p is hyperbolic for F if and only if it is a non degenerate critical point of V.

Proof Let $p \in \mathbf{Eq}(F) \cap \dot{\Delta}$. Set $L = L(p)$ and recall that $L^T : T\Delta \mapsto T\Delta$ is defined by $L^T h = hL$. Then, by Lemma 8 in the Appendix, $-L^T$ is a definite positive operator for the scalar product on $T\Delta$ defined by $\langle u, v\rangle_{1/p} = \sum_i u_i v_i \frac{1}{p_i}$. Define now $g_0(p)$ by

$$g_0(p)(u, v) = -\langle (L^T)^{-1} u, v\rangle_{\frac{1}{p}}. \tag{21}$$

Using (19) and (20) it comes that for all $u, v \in T\Delta$

$$\begin{aligned} g_0(p)(DF(p)u, v) &= -\langle (I - D\pi(p))u, v\rangle_{1/p} \\ &= -[\alpha(p)s'(1)]^{-1}\langle Hess(V)(p)u, v\rangle. \end{aligned}$$

Replacing $g_0(p)$ by $\alpha(p)s'(1)g_0(p)$ proves the result. ■

A useful consequence of this later proposition is that it is usually much easier to verify non degeneracy of equilibria rather than hyperbolicity. Here is an illustration:

Example 6 (Gibbs Systems, 4) Consider the symmetric Gibbs model analyzed in [10] (see Remark 5 and Example 7). We suppose that the symmetric matrix $U = (U_{ij})$ is given and we treat $U^0 = (U_i^0)_{i\in S}$ and β as parameters. Let $\Xi_{rev}(U^0)$ denote the set of maps

$$\mathbb{R}^+ \times \Delta \mapsto T\Delta,$$

$$(\beta, x) \mapsto F_\beta(x) = xL_\beta(x)$$

such that L_β verifies assumption 2, is C^1 in x, and $L_\beta(x)$ is reversible with respect to $\pi_\beta(x)$ where

$$\pi_{\beta,i}(x) = \frac{e^{-U_i^0 - \beta\sum_j U_{ij}x_j}}{Z(x)}.$$

Proposition 4 *There exists an open and dense set $\mathcal{G}^0 \subset \mathbb{R}^n$ such that for all $U^0 \in \mathcal{G}^0$ and $F \in \Xi_{rev}(U^0)$*

(i) The set $\{(x, \beta) \in \Delta \times \mathbb{R}_\cdot^+ : F_\beta(x) = 0\}$ is a C^∞ one dimensional submanifold,
(ii) There exists an open dense set $\mathcal{B}^0 \subset \mathbb{R}^+$ containing 0 such that for all $\beta \in \mathcal{B}^0$ equilibria of F_β are hyperbolic.

Proof Let $H : \dot{\Delta} \times \mathbb{R}^n \times \mathbb{R}^+ \mapsto T\Delta$ be defined by $H(x, U^0, \beta) = \nabla V_{U^0,\beta}(x)$ where $V_{U^0,\beta}$ is given by (15). Since $\frac{\partial H}{\partial U^0}(x, U^0, \beta)$ is the identity map, H is a submersion. Hence, by Thom's parametrized transversality Theorem (see [12], Chapter 3), there exists an open and dense set $\mathcal{G}^0 \in \mathbb{R}^n$ such that for all $U^0 \in \mathcal{G}^0, (x, \beta) \mapsto H(x, U^0, \beta)$ is a submersion. This proves (*i*). By the same theorem, for all $\beta \in \mathcal{B}^0$ with $\mathcal{B}^0$ open and dense in $\mathbb{R}^+, x \mapsto H(x, U^0, \beta)$ is a submersion, meaning that critical points of $V_{U^0,\beta}$ are nondegenerate. By Proposition 3, equilibria of F_β are hyperbolic. ■

Remark 9 Other genericity results can be proved, if one fix U^0 or β and treat U as a parameter. Compare to the proof of Theorem 2.10 in [4] in an infinite dimensional setting.

6.1 Equilibria of Potential Games

Consider a population game with C^1 payoff function $U : \Delta \mapsto \mathbb{R}^n$. Recall that the game is called a potential game, provided $U_i(x) = -\frac{\partial W}{\partial x_i}(x)$ for all $x \in \Delta$ and some *potential* $W : \mathbb{R}^n \mapsto \mathbb{R}$.

Point $x^* \in \Delta$ is called a *Nash equilibrium* of U if, given the population state x^*, every agent has interest to play the mixed strategy x^*. That is

$$\forall i \in \{1,\ldots,n\}\ U_i(x^*) \leq \langle U(x^*), x^* \rangle \tag{22}$$

Let

$$Supp(x^*) = \{i \in \{1,\ldots,n\} : x_i^* > 0\}.$$

It follows from (22) that

$$\forall i \in Supp(x^*)\ U_i(x^*) = \langle U(x^*), x^* \rangle.$$

We let $\mathbf{NE}(U)$ denote the set of Nash equilibria of U. For all $\beta \geq 0$ and $x \in \Delta$ we let $\pi_\beta(x) \in \Delta$ be defined as

$$\pi_{\beta,i}(x) = \frac{e^{\beta U_i(x)}}{\sum_j e^{\beta U_j(x)}},\ i = 1,\ldots,n \tag{23}$$

and we let $\chi(\beta, U)$ (respectively, $\chi_{rev}(\beta, U)$) denote the set of all vector fields having the form given by (1) where $L(x)$ is C^1 in x, irreducible and admits $\pi_\beta(x)$ as invariant (respectively reversible) probability. Recall (see Eq. (9)) that

$$F_{\pi_\beta} = -Id + \pi_\beta.$$

Our aim here is to describe $\mathbf{Eq}(F)$ for $F \in \chi(\beta, U)$ in term of $\mathbf{NE}(U)$ for large β, with a particular emphasis on potential games. Some of the results here are similar to the results obtained in Benaïm and Hirsch (n 2 coordinative games, 2000, unpublished manuscript) for $n \times 2$ pseudo games.

Proposition 5 *Let $\mathcal{N}$ be a neighborhood of* $\mathbf{NE}(U)$. *There exists $\beta_0 \geq 0$ such that for all $\beta \geq \beta_0$ and $F \in \chi(\beta, U)$*

$$\mathbf{Eq}(F) \subset \mathcal{N}$$

Proof Equilibria of F coincide with equilibria of F_{π_β}. Let $x(\beta)$ be such an equilibrium. Then for all i, j

$$\frac{\log(x_i(\beta)) - \log(x_j(\beta))}{\beta} = U_i(x(\beta)) - U_j(x(\beta)).$$

Thus for every limit point $x^* = \lim_{\beta_k \to \infty} x(\beta_k)$ it follows that

$$U_i(x^*) = U_j(x^*)$$

if $i, j \in Supp(x^*)$ and

$$U_i(x^*) \geq U_j(x^*)$$

if $i \notin Supp(x^*)$ and $j \in Supp(x^*)$. ■

Remark 10 Note that Proposition 5 only requires the continuity of U.

We shall now prove some converse results.

A Nash equilibrium x^* is called *pure* if $Supp(x^*)$ has cardinal 1 and *mixed* otherwise. It is called *strict* if inequality (22) is strict for all $i \notin Supp(x^*)$.

Theorem 5 *Let x^* be a pure strict Nash equilibrium and $\mathcal{N}$ a (sufficiently small) neighborhood of x^*. Then, there exists $\beta_0 > 0$ such that for all $\beta \geq \beta_0$ and $F \in \chi(\beta, U)$*

(i) $\mathbf{Eq}(F) \cap \mathcal{N} = \{x^*_\beta\}$
(ii) Equilibrium x^*_β is linearly stable for F_{π_β}.
(iii) Assume furthermore that the game is a potential game. Then x^*_β is linearly stable for F under one of the following conditions:

(a) L (hence F) is C^2 and x^*_β is hyperbolic for F, or
(b) $F \in \chi_{rev}(\beta, U)$.

Proof Suppose without loss of generality that $x^*_1 = 1$ and $x^*_i = 0$ for $i \neq 1$.

Set $R_{ij} = U_j - U_i$. By assumption and continuity, there exists $\delta > 0, \alpha > 0$ such that for all $x \in B(x^*, \alpha) = \{x \in \Delta : \|x - x^*\| \leq \alpha\}$,

$$R_{i1}(x) \geq \delta \text{ for } i > 1;$$

$$\|R_{ij}(x)\| \geq \delta \text{ if } R_{ij}(x^*) \neq 0$$

and

$$\|R_{ij}(x)\| \leq \delta \text{ if } R_{ij}(x^*) = 0.$$

Thus

$$1 \geq \pi_{\beta,1}(x) = (1 + \sum_{i>1} e^{-\beta R_{i1}(x)})^{-1} \geq (1 + (n-1)e^{-\beta\delta})^{-1}.$$

This implies that π_β maps $B(x^*, \alpha)$ into itself for β large enough. By Brouwer's Theorem, it then admits a fixed point x^*_β. To prove uniqueness and assertion *(ii)* it suffices to prove that π_β restricted to $B(x^*, \alpha)$ is a contraction. From the expression $\pi_{\beta,i} = (\sum_j e^{\beta R_{ij}})^{-1}$, we get

$$\frac{\partial \pi_{\beta,i}}{\partial x_m} = -\sum_j [\beta e^{\beta R_{ij}} (\sum_k e^{\beta R_{ik}})^{-2} \frac{\partial R_{ij}}{\partial x_m}] := \sum_j D_{ij} = \sum_{j \neq i} D_{ij}.$$

Let $j \neq i$. If $R_{ij}(x^*) \neq 0$

$$|D_{ij}| \leq \beta e^{\beta R_{ij}} (1 + e^{\beta R_{ij}})^{-2} \leq \beta \min(e^{\beta R_{ij}}, e^{-\beta R_{ij}}) \leq \beta e^{-\beta \delta}.$$

If $R_{ij}(x^*) = 0$. Then $i \neq 1$ and

$$|D_{ij}| \leq \beta e^{\beta R_{ij}} (e^{\beta R_{i1}})^{-2} = \beta e^{\beta (R_{ij} - 2R_{i1})} \leq \beta e^{-\beta \delta}$$

These inequalities show that $\|D\pi_\beta(x)\| < 1$ for all $x \in B(x^*, \alpha)$ and β large enough, proving uniqueness of the equilibrium as well as assertion (*ii*). The last assertion follows from Propositions 2 and 3. ■

A Nash equilibrium $x*$ is called *fully mixed* if $Supp(x^*) = \{1, \ldots, n\}$ and *partially mixed* if $1 < card(Supp(x^*)) < n$.

A fully mixed Nash equilibrium is called *non degenerate* if for all $u \in T\Delta$

$$\left[\forall w \in T\Delta \ \langle DU(x^*)u, w\rangle = 0\right] \Rightarrow u = 0.$$

Let

$$T\Delta(x^*) = \{u \in T\Delta : \ u_i = 0 \text{ for } i \notin Supp(x^*)\}.$$

A partially mixed equilibrium x^* is called non degenerate if for all $u \in T\Delta(x^*)$

$$\left[\forall w \in T\Delta(x^*) \ \langle DU(x^*)u, w\rangle = 0\right] \Rightarrow u = 0,$$

Lemma 5 *Let $x^* \in \Delta$ be a mixed equilibria. Assume that $Supp(x^*) = \{1, \ldots, r\}$ for some $1 < r \leq n$ and set*

$$x^* = (q_1, \ldots, q_{r-1}, 1 - \sum_{i=1}^{r-1} q_i, 0, \ldots, 0).$$

Let, for $i = 1, \ldots, r-1$,

$$h_i^r(x_1, \ldots, x_{r-1}, y_1, \ldots, y_{n-r}) = U_i(x_1, \ldots, x_{r-1}, 1 - \sum_{i=1}^{r-1} x_i - \sum_{i=1}^{n-r} y_i, y_1, \ldots, y_{n-r})$$
$$- U_r(x_1, \ldots, x_{r-1}, 1 - \sum_{i=1}^{r-1} x_i - \sum_{i=1}^{n-r} y_i, y_1, \ldots, y_{n-r})$$

Then x^* is non degenerate if and only if the $(r-1) \times (r-1)$ matrix

$$\left[\frac{\partial h_i^r}{\partial x_j}((q,0))\right]_{i,j=1,\ldots r-1}$$

is invertible.

Proof One has

$$\frac{\partial h_i^r}{\partial x_j}(q,0) = (\frac{\partial U_i}{\partial x_j}(x^*) - \frac{\partial U_i}{\partial x_r}(x^*)) - (\frac{\partial U_r}{\partial x_j}(x^*) - \frac{\partial U_r}{\partial x_r}(x^*)).$$

Let

$$v = (v_1, \ldots, v_{r-1}, -\sum_{i=1}^{r-1} v_i, 0, \ldots, 0) \in T\Delta(x^*)$$

and

$$w = (w_1, \ldots, w_{r-1}, -\sum_{i=1}^{r-1} w_i, 0, \ldots, 0) \in T\Delta(x^*).$$

Then it is easily seen that

$$\sum_{i=1}^{r-1}\sum_{j=1}^{r-1} \frac{\partial h_i^r}{\partial x_j}(q) v_i w_j = \langle DU(x^*)v, w\rangle.$$

This proves that x^* is non degenerate if and only if $\left[\frac{\partial h_i^r}{\partial x_j}((q,0))\right]_{i,j=1,\ldots r-1}$ is invertible. ■

Theorem 6 *Let x^* be a non degenerate fully mixed Nash equilibrium for U and $\mathcal{N}$ a (sufficiently small) neighborhood of x^*. Then, there exists $\beta_0 > 0$ such that for all $\beta \geq \beta_0$ and $F \in \chi(\beta, U)$*

$$\mathbf{Eq}(F) \cap \mathcal{N} = \{x_\beta^*\}.$$

Assume furthermore that the game is a potential game with potential W. Then x_β^* is hyperbolic for F and its unstable manifold (for F) has dimension $\mathbf{Ind}(x^*, W|_\Delta)$ under one of the following conditions:

(a) L (hence F) is C^2 and x_β^* is hyperbolic for F, or
(b) $F \in \chi_{rev}(\beta, U)$.

Proof Set $T = 1/\beta$. Equilibria of F_{π_β} are given by the set of equations

$$T(\log(x_i) - \log(x_n)) = U_i(x) - U_n(x), i = 1, \dots, n-1$$

or, with the notation of Lemma 5,

$$T(\log(x_i) - \log(1 - \sum_{i=1}^{n-1} x_i)) = h_i^n(x_1, \dots, x_{n-1}), i = 1, \dots, n-1. \tag{24}$$

Write $x^* = (q_1, \dots, q_{n-1}, 1 - \sum_{i=1}^{n-1} q_i)$. For $T = 0$, $q = (q_1, \dots, q_{n-1})$ is solution to (24). Hence, by the implicit function theorem (which hypothesis is fulfilled by the non degeneracy of x^* and Lemma 5) there exists $\alpha_0 > 0$, a neighborhood O of q in $(\mathbb{R}_+^*)^{n-1}$ and a C^1 map $T \in] - \alpha_0, \alpha_0[\mapsto q(T) \in O$ such that $(T, q(T))$ is the unique solution to (24) in $] - \alpha_0, \alpha_0[\times O$. This proves the first assertion of the theorem with $\beta_0 > 1/\alpha_0$ and $x_\beta^* = (q(1/\beta), 1 - \sum_{i=1}^{n-1} q_i(1/\beta))$.

In case, the game is a potential game with potential W, F is gradient-like with Lyapounov function V_β given by (16). Since x^* is fully mixed, $\frac{1}{q_i} < \infty$ so that $\|\frac{1}{\beta} Hess V_\beta(x_\beta^*) - Hess W(x^*)\| \to 0$ as $\beta \to \infty$. In particular, for β large enough $Hess V_\beta(x_\beta^*)$ is non degenerate, because x^* is non degenerate. The last assertion then follows from Propositions 2 and 3. ■

Theorem 7 *Let x^* be a strict and non degenerate partially mixed Nash equilibrium for U which support has cardinal $1 < r < n$. Let $\mathcal{N}$ be a (sufficiently small) neighborhood of x^*. Then, there exists $\beta_0 > 0$ such that for all $\beta \geq \beta_0$ and $F \in \chi(\beta, U)$*

$$\mathbf{Eq}(F) \cap \mathcal{N} = \{x_\beta^*\}.$$

Assume furthermore that the game is a potential game with potential W and that one of the following conditions hold:

(a) L (hence F) is C^2 and x_β^* is hyperbolic for F, or
(b) $F \in \chi_{rev}(\beta, U)$.

Then x_β^* is hyperbolic and

$$k \leq \dim(E^u_{x_\beta^*}) \leq \min(n - r + k, r - 1).$$

with $k = \mathbf{Ind}(x^*, W|_{\Delta(x^*)})$ *and* $\dim(E^u_{x_\beta^*})$ *stands for the dimension of the unstable manifold (for F).*

Proof Assume without loss of generality that $Supp(x^*) = \{1, \dots, r\}$ and set $x^* = (q_1, \dots, q_{r-1}, 1 - \sum_{i=1}^{r-1} q_i, 0, \dots, 0)$. Write every element of Δ as $(x_1, \dots, x_{r-1}, 1 - \sum_{i=1}^{r-1} x_i - \sum_{i=1}^{n-r} y_i, y_1, \dots y_{n-r})$ and set $x = (x_1, \dots, x_{r-1}), y = (y_1, \dots, y_{n-r})$. Thus, with $\beta = 1/T$, equilibria of F_{π_β} are given by the following system of equations:

$$T(\log(x_i) - \log(1 - \sum_{i=1}^{r-1} x_i - \sum_{i=1}^{n-r} y_i)) = h_i^r(x, y), i = 1, \dots r - 1 \tag{25}$$

and

$$T(\log(y_i) - \log(1 - \sum_{i=1}^{r-1} x_i - \sum_{i=1}^{n-r} y_i)) = h_{i+r}^r(x, y), i = 1 \dots n - r \tag{26}$$

where h_i^r is defined in Lemma 5. The triplet $(T = 0, x = q, y = 0)$ is solution to (25). Thus by the non degeneracy hypothesis and the implicit function theorem, there exists a smooth map

$$\hat{x} : \mathcal{O} \mapsto \mathcal{V}, (T, y) \mapsto \hat{x}(T, y)$$

where $\mathcal{O}$ is a neighborhood of $(0, 0)$ in $\mathbb{R} \times \mathbb{R}^{n-r}$ and $\mathcal{V}$ a neighborhood of q in $\mathbb{R}^{r-1}$ such that $(T, \hat{x}(T, y), y)$ is solution to (25). Recall that $0 < \sum_{i=1}^{r-1} q_i < 1$ and $h_{i+r}^r(q, 0) < 0$ for all $i = 1, \dots, n - r$ (because x^* is strict). Thus, by choosing $\mathcal{O}$ small enough we can furthermore ensure that

$$0 < 1 - \sum_{i=1}^{r-1} \hat{x}_i(T, y) - \sum_{i=1}^{n-r} y_i < 1 \tag{27}$$

and

$$h_{i+r}^r(\hat{x}(T, y), y) \leq -\delta < 0, i = 1 \dots n - r \tag{28}$$

for all $(T, y) \in \mathcal{O}$.

Now replacing x by $\hat{x}(T, y)$ in (26) leads to

$$y_i = G_i(T, y), i = 1 \dots n - r$$

where

$$G_i(T, y) = (1 - \sum_{i=1}^{r-1} \hat{x}_i(T, y) - \sum_{i=1}^{n-r} y_i) \exp(\frac{1}{T} h_{i+r}^r(\hat{x}(T, y), y)).$$

Using (27) and (28) we see that α small enough and $T \leq \frac{\log(1/\alpha)}{\delta}$ $G(T, \cdot)$ maps $\{y \in \mathbb{R}^{n-r} : 0 \leq y_i \leq \alpha\}$ into itself. By Brouwer's fixed point theorem, $G(T, \cdot)$ admits a fixed point $\hat{y}(T)$. Furthermore, $\|D_y G(T, y)\| \leq \frac{C}{T} e^{-\delta/T}$ for some constant C making $G(T, \cdot)$ a contraction. This implies that $\hat{y}(T)$ is unique. Finally define x_β^* by $x_{\beta,i}^* = \hat{x}_i(T, \hat{y}(T))$ for $1 \leq i < r$ and $x_{\beta,i+r}^* = \hat{y}_i(T)$ for $1 \leq 1 \leq n - r$.

We now prove the last assertions. By assumption, $T\Delta(x^*)$ admits a decomposition $T\Delta(x^*) = E_+ \oplus E_-$ with $\langle Hess(W)(x^*)u, u\rangle > 0$ (respectively < 0) for all $u \in E_+$ (respectively E_-) and $u \neq 0$.

Set $T\Delta_s(x^*) = \{u \in T\Delta : u_1 = \ldots = u_r = 0\}$. Then

$$T\Delta = E_+ \oplus E_- \oplus T\Delta_s(x^*).$$

Let now V_β be the Lyapounov function given (16). Then for all $u \in T\Delta$

$$Q_\beta(u) := \langle\frac{1}{\beta}Hess(V_\beta)(x_\beta^*)u, u\rangle = \langle HessW(x_\beta^*)u, u\rangle + \frac{1}{\beta}\sum_i \frac{1}{x_{\beta,i}^*}u_i^2.$$

The construction of x_β^* shows that $\frac{1}{\beta}\frac{1}{x_{\beta,i}^*} \to 0$ for $i \leq r$ and $\frac{1}{\beta}\frac{1}{x_{\beta,i}^*} \to \infty$ for $i > r$ when $\beta \to \infty$. Thus, for β large enough, Q_β is non degenerate, definite positive on E_+ and $T\Delta_s(x^*)$, and definite negative on E_-.

This implies that its index is bounded below by $k = \dim(E_-)$ and above by $\min(r-1, n-r-k)$. This index equals the dimension of the stable manifold by Proposition 2. Under the reversibility assumption hyperbolicity follows from Proposition 3. ■

7 Reversibility and Gradient Structure

Recall that an irreducible rate matrix L is called *reversible* with respect to $\pi \in \dot{\Delta}$ is $\pi_i L_{ij} = \pi_j L_{ji}$. In this case π is the (unique) invariant probability of L. Here we will consider gradient properties of (1) under the assumption that $L(x)$ is reversible.

A $C^k, k \geq 0$ (Riemannian) metric on $\dot{\Delta}$ (or Δ) is a C^k map g such that for each $x \in \Delta$ $g(x) : T\Delta \times T\Delta \mapsto \mathbb{R}$ is a definite positive bilinear form. Given a C^1 map $V : \dot{\Delta} \mapsto \mathbb{R}$ we let $grad_g V$ denote the gradient vector field of V with respect to g. That is

$$g(x)(grad_g V(x), u) = \langle\nabla V(x), u\rangle$$

for all $u \in T\Delta$.

Proposition 6 *Assume that for all $x \in \dot{\Delta}$ $L(x)$ is reversible with respect to $\pi(x)$ and assume that the map $h : \dot{\Delta} \mapsto \mathbb{R}^n$, defined by*

$$h(x) = \frac{x}{\pi(x)}$$

is a α-quasigradient. Then there exists a metric g on $\dot{\Delta}$ such that for all $x \in \dot{\Delta}$ $F(x) = -grad_g V(x)$. If L and α are C^k then g is C^k.

Proof The proof is similar to the proof of Proposition 3. Let $A(x) : T\Delta \mapsto T\Delta$ be defined by $A(x)h = -hL(x)$. Then $A(x)$ and $L(x)$ are conjugate by the relation $\pi(x)L(x)h = A(x)\pi(x)h$ and $A(x)$ is a definite positive operator for the scalar product on $T\Delta$ defined by $\langle u, v\rangle_{1/\pi(x)} = \sum_i u_i v_i \frac{1}{\pi_i(x)}$. Define now a Riemannian metric on $T\Delta$ by

$$g_0(x)(u, v) = \langle A(x)^{-1}u, v\rangle_{\frac{1}{\pi(x)}}. \tag{29}$$

Since $F(x) = xL(x) = (x - \pi(x))L(x) = A(x)(-x + \pi(x))$, we get

$$g_0(x)(F(x), u) = -\langle \frac{x}{\pi(x)} - 1, u\rangle = -\langle \frac{x}{\pi(x)}, u\rangle.$$

If $x \mapsto \frac{x}{\pi(x)}$ is a quasi gradient, this makes F a gradient for the metric $g(x) = \alpha(x)g_0(x)$. ■

Example 7 Suppose that $L(x)$ is reversible with respect to π, independent on x. Then $x \mapsto \frac{x}{\pi}$ is the gradient of the χ^2 function $V(x) = \sum_i(\frac{x_i}{\pi_i} - 1)^2\pi_i$. Hence $F(x) = -grad_g V(x)$ for some metric g.

Under the weaker assumption that $x \mapsto s(\frac{x}{\pi(x)})$ is a quasi-gradient for some strictly increasing function s (see Theorem 3) it is no longer true that F is a gradient, but it can be approximated by a gradient. The next Lemma is the key tool. Its proof is identical to the proof of Proposition 3.

Lemma 6 *Assume that assumptions of Theorem 3 hold and that for all $x \in \dot{\Delta}$, $L(x)$ is reversible with respect to $\pi(x)$. Then there exists a metric g_0 on $\dot{\Delta}$ such that for $p \in$ **Eq**$(F) \cap \dot{\Delta}$ and $u, v \in T\Delta$*

$$g_0(p)(DF(p)u, v) = -\langle Hess(V)(p)u, v\rangle.$$

If, furthermore, L and α (in Eq. (13)) are C^k then g_0 is C^k.

Theorem 8 *Assume that*

(a) Assumptions of Theorem 3 hold with s, α and L $C^k, k \geq 2$,
(b) For all $x \in \dot{\Delta}$ $L(x)$ is reversible with respect to $\pi(x)$,
(c) **Eq**$(F) \cap \dot{\Delta}$ *is finite.*

Then for every neighborhood $\mathcal{U}$ of **Eq**$(F) \cap \dot{\Delta}$ *and every $\varepsilon > 0$ there exists a C^k metric g on $\dot{\Delta}$ such that*

(i) $-grad_g V = F$ on $\dot{\Delta} \setminus \mathcal{U}$.
(ii) $\| - grad_g V - F\|_{C^1,\mathcal{U}} \leq \varepsilon$ where

$$\|G\|_{C^1,\mathcal{U}} = \sup_{x\in\mathcal{U}} \|G(x)\| + \|DG(x)\|.$$

Proof Let $\mathcal{E} = \dot{\Delta} \cap \mathbf{Eq}(F)$, $v(x) = d(x, \mathcal{E}) = \min_{p\in\mathcal{E}} \|x - p\|$ and let $\psi : \mathbb{R}^+ \mapsto [0, 1]$ be a C^∞ function which is 0 on $[0, 1]$, 1 on $[3, \infty[$ and such that $0 \le \psi' \le 1$. Fix $\varepsilon > 0$ and let $\lambda(x) = \psi(\frac{v(x)}{\varepsilon})$, $G_0 = -grad_{g_0} V$ where g_0 is given by Lemma 6 and

$$G(x) = (1 - \lambda(x))G_0(x) + \lambda(x)F(x).$$

Since for all $p \in \mathcal{E}, F(p) - G_0(p) = DF(p) - DG_0(p) = 0$ there exists a constant $C > 0$ such that

$$\|G_0(x) - F(x)\| \le Cv(x)^2, \ \|DG_0(x) - DF(x)\| \le Cv(x).$$

Thus

$$\|G(x) - F(x)\| = (1 - \lambda(x))\|G_0(x) - F(x)\| \le C(1 - \lambda(x))v(x)^2 \le C\varepsilon^2$$

and

$$\|DG(x) - DF(x)\| = \|(1 - \lambda(x))(DG_0(x) - DF(x)) + \langle \nabla\lambda(x), G_0(x) - F(x)\rangle\|$$

$$\le C((1 - \lambda(x))v(x) + \frac{1}{\varepsilon}v(x)^2) \le C\varepsilon.$$

This shows that G is a C^1 approximation of F which coincides with F on $\{v(x) \ge 3\varepsilon\}$ and with G_0 on $\{v(x) \le \varepsilon\}$. Furthermore,

$$\langle \nabla V(x), G(x)\rangle = -(1 - \lambda(x))g_0(x)(G_0(x), G_0(x)) + \lambda(x)\langle \nabla V(x), F(x)\rangle \le 0$$

with equality if and only if $x \in \mathcal{E}$.

Now, for all $x \in \dot{\Delta} \setminus \mathcal{E}$

$$T\Delta = \nabla V(x)^\perp \oplus \mathbb{R}G(x)$$

and the splitting is smooth in x. Hence $u \in T\Delta$ can be uniquely written as $u = P_x(u) + t_x(u)G(x)$ with $t_x(u) \in \mathbb{R}$ and $P_x(u) \in \nabla V(x)^\perp$. Let g be the metric on $\dot{\Delta} \setminus \mathcal{E}$ defined by

$$g(x)(u, v) = g_0(P_x(u), P_x(v)) + t_x(u)t_x(v)g_0(x)(G_0(x), G(x)).$$

Then g coincides with g_0 on $\{0 < x < v(x) < \varepsilon\}$ so that g can be extended to a C^2 metric on $\dot{\Delta}$. By construction of G and g, $G = -grad_g V$. ■

8 Questions of Structural Stability

Let $C^k_{pos}(\Delta, T\Delta)$ denote the set of C^k vector fields $F : \Delta \mapsto T\Delta$ leaving Δ positively invariant.

Two elements $F, G \in C^k_{pos}(\Delta, T\Delta)$ are said *topologically equivalent* if there exists a homeomorphism $h : \Delta \mapsto \Delta$ which takes orbits of F to orbits of G preserving their orientation. A set $\chi \subset C^k_{pos}(\Delta, T\Delta)$ is said *structurally stable* if all its elements are topologically equivalents.

Let $\pi : \Delta \mapsto \dot{\Delta}$ be a smooth function. Assume that π verifies the assumption of Theorem 3 and that F_π has non degenerate equilibria. Let $\chi_{\pi, rev}$ denote the convex set of vector fields having the form given by (the right hand side of) (1), where for each $x \in \Delta$, $L(x)$ is irreducible and reversible with respect to $\pi(x)$. By Theorem 3, Proposition 3 and Theorem 8 all the elements of $\chi_{\pi, rev}$ have the same strict Lyapounov function V, hyperbolic equilibria (given by the critical points of V) and are C^1 close to $-grad_g V$ for some metric g. We may then wonder wether $\chi_{\pi, rev}$ is structurally stable. The following construction shows that this is not the case.

8.1 Potential Games are not Structurally Stable

Here Δ stands for the two-dimensional simplex in $\mathbb{R}^3$. Let

$$\tilde{\Delta} = \{(y_1, y_2) \in \mathbb{R}^2 \ : y_1, y_2 \geq 0, \ y_1 + y_2 \leq 1\}$$

and $\jmath : \mathbb{R}^3 \mapsto \mathbb{R}^2$ be the projection defined by $\jmath(x_1, x_2, x_3) = (x_1, x_2)$. Note that $\jmath$ maps Δ homeomorphically onto $\tilde{\Delta}$.

Let $\tilde{W} : \mathbb{R}^2 \mapsto \mathbb{R}$ be a smooth function. Assume that

(a) $-\nabla \tilde{W}$ points inward $\tilde{\Delta}$ on $\partial \tilde{\Delta}$;
(b) The critical set $crit(\tilde{W}) = \{y \in \tilde{\Delta} : \nabla \tilde{W}(y) = 0\}$ consist of (finitely many) non degenerate points,
(c) For all $u \in \mathbb{R}$

$$\frac{\partial \tilde{W}}{\partial y_1}(u, u) = \frac{\partial \tilde{W}}{\partial y_2}(u, u).$$

In particular, the diagonal $D(\tilde{\Delta}) = \{(y_1, y_2) \in \tilde{\Delta} : y_1 = y_2\}$ is positively invariant under the dynamics

$$\dot{y} = -\nabla \tilde{W}(y) \tag{30}$$

(c) There is a saddle connection contained in $D(\tilde{\Delta})$, meaning that there are two saddle points of $\tilde{W}$ $s^1, s^2 \in D(\tilde{\Delta})$ and some (hence every) point $y \in]s^1, s^2[$ which α limit set under (30) is s^1 and omega limit set is s^2.

It is not hard to construct such a map.

Let $W : \mathbb{R}^3 \mapsto \mathbb{R}$ be defined by $W = \tilde{W} \circ \jmath$.

Consider now the 3-strategies potential game associated to W. Payoffs are then defined by

$$U_i(x) = -\frac{\partial \tilde{W}}{\partial x_i}(x_1, x_2), i = 1, 2 \text{ and } U_3(x) = 0.$$

Using the notation of Sect. 6.1, record that $F_{\pi_\beta} = -Id + \pi_\beta$ where π_β is defined by (23), and $\chi_{rev}(\beta, U)$ is the set of vector fields given by (1) with $L(x)$ irreducible and reversible with respect to $\pi_\beta(x)$.

Proposition 7 *For all $\beta > 0$ sufficiently large, there exists $F \in \chi_{rev}(\beta, U)$ (which can be chosen C^1 close to F_{π_β}) which is not equivalent to F_{π_β}.*

Proof of Proposition 7

By definition of Nash equilibria (see Sect. 6.1) and condition (a) above, Nash equilibria of U are fully mixed and coincide with critical points of $\tilde{W}$:

$$crit(\tilde{W}) = \jmath(\mathbf{NE}(U)).$$

Lemma 7 *For all $\varepsilon > 0$ there exists $\beta_0 > 0$ such that for all $\beta \geq \beta_0$ and $F \in \chi_{rev}(\beta, U)$ there is a one to one map*

$$p \in crit(\tilde{W}) \mapsto p_\beta \in \mathbf{Eq}(F),$$

such that

(i) $\|p - \jmath(p_\beta)\| \leq \varepsilon$,

(ii) The unstable (respectively stable) manifold of p_β has dimension $\mathbf{Ind}(p, \tilde{W},)$ *(resp.* $2 - \mathbf{Ind}(p, \tilde{W})$*). In particular, s^1_β and s^2_β are saddle points.*

(iii) $p \in D(\tilde{\Delta}) \Leftrightarrow \jmath(p_\beta) \in D(\tilde{\Delta})$

(iv) Under the dynamics induced by F_{π_β}, the interval $[s^1_\beta, s^2_\beta]$ is invariant and for some (hence all) $q \in]s^1_\beta, s^2_\beta[$ the alpha limit (respectively omega limit) set of q equals s^1_β (respectively s^2_β).

Proof Assertions (i) and (ii) this follows from Propositions 5 and 6.

On $\jmath^{-1}(D(\tilde{\Delta})) = \{(x_1, x_1, 1 - 2x_1)\}$ equilibria of F_{π_β} are given by the implicit equation $T(\log(x_1) - \log(1 - 2x_1)) = U_1(x_1, x_1)$ where $T = 1/\beta$. Solutions for $T = 0$ coincide with $\jmath^{-1}(D(\tilde{\Delta}) \cap crit(\tilde{W}))$. For $T > 0$ and small enough, assertion (iii) then follows from the implicit function theorem.

By condition (c), $\frac{\partial \tilde{W}}{\partial x_1} = \frac{\partial \tilde{W}}{\partial x_2}$ on $D(\tilde{\Delta})$. Thus $U_1(x) = U_2(x)$ (hence $F_{\pi_{\beta,1}}(x) = F_{\pi_{\beta,2}}(x)$) on $J^{-1}(D(\tilde{\Delta}))$ proving invariance of $[s_\beta^1, s_\beta^2] \subset J^{-1}(D(\tilde{\Delta}))$. Assertion (iv) follows since, by (iii), there are no equilibria in $]s_\beta^1, s_\beta^2[$. ■

We now construct $F \in \chi_{rev}(\beta, U)$. Let $L(x)$ be the rate matrix defined for $i \neq j$ by

$$L_{ij}(x) = \pi_{\beta,j}(x) \text{ if } i,j \notin \{1,3\}$$

$$L_{13}(x) = (1 + a(x))\pi_{\beta,3}(x) \text{ and } L_{31}(x) = (1 + a(x))\pi_{\beta,1}(x)$$

where $a : \Delta \mapsto \mathbb{R}^+$ is a smooth function to be defined below. Then Eq. (1) reads

$$\begin{aligned}
\dot{x_1} &= (x_2\pi_{\beta,1}(x) - x_1\pi_{\beta,2}(x)) + (x_3\pi_{\beta,1}(x) - x_1\pi_{\beta,3}(x))(1 + a(x)), \\
\dot{x_2} &= (x_1\pi_{\beta,2}(x) - x_2\pi_{\beta,1}(x)) + (x_3\pi_{\beta,2}(x) - x_2\pi_{\beta,3}(x)), \\
\dot{x_3} &= -\dot{x_1} - \dot{x_2}.
\end{aligned}$$

Thus, on $x_1 = x_2$,

$$\begin{aligned}
\dot{x_1} - \dot{x_2} &= [x_3\pi_{\beta,1}(x) - x_1\pi_{\beta,3}(x)]a(x) \\
&= \frac{a(x)}{Z(x)}(x_3 e^{\beta U_1(x)} - x_1).
\end{aligned}$$

The map $x \mapsto x_3 e^{\beta U_1(x)} - x_1$ vanishes at points s_β^1, s_β^2 and has a constant sign over $[s_\beta^1, s_\beta^2]$ (for otherwise there would exists an equilibrium for F in $]s_\beta^1, s_\beta^2[$ contradicting Lemma 7). Let $p = (s_\beta^1 + s_\beta^2)/2$ and B_η be the Euclidean open ball with center p and radius η. Choose η small enough so that

(i) $B_\eta \cap [s_\beta^1, s_\beta^2] =]q^1, q^2[$ with $s_\beta^1 < q^1 < q^2 < s_\beta^2$ where $<$ stands for the natural ordering on $[s_\beta^1, s_\beta^2]$.
(ii) $x \mapsto x_3\pi_{\beta,1}(x) - x_1\pi_{\beta,3}(x)$ has constant sign on B_η.

Let $x \mapsto a(x)$ be such $a = 0$ on $\Delta \setminus B_\eta$, $a > 0$ on B_η and $0 \leq a \leq \eta$ on Δ. Then, the alpha limit set of q^1 equals s_β^1, for both F and F_{π_β} but since $\dot{x_1} - \dot{x_2}$ doesn't vanish on B_η the trajectory through q^1 exits B_η at a point $\neq q^2$ and, consequently, the omega limit set of q^1 for F is distinct from s_β^2. This proves that F and F_{π_β} are not equivalent.

8.2 *Open Question*

The preceding construction shows that $\chi_{rev}(\beta, U)$ is not structurally stable for an arbitrary potential game but this might be the case for particular examples. Consider for example the Gibbs model described in Remark 5. For $U^0 \in \mathbb{R}^n$ and $U = (U_{ij})$ symmetric, let $\chi_{rev}(\beta, U^0, U)$ be the set of C^1 vector field given by (1) with $L(x)$ irreducible and reversible with respect to the Gibbs measure (14).

Question

For generic (U^0, U) and β large enough, is $\chi_{rev}(\beta, U^0, U)$ structurally stable ?

Appendix

Let L be an irreducible rate matrix and $\pi \in \dot{\Delta}$ denote the invariant probability of L. That is the unique solution (in Δ) of $\pi L = 0$. For all $f, g \in \mathbb{R}^n$ we let

$$\langle f, g\rangle = \sum_i f_i g_i,\ \langle f, g\rangle_\pi = \sum_i f_i g_i \pi_i \text{ and } \langle f, g\rangle_{1/\pi} = \sum_i f_i g_i \frac{1}{\pi_i}.$$

The *Dirichlet form* of L is the map $\mathcal{E} : \mathbb{R}^n \mapsto \mathbb{R}_+$ defined as

$$\mathcal{E}(f) = -\langle f, Lf\rangle_\pi = \frac{1}{2}\sum_{i,j}(f_i - f_j)^2 L_{ij}\pi_i.$$

By irreducibility, $\mathcal{E}(f) > 0$ unless f is constant, and the *spectral gap*

$$\lambda = \sup\{\mathcal{E}(f) :\ \langle f, 1\rangle_\pi = 0, \langle f, f\rangle_\pi = 1\}$$

is positive. We let L^* be the irreducible rate matrix defined by

$$L^*_{ij} = \frac{\pi_j L_{ji}}{\pi_i}.$$

Note that L^* admits π as invariant probability and that L^* is the adjoint of L for $\langle , \rangle_\pi$.

We let $L^T : T\Delta \mapsto T\Delta$ be defined by

$$L^T h = hL.$$

Finally recall that for all $f \in \mathbb{R}^n$ $\frac{f}{\pi}$ stands for the vector defined by $(\frac{f}{\pi})_i = \frac{f_i}{\pi_i}, i = 1 \ldots n$.

Lemma 8 *For all $u, v \in T\Delta$*

$$\langle L^T u, v\rangle_{1/\pi} = \langle L^*(\frac{u}{\pi}), \frac{v}{\pi}\rangle\rangle_\pi$$

In particular L^T is invertible and L^T is a definite negative operator for $\langle , \rangle_{\frac{1}{\pi}}$ whenever L is reversible with respect to π.

Proof The first assertion follows from elementary algebra. For the second, note that $\langle L^T u, u\rangle_{1/\pi} = -\mathcal{E}(\frac{u}{\pi})$. Thus, by irreducibility,

$$\langle L^T u, u\rangle_{1/\pi} < 0$$

unless $u = 0$. ■

Proof of Lemma 4

Given $f \in \mathbb{R}^n$ we write $f \geq 0$ if $f_i \geq 0$ for all i. We let $1 \in \mathbb{R}^n$ denote the vector which components are all equal to 1. For all $t \geq 0$ we let $P_t = e^{tL}$. Since L is a rate matrix, (P_t) is a Markov semigroup meaning that $P_t f \geq 0$ for all $f \in \mathbb{R}^n$ with $f \geq 0$ and $P_t 1 = 1$.

Lemma 9 *Let $I \subset \mathbb{R}$ be an open interval and $S : I \mapsto \mathbb{R}$ a C^2 function such that $S''(t) \geq \alpha > 0$. Let $f \in \mathbb{R}^n$ be such that $f_i \in I$ for all i. Then*

$$\frac{d}{dt}\langle S(P_t f), 1\rangle_\pi|_{t=0} \leq -\alpha\mathcal{E}(f).$$

Proof For all $u, v \in I$ $S(v) - S(u) - S'(u)(v-u) \geq \alpha/2(v-u)^2$. Hence for all i, j

$$S(f_j) - S((P_t f)_i) - S'((P_t f)_i)(f_j - (P_t f)_i) \geq \alpha/2(f_j - (P_t f)_i)^2.$$

Applying P_t to this inequality gives

$$P_t(Sf)_i - S((P_t f)_i) \geq \alpha/2 P_t(f_i - (P_t f)_i)^2) = \alpha/2(P_t f_i^2 - (P_t f)_i^2)$$

Hence

$$P_t(Sf) - S((P_t f)) \geq \alpha/2 P_t(f - (P_t f))^2) = \alpha/2(P_t f^2 - (P_t f)^2).$$

Therefore, using the fact that $\langle P_t g, 1\rangle_\pi = \langle g, 1\rangle_\pi$ leads to

$$\langle Sf - S(P_t f), 1\rangle_\pi \geq \alpha\langle f^2 - (P_t f)^2, 1\rangle_\pi.$$

Dividing by t and letting $t \to 0$ leads to the desired inequality. ■

Let $S :]0, \infty[\mapsto \mathbb{R}$ be a C^2 function with positive second derivative. Let $H_\pi^S : \Delta \mapsto \mathbb{R}$ be the map defined by

$$H_\pi^S(x) = \sum_i \pi_i S(\frac{x_i}{\pi_i}).$$

Corollary 4 *For all $x \in \Delta$*

$$\langle \nabla H_\pi^S(x), xL\rangle \leq -\alpha\lambda Var_\pi(f)$$

where $f_i = \frac{x_i}{\pi_i}$

Proof For $x \in \Delta$ let $x(t) = xe^{tL}, f_i = \frac{x_i}{\pi_i}, f_i(t) = \frac{x_i(t)}{\pi_i}$ and $P_t^* g = e^{tL^*} g$. Note that $P_t^*)$ is the adjoint of P_t with respect to $\langle , \rangle_\pi$.

For all $g \in \mathbb{R}^n$, $\langle x(t), g\rangle = \langle x, P_t g\rangle = \langle f, P_t g\rangle_\pi = \langle P_t^* f, g\rangle_\pi$ so that $f(t) = P_t^* f$. Hence by the preceding lemma applied to L^* it follows that

$$\langle \nabla H_\pi^S(x), xL\rangle = \frac{d}{dt}\langle S(P_t^* f), 1\rangle_\pi|_{t=0} \leq -\alpha\mathcal{E}(f) \leq -\alpha\lambda Var_\pi(f)$$

where $\alpha = \min_i S''(\frac{x_i}{\pi_i}) > 0$. ■

We now prove the Lemma. Set $S(t) = \int_1^t s(u)du$. Then for all $u \in T\Delta$

$$\langle \nabla H_\pi^S(x), u\rangle = \sum_i u_i s(\frac{x_i}{\pi_i})$$

and the results follows from Corollary 4.

Acknowledgements This work was supported by the SNF grant $200020_1 49871/1$ I would like to thank J. B Bardet, F. Malrieu, M. W Hirsch, J. Hofbauer, J. Robbin, B. Sandholm, S. Sorin, P. A Zitt for numerous discussions on topics related to this paper.

References

1. Benaïm, M.: Vertex-reinforced random walks and a conjecture of Pemantle. Ann. Probab. **25**(1), 361–392 (1997)
2. Benaïm, M.: Dynamics of Stochastic Approximation Algorithms, Séminaire de Probabilités, XXXIII. Lecture Notes in Mathematics, vol. 1709, pp. 1–68. Springer, Berlin (1999)

3. Benaïm, M., Hirsch, M.W.: Asymptotic pseudotrajectories and chain recurrent flows, with applications. J. Dyn. Diff. Equ. **8**(1), 141–176 (1996). MR 1388167 (97d:58165)
4. Benaïm, M., Raimond, O.: Self interacting diffusions iii: symmetric interactions. Ann. Probab. **33**(5), 1716–1759 (2005)
5. Benaïm, M., Weibull, J.: Deterministic approximation of stochastic evolution in games. Econometrica **71**(3), 873–903 (2003)
6. Benaïm, M., Benjamini, O., Chen, J., Lima, Y.: A generalized polya's urn with graph based interactions. Rand. Struct. Algorithms **46**(4), 614–634 (2015)
7. Benaïm, M., Raimond, O., Schapira, B.: Strongly reinforced vertex-reinforced random walks on the complete graph. ALEA. Latin Am. J. Probab. Math. Stat. **10**(2), 767–782 (2013)
8. Chen, J.: Lucas, C.: Generalized polya's urn: convergence at linearity (2013, preprint) [arXiv:1306.5465]
9. Chill, R., Haraux, A., Ali-Jendoubi, M., Applications of the lojasiewicz simon, gradient inequality to gradient-like evolution equations. Anal. Appl. **7**(4), 351–372 (2009)
10. Dupuis, P., Fisher, M.: On the construction of Lyapunov functions for nonlinear Markov processes via relative entropy. Lefschetz Center for Dynamical Systems (2011, preprint)
11. Freildin, M., Wentzell, A.D.: Random Perturbations of Dynamical Sytems, 3rd edn. Springer, Heidelberg (2012)
12. Hirsch, M.W.: Differential Topology, vol. 33. Springer, New York (1976)
13. Hofbauer, J., Sigmund, K.: Evolutionary Games and Population Dynamics. Cambridge University Press, Cambridge (1998)
14. Hofbauer, J., Sorin, S., Viossat, Y.: Time average replicator and best reply dynamics. Math. Oper. Res. **34**(2), 263–269 (2009)
15. Kurtz, T.G.: Solutions of ordinary differential equations as limits of pure jump markov processes. J. Appl. Probab. **7**, 49–58 (1970)
16. Lojasiewicz, S.: Une propriété topologique des sous-ensembles analytiques réels. Les Équations aux Dérivées Partielles, pp. 87–89. Éditions du C.N.R.S, Paris (1963)
17. Merlet, B., Nguyen, T.N.: Convergence to equilibrium for discretizations of gradient-like flows on riemannian manifolds. Differ. Integr. Equ. **26**(5–6), 571–602 (2013)
18. Palis, J., de Melo, W.: Geometric Theory of Dynamical Systems. Springer, New York (1980)
19. Pemantle, R.: Vertex-reinforced random walk. Probab. Theory Relat. Fields **1** 117–136 (1992)
20. Pemantle, R.: A survey of random processes with reinforcement. Probab. Surv. **4**, 1–79 (2007). MR 2282181 (2007k:60230)
21. Robinson, C.: Dynamical Systems: Stability, Symbolic Dynamics and Chaos, 2nd edn. CRC Press, Boca Raton (1999)
22. Sandholm, W.H.: Population Games and Evolutionary Dynamics. MIT, Cambridge (2010)
23. Sandholm, W.H.: Population games and deterministic evolutionary dynamics. In: Young, H.P., Zamir, S. (eds.) Handbook of Game Theory, vol. 4, pp. 703–775. North Holland (2015)
24. van der Hofstad, R., Holmes, M., Kuznetsov, A., Ruszel, W.: Strongly reinforced polya urns with graph-based competition (2014, preprint) [arXiv:1406.0449]

Wave Interaction with Floating Bodies in a Stratified Multilayered Fluid

Filipe S. Cal, Gonçalo A.S. Dias, and Juha H. Videman

Abstract We derive from first principles the dynamical equations that govern the interaction of small-amplitude water waves with freely floating obstacles in a stratified multilayer fluid. Focusing on two-layer fluids, we present the equations in an easily manageable matrix form, write down conditions for the stability of equilibrium and, by limiting ourselves to time-harmonic motions, recast the problem as a spectral boundary-value problem composed of a differential equation and an algebraic system, coupled through boundary conditions. Proceeding with a suitable variational and operator formulation, we present an elimination scheme that simplifies the system to a linear spectral problem for a self-adjoint operator in a Hilbert space. Under symmetry assumptions on the geometry of the fluid domain, we derive a sufficient condition guaranteeing the existence of trapped modes in a two-layer fluid channel.

1 Introduction

Given the recent growing interest in the interaction of water waves with freely floating structures, cf. [7, 10–12, 16, 19, 20, 22, 23], it is natural to start thinking beyond homogeneous constant-density fluids. Now, while the equations governing the fluid and wave motion in stratified media are well-known (cf. [9]), the equations that couple this motion with the motion of a freely floating body have been derived for a homogeneous fluid only, see John [8] and Mei et al. [17].

As the simplest generalization of a constant-density fluid, we may consider a fluid that lies in several homogeneous layers of uniform but distinct densities. This situation is likely to occur, e.g., in fjords, channels, rivers and estuaries, being, therefore, of great practical interest. Besides, it is often used as an approximation of a continuously stratified fluid which itself does not maintain irrotational flow.

In this chapter, we set down the equations coupling the motion of a freely floating structure to the wave motion in a multilayer fluid. The equations are derived from the

F.S. Cal (✉) • G.A.S. Dias • J.H. Videman
CAMGSD/Departamento de Matemática, Instituto Superior Técnico, Universidade de Lisboa, Av. Rovisco Pais 1, 1049-001 Lisboa, Portugal
e-mail: fcal@adm.isel.pt; goncalo.dias@tecnico.ulisboa.pt; jvideman@math.tecnico.ulisboa.pt

J.-P. Bourguignon et al. (eds.), *Dynamics, Games and Science*, CIM Series in Mathematical Sciences 1, DOI 10.1007/978-3-319-16118-1_9

fundamental principles of physics under the assumption that the motion is of small amplitude near the equilibrium position. Moreover, conditions ensuring stability of the equilibrium position are inferred from energy considerations. For expediency, we establish the equations and stability conditions for a two-layer fluid and only outline the generalization to multilayer fluids. We also express the equations in a non-dimensional form and, considering time-harmonic motions, recast the problem in the frequency domain as a coupled spectral boundary-value problem consisting of a differential equation and an algebraic system. Following an argument set forth in [22], we then proceed to rewrite the original quadratic eigenvalue problem as a linear one and present, as an example, a trapping condition in a two-layer fluid channel.

Built upon the general framework laid down here, one can now investigate the wave/structure interaction in multilayer fluids, see [3, 4] for numerous examples of floating bodies supporting trapped modes. Some of the results presented in this review paper have been published in a more concise form in [2].

This chapter is organised as follows. In Sect. 2 we introduce our notation and derive the kinematic and dynamic (zeroth and first order) conditions on the wetted surface of a freely floating body partially immersed in a two-layer fluid. We restate the equations in a matrix form, for two-dimensional motions and for totally submerged obstacles. In Sect. 3, we investigate the stability of equilibrium of a freely floating body in a two-layer fluid and in Sect. 4 generalize the results to a stratified multilayer fluid. In Sect. 5, we formulate the problem for a time-harmonic motion, rephrase it in a variational and operator form, and present a scheme which reduces the problem to a linear pencil $\boldsymbol{B} - \omega \boldsymbol{D}$. Making symmetry assumptions on the fluid motion and on the movements of the freely-floating body, we are able to guarantee that the self-adjoint operator $\boldsymbol{B}$ is positive definite and, consequently, rewrite the problem in normal form $\boldsymbol{M} - \mu \boldsymbol{I}$. Finally, we derive a condition that guarantees the existence of trapped modes in a two-layer fluid channel.

2 Equations of Motion for a Two-Layer Fluid

2.1 *Equations of Motion in the Absence of Obstacles*

Consider the mechanical system consisting of an incompressible inviscid heavy fluid occupying two homogeneous immiscible layers, one on top of the other, and of a rigid body floating freely in it, i.e. the only external force acting on the totally or partially immersed body is gravity. Assume that the constant density of the lower layer is greater than the one in the upper layer ($\rho_2 > \rho_1 > 0$) and that the flow in each layer is irrotational (cf. Lamb [14]).

The fluid domain extends to infinity in horizontal directions, but has finite depth, being bounded from below by a horizontal rigid bottom and from above by a free surface. A Cartesian coordinate system is chosen in such a way that when the fluid is at rest the (x, y)−plane coincides with the free surface and the interface is a horizontal surface at $z = -h$.

Fig. 1 A freely floating body in a two-layer fluid

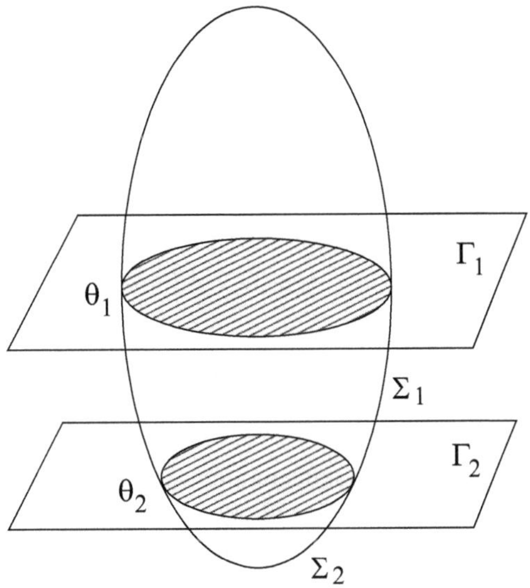

The upper and lower fluid layers are denoted by $\Pi_1 = \mathbb{R}^2 \times (-h, 0)$ and $\Pi_2 = \mathbb{R}^2 \times (-h_b, -h)$, $h_b, h \in \mathbb{R}^+$ $(h_b > h)$. Let B be a bounded connected open subset of $\mathbb{R}^3$ representing a partially immersed rigid body (in its equilibrium state). Within the upper and lower fluid layers, we introduce the open sets $\Theta_1 = B \cap \Pi_1$ and $\Theta_2 = B \cap \Pi_2$ corresponding to the submerged parts of the body B and assume that the fluid regions

$$\Omega_1 = \Pi_1 \setminus \overline{\Theta_1}, \qquad \Omega_2 = \Pi_2 \setminus \overline{\Theta_2}$$

are Lipschitz domains so that the normal vector is defined almost everywhere on $\partial\Omega_1$ and $\partial\Omega_2$. We also define (see Fig. 1) the unpierced parts of the free surface and the interface (at their rest position) by Γ_1 and Γ_2, i.e.,

$$\Gamma_1 = \{(x, y, z) \in \partial\Omega_1 : z = 0\}, \qquad \Gamma_2 = \{(x, y, z) \in \partial\Omega_1 \cap \partial\Omega_2 : z = -h\},$$

and the rigid bottom by

$$\Gamma_b = \{(x, y, z) \in \partial\Omega_2 : z = -h_b\}.$$

The Eulerian description of the fluid motion is given by the vector field $\mathbf{u}^{(j)}(x, y, z, t)$, that is, the fluid velocity in layer j at a fixed position $(x, y, z) \in \Omega_j$, $j = 1, 2$, at an instant $t > 0$. Assuming the fluid to be inviscid and ignoring surface tension, the equations of motion can be derived layer-wise from the basic conservation laws (see Lamb [14] for details). The conservation of mass implies the continuity equation

$$(\rho_j)_t + \nabla \cdot (\rho_j \mathbf{u}^{(j)}) = 0 \quad \text{in} \quad \Omega_j, \quad j = 1, 2.$$

Under the assumption of constant-density at each layer, the previous equation reduces to

$$\nabla \cdot \mathbf{u}^{(j)} = 0 \quad \text{in} \quad \Omega_j\,, \quad j = 1,2. \tag{1}$$

The conservation of linear momentum leads to the (layer-wise) Euler equations:

$$\mathbf{u}_t^{(j)} + \mathbf{u}^{(j)} \cdot \nabla \mathbf{u}^{(j)} = -\frac{\nabla P^{(j)}}{\rho_j} + \mathbf{g} \quad \text{in} \quad \Omega_j\,, \quad j = 1,2. \tag{2}$$

Here $P^{(j)}$ is the mechanical mean pressure at point (x, y, z) in layer j at time t, and $\mathbf{g} = (0, 0, -g)$ is the gravitational force per unit volume (g denotes the acceleration due to gravity).

Assuming that the motion stays irrotational at all times, the velocity vectors $\mathbf{u}^{(j)}$ can be expressed in simply connected domains Ω_j as gradients of (scalar) velocity potentials $\Phi^{(1)}(x, y, z, t)$ and $\Phi^{(2)}(x, y, z, t)$, i.e.

$$\mathbf{u}^{(j)} = \nabla \Phi^{(j)} \quad \text{in} \quad \Omega_j\,, \quad j = 1,2. \tag{3}$$

From (1) it then follows that the potentials $\Phi^{(j)}$ satisfy the Laplace equation in their respective domains at every time instant t.

Next, assuming that the wave motion is of small amplitude, we may linearise the equations of motion (see John [8] and Mei et al. [17]) and thus make the following ansatz for the velocity potentials

$$\Phi^{(j)}(x, y, z, t) = \epsilon\, \Phi_1^{(j)}(x, y, z, t) + \epsilon^2 \Phi_2^{(j)}(x, y, z, t) + \ldots\,, \quad j = 1, 2\,,$$

where $\epsilon > 0$ is a small parameter. The first order velocity potentials $\Phi_1^{(1)}$ and $\Phi_1^{(2)}$, the functions we are interested in, satisfy the Laplace equation in their domains, i.e.,

$$\rho_j\, \Delta \Phi_1^{(j)} = 0 \quad \text{in} \quad \Omega_j\,, \quad j = 1, 2\,.$$

We also assume that the function $\eta(x, y, t)$ which describes the vertical position of the free surface at time t, can be expanded in powers of ϵ as

$$\eta(x, y, t) = \eta_0(x, y) + \epsilon\, \eta_1(x, y, t) + \epsilon^2 \eta_2(x, y, t) + \ldots\,.$$

Note that the free surface at its rest position, defined by equation $z = \eta_0(x, y)$, cannot depend on t. The kinematic boundary condition at the free surface requires that

$$\Phi_x^{(1)} \eta_x + \Phi_y^{(1)} \eta_y + \eta_t = \Phi_z^{(1)} \quad \text{on} \quad z = \eta(x, y, t)\,.$$

Substituting the expansions of $\Phi^{(1)}$ and η into the previous equation, gives at the zeroth order

$$\left(\eta_0\right)_t = 0$$

and at the first order

$$\left(\Phi_1^{(1)}\right)_x\left(\eta_0\right)_x + \left(\Phi_1^{(1)}\right)_y\left(\eta_0\right)_y + \left(\eta_1\right)_t = \left(\Phi_1^{(1)}\right)_z \tag{4}$$

on the free surface. From Eqs. (2) and (3) one derives the Bernoulli equation

$$\Phi_t^{(j)} + \frac{1}{2}|\nabla\Phi^{(j)}|^2 + gz = -\frac{P^{(j)}}{\rho_j} + C_j \quad \text{in} \quad \overline{\Omega_j}\,, \quad j = 1, 2, \tag{5}$$

where C_j is an arbitrary function of t. The last term on the left-hand side is the hydrostatic contribution, whereas the rest is the hydrodynamic contribution to the total pressure. Choosing $C_1 = \frac{P_0}{\rho_1}$, where P_0 is the constant atmospheric pressure, the Bernoulli equation (5) leads to the dynamic boundary condition on the free surface

$$g\,\eta + \Phi_t^{(1)} + \frac{1}{2}\left(\left(\Phi_x^{(1)}\right)^2 + \left(\Phi_y^{(1)}\right)^2 + \left(\Phi_z^{(1)}\right)^2\right) = 0 \quad \text{on} \quad z = \eta(x, y, t)\,,$$

Similarly, substituting the expansions of $\Phi^{(1)}$ and η into previous equation, we obtain the zeroth order dynamic boundary condition

$$\eta_0 = 0$$

and the first order dynamic boundary condition

$$g\,\eta_1 + \left(\Phi_1^{(1)}\right)_t = 0\,, \tag{6}$$

both valid on Γ_1. Thus, we can see that the free surface is horizontal in its rest position and, eliminating η_1 between (4) and (6), we find the classical linearised kinematic/dynamic boundary condition

$$\left(\Phi_1^{(1)}\right)_{tt} + g\left(\Phi_1^{(1)}\right)_z = 0 \quad \text{on} \quad \Gamma_1\,.$$

On the interface between the two fluid layers (at its rest position), we have the following linearised transmission conditions

$$\rho_1\left(\left(\Phi_1^{(1)}\right)_{tt} + g\left(\Phi_1^{(1)}\right)_z\right) = \rho_2\left(\left(\Phi_1^{(2)}\right)_{tt} + g\left(\Phi_1^{(2)}\right)_z\right) \text{ and } \left(\Phi_1^{(1)}\right)_z = \left(\Phi_1^{(2)}\right)_z \text{ on } \Gamma_2\,.$$

On the rigid bottom, we impose the Neumann boundary condition (no normal flow)

$$\left(\Phi_1^{(2)}\right)_n = 0 \quad \text{on} \quad \Gamma_b\,.$$

Collecting the previous equations, we obtain the system (cf. Lamb [14])

$$\Delta\Phi_1^{(1)} = 0 \quad \text{in} \quad \Omega_1 \tag{7}$$

$$\Delta\Phi_1^{(2)} = 0 \quad \text{in} \quad \Omega_2 \tag{8}$$

$$\left(\Phi_1^{(1)}\right)_{tt} + g\left(\Phi_1^{(1)}\right)_z = 0 \quad \text{on} \quad \Gamma_1 \tag{9}$$

$$\rho_1\left(\left(\Phi_1^{(1)}\right)_{tt} + g\left(\Phi_1^{(1)}\right)_z\right) = \rho_2\left(\left(\Phi_1^{(2)}\right)_{tt} + g\left(\Phi_1^{(2)}\right)_z\right) \qquad \text{on} \quad \Gamma_2 \tag{10}$$

$$\left(\Phi_1^{(1)}\right)_z = \left(\Phi_1^{(2)}\right)_z \quad \text{on} \quad \Gamma_2 \tag{11}$$

$$\left(\Phi_1^{(2)}\right)_z = 0 \quad \text{on} \quad \Gamma_b \tag{12}$$

2.2 Boundary Condition on the Surface of the Floating Body

Let σ_1 and σ_2 be the wetted surfaces of the partially immersed rigid body B belonging, at each instant of time t, to the upper layer and to the lower layer, respectively, i.e.,

$$\sigma_j = \{(x, y, z) \in \Pi_j : z = f(x, y, t)\}\,, \quad j = 1, 2\,,$$

where f is a sufficiently smooth real-valued function satisfying the ansatz

$$f(x, y, t) = f_0(x, y) + \epsilon f_1(x, y, t) + \epsilon^2 f_2(x, y, t) + \dots\,. \tag{13}$$

The equilibrium counterparts of σ_1 and σ_2 are defined by

$$\Sigma_1 = \{(x, y, z) : z = f_0(x, y);\ -h < z < 0\}\,,$$
$$\Sigma_2 = \{(x, y, z) : z = f_0(x, y);\ z < -h\}\,.$$

Let $\boldsymbol{X}(t) = (X(t), Y(t), Z(t))$ be the vector position of the centre of rotation (coinciding with the centre of mass) of the body B at time instant t, expanded as

$$\boldsymbol{X}(t) = \boldsymbol{X}_0 + \epsilon\boldsymbol{X}_1(t) + \epsilon^2\boldsymbol{X}_2(t) + \dots\,, \tag{14}$$

where $\boldsymbol{X}_0 = (X_0, Y_0, Z_0)$ is the rest position of the centre of mass of B. Note that for our purposes here any other choice of rotation centre would increase complexity needlessly (see Mei et al. [17]). Moreover, let $\bar{\boldsymbol{x}} = (\bar{x}, \bar{y}, \bar{z})$ denote a time-dependent coordinate system attached to the body and chosen so that $\bar{\boldsymbol{x}} = \boldsymbol{x}$ for any point in the body (or the fluid) when the system is at rest. Points in the coordinate systems $\mathcal{O}_{x,y,z}$ and $\mathcal{O}_{\bar{x},\bar{y},\bar{z}}$ are related, up to the first order, by

$$\bar{\boldsymbol{x}} = \boldsymbol{x} - \epsilon\big(\boldsymbol{X}_1 + \boldsymbol{\theta}_1 \times (\boldsymbol{x} - \boldsymbol{X}_0)\big) + O(\epsilon^2) , \tag{15}$$

where we have the angular position of the body denoted by $\boldsymbol{\theta}(t) = \epsilon\boldsymbol{\theta}_1(t) + O(\epsilon)$, with $\epsilon\boldsymbol{\theta}_1 = (\epsilon\alpha, \epsilon\beta, \epsilon\gamma)$ denoting the angles of rotation of the body with respect to the lines going through the centre of mass $\boldsymbol{X}_0$ parallel to the x,y and z-axis, respectively, for order ϵ.

Since $\bar{z} = f_0(\bar{x}, \bar{y})$ when the system is at rest, expanding f_0 in Taylor series with respect to (x, y), and using relations (15), from Eq. (13) we obtain at first order ϵ

$$\begin{aligned} f_1 &= Z_1 + \alpha\big(y - Y_0\big) - \beta\big(x - X_0\big) \\ &\quad - \big(f_0\big)_x\Big(X_1 + \beta\big(z - Z_0\big) - \gamma\big(y - Y_0\big)\Big) \\ &\quad - \big(f_0\big)_y\Big(Y_1 + \gamma\big(x - X_0\big) - \alpha\big(z - Z_0\big)\Big) . \end{aligned} \tag{16}$$

Continuity of normal velocity requires that

$$\Phi^{(j)}_x f_x + \Phi^{(j)}_y f_y + f_t = \Phi^{(j)}_z \quad \text{on} \quad \sigma_j\,, \quad j = 1, 2\,. \tag{17}$$

On substituting the expansions of $\Phi^{(j)}$ and f, we obtain at the first order

$$\big(\Phi^{(j)}_1\big)_x\big(f_0\big)_x + \big(\Phi^{(j)}_1\big)_y\big(f_0\big)_y + \big(f_1\big)_t = \big(\Phi^{(j)}_1\big)_z, \quad j = 1, 2\,.$$

Recalling (16), we can write the linearised kinematic boundary condition at Σ_j, j=1,2, omitting the subindices of the first order velocity potentials, as

$$\Phi^{(j)}_n = \big((\boldsymbol{X}_1)_t + (\boldsymbol{\theta}_1)_t \times (\boldsymbol{x} - \boldsymbol{X}_0)\big) \cdot \boldsymbol{n}\,, \quad j = 1, 2\,, \tag{18}$$

where $\boldsymbol{n} = (-(f_0)_x, -(f_0)_y, 1)$ is the unit normal vector to Σ_j pointing into B. Alternatively (see Nazarov and Videman [22]), we can introduce a vector $\boldsymbol{a} \in \mathbb{R}^6$, describing the translational displacements (components a_j, $j = 1, 2, 4$) of the mass centre of the body and the angular displacements (components a_j, $j = 3, 5, 6$) of the body about the axes passing through the centre of mass, by

$$\boldsymbol{a} = \big(X_1, Y_1, \gamma, Z_1, \alpha, \beta\big) , \tag{19}$$

and define a matrix of rigid-body motions $D(\boldsymbol{x}) \in \mathbb{R}^{3\times 6}$ by

$$D(\boldsymbol{x}) = \begin{bmatrix} 1 & 0 & -y & 0 & 0 & z \\ 0 & 1 & x & 0 & -z & 0 \\ 0 & 0 & 0 & 1 & y & -x \end{bmatrix} . \tag{20}$$

Consequently, the equations in (18) take the form

$$\Phi_n^{(j)} = \boldsymbol{n}^T D(\boldsymbol{x} - \boldsymbol{X}_0)\,\boldsymbol{a}_t \quad \text{on} \quad \Sigma_j\,, \quad j = 1, 2\,. \tag{21}$$

Remark 1 The kinematic equation at the boundary of the rigid body can be written locally without necessarily taking into consideration the entire body surface. Considering a neighbourhood of a certain point $\boldsymbol{x}_0$ on the surface σ_j, where one can define a local coordinate system (ξ_1, ξ_2, ζ), with the ζ axis normal to the surface at $\boldsymbol{x}_0$, pointing into the body, one can write the equation analog of (17) in local coordinates (cf. Kuznetsov et al. [13], p. 9). This allows us to generalise the treatment given in this work to surfaces that cannot be described globally as $z = f(x, y, t)$, but only locally as the graph of a function of ξ_1, ξ_2, and t.

2.3 Translational Dynamics of the Floating Body

The equations governing the translational motion of a rigid body, totally or partially submerged in an inviscid fluid (the only forces acting on the body's surface are due to the fluid pressure), and freely floating (buoyancy and gravity balance each other) result from the Newton's second law of motion or, equivalently, from the balance of linear momentum. These equations can be written in the form

$$I^B \boldsymbol{X}_{tt} = \sum_{j=1}^{2} \int_{\sigma_j} P^{(j)}\, \boldsymbol{n}\, \mathrm{d}s - I^B g\, \boldsymbol{e}_3\,, \tag{22}$$

where I^B is the total mass of the body, $\boldsymbol{X}$ is the instantaneous position of the centre of mass of B and $\boldsymbol{n} = (-f_x, -f_y, 1)$ is the normal vector to σ_j pointing into B. Moreover, $P^{(1)}$ is the pressure acting on the surface of the body in the upper layer which, according to the linearised Bernoulli equation, can be written as

$$P^{(1)} = -\rho_1\, g f - \epsilon\, \rho_1 \Phi_t^{(1)} + O(\epsilon^2)\,. \tag{23}$$

In the lower layer, the pressure exerted on the body is given by

$$P^{(2)} = \rho_1\, g\, h - \rho_2\, g\, (f + h) - \epsilon\, \rho_2 \Phi_t^{(2)} + O(\epsilon^2)\,. \tag{24}$$

Substituting (23) and (24) into (22) and expanding $\boldsymbol{X}$ and f in ϵ, results in

$$\begin{aligned} \epsilon\, I^B (\boldsymbol{X}_1)_{tt} = &- \int_{\sigma_1} \rho_1\, g\, (f_0 + \epsilon f_1)\, \boldsymbol{n}\, \mathrm{d}s + \int_{\sigma_2} (\rho_1\, g\, h - \rho_2\, g\, (f_0 + \epsilon f_1 + h))\, \boldsymbol{n}\, \mathrm{d}s \\ &- \epsilon \sum_{j=1}^{2} \rho_j \int_{\sigma_j} \Phi_t^{(j)}\, \boldsymbol{n}\, \mathrm{d}s - I^B g\, \boldsymbol{e}_3 + O(\epsilon^2)\,, \end{aligned} \tag{25}$$

where the left hand side has no zeroth order terms, since the rest position $\boldsymbol{X}_0$ of the centre of mass is constant in time. On the right hand side, we have both zeroth and first order hydrostatic and hydrodynamic terms. All higher order terms are included in $O(\epsilon^2)$.

Selecting the vertical component, we obtain

$$\epsilon I^B (Z_1)_{tt} = -\rho_1 \, g \int_{\vartheta_1} (f_0 + \epsilon f_1) \; \mathrm{d}x\mathrm{d}y - (\rho_2 - \rho_1) \, g \int_{\vartheta_2} (f_0 + \epsilon f_1 + h) \; \mathrm{d}x\mathrm{d}y$$
$$- \epsilon \sum_{j=1}^{2} \rho_j \int_{\sigma_j} \Phi_t^{(j)} \, n_3 \, \mathrm{d}s - I^B g + O(\epsilon^2) \, , \tag{26}$$

where ϑ_1 and ϑ_2 are the projections of the cross-sectional areas of the parts of the body piercing the free surface and the interface, respectively, onto a horizontal plane, at each instant in time.

The equilibrium counterparts of the ϑ_j are the θ_j. These θ_j are the cross-sectional areas of the parts of the body piercing the free surface and the interface at the same horizontal plane, when the body is at its rest position. Given that the measure of ϑ_1 differs from the measure of θ_1 by $O(\epsilon)$, and since f_0 is of order $O(\epsilon)$ in the region where ϑ_1 and θ_1 differ from each other (note that $f_0 = 0$ at the boundary of θ_1), we obtain

$$\int_{\vartheta_1} (f_0 + \epsilon f_1) \; \mathrm{d}x\mathrm{d}y = \int_{\theta_1} (f_0 + \epsilon f_1) \; \mathrm{d}x\mathrm{d}y + O(\epsilon^2) \, .$$

A similar reasoning can be made for ϑ_2 and θ_2, with $f_0 + h = 0$ at the boundary of θ_2.

Expressing also σ_j in terms of its equilibrium counterpart $\Sigma_j, j = 1, 2$, plus higher order terms in ϵ, the vertical component of the linear momentum balance becomes

$$\epsilon I^B (Z_1)_{tt} = -\rho_1 \, g \int_{\theta_1} (f_0 + \epsilon f_1) \; \mathrm{d}x\mathrm{d}y - (\rho_2 - \rho_1) \, g \int_{\theta_2} (f_0 + \epsilon f_1 + h) \; \mathrm{d}x\mathrm{d}y$$
$$- \epsilon \sum_{j=1}^{2} \rho_j \int_{\Sigma_j} \Phi_t^{(j)} \, n_3 \, \mathrm{d}s - I^B g + O(\epsilon^2) \, . \tag{27}$$

From the zeroth order, writing $f_0 = -\int_{f_0}^{0} \mathrm{d}z$ and $f_0 + h = -\int_{f_0}^{-h} \mathrm{d}z$, and defining the momenta

$$I^{\Theta_j} = \int_{\Theta_j} \mathrm{d}\boldsymbol{x} \, , \quad j = 1, 2 \, ,$$

where Θ_1 and Θ_2 are the submerged parts of the body in its equilibrium state, in the upper and lower layer, respectively, we obtain

$$I^B g = \rho_1 \, g \, I^{\Theta_1} + \rho_2 \, g \, I^{\Theta_2} \,, \tag{28}$$

which is nothing but the Archimedes' principle of flotation.

At the first order in ϵ, the vertical component satisfies the equation

$$I^B (Z_1)_{tt} = -\sum_{j=1}^{2} \rho_j \int_{\Sigma_j} \Phi_t^{(j)} \, n_3 \, \mathrm{d}s - \rho_1 \, g \int_{\theta_1} f_1 \, \mathrm{d}x\mathrm{d}y - (\rho_2 - \rho_1) \, g \int_{\theta_2} f_1 \, \mathrm{d}x\mathrm{d}y \,. \tag{29}$$

Now, f_0 is constant at the boundary of θ_1, as well as at the boundary of θ_2, so that

$$\int_{\theta_j} (f_0)_x \, \mathrm{d}x\mathrm{d}y = \int_{\theta_j} (f_0)_y \, \mathrm{d}x\mathrm{d}y = 0 \,, \quad j = 1, 2 \,.$$

Hence, substituting Eq. (16) into (29) gives

$$\begin{aligned} I^B (Z_1)_{tt} &= -\sum_{j=1}^{2} \rho_j \int_{\Sigma_j} \Phi_t^{(j)} \, n_3 \, \mathrm{d}s \\ &\quad - \rho_1 \, g \left(\alpha I_y^{\theta_1} - \beta I_x^{\theta_1} + Z_1 I^{\theta_1} \right) - (\rho_2 - \rho_1) \, g \left(\alpha I_y^{\theta_2} - \beta I_x^{\theta_2} + Z_1 I^{\theta_2} \right) , \end{aligned} \tag{30}$$

where

$$I^{\theta_j} = \int_{\theta_j} \mathrm{d}x\mathrm{d}y \,, \quad I_x^{\theta_j} = \int_{\theta_j} (x - X_0) \, \mathrm{d}x\mathrm{d}y \,, \quad I_y^{\theta_j} = \int_{\theta_j} (y - Y_0) \, \mathrm{d}x\mathrm{d}y \,, \quad j = 1, 2 \,.$$

Let us next consider the x-component of the equation of motion (25)

$$\begin{aligned} \epsilon I^B (X_1)_{tt} &= -\rho_1 \, g \int_{\sigma_1} (f_0 + \epsilon f_1) n_1 \, \mathrm{d}s - \int_{\sigma_2} \left(\rho_2 \, g \, (f_0 + \epsilon f_1 + h) - \rho_1 \, g \, h \right) n_1 \, \mathrm{d}s \\ &\quad - \epsilon \sum_{j=1}^{2} \rho_j \int_{\sigma_j} \Phi_t^{(j)} \, n_1 \, \mathrm{d}s + O(\epsilon^2) \,. \end{aligned} \tag{31}$$

Reasoning as above, we can replace σ_j by Σ_j. Integrating by parts and observing that $f_0 = 0$ and $f_0 + h = 0$ at the boundaries of θ_1 and θ_2, respectively, we get rid of the hydrostatic terms and obtain, for the x-component,

$$\epsilon I^B (X_1)_{tt} = -\epsilon \sum_{j=1}^{2} \rho_j \int_{\Sigma_j} \Phi_t^{(j)} \, n_1 \, \mathrm{d}s + O(\epsilon^2) \,. \tag{32}$$

Similarly, for the y-component

$$\epsilon I^B (Y_1)_{tt} = -\epsilon \sum_{j=1}^{2} \rho_j \int_{\Sigma_j} \Phi_t^{(j)}\, n_2 \,\mathrm{d}s + O(\epsilon^2)\,. \tag{33}$$

Note that at the zeroth order the previous two equations are trivially satisfied.

Collecting Eqs. (30), (32) and (33), the linearised form of the equation of motion (22) is written as

$$I^B (\boldsymbol{X}_1)_{tt} = -\sum_{j=1}^{2} \rho_j \int_{\Sigma_j} \Phi_t^{(j)}\, \boldsymbol{n} \,\mathrm{d}s$$
$$- \Big(\rho_1\, g \left(\alpha I_y^{\theta_1} - \beta I_x^{\theta_1} + Z_1 I^{\theta_1}\right) + (\rho_2 - \rho_1)\, g \left(\alpha I_y^{\theta_2} - \beta I_x^{\theta_2} + Z_1 I^{\theta_2}\right)\Big)\boldsymbol{e}_3\,. \tag{34}$$

2.4 Rotational Dynamics of the Floating Body

The rotational motion of a rigid body is governed by an equation where the body's angular acceleration times its moments of inertia equal the resultant of the moment of forces applied. In our case, the moments acting on the body arise only from the fluid pressure because the body's centre of mass is assumed to coincide with its centre of rotation. Therefore,

$$I\boldsymbol{\theta}_{tt} = \sum_{j=1}^{2} \int_{\sigma_j} (\boldsymbol{x} - \boldsymbol{X}) \times P^{(j)}\, \boldsymbol{n}\, \mathrm{d}s\,, \tag{35}$$

with I being the inertia tensor defined through

$$I = \begin{bmatrix} I_{yy}^B + I_{zz}^B & -I_{xy}^B & -I_{xz}^B \\ -I_{yx}^B & I_{xx}^B + I_{zz}^B & -I_{yz}^B \\ -I_{zx}^B & -I_{zy}^B & I_{xx}^B + I_{yy}^B \end{bmatrix}$$

and $\boldsymbol{\theta}_{tt}$ the angular acceleration. In the inertia tensor matrix we have

$$I_{xx}^B = \int_B (x - X_0)^2 \,\mathrm{dm}\,, \qquad I_{xy}^B = \int_B (x - X_0)(y - Y_0)\,\mathrm{dm}\,,$$

with $\mathrm{dm} = \rho_B(\boldsymbol{x})\mathrm{d}x\mathrm{d}y\mathrm{d}z$ denoting the mass element of the body, and $\rho_B(\boldsymbol{x})$ its density distribution, similarly for I_{yy}^B, I_{zz}^B, I_{xz}^B and I_{yz}^B (see, e.g., Landau and Lifshitz [15], Sect. 32). Note that, by definition, $I_x^B = I_y^B = I_z^B = 0$.

Let us derive from (35) the equations of equilibrium for the body at rest (zeroth order equation in ϵ) and the equations of motion for small-amplitude motions (first order in ϵ). Using expansions (14), (23) and (24), the zeroth order equation becomes

$$0 = -\rho_1 \, g \int_{\Sigma_1} (\boldsymbol{x} - \boldsymbol{X}_0) \times \boldsymbol{n} f_0 \, \mathrm{d}s - g \int_{\Sigma_2} (\boldsymbol{x} - \boldsymbol{X}_0) \times \boldsymbol{n} \left(\rho_2 \, (f_0 + h) - \rho_1 \, h\right) \mathrm{d}s \,,$$

where $\boldsymbol{n} = \big(-(f_0)_x, -(f_0)_y, 1\big)$ is the normal vector to Σ_j pointing into B. For the x-component, it follows

$$\begin{aligned} 0 = & -\rho_1 \, g \int_{\theta_1} (y - Y_0) f_0 \, \mathrm{d}x\mathrm{d}y - \rho_1 \, g \int_{\theta_1} (f_0 - Z_0) f_0 \, (f_0)_y \, \mathrm{d}x\mathrm{d}y \\ & -(\rho_2 - \rho_1) \, g \int_{\theta_2} (y - Y_0) \, (f_0 + h) \, \mathrm{d}x\mathrm{d}y \\ & -(\rho_2 - \rho_1) \, g \int_{\theta_2} (f_0 - Z_0) \, (f_0 + h) \, (f_0)_y \, \mathrm{d}x\mathrm{d}y \,. \end{aligned} \tag{36}$$

Since f_0 vanishes at the boundary of θ_1 and $f_0 + h$ at the boundary of θ_2, the second and the fourth terms integrate to zero. Therefore, defining

$$I_x^{\Theta_j} = \int_{\Theta_j} (x - X_0) \, \mathrm{d}\boldsymbol{x} \,, \quad I_y^{\Theta_j} = \int_{\Theta_j} (y - Y_0) \, \mathrm{d}\boldsymbol{x} \,, \quad j = 1, 2 \,,$$

where the $\Theta_j, j = 1, 2$, denote the submerged parts of the body at rest in each of the layers, and writing $f_0 = -\int_{f_0}^{0} \mathrm{d}z$ and $f_0 + h = -\int_{f_0}^{-h} \mathrm{d}z$, Eq. (36) reads as

$$\rho_1 \, g \, I_y^{\Theta_1} + \rho_2 \, g \, I_y^{\Theta_2} = 0 \,. \tag{37}$$

For the y-component, a similar reasoning results in

$$\rho_1 \, g \, I_x^{\Theta_1} + \rho_2 \, g \, I_x^{\Theta_2} = 0 \,. \tag{38}$$

The zeroth order equation for the z-component is trivially satisfied.

Moving on to the next order in ϵ, we first notice that the left-hand side of (35) can be written as

$$\left[(I_{yy}^B + I_{zz}^B)\alpha_{tt} - I_{xy}^B \beta_{tt} - I_{xz}^B \gamma_{tt}\right] , \tag{39}$$

$$\left[(I_{zz}^B + I_{xx}^B)\beta_{tt} - I_{yz}^B \gamma_{tt} - I_{xy}^B \alpha_{tt}\right] , \tag{40}$$

$$\left[(I_{xx}^B + I_{yy}^B)\gamma_{tt} - I_{xz}^B \alpha_{tt} - I_{yz}^B \beta_{tt}\right] . \tag{41}$$

Now, the right hand side of (35) can be divided into two parts. The first term corresponds to the hydrodynamic torque and is given by

$$-\epsilon \sum_{j=1}^{2} \int_{\Sigma_j} \rho_j \Phi_t^{(j)} \, (\boldsymbol{x} - \boldsymbol{X}_0) \times \boldsymbol{n} \, \mathrm{d}s + O(\epsilon^2) \, .$$

The other terms, representing the buoyancy torque, read as

$$-\rho_1 \, g \int_{\sigma_1} f \, (\boldsymbol{x} - \boldsymbol{X}) \times \boldsymbol{n} \, \mathrm{d}s - g \int_{\sigma_2} \big(\rho_2 \, (f + h) - \rho_1 \, h\big) \, (\boldsymbol{x} - \boldsymbol{X}) \times \boldsymbol{n} \, \mathrm{d}s \, ,$$

where $\boldsymbol{n} = \big(-f_x, -f_y, 1\big)$.

Let us look more closely at the buoyancy torque. Writing out the cross products and recalling that $f = \epsilon f_1 + O(\epsilon^2)$ at the boundary of θ_1 and that $f + h = \epsilon f_1 + O(\epsilon^2)$ at the boundary of θ_2, we see that the terms multiplied by f_x or f_y can be integrated in x or in y and, thus, are $O(\epsilon^2)$. Therefore, taking Eqs. (13) and (14) into account and subtracting the buoyancy torque terms of order zero in ϵ, yields for the x component

$$\begin{aligned} &-\epsilon \, \rho_1 \, g \int_{\theta_1} (f_1 (y - Y_0) - f_0 Y_1) \, \mathrm{d}x\mathrm{d}y \\ &-\epsilon \, (\rho_2 - \rho_1) \, g \int_{\theta_2} (f_1 (y - Y_0) - (f_0 + h) Y_1) \, \mathrm{d}x\mathrm{d}y + O(\epsilon^2) \, . \end{aligned}$$

Introducing now the cross-section second-order moments, defined by

$$I_{xx}^{\theta_j} = \int_{\theta_j} (x - X_0)^2 \, \mathrm{d}x\mathrm{d}y \, , \qquad I_{xy}^{\theta_j} = \int_{\theta_j} (x - X_0)(y - Y_0) \, \mathrm{d}x\mathrm{d}y \, , \quad j = 1, 2,$$

and similarly for $I_{yy}^{\theta_j}$, $I_{zz}^{\theta_j}$, $I_{xz}^{\theta_j}$ and $I_{yz}^{\theta_j}$, substituting f_1 from (16), writing $f_0 = -\int_{f_0}^{0} \mathrm{d}z$ and $f_0 + h = -\int_{f_0}^{-h} \mathrm{d}z$, and collecting the first order terms in ϵ, yields for the x-component

$$\begin{aligned} &(I_{yy}^B + I_{zz}^B)\alpha_{tt} - I_{xy}^B \beta_{tt} - I_{xz}^B \gamma_{tt} \\ &= -\sum_{j=1}^{2} \rho_j \int_{\Sigma_j} \Phi_t^{(j)} \big(-(z - Z_0) n_2 + (y - Y_0) n_3\big) \, \mathrm{d}s \\ &\quad -\rho_1 \, g \, \big[Z_1 \, I_y^{\theta_1} + \alpha I_{yy}^{\theta_1} - \beta I_{xy}^{\theta_1} - \gamma I_x^{\Theta_1} + \alpha I_z^{\Theta_1}\big] \\ &\quad -g \, \big[(\rho_2 - \rho_1) \, \big(Z_1 \, I_y^{\theta_2} + \alpha I_{yy}^{\theta_2} - \beta I_{xy}^{\theta_2}\big) + \rho_2 \, \big(-\gamma I_x^{\Theta_2} + \alpha I_z^{\Theta_2}\big)\big] \, . \end{aligned} \tag{42}$$

The corresponding equations for the y- and z-component read as

$$
\begin{aligned}
&(I^B_{zz} + I^B_{xx})\beta_{tt} - I^B_{yz}\gamma_{tt} - I^B_{xy}\alpha_{tt} \\
&= -\sum_{j=1}^{2} \rho_j \int_{\Sigma_j} \Phi_t^{(j)} \big((z - Z_0)n_1 - (x - X_0)n_3\big)\, \mathrm{d}s \\
&+\rho_1\, g\left[Z_1\, I_x^{\theta_1} + \alpha I_{xy}^{\theta_1} - \beta I_{xx}^{\theta_1} + \gamma I_y^{\Theta_1} - \beta I_z^{\Theta_1}\right] \\
&+g\left[(\rho_2 - \rho_1)\left(Z_1\, I_x^{\theta_2} + \alpha I_{xy}^{\theta_2} - \beta I_{xx}^{\theta_2}\right) + \rho_2\left(\gamma I_y^{\Theta_2} - \beta I_z^{\Theta_2}\right)\right] ,
\end{aligned}
\tag{43}
$$

and

$$
\begin{aligned}
&(I^B_{xx} + I^B_{yy})\gamma_{tt} - I^B_{xz}\alpha_{tt} - I^B_{yz}\beta_{tt} \\
&= -\sum_{j=1}^{2} \rho_j \int_{\Sigma_j} \Phi_t^{(j)} \big((y - Y_0)n_1 - (x - X_0)n_2\big)\, \mathrm{d}s\, .
\end{aligned}
\tag{44}
$$

2.5 Equations of Motion in Matrix Form

We can now summarise the linear system of dynamic equations for floating bodies

$$
M\boldsymbol{a}_{tt} = -\sum_{j=1}^{2} \rho_j \int_{\Sigma_j} \Phi_t^{(j)}\, D(\boldsymbol{x} - \boldsymbol{X}_0)^T \boldsymbol{n}\, \mathrm{d}s - gK\boldsymbol{a}\, , \tag{45}
$$

where $\boldsymbol{n} \in \mathbb{R}^3$ is the unit normal vector to Σ_j pointing into B and $\boldsymbol{a} \in \mathbb{R}^6$ is the generalised vector, defined in (19), describing the translational displacements of the centre of mass of the body (components a_j, $j = 1, 2, 4$) as well as the angular displacements about the axis through the centre of mass $\boldsymbol{X}_0$ (components a_j, $j = 3, 5, 6$). Moreover, $D \in \mathbb{R}^{3\times 6}$ is the matrix of rigid body displacements defined in (20) and $M \in \mathbb{R}^{6\times 6}$ is the mass matrix given by

$$
M = \int_B D(\boldsymbol{x} - \boldsymbol{X}_0)^T D(\boldsymbol{x} - \boldsymbol{X}_0)\, \rho_B(\boldsymbol{x})\, \mathrm{d}x\mathrm{d}y\mathrm{d}z\, .
$$

The mass matrix is a Gram matrix, thus, in particular, symmetric and positive definite. Recalling that all first-order moments of inertia of B vanish in view of the definition of $\boldsymbol{X}_0$, we obtain

$$M = \begin{bmatrix} I^B & 0 & 0 & 0 & 0 & 0 \\ 0 & I^B & 0 & 0 & 0 & 0 \\ 0 & 0 & I^B_{xx} + I^B_{yy} & 0 & -I^B_{xz} & -I^B_{yz} \\ 0 & 0 & 0 & I^B & 0 & 0 \\ 0 & 0 & -I^B_{xz} & 0 & I^B_{yy} + I^B_{zz} & -I^B_{xy} \\ 0 & 0 & -I^B_{yz} & 0 & -I^B_{xy} & I^B_{xx} + I^B_{zz} \end{bmatrix} . \tag{46}$$

The buoyancy matrix $K \in \mathbb{R}^{6\times 6}$ can be defined block-wise as

$$K = \begin{bmatrix} \mathbb{O}_3 & \mathbb{O}_3 \\ \mathbb{O}_3 & K' \end{bmatrix} , \tag{47}$$

with $\mathbb{O}_3$ denoting the 3×3 null matrix and $K' \in \mathbb{R}^{3\times 3}$ a matrix of the form

$$\begin{aligned} K' &= K^{\theta} + K^{\Theta} , \\ K^{\theta} &= \rho_1 \int_{\theta_1} d(\boldsymbol{x} - \boldsymbol{X}_0)^T d(\boldsymbol{x} - \boldsymbol{X}_0)\,\mathrm{d}x\mathrm{d}y \\ &\quad +(\rho_2 - \rho_1) \int_{\theta_2} d(\boldsymbol{x} - \boldsymbol{X}_0)^T d(\boldsymbol{x} - \boldsymbol{X}_0)\,\mathrm{d}x\mathrm{d}y, \\ K^{\Theta} &= \mathrm{diag}\{0, \rho_1 I_z^{\Theta_1} + \rho_2 I_z^{\Theta_2}, \rho_1 I_z^{\Theta_1} + \rho_2 I_z^{\Theta_2}\}, \end{aligned}$$

where $d(\boldsymbol{x}) = (1, y, -x)$ and where we have used the zeroth order balance equations (37) and (38). Note that the part K^{θ} of the buoyancy matrix is symmetric and positive definite as a sum of two 3×3 Gram matrices. Recalling the definitions of the area moments, we can finally write the matrix K' as

$$\begin{bmatrix} \begin{pmatrix} \rho_1 I^{\theta_1} \\ +(\rho_2 - \rho_1) I^{\theta_2} \end{pmatrix} & \begin{pmatrix} \rho_1 I_y^{\theta_1} \\ +(\rho_2 - \rho_1) I_y^{\theta_2} \end{pmatrix} & \begin{pmatrix} -\rho_1 I_x^{\theta_1} \\ -(\rho_2 - \rho_1) I_x^{\theta_2} \end{pmatrix} \\ \begin{pmatrix} \rho_1 I_y^{\theta_1} \\ +(\rho_2 - \rho_1) I_y^{\theta_2} \end{pmatrix} & \begin{pmatrix} \rho_1 (I_{yy}^{\theta_1} + I_z^{\Theta_1}) \\ +(\rho_2 - \rho_1) I_{yy}^{\theta_2} + \rho_2 I_z^{\Theta_2} \end{pmatrix} & \begin{pmatrix} -\rho_1 I_{xy}^{\theta_1} \\ -(\rho_2 - \rho_1) I_{xy}^{\theta_2} \end{pmatrix} \\ \begin{pmatrix} -\rho_1 I_x^{\theta_1} \\ -(\rho_2 - \rho_1) I_x^{\theta_2} \end{pmatrix} & \begin{pmatrix} -\rho_1 I_{xy}^{\theta_1} \\ -(\rho_2 - \rho_1) I_{xy}^{\theta_2} \end{pmatrix} & \begin{pmatrix} \rho_1 (I_{xx}^{\theta_1} + I_z^{\Theta_1}) \\ +(\rho_2 - \rho_1) I_{xx}^{\theta_2} + \rho_2 I_z^{\Theta_2} \end{pmatrix} \end{bmatrix} . \tag{48}$$

2.6 Two-Dimensional Motion

The two-dimensional motion, say in the xz-plane is described by the velocity potentials $\Phi^{(1)}(x,z,t)$, in the upper layer, and $\Phi^{(2)}(x,z,t)$, in the lower layer, and by three rigid-body motions. The zeroth order equations can be written as

$$I^B g = \rho_1\, g\, I^{\Theta_1} + \rho_2\, g\, I^{\Theta_2}\,, \qquad \rho_1\, g\, I_x^{\Theta_1} + \rho_2\, g\, I_x^{\Theta_2} = 0\,,$$

where I^B denotes the total mass of the body, and

$$I^{\Theta_j} = \int_{\Theta_j} \mathrm{d}x\mathrm{d}z\,, \quad I^{\theta_j} = \int_{\theta_j} \mathrm{d}x\,, \quad j = 1,2\,.$$

Above, Θ_1 and Θ_2 stand for the submerged parts of the body B (in its rest state) in the upper and lower layer, and θ_1 and θ_2 are the segments of the free surface and interface pierced by the body and the free surface, respectively.

It is easy to see that the first-order equations of motion read as

$$M\boldsymbol{a}_{tt} = -\sum_{j=1}^{2} \rho_j \int_{\Sigma_j} \Phi_t^{(j)}\, D(x - X_0, z - Z_0)^T \boldsymbol{n}\, \mathrm{d}s - gK\boldsymbol{a}\,, \tag{49}$$

where Σ_1 and Σ_2 denote the wetted surfaces of the body in the upper and lower layer, respectively, and $\boldsymbol{n} = (n_1, n_3)$ is the vector normal to Σ_j and pointing into B. The vector $\boldsymbol{a} = (X_1, Z_1, \beta)$ represents the rigid-body motions (two translational displacements and the rotation about the axis passing through the centre of mass and perpendicular to the xz-plane) and $D \in \mathbb{R}^{2\times 3}$ is a matrix defined by

$$D(x,z) = \begin{bmatrix} 1 & 0 & z \\ 0 & 1 & -x \end{bmatrix}. \tag{50}$$

The mass matrix $M \in \mathbb{R}^{3\times 3}$ is given by

$$M = \begin{bmatrix} I^B & 0 & 0 \\ 0 & I^B & 0 \\ 0 & 0 & I_{xx}^B + I_{zz}^B \end{bmatrix},$$

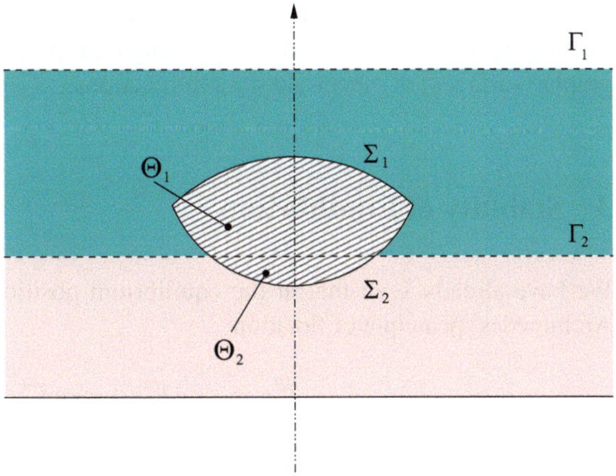

Fig. 2 A totally submerged body

and the buoyancy matrix $K \in \mathbb{R}^{3\times 3}$ takes the form

$$K = \begin{bmatrix} 0 & 0 & 0 \\ 0 & \begin{pmatrix} \rho_1 I^{\theta_1} \\ +(\rho_2 - \rho_1) I^{\theta_2} \end{pmatrix} & \begin{pmatrix} -\rho_1 I_x^{\theta_1} \\ -(\rho_2 - \rho_1) I_x^{\theta_2} \end{pmatrix} \\ 0 & \begin{pmatrix} -\rho_1 I_x^{\theta_1} \\ -(\rho_2 - \rho_1) I_x^{\theta_2} \end{pmatrix} & \begin{pmatrix} \rho_1 (I_{xx}^{\theta_1} + I_z^{\Theta_1}) \\ +(\rho_2 - \rho_1) I_{xx}^{\theta_2} + \rho_2 I_z^{\Theta_2} \end{pmatrix} \end{bmatrix}.$$

2.7 *Equations of Motion for a Totally Submerged Body*

The contour of a totally submerged body can be defined by gluing the graphs of two (or more) functions of the form $f(x, y)$, see Fig. 2 for a simple example. It is straightforward to show that the equations of motions are the same as before except for the part K' of the buoyancy matrix, see (48), which now writes as

$$K' = \begin{bmatrix} (\rho_2 - \rho_1) I^{\theta_2} & (\rho_2 - \rho_1) I_y^{\theta_2} & -(\rho_2 - \rho_1) I_x^{\theta_2} \\ (\rho_2 - \rho_1) I_y^{\theta_2} & \begin{pmatrix} (\rho_2 - \rho_1) I_{yy}^{\theta_2} \\ +\rho_1 I_z^{\Theta_1} + \rho_2 I_z^{\Theta_2} \end{pmatrix} & -(\rho_2 - \rho_1) I_{xy}^{\theta_2} \\ -(\rho_2 - \rho_1) I_x^{\theta_2} & -(\rho_2 - \rho_1) I_{xy}^{\theta_2} & \begin{pmatrix} (\rho_2 - \rho_1) I_{xx}^{\theta_2} \\ +\rho_1 I_z^{\Theta_1} + \rho_2 I_z^{\Theta_2} \end{pmatrix} \end{bmatrix}.$$

At the same time, we end up showing that all previous equations are valid for freely floating bodies defined not only by a single graph $z = f(x, y)$ but by a union of graphs (surfaces) of piecewise smooth functions.

3 Stability of Equilibrium

We have already seen that at the equilibrium position the body must satisfy the Archimedes' principle of flotation

$$I^B g = \rho_1 \, g \, I^{\Theta_1} + \rho_2 \, g \, I^{\Theta_2} \, ,$$

that is, the upward buoyancy force acting on the body equals the weight of the displaced fluid. Defining $W_j = \rho_j \, g \, I^{\Theta_j}$ as the weight of the displaced fluid in layer j and letting

$$\boldsymbol{X}_F^j = (I^{\Theta_j})^{-1} \int_{\Theta_j} \boldsymbol{x} \, \mathrm{d}\boldsymbol{x} \, , \quad j = 1, 2 \, ,$$

be the centres of buoyancy of the parts of the body submerged in the upper layer ($j = 1$) and the lower layer ($j = 2$), we define the centre of buoyancy of the entire submerged part of the body by

$$\boldsymbol{X}_F = \frac{W_1 \boldsymbol{X}_F^1 + W_2 \boldsymbol{X}_F^2}{W_1 + W_2} \, .$$

Thus, the balance equations (37) and (38) can be written as

$$X_F - X_0 = 0 \, , \qquad Y_F - Y_0 = 0 \, . \tag{51}$$

These equations confirm that, in the equilibrium position, the centre of buoyancy of the submerged part of the body must lie on the same vertical line as the centre of mass of the entire body.

Let us now analyse the stability of equilibrium. The kinetic energy of the coupled system is given by

$$T = \sum_{j=1}^{2} \int_{\Omega_j} \rho_j \frac{1}{2} |\nabla \Phi^{(j)}|^2 \, \mathrm{d}\boldsymbol{x} + \frac{1}{2} \boldsymbol{a}_t^T \boldsymbol{M} \boldsymbol{a}_t$$

and its potential energy by

$$V = \frac{1}{2} \rho_1 g \int_{\Gamma_1} \eta^2 \mathrm{d}s + \frac{1}{2} (\rho_2 - \rho_1) g \int_{\Gamma_2} \zeta^2 \mathrm{d}s + \frac{1}{2} g \, \boldsymbol{a}^T \boldsymbol{K} \boldsymbol{a} \, ,$$

where $\eta(x, y, t)$ and $\zeta(x, y, t)$ describe the vertical positions of the free surface and the interface. The integrals over the free surface Γ_1 and the interface Γ_2 correspond to the potential energy of the fluid and the term defined by the buoyancy matrix K to the potential energy of the body.

The stability analysis consists in studying the properties of the Hessian of the potential energy in the equilibrium configuration of the system. Considering the potential energy as a function of η, ζ and $\boldsymbol{a}$, we can guarantee stability if the Hessian is positive semi-definite. The Hessian matrix is an 8×8 block-diagonal matrix H such that

$$\det H = \det H_{\text{fluid}} \det K = \det H_{\text{fluid}} \det \mathbb{O}_3 \det K' = 0\,,$$

where the Hessian of the fluid parcel is given by

$$H_{\text{fluid}} = g\,\text{diag}\left\{\rho_1 \text{meas}(\Gamma_1),\ (\rho_2 - \rho_1)\text{meas}(\Gamma_2)\right\}\,.$$

Hence, if the matrices H_{fluid} and K' are positive definite, the system is stable. The first condition requires that $\rho_1 > 0$ and $\rho_2 > \rho_1$, which translates to gravitational stability. The second condition corresponds to fluctuation stability. Recall that the first three components of the vector $\boldsymbol{a}$ (horizontal translations and rotation about the vertical axis) do not influence buoyancy nor potential energy.

Recalling the form of the floating matrix, cf. (47) and (48), we readily obtain

$$\begin{aligned}
\frac{1}{2} g\,\boldsymbol{a}^T K \boldsymbol{a} &= \frac{1}{2} g \left[Z_1\ \alpha\ \beta\right] K' \left[Z_1\ \alpha\ \beta\right]^T \\
&= \rho_1 \left(\alpha^2 I_{yy}^{\theta_1} - 2\alpha\beta I_{xy}^{\theta_1} + \beta^2 I_{xx}^{\theta_1}\right) + (\rho_2 - \rho_1)\left(\alpha^2 I_{yy}^{\theta_2} - 2\alpha\beta I_{xy}^{\theta_2} + \beta^2 I_{xx}^{\theta_2}\right) \\
&\quad + (\alpha^2 + \beta^2)\left(\rho_1 I_z^{\Theta_1} + \rho_2 I_z^{\Theta_2}\right) \\
&\quad + \rho_1 I^{\theta_1}\left(Z_1 - \frac{\beta I_x^{\theta_1} - \alpha I_y^{\theta_1}}{I^{\theta_1}}\right)^2 + (\rho_2 - \rho_1) I^{\theta_2}\left(Z_1 - \frac{\beta I_x^{\theta_2} - \alpha I_y^{\theta_2}}{I^{\theta_2}}\right)^2 \\
&\quad - \rho_1 \frac{\left(\alpha I_y^{\theta_1} - \beta I_x^{\theta_1}\right)^2}{I^{\theta_1}} - (\rho_2 - \rho_1)\frac{\left(\alpha I_y^{\theta_2} - \beta I_x^{\theta_2}\right)^2}{I^{\theta_2}}\,.
\end{aligned}$$

Note that the matrix K' becomes singular if the body crosses neither the free surface nor the interface (see Eq. (48)). Hence, assuming that at least one of the surfaces is pierced, the matrix K' is positive definite if

$$\begin{aligned}
&\rho_1 \left(\alpha^2 I_{yy}^{\theta_1} - 2\alpha\beta I_{xy}^{\theta_1} + \beta^2 I_{xx}^{\theta_1}\right) + (\rho_2 - \rho_1)\left(\alpha^2 I_{yy}^{\theta_2} - 2\alpha\beta I_{xy}^{\theta_2} + \beta^2 I_{xx}^{\theta_2}\right) \\
&+ (\alpha^2 + \beta^2)\left(\rho_1 I_z^{\Theta_1} + \rho_2 I_z^{\Theta_2}\right) \\
&- \rho_1 \frac{\left(\alpha I_y^{\theta_1} - \beta I_x^{\theta_1}\right)^2}{I^{\theta_1}} - (\rho_2 - \rho_1)\frac{\left(\alpha I_y^{\theta_2} - \beta I_x^{\theta_2}\right)^2}{I^{\theta_2}} > 0\,,
\end{aligned} \qquad (52)$$

for all $(\alpha, \beta) \in \mathbb{R}^2 \backslash \{(0,0)\}$, then the matrix K'. Now, let us define

$$\left(X_A^j, Y_A^j\right) = (I^{\theta_j})^{-1} \left(\int_{\theta_j} x \,\mathrm{d}x\mathrm{d}y, \int_{\theta_j} y \,\mathrm{d}x\mathrm{d}y \right), \quad j = 1, 2,$$

and introduce second order moments of inertia of θ_j, with respect to their area centres, by

$$\tilde{I}_{xx}^{\theta_j} = \int_{\theta_j} (x - X_A^j)^2 \mathrm{d}x\mathrm{d}y\,, \quad \tilde{I}_{yy}^{\theta_j} = \int_{\theta_j} (y - Y_A^j)^2 \mathrm{d}x\mathrm{d}y\,, \quad \tilde{I}_{xy}^{\theta_j} = \int_{\theta_j} (x - X_A^j)(y - Y_A^j)\, \mathrm{d}x\mathrm{d}y\,.$$

Given that

$$I_{xx}^{\theta_j} - \frac{\left(I_x^{\theta_j}\right)^2}{I^{\theta_j}} = \tilde{I}_{xx}^{\theta_j}\,, \qquad I_{yy}^{\theta_j} - \frac{\left(I_y^{\theta_j}\right)^2}{I^{\theta_j}} = \tilde{I}_{yy}^{\theta_j}\,, \qquad I_{xy}^{\theta_j} - \frac{I_x^{\theta_j} I_y^{\theta_j}}{I^{\theta_j}} = \tilde{I}_{xy}^{\theta_j},$$

condition (52) can be written as

$$\begin{aligned}
&\rho_1 \left(\alpha^2 \tilde{I}_{yy}^{\theta_1} + \beta^2 \tilde{I}_{xx}^{\theta_1} - 2\alpha\beta \tilde{I}_{xy}^{\theta_1}\right) + (\rho_2 - \rho_1)\left(\alpha^2 \tilde{I}_{yy}^{\theta_2} + \beta^2 \tilde{I}_{xx}^{\theta_2} - 2\alpha\beta \tilde{I}_{xy}^{\theta_2}\right) \\
&\quad + (\alpha^2 + \beta^2)\left(\rho_1 I_z^{\Theta_1} + \rho_2 I_z^{\Theta_2}\right) > 0\,.
\end{aligned}$$

This inequality is valid if and only if it holds for $(\alpha, \beta) = (\cos\zeta, \sin\zeta)$ $\forall \zeta \in [0, \pi[$, i.e.

$$\begin{aligned}
&\rho_1 \left(\cos^2\zeta\, \tilde{I}_{yy}^{\theta_1} + \sin^2\zeta\, \tilde{I}_{xx}^{\theta_1} - \sin(2\zeta)\, \tilde{I}_{xy}^{\theta_1}\right) \\
&+ (\rho_2 - \rho_1)\left(\cos^2\zeta\, \tilde{I}_{yy}^{\theta_2} + \sin^2\zeta\, \tilde{I}_{xx}^{\theta_2} - \sin(2\zeta)\, \tilde{I}_{xy}^{\theta_2}\right) \\
&+ \rho_1 I_z^{\Theta_1} + \rho_2 I_z^{\Theta_2} > 0 \quad \forall \zeta \in [0, \pi[\,.
\end{aligned}$$

Defining a second moment of inertia with respect to a horizontal line rotated counterclockwise by an angle ζ relative to the x-axis and passing through the area centre of θ_j by

$$\tilde{I}_{\zeta\zeta}^{\theta_j} = \int_{\theta_j} \left(-(x - X_A^j)\sin\zeta + (y - Y_A^j)\cos\zeta\right)^2 \mathrm{d}x\mathrm{d}y\,,$$

the sufficient condition for the matrix K' to be positive definite becomes

$$\rho_1 \tilde{I}_{\zeta\zeta}^{\theta_1} + (\rho_2 - \rho_1)\tilde{I}_{\zeta\zeta}^{\theta_2} + \rho_1 I_z^{\Theta_1} + \rho_2 I_z^{\Theta_2} > 0\,, \tag{53}$$

or, equivalently,

$$\rho_1\left(\tilde{I}^{\theta_1}_{\zeta\zeta}-\tilde{I}^{\theta_2}_{\zeta\zeta}+I^{\Theta_1}(Z^1_F-Z_0)\right)+\rho_2\left(\tilde{I}^{\theta_2}_{\zeta\zeta}+I^{\Theta_2}(Z^2_F-Z_0)\right)>0\quad\forall\zeta\in[0,2\pi[\ .$$

Finally, in view of Archimedes' principle, the stability condition (53) takes the form

$$Z_0-Z_F<\frac{\rho_1\,\tilde{I}^{\theta_1}_{\zeta\zeta}+(\rho_2-\rho_1)\,\tilde{I}^{\theta_2}_{\zeta\zeta}}{I^B}\qquad\forall\zeta\in[0,\pi[\ .\tag{54}$$

This condition generalises the classical condition of stability of equilibrium (see Euler [5], John [8]) to two-layer fluids. It guarantees in particular that if the centre of buoyancy of the body is below its centre of mass, i.e. $Z_0>Z_F$, then the configuration is stable if the distance between the two centres is small enough or one of the area moments $\tilde{I}^{\theta_j}_{\zeta\zeta}$ large enough.

In the two-dimensional case, the sufficient condition (53) simplifies to

$$\rho_1\,\tilde{I}^{\theta_1}_{xx}+(\rho_2-\rho_1)\,\tilde{I}^{\theta_2}_{xx}+\rho_1\,I^{\Theta_1}_z+\rho_2\,I^{\Theta_2}_z>0\ .\tag{55}$$

4 Stratified Fluid

The generalization to the multilayer $n>2$ case is quite straightforward. The Archimedes' principle of flotation becomes

$$I^B g=\sum_{j=1}^{n}\rho_j\,g\,I^{\Theta_j}\ .$$

Defining $W_j=\rho_j\,g\,I^{\Theta_j}$ as the weight of the displaced fluid in layer j and letting

$$\boldsymbol{X}^j_F=(I^{\Theta_j})^{-1}\int_{\Theta_j}\boldsymbol{x}\,\mathrm{d}\boldsymbol{x}\ ,\quad j=1,\ldots,n\,,$$

be the centre of buoyancy of the part of the body submerged in the jth layer, we define the coordinates of the centre of buoyancy of the entire body by the weighted average of these centres,

$$\boldsymbol{X}_F=\frac{\sum_{j=1}^{n}W_j\boldsymbol{X}^j_F}{\sum_{j=1}^{n}W_j}\ .$$

The balance equations for the x- and y-component are now

$$\sum_{j=1}^{n} \rho_j \, g \, I_y^{\Theta_j} = 0 \quad \text{and} \quad \sum_{j=1}^{n} \rho_j \, g \, I_x^{\Theta_j} = 0 \, .$$

Using the definition of the centre of buoyancy of the entire body, the above balance equations reduce to

$$X_F - X_0 = 0 \quad \text{and} \quad Y_F - Y_0 = 0 \, .$$

The linear system of dynamic equations for a body floating in n layers is

$$\boldsymbol{M}\boldsymbol{a}_{tt} = -\sum_{j=1}^{n} \rho_j \int_{\Sigma_j} \Phi_t^{(j)} \, D(\boldsymbol{x} - \boldsymbol{X}_0)^T \boldsymbol{n} \, \mathrm{d}s - gK\boldsymbol{a} \, ,$$

where the only difference with respect to the case $n = 2$ case lies in the new matrix $K' = K^{\theta} + K^{\Theta}$, for which we define

$$K^{\theta} = \sum_{j=1}^{n} (\rho_j - \rho_{j-1}) \int_{\theta_j} d(\boldsymbol{x} - \boldsymbol{X}_0)^T d(\boldsymbol{x} - \boldsymbol{X}_0) \, \mathrm{d}x\mathrm{d}y \, ,$$

$$K^{\Theta} = \mathrm{diag}\{0, \sum_{j=1}^{n} \rho_j I_z^{\Theta_j}, \sum_{j=1}^{n} \rho_j I_z^{\Theta_j}\} \, ,$$

and where $\rho_0 = 0$, which means that the air above the free surface exerts negligible pressure on the fluid layers and the floating body below.

Concerning the stability of the equilibrium, matrix K' is positive definite if

$$\sum_{j=1}^{n} (\rho_j - \rho_{j-1}) \, \tilde{I}_{\zeta\zeta}^{\theta_j} + \sum_{j=1}^{n} \rho_j \, I_z^{\Theta_j} > 0 \, ,$$

with the obvious definitions. Equivalently,

$$\sum_{j=1}^{n} (\rho_j - \rho_{j-1}) \, \tilde{I}_{\zeta\zeta}^{\theta_j} + \sum_{j=1}^{n} \rho_j \, I^{\Theta_j} (Z_F^j - Z_0) > 0 \quad \forall \zeta \in [0, \pi[\, ,$$

or, in view of Archimedes' principle,

$$Z_0 - Z_F < \frac{\sum_{j=1}^{n} (\rho_j - \rho_{j-1}) \, \tilde{I}_{\zeta\zeta}^{\theta_j}}{I^B} \qquad \forall \zeta \in [0, \pi[\, .$$

5 Time-Harmonic Motion

Assuming that the motion of the coupled system is time harmonic with frequency ω, we may express the (first order) velocity potentials $\Phi^{(j)}$, $j = 1, 2$, and the displacement vector $\boldsymbol{a}$ as

$$(\Phi^{(j)}(x, y, z, t), \boldsymbol{a}) = \mathrm{Re}\left(\mathrm{e}^{-\mathrm{i}\omega t}(\varphi^{(j)}(x, y, z), \boldsymbol{\alpha})\right) , \quad j = 1, 2$$

and write Laplace equations (7)–(8), kinematic/dynamic boundary condition on the free surface (9), linearised transmission conditions on the interface (10)–(11), Neumann boundary conditions on the bottom (12), kinematic boundary conditions (21) and dynamic boundary conditions (45) on the surface of the floating body as the following spectral boundary-value problem for the eigenpair $\left((\varphi^{(1)}, \varphi^{(2)}, \boldsymbol{\alpha}), \omega\right)$

$$\rho_j \, \Delta\varphi^{(j)} = 0 \quad \text{in} \quad \Omega_j , \quad j = 1, 2 , \tag{56}$$

$$\varphi_z^{(1)} = g^{-1}\omega^2\varphi^{(1)} \quad \text{on} \quad \Gamma_1 , \tag{57}$$

$$\rho_1\left(\varphi_z^{(1)} - g^{-1}\omega^2\varphi^{(1)}\right) = \rho_2\left(\varphi_z^{(2)} - g^{-1}\omega^2\varphi^{(2)}\right) \quad \text{and} \quad \varphi_z^{(1)} = \varphi_z^{(2)} \quad \text{on} \quad \Gamma_2 , \tag{58}$$

$$\varphi_n^{(2)} = 0 \quad \text{on} \quad \Gamma_b , \tag{59}$$

$$\varphi_n^{(j)} = -\mathrm{i}\omega \boldsymbol{n}^T D(\boldsymbol{x} - \boldsymbol{X}_0)\boldsymbol{\alpha} \quad \text{on} \quad \Sigma_j , \quad j = 1, 2 , \tag{60}$$

$$\omega^2 M\boldsymbol{\alpha} = -\mathrm{i}\omega \sum_{j=1}^{2} \rho_j \int_{\Sigma_j} \varphi^{(j)} D(\boldsymbol{x} - \boldsymbol{X}_0)^T \boldsymbol{n} \, \mathrm{d}s + gK\boldsymbol{\alpha} , \tag{61}$$

with the composite eigenvector $(\varphi^{(1)}, \varphi^{(2)}, \boldsymbol{\alpha})$ consisting of two scalar functions $\varphi^{(1)}$ and $\varphi^{(2)}$, and a number vector $\boldsymbol{\alpha}$.

Problem (56)–(61) should be complemented with suitable radiation conditions at infinity which we omit here since we do not wish to limit neither the fluid domain nor the type of solutions satisfying the equations (see Kuznetsov et al. [13] for a discussion on radiation conditions). However, since we are ultimately interested in studying solutions with finite energy, all results that follow are based upon the assumption that problem (56)–(61) admits a solution $(\varphi^{(1)}, \varphi^{(2)}, \boldsymbol{\alpha}) \in H^1(\Omega_1) \times H^1(\Omega_2) \times \mathbb{C}^6$.

Remark 2 Equations (56)–(61) have been written for a single freely floating body although it is easy to consider the multi-body case. In fact, one only needs to couple the fluid motion (in both layers) with every floating structure via boundary conditions of the form (60) and consider, for each body, an equation of motion such as (61). Essentially, this just increases the dimension of the algebraic part of the system of Eqs. (56)–(61).

It is now possible to rewrite Eqs. (56)–(61) in a dimensionless form. For that we have to redefine coordinates and unknowns through the following transformations:

$$\tilde{\boldsymbol{x}} = \frac{\boldsymbol{x}}{d}, \quad \tilde{\varphi}(\tilde{\boldsymbol{x}}) = \frac{\varphi(\boldsymbol{x})}{\sqrt{d^3 g}}, \qquad \tilde{\boldsymbol{\alpha}} = G\frac{\boldsymbol{\alpha}}{d}, \tag{62}$$

using for characteristic length the depth of the lower fluid layer $d = h_b - h$, and defining the matrix $G = \operatorname{diag}\{1, 1, d, 1, d, d\}$, cf. [12]. In this regard, the domains are transformed from their original description to this new metric with respect to d, thus being denoted by $\tilde{\Omega}_j$, $\tilde{\Gamma}_j$, $\tilde{\Gamma}_b$, and $\tilde{\Sigma}_j$. Furthermore, the new nabla operator is given by $\tilde{\nabla} = (\partial_{\tilde{x}}, \partial_{\tilde{y}}, \partial_{\tilde{z}})$; the new normal vector to $\tilde{\Sigma}_j$ is denoted by $\tilde{\boldsymbol{n}}$; the propagation frequency is now $\tilde{\omega} = \omega\sqrt{d/g}$, making the new spectral parameter $\tilde{\omega}^2$. Finally, the quantities represented by matrices in Eqs. (56)–(61) become

$$\tilde{D}_0 = D_0 G^{-1}, \qquad \tilde{M} = \rho_2^{-1} d^{-3} G^{-1} M G^{-1}, \qquad \tilde{K} = \rho_2^{-1} d^{-2} G^{-1} K G^{-1}, \tag{63}$$

where G^{-1} is the inverse of G defined above. Furthermore, we will reparameterize the masses by dividing all densities by ρ_2, thus defining $\rho = \rho_1/\rho_2$ and $\tilde{\rho}_B = \rho_B/\rho_2$. These definitions transform (56)–(61) into the following non-dimensional spectral-value problem:

$$\rho^{2-j}\,\tilde{\Delta}\tilde{\varphi}^{(j)} = 0 \quad \text{in} \quad \tilde{\Omega}_j, \quad j = 1, 2, \tag{64}$$

$$\tilde{\varphi}^{(1)}_{\tilde{z}} = \tilde{\omega}^2 \tilde{\varphi}^{(1)} \quad \text{on} \quad \tilde{\Gamma}_1, \tag{65}$$

$$\rho\big(\tilde{\varphi}^{(1)}_{\tilde{z}} - \tilde{\omega}^2 \tilde{\varphi}^{(1)}\big) = \big(\tilde{\varphi}^{(2)}_{\tilde{z}} - \tilde{\omega}^2 \tilde{\varphi}^{(2)}\big) \quad \text{and} \quad \tilde{\varphi}^{(1)}_{\tilde{z}} = \tilde{\varphi}^{(2)}_{\tilde{z}} \quad \text{on} \quad \tilde{\Gamma}_2, \tag{66}$$

$$\tilde{\varphi}^{(2)}_{\tilde{n}} = 0 \quad \text{on} \quad \tilde{\Gamma}_b, \tag{67}$$

$$\tilde{\varphi}^{(j)}_{\tilde{n}} = -\mathrm{i}\tilde{\omega}\tilde{\boldsymbol{n}}^T \tilde{D}(\tilde{\boldsymbol{x}} - \tilde{\boldsymbol{X}}_0)\tilde{\boldsymbol{\alpha}} \quad \text{on} \quad \tilde{\Sigma}_j, \quad j = 1, 2, \tag{68}$$

$$\tilde{\omega}^2 \tilde{M}\tilde{\boldsymbol{\alpha}} = -\mathrm{i}\tilde{\omega}\sum_{j=1}^{2} \rho^{2-j} \int_{\tilde{\Sigma}_j} \tilde{\varphi}^{(j)} \tilde{D}(\tilde{\boldsymbol{x}} - \tilde{\boldsymbol{X}}_0)^T \tilde{\boldsymbol{n}}\, \mathrm{d}\tilde{s} + \tilde{K}\tilde{\boldsymbol{\alpha}}, \tag{69}$$

where we have set $\rho = \rho_1/\rho_2$. The following analysis is based on the non-dimensional description even though the tildes will not appear anymore.

5.1 *Variational and Operator Formulation*

Let us derive a variational formulation for problem (56)–(61) (see Nazarov and Videman [21, 22] for similar computations). We start by multiplying the Laplace equations by test functions $\psi^{(j)} \in C^\infty(\overline{\Omega_j})$, where $C^\infty(\overline{\Omega_j})$ denotes the set of restritions to Ω_j of functions in $C_c^\infty(\mathbb{R}^3)$. Integrating by parts and using the boundary condition (57) on the free surface, the boundary condition (59) on the bottom, the transmission conditions at the interface (58), and the kinematic boundary condition on the wetted surface of the body, we obtain

$$\rho \int_{\Omega_1} \nabla\varphi^{(1)} \cdot \overline{\nabla\psi^{(1)}}\,\mathrm{d}\boldsymbol{x} = \rho\omega^2 \int_{\Gamma_1} \varphi^{(1)}\,\overline{\psi^{(1)}}\,\mathrm{d}x\mathrm{d}y - \mathrm{i}\omega\rho \int_{\Sigma_1} \boldsymbol{n}^T D(\boldsymbol{x}-\boldsymbol{X}_0)\boldsymbol{\alpha}\,\overline{\psi^{(1)}}\,\mathrm{d}s$$
$$- \rho\frac{\omega^2}{(1-\rho)} \int_{\Gamma_2} (\varphi^{(2)} - \rho\varphi^{(1)})\overline{\psi^{(1)}}\,\mathrm{d}x\mathrm{d}y\,,$$
$$\int_{\Omega_2} \nabla\varphi^{(2)} \cdot \overline{\nabla\psi^{(2)}}\,\mathrm{d}\boldsymbol{x} = -\mathrm{i}\omega \int_{\Sigma_2} \boldsymbol{n}^T D(\boldsymbol{x}-\boldsymbol{X}_0)\boldsymbol{\alpha}\,\overline{\psi^{(2)}}\,\mathrm{d}s$$
$$+ \frac{\omega^2}{(1-\rho)} \int_{\Gamma_2} \left(\varphi^{(2)} - \rho\varphi^{(1)}\right)\overline{\psi^{(2)}}\,\mathrm{d}x\mathrm{d}y\,.$$

Summing the previous equations, gives

$$\sum_{j=1}^{2} \rho^{2-j}\left(\nabla\varphi^{(j)}, \nabla\psi^{(j)}\right)_{\Omega_j} + \sum_{j=1}^{2} \mathrm{i}\omega\rho^{2-j}\left(\boldsymbol{n}^T D(\boldsymbol{x}-\boldsymbol{X}_0)\boldsymbol{\alpha}, \psi^{(j)}\right)_{\Sigma_j}$$
$$= \omega^2 \left(\rho\left(\varphi^{(1)}, \psi^{(1)}\right)_{\Gamma_1} + \frac{1}{1-\rho}\left(\varphi^{(2)} - \rho\varphi^{(1)}, \psi^{(2)} - \rho\,\psi^{(1)}\right)_{\Gamma_2}\right) \tag{70}$$

where $(\cdot,\cdot)_{\Omega_j}$, $(\cdot,\cdot)_{\Sigma_j}$ and $(\cdot,\cdot)_{\Gamma_j}$ denote the usual scalar products in $\left[L_2(\Omega_j)\right]^3$, $L_2(\Sigma_j)$ and $L_2(\Gamma_j)$, respectively. On the other hand, taking the (complex) inner product between Eq. (61) and a vector $\boldsymbol{\beta} \in \mathbb{C}^6$, results in

$$(K\boldsymbol{\alpha}, \boldsymbol{\beta})_{\mathbb{C}^6} - \sum_{j=1}^{2} \mathrm{i}\omega\rho^{2-j}\left(\varphi^{(j)}, \boldsymbol{n}^T D(\boldsymbol{x}-\boldsymbol{X}_0)\boldsymbol{\beta}\right)_{\Sigma_j} = \omega^2 (M\boldsymbol{\alpha}, \boldsymbol{\beta})_{\mathbb{C}^6}\,. \tag{71}$$

The variational formulation for problem (56)–(61) thus consists in finding a non-trivial $(\varphi^{(1)}, \varphi^{(2)}, \boldsymbol{\alpha}) \in H^1(\Omega_1) \times H^1(\Omega_2) \times \mathbb{C}^6$ and $\omega \in \mathbb{C}$, such that Eqs. (70) and (71) are satisfied for all $(\psi^1, \psi^2, \boldsymbol{\beta}) \in H^1(\Omega_1) \times H^1(\Omega_2) \times \mathbb{C}^6$.

Definition 1 A non-trivial solution $(\varphi^{(1)}, \varphi^{(2)}, \boldsymbol{\alpha}) \in H^1(\Omega_1) \times H^1(\Omega_2) \times \mathbb{C}^6$ of problem (70)–(71) is called a *trapped mode*; the corresponding value of ω is referred to as a *trapping frequency*.

Choosing $\boldsymbol{\beta} = \boldsymbol{\alpha}$ in (71) and $\psi = \varphi$ in (70), proves the following result.

Proposition 1 (Equipartition of Energy) *Let* $(\varphi^{(1)}, \varphi^{(2)}, \boldsymbol{\alpha}) \in H^1(\Omega_1)\times H^1(\Omega_2)\times \mathbb{C}^6$ *be a solution to problem* (70)–(71)*. Then the following equality holds*

$$\rho\|\nabla\varphi^{(1)}\|^2_{L^2(\Omega_1)} + \|\nabla\varphi^{(2)}\|^2_{L^2(\Omega_2)} + \omega^2\,(\boldsymbol{\alpha}, M\boldsymbol{\alpha})_{\mathbb{C}^6}$$
$$= \omega^2\left(\rho\|\varphi^{(1)}\|^2_{L^2(\Gamma_1)} + \frac{1}{1-\rho}\|\,\varphi^{(2)} - \rho\,\varphi^{(1)}\|^2_{L^2(\Gamma^2)}\right) + (\boldsymbol{\alpha}, K\boldsymbol{\alpha})_{\mathbb{C}^6}\ .$$

where the left-hand side represents the kinetic energy and the right-hand side the potential energy of the coupled system.

Let H be the Hilbert space composed of elements $\varphi = (\varphi^{(1)}, \varphi^{(2)}) \in H^1(\Omega_1) \times H^1(\Omega_2)$ and equipped with the scalar product

$$\langle\varphi, \psi\rangle = \sum_{j=1}^{2}\rho^{2-j}\big(\nabla\varphi^{(j)}, \nabla\psi^{(j)}\big)_{\Omega_j} + \sum_{j=1}^{2}\rho^{2-j}\big(\varphi^{(j)}, \psi^{(j)}\big)_{\Gamma_j}\ .$$

and the associated norm $\|\cdot\| = \langle\cdot,\cdot\rangle^{\frac{1}{2}}$. We introduce linear operators A, T by

$$\langle \boldsymbol{A}\varphi, \psi\rangle = \rho\big(\nabla\varphi^{(1)}, \nabla\psi^{(1)}\big)_{\Omega_1} + \big(\nabla\varphi^{(2)}, \nabla\psi^{(2)}\big)_{\Omega_2} \quad \forall\,\varphi,\,\psi \in H\ ,$$
$$\langle \boldsymbol{T}\varphi, \psi\rangle = \rho\big(\varphi^{(1)}, \psi^{(1)}\big)_{\Gamma_1} + \frac{1}{1-\rho}\big(\varphi^{(2)} - \rho\varphi^{(1)}, \psi^{(2)} - \rho\psi^{(1)}\big)_{\Gamma_2} \quad \forall\,\varphi,\,\psi \in H\ .$$

The operators $\boldsymbol{A} : H \to H$ and $\boldsymbol{T} : H \to H$ are positive, continuous and self-adjoint. We also define the operator $\boldsymbol{S} : H \to \mathbb{C}^6$ through

$$\langle\boldsymbol{\alpha}, \boldsymbol{S}\psi\rangle = \left(\boldsymbol{\alpha}, \rho\int_{\Sigma_1} D(\boldsymbol{x}-\boldsymbol{X}_0)^T\boldsymbol{n}\psi^{(1)}\,\mathrm{d}s + \int_{\Sigma_2} D(\boldsymbol{x}-\boldsymbol{X}_0)^T\boldsymbol{n}\psi^{(2)}\,\mathrm{d}s\right)_{\mathbb{C}^6}\ ,$$

for all $\boldsymbol{\alpha} \in \mathbb{C}^6$, $\psi \in H$. Note that the operator $\boldsymbol{S}$ is compact given that the boundary of the floating body is a compact set and, consequently, $H^1(\Omega_j)$ is compactly embedded into $L^2(\Sigma_j)$.

Problem (70)–(71) can now be written as

$$\langle\boldsymbol{A}\varphi, \psi\rangle + \mathrm{i}\omega\langle\boldsymbol{S}^*\boldsymbol{\alpha}, \psi\rangle = \omega^2\langle\boldsymbol{T}\varphi, \psi\rangle\ , \tag{72}$$

$$(K\boldsymbol{\alpha}, \boldsymbol{\beta})_{\mathbb{C}^6} - \mathrm{i}\omega\langle\boldsymbol{\beta}, \boldsymbol{S}\varphi\rangle = \omega^2(M\boldsymbol{\alpha}, \boldsymbol{\beta})_{\mathbb{C}^6}\ , \tag{73}$$

for all $(\psi, \boldsymbol{\beta}) \in H \times \mathbb{C}^6$, where $\boldsymbol{S}^*$ is the adjoint operator of $\boldsymbol{S}$, defined through

$$\langle\boldsymbol{S}^*\boldsymbol{\alpha}, \psi\rangle = \rho\big(\boldsymbol{n}^T D(\boldsymbol{x}-\boldsymbol{X}_0)\boldsymbol{\alpha}, \psi^{(1)}\big)_{\Sigma_1} + \big(\boldsymbol{n}^T D(\boldsymbol{x}-\boldsymbol{X}_0)\boldsymbol{\alpha}, \psi^{(2)}\big)_{\Sigma_2}\ ,$$

for all $\boldsymbol{\alpha} \in \mathbb{C}^6$, $\psi \in H$.

5.2 Reduction Scheme

Unlike the trapping of water waves by fixed obstacles, the interaction of time-harmonic waves with freely floating objects gives rise to a quadratic operator pencil. Following Nazarov and Videman [22], we present a scheme that reduces the quadratic pencil to a linear spectral problem (linear pencil) for a self-adjoint operator in a Hilbert space. For any $\omega \neq 0$, problem (72)–(73) can formally be reduced to a linear problem by defining $\xi = \omega \boldsymbol{T}^{\frac{1}{2}}\varphi$, $\boldsymbol{\eta} = \omega M\boldsymbol{\alpha}$ and $\mathbf{X} = (\varphi, \xi, \boldsymbol{\alpha}, \boldsymbol{\eta})$. This results in the system

$$\begin{bmatrix} A & 0 & 0 & 0 \\ 0 & \boldsymbol{I} & 0 & 0 \\ 0 & 0 & K & 0 \\ 0 & 0 & 0 & N \end{bmatrix} \mathbf{X} = \omega \begin{bmatrix} 0 & \boldsymbol{T}^{\frac{1}{2}} & -i\boldsymbol{S}^* & 0 \\ \boldsymbol{T}^{\frac{1}{2}} & 0 & 0 & 0 \\ i\boldsymbol{S} & 0 & 0 & \mathbb{I}_6 \\ 0 & 0 & \mathbb{I}_6 & 0 \end{bmatrix} \mathbf{X}, \tag{74}$$

where $\boldsymbol{T}^{\frac{1}{2}}$, the operator square root of $\boldsymbol{T}$, is a continuous self-adjoint operator in H, $N = M^{-1}$ is a symmetric and positive definite matrix, $\boldsymbol{I}$ is the identity operator in H and $\mathbb{I}_6$ is the 6×6 identity matrix. Note that the spectral parameter ω appears only linearly.

Remark 3 If $((\varphi, \xi, \boldsymbol{\alpha}, \boldsymbol{\eta}), \omega)$ is a solution to (74) then $((\varphi, -\xi, -\boldsymbol{\alpha}, \boldsymbol{\eta}), -\omega)$ solves the same problem. It thus suffices to consider positive values of ω.

The matrix on the left-hand side in (74) is necessarily singular because the buoyancy matrix K is singular. To deal with the singular part of the matrix K, we need to rewrite the previous system (74) in an equivalent form where the first three components of $\boldsymbol{\alpha}$ are eliminated, removing the rigid body movements not influenced by buoyancy. To this end, we decompose the vectors $\boldsymbol{\alpha}$ and $\boldsymbol{\eta}$ as

$$\boldsymbol{\alpha} = \begin{pmatrix} \boldsymbol{\alpha}_\circ \\ \boldsymbol{\alpha}_\bullet \end{pmatrix}, \quad \boldsymbol{\eta} = \begin{pmatrix} \boldsymbol{\eta}_\circ \\ \boldsymbol{\eta}_\bullet \end{pmatrix}, \quad \boldsymbol{\alpha}_\circ, \boldsymbol{\alpha}_\bullet, \boldsymbol{\eta}_\circ, \boldsymbol{\eta}_\bullet \in \mathbb{C}^3,$$

where $\boldsymbol{\alpha}_\circ = (\alpha_1, \alpha_2, \alpha_3)$, $\boldsymbol{\alpha}_\bullet = (\alpha_4, \alpha_5, \alpha_6)$, similarly for $\boldsymbol{\eta}_\circ, \boldsymbol{\eta}_\bullet$, and write the matrix $N = M^{-1}$ block-wise as

$$N = \begin{bmatrix} N_{\circ\circ} & N_{\circ\bullet} \\ N_{\bullet\circ} & N_{\bullet\bullet} \end{bmatrix},$$

where all blocks are 3×3 matrices. Moreover, we introduce the operators $\boldsymbol{S}_\circ : H \to \mathbb{C}^3$ and $\boldsymbol{S}_\bullet : H \to \mathbb{C}^3$ as compositions of projections from $\mathbb{C}^6$ into $\mathbb{C}^3$ and $\boldsymbol{S}$, that is, $\boldsymbol{S}_\circ$ ($\boldsymbol{S}_\bullet$) returns the first (last) three components of the image of $\boldsymbol{S}$, respectively. Now, consider system (74) and write the first three lines of the third row as

$$\mathbf{0} = \omega\left(\mathrm{i}\boldsymbol{S}_\circ\varphi + \boldsymbol{\eta}_\circ\right). \tag{75}$$

Decomposing the fourth row of system (74) as

$$N_{\circ\circ}\boldsymbol{\eta}_\circ + N_{\circ\bullet}\boldsymbol{\eta}_\bullet = \omega\boldsymbol{\alpha}_\circ\,,$$
$$N_{\bullet\circ}\boldsymbol{\eta}_\circ + N_{\bullet\bullet}\boldsymbol{\eta}_\bullet = \omega\boldsymbol{\alpha}_\bullet\,,$$

and using (75), we obtain

$$\boldsymbol{S}_\circ^* N_{\circ\circ}\boldsymbol{S}_\circ\varphi + \mathrm{i}\boldsymbol{S}_\circ^* N_{\circ\bullet}\boldsymbol{\eta}_\bullet = \mathrm{i}\omega\boldsymbol{S}_\circ^*\boldsymbol{\alpha}_\circ\,,$$
$$-\mathrm{i}N_{\bullet\circ}\boldsymbol{S}_\circ\varphi + N_{\bullet\bullet}\boldsymbol{\eta}_\bullet = \omega\boldsymbol{\alpha}_\bullet\,.$$

Defining a truncated eigenvector $\mathbf{X}_\bullet = (\varphi, \xi, \boldsymbol{\alpha}_\bullet, \boldsymbol{\eta}_\bullet) \in H \times H \times \mathbb{C}^3 \times \mathbb{C}^3$ and considering the previous equalities, system (74) can be written as

$$\boldsymbol{B}\mathbf{X}_\bullet = \omega\boldsymbol{D}\mathbf{X}_\bullet\,, \tag{76}$$

where the matrix operators $\boldsymbol{B}$ and $\boldsymbol{D}$ take the form

$$\boldsymbol{B} = \begin{bmatrix} A + \boldsymbol{S}_\circ^* N_{\circ\circ}\boldsymbol{S}_\circ & 0 & 0 & \mathrm{i}\boldsymbol{S}_\circ^* N_{\circ\bullet} \\ 0 & \boldsymbol{I} & 0 & 0 \\ 0 & 0 & K' & 0 \\ -\mathrm{i}N_{\bullet\circ}\boldsymbol{S}_\circ & 0 & 0 & N_{\bullet\bullet} \end{bmatrix}, \quad \boldsymbol{D} = \begin{bmatrix} 0 & \boldsymbol{T}^{\frac{1}{2}} & -\mathrm{i}\boldsymbol{S}_\bullet^* & 0 \\ \boldsymbol{T}^{\frac{1}{2}} & 0 & 0 & 0 \\ \mathrm{i}\boldsymbol{S}_\bullet & 0 & 0 & \mathbb{I}_3 \\ 0 & 0 & \mathbb{I}_3 & 0 \end{bmatrix}.$$

The operators $\boldsymbol{B}$ and $\boldsymbol{D}$ are both continuous and self-adjoint, and the operator $\boldsymbol{B}$ is positive because

$$\begin{aligned}\langle \boldsymbol{B}\mathbf{X}_\bullet, \mathbf{X}_\bullet\rangle &= \langle A\varphi, \varphi\rangle + \langle \xi, \xi\rangle + \left(K'\boldsymbol{\alpha}_\bullet, \boldsymbol{\alpha}_\bullet\right)_{\mathbb{C}^3} + \left(N\begin{pmatrix}\mathrm{i}\boldsymbol{S}_\circ\varphi \\ \boldsymbol{\eta}_\bullet\end{pmatrix}, \begin{pmatrix}\mathrm{i}\boldsymbol{S}_\circ\varphi \\ \boldsymbol{\eta}_\bullet\end{pmatrix}\right)_{\mathbb{C}^6} \\ &\geq \langle A\varphi, \varphi\rangle + \|\xi\|_H^2 + \left(K'\boldsymbol{\alpha}_\bullet, \boldsymbol{\alpha}_\bullet\right)_{\mathbb{C}^3} + C_N\left(\|\boldsymbol{S}_\circ\varphi\|_{\mathbb{C}^3}^2 + \|\boldsymbol{\eta}_\bullet\|_{\mathbb{C}^3}^2\right)\,,\end{aligned}$$

where C_N denote some positive constant due to the positive definiteness of N. Recalling that K' is positive definite, we conclude that the operator $\boldsymbol{B}$ is positive.

Having determined a solution $((\varphi, \xi, \boldsymbol{\alpha}_\bullet, \boldsymbol{\eta}_\bullet), \omega)$ to the linear operator pencil (76), we conclude that $((\varphi, \xi, \boldsymbol{\alpha}, \boldsymbol{\eta}), \omega)$ solves problem (74), with

$$\boldsymbol{\alpha} = \begin{pmatrix} \omega^{-1}(-\mathrm{i}N_{\circ\circ}\boldsymbol{S}_\circ\varphi + N_{\circ\bullet}\boldsymbol{\eta}_\bullet) \\ \boldsymbol{\alpha}_\bullet \end{pmatrix}, \quad \boldsymbol{\eta} = \begin{pmatrix} -\mathrm{i}\boldsymbol{S}_\circ\varphi \\ \boldsymbol{\eta}_\bullet \end{pmatrix}.$$

Moreover, $((\varphi, \boldsymbol{\alpha}), \omega)$ is a non-trivial solution to problem (72)–(73).

5.3 Condition for the Existence of Trapped Modes

Assuming now that the coupled motion of the system takes place in an open channel bounded laterally by rigid walls. To this end, we redefine the upper and lower fluid layers as

$$\Pi_1 = \{(x,y,z) \in \mathbb{R}^3 : x \in (-\infty,\infty),\ y \in (-l,l),\ z \in (-h,0)\}$$

$$\Pi_2 = \{(x,y,z) \in \mathbb{R}^3 : x \in (-\infty,\infty),\ y \in (-l,l),\ z \in (-h_b,-h)\}$$

where l denotes half of the (non-dimensionalised) distance between the vertical walls, and h and h_b denote the (non-dimensionalised) depths.

Using the ideas of Evans et al. [6], we assume that the body is symmetric with respect to the centreplane $\{y = 0\}$ and impose symmetry conditions on the fluid motion and rigid body movements. We introduce the following subspace of anti-symmetric functions

$$H_0 = \left\{\varphi \in H : \varphi^{(j)}(x,-y,z) = -\varphi^{(j)}(x,y,z)\,,\ j = 1,2\right\}$$

and restrict the body movements to swaying, rolling and yawing, i.e.,

$$\alpha_1 = \alpha_4 = \alpha_6 = 0\,.$$

In this way we create a new spectral problem whose continuous spectrum is the set $(-\infty,-\omega_\dagger] \cup [\omega_\dagger,\infty)$, where $\omega_\dagger$ is a positive cut-off value that leaves room for a discrete non-empty spectrum belonging to the interval $(-\omega_\dagger,\omega_\dagger)$, cf. [18]. At the same time, the positive operator $\boldsymbol{B}$ in Eq. (76) becomes positive definite and thus, there is a self-adjoint, positive definite operator $\boldsymbol{B}^{\frac{1}{2}}$, which is the positive square root of $\boldsymbol{B}$. Defining $\mathbf{Y} = \boldsymbol{B}^{\frac{1}{2}}\mathbf{X}_\bullet$ and $\mu = 1/\omega$, the spectral problem (76) can be written as

$$\boldsymbol{M}\mathbf{Y} = \mu\mathbf{Y}\,, \tag{77}$$

with the new self-adjoint operator $\boldsymbol{M} = \boldsymbol{B}^{-\frac{1}{2}}\boldsymbol{D}\boldsymbol{B}^{-\frac{1}{2}}$. The continuous spectrum of the operator $\boldsymbol{M}$ is $[-\mu_\dagger,\ 0) \cup (0,\ \mu_\dagger]$, with the obvious identification $\mu_\dagger = 1/\omega_\dagger$. Since $\mu = 0$ is an eigenvalue of infinite multiplicity, it belongs to the essential spectrum $[-\mu_\dagger,\ \mu_\dagger]$ of the operator $\boldsymbol{M}$, but does not influence the spectrum of the original problem. For the discrete spectrum of $\boldsymbol{M}$, there are two possibilities: either the norm of the operator concides with $\mu_\dagger$, so that the discrete spectrum is empty, or the norm is greater then $\mu_\dagger$, and the discrete spectrum of $\boldsymbol{M}$ is non-empty since the norm belongs to its spectrum. Hence, if

$$\|\boldsymbol{M}\| = \sup_{\mathbf{Y}\neq 0} \frac{|\langle \boldsymbol{M}\mathbf{Y},\ \mathbf{Y}\rangle|}{\langle \mathbf{Y},\ \mathbf{Y}\rangle} > \mu_\dagger \tag{78}$$

then the discrete spectrum of the operator $\boldsymbol{M}$ contains at least one eigenvalue $\mu > \mu_\dagger$. By the definition of the square root of $\boldsymbol{B}$, inequality (78) can be written as

$$\sup_{\mathbf{X}_\bullet \neq 0} \frac{|\langle \boldsymbol{D}\mathbf{X}_\bullet, \mathbf{X}_\bullet \rangle|}{\langle \boldsymbol{B}\mathbf{X}_\bullet, \mathbf{X}_\bullet \rangle} > \frac{1}{\omega_\dagger}. \tag{79}$$

This is a sufficient condition for the existence of a trapped mode. Since, in general, there is no hope to calculate the norm of the operator $\boldsymbol{M}$, we will rewrite the condition choosing a particular test function.

Consider a function $\varphi_\epsilon \in H_0$ defined by $\varphi_\epsilon(\boldsymbol{x}) = \mathrm{e}^{-\epsilon|x|}\phi_\dagger(y,z)$, where $\phi_\dagger = (\phi_\dagger^{(1)}, \phi_\dagger^{(2)})$ is the non-trivial solution of the problem in the absence of bodies corresponding to the cutoff value $\lambda_\dagger = \omega_\dagger^2$, defined by (see Cal et al. [4])

$$\phi_\dagger^{(1)}(y,z) = \sin\left(\frac{\pi}{2l}y\right)\left(\mathrm{e}^{\frac{\pi}{2l}(z+h)} + \frac{\frac{\pi}{2l} - \lambda_\dagger}{\frac{\pi}{2l} + \lambda_\dagger}\,\mathrm{e}^{\frac{\pi}{l}h}\,\mathrm{e}^{-\frac{\pi}{2l}(z+h)}\right),$$

$$\phi_\dagger^{(2)}(y,z) = \sin\left(\frac{\pi}{2l}y\right)\left(1 - \frac{\frac{\pi}{2l} - \lambda_\dagger}{\frac{\pi}{2l} + \lambda_\dagger}\,\mathrm{e}^{\frac{\pi}{l}h}\right)\operatorname{csch}\left(\frac{\pi}{2l}\right)\cosh\left(\frac{\pi}{2}(z + h_b)\right) \tag{80}$$

and $\epsilon \ll 1$ is a small positive parameter. Defining the following test function

$$\mathbf{X}_\epsilon = (\varphi_\epsilon, \xi_\epsilon, \mathbf{0}, \mathbf{0}), \quad \xi_\epsilon = \omega_\dagger \boldsymbol{T}^{\frac{1}{2}}\varphi_\epsilon,$$

we obtain

$$\langle \boldsymbol{B}\mathbf{X}_\epsilon, \mathbf{X}_\epsilon \rangle = \langle \varphi_\epsilon, \varphi_\epsilon \rangle + \langle \xi_\epsilon, \xi_\epsilon \rangle + \left(N_{\circ\circ}\boldsymbol{S}_\circ\phi_\dagger, \boldsymbol{S}_\circ\phi_\dagger\right)_{\mathbb{C}^3}.$$

Computing

$$\langle \boldsymbol{D}\mathbf{X}_\epsilon, \mathbf{X}_\epsilon \rangle = 2\,\mathrm{Re}\langle \boldsymbol{T}^{\frac{1}{2}}\xi_\epsilon, \varphi_\epsilon \rangle$$

and observing that $\langle \boldsymbol{T}\varphi_\epsilon, \varphi_\epsilon \rangle = O(\epsilon^{-1})$, cf. Cal et al. [1], we see that

$$\langle \boldsymbol{D}\mathbf{X}_\epsilon, \mathbf{X}_\epsilon \rangle = 2\,\omega_\dagger \langle \boldsymbol{T}\varphi_\epsilon, \varphi_\epsilon \rangle$$

is positive for sufficiently small $\epsilon > 0$. Hence, the sufficient condition (79) is satisfied if

$$\omega_\dagger \langle \boldsymbol{D}\mathbf{X}_\epsilon, \mathbf{X}_\epsilon \rangle - \langle \boldsymbol{B}\mathbf{X}_\epsilon, \mathbf{X}_\epsilon \rangle > 0. \tag{81}$$

Recalling the identity (see Cal et al. [1])

$$\rho\big(\nabla\varphi_\epsilon^{(1)},\nabla\varphi_\epsilon^{(1)}\big)_{\Pi_1}+\big(\nabla\varphi_\epsilon^{(2)},\nabla\varphi_\epsilon^{(2)}\big)_{\Pi_2}=\omega_\dagger^2\,\rho\big(\varphi_\epsilon^{(1)},\varphi_\epsilon^{(1)}\big)_{\Upsilon^1}$$
$$+\,\omega_\dagger^2\frac{1}{1-\rho}\big(\varphi_\epsilon^{(2)}-\rho\varphi_\epsilon^{(1)},\rho\varphi_\epsilon^{(2)}-\rho\varphi_\epsilon^{(1)}\big)_{\Upsilon^2}+O(\epsilon)\,,$$

where

$$\Upsilon_1=\big\{(x,y,z)\in\partial\Pi_1\ :z=0\big\}\,,\qquad \Upsilon_2=\big\{(x,y,z)\in\partial\Pi_2\ :z=-h\big\}\,,$$

and noting that $e^{-\epsilon|x|}=1+O(\epsilon)$ in any compact set, yields the following equality

$$\omega_\dagger\,\langle \boldsymbol{D}\mathbf{X}_\epsilon,\mathbf{X}_\epsilon\rangle-\langle \boldsymbol{B}\mathbf{X}_\epsilon,\mathbf{X}_\epsilon\rangle=\rho\big(\nabla\phi_\dagger^{(1)},\nabla\phi_\dagger^{(1)}\big)_{\Theta_1}+\big(\nabla\phi_\dagger^{(2)},\nabla\phi_\dagger^{(2)}\big)_{\Theta_2}$$
$$-\,\omega_\dagger^2\rho\big(\phi_\dagger^{(1)},\phi_\dagger^{(1)}\big)_{\theta_1}-\omega_\dagger^2\frac{1}{1-\rho}\big(\phi_\dagger^{(2)}-\rho\phi_\dagger^{(1)},\phi_\dagger^{(2)}-\rho\phi_\dagger^{(1)}\big)_{\theta_2}$$
$$-\,\big(N_{\circ\circ}\boldsymbol{S}_\circ\phi_\dagger,\,\boldsymbol{S}_\circ\phi_\dagger\big)_{\mathbb{C}^3}+O(\epsilon)\,.$$

Therefore, by choosing a trial function $\mathbf{X}_\epsilon$, with small enough $\epsilon>0$, we have come up with the following sufficient condition for the existence of a trapped mode for problem (72)–(73) (subject to the symmetry assumptions)

$$\rho\big(\nabla\phi_\dagger^{(1)},\nabla\phi_\dagger^{(1)}\big)_{\Theta_1}+\big(\nabla\phi_\dagger^{(2)},\nabla\phi_\dagger^{(2)}\big)_{\Theta_2}-\omega_\dagger^2\rho\big(\phi_\dagger^{(1)},\phi_\dagger^{(1)}\big)_{\theta_1}$$
$$-\,\frac{\omega_\dagger^2}{1-\rho}\big(\phi_\dagger^{(2)}-\rho\phi_\dagger^{(1)},\phi_\dagger^{(2)}-\rho\phi_\dagger^{(1)}\big)_{\theta_2}-\big(N_{\circ\circ}\boldsymbol{S}_\circ\phi_\dagger,\,\boldsymbol{S}_\circ\phi_\dagger\big)_{\mathbb{C}^3}>0\,. \tag{82}$$

We have proved the following statement

Theorem 1 *Assume that the obstacle is symmetric with respect to the centreplane* $\{y=0\}$ *of the channel. If the inequality* (82) *holds, then problem* (70)–(71) *admits a trapped mode* $(\varphi^{(1)},\varphi^{(2)},\boldsymbol{\alpha})\in H^1(\Omega_1)\times H^1(\Omega_2)\times\mathbb{C}^6$ *corresponding to a trapping frequency* $\omega<\omega_\dagger$.

6 Conclusions

One of the main challenges in addressing the problem of wave trapping by freely floating obstacles in its full generality is the singular structure of the floating matrix, recall that the surge, sway and yaw motions are all unaffected by the buoyancy forces. Besides, if the body is totally submerged it can only gain a stable, freely floating position if it is bottom-heavy (and weighs as much as the liquid it displaces), i.e. the centre of buoyancy must lie above the centre of mass. In a multilayer fluid, this situation becomes more interesting, although not necessarily simpler, because of the density differences across the interfaces between the fluid layers. Now also a homogeneous body can be fully submerged and have a stable equilibrium position. Moreover, the structure can interact with both surface and internal (interface) gravity waves.

In this vein, we have recently shown that a totally submerged body not piercing the interface generates trapped modes, see [3]. The result was shown for any stable, that means bottom-heavy, floating cuboid but generalization to other objects is straightforward. Using an algebraic test function, we also proved the existence of trapped modes for objects piercing the free-surface and the interface. Moreover, in [4] we gave examples of periodic arrays of freely-floating structures supporting trapped waves and studied how the problem parameters (density ratio, obstacle dimensions, layer depths and radian frequency) influence the trapping condition.

Acknowledgements We would like to thank Professor S.A. Nazarov for several fruitful discussions and suggestions regarding this work. G.A.S.D. was supported by a FCT (Fundação para a Ciência e a Tecnologia) fellowship SFRH/BPD/70578/2010. J.H.V was partially supported by the projects UTA-CMU/MAT/0007/2009 and PTDC/MAT-CAL/0749/2012.

References

1. Cal, F.S., Dias, G.A.S., Videman, J.H.: Existence of trapped modes along periodic structures in a two-layer fluid. Q. J. Mech. Appl. Math. **65**(2), 273–293 (2012)
2. Cal, F.S., Dias, G.A.S., Nazarov, S.A., Videman, J.H.: Linearised theory for surface and interfacial waves interacting with freely floating bodies in a two-layer fluid. Z. Angew. Math. Phys. **66**(2), 417–432 (2014)
3. Cal, F.S., Dias, G.A.S., Videman, J.H.: Trapped modes around freely floating bodies in a two layer fluid channel. Proc. R. Soc. A **470**, 20140396 (2014)
4. Cal, F., Dias, G., Videman, J.: Trapped waves along freely floating structures in two-layer fluids. Int. J. Numer. Anal. Model. Ser. B **5**(4), 400–413 (2014)
5. Euler, L.: Théorie complète de la construction et de la manoeuvre des vaisseaux mise à la portée des ceux, qui s'appliquent à la navigation. Imperial Academy of Sciences, St.-Petersburg (1773)
6. Evans, D.V., Levitin, M., Vassilev, D.: Existence theorems for trapped modes. J. Fluid Mech. **261**, 21–31 (1994)
7. Fitzgerald, C.J., McIver, P.: Passive trapped modes in the water-wave problem for a floating structure. J. Fluid Mech. **657**, 456–477 (2010) doi:10.1017/S0022112010001503. http://journals.cambridge.org/article_S0022112010001503
8. John, F.: On the motion of floating bodies I. Commun. Pure Appl. Math. **2**(1), 13–57 (1949). doi:10.1002/cpa.3160020102. http://dx.doi.org/10.1002/cpa.3160020102
9. Kundu, P.K., Cohen, I.M., Dowling, D.R.: Fluid Mechanics, 5th edn. Academic Press, Boston (2012)
10. Kuznetsov, N.: On the problem of time-harmonic water waves in the presence of a freely floating structure. Algebra Anal. **226**, 185–199 (2009). doi:http://dx.doi.org/10.1090/S1061-0022-2011-01179-3 [Transl. St.-Petersburg Math. J. **226**, 985–995 (2011)]

11. Kuznetsov, N., Motygin, O.: On the coupled time-harmonic motion of water and a body freely floating in it. J. Fluid Mech. **679**, 616–627 (2011). doi:10.1017/jfm.2011.161. http://journals.cambridge.org/article_S0022112011001613
12. Kuznetsov, N., Motygin, O.: On the coupled time-harmonic motion of deep water and a freely floating body: trapped modes and uniqueness theorems. J. Fluid Mech. **703**, 142–162 (2012). doi:10.1017/jfm.2012.202. URL http://journals.cambridge.org/article_S0022112012002029
13. Kuznetsov, N., Maz'ya, V., Vainberg, B.: Linear Water Waves: A Mathematical Approach. Cambridge University Press, Cambridge (2002)
14. Lamb, H.: Hydrodynamics, 6th edn. Dover, New York (1945)
15. Landau, L.D., Lifshitz, E.M.: Course of Theoretical Physics, 2nd edn., vol. 1. Pergamon Press, New York (1969)
16. McIver, P., McIver, M.: Trapped modes in the water-wave problem for a freely floating structure. J. Fluid Mech. **558**, 53–67 (2006). doi:10.1017/S0022112006009803. http://journals.cambridge.org/article_S0022112006009803
17. Mei, C.C., Stiassnie, M., Yue, D.K.P.: Theory and Applications of Ocean Surface Waves. Part 1: Linear Aspects. World Scientific, Singapore (2005)
18. Nazarov, S.A.: Sufficient conditions for the existence of trapped modes in problems of the linear theory of surface waves. Zap. Nauchn. Sem. S.-Peterburg. Otdel. Mat. Inst. Steklov. **369**, 202–223 (2009) [Transl. J. Math. Sci. **167**(5), 713–725 (2010)]
19. Nazarov, S.A.: Incomplete comparison principle in problems about surface waves trapped by fixed and freely floating bodies. Probl. Mat. Analiz. **56**, 83–114 (2011). doi:10.1007/s10958-011-0349-z. http://dx.doi.org/10.1007/s10958-011-0349-z [Transl. J. Math. Sci. (N.Y.) **175**(3), 309–348 (2011)]
20. Nazarov, S.A.: Concentration of frequencies of trapped waves in problems on freely floating bodies. Mat. Sbornik. **203**(9), 41–66 (2012). doi:10.1070/SM2012v203n09ABEH004264. http://dx.doi.org/10.1070/SM2012v203n09ABEH004264
21. Nazarov, S.A., Videman, J.H.: A sufficient condition for the existence of trapped modes for oblique waves in a two-layer fluid. Proc. R Soc. A **465**(2112), 3799–3816 (2009)
22. Nazarov, S.A., Videman, J.H.: Trapping of water waves by freely floating structures in a channel. Proc. R Soc. A **467**(2136), 3613–3632 (2011). doi:10.1098/rspa.2011.0288
23. Porter, R., Evans, D.V.: Examples of trapped modes in the presence of freely floating structures. J. Fluid Mech. **606**, 189–208 (2008)

[illegible] Bodies [illegible] Water Waves [illegible]

11. Kuznetsov, N., Maz'ya, V.: On the [illegible]
[illegible]
12. [illegible]
13. Kuznetsov, N., [illegible] Cambridge University Press, Cambridge (2002)
14. Lamb, H.: Hydrodynamics [illegible] (1945)
15. [illegible]
16. [illegible]
17. [illegible]
18. [illegible]
19. [illegible]
20. [illegible]
21. [illegible]
22. [illegible]

Shannon Switching Game and Directed Variants

A.P. Cláudio, S. Fonseca, L. Sequeira, and I.P. Silva

Dedicated to the memory of Yahya Ould Hamidoune (1947–2011) and Michel las Vergnas (1941–2013)

Abstract Shannon's switching game is a combinatorial game invented by C. Shannon circa 1955 as a simple model for breakdown repair of the connectivity of a network. The game was completely solved by A. Lehman, shortly after, in what is considered the first application of matroid theory. In the middle 1980s Y. O. Hamidoune and M. Las Vergnas introduced and solved directed versions of the game for graphs considering their generalization to oriented matroids. We do a brief review of the main results and conjectures of the directed case.

1 Introduction

In the middle 1950s C. Shannon invented the following game: given a connected graph G with two distinguished vertices x and y, two players JOIN and CUT choose alternately one unplayed edge. JOIN reinforces the chosen edge that becomes invulnerable and CUT deletes the chosen edge. JOIN wins if he succeeds in making invulnerable the edges of a path connecting x to y. CUT wins otherwise.

Shannon's game, like Hex, invented earlier by Piet Hein are examples of two player games where the objective of one or both players is to connect or keep connected some subset of nodes of a network.

A.P. Cláudio • S. Fonseca, Collaborators of BioISI
FCUL, Campo Grande, Edificio C6, Piso 3, 1749-016 Lisboa, Portugal
e-mail: apc@di.fc.ul.pt; sashaafm@gmail.com

L. Sequeira, Collaborator of CEMAT
FCUL, Campo Grande, Edificio C6, Piso 2, 1749-016 Lisboa, Portugal
e-mail: lfsequeira@fc.ul.pt

I.P. da Silva (✉), Collaborator of CMAF, Member of CFCUL
FCUL, Campo Grande, Edificio C6, Piso 2, 1749-016 Lisboa, Portugal
e-mail: ipsilva@fc.ul.pt

J.-P. Bourguignon et al. (eds.), *Dynamics, Games and Science*, CIM Series in Mathematical Sciences 1, DOI 10.1007/978-3-319-16118-1_10

Mathematically both games are well known combinatorial games where J. Nash's stealing argument applies [2]. In the case of the Shannon switching game Nash's argument guarantees that if one of the players, JOIN or CUT, has a winning strategy playing second then the same strategy applied to any fictitious first move of its opponent, guarantees that he, JOIN, respectively CUT, playing first will win the game too.

The game was completely solved by Lehman in 1964 [15] and is a classical game in combinatorics and its applications [2, 5, 11, 16].

Less known are the results and conjectures concerning the directed versions of Shannon's switching game for graphs and oriented matroids that were introduced and studied by Hamidoune and Las Vergnas in [12, 13]. The detailed presentation and discussion of these directed versions in [12] makes it clear that the directed versions are considerably more difficult to analyse.

In general all these games were inspired and have direct applications in engineering problems and network analysis of electrical networks, optical networks, more generally, communication networks and information theory.

In what follows we recall Lehman's results on Shannon's game, and then briefly review the main results and conjecture of the directed case.

We have implemented a computational prototype of two games: TREE and ARBORESCENCE whose analysis, as the reader will see, is the key for solving respectively, the undirected and the directed, Shannon switching games. In both games the human player plays as JOIN, starts second, and has a winning strategy. However at his first mistake, the computer is expected to take the lead and win. The interested reader may download the games from [9]: http://shlvgraphgame.fc.ul.pt/.

2 Solution of Shannon's Classical Game

2.1 *Shannon's Switching Game* $(G; e)$

Shannon's switching game $(G; e)$ is a two player game played on a connected graph G with a distinguished edge $e = \{x, y\}$, $x \neq y$, not subject to play. Two players, JOIN and CUT, play alternately choosing one unplayed edge of the graph. JOIN reinforces the chosen edge, CUT deletes the chosen edge.

JOIN wins if he succeeds in reinforcing the edges of a path connecting x to y. CUT wins otherwise.

Shannon's switching games $(G; e)$ are combinatorial games that fall into one of the following three classes:

A JOIN GAME, if JOIN playing second has a winning strategy, a CUT GAME, if CUT playing second has a winning strategy or a NEUTRAL GAME, when the first player has a winning strategy (see Fig. 1).

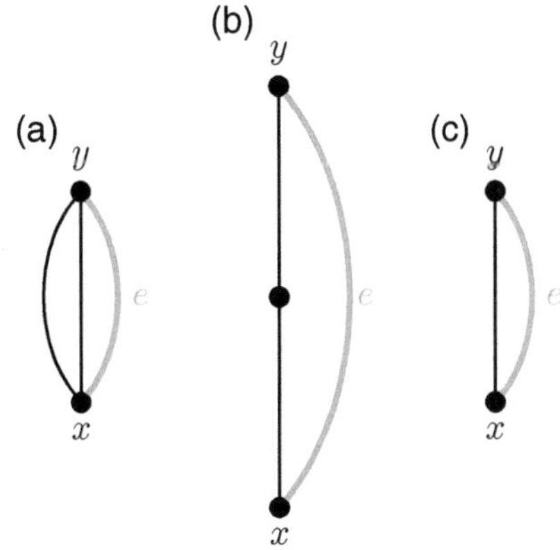

Fig. 1 (**a**) JOIN game; (**b**) CUT game; (**c**) NEUTRAL game

The game was solved by A. Lehman who characterized each class of games. The clues of Lehman's solution are the following:

1. The observation that the game depends exclusively on the matroid of cycles of the graph (see Definition 1).

   This observation enlightens the duality between the objectives of JOIN and CUT simplifying, on one hand, the analysis of the game and showing simultaneously, that the natural context of the game is matroids, rather than graphs or graphic matroids.
2. The strategies and characterization of Shannon's game are derived from the strategies of an associated game: the TREE GAME on a particular kind of graph: blocks (see Definition 2).

   Constructive and algorithmic aspects of Lehman's results for graphs were considered by Bruno and Weinberg in [6] exploring the notion of principal partition of a graph introduced by Kishi and Kajitani [14] that they have generalized to matroids in [7].

In the next paragraphs we sketch the main results and strategies, starting with the description of the associated game: TREE.

We assume the reader is acquainted with the basic notions of graph theory. Good references are [1, 11]. In the next paragraph we recall some terminology and notation.

Notation Given a graph G, we denote by $V(G)$ the set of its vertices and by $E(G)$ the set of its edges. A subgraph G' of G is any graph satisfying the conditions $V(G') \subseteq V(G)$ and $E(G') \subseteq E(G)$. For every edge e, $G \setminus e$ denotes the *subgraph obtained from* G *deleting* e, and G/e the *subgraph obtained from* G *contracting* e. We recall that the contraction G/e is the graph obtained from G identifying the vertices of the edge e and then eliminating e.

A *tree* T is a connected graph with minimum number of edges. The numbers of vertices and edges of a tree are related by: $|E(T)| = |V(T)| - 1$. A *forest* is a graph whose connected components are trees.

Given a connected graph G, a *spanning tree* T *of* G is a maximal tree that is a subgraph of G. One has $V(T) = V(G)$ and $|E(T)| = |V(G)| - 1$.

Definition 1 (Matroid of Cycles of a Connected Graph) The *matroid of cycles of a connected graph* G is the pair $M(G) = (E(G), \mathscr{B})$ where $\mathscr{B} \subseteq 2^{E(G)}$, the family of bases of the matroid, is defined as: $\mathscr{B} := \{E(T) : T \text{ is a spanning tree of } G\}$. The number of edges, $|V(G)| - 1$, of a (any) base is the rank of the cycle matroid.

The *circuits* of the matroid $M(G)$ are the sets of edges of a minimal closed path of the graph. A cocircuit of $M(G)$ is a minimal subset of edges whose removal disconnects the graph.

One says that *an edge* e *is spanned by a subset of edges* $A \subseteq E(G)$ if there is a circuit C of $M(G)$ such that $e \in C \subseteq A \cup e$. Clearly, given a connected graph G and a spanning tree T of G every edge $e \notin E(T)$ is spanned by $E(T)$ or simply, by T. Moreover, there is a unique circuit $C(T; e)$ such that $e \in C(T; e) \subseteq E(T) \cup e$.

A *graphic matroid* is the matroid of cycles of some graph.

For an introduction to matroids the reader may consult [18] or [17].

2.2 *TREE*

Like Shannon's switching game TREE is a two person game played on a connected graph G. The two players, JOIN and CUT, play like in Shannon's game: alternately, JOIN reinforcing one unplayed edge, CUT deleting one edge. The objective of JOIN is to reinforce the edges of a connected spanning tree of G. If he does not succeed, CUT wins.

Notice that CUT wins if he deletes the edges of a cocircuit of the graph G, equivalently, a cocircuit of the graphic matroid of G.

The analysis of this game depends exclusively on the notion of pair of maximally distant spanning trees of the graph G, resp. pair of maximally distant bases in the case of a matroid. In other words, the game depends on the matroid union $M(G) \bigvee M(G)$, with $M(G)$ the cycle matroid of the graph.

Definition 2 (Blocks and Maximally Distant Spanning Trees) A block is a matroid that is the union of two disjoint bases. A connected graph G is a (connected) block if $E(G) = E(T_1) \uplus E(T_2)$ with T_1, T_2 two edge-disjoint spanning trees of G.

A *pair of maximally distant spanning trees of a connected graph* G is a pair of spanning trees (T_1, T_2) that maximizes the cardinal $|E(T) \cup E(T')|$ with (T, T') a pair of spanning trees of the graph. A connected graph G is a (connected) block if and only every pair (T_1, T_2) of maximally distant spanning trees satisfies the equalities: $|E(T_1) \cup E(T_2)| = |E(T_1)| + |E(T_2)| = 2|E(T_1)|$.

All the strategies to play TREE and Shannon's switching game follow from the proof of the next Lemma:

Lemma 1 *The TREE game on a connected block is a JOIN game.*

The proof describes a recursive winning strategy for JOIN playing second.

Strategy for JOIN winning TREE on a connected block, playing second

(1) Let $G = T_1 \uplus T_2$ be a connected block, with set of edges $E(G) = E(T_1) \uplus E(T_2)$, T_1, T_2 two edge disjoint spanning trees of G.

(2) Denote by c the edge deleted by CUT in his first move. Assume w.l.o.g. that $c \in E(T_1)$.

Notice that once c is deleted $T_1 \setminus c$ has exactly two connected components, say R and S.

(3) JOIN then responds reinforcing(contracting) any edge $j \in E(T_2)$ that has one vertex in R and the other in S, equivalently, j is any edge of T_2 such that *the unique circuit -* $C(T_1; j)$*—contained in* $E(T_1) \cup j$ contains c.

For every such j, the graph $G_1 = G \setminus c/j$ is a connected block whose rank is $rank(G) - 1$. The game continues with G replaced by G_1.

Figure 2, below, illustrates the strategy.

From the above strategy one concludes that the game TREE is a JOIN game for any connected graph G containing a spanning block. Notice that when the graph strictly contains a spanning block $T_1 \cup T_2$, CUT may delete an edge outside $E(T_1) \cup E(T_2)$. In this case JOIN imagines a possible move for CUT in the block and replies to that fictitious move of CUT.

If a graph G does not contain a spanning block then every pair (T_1, T_2) of maximally distant spanning trees of G satisfies one of the following two conditions: either (i) $|E(T_1) \cap E(T_2)| = 1$ or (ii) $|E(T_1) \cap E(T_2)| \geq 2$. It is not hard to derive from the proof of the lemma winning strategies for the first player, JOIN or CUT, in case (i), as well as winning strategies for CUT playing second in case (ii).

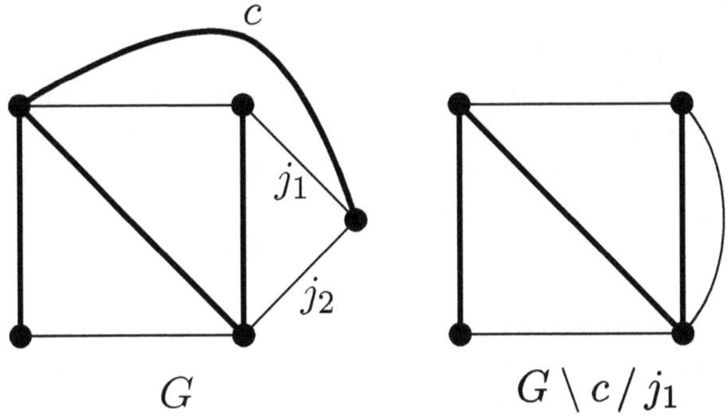

Fig. 2 JOIN possible responses to CUT playing c

Notice also that the above algorithm gives the strategy for JOIN playing second in Shannon's switching game when the distinguished edge *e* is spanned by a (connected) block not containing it. The fact that JOIN can reinforce a spanning tree of such a block will certainly guarantee that JOIN reinforces the edges of a circuit contained in that block union *e*, in other words that block contains a circuit of the graph, broken at *e*.

The next theorem solves TREE. Theorem 2, a direct consequence of Theorem 1, solves Shannon's switching Game.

Theorem 1 (Classification of TREE [15], See Also [12]) *Let G be a connected graph. Consider a pair* (T_1, T_2) *of maximally distant spanning trees of G.*

1. *If* $|E(T_1) \cap E(T_2)|$ *the TREE game is a JOIN game.*
2. *If* $|E(T_1) \cap E(T_2)|$ *the TREE game is a NEUTRAL game.*
3. *If* $|E(T_1) \cap E(T_2)|$ *the TREE game is a CUT game.*

Theorem 2 (Classification of Shannon's Switching game [15], See Also [12]) *A Shannon's switching Game* (G, e) *is:*

1. *a JOIN game if and only if G contains a block spanning, but not containing e;*
2. *a NEUTRAL game if and only if G contains a block spanning e and every such block contains e;*
3. *a CUT game if and only if there is no block spanning e.*

3 Hamidoune-Las Vergnas Directed Switching Games

In the 1980s Hamidoune and Las Vergnas [12] studied generalizations of Lehman's results to oriented matroids. They defined two types of directed variants of Shannon's game: in the first type CUT plays as in the original game, deleting edges, but JOIN plays differently. Each move of JOIN consists in directing one edge of the graph. In the second type of variants both players direct edges.

The objectives of JOIN are then translated in terms of directed paths, positive directed circuits or cocircuits.

Concerning the games of the second type, very little is known about them. Even their classification is not clear. In contrast with the undirected case, where a player having a winning strategy playing second has a winning strategy playing first, there are examples [12] of games of the second type where J. Nash stealing argument does not apply, games where a player has a winning strategy playing second but looses when playing first.

Concerning the directed games of the first type, Hamidoune and Las Vergnas proved that, in this case, the classification of the directed and undirected Shannon's switching games are exactly the same. Their proofs for the directed case, follow the same ideas of the non-oriented case, however the strategies described are more elaborate and not generalizable to oriented matroids in general.

The complete characterization of the winning positions, which is known from [13] only for blocks, requires in that case, and in contrast with the undirected case, the analysis of all previous moves evidentiating the complexity of the analysis of the directed case.

In the next paragraphs we briefly review the main results of Hamidoune and Las Vergnas.

In what follows, we shall call the JOIN player in the directed games d-JOIN. Also the directed Shannon's switching game, introduced in [12], will be called Hamidoune-Las Vergnas switching game.

3.1 *Hamidoune-Las Vergnas Switching Game* $(G;\mathbf{e})$

Let $(G;\mathbf{e})$ be a graph (or a digraph) with a distinguished arc $\mathbf{e} = (x, y)$, *not subject to play*. Two players d-JOIN and CUT play alternately choosing one unplayed edge of the graph. d-JOIN directs the chosen edge and CUT deletes the chosen edge. d-JOIN wins if he succeeds in directing a path from x to y. CUT wins otherwise.

It is clear that we can import several results from Lehman's study of the undirected case namely that a winning strategy for CUT in the associated undirected game $(G; e)$ must be a winning strategy for CUT in the directed game $(G;\mathbf{e})$. The, non obvious question is then whether the existence of a winning strategy for JOIN in Shannon's Game implies the existence of a strategy for d-JOIN in the corresponding Hamidoune-Las Vergnas Game or not.

Hamidoune and Las Vergnas, in [12] and [13], prove that this is the case. Although the winning strategies for the undirected game can not, in general, be adapted do the directed case they follow a similar approach, starting with the analysis of a directed TREE game, ARBORESCENCE, on a connected block.

3.2 *ARBORESCENCE* $(G;x)$

ARBORESCENCE is a game played on a connected graph G with a distinguished vertex x. The two players CUT and d-JOIN play as in Hamidoune-Las Vergnas directed switching game. In this case the objective of d-JOIN is to direct the edges of an arborescence rooted at x.

Lemma 2 *ARBORESCENCE on a connected block is a d-JOIN game.*

3.2.1 Hamidoune-Las Vergnas Strategy for d-JOIN Playing Second on a Connected Block [13]

This is the crucial step of the Hamidoune-Las Vergnas results. We use the short proof in [13] that works by induction on the number of edges of the block.

Let G be a connected block. If $|E(G)| = 2$ obvious. If $|E(G)| > 2$ and CUT deletes an edge c then either there is a nonempty connected block X of $G \setminus c$ incident to x or not.

In the first case G', the subgraph induced by X, and $G'' := G/X$ are both connected blocks. By induction d-JOIN has winning strategies, Σ' and Σ'', for winning playing second, resp. in G' and in G''. The strategy of d-JOIN in G is the following: if the move c of CUT falls into X he responds directing an edge j given by strategy Σ', if the move c falls into $E(G) \setminus X$ he answers directing an edge according to strategy Σ''.

In the second case, G itself is the only connected block incident to x. In this case d-JOIN must respond to the move c of CUT by directing one edge j of $G \setminus c$ incident to x (outgoing from x).

The game restarts with G replaced by $G_1 := G \setminus c/j$.

Example 1 In this example we play ARBORESCENCE in the connected block represented in the next Fig. 3a. The player *d*-JOIN plays second and wins using the strategy described in the proof of the Lemma. The *i*-th moves of CUT is denoted c_i and the *i*-th move of *d*-JOIN is denoted j_i.

The first move of CUT is $c_1 = 8$. Since the smallest block incident to x is G itself, JOIN responds directing one edge incident to x, outgoing from x. *d*-JOIN played the edge $j_1 = 7$ in the first figure. The game continues in $G1 := G \setminus c_1/j_1$ represented in Fig. 3b.

CUT's second move is $c_2 = 2$. Now there is a connected block incident to x contained in $G_1 \setminus c_2$, namely the block with edges $X = \{1, 5\}$. According to the above strategy *d*-JOIN plays in $G_1'' := G_1/X$. Since the edge $c_2 = 2$ is contained in the block $X_1 = \{c_2 = 2, 6\}$ of G_1', incident to x, according to the above strategy *d*-JOIN answers $j_2 = 6$.

The game continues in the graph $G_2 = G1 \setminus c_2/j_2$, represented in Fig. 4a.

The reader may easily see from G_2 that, according to the strategy defined, the answer to CUT's move $c_3 = 3$ must be $j_3 = 4$. The last Fig. 4b, represents $G_3 = G_2 \setminus c_3/j_3$, and it is clear that JOIN's answer to $c_4 = 1$ must be $j_4 = 5$.

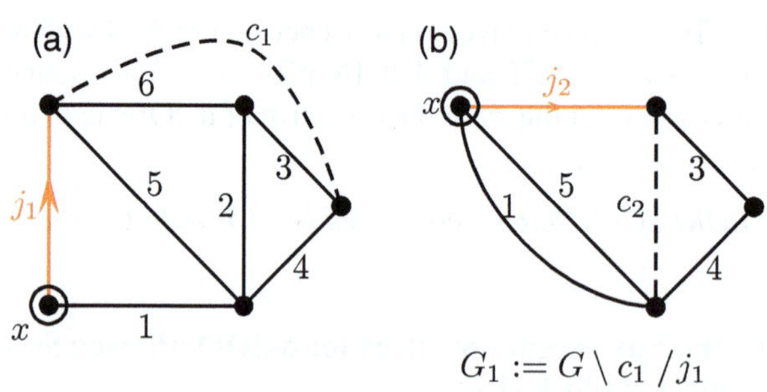

Fig. 3 Game played on a block with d-JOIN answering with Hamidoune-Las Vergnas strategy

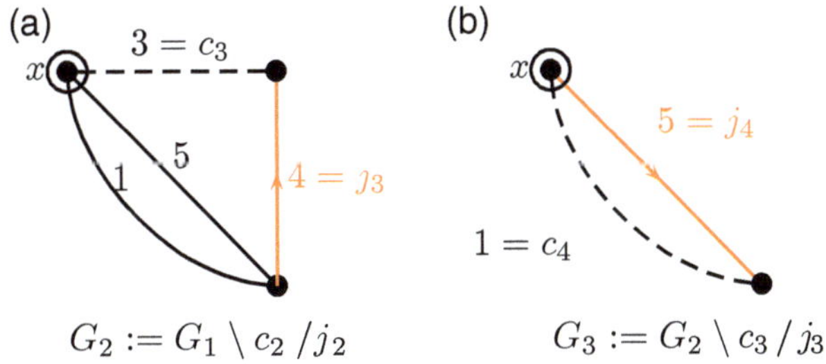

Fig. 4 Game played on a block with d-JOIN answering with Hamidoune-Las Vergnas strategy

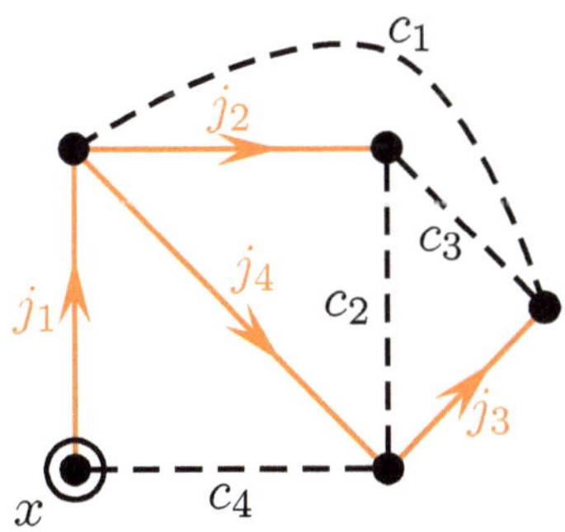

Fig. 5 The previous ARBORESCENCE game in the original graph

In Fig. 5 we have represented all the moves of the previous game in the original graph G.

The next Theorem is a direct consequence of the above lemma:

Theorem 3 (Classification of ARBORESCENCE and Hamidoune-Las Vergnas Switching Game [12, 13])

1. *ARBORESCENCE in a connected graph G with root x is a d-JOIN game (resp. a CUT game or a NEUTRAL game) if and only if TREE is a JOIN game on G (resp. a CUT game or a NEUTRAL game on G).*
2. *A Hamidoune-Las Vergnas switching game* $(G; \mathbf{e})$ *is a d-JOIN game (resp. a CUT game or a NEUTRAL game) if and only if the corresponding undirected switching game is a JOIN game on G (resp. a CUT game or a NEUTRAL game on G).*

We point out that from a computational point of view all the algorithms involved in evaluating a position and playing strategically any one of the unoriented games, TREE or Shannon's switching game, are polynomial time algorithms. For the directed games, ARBORESCENCE and Hamidoune-Las Vergnas switching game, this is not known except in very particular cases [12]. The reason being that there is no general theorem characterizing all the winning positions at some point of the game.

Since TREE and Shannon switching games have natural generalized versions in terms of matroids, one would expect that Hamidoune-Las Vergnas switching game and ARBORESCENCE had a natural generalized version in terms of oriented matroids. Although ARBORESCENCE, at least so far, has no natural generalization to oriented matroids Hamidoune-Las Vergnas switching game does have. In the next paragraph we consider the generalization of both switching games to configurations of (real) vectors a particular case of matroids and oriented matroids.

The interested reader may consult [3] for an introduction to oriented matroids.

3.3 *Shannon Switching Game, TREE and Hamidoune-Las Vergnas Switching Game on Configurations of Vectors*

Shannon switching game $(V; e)$ *and TREE on a configuration of vectors* are particular cases of the generalized versions of the games for matroids. The board instead of a graph is now a (finite) configuration of vectors V in some linear space over an *arbitrary field*. In both games the two players JOIN and CUT choose alternately one unplayed vector of V, CUT deletes it, JOIN reinforces/marks it.

In Shannon's switching game (V, e), where e is a distinguished vector not subject to play, JOIN wins if he succeeds in marking a set of vectors whose linear span contains e. CUT wins otherwise.

In TREE on a configuration of vectors V, JOIN wins if he succeeds in marking a base of the linear span of V.

Lehman's results [15] on Shannon's switching game and TREE, namely Theorems 1 and 2 of the last section, hold for the more general games on matroids and therefore on configurations of vectors which correspond to coordinatizable matroids (see [17, 18]). We recall that a matroid is a block if it the union of two disjoint bases. In terms of coordinatizable matroids or configurations vectors this translates as: a configuration V of vectors of some linear space is a block iff V it is the union of two disjoint bases of the linear span of V.

Concerning the directed versions of these games, Hamidoune-Las Vergnas switching game and ARBORESCENCE, only the first has a generalization to configurations of vectors necessarily over an ordered field which may be assumed to be the reals.

3.3.1 Hamidoune-Las Vergnas Switching Game $(V; e)$ on a Configuration of Vectors

Let $(V; e)$ be a configuration of vectors of $\mathbb{R}^d$, e a distinguished vector not subject to play. The two players CUT and d-JOIN choose alternately one unplayed vector. CUT deletes it and d-JOIN decides if he leaves the vector or if he replaces it by its opposite before marking the chosen one.

The objective of d-JOIN is to capture the distinguished vector e inside the positive cone spanned by his marked vectors.

We recall that the following conjecture of Hamidoune and Las Vergnas is open even for configurations of real vectors:

Conjecture 1 (Hamidoune-Las Vergnas [12]) The classification of a directed switching game $(M; e)$ on an oriented matroid M is the same as the classification of the associated undirected game $(\underline{M}; e)$ on the underlying matroid $\underline{M}$. More precisely, a directed switching game $(M; e)$ is:

1. a JOIN game if and only if there is a block of $\underline{M}$ spanning but not containing e.
2. a CUT game if and only if there is a block of $\underline{M}^*$ spanning but not containing e.
3. a NEUTRAL game if and only if both $\underline{M}$ and $\underline{M}^*$ have blocks containing e.

The structure of the positive cones spanned by a configuration of vectors is encoded by the associated oriented matroid.

In the case of configurations of vectors corresponding to graphs, and since graphic matroids have a unique class of orientations [4] or [3], the structure of positive cones is actually encoded in the underlying matroid.

The opposite situation occurs on configurations of vectors admitting many orientation classes. This is the case of the uniform matroids, $U_{n,d}$ corresponding to configurations of n vectors in general position in $\mathbb{R}^d$ (every d-subset of vectors is a base of $\mathbb{R}^d$). An important step would be to understand Hamidoune-Las Vergnas conjecture in this class of oriented matroids.

Notice that because Lehman's results hold for matroids, in order to prove the conjecture for uniform matroids one only needs to prove that d-JOIN has a winning strategy playing first when playing in a block, i.e. in an oriented uniform matroid $U_{2d,d}$.

Example 2 Shannon and Hamidoune-Las Vergnas switching games on a configuration of vectors that is a block.

The next figure, Figure 6, represents a configuration $(V; e)$ of four vectors in general position of $\mathbb{R}^2$. The corresponding matroid and oriented matroid are $U_{4,2}$ and one of its orientations. This is a first example which is not included in Theorem 3, because $U_{4,2}$ is not the matroid of cycles of a graph [17, 18] (Fig. 6).

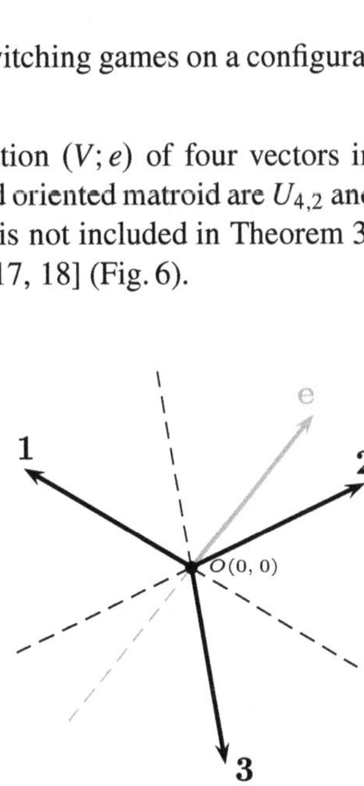

Fig. 6 A configuration $(V; e)$ of 4 vectors in general position in $\mathbb{R}^2$

Analysis of the Shannon Switching Game $(V; e)$

In this game *JOIN playing first wins, independently of how he plays*. In fact, JOIN, starting first will mark two vectors $j_1, j_2 \in \{\mathbf{1}, \mathbf{2}, \mathbf{3}\}$. Every subset of two vectors of $\{\mathbf{1}, \mathbf{2}, \mathbf{3}\}$ is an (unordered) basis of $\mathbb{R}^2$ and therefore $\{j_1, j_2, e\}$ contains (actually is) a minimal linearly dependent subset of V, i.e. a circuit of the matroid $U_{4,2}$ containing e.

Analysis of the Hamidoune-Las Vergnas Switching Game $(V; e)$

In this game *d-JOIN playing first has a winning strategy*. In contrast with the non oriented case d-JOIN must define carefully his strategy. Notice that if in his first move d-JOIN marks one of the vectors **1** or its opposite, $-\mathbf{1}$, he may loose. In fact, if he marks **1**, CUT eliminates **2** and e does not belong to $poscone(\{\mathbf{1}, \mathbf{3}\})$ nor to $poscone(\{\mathbf{1}, -\mathbf{3}\})$ so d-JOIN loses. Similarly if he plays $-\mathbf{1}$, CUT wins responding **3**.

The winning strategy for d-JOIN consists in playing in the first move either **2** or $-\mathbf{3}$. Only playing in this way he guarantees that he has a good answer to every possible move of CUT.

Recently, Chatelain and Ramirez [8], extending previous results of Forge and Vielleiribière [10], proved that d-JOIN has a winning strategy playing first in the orientations of $U_{2d,d}$ that are obtained as unions of rank 1/or rank 2 uniform oriented matroids. We point out that the oriented matroids arising in this way all correspond to vector configurations of vectors in general position in some $\mathbb{R}^d$ and do not cover all the possible orientations of $U_{2d,d}$.

The Hamidoune-Las Vergnas conjecture, for oriented matroids, even for configurations of vectors, remains open.

Acknowledgements This work was partially supported by Fundação para a Ciência e a Tecnologia, PEst-OE/MAT/UI0209/2013.

References

1. Berge, C.: Graphs. North-Holland Mathematical Library, vol. 6. North-Holland, Amsterdam (1989)
2. Berlekamp, E.R., Conway, J.H., Guy, R.K.: Winning Ways for Your Mathematical Plays, Volume II, 3rd ed. Academic Press, London (1985)
3. Björner, A., Las Vergnas, M., Sturmfels, B., White, N., Ziegler, G.: Oriented Matroids. Encyclopedia of Mathematics and Its Applications, vol. 46, 2nd edn. Cambridge University Press, Cambridge (1999)
4. Bland, R., Las Vergnas, M.: Orientability of matroids. J. Combin. Theory, Ser. B **24**, 94–123 (1978)
5. Brualdi, R.: Introductory Combinatorics. North-Holland, Amsterdam (1977)

6. Bruno, J., Weinberg, L.: A constructive graph theoretic solution of the Shannon switching game. IEEE Trans. Circuit Theory **CT-17**(1), 74–81 (1970)
7. Bruno, J., Weinberg, L.: The principal minors of a matroid. Linear Algebra Appl. **4**, 17–54 (1971)
8. Chatelain, V., Ramirez Alfonsin, J.L.: The switching game on unions of oriented matroids. Eur. J. Combin. **33**(2), 215–219 (2012)
9. Cláudio, A.P., Fonseca, S., Sequeira, L., Silva, I.P.: Implementations of TREE and ARBORESCENCE: http://shlvgraphgame.fc.ul.pt/ (2014)
10. Forge, D., Vieilleribière, A.: The directed switching game on Lawrence Oriented matroids. Eur. J. Combin. **30**(8), 1833–1834 (2009)
11. Gross, J.L., Yellen, J., Zhang, P. (eds.): Handbook of Graph Theory, 2nd ed, CRC Press 2004.
12. Hamidoune, Y.O., Las Vergnas, M.: Directed switching games on graphs and matroids. J. Combin. Theory, Ser. B **40**, 237–269 (1986)
13. Hamidoune, Y.O., Las Vergnas, M.: Directed switching games II. Discret. Math. **165/166**, 397–402 (1997)
14. Kishi, G., Kajitani, Y.: On maximally distant trees. In: Proceedings of 5th Allerton Conference on Circuits and Systems Theory, pp. 635–643 (1967)
15. Lehman, A.: A Solution to the Shannon switching game. J. Soc. Ind. Appl. Math. **12**, 687–725 (1964)
16. Novak, L., Gibbons, A.: Hybrid Graph Theory and Network Analysis. Cambridge Tracts in Theoretical Computer Science, vol. 49. Cambridge University Press, Cambridge (1999)
17. Oxley, J.: Matroid Theory, 2nd ed. Oxford University Press, New York (2011)
18. Welsh, D.: Matroid Theory. Academic Press, London (1976)

6. Bruno, J., Weinberg, L.: A constructive graph-theoretic solution of the Shannon switching game. IEEE Trans. Circuit Theory CT-17(1), 74–81 (1970)
7. Brown, J., Weinberg, L.: The principal minors of a matroid. Linear Algebra Appl. 4, 33 (1971)
8. Chatelain, V., Ramírez Alfonsín, J.L.: The switching game on unions of oriented matroids. Eur. J. Combin. 33(2), 215 (2012)
9. Cibulka, J., Fonseca, S., Sequeira, L.: Solvers Implementations of The Shannon Switching Game. http://blognaphgame.blogspot.com
10. Forge, D., Vieilleribière, A.: The directed switching game on Lawrence Oriented matroids. Eur. J. Combin. 30(8), 1833–1834 (2009)
11. Gross, J.L., Yellen, J., Zhang, P. (eds.): Handbook of Graph Theory 2nd edn. CRC Press (2013)
12. Hamidoune, Y.O., Las Vergnas, M.: Directed switching games on graphs and matroids. J. Combin. Theory Ser. B 40, 237–269 (1986)
13. Hamidoune, Y.O., Las Vergnas, M.: Directed switching games II. Discrete Math. 165/166, 395 (1997)
14. Kishi, G., Kajitani, Y.: On maximally distant trees. In: Proceedings of 5th Allerton Conference on Circuits and System Theory, pp. 635–643 (1967)
15. Lehman, A.: A Solution to the Shannon switching game. J. Soc. Ind. Appl. Math. 12, 687–725 (1964)
16. [illegible]
17. [illegible]
18. Welsh, D.J.A.: Matroid Theory. Academic Press, London (1976)

A Proposal to Measure the Functional Efficiency of Futures Markets

Meliyara Consuegra and Javier García-Verdugo

Abstract This paper presents a method to measure the functional efficiency of futures markets in terms of social welfare using a standard futures market structural model. Employing the concept of social surplus, it can be shown that the error committed when using futures prices to estimate spot prices in the future results in a welfare loss caused by the erroneous allocation of resources. Therefore, the social welfare associated with the presence of futures markets can be measured using a social loss (SL) statistic and its components. The results confirm the consistency and robustness of the method. Finally, several practical uses for the SL statistic are suggested.

1 Introduction

In the second half of the twentieth century, futures markets received increasing attention from academics, governments and companies in general. An important part of the research has focused in the efficiency of futures markets from the perspective of the Efficient Market Hypothesis (EMH). This paper aims to study the functional efficiency of these markets. When we talk about *functional efficiency,* we refer to the efficiency with which futures markets perform the functions of price risk transfer and price discovery. Regarding the transfer of price risks, participants seek to protect themselves from the variability of prices, and the efficiency of the hedging instrument depends on the relative variation between futures contract prices and the prices in the physical market. Price discovery refers to the fact that each participant in the futures markets acts using all available information and their own estimates about future price changes. In this paper the functional efficiency of futures markets is assessed estimating the social loss derived from allocation errors that are committed when the prices of futures contracts are used as estimators for prices in the physical markets.

The outline of this paper is as follows. The second section reviews all the previous research on this topic. The next section presents the basic model, while

M. Consuegra (✉) • J. García-Verdugo
Department of Applied Economics, UNED, Madrid, Spain
e-mail: mconsuegra@cee.uned.es; fgarcia-verdugo@cee.uned.es

J.-P. Bourguignon et al. (eds.), *Dynamics, Games and Science*, CIM Series in Mathematical Sciences 1, DOI 10.1007/978-3-319-16118-1_11

section four develops the theoretical and empirical indicators for the quantification of social welfare loss in futures markets. The fifth section presents the application of the model through some examples. The last section concludes and proposes other directions for future research.

2 Previous Research

Different papers such as Peroni and McNown [16], Switzer and El-Khoury [24], Maslyuk and Smyth [13], Kawamoto and Hamori [10] and Stevens [23] have focused on the efficiency of energy futures markets from a theoretical perspective. Other authors such as Chowdhury [6] and Timmermann and Granger [25] studied the efficiency in futures markets in general. The paper by Kawamoto and Hamori [10] is closest to our approach in the sense that they also used a sample of futures contracts with different maturities. A crucial difference is that they test the EMH, which assumes that the information of all past prices is reflected in today's prices. Stevens [23] found that the WTI futures market[1] is shown to be inefficient[2] according to the weighted least squares (WLS) and the trimmed least squares (TLS) tests, but efficient when the ordinary least squares (OLS) test is used. There are similar papers that studied the efficiency of futures contracts in other markets. McKenzie and Holt [14] tested the market efficiency and unbiasedness in agricultural futures markets and Wang and Ke [27] studied the efficiency in Chinese agricultural futures markets.

All these studies are only able to produce dichotomic results on the existence of efficiency, showing their limitation to examine the efficiency of futures markets. Contrary to the literature reviewed in the previous paragraph, this paper employs a generic structural model of futures markets in which efficiency is measured with an indicator that evaluates the functional efficiency of futures markets in terms of social welfare. This model is useful for different types of commodities, such as metals, agricultural and energy.

The model used in this paper was originally developed by Stein [18, 19, 21]. This model has been used by several authors along the years, which is a proof of its usefulness: Brooks [4] and Stein [22] studied the financial futures markets, Hong [8] applied the model to the non-financial futures, Avsar and Goss [1] studied the informational efficiency of electricity futures contracts, Pennings and Garcia [15] examined the determinants of heterogeneity in hedging behavior in a concomitant mixture regression framework, and García-Verdugo and Consuegra [7] focused on energy futures contracts. Different versions of Stein's model have been used by Kawai [9] and Turnovsky [26] who studied the spot and futures prices of non-

[1]The crude oil price is that of West Texas Intermediate traded on the New York mercantile exchange.

[2]Inefficient in the sense of the EMH.

storable and storable commodities respectively; Bond [3] examined the effects of supply and interest rate shocks in commodity futures markets; Pindyck and Rotemberg [17] who proved that the prices of largely unrelated raw commodities have a persistent tendency to move together, and Chari and Jagannathan who studied the volatility of prices with the introduction of futures markets [5].

Stein's model is based on the optimization of individual decisions made by different market participants and has several useful features. First, it explains theoretically which variables determine equilibrium prices, equilibrium open interest and the variability of prices. Second, it can incorporate exogenous and endogenous expectations, as well as participants with different forecasting abilities. Above all, it can be used to analyse the ex-post contribution of futures markets to social welfare through the optimal inter-temporal allocation of resources, which is the main reason why it was selected to be used in this paper.

3 The Basic Model

The basic model has two periods. In period t producers and consumers decide the proportion of their commercial positions to be hedged with futures, and speculators make their investment decisions. In period $t + 1$ exchanges are made in the physical market and open positions in the futures markets are canceled. Commercial participants are attracted to futures markets by the possibility of protecting themselves from price risks. On the contrary, the variability of these same prices attracts speculators and determines their corresponding level of expected benefits. As a result of the participant's hedging, an optimal level of production as well as optimal positions in the futures market are obtained. In this section we present the mathematical expressions of the participants' positions, supply and demand functions, and futures and spot prices.

In general, it can be stated that commercial participants are attracted to futures markets by the possibility of protecting themselves from price risks. On the other hand, the variability of these same prices attracts speculators and determines their corresponding level of expected benefits. As a result of the participants' hedging or speculative decisions, an optimal level of production and an optimal position in the futures market is obtained. Since Stein's model is based on the optimization of individual decisions made by different market participants, we start from the expressions that summarize the determinants of the open positions of commercial firms with sale price uncertainty, commercial firms with purchase price uncertainty and speculators [21].

It can be shown that the open position $x(t)$ for each commercial firm with sale price uncertainty is given by:

$$x(t) = \frac{q_{t+1}(t)}{\alpha(1 - r^2)\operatorname{var} p + c} - \frac{E_1 p(t+1;t) - q_{t+1}(t)}{\alpha r^2 \operatorname{var} p} \tag{1}$$

Similarly, the open position $y(t)$ for each commercial firm with purchase price uncertainty is given by:

$$y(t) = \frac{a - q_{t+1}(t)}{\alpha(1 - r^2)\mathrm{var}\,p + b} + \frac{E_2p(t+1;t) - q_{t+1}(t)}{\alpha r^2 \mathrm{var}\,p} \tag{2}$$

Finally, the open position $z(t)$ for each speculator is given by:

$$z(t) = \frac{E_s p(t+1;t) - q_{t+1}(t)}{\beta\,\mathrm{var}\,p} \tag{3}$$

The price of the future contract in t is denoted by $q_{t+1}(t)$ in the three equations. E_1p and E_2p are the expected prices by each type of commercial firms. Assuming that all commercial firms have identical expectations, $E_1p = E_2p = E_cp$. β and α are the risk aversion parameters, var$p$ is the price risk, r^2 will referred to as the quality of the hedging instrument,[3] b is the slope of the individual demand functions and c is the slope of the individual supply functions of the commercial participants.

From (1)–(3) the aggregated open positions are obtained. Therefore, the open position X for n_1 commercial firms of the first type is represented in Eq. (4), where $x_i(t)$ are the sales (+) or purchases (-) at the time t of futures contracts maturing at $t+1$. The open position Y for n_2 commercial firms of the second type can be aggregated as (5), where $y_i(t) = -x_i(t)$. Accordingly, y_i represents the purchases (+) and sales (-) at the time t of futures contracts maturiting at $t+1$. And the open position Z for n_s speculators is denoted by (6). The total demand of futures by speculators $Z(t)$ is the sum of the long (+) or short (-) position of futures speculators, z_i.

$$X(t) = \sum_{i=1}^{n_1} x_i(t) = n_1[x(t)] \tag{4}$$

$$Y(t) = \sum_{i=1}^{n_2} y_i(t) = n_2[y(t)] \tag{5}$$

$$Z(t) = \sum_{i=1}^{n_s} z_i(t) = n_s[z(t)] \tag{6}$$

By subtracting $X(t)$ minus $Y(t)$ we obtain the supply function of futures contracts by commercials, denoted by C in Eq. (7). And the demand function $S = Z(t)$ is a

[3] r is the correlation between the price of the commodity relevant for the commercial firm and the price of the standardized commodity defined in the futures contract.

result of the open position of speculators.

$$C = X(t) - Y(t) = w_c[q_{t+1}(t) - E_c p] + v_1 q_{t+1}(t) - v_2[a - q_{t+1}(t)] \tag{7}$$

$$S = Z(t) = w_s[E_s p - q_{t+1}(t)] \tag{8}$$

The speculative coefficients $w_c = \frac{n_1+n_2}{\alpha r^2 \text{varp}}$ and $w_s = \frac{n_s}{\beta \text{varp}}$ determine the magnitude of speculation by commercial firms and speculators respectively. Accordingly, the higher the risk aversion of the participants, the lower the speculative component of the commercial firms' futures position and the lower the speculators' position. On the other hand, the larger the number of participants, the higher the speculative futures position of firms and speculators. The output and input coefficients $v_1 = \frac{n_1}{\alpha(1-r^2)\text{varp}+c}$ and $v_2 = \frac{n_2}{\alpha(1-r^2)\text{varp}+b}$ represent the magnitude of hedging by commercials entities with uncertainty in sales and purchase prices respectively. Assuming that all the participants are rational $E_c p \sim E_s p \sim E_m p$, which is the subjective expected price in t for period $t + 1$ by commercial participants and speculators. If $q_{t+1}(t) = E_c p$, the open positions in the futures markets of commercials entities will not depend on w_c, and it will only depend on production and demand parameters. The net hedging pressure (h) is the excess supply of futures contracts by commercials when the futures price is equal to their expected price, so $h = v_1 q_{t+1}(t) - v_2[a - q_{t+1}(t)]$.

The literature on commodities futures markets traditionally assumes that speculative transactions result in net long speculative positions. Accordingly, the only commercial participants included in Stein's model are assumed to hold a net short position in futures contracts, i.e. they are sellers hedging against the risk of falling prices. In the model, futures prices determine production, while consumption exogenously equals production.[4]

The market equilibrium is obtained when the supply function (7) equals the demand function (8). Then, the prices of the futures contracts are obtained:

$$q_{t+1}(t) = (1 - \delta)E_m p(t + 1; t), \text{ where } \delta = \frac{h/w}{E_m p} \tag{9}$$

The term w is the sum of the coefficients w_c and w_s. The parameter δ reflects the sufficiency or insufficiency of speculation to satisfy the need for commercial hedging or hedging pressure. At the same time, if the quality of the hedging instrument is assumed to be perfect ($r^2 = 1$), the equilibrium of the goods market determines (10):

$$q_{t+1}(t) = cS(t + 1) \tag{10}$$

[4]Futures prices collect the intertemporal allocation of resources for production and not for consumption, but welfare in the model does not vary significantly.

Equation (11) is the market supply function since $cS(t+1)$ is the aggregate marginal production cost when marginal costs of individual commercial participants are assumed to be linear. Using (9), the supply equation can be written in terms of the participants' subjective expectation of the commodity spot price:

$$O \equiv E_m p(t+1;t) = \frac{c}{1-\delta} S(t+1) \quad (11)$$

On the other hand, since consumption exogenously equals production in the model, the market demand function can be given as:

$$D(t+1) \equiv p^*(t+1) = u^*(t+1) - bS(t+1) \quad (12)$$

Considering that $u^*(t+1)$ is a random parameter relevant to the second period, demand $D(t+1)$ follows an unknown probability distribution and $p^*(t+1)$ is the random spot price in $t+1$. $ED(t+1)$ is the objective expectation of demand, and its expression is:

$$ED(t+1) \equiv Ep(t+1) = Eu(t+1) - bS(t+1) \quad (13)$$

The expected value of the spot price in (13) is denoted by $Ep(t+1)$. The difference between value D and its objective expectation of demand ED is known as the inevitable error $\varepsilon(t+1)$. This error is due to the unpredictable variation of the random parameter u^* around its expected value $Eu(t+1)$. This difference can be shown to be equal to $\varepsilon(t+1) = p(t+1) - Ep(t+1)$, and is represented in Fig. 1 by the segment CB.

Lacking the capacity for perfect forecast, companies can only attempt to predict the value of the objective expectation of demand ED. Their actual estimate of ED is known as the subjective expectation of demand E_mD.

$$E_mD(t+1) \equiv E_mp(t+1) = Eu(t+1) + y_m(t) - bS(t+1) \quad (14)$$

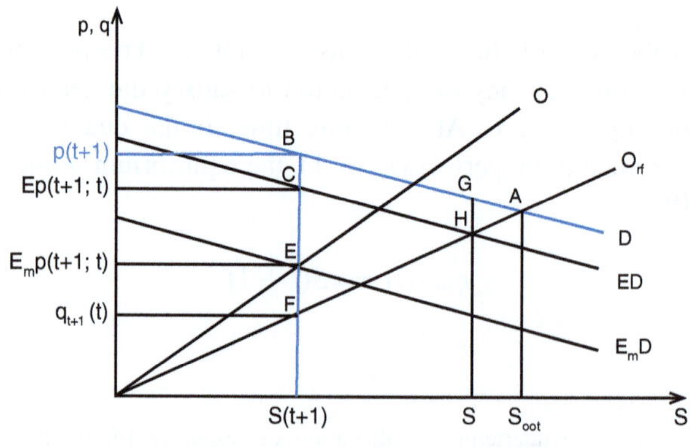

Fig. 1 Social loss associated with the suboptimal allocation of resources

The difference between both expectations of the demand function is known as the Bayesian error $y_m(t)$, represented in Fig. 1 as the segment EC. It can be shown that $y_m(t) = Ep(t+1) - E_mp(t+1;t)$. For convenience, Fig. 1 represents $D > ED > EmD$, but it need not be so. The intersection of the supply curve with E_mD determines the subjective expectation of the equilibrium price and the amount to be produced in t. In turn, $E_mp(t+1;t)$ and the risk premium determines the futures price$q_{t+1}(t)$, as is shown in Eq. (10). Eventually, the production $S(t+1)$ reaches the market in $t+1$ and is faced with the actual demand curve D, which results in $p(t+1)$. The consideration or not of a risk premium (embodied in the parameter δ) affects the futures price because it shifts the supply function, as can be seen from Eq. (12): the actual supply curve O in Fig. 1 shifts to the left of the risk-free supply function O_{rf} because a positive risk premium is assumed. Therefore the distance between these two curves (EF in Fig. 1) represents the risk premium, given by $E_mp - q_{t+1}(t)$, that can be considered the third source of forecast error. Let us take a step back to put these concepts in perspective. Stein [20] identified two types of social loss in the forward markets: the avoidable and the unavoidable. The unavoidable error represents the difference between the market equilibrium price and the expected equilibrium price. The avoidable error is the gap between the expected equilibrium price and the forward price. This model was expanded by Stein [21] for futures markets. In this paper he identified three types of errors: the inevitable error, the Bayesian error and the risk premium. For futures markets the unavoidable error is called inevitable; the Bayesian error occurs because of the difference between the subjective and objective expected price of the contract; finally, the risk premium occurs when expected demand exceeds marginal cost. The inevitable and Bayesian errors plus the risk premium comprise the total forecast error:

$$p(t+1) - q_{t+1}(t) = [p(t+1) - Ep] + [Ep - E_mp] + [E_mp - q_{t+1}(t)] \tag{15}$$

4 The Empirical Model

As we saw in Fig. 1 $S(t+1)$ is the volume of output that reaches the market at $t+1$, while S_{opt} is the optimal volume of production, which is obtained from the intersection of O_{rf} and actual demand D. Following Stein [21], we assume that the loss of social welfare—or simply social loss—is the triangular area ABF between the effective demand curve and the marginal cost curve, O_{rf}, between the actual output $S(t+1)$ and the perfect-foresight equilibrium output S_{opt}, while the triangle AGH represents the inevitable social loss caused by the random parameter $u^*(t+1)$ included in D (see Eq. (12)). Therefore, total social loss can be represented with the expression:

$$L(t+1) = \frac{1}{2}[p(t+1) - q_{t+1}(t)].[S_{opt} - S(t+1)] \tag{16}$$

This value is the product of the price forecast error and the deviation of production $S(t+1)$ from the optimal value S_{opt}. Since the production deviation depends on the price deviation, the loss of social welfare can be rewritten as:

$$L(t+1) = K\,[p(t+1) - q_{t+1}(t)]^2 \text{ where } K = \frac{1}{2(b+c)} \tag{17}$$

Therefore, ex-post social loss $L(t+1)$ is a multiple K of the square of the price deviation between the subsequently realized cash price $p(t+1)$ and the futures price. Stein [21] defines the social loss statistic SL as the ratio of the social loss $L(t+1)$ to the minimum or inevitable social lossL_0. Using Eq. (17), it can be seen that the expected social loss $E[L(t+1)]$ is equal to the constant K times the mean squared error, MSE, of the price forecast for t+1. On the other hand, the expectation of the inevitable social loss $E(L_0) = EK[\varepsilon(t+1)]^2$ can be written as K times MSE_0. Therefore, the value of K is not needed to compute the SL statistic for the estimation of social welfare loss:

$$SL = \frac{E[L(t+1)]}{E(L_0)} = \frac{EK[p(t+1) - q_{t+1}(t)]^2}{EK[\varepsilon(t+1)]^2} = \frac{MSE(t+1)}{MSE_0} \tag{18}$$

At the same time that we move from two periods to a more realistic k periods, we define the empirical equivalent of $MSE(k)$ as:

$$MSE(k) = \frac{1}{n}\sum_{t=1}^{n}[\ln p(t+1) - \ln q_{t+k}(t)]^2 \tag{19}$$

where n is the number of observations in the data. $MSE(k)$ is the mean squared error derived from the estimation of the spot price in period $t+k$ using the price in t of the futures contract which expires k periods later. On the other hand, $MSE(1)$ will be used as an empirical proxy of the unobservable minimum expected social loss MSE_0. Thus the last term in (18) can be rewritten as

$$SL = \frac{MSE}{MSE_0} = \frac{E[p(t+1) - q_{t+1}(t)]^2}{E[\varepsilon(t+1)]^2} \tag{20}$$

In a similar way, the decomposition of the forecast error in (15) contains only theoretical variables that could only be used if observable proxies are found. Fortunately, an alternative method provides an empirical equivalent for the decomposition of $MSE(k)$ according to the type of error:

$$MSE(k) = E[p(t+k) - q_{t+k}(t)]^2 = (\overline{p} - \overline{q})^2 + [\sigma_q(1-d)]^2 + (1-r^2)\sigma_p^2 \tag{21}$$

where $\bar{p}$ and $\bar{q}$ are the means of spot and futures prices during the relevant period, σ_p^2 is the variance of spot prices, σ_q is the standard deviation of futures prices, d is the regression coefficient of p over q, and r is the correlation coefficient[5] between p and q. The first term is the part of *MSE* which is derived from the difference between the mean values of spot and future prices, the second is due to the risk premium which separates the value of d from the unit, and the third is a composition of the inevitable and Bayesian errors. Now, the empirical approximation to $SL(k)$ in (20) includes squared terms that exaggerate the absolute differences between the values of the statistic and reduce the informative content of the computed mean of forecast deviations. Following the method applied by Ma [12] in his efficiency contrasts, the squared root of the mean squared error can be used as an alternative:

$$RMSE(k) = \left[\frac{1}{n}\sum_{t=1}^{n}[\ln p(t+1) - \ln q_{t+k}(t)]^2\right]^{\frac{1}{2}} \tag{22}$$

so that:

$$SL(k) = \frac{RMSE(k)}{RMSE(1)} \tag{23}$$

4.1 Applications of the Model

In this section we are going to apply the model to evaluate the functional efficiency of certain futures markets. Five energy futures markets and three food futures markets with monthly data from 1992 to 2012 were use for the analysis. The futures contracts selected for the empirical analysis are traded in the Intercontinental Exchange (ICE) of London, the Chicago Mercantile Exchange (CME) and the Chicago Board of Trade (CBOT), which is part of the CME Group. Eight products and six maturities for each futures contract were selected: crude Brent and diesel from ICE; WTI (West Texas Intermediate) crude oil, heating oil, gasoline, and natural gas from CME; and corn, wheat and soybean that are traded in CBOT.

Following Kumar's [11] approach, we used futures prices corresponding to the last trading day of each month during the period of study. Kumar tested the hypothesis that the last futures price of each month contains all relevant information up to that moment, which is why those prices should be more accurate in predicting prices in the future. He concluded that price predictions made during the last trading day were superior to those obtained with alternative methods.

Table 1 shows the SL statistic for every product during the period 1992–2012. Since lower SL values represent a higher functional efficiency of the market, the products are organized accordingly with the most efficient at the top. Thus, in terms

[5]Note that these two variables are different from those that determine the quality of the hedging instrument in the basic model.

Table 1 SL values for the period 1992–2012

	SL(2)	SL(3)	SL(4)	SL(5)	SL(6)
Heating oil	1.18	1.42	1.63	1.84	2.02
Natural gas	1.32	1.49	1.67	1.82	1.92
Brent crude oil	1.59	2.02	2.37	2.67	2.93
WTI crude oil	1.62	2.07	2.46	2.80	3.08
Gasoline	1.75	2.25	2.52	2.77	2.99
Soybean	1.80	2.29	2.74	3.09	3.50
Wheat	1.81	2.32	2.81	3.41	3.90
Gasoil	1.87	2.46	2.98	3.41	3.82
Corn	1.89	2.57	3.21	3.69	4.02

of social welfare, heating oil was the most efficient futures market for maturities 2 through 4, natural gas was the most efficient for maturities 5 and 6, and corn was the least efficient for every maturity. It should be highlighted that, in terms of social loss, agricultural futures contracts generally perform worse than energy futures. However, the specific ranking varies somewhat as k increases. Broadly speaking, the futures contracts whose associated SL values increased most with the time to maturity are usually those that fare worse when considering the absolute SL values and vice versa.

More interesting is to compare the evolution of functional efficiency in each market over time. In Table 2 data are divided in three periods: period 1: 1992–1996, period 2: 1997–2006 and period 3: 2007–2012. In 1992–1996 the most and least efficient markets were the same as throughout the total period; in 1997–2006 the least efficient market of the group was gasoline while heating oil remained the most efficient; in period 3 natural gas was more efficient than heating oil while the wheat futures market was the least efficient for every maturity. Between the period 1 and 2, the corn market reduced its SL value for every maturity, moving from the least efficient to the sixth most efficient market in the group, remaining at this level of efficiency in period 3. On average, between the sub-periods 1992–1996 and 2007–2012, heating oil markets, diesel, and natural gas presented the greatest increases in social loss in the past two sub-periods (32.5, 31.5 and 30.6 % respectively), while the gasoline market presented a reduction in social loss almost as great as the increase presented in the other two markets (−32.1 %). Crude WTI and Brent Crude showed positive variations of 15.4 and 16.4 % respectively, much less than that of natural gas and diesel.

Observing the overtime rate of variation of the SL statistic, we found that between the first and the last period, heating oil, wheat and gasoil futures presented the greatest increases in SL values for every k (on average 46.9, 40.2 and 38.6 % respectively), while the corn market presented the largest reduction in social loss (−32.3 %) followed by gasoline with a reduction 22.4 %. Excluding corn and gasoline, every product reduced its efficiency, which means a decrease in social welfare. Corn futures markets significantly increased their efficiency between the period 1 and 2. Between period 2 and 3 this market maintained its efficiency level

Table 2 SL values for three periods

1: 1992–1996	SL(2)	SL(3)	SL(4)	SL(5)	SL(6)
Heating oil	1.12	1.29	1.44	1.59	1.75
Natural gas	1.27	1.41	1.43	1.45	1.49
Soybean	1.58	1.76	1.94	2.28	2.78
Brent crude oil	1.59	2.06	2.34	2.53	2.65
WTI crude oil	1.64	2.13	2.37	2.57	2.68
Gasoil	1.75	2.23	2.57	2.80	2.94
Wheat	1.81	2.35	2.81	3.21	3.56
Gasoline	1.92	2.53	2.93	3.27	3.64
Corn	2.56	3.55	4.57	5.46	6.03
2: 1997–2006	**SL(2)**	**SL(3)**	**SL(4)**	**SL(5)**	**SL(6)**
Heating oil	1.14	1.30	1.38	1.44	1.48
Natural gas	1.34	1.48	1.65	1.77	1.83
Brent crude oil	1.53	1.85	2.14	2.40	2.67
WTI crude oil	1.58	1.94	2.25	2.54	2.85
Wheat	1.70	2.06	2.51	3.24	3.63
Corn	1.81	2.42	2.97	3.37	3.66
Gasoil	1.83	2.32	2.69	3.03	3.38
Soybean	1.94	2.48	2.97	3.38	3.85
Gasoline	2.14	2.70	2.90	3.10	3.33
3: 2007–2012	**SL(2)**	**SL(3)**	**SL(4)**	**SL(5)**	**SL(6)**
Natural gas	1.36	1.64	1.97	2.27	2.49
Heating oil	1.37	1.78	2.18	2.55	2.85
Gasoline	1.52	1.97	2.29	2.55	2.74
WTI crude oil	1.66	2.21	2.73	3.13	3.42
Brent crude oil	1.70	2.25	2.70	3.07	3.35
Soybean	1.73	2.26	2.74	3.05	3.39
Corn	1.81	2.48	3.10	3.57	3.89
Gasoil	2.00	2.80	3.60	4.23	4.80
Wheat	2.22	3.03	3.77	4.74	5.98

for $k = 2$ and reduced it in a small amount for the other maturities. Gasoline futures markets contracts considerably increase its efficiency overtime, specially between period 1 and 3. In general, between the first and the third period, corn and gasoline futures contracts increased their functional efficiency while that of the other studied contracts reduced.

As expected, for every sub-period as well as for the whole period 1992–2012, SL values increase for every product with the distance to contract maturity, showing that futures prices see their capacity for prediction reduced when k increases. It should be noted again that agricultural futures markets showed worse results than energy futures markets in terms of social loss. This result can be explained by the dependence of the SL value on the forecast error, which is usually higher in the food sector than in the energy industry. Food products are more perishable than energy products, naturals disasters and climate variations affect them more, and the influence on crops of other variables that are quite hard to forecast [2] increase the probability of a higher forecast error and of a corresponding higher SL value.

5 Conclusions

This paper has presented a useful and simple measurement of the functional efficiency of futures markets. The SL statistic is shown to be a consistent indicator that can be used to quantitatively estimate social losses associated with the use of futures markets for spot price forecasting by using concepts and tools related to social surplus theory. The SL statistic computed for several energy and agricultural futures and maturities show that this indicator can be used to compare the relative behaviour of different markets and to analyse the evolution of their functional efficiency over time.

A great deal of research is still needed in this area. One direction of advance would be to explain the SL statistic and its evolution over time using the variations in futures markets indicators such as open interest, trading volume and commodity price volatility. These indicators were only taken into account in this paper as criteria for selecting the contracts to be considered, since the energy and food futures that were chosen were those with higher open interest and trade volume. Another direction of research would be the comparison between the values of the SL statistic associated with futures markets with the values of the indicator computed for forward contracts traded in the physical market, as was suggested by Stein [20]. Finally, a natural expansion of this research would be to apply this quantification method to other groups of commodities, such as other agricultural futures, metals and financial products such as equities, bonds and currencies.

References

1. Avsar, S.G., Goss, B.A.: Forecast errors and efficiency in the us electricity futures market. Aust. Econ. Pap. **40**(4), 479–499 (2001)
2. Baffes, J., Dennis, A.: Long-term drivers of food prices. Policy Research Working Paper, vol. 6455, pp. 1–35 (2013)
3. Bond, G.E.: The effects of supply and interest rate shocks in commodity futures markets. Am. J. Agric. Econ. **66**(3), 294–301 (1984)
4. Brooks, R.D.: A social loss approach to testing the efficiency of Australian financial futures. Monash University Department of Economics, Working Paper (1989)
5. Chari, V.V., Jagannathan, R.: The simple analytics of commodity futures markets: do they stabilize prices? Do they raise welfare? Federal Reserve Bank of Minneapolis. Q. Rev. **14**, 1–13 (1990)
6. Chowdhury, A.R.: Futures market efficiency: evidence from cointegration tests. J. Futur. Mark. **11**(5), 577–589 (1991)

7. García-Verdugo, J., Consuegra, M.: Estimating functional efficiency in energy futures markets. Econ. Bus. Lett. **2**(3), 105–115 (2013)
8. Hong, B.G.: Speculation and market performance. Ph.D. thesis, Brown University (1989)
9. Kawai, M.: Spot and futures prices of nonstorable commodities under rational expectations. Q. J. Econ. **98**(2), 235–254 (1983)
10. Kawamoto, K., Hamori, S.: Market efficiency among futures with different maturities: evidence from the crude oil futures. J. Futur. Mark. **31**(5), 487–501 (2011)
11. Kumar, M.S.: Forecasting accuracy of crude oil futures prices. International Monetary Fund Working Paper (1991)
12. Ma, C.W.: Forecasting efficiency of energy futures prices. J. Futur. Mark. **5**, 393–419 (1989)
13. Maslyuk, S., Smyth, R.: Cointegration between oil spot and futures prices of the same and different grades in the presence of structural change. Energy Policy **37**, 1687–1693 (2009)
14. McKenzie, A.M., Holt, M.T.: Market efficiency in agricultural futures markets. Appl. Econ. **34**(12), 1519–1532 (2002)
15. Pennings, J.M., Garcia, P.: Risk and hedging behavior: the role and determinants of latent heterogeneity. J. Financ. Res. **XXXIII**(4), 373–401 (2010)
16. Peroni, E., McNown, R.: Noninformative and informative tests of efficiency in three energy futures markets. J. Futur. Mark. **18**(8), 939–964 (1998)
17. Pindick, R.S., Rotemberg, J.J.: The excess co-movement of commodity prices. NBER Working Paper Series, vol. 2671 (1988)
18. Stein, J.: The simultaneous determination of spot and futures prices. Am. Econ. Rev. **51**, 1012–1025 (1961)
19. Stein, J.: Spot, forward and futures. Res. Financ. **1**, 225–310 (1979)
20. Stein, J.: Speculative price: Economic welfare and the idiot of chance. Rev. Econ. Stat. **63**(2), 565–583 (1981)
21. Stein, J.: The Economics of Futures Markets. Basil Blackwell, Oxford (1986)
22. Stein, J.: An evaluation of the performance of speculative markets. Commodity Futures Financ. Mark. **21**, 153–178 (1991)
23. Stevens, J.: Testing the efficiency of futures market for crude oil using weighted least squares. Appl. Econ. Lett. **20**(18), 1611–1613 (2013)
24. Switzer, L.N., El-Khoury, M.: Extreme volatility, speculative efficiency, and the hedging effectiveness of the oil futures markets. J. Futur. Mark. **27**, 61–84 (2007)
25. Timmermann, A., Granger, C.W.: Efficient market hypothesis and forecasting. Int. J. Forecast. **20**, 15–27 (2004)
26. Turnovsky, S.J.: The determinants of spot and futures prices with storable commodities. Econometrica **51**, 1363–1387 (1983)
27. Wang, H.H., Ke, B.: Efficiency tests of agricultural commodity futures markets in China. Aust. J. Agric. Resour. Econ. **49**, 125–141 (2005)

On the Fundamental Bifurcation Theorem for Semelparous Leslie Models

J.M. Cushing

Abstract This brief survey of nonlinear Leslie models focuses on the fundamental bifurcation that occurs when the extinction equilibrium destabilizes as R_0 increases through 1. Of particular interest is the bifurcation that occurs when only the oldest age class is reproductive, in which case the Leslie projection matrix is not primitive. This case is distinguished by the invariance of the boundary of the positive cone on which orbits contain temporally synchronized, missing age classes and by the bifurcation of oscillatory attractors, lying on the boundary of the positive cone, in addition to the bifurcation of positive equilibria. The lack of primitivity of the Leslie projection matrix, while seemingly only a mathematically technicality, corresponds to a fundamental life history strategy in population dynamics, namely, semelparity (when individuals have one reproductive event before dying). The study of semelparous Leslie models was historically motivated by the synchronized outbreak cycles of periodical insects, the most famous being the long-lived cicadas (C. *magicicada* spp).

1 Introduction

Many mathematical models used to describe the dynamics of biological populations aggregate all individuals into a single state variable, such as population numbers, densities, biomass, etc. Structured population dynamics allows for differences among individuals by means of some designated characteristics. As models for the dynamics of structured populations, matrix models describe discrete time dynamical systems which advance a distribution vector $x = \mathrm{col}\,(x_i)$ of numbers (or densities) x_i of individuals, assigned to a finite collection of (say m) specified classes, forward in time by a multiplication by a projection matrix P [4]. Typically the classification scheme is based on characteristics such as chronological age, a physiological trait (size, weight, etc.), life history stages (juvenile, adult, quiescent, etc.), the state of health (disease susceptible, infected, etc.), spatial location, and so on. Historically

J.M. Cushing (✉)
Department of Mathematics & The Interdisciplinary Program in Applied Mathematics, University of Arizona, 617 N. Santa Rita, Tucson, AZ 85721, USA
e-mail: cushing@math.arizona.edu

J.-P. Bourguignon et al. (eds.), *Dynamics, Games and Science*, CIM Series in Mathematical Sciences 1, DOI 10.1007/978-3-319-16118-1_12

the first influential use of matrix models can be found in the seminal work of P.H. Leslie who studied populations structured by age [47, 48].

The projection matrix P has nonnegative entries that describe transition probabilities of individuals between classes and their mortality and fecundity rates. If these vital rates remain constant in time, then the resulting dynamic system is linear. Assuming no other processes (such as immigration or emigration, harvesting or seeding, etc.), the sequence of population densities is $x(t) = P^t x(0)$, $x(0) \geq 0$, and the study of this sequence is a beautiful application of Perron-Frobenius theory. This classic theory is applicable to the nonnegative matrix P when it is irreducible, an assumption generally made in applications to population dynamics. This amounts to requiring that there is a path that, in time, connects any two classes (through transitions or births). The vector $x(0) = 0$ remains fixed in time, a fixed point we refer to as the *extinction equilibrium*. Extinction is obviously of fundamental important in population dynamics; thus the stability of the extinction equilibrium is of basic importance in mathematical models. If the dominant eigenvalue r (the spectral radius) of P is less than 1, then the extinction equilibrium is globally attracting. If $r > 1$ then the extinction equilibrium unstable (and repeller for $x(0) \geq 0$) and $x(t)$ grows exponentially without bound. If $r = 1$, there exist bounded non-extinction states, including equilibria given by constant multiples of the positive Perron eigenvector $v > 0$ of P associated with $r = 1$ (and, if P is not primitive, there can be other bounded dynamics such as periodic cycles [1, 36]). Thus, the destabilization of the extinction state at $r = 1$ results in a bifurcation phenomenon which creates bounded non-extinction states, but in this linear case only non-generically at exactly $r = 1$. We say this bifurcation is vertical and the spectrum associated with non-extinction states is a point spectrum.

Density-dependence is a term used in population dynamics to describe the situation when vital rates of a population depend on population density. For a matrix model this means $P = P(x)$ and the resulting discrete time dynamical system becomes nonlinear. That the extinction equilibrium $x = 0$ loses stability as r increases through 1, where now r is the dominant eigenvalue of the inherent (density free) projection matrix $P(0)$, is a consequence of the linearization principle for maps [34]. The nature of the bifurcation that results, at least in a neighborhood of the bifurcation point $(r, x) = (1, 0)$, is well-known provided $P(x)$ is primitive. By primitive is meant that $P(x)$ is nonnegative, irreducible and has a *strictly* dominant eigenvalue $r(x)$ (equivalently that some integer power $P^n(x)$ is positive). In this case, a continuum of positive equilibria bifurcates from $x = 0$ as r is increased through 1 whose stability depends on the direction of bifurcation (at least in a neighborhood of the bifurcation point): they are (locally asymptotically) stable if the bifurcation is forward (i.e. they correspond to $r > 1$) and unstable if it is backward (they correspond to $r < 1$) [8, 11]. Thus, for nonlinear models the bifurcation is not vertical and the spectrum is a continuum, unlike linear matrix models. This fundamental bifurcation result is described in more detail for nonlinear Leslie age-structured matrix models in Sect. 3.

The primitivity assumption, i.e. that the dominant eigenvalue r be a strictly dominant eigenvalue of $P(0)$, might seem a minor technicality in a rigorously stated mathematical theorem. Indeed, strict dominance is not needed for the nonlinear bifurcation results described in the previous paragraph but for one crucial exception, namely, that the direction of bifurcation determines the stability of the bifurcation. This is no longer true (in general) if $P(0)$ is imprimitive. The mathematical reason is that destabilization of the extinction equilibrium occurs not solely because the real, dominant eigenvalue r leaves the unit circle in the complex plane, but because other eigenvalues simultaneously leave the unit circle. This occurrence also leads to other possible bifurcation phenomenon from $x = 0$ at $r = 1$.

These mathematical details are not insignificant with regard to applications to structured population dynamics. The semelparous Leslie model discussed in Sect. 3 is, as we will see, an example possessing an imprimitive projection matrix. A population is semelparous if individuals have only one reproductive event before death. This life history strategy is used by numerous species across many taxa, including species of insects, arachnids, molluscs, and a few species of reptiles, amphibians, and marsupials, and many species plants (for which the strategy is also known as monocarpy). Perhaps the most famous examples are certain species of cicadas and salmon and, of course, annual plants. The opposite life history strategy of multiple reproductive events before death is called iteroparity. Semelparity and iteroparity, along with traits such as the timing of reproduction, resource allocation trade-offs, and number or size of offspring, play central roles in studies of life history strategies (see for example [60, 63]). A study of matrix models with imprimitive projection matrices is, therefore, of more than just mathematical interest.

Historically, the hallmark example of a matrix model with imprimitive projection matrix is the semelparous Leslie (age structured) model. This interest was particularly stimulated by studies of cicada population dynamics that utilized Leslie matrices [2, 3] and whose periodic, synchronized outbreaks have long fascinated biologists. In Sect. 3 we survey some recent results for this model with regard to the fundamental bifurcation that occurs when the extinction equilibrium is destabilized as r increases through 1. As we will see, certain basic features of the bifurcation are known in general, but a full understanding of the bifurcation has not yet been obtained except in lower dimensions $m = 2$ and 3. The complexity of the dynamic possibilities rapidly increases with the dimension m and a full accounting of the possibilities might be attainable only for specialized models. (The same conclusion, using methods other than those described in this paper, was reached in [30].) The dimension m can be thought of as the maturation time for individuals in the population, which for the long lived periodic cicada *Cicadidae Magicicada* (which in fact is the longest lived insect known) is 13 or 17 years ($m = 13$ or 17 in the Leslie model). This provides one stimulus for further study of higher dimensional semelparous Leslie models.

2 Preliminaries

Denote m-dimensional Euclidean space by R^m and its positive cone by

$$R_+^m \overset{\circ}{=} \{x = \mathrm{col}\,(x_i) \in R^m \mid x_i > 0\}\,.$$

The closure and boundary of R_+^m are denoted by $\bar{R}_+^m$ and $\partial R_+^m = \bar{R}_+^m \backslash R_+^m$ respectively. We consider discrete time dynamical systems defined by matrix multiplication

$$x\,(0) = x_0 \in R_+^m$$
$$x\,(t+1) = Px\,(t) \;\; \text{for } t = 1, 2, \ldots$$

where the $m \times m$ matrix is called the projection P matrix. In population dynamic models, P generally has an additive decomposition

$$P = F + T$$

where F and T are the fertility and transition matrices respectively. Specifically

$$F = \left(f_{ij}\right), \qquad T = \left(s_{ij}\right)$$

$$f_{ij} \geq 0, \quad 0 \leq s_{ij} \leq 1, \quad \Sigma_{i=1}^{m} s_{ij} \leq 1 \text{ for } i, j = 1, 2, \ldots, m$$

where f_{ij} is the per unit number of i-class offspring produced by a j-class individual during a time unit that survive to the end of the time unit.

An example is the (extended) Leslie matrix model based on age classes when the census time interval for t is equal to the length of the age classes. We denote the projection matrix for a Leslie model by $L = F + T$ where

$$F = \begin{pmatrix} 0 & 0 & \cdots & 0 & s_m \\ 0 & 0 & \cdots & 0 & 0 \\ 0 & 0 & \cdots & 0 & 0 \\ \vdots & \vdots & & \vdots & \vdots \\ 0 & 0 & \cdots & 0 & 0 \end{pmatrix}, \quad T = \begin{pmatrix} 0 & 0 & \cdots & 0 & 0 \\ s_1 & 0 & \cdots & 0 & 0 \\ 0 & s_2 & \cdots & 0 & 0 \\ \vdots & \vdots & & \vdots & \vdots \\ 0 & 0 & \cdots & s_{m-1} & s_a \end{pmatrix}$$

$$0 < s_i \leq 1 \text{ for } i = 1, 2, \ldots, m-1 \tag{1}$$
$$s_m > 0, \quad 0 \leq s_a < 1.$$

Here the $i = 1, 2, \ldots, m-1$ classes consist of juveniles (non-reproducing) individuals and the adult class x_m is not structured. The number s_a is the fraction of adults who survive a time unit (and hence reproduce again). Assuming population counts are made at the end of each time step, the quantity s_m is the number of newborns produced per adult during a time unit that survive to the census time. In this way s_m contains both adult reproduction and newborn survivorship characteristics.

The famous Perron-Frobenius Theorem applies to the (nonnegative and irreducible) Leslie projection matrix L. Therefore, its spectral radius $r = \rho[L]$ is positive and is a simple eigenvalue of L with positive eigenvector

$$v(r) = \begin{pmatrix} p_1 \\ \vdots \\ \frac{p_i}{r^{i-1}} \\ \vdots \\ \frac{p_{m-1}}{r^{m-2}} \\ \frac{r}{s_m} \end{pmatrix} \tag{2}$$

where

$$p_i = \prod_{n=1}^{i-1} s_n \text{ for } i = 2, 3, \cdots, m$$

is the probability of living to age i. For later notational convenience we define $p_1 = 1$. The dominant eigenvalue r satisfies the characteristic equation.

$$r^m - s_a r^{m-1} - p_m s_m = 0.$$

Moreover, no other eigenvalue has larger absolute value nor has a nonnegative right or left eigenvector.

If the population is iteroparous, i.e. $s_a > 0$, then the Leslie matrix L is primitive. That is to say r strictly dominates all other eigenvalues. A bifurcation of equilibria (fixed points) occurs at $r = 1$. The equilibrium $x(t) \equiv 0$ is a global attractor if $r < 1$ and is a repeller if $r > 1$. At $r = 1$ there is a continuum (of global extent) of positive equilibria, namely the positive scalar multiples of $v = v(1)$. This is a vertical transcritical bifurcation whose spectrum is a single point $r = 1$. At $r = 1$ all orbits with $x_0 \in R^m_+ \setminus \{0\}$ approach a multiple of v as $t \to +\infty$ and in this sense the bifurcating branch of positive equilibria is stable.

In these assertions we can replace r by the quantity

$$R_0 = s_m \frac{p_m}{1 - s_a}. \tag{3}$$

This follows from a basic theorem in [23] that guarantees r and R_0 (in general matrix models) equal 1 simultaneously or always lie on the same side of 1. Also see [8, 11, 15, 51]. This allows for a stability determination by a simple calculation from the entries of L (no algebraic formula exists for r, in general). Biologically R_0 is the expected number of newborns per newborn over the course of its lifetime and is called the *net reproductive number* (or rate).

If the population is semelparous, i.e. $s_a = 0$, the Leslie matrix L is imprimitive. This is because r is no longer *strictly* dominant. Indeed, the eigenvalues of L are $\lambda = ru_k$ where

$$u_k \stackrel{\circ}{=} \exp\left(\frac{2\pi\,(k-1)}{m}i\right), \quad k = 1, 2, \ldots, m$$

are the mth roots of unity. In this case,

$$r = R_0^{1/m}, \quad R_0 = \prod_{i=1}^{m} s_i. \tag{4}$$

The vertical bifurcation of positive equilibria still occurs at $r = 1$, as in the primitive case, but it is no longer stable in the sense that all orbits with $x_0 \in R_+^m \setminus \{0\}$ approach a multiple of v as $t \to +\infty$. At $r = 1$ there also exists a continuum of periodic cycles. These cycles have a special form. Because $L^m = \text{diag}\,(R_0)$ and hence $L^m = I$ when $r = 1$ all points $x_0 \in R_+^m$ produce an m-cycle, i.e., a periodic orbit of period m (although m might not be the minimal period). This includes so called synchronous cycles, which are periodic orbits lying on the boundary ∂R_+^m of the positive cone. The boundary ∂R_+^m is straightforwardly seen to be forward invariant since a zero component in x_0 advances one entry (modulo m) at each time step. Similarly, positive components advance one entry at each step. These cycles have the same number of missing age classes (and positive age classes) at each point and they sequentially move between the coordinate hyperplanes. At the extreme are single class m-cycles in which only one entry is positive at each point of the cycle. In the case of semelparity we see, then, that continua of such so-called *synchronous m-cycles* also exist when $r = 1$. These synchronous cycles for the linear case can be the source of synchronous oscillations in nonlinear matrix models, to which we next turn our attention.

3 Nonlinear Leslie Matrix Models

The linear Leslie model predicts either extinction when $r < 1$ or unbounded (exponentially) unbounded growth when $r > 1$. Bounded population persistence can only occur at $r = 1$, the point spectrum of the bifurcating branch of equilibria and/or periodic cycles. The ecological notion of density dependence, i.e. the dependence of the components in F and T on population density, allows for population self regulation and bounded persistence on a spectrum of r values of positive measure (for example, all $r > 1$). This assumption results in a nonlinear matrix model of the form

$$x(0) = x_0 \in R_+^m \tag{5}$$
$$x(t+1) = L(x)\,x(t) \text{ for } t = 1, 2, \ldots$$

in which the age-specific entries in the fertility and transition matrices may depend on population density

$$L(x) = F(x) + T(x).$$

$$F(x) = \begin{pmatrix} 0\ 0 \cdots 0 & s_m\sigma_m(x) \\ 0\ 0 \cdots 0 & 0 \\ 0\ 0 \cdots 0 & 0 \\ \vdots\ \vdots \quad \vdots & \vdots \\ 0\ 0 \cdots 0 & 0 \end{pmatrix}$$

$$T(x) = \begin{pmatrix} 0 & 0 & \cdots & 0 & 0 \\ s_1\sigma_1(x) & 0 & \cdots & 0 & 0 \\ 0 & s_2\sigma_2(x) & \cdots & 0 & 0 \\ \vdots & \vdots & & \vdots & \vdots \\ 0 & 0 & \cdots & s_{m-1}\sigma_{m-1}(x) & s_a\sigma_a(x) \end{pmatrix}$$

where the fertility and survivorship parameters s_i (satisfying (1)) have been modified by density dependent, multiplicative factors $\sigma_i(x)$ normalized so that

$$\sigma_i(0) = 1 \text{ for all } i = 1, 2, \cdots, m \text{ and } i = a. \tag{6}$$

In this way, the s_i are the *inherent* fertility and survivorship rates, by which we mean the rates in the absence of density effects. We refer to r and R_0 given by (3) and (4) as the *inherent population growth rate* and the *inherent net reproductive number* respectively. We must supply the multiplicative factors σ_i with some mathematical properties, which we do by assuming the following.

A1: $\sigma_i \in C^2\left(D, \bar{R}_+^m\right)$ where D is an open set in R^m containing $\bar{R}_+^m$. In addition to the normalizations (6) we require, for $x \in \bar{R}_+^m$, that $s_m\sigma_m(x) > 0$ and $0 < s_i\sigma_i(x) \leq 1$ for $i = 1, 2, \ldots, m-1$ and $i = a$.

We denote partial derivatives by

$$\partial_j\sigma_i(x) \stackrel{\circ}{=} \frac{\partial\sigma_i(x)}{\partial x_j}$$

and introduce the notation

$$\partial_j^0 \sigma_i \triangleq \left.\frac{\partial \sigma_i(x)}{\partial x_j}\right|_{x=0}.$$

We denote the (row vector) gradient of σ_i with respect to x evaluated at $x = 0$ by

$$\nabla^0 \sigma_i \triangleq \left(\partial_1^0 \sigma_i \cdots \partial_j^0 \sigma_i \cdots \partial_m^0 \sigma_i \right).$$

A negative derivative $\partial_j \sigma_m(x) < 0$ represents a *negative feedback* on fertility with respect to an increase in the density of the jth age class when the population age distribution is x. A positive derivative represents a *positive feedback* (a so-called *component Allee effect*). Similarly for other survivorship factors $\sigma_i(x)$.

Functions commonly used by modelers for negative feedback factors include the rational function

$$\frac{1}{1 + \Sigma_{i=1}^m c_i x_i} \tag{7}$$

often referred to as a Leslie-Gower [49] (or Beverton-Holt or discrete Lotka-Volterra nonlinearity) and exponential function

$$\exp\left(-\Sigma_{i=1}^m c_i x_i\right) \tag{8}$$

(often called the Ricker model [57]). Functions that have been used for positive feedback include [7]

$$x_j \frac{1}{1 + \Sigma_{i=1}^m c_i x_i} \text{ or } x_j \exp\left(-\Sigma_{i=1}^m c_i x_i\right).$$

A basic biological question concerns the extinction or persistence of a population, which mathematically concerns the stability properties of the extinction equilibrium $x = 0$. The Jacobian of (5) at $x = 0$ is $L(0)$, which is the linear Leslie matrix in Sect. 1 with dominant eigenvalue r. The Linearization Principle [34] implies $x = 0$ is stable or unstable if $r < 1$ or $r > 1$ respectively, or mathematically more conveniently if $R_0 < 1$ or $R_0 > 1$.

Theorem 1 *Assume* A1. *If $R_0 < 1$ (equivalently $r < 1$) then the extinction equilibrium $x = 0$ is (locally asymptotically) stable. If $R_0 > 1$ (equivalently $r > 1$) then the extinction equilibrium is unstable.*

Theorems from persistence theory add to the dynamics near $x = 0$ when $r > 1$. Define

$$|x| \triangleq \sum_{i=1}^m |x_i|.$$

The nonlinear Leslie model (5) is *uniformly persistent* with respect to $x = 0$ if there exists a $\delta > 0$ such that $\liminf_{t\to+\infty} |x(t)| > \delta$ for all $x(0) \in R^m_+\backslash\{0\}$. It is *dissipative* on $\bar{R}^m_+$ if there is a compact subset of $\bar{R}^m_+$ into which all orbits in $\bar{R}^m_+$ enter and remain after a finite number of time steps. If (5) is both uniformly persistent with respect to $x = 0$ and dissipative, then it is called *permanent* on $\bar{R}^m_+$. That is to say, there exist constants $\delta_1, \delta_2 > 0$ such that

$$\delta_1 < \lim_{t\to+\infty}\inf |x(t)| \leq \lim_{t\to+\infty}\sup |x(t)| < \delta_2$$

for all $x(0) \in R^m_+\backslash\{0\}$. The following theorem is proved in [46] (also see Theorem 1.2.1 in [8] and, for the semelparous case $s_a = 0$, Proposition 3.3 in [45]).

Theorem 2 *Assume A1. If the nonlinear Leslie model (5) is dissipative on $\bar{R}^m_+$, then for $R_0 > 1$ it is permanent with respect to the extinction equilibrium $x = 0$ on $\bar{R}^m_+$.*

Under the conditions of this theorem not only is $x = 0$ unstable when $R_0 > 1$ (equivalently $r > 1$), but no orbit in $R^m_+\backslash\{0\}$ leads to extinction.

A sufficient condition for dissipativity is that there exists a number $k_0 > 0$ such that

$$\sigma_m(x)\,x_m,\ \sigma_a(x)\,x_m \leq k_0 \text{ for } x \in R^m_+. \tag{9}$$

These inequalities mean that the adult class self-regulates its vital rates. To see this we note, from the first component of the Leslie model (5), that

$$0 \leq x_1(t+1) = b\sigma_m(x(t))\,x_m(t) \leq bk_0$$

for all $t \geq 0$ from which follow the inequalities

$$0 \leq x_i(t+1) = s_{i-1}\sigma_{i-1}(x(t))\,x_{i-1}(t) \leq bk_0$$

for all $t \geq i-1$ and $i = 1, 2, \cdots, m-1$. Finally

$$0 \leq x_m(t+1) = s_{m-1}\sigma_{m-1}(x(t))\,x_{i-1}(t) + s_a\sigma_a(x(t))\,x_m(t) \leq bk_0 + s_ak_0$$

for all $t \geq m-1$. Thus, all orbits in $\bar{R}^m_+$ lie and remain in the rectangular region

$$B \overset{\circ}{=} \left\{x \in \bar{R}^m_+ \mid 0 \leq x_i \leq bk_0 \text{ for } i = 1, 2, \cdots, m-1 \text{ and } 0 \leq x_m \leq bk_0 + s_ak_0\right\}$$

after $m-1$ time steps.

From the destabilization of the equilibrium $x \equiv 0$, as caused by an eigenvalue of the Jacobian increasing through 1 as R_0 (and hence r) increases through 1, we expect that a branch of non-zero equilibria will (transcritically) bifurcate from $x \equiv 0$ at $R_0 = 1$. That is to say, we expect there will exist a continuum of pairs (R_0, x) for which x is a nonzero equilibrium of (5) whose closure contains the bifurcation point

$(1, 0)$. Theorems from bifurcation theory can be applied to validate this assertion. One way to do this is to write the equilibrium equation of (5), namely, the algebraic equation

$$x = L(x)\,x \tag{10}$$

as

$$(I - T(x))\,x = R_0 \Phi(x)\,x \tag{11}$$

where

$$\Phi(x) = \begin{pmatrix} 0 & 0 & \cdots & 0 & \frac{1-s_a}{p_m}\sigma_m(x) \\ 0 & 0 & \cdots & 0 & 0 \\ 0 & 0 & \cdots & 0 & 0 \\ \vdots & \vdots & & \vdots & \vdots \\ 0 & 0 & \cdots & 0 & 0 \end{pmatrix}.$$

Note that $s_a\sigma_a(x) < 1$ implies $I - T(x)$ has a nonnegative inverse

$$\begin{pmatrix} 1 & 0 & \cdots & 0 & 0 & 0 \\ s_1\sigma_1(x) & 1 & \cdots & 0 & 0 & 0 \\ s_1\sigma_1(x)\,s_2\sigma_2(x) & s_2\sigma_2(x) & \cdots & 0 & 0 & 0 \\ \vdots & \vdots & & \vdots & \vdots & \vdots \\ \Pi_{j=1}^{m-2} s_j\sigma_j(x) & \Pi_{j=2}^{m-2} s_j\sigma_j(x) & \cdots & s_{m-2}\sigma_{m-2}(x) & 1 & 0 \\ \frac{\Pi_{j=1}^{m-1} s_j\sigma_j(x)}{1-s_a\sigma_a(x)} & \frac{\Pi_{j=2}^{m-1} s_j\sigma_j(x)}{1-s_a\sigma_a(x)} & \cdots & \frac{\Pi_{j=m-2}^{m-1} s_j\sigma_j(x)}{1-s_a\sigma_a(x)} & \frac{s_{m-1}\sigma_{m-1}(x)}{1-s_a\sigma_a(x)} & \frac{1}{1-s_a\sigma_a(x)} \end{pmatrix}.$$

We write the equilibrium equation (11) equivalently as

$$x = R_0 M(x)\,x$$

where

$$M(x) \mathrel{\mathring{=}} (I - T(x))^{-1}\,\Phi(x) = \begin{pmatrix} 0 & 0 & \cdots & 0 & \frac{1-s_a}{p_m} \\ \vdots & \vdots & & \vdots & \vdots \\ 0 & 0 & \cdots & 0 & p_j\frac{1-s_a}{p_m} \\ \vdots & \vdots & & \vdots & \vdots \\ 0 & 0 & \cdots & 0 & 1 \end{pmatrix}$$

which has the abstract form

$$x = R_0 M(0)\, x + R_0 h(x) \tag{12}$$

where

$$|R_0 h(x)| = O\left(|x|^2\right) \text{ near } x = 0$$

uniformly in R_0 on compact intervals.

Corresponding to a (nonzero, positive or negative) solution of Eq. (12) we refer to a *(nonzero, positive or negative) equilibrium pair* (R_0, x). Equation (12) has the form of the nonlinear eigenvalue problem whose bifurcation properties are studied in [56] (also see [40]). A characteristic value of a matrix is the reciprocal of an eigenvalue. Note that $M(0)$ has one characteristic value, namely, 1 and that it is geometrically simple. Using Theorem 1.20 in [56] (see the Appendix) we find that there exists a continuum $\mathscr{C}_+^e$ of positive equilibrium pairs (R_0, x) that contains $(1, 0)$ and is unbounded in $R_+^1 \times R_+^m$. (See the Appendix for further details.) It follows that either the *spectrum* of $\mathscr{C}_+^e$

$$\mathscr{S} \mathrel{\hat{=}} \left\{R_0 |\, (R_0, x) \in \mathscr{C}_+^e\right\} \subset R_+$$

is unbounded in R_+ or the *range* of $\mathscr{C}_+^e$

$$\mathscr{R} \mathrel{\hat{=}} \left\{x |\, (R_0, x) \in \mathscr{C}_+^e\right\} \subset R_+^m$$

of $\mathscr{C}_+^e$ is unbounded in R_+^m or both. Both are continua.

Perturbation methods in classic bifurcation theory (e.g. Lyapunov-Schmidt techniques) allow for a parameterization of the bifurcating continuum $\mathscr{C}$ of positive equilibria near the bifurcation point $(R_0, x) = (1, 0)$. The result is contained in the following theorem (see [8] for details). Let $v = v(1)$, i.e.

$$v = \begin{pmatrix} p_1 \\ \vdots \\ p_i \\ \vdots \\ p_{m-1} \\ \frac{1}{s_m} \end{pmatrix}$$

[see (2)].

Theorem 3 *Assume A1 and $a_1 \neq 0$ where*

$$a_1 \mathrel{\hat{=}} \Sigma_{i=1}^m \nabla^0 \sigma_i\, v + \frac{s_m}{1 - s_a} \nabla^0 \sigma_a\, v. \tag{13}$$

The nonlinear Leslie model (5) has is a continuum $\mathscr{C}_+^e$ of positive equilibrium pairs which bifurcates from $(R_0, x) = (1, 0)$, is unbounded in $R_+ \times R_+^m$ and, near the bifurcation point $(R_0, x) = (1, 0)$, has the parameterization

$$x = -\frac{1}{a_1} v\varepsilon + \eta\varepsilon^2 + O\left(\varepsilon^3\right), \qquad R_0 = 1 + \varepsilon \tag{14}$$

for $\varepsilon \gtrapprox 0$ if $a_1 < 0$ and $\varepsilon \lessapprox 0$ if $a_1 > 0$.

Definition 1 The bifurcation of the continuum $\mathscr{C}_+^e$ is forward (to the right or super-critical) if $(R_0, x) \in \mathscr{C}_+^e$ implies $R_0 > 1$ in a neighborhood of the bifurcation point $(1, 0)$. The bifurcation of the continuum $\mathscr{C}_+^e$ is backward (to the left or sub-critical) if $(R_0, x) \in \mathscr{C}_+^e$ implies $R_0 < 1$ in a neighborhood of the bifurcation point $(1, 0)$.

Note that the *direction of bifurcation* of $\mathscr{C}_+^e$ is determined by the sign of the quantity a_1 (if it is nonzero). If $a_1 < 0$ (which is certainly the case if there are no positive feedback components at $x = 0$) then $\mathscr{C}_+^e$ bifurcates forward. On the other hand, if $a_1 > 0$ then the bifurcation is backward. An inspection of the formula for a_1 shows that the latter case requires positive feedback density effects at low population densities (i.e. component Allee effects [7]) and these must be of sufficient magnitude if negative feedback components are also present.

The set of R_0 values for which the nonlinear Leslie model (5) has a positive equilibrium is of obvious interest applications. This set includes the spectrum $\mathscr{S}$ of the bifurcating continuum $\mathscr{C}_+^e$. $\mathscr{S}$ is a continuum, i.e. is an interval in R_+, whose closure contains 1. In an exceptional case, $\mathscr{S}$ could be the singleton set $\{1\}$ as it is in the linear case. However, more generally $\mathscr{S}$ is an interval of real numbers of positive measure. This is certainly the case if $a_1 \neq 0$ since, in that case, the bifurcation at $(1, 0)$ is not vertical by Theorem 3.

When can we expect there to exist a positive equilibrium for *all* values of $R_0 > 1$?

Corollary 1 ([17]) *Assume A1. If there exists a function $k : R_+ \to R_+$, bounded on bounded sets in R_+, such that*

$$|x| \leq k\left(R_0\right) \text{ for all } \left(R_0, x\right) \in \mathscr{C}_+^e, \tag{15}$$

then the spectrum $\mathscr{S} \subset R_+$ of $\mathscr{C}_+^e$ is unbounded and there exists (at least one) positive equilibrium for each $R_0 > 1$.

This corollary follows because (15) implies that a bounded spectrum $\mathscr{S}$ would imply a bounded range $\mathscr{R}$, in contradiction to $\mathscr{C}_+^e$ being unbounded.

As an example, suppose the inequalities (9) hold. We showed above that all orbits eventually lie and remain in the compact region B. It follows that any equilibrium must lie in this region and therefore (15) holds with

$$k\left(R_0\right) \stackrel{\circ}{=} m \frac{R_0}{\Pi_{j=1}^{m-1} s_j} k_0 + s_a k_0.$$

Note: A pair $(R_0, x) \in \mathscr{C}_+^e$ corresponds to a positive equilibrium $x \in R_+^m$ of the nonlinear Leslie model for parameters s_i that yield the corresponding R_0 value. The inherent projection matrix $L(0)$ of that model has a dominant eigenvalue r and we can, therefore, associate with each pair in $\mathscr{C}_+^e$ a positive equilibrium pair (r, x). If the R_0 spectrum $\mathscr{S}$ of $\mathscr{C}_+^e$ is unbounded, it follows from a theorem of Li and Schneider [51] that the spectrum of r values obtained from the corresponding equilibrium pairs (r, x) is also unbounded. ■

When might there be positive equilibria for $R_0 < 1$? This will certainly be the case when $a_1 > 0$ and the continuum $\mathscr{C}_+^e$ bifurcates backward at $(1, 0)$. As we observe from the formula (13) for a_1, this requires the presence of component Allee effects of sufficient magnitude. If $a_1 < 0$ then there are no positive equilibria for $R_0 < 1$ in a neighborhood of the bifurcation point $(1, 0)$, but this does preclude the possibility of equilibrium pairs (R_0, x) from the continuum $\mathscr{C}_+^e$ outside a neighborhood of $(1, 0)$ for which $R_0 < 1$. One case in which this can be ruled out altogether is when

$$\sigma_i(x), \ \sigma_a(x) \leq 1 \text{ for } x \in \bar{R}_+^m \tag{16}$$

(which disallows any component Allee effects near $x = 0$). In this case we obtain from equilibrium equation (10) the inequality

$$0 \leq x = L(x) \leq L(0)x.$$

If $R_0 < 1$ then $r < 1$ [23] and all orbits of

$$y_0 = x_0 \in R_+^m$$
$$y(t+1) = L(0)\, y(t)$$

satisfy $\lim_{t++\infty} y(t) = 0$, By a straightforward comparison argument, the orbits of the nonlinear Leslie model (5) satisfy $0 \leq x(t) \leq y(t)$ and hence also tend to the origin.

Corollary 2 *Assume A1 and (16). Then $R_0 < 1$ (equivalently $r < 1$) implies that the extinction equilibrium $x = 0$ is globally asymptotically stable.*

Combining these results we have the following result.

Corollary 3 *Assume A1, the adult self regulation assumption (9), and that the inequalities (16) hold. Then for the nonlinear Leslie model (5) we have that:*

- *the extinction equilibrium $x = 0$ is globally asymptotically stable for $R_0 < 1$;*
- *the model is permanent with respect to $x = 0$ for $R_0 > 1$;*
- *there exists at least one positive equilibrium for all values of $R_0 > 1$.*

Example sub-models for the vital rates $\sigma_i(x)$ and $\sigma_a(x)$ that satisfy (16) are the Leslie-Gower (7) and the Ricker (8) functions. The adult self regulation inequalities (9) are satisfied if $c_m > 0$, and hence all the conclusions in Corollary 3 hold for models built using any combinations of these familiar nonlinearities.

We have not yet taken up the question of the stability or instability of the equilibria from the bifurcating continuum $\mathscr{C}_+^e$. In general such equilibria can be either stable or unstable, depending on the specifics of the nonlinearities used in the model. Some general conclusions can be made, however, in the neighborhood of the bifurcation point $(R_0, x) = (1, 0)$. Tractability of this question is obtained from the parameterization (14) of $\mathscr{C}_+^e$ near $(1, 0)$. This parameterization allows for a parameterization of the Jacobian of (5) evaluated at the positive equilibrium and, in turn, a parameterization of the eigenvalues $\lambda = \lambda(\varepsilon)$ of the Jacobian.

If the population is iteroparous, i.e. if $s_a > 0$, then the Jacobian at the bifurcation point has a strictly dominant eigenvalue of 1. Thus, $\lambda_1(0) = 1$ and all other eigenvalues lie inside the unit circle. This means, in this case, that these eigenvalues will remain inside the complex unit circle for ε small (by continuity) and the stability of the bifurcating positive equilibria can be determined by the eigenvalue $\lambda_1(\varepsilon) = 1 + \lambda_1'(0)\varepsilon + O(\varepsilon^2)$ alone. A calculation of $\lambda_1'(0)$ can be made by perturbation or calculus methods and the result is (see Lemma 1.2.2 in [8] or, in a more general abstract setting, see the exchange of stability principle for transcritical bifurcations in [40])

$$\lambda_1(\varepsilon) = 1 - ca_1\varepsilon + O(\varepsilon^2)$$

for a positive constant $c > 0$. This leads to the following stability result for the equilibrium on the bifurcating continuum $\mathscr{C}_+^e$ in a neighborhood of the bifurcation point.

Definition 2 The bifurcation of $\mathscr{C}_+^e$ at $R_0 = 1$ is called stable (unstable) if, in a neighborhood of $(R_0, x) = (1, 0)$, the positive equilibria from the range of the continuum $\mathscr{C}_+^e$ are locally asymptotically stable (unstable).

Theorem 4 *Assume A1 and $s_a > 0$. If $a_1 < 0$ then the bifurcation of the continuum $\mathscr{C}_+^e$ of positive equilibrium pairs of the nonlinear, iteroparous Leslie model (5) at $(R_0, x) = (1, 0)$ is forward and stable. If $a_1 > 0$ it is backward and unstable.*

Taken together Theorems 3 and 4 constitute a fundamental bifurcation theorem for the iteroparous nonlinear Leslie model (5) that guarantees the occurrence of a transcritical bifurcation of positive equilibria at the destabilization of the extinction equilibrium when R_0 increases through 1 and the fact that the stability or instability of the bifurcation depends on the direction of bifurcation.

Example 1 The $m = 3$ stage nonlinear Leslie model with projection matrix

$$L(x) = \begin{pmatrix} 0 & 0 & be^{-c_1x_1(t) - c_3x_3(t)} \\ 1 - \mu_l & 0 & 0 \\ 0 & (1 - \mu_p)\, e^{-c_2x_3(t)} & 1 - \mu_a \end{pmatrix} \tag{17}$$

$$b > 0, 0 < \mu_l,\ \mu_p,\ \mu_a < 1 \text{ and } c_i > 0.$$

(known as the LPA model) was extensively used over a period of several decades in numerous experimental studies of nonlinear dynamics involving the insect *Tribolium castaneum* (aka flour beetles). See [24] and [6]. The three stages represent larval, pupal and adult stages in this insect and the unit of time is 2 weeks. This matrix model has the form (5) with parameters

$$s_3 = b,\ s_1 = 1 - \mu_l,\ s_2 = 1 - \mu_p,\ s_a = 1 - \mu_a$$

$$R_0 = b\frac{(1-\mu_l)\left(1-\mu_p\right)}{\mu_a}$$

and Ricker-type nonlinearities (8):

$$\sigma_1(x) \equiv 1, \quad \sigma_2(x) = e^{-c_2 x_3(t)}, \quad \sigma_3(x) = e^{-c_1 x_1(t) - c_3 x_3(t)}, \quad \sigma_a(x) \equiv 1$$

for which assumption A1 holds with $D = R^3$. There are no positive feedbacks and, indeed, the inequalities (16) hold. The population is iteroparous ($s_a > 0$). From these observations we conclude from Theorem 3, Corollary 2 and Theorem 4 that the extinction equilibrium is globally asymptotically stable for $R_0 < 1$, that the population permanent when $R_0 > 1$, and the bifurcation of positive equilibria at $R_0 = 1$ is forward and stable.

Although the inequalities (9) do not both hold ($\sigma_3(x)\,x_3$ is bounded for $x \in R^3_+$, but $s_a\sigma_a(x)\,x_3$ is not), an observation of the components of the equilibrium equations yields, for $x \in R^3_+$, the inequalities

$$0 \le x_1 \le b\frac{1}{c_1 e}$$

$$0 \le x_2 \le (1-\mu_l)\, b\frac{1}{c_1 e}$$

$$0 \le x_3 \le \left(1-\mu_p\right)(1-\mu_l)\, b\frac{1}{c_1 e} + (1-\mu_a)\, x_3$$

the latter of which implies

$$0 \le x_3 \le \frac{\left(1-\mu_p\right)(1-\mu_l)}{\mu_a} b\frac{1}{c_1 e}.$$

A summation shows that (15) holds with

$$k(R_0) = \frac{\mu_a + \mu_a(1-\mu_l) + (1-\mu_l)\left(1-\mu_p\right)}{(1-\mu_l)\left(1-\mu_p\right) c_1 e} R_0.$$

It follows from Corollary 1 that there exists at least one positive equilibrium for all values of $R_0 > 1$.

The stability properties of the bifurcating positive equilibria, in a neighborhood of the bifurcation point, given in Theorem 4 might or might not persist globally along the continuum $\mathscr{C}_+^e$. As is well known for nonlinear maps further bifurcations (numerous types), and even routes-to-chaos, can occur as R_0 is increased. This can indeed happen for the LPA model in Example 1, which formed the basis of the nonlinear studies described in [6, 24].

Strong Allee effects have been of increasing interest in theoretical ecology during the last couple of decades [7]. This is a dynamic scenario in which there exist multiple (nonnegative) attractors one of which is the extinction equilibrium, a scenario which in matrix models can only occur if $R_0 \leq 1$. One common way that a strong Allee effect arises in models is when a backward bifurcation occurs at $R_0 = 1$ and the spectrum $\mathscr{S}$ of $\mathscr{C}_+^e$ is infinite. In this case, $R_0 = 1$ necessarily lies in the spectrum $\mathscr{S}$ which would imply the existence of a positive equilibrium for R_0 at and near 1 and, in particular for $R_0 \lessapprox 1$. This occurs, for example if $a_1 > 0$ and the bound (15) hold (Theorem 3 and Corollary 1).

Geometrically, one can think of the backward bifurcating continuum $\mathscr{C}_+^e$ as "turning around" at a point $(R_0^*, x^*) \in R_+^1 \times R_+^m$ (usually at a saddle-bifurcation) so as to have an infinite spectrum (or so as to at least include $R_0 = 1$). Thus, for $R_0 \lessapprox 1$ the extinction equilibrium $x = 0$ is stable and there exist (at least) two other positive equilibria, one of which (near the bifurcation point) is unstable and the other on $\mathscr{C}_+^e$ is potentially stable.

The turning point of $\mathscr{C}_+^e$ usually occurs as a saddle-node (blue-sky) bifurcation which creates stable positive equilibria (and hence a strong Allee effect involving equilibria for at least $R_0 \gtrapprox R_0^*$). If this stability of the positive equilibria persists along the continuum $\mathscr{C}_+^e$ until $R_0 = 1$, then a strong Allee effect involving equilibria occurs for $R_0 \lessapprox 1$. It can happen, however, that the stable positive equilibria created by the saddle-node bifurcation lose their stability at a spectrum point $R_0 < 1$, say by a period doubling or Neimark-Sacker bifurcation. In this case, a strong Allee effect occurs for $R_0 \lessapprox 1$ that involve non-equilibrium attractors. For examples, see [17].

In this scenario, a strong Allee effect provides the possibility of population survival when environmental conditions degrade so as to produce $R_0 < 1$. It requires a backward bifurcation which, in turn, requires sufficiently strong positive feedbacks (component Allee effects) at low population densities. The caveat is, of course, that the population must remain out of the basin of attraction of the extinction state (the Allee basin).

4 Nonlinear Semelparous Leslie Models

All the theorems and corollaries in Sect. 3 are valid for the semelparous ($s_a = 0$) Leslie model

$$\begin{aligned} x(0) &= x_0 \in R_+^m \\ x(t+1) &= L(x)\, x(t) \quad \text{for } t = 1, 2, \ldots \end{aligned} \tag{18}$$

$$L(x) = T \begin{pmatrix} 0 & 0 & \cdots & 0 & s_m\sigma_m(x) \\ s_1\sigma_1(x) & 0 & \cdots & 0 & 0 \\ 0 & s_2\sigma_2(x) & \cdots & 0 & 0 \\ \vdots & \vdots & & \vdots & \vdots \\ 0 & 0 & \cdots & s_{m-1}\sigma_{m-1}(x) & 0 \end{pmatrix}$$

with the exception of Theorem 4. For the semelparous Leslie model, as we will see, the stability of the bifurcating positive equilibria does not depend solely on the direction of bifurcation.

4.1 Bifurcating Equilibria

The mathematical reason underlying the failure of the direction of bifurcation to sufficiently determine the stability of the bifurcating continuum $\mathscr{C}_+^e$ of positive equilibria at $R_0 = 1$ for semelparous Leslie models is the imprimitivity of the inherent projection matrix $L(0)$. The destabilization of the extinction equilibrium $x = 0$ in this case is not caused by the dominant eigenvalue $L(0)$ alone leaving the complex unit circle, but by all m eigenvalues simultaneously leaving the unit circle (at the mth roots of unity) as R_0 increases through 1. As a consequence of this, when analyzing the parameterized branch of positive equilibria, as outlined in the paragraph preceding Theorem 4, one needs to calculate expansions of all m eigenvalues of the Jacobian in order to see whether they all move into the complex unit disk or whether at least one moves out of the circle as one follows the bifurcating branch of positive equilibria. These calculations are carried out in [18] where conditions for stability and instability are obtained that involve quantities in addition to a_1. Define

$$a_k \triangleq \sum_{n=1}^{m}\sum_{j=1}^{m} p_j\partial_j^0\sigma_n \operatorname{Re} u_k^{n-j} \text{ for } k = 1, 2, \ldots, m_{1/2} + 1 \tag{19}$$

where

$$m_{1/2} \triangleq \begin{cases} \frac{m}{2} & \text{if } m \text{ is even} \\ \frac{m-1}{2} & \text{if } m \text{ is odd.} \end{cases}$$

Since $u_1 = 1$ the definition of a_1 is consistent with that in Sect. 3 when $s_a = 0$.

The following theorem provides stability and instability criteria for the bifurcating positive equilibria, in a neighborhood of the bifurcation point, in relation to the direction of bifurcation. It follows from the expansion

$$|\lambda_k| = 1 - \frac{1}{m}\frac{a_k}{a_1}\varepsilon + O\left(\varepsilon^2\right), \quad \varepsilon = R_0 - 1 \tag{20}$$

of the eigenvalue magnitudes for the Jacobian evaluated at the bifurcating equilibria near the bifurcation point, as calculated in [18].

Theorem 5 ([18]) *Assume A1. Let $\mathscr{C}_+^e$ be the unbounded continuum of positive equilibrium pairs that bifurcates from $(R_0, x) = (1, 0)$ as given in Theorem 3.*

(a) If $a_1 < 0$ then the bifurcation of $\mathscr{C}_+^e$ at $R_0 = 1$ is forward. If $a_k < 0$ for all $k = 1, 2, \ldots, m_{1/2} + 1$ then the bifurcation of $\mathscr{C}_+^e$ is stable. If at least one of these $a_k > 0$, then the bifurcation of $\mathscr{C}_+^e$ is unstable.
(b) If $a_1 > 0$ then the bifurcation $\mathscr{C}_+^e$ at $R_0 = 1$ is backward and unstable.

Generically, in the sense that $a_1 \neq 0$, Theorem 5(b) implies the backward bifurcation of positive equilibria at $R_0 = 1$ $(a_1 > 0)$ is always unstable.[1] However, Theorem 5(a) shows that a forward bifurcation is not necessarily stable, in contrast to the iteroparous case in Theorem 4. While analytically rather clear-cut, the stability criteria for a forward bifurcation, namely, that all a_k be negative, does not lend itself to an immediately obvious biological interpretation. They have to do with the relationship between the effects of density on vital rates among individuals within the same age class and those among individuals of different age classes.

For example, suppose there are no density effects between age classes; that is to say, suppose σ_i does not depend on x_j for all $j \neq i$. Then $\partial_j^0 \sigma_n = 0$ for all $j \neq n$ and $a_k = a_1$ for all k. It follows from Theorem 5 that a forward bifurcation is stable. More generally, write

$$a_k = a_1 + \sum_{n=1}^{m} \sum_{j=1}^{m} p_j \partial_j^0 \sigma_n \left(\operatorname{Re} u_k^{n-j} - 1 \right)$$

and note that the double sum on the right side contains no within-class density effects, i.e. no derivatives $\partial_j^0 \sigma_n$ with $j = n$. Thus, if $a_1 < 0$ and the magnitudes of all between-class density effects are sufficiently small, then $a_k < 0$ for all k.

Corollary 4 *If $a_1 < 0$ and between-class density effects are weak, i.e. $\left| \partial_j^0 \sigma_n \right|$ are sufficiently small for all $j \neq n$, then the bifurcation of $\mathscr{C}_+^e$ at $R_0 = 1$ is forward and stable for the semelparous model (18).*

As we will see in Sect. 4.2, when between-class density effects become significant, the stability of the bifurcating branch of positive equilibria can be lost. For a further analysis of the relationship between between-class and within-class density effects, the direction of bifurcation, and the stability properties of the bifurcating positive equilibria see [18].

Note that Corollary 1 concerning an unbounded spectrum and the sufficiency of (9) for an unbounded spectrum both hold for the semelparous model (18).

[1] This corrects an error in Theorem 4.1 of [10].

4.2 Bifurcating Synchronous Cycles

To investigate further the nature of the bifurcation at $R_0 = 1$ for the nonlinear semelparous Leslie model (18), we begin with the $m = 2$ dimensional case

$$L(x) = \begin{pmatrix} 0 & s_2\sigma_2(x_1, x_2) \\ s_1\sigma_1(x_1, x_2) & 0 \end{pmatrix}.$$

This semelparous, juvenile-adult model has been extensively studied by several authors [19–22, 32, 33, 55, 64, 66] (also see [54]).

If one begins with a population of only $x_1 > 0$ juveniles, then the resulting orbit

$$x(0) = \begin{pmatrix} x_1 \\ 0 \end{pmatrix}, \quad x(1) = \begin{pmatrix} 0 \\ s_1\sigma_1(x_1, 0)\, x_1 \end{pmatrix},$$

$$x(2) = \begin{pmatrix} R_0\sigma_2(0, s_1\sigma_1(x_1, 0)\, x_1)\, \sigma_1(x_1, 0)\, x_1 \\ 0 \end{pmatrix}, \cdots$$

sequentially visits the positive coordinate axes. This shows that the boundary of the positive cone is invariant and has a dynamic that can be understood by an analysis of the one-dimensional composite map

$$x_1(t+1) = R_0\bar{\sigma}(x_1(t))\, x_1(t) \tag{21}$$

$$\bar{\sigma}(x_1) \overset{\circ}{=} \sigma_2(0, s_1\sigma_1(x_1, 0)\, x_1)\, \sigma_1(x_1, 0), \quad R_0 = s_1 s_2$$

which describes the dynamics of every other point (the juvenile component) of the orbit. An equilibrium of this composite map corresponds to an orbit of period 2 of the semelparous model. This cycle is an example of a *single-class synchronous 2-cycle* by which is meant a periodic cycle in which the age classes are synchronized in a way that they are temporally separated and that only one class is present at each point in time.

One-dimensional maps, such as (21), have been well studied and there is a large literature, and a large quantity of analytic methods, available for their analysis. For example, one approach is to view (21) as an $m = 1$ dimensional matrix model to which we can apply the equilibrium bifurcation theorems in Sect. 3. Or, more straightforwardly, one can investigate the equilibrium equation for positive solution pairs (R_0, x_1), which obviously are defined by the equation

$$1 = R_0\bar{\sigma}(x_1).$$

Noting that $\bar{\sigma}(x_1)$ is positive valued for $x_1 \geq 0$ and that $\bar{\sigma}(0) = 1$, we see that the pairs $(R_0, x_1) = (1/\bar{\sigma}(x_1), x_1)$ for $x_1 \in R^1_+$ define a continuum of equilibrium pairs that bifurcates from $(1, 0)$ at $R_0 = 1$ (whose range is R^1_+) and whose direction of bifurcation is forward and stable or backward and unstable if

$$\partial_1^0 \bar{\sigma} < 0 \text{ (or } \partial_1^0 \bar{\sigma} > 0)$$

respectively. A calculation shows $\partial_1^0 \bar{\sigma} = c_w$ where

$$c_w \triangleq \partial_1^0 \sigma_1 + s_1 \partial_2^0 \sigma_2.$$

Each of these equilibrium pairs (R_0, x_1) corresponds to a single-class synchronous 2-cycle of the $m = 2$ of the dimensional semelparous Leslie model (18) as defined by the two points

$$\begin{pmatrix} x_1(1) \\ x_2(1) \end{pmatrix} = \begin{pmatrix} x_1 \\ 0 \end{pmatrix}, \quad \begin{pmatrix} x_1(2) \\ x_2(2) \end{pmatrix} = \begin{pmatrix} 0 \\ s_1 \sigma_1(x_1, 0) x_1 \end{pmatrix}$$

of the cycle. Identify this cycle by its pair of positive components $x_1(1), x_2(2)$ (which are the two cohort densities that temporally alternate) and denote the corresponding *single-class synchronous 2-cycle pair* by

$$(R_0, [x_1(1), x_2(2)]) \in R_+^1 \times R_+^2$$

where for every $x_1 \in R_+^1$

$$R_0 = \frac{1}{\bar{\sigma}_1(x_1)}, \quad x(1) = x_1, \quad x(2) = s_1 \sigma_1(x_1, 0) x_1.$$

The continuum of equilibrium pairs (R_0, x_1) of (21) produces a continuum $\mathscr{C}_+^2$ of these single-class synchronous 2-cycle pairs of the semelparous Leslie model (18). This continuum $\mathscr{C}_+^2$ bifurcates from $(1, 0, 0)$ at $R_0 = 1$ and it is a forward bifurcation if $c_w < 0$ and a backward bifurcation if $c_w > 0$.

That the stability of the $x(1) = x_1$ component as an equilibrium of the composite map (21) depends on the direction of bifurcation only tells us about the stability of the single-class 2-cycles with respect to the dynamics on the boundary ∂R_+^2. The stability or instability of these cycles as cycles of the semelparous Leslie model (18) on $\bar{R}_+^2$ requires further analysis. This stability analysis involves the eigenvalues μ_1, μ_2 of the Jacobian of the composite map (which is the product of the Leslie model's Jacobian evaluated at the two points of the 2-cycle). Making use of the parameterizations

$$\begin{pmatrix} x_1 \\ 0 \end{pmatrix} = \begin{pmatrix} -\frac{1}{c_w} \\ 0 \end{pmatrix} \varepsilon + O(\varepsilon^2), \quad \begin{pmatrix} 0 \\ s_1 \sigma_1(x_1, 0) x_1 \end{pmatrix} = \begin{pmatrix} 0 \\ -\frac{s_2}{c_w} \end{pmatrix} \varepsilon + O(\varepsilon^2),$$

$$R_0 = 1 + \varepsilon$$

of the synchronous 2-cycle points, one can calculate the parameterizations

$$\mu_1 = 1 - \varepsilon + O\left(\varepsilon^2\right), \quad \mu_2 = 1 + \frac{c_w - c_b}{c_w}\varepsilon + O\left(\varepsilon^2\right)$$

of these eigenvalues, where

$$c_b \stackrel{\circ}{=} a_1 - c_w = \partial_1^0 \sigma_2 + s_1 \partial_2^0 \sigma_1$$

measures the between-class density effects. Accompanying these bifurcating, single-class synchronous 2-cycles, are the bifurcating positive equilibria, which have, together with eigenvalues of the associated Jacobian, the expansions

$$\begin{pmatrix} x_1 \\ x_2 \end{pmatrix} = \begin{pmatrix} -\frac{1}{c_w + c_b} \\ -\frac{s_1}{c_w + c_b} \end{pmatrix} \varepsilon + O\left(\varepsilon^2\right), \quad R_0 = 1 + \varepsilon$$

$$\lambda_1 = 1 - \frac{1}{2}\varepsilon + O\left(\varepsilon^2\right), \quad \lambda_2 = -1 + \frac{c_w - c_b}{c_w + c_b}\varepsilon + O\left(\varepsilon^2\right).$$

From these expansions one can sort out the direction of bifurcation and stability properties of both the positive equilibria and synchronous 2-cycles. The results are summarized in the following theorem.

Theorem 6 ([22]) *Consider the $m = 2$ dimensional, semelparous Leslie model (18) and assume A1 holds. Also assume that $a_1 = c_w + c_b \neq 0$ and $c_w \neq 0$. Then a bifurcation of unbounded continua $\mathscr{C}_+^e$ and $\mathscr{C}_+^2$ of positive equilibrium pairs and single-class, synchronous 2-cycles respectively occurs at $R_0 = 1$.*

(a) If $c_w + c_b < 0$ then the bifurcation of $\mathscr{C}_+^e$ at $R_0 = 1$ is forward. The bifurcation is stable if $c_w - c_b < 0$ and unstable if $c_w - c_b > 0$.
(b) If $c_w + c_b > 0$ then the bifurcation of $\mathscr{C}_+^e$ at $R_0 = 1$ is backward and unstable.
(c) If $c_w < 0$ then the bifurcation of $\mathscr{C}_+^2$ at $R_0 = 1$ is forward. The bifurcation is stable if $c_w - c_b > 0$ and unstable if $c_w - c_b < 0$.
(d) If $c_w > 0$ then the bifurcation of $\mathscr{C}_+^2$ at $R_0 = 1$ is backward and unstable.

That a forward bifurcation in a nonlinear semelparous Leslie model (5) is not necessarily stable can be seen by cases (a) and (c). Also note that the positive equilibria and the synchronous 2-cycles can bifurcate in opposite directions, since $c_w + c_b$ and c_w do not necessarily have the same signs. In any case, however, a backward bifurcation is always unstable. Also note that it is never the case that both bifurcating continua are stable, although it is possible that both are unstable.

A natural question to ask is whether Theorem 6 can be extended in some manner to higher dimensional semelparous Leslie models with $m \geq 3$. The properties of the $\mathscr{C}_+^e$ equilibrium bifurcation for $m \geq 3$, namely its occurrence, direction of bifurcation and stability properties, are described by Theorem 5. We next turn our attention to the bifurcation of single-class, synchronous m-cycles at $R_0 = 1$.

Periodic m-cycles are fixed points of the m-fold composite map obtained from (18) whose components satisfy equations of the form

$$\begin{aligned} x_1 &= R_0\bar{\sigma}_1(x)\,x_1 \\ x_2 &= R_0\bar{\sigma}_2(x)\,x_2 \\ &\vdots \\ x_m &= R_0\bar{\sigma}_m(x)\,x_m \end{aligned} \tag{22}$$

where $\bar{\sigma}_i(x)$ is a composite made from the coefficients $\sigma_i(x)$ of the Leslie projection matrix $L(x)$. Single class synchronous cycles correspond to fixed points with $x_j = 0$ for $j = 2, 3, \ldots, m$, and where $x_1 > 0$ is a positive fixed point of the one dimensional map

$$x_1 = R_0\bar{\sigma}(x_1)\,x_1 \tag{23}$$

with

$$\bar{\sigma}(x_1) \overset{\circ}{=} \bar{\sigma}_1(x)\big|_{x_j=0,\,j\neq 1} > 0, \qquad \bar{\sigma}(0) = 1.$$

This one dimensional map can be treated in the manner that we treated the map (21) when $m = 2$. There is a continuum of equilibrium pairs (R_0, x_1) that bifurcates forward (backward) from $(1, 0)$ if $c_w < 0$ (or $c_w > 0$) where

$$c_w = \sum_{n=1}^{m} p_n \partial_n^0 \sigma_n. \tag{24}$$

Each equilibrium pair corresponds to a single-class, synchronous m-cycle of the semelparous Leslie model (18) and the bifurcating continuum of equilibrium pairs produces a continuum $\mathscr{C}_+^m$ of single-class synchronous m-cycle pairs

$$(R_0, [x_1(1), x_2(2), \ldots, x_m(m)]) \in R_+^1 \times R_+^m$$

of the semelparous Leslie model (18). This continuum $\mathscr{C}_+^m$ bifurcates from $(1, [0, \ldots, 0])$ at $R_0 = 1$ and it is a forward bifurcation if $c_w < 0$ and a backward bifurcation if $c_w > 0$.

Since positive fixed points of (23) satisfy the equation

$$1 = R_0\bar{\sigma}(x_1)$$

the range of the continuum (i.e. of the set of x_1 values obtained from the continuum) is the half line R_+^1, since there is a unique $R_0 = 1/\bar{\sigma}(x_1)$ for each $x_1 \in R_+^1$. With regard to the spectrum of R_0 values from the continuum $\mathscr{C}_+^m$, a little thought about the composite $\bar{\sigma}(x_1)$ reveals that the adult regulation assumption (9) on σ_m implies $\bar{\sigma}(x_1)x_1$ is bounded for $x_1 \geq 0$. This in turn implies the spectrum of R_0 on the continuum is unbounded (and hence is a half line in R_+^1).

We summarize these results in the following theorem.

Theorem 7 *Assume A1 and $c_w \neq 0$. A continuum $\mathscr{C}_+^m$ of single-class m-cycles for the semelparous Leslie model (18) bifurcates from the origin $x = 0$ at $R_0 = 1$. If $c_w < 0$ the bifurcation is forward. If $c_w > 0$ the bifurcation is backward. The range of the continuum, i.e. the set of first class cohort densities x_1 from the cycles, is the half line R_+^1. If $\sigma_m(x)$ satisfies (9), then the spectrum of R_0 values is infinite (and hence is a positive half line whose closure contains 1).*

By Theorem 3 the direction of the bifurcation at $R_0 = 1$ of the continuum $\mathscr{C}_+^e$ of positive equilibria is determined by the sign of

$$a_1 = \Sigma_{n=1}^m \nabla^0 \sigma_n \, v, \quad v = \text{col}(p_i).$$

This quantity involves both between-class and within-class density effects $\partial_j^0 \sigma_n$. On the other hand, by Theorem 7 the direction of bifurcation at $R_0 = 1$ of the continuum $\mathscr{C}_+^m$ of single class, synchronous m-cycles determined by the sign of the quantity c_w given by (24), which involves only within-class density effects $\partial_n^0 \sigma_n$. Since these quantities can have different signs, it follows that the two continuum can bifurcate in opposite directions.

Theorem 7 provides the existence of a bifurcating continuum of single-class, synchronous m- cycles . General stability and instability criteria for these cycles have yet to be obtained for dimensions $m \geq 3$ and this remains a challenging open problem. Some results are known, however, under special assumptions.

4.3 Negative Feedback Only

Most of the literature on nonlinear Leslie matrix focusses on the case of negative feedback density effects (negative derivatives $\partial_j^0 \sigma_n$) and the absence of positive density effects (no positive derivatives $\partial_j^0 \sigma_n$, i.e. no component Allee effects). Assume

A2: $\partial_j^0 \sigma_n \leq 0$ for all $1 \leq j, n \leq m$ with at least one $\partial_n^0 \sigma_n < 0$.

Then $a_1 < 0$ and $c_w < 0$ (and $c_b \leq 0$) and, by Theorems 3 and 7, both the bifurcating continua of positive equilibria and single-class m-cycle are forward.

In the $m = 2$ dimensional case we have the following corollary of Theorem 6.

Corollary 5 *Assume $m = 2$ and that A1 and A2 hold. The bifurcation at $R_0 = 1$ of the continua $\mathscr{C}_+^e$ and $\mathscr{C}_+^2$ of positive equilibria and single-class, synchronous 2-cycles of semelparous Leslie model (18) are both forward and the following two*

alternatives hold:

(a) If $c_w < c_b$ then the bifurcation of $\mathscr{C}_+^e$ is stable and the bifurcation of $\mathscr{C}_+^2$ is unstable;
(b) If $c_w > c_b$ then the bifurcation of $\mathscr{C}_+^e$ is unstable and the bifurcation of $\mathscr{C}_+^2$ is stable.

The two alternatives in Corollary 5 describe a *dynamic dichotomy* at the bifurcation point $R_0 = 1$: either the bifurcating positive equilibria or bifurcating the synchronous 2-cycles are stable, but not both. Which bifurcating branch is stable depends on the relative strength of between-class and in-class density effects, which we can express in terms of the ratio

$$\rho \triangleq \frac{c_b}{c_w}. \tag{25}$$

If $\rho < 1$ then within-class effects out weigh between-class effects and the population equilibrates. On the other hand, if $\rho > 1$ then between-class effects out weigh between-class effects and the population tends towards synchronous 2-cycle oscillations in which the juveniles and adults are temporally separated.

It is interesting to notice that this dynamic dichotomy, in the case $m = 2$, bears a similarity with the classic two species competitive exclusion principle, except that in the case of semelparous Leslie populations the two dynamic outcomes have to do with age classes within a single population and not the presence or absence of species. (Mathematically, this relates to the fact that the composite of the semelparous model has the same mathematical form as a two species competition model.)

The dynamic dichotomy between the bifurcating positive equilibria and the single-class, synchronous 2-cycles described in Corollary 5 for $m = 2$ does not hold in higher dimensions $m \geq 3$. This can be seen, for example, in studies of the $m = 3$ dimensional case [9, 12, 18]. As we will see in Theorem 9, however, when $m = 3$ there is a dynamic dichotomy between the stability of the bifurcating positive equilibria and the boundary of the positive cone as an attractor or repeller.

The boundary ∂R_+^m of the positive cone is an *attractor* if there exists an open neighborhood $U \subset R_+^m$ of ∂R_+^m (in the relative topology of R^3) such that orbits with initial conditions in U have ω-limit sets in ∂R_+^m. The boundary ∂R_+^m of the positive cone is a *repeller* if there exists a neighborhood $U \subset R_+^m$ of ∂R_+^m such that for the orbit from each initial condition not in ∂R_+^m there exists a time $T > 0$ such that the orbit lies outside of U for all $t \geq T$.

The use of average Lyapunov functions for the study of nonlinear Leslie models was pioneered by Kon et al. [43, 45, 46]. This approach leads to criteria for the attracting and repelling properties of the boundary ∂R_+^m that require the calculation the maxima and minima of quantities taken along all orbits on ∂R_+^m (e.g. see Theorem 4.1 in [45]). This obviously requires some knowledge of the dynamics on ∂R_+^m, which in general can be complicated. Near the bifurcation point, however, the dynamics on ∂R_+^m is usually simpler and, in fact, usually involves attracting (synchronous) cycles. Here, we will restrict attention to this case and assume all orbits on ∂R^m_+ approach a cycle.

Theorem 8 ([18]) *In addition to A1 and A2, suppose the follow two assumptions hold for the semelparous model (18):*

A3: $\sigma_i(x)\, x_i$ is bounded for $x \in \bar{R}^m_+$ for all $1 \leq i \leq m$;
A4. every orbit on ∂R^m_+ approaches a synchronous cycle as $t \to +\infty$.

The boundary ∂R^m_+ of the positive cone is an attractor (repeller) if, for every periodic cycle $c(j)$ on ∂R^m_+, the quantity

$$\theta \triangleq \sum_{j=1}^{p} \ln \left(R_0 \prod_{n=1}^{m} \sigma_n(c(j)) \right) \tag{26}$$

is negative (positive). Here p is the period of the cycle.

For dimension $m = 2$ we saw that the dynamics on ∂R^m_+ were described by a (composite) one-dimensional map, which permitted us to address the assumption A4 near $R_0 = 1$. In higher dimensions, however, it is possible for an increasing number of different types of synchronous cycles to arise at bifurcation, namely *k-class synchronous cycles* with k positive entries at each time step. These cycles correspond to fixed points of the composite map (22) with k positive and $m - k$ zero entries. It is an open problem to obtain conditions under which the bifurcation of k-class synchronous cycles occurs at $R_0 = 1$, although one approach that uses Eq. (22) is clear. Select any subset of k of Eq. (22), set $x_i = 0$ for all other subscripts, and study the resulting system of equations using the bifurcation methods used above for single-class cycles.

Under assumption A2, the bifurcation of the single-class m-cycles is forward. Therefore, for $R_0 \gtrapprox 1$ it is necessary, in order to apply Theorem 8, to calculate θ for the bifurcating single-class cycle. Since a parameterization of the cycle is possible by perturbation methods, one can calculate an approximation of θ for $R_0 \gtrapprox 1$, which turns out to be

$$\theta = \left(m - 1 - \sum_{q=1}^{m-1} \rho_q \right) \varepsilon + O\left(\varepsilon^2\right)$$

where

$$\rho_q = \frac{\sum_{i=1}^{m} p_i \partial_i^0 \sigma_{i+q}}{\sum_{i=1}^{m} p_i \partial_i^0 \sigma_i}, \quad q = 1, 2, \ldots, m-1 \tag{27}$$

(subscripts on σ_i are calculated modulo m) [18]. Theorem 8 gives the following result.

Corollary 6 *Assume A1, A2 and A3 hold for the semelparous model (18). Assume for $R_0 \gtrapprox 1$ that all boundary orbits tend to the single-class m-cycle as $t \to +\infty$.*

Then for $R_0 \gtrapprox 1$

$$\sum_{q=1}^{m-1} \rho_q > m-1 \text{ implies } \partial R_+^m \text{ is an attractor}$$

$$\sum_{q=1}^{m-1} \rho_q < m-1 \text{ implies } \partial R_+^m \text{ is a repeller.}$$

The ratios ρ_q are measures of the relative strength of between-class density effects in comparison to with-in class density effects. The denominator of ρ_q in (27) is a measure of within-class competition (at low population densities) as based on the derivatives $\partial_i^0 \sigma_i$. In the numerator, the derivative $\partial_i^0 \sigma_{i+q}$ measures the density effect that i^{th} age class has on the survivorship of age class $i+q$ modulo m. This means that the numerator of the ratio ρ_q is a measure of the density effects among these selected (but not all) unidirectional pairings of age classes. Thus, Corollary 6 generalizes the conclusion stated after Corollary 5 for $m = 2$, namely, that weak density effects between age classes promotes stabilization with all age classes present, while strong density effects between age classes promotes synchronized oscillations with missing age cohorts.

Suppose we strengthen the local monotonicity assumption A2 to the following global assumptions, which are satisfied, for example, by the Leslie-Gower type nonlinearities (7).

A4: $\partial_i \sigma_j (x) < 0$ and $\partial_i (\sigma_i (x) x_i) > 0$ for $x \in \bar{R}_+^m$.

In the $m = 2$ dimensional case, Eq. (22) reduce to a single (one dimensional) map

$$x_1 = R_0 \bar{\sigma}_1 (x_1) x_1$$

which by A4 is a monotone map and, as a result, all orbits equilibrate as $t \to +\infty$. For $R_0 > 0$ assumption A4 also implies $x = 0$ is a repeller and there exists a unique positive fixed point $x_1 > 0$. All this goes to show that all boundary orbits tend to the single-class 2-cycle when $R_0 > 1$. Corollary 6 implies the boundary ∂R_+^2 is an attractor or repeller when ρ_1 is greater than or less than 1. Here ρ_1 is identical to ρ in (25) and this result provides an enhancement of Corollary 5.

Unlike the case $m = 2$, however, the monotonicity assumptions A4 do not guarantee that all boundary orbits tend to the single-class synchronous 3-cycle when $m = 3$. When $m = 3$ Eq. (22) defining the boundary cycles is a planar map, which under A4, is strictly competitive on $\bar{R}_+^2$ and strongly competitive on R_+^2 to which the powerful theory of planar monotone maps can be applied (e.g. Proposition 2.1 in [62]). The result is that if $R_0 > 1$ then all orbits converge to an equilibrium in $\bar{R}_+^2$, specifically to a non-negative equilibrium lying on a positive axis ∂R_+^2 or possibly a positive equilibrium in R_+^2. These fixed points correspond to a single-class synchronous 3-cycle and a two-class, synchronous 3-cycle respectively. Under our working assumptions we know that a single-class 3-cycle exists for $R_0 \gtrapprox 1$. Using bifurcation theory it can be shown that positive two-class synchronous

3-cycles also bifurcate from the origin at $R_0 = 1$ if both $\rho_i > 1$ or both $\rho_i < 1$ [18]. A parameterization of these bifurcating two-class 3-cycles near $R_0 = 1$ leads to the expansion

$$\theta = \frac{\rho_1 + \rho_2 - \rho_1^2 - \rho_2^2 + \rho_1\rho_2 - 1}{\rho_1\rho_2 - 1}\varepsilon + O\left(\varepsilon^2\right)$$

which, coupled with the expansion (27), with $m = 3$, for the single-class 3-cycles lead to the following result.

Theorem 9 *Assume A1, A2 and A3 hold for the semelparous model (18). For $R_0 \gtrapprox 1$ we have the following alternatives for the cases $m = 2$ and $m = 3$.*

Suppose $m = 2$. *If $\rho_1 < 1$ then the bifurcating positive equilibria are stable and the boundary ∂R_+^2 is a repeller. If $\rho_1 > 1$, the bifurcating positive equilibria are unstable and ∂R_+^2 is an attractor.*

Suppose $m = 3$. *If $\rho_1 + \rho_2 < 2$ then the bifurcating positive equilibria are stable and the boundary ∂R_+^3 is a repeller. If $\rho_1 + \rho_2 > 2$ then the bifurcating positive equilibria are unstable and ∂R_+^3 is an attractor.*

This theorem describes, for $R_0 \gtrapprox 1$, a dynamic dichotomy between the boundary of the positive cone and a positive equilibrium. By this is meant, roughly speaking, that strong within-class (negative feedback) density effects, measured by the ratios ρ_i, promotes equilibration with over-lapping age classes while strong between-class density effects destabilize this equilibration and promotes oscillations with missing age classes present at each time step. In the latter case (i.e. when $\rho_1 + \rho_2 > 2$) for the $m = 3$, an orbit within the positive cone does not necessarily approach a synchronous cycle, or a periodic oscillation of any kind, even though it approaches the boundary of the cone on which orbits do approach the single-class 3-cycle. If, in addition to $\rho_1 + \rho_2 > 2$, one of the ratios ρ_i is less than 1, then the single-class synchronous 3-cycle is also unstable, namely it is a saddle (it is stable if both $\rho_i > 1$) [12]. In this case orbits in the positive cone that approach the boundary, approach a *cycle chain* on the boundary. This invariant set consists of the three phases of the synchronous 3-cycle together with heteroclinic connections among them. See Fig. 1. Other types of cycle chains (ones that also contain 2-cycle synchronous cycles and their phases) can also bifurcate from the origin, under different circumstances. For a list of the possibilities when $m = 3$ see [12]. Since $\rho_1 + \rho_2 > 2$ also implies that the positive equilibrium is unstable, we see that the dynamic dichotomy in the case $m = 3$ is not between the positive equilibrium and the single-class synchronous cycle, unlike the case $m = 2$.

In general, for dimensions $m \geq 4$ a description of the dynamic alternatives near the bifurcation point $R_0 = 1$ remains an open problem. It is not known in general if a dynamic dichotomy exists between the two alternatives of a stable positive equilibrium and an attracting boundary ∂R_+^m. Even if the boundary is known to be an attractor (from inside the positive cone), an understanding of the dynamics on the boundary is complicated by the possibility of many types of synchronous

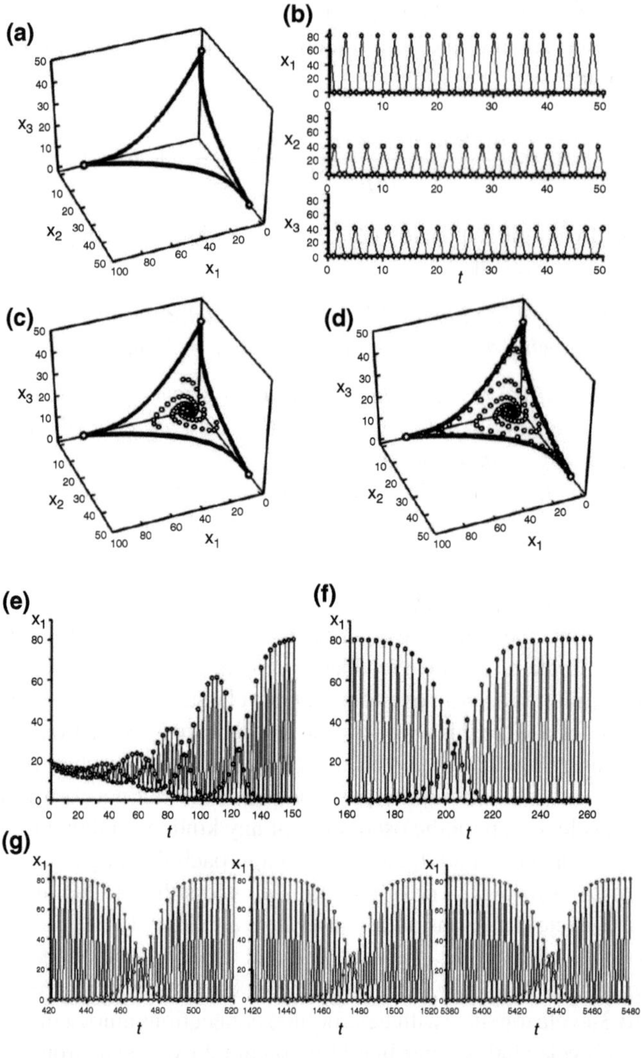

Fig. 1 These plots show orbits from the semelparous LPA model (17) with coefficients $\mu_a = 1$ and $b = 3$, $\mu_l = 0.5$, $\mu_p = 0$, $c_1 = c_3 = 0.1$, $c_2 = 0.2$. We calculate $\rho_1 = 0$ and $\rho_2 = 5$ and conclude that both continua of bifurcating positive equilibria and single-class synchronous cycles are unstable, but that the boundary ∂R^3_+ is an attractor. Note that $R_0 = 1.5$. (**a**) The *open circles* are the points of the single-class 3-cycle which are temporally visited counter-clockwise. Also shown are points on heteroclinic connecting orbits lying in the coordinate planes (which temporally are visited sequentially). (**b**) The time series of the single-class 3-cycle showing the synchrony of the age classes. (**c**) An orbit starting near the unstable positive orbits displays a spiral departure from the equilibrium. (**d**) This plot shows how the orbit approaches ∂R^3_+, specifically the cycle chain shown in (**a**). (**e**) The x_1 component of this orbit appears to approach a period three oscillations by $t = 160$. (**f**) However, this component departs from this oscillation and undergoes a phase shift at around $t = 200$, after which it returns to a period three oscillation, but with its phase shifted by one time unit. (**g**) Three more such phase shifts are shown. They occur infinitely often, increasingly further apart, creating infinitely many longer and longer episodes of (near) single-class oscillations of period three.

m-cycles and connecting heteroclinics, i.e. types of cycle chains (to say nothing of other possible bifurcating invariant sets lying on the boundary). The dynamic complexity can greatly increase with the dimension $m \geq 4$, and it is likely that a complete and general description of the bifurcation at $R_0 = 1$ will not be possible for higher dimensions [30].

4.4 *Backward Bifurcations*

As mentioned at the end of Sect. 3 a strong Allee effect, that is to say, the occurrence of multiple attractors, one of which is the extinction equilibrium and another which is a non-extinction attractor, often (if not usually) arises in population models from a backward bifurcation. For nonlinear matrix models, a backward bifurcating continuum of positive equilibria (necessarily unstable for $R_0 \lessapprox 1$) can "turn around" at a saddle-node bifurcation point $R_0^* < 1$, creating stable positive equilibria for $R_0 < 1$ when the extinction state is also stable. For the semelparous Leslie model (18) opportunities for a strong Allee effect arise from such an occurrence for both the continuum of positive equilibria and the continuum of single-class synchronous cycles (or more complicated cycle chains).

For example, we see from Theorem 6 that several bifurcation scenarios at $R_0 = 1$ are possible for the $m = 2$ dimensional case, due to the fact that the continua $\mathscr{C}_+^e$ and $\mathscr{C}_+^e$ can bifurcate in the same or different directions. Figure 2 shows orbits calculated from the semelparous Leslie model (18) with $m = 2$ and

$$\sigma_1(x) = \frac{1 + \alpha x_2}{1 + c_{21}x_1 + c_{22}x_2}, \quad \sigma_2(x) = \frac{1}{1 + c_{11}x_1 + c_{12}x_2}. \tag{28}$$

The parameter values chosen in Fig. 2, together with Theorem 6, imply that both continua $\mathscr{C}_+^e$ and $\mathscr{C}_+^2$ bifurcate backward. The result is a strong Allee effect in this model in which there are two non-extinction attractors for values of $R_0 < 1$, namely a positive equilibrium and a single-class, synchronous 2-cycle, as well as a stable extinction equilibrium. Among other things, this example shows that the dynamic dichotomy that occurs in the $m = 2$ dimensional semelparous Leslie model described in Corollary 5 does not occur when both continua bifurcate backwards.

5 Concluding Remarks

I have focussed in this paper on the dynamics of m dimensional, nonlinear semelparous Leslie models that arise from bifurcations that occur at $R_0 = 1$ due to the loss of stability of the extinction equilibrium. Under quite general conditions, methods from modern and classic bifurcation theory establish the existence of two basic continua that bifurcate from the extinction equilibrium at $R_0 = 1$, one $\mathscr{C}_+^e$

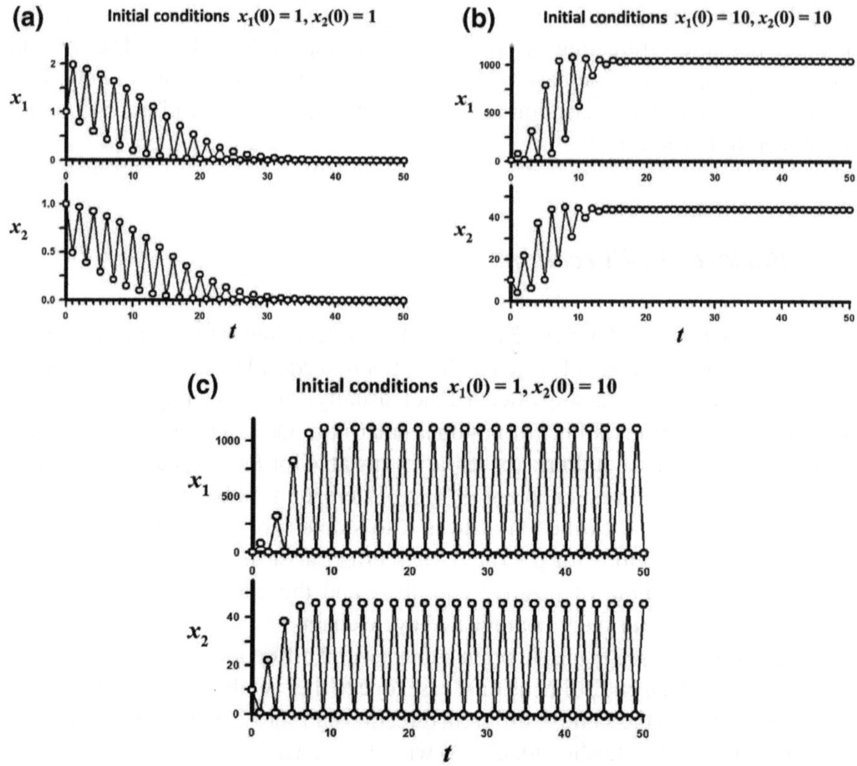

Fig. 2 These plots show three example orbits for the $m = 2$ semelparous Leslie model (18) with $\sigma_1(x)$ and $\sigma_2(x)$ given by (28) with parameter values $c_{11} = c_{12} = c_{22} = 0.01$, $c_{21} = 0$, $\alpha = 0.6$, $s_1 = 0.5$, and $b = 1.25$. These values imply $R_0 = 0.625 < 1$ and $c_w = -0.015$, $c_b = 0.295$, $c_w + c_b = 0.280$, and $c_w - c_b = -0.310$. By Theorem 6 both continua bifurcate backwards, which accounts for a strong Allee effect with two non-extinction attractors in the presence of an attracting extinction equilibrium. Some initial conditions lead to extinction as in plot (**a**), some to a positive equilibrium as in plot (**b**), and some to a single-class synchronous 2-cycle as in plot (**c**)

consisting of positive equilibrium pairs and the other $\mathscr{C}_+^m$ consisting of single-class, synchronous m-cycle pairs. These continua have a global extent in the sense that they are unbounded, i.e. either their spectrum or range is unbounded. The directions of bifurcation, in a neighborhood of the bifurcation point $(R_0, x) = (1, 0)$, can be determined from the signs of the quantity a_1 in the case of $\mathscr{C}_+^e$ and c_w in the case of $\mathscr{C}_+^m$. Both a_1 and c_w are linear combinations of the derivatives $\partial_j^0 \sigma_i$ (the sensitivities of the Leslie matrix entries σ_i to changes in low age-class densities). In models with no positive feedback density effects (i.e. no derivative $\partial_j^0 \sigma_i > 0$ or, in other words, no component Allee effects) the bifurcation of both continua is forward. Positive feedback effects can, if of sufficient magnitude, lead to backward bifurcations (which can, in turn, lead to strong Allee effects).

The question of what attractors arise from these bifurcations is a difficult one and has not been fully resolved in general, except in lower dimensions. The two dimensional case $m = 2$ is well understood (Theorem 6): backward bifurcations are unstable and, in the case of forward bifurcations, one but not both bifurcations of $\mathscr{C}_+^e$ and $\mathscr{C}_+^e$ is stable. The case $m = 3$ is also well-understood, at least in the absence of positive feedback terms, in that there is a dynamic dichotomy between $\mathscr{C}_+^e$ and ∂R_+^3 (Theorem 9). In this case, the attractor is not necessarily a positive equilibrium or a single-class synchronous 3-cycle, but can be a cycle chain consisting of the three phases of the 3-cycle connected by heteroclinic orbits lying on ∂R_+^3. This anticipates the complexity of the dynamics on ∂R_+^m, in particular the number and type of synchronous cycles, that can occur for higher dimensions m. The boundary dynamics significantly influence the dynamics and type of attractors for the semelparous Leslie model (5) on $\bar{R}_+^m$. It seems unlikely that a thorough accounting of the possibilities is possible for higher dimensions $m \geq 4$ without specialized and simplifying assumptions on the model. Using a different approach to the study of the dynamics of the Leslie model (5)—one based on a formal limiting procedure and comparison with associated differential equation models—Diekmann and van Gils come to the same conclusion, even with the specialized assumption they make that all density dependence is through a single weighted population size [30].

There is a large literature that investigates the dynamics of semelparous Leslie models from other points of view that do not restrict attention to a neighborhood of the bifurcation at $R_0 = 1$. These studies generally restrict the nonlinearities in the model in some way or another. Common assumptions include limiting density dependence to a few or even just one age class [50, 53, 61, 66], assuming density effects are through a dependence on one weighted population size $w = \Sigma_{j=1}^m w_j x_j$ (so that $\sigma_i = \sigma_i(w)$ for all i) [25–27, 29–31, 65–68], use of specific nonlinearities such as Leslie-Gower (7) [49] or Ricker (8) types [57], and an hierarchical structure to density dependence in which the vital rates of an age class depend only on densities of older (or younger) age classes [16, 37, 38]. These studies often are done with an eye towards the possibility of positive (non-synchronous) periodic cycles, invariant loops, and chaotic attractors. Given that one-dimensional maps $m = 1$ can, as is well known, exhibit complex dynamics, it is certainly to be expected that such attractors will occur in nonlinear Leslie models of dimension $m \geq 2$. They generally arise when R_0 is increased and a destabilization of the positive equilibria on $\mathscr{C}_+^e$ occurs by means of a period-doubling or Neimark-Sacker (discrete Hopf) bifurcation, and subsequently by destabilizations of non-equilibrium attractors, all of which can lead to a so-called route-to-chaos. For semelparous Leslie models, since both the boundary ∂R_+^m and the interior R_+^m of the positive cone are invariant, it is possible for such bifurcation scenarios to occur in the interior and on the boundary positive cone. Thus, one can see complicated attractors on ∂R_+^m and/or in R_+^m. Biologically the former are distinguished by always having a missing age class while the latter never have a missing age class.

A theme that arises from the study of the semelparous Leslie model (5) is that strong competition (negative feedback density effects) between age-classes (relative to within class competition) promotes synchronous oscillations. This is viewed as a kind of competitive exclusion principles among age-classes (in analogy to the competitive exclusion principle among different species). This idea forms one of the principle hypotheses offered to explain the synchronized, recurrent outbreaks of periodical insects, the periodical cicadas being the most famous example. Another competing hypothesis is predator saturation: by synchronizing their emergence adults overwhelm predators by their number and thereby assure successful reproduction of at least a fraction of their number (for a discussion see [52, 69]). Bulmer and Bencke [2, 3] concluded from their seminal model studies, in which predation and fungal infections of adult were included in a semelparous Leslie model, that between-class competitive effects (particularly in the youngest age classes) are the primary cause of synchronized cohort oscillations. Although there is some evidence of such competition among cicada nymphs [39], as they struggle for feeding locations on tree roots, it is difficult to obtain observational data about the interactions among age classes of nymphs.

There is, however, some striking experimental evidence of the phenomenon of competition induced synchronization of age classes. Decades long experimental studies of nonlinear dynamics conducted with *Tribolium castaneum* (flour beetles), reported in [5, 24, 28, 41], were not designed to study synchronized oscillations in a semelparous species. Indeed, *T. castaneum* is not naturally semelparous. However, the experimental protocol used in the keystone study of dynamic bifurcations and routes-to-chaos in effect made the experimental cultures of *T. castaneum* semelparous by imposing high adult mortality. (Theoretically $\mu_a = 0.96$ in the LPA model (17), although in practice 100 % mortality was often imposed during the long term study.) With the cultures placed into an essentially semelparous life history, between-class density effects were increased in a sequence of replicated cultures (specifically, c_2 was increased from 0 to 1 in (17). The goal of that experiment was to document a sequence of bifurcations and their resulting complicated attractors (including chaotic attractors) that were predicted to occur by the LPA model (17). For our purposes here, we point out that at the lowest level of between-class competition, $c_2 = 0$, there was observed an equilibrium state with all age-class present and at the highest level, $c_2 = 1$, single-class synchronous 3-cycles were observed. This is in agreement with the principle that strong between-class competition promotes synchronized oscillations. See Fig. 3. Furthermore, attracting cycle chains, as illustrated in Fig. 1b, offer a deterministic explanation of experimentally observed phase shifts in the synchronous 3-cycles (which were explained in [35] by stochastic jumps among the basins of attractions of the three phases).

In addition to many unanswered questions concerning the nature of the primary bifurcation at $R_0 = 1$ for the semelparous Leslie model in higher dimensions, there are extensions and elaborations of the model that also present interesting challenges. For example, the role of evolution in determining semelparity or iteroparity has long been of interest in life history theory [60, 63]. In this regard, the fundamental bifurcation at $R_0 = 1$ has been studied for evolutionary versions of matrix models primitive projection matrices [13, 14] and for semelparous Leslie models of dimension $m = 2$ [22]. As already mentioned, models that include the effects of

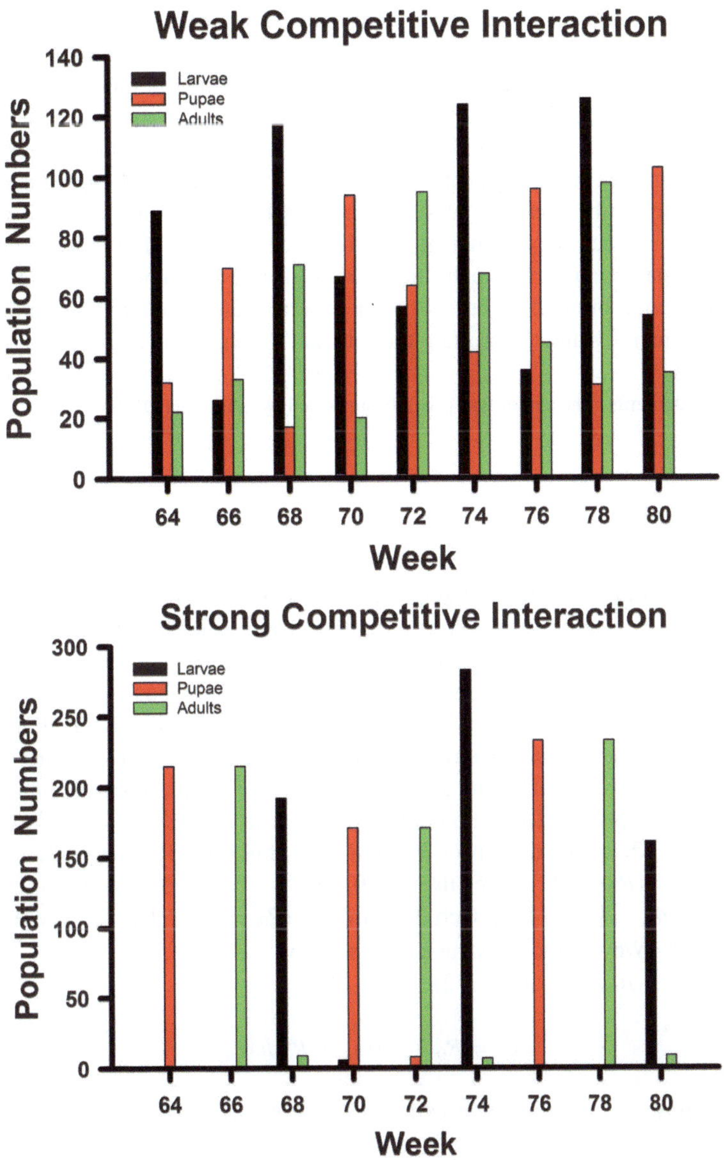

Fig. 3 Experimental evidence for single-class cycles induced by inter-class competition. The graphs show the age class histograms (at the end of 80 weeks when transients were dissipated) in the experimental treatments that underwent the weakest and strongest cannibalism rates. The former treatment shows overlapping age classes (larvae, pupae, adults), while the latter shows single cohorts of non-overlapping age classes. (The larval and pupal stages for *T. castaneum* are 2 weeks in length, which is the time unit used in the LPA model (17))

predation (and parasitism) have been formulated and studied [2, 3, 44], although not specifically with the bifurcation at $R_0 = 1$ in mind. The fundamental bifurcation at $R_0 = 1$ has also been studied for spatial versions of matrix models, but only for primitive projection matrices [58, 59].

The semelparous Leslie model is a notable and important example of a matrix population model that has an imprimitive projection matrix. A signature feature of this model is the presence of synchronous orbits, i.e. orbits with missing age classes. It has been shown that no matrix model, of any kind, with a primitive projection matrix can have synchronous orbits [42]. It would be interesting to study the fundamental bifurcation at $r = 1$ for general matrix models with imprimitive projection matrices and to ascertain the role that synchronous cycles play.

Acknowledgements The author was partially supported by NSF grant DMS 0917435.

Appendix

Theorem 1.20 in [56] implies the existence of two, globally distinct continua $\mathscr{C}_+^e$ and $\mathscr{C}_-^e$ of nonzero equilibrium pairs each of which satisfies the two alternatives, i.e. is unbounded in $R^1 \times R^m$ or contains a point $(\lambda, 0)$ where $\lambda \neq 1$ is a characteristic value of $M(0)$.[2] In a neighborhood of $(1,0)$, $\mathscr{C}_+^e$ and $\mathscr{C}_-^e$ consist of positive and negative equilibrium pairs respectively. Since second alternative is ruled out by the fact that $M(0)$ has no characteristic value other than 1, $\mathscr{C}_+^e$ and $\mathscr{C}_-^e$ are globally distinct distinct continua that are unbounded in $R \times R^m$. For purposes of contradiction we assume the unbounded continuum C_+^e, which in a neighborhood of $(1,0)$ lies in $R_+^1 \times R_+^m$, does not remain in $R_+^1 \times R_+^m$. In this case, it must contain a point $(R_0^*, x^*) \in \partial (R_+ \times R_+^m)$ other than $(1,0)$ and we can find a sequence of points $(R_{0n}, x_n) \in \mathscr{C}_+^e \cap (R_+ \times R_+^m)$ such that $\lim_{n\to\infty} (R_{0n}, x_n) = (R_0^*, x^*)$ where $R_0^* \geq 0$ and $x^* \in \bar{R}_+^m$. We want to arrive at a contradiction.

The points (R_{0n}, x_n) satisfy (12)

$$x_n = R_{0n} M(0)\, x_n + R_{0n} h(x_n). \tag{29}$$

First, suppose $x^* = 0$. We can extract a subsequence from the sequence of unit vectors

$$u_n = \frac{x_n}{|x_n|} \in R_+^m$$

[2] To apply Theorem 1.20 in [56] we extend the domain of the $\sigma_i(x)$ to R^m by re-defining them smoothly outside of the closure $\bar{R}_+^m$ of the positive cone. This is possible by assumption A1.

that converges to a nonnegative unit vector u :

$$\lim_{n\to\infty} u_n = u \in \bar{R}_+$$

Passing to the limit in

$$\frac{x_n}{|x_n|} = R_{0n}M(0)\frac{x_n}{|x_n|} + R_{0n}\frac{h(x_n)}{|x_n|}.$$

we obtain $u = R_0^* M(0)u$. This leads to an immediate contradiction if $R_0^* = 0$. If $R_0^* \neq 0$ then since the only characteristic value of $M(0)$ is 1 we obtain another contradiction, namely, $R_0^* = 1$. Having ruled out $x^* = 0$, we conclude that $x^* \in \partial R_+^m \backslash \{0\}$. Passing to the limit in Eq. (29) we conclude that x^* is an equilibrium of the nonlinear Leslie model (with $R_0 = R_0^*$). However, an inspection of components of the equilibrium equation (10) shows that if one component equals 0 then all components equal 0, i.e. $x^* = 0$. This is a contradiction to $x^* \in \partial R_+^m \backslash \{0\}$.

References

1. Allen, L.J.S.: A density-dependent Leslie matrix model. Math. Biosci. **96**(2), 179–187 (1989)
2. Behncke, H.: Periodical cicadas. J. Math. Biol. **40**, 423–431 (2000)
3. Bulmer, M.G.: Periodical insects. Am. Nat. **111**, 1099–1117 (1977)
4. Caswell, H.: Matrix Population Models: Construction, Analysis and Interpretation, 2nd edn. Sinauer Associates, Inc., Sunderland, MA (2001)
5. Costantino, R.F., Desharnais, R.A., Cushing, J.M., Dennis, B.: Chaotic dynamics in an insect population. Science **275**, 389–391 (1997)
6. Costantino, R.F., Desharnais, R.A., Cushing, J.M., Dennis, B., Henson, S.M., King, A.A.: The flour beetle *Tribolium* as an effective tool of discovery. Adv. Ecol. Res. **37**, 101–141 (2005)
7. Courchamp, F., Berec, L., Gascoigne, J.: Allee Effects in Ecology and Conservation. Oxford University Press, Oxford (2008)
8. Cushing, J.M.: An Introduction to Structured Population Dynamics. Conference Series in Applied Mathematics, vol. 71. SIAM, Philadelphia (1998)
9. Cushing, J.M.: Cycle chains and the LPA model. J. Differ. Equ. Appl. **9**, 655–670 (2003)
10. Cushing, J.M.: Nonlinear semelparous Leslie models. Math. Biosci. Eng. **3**(1), 17–36 (2006)
11. Cushing, J.M.: Matrix models and population dynamics. In: Mark Lewis, A.J.C., James, P.K., Philip, K.M. (eds.) Mathematical Biology. IAS/Park City Mathematics Series, vol. 14, pp. 47–150. American Mathematical Society, Providence (2009)
12. Cushing, J.M.: Three stage semelparous Leslie models. J. Math. Biol. **59**, 75–104 (2009)
13. Cushing, J.M.: A bifurcation theorem for Darwinian matrix models. Nonlinear Stud. **17**(1), 1–13 (2010)
14. Cushing, J.M.: On the dynamics of a class of Darwinian matrix models. Nonlinear Dyn. Syst. Theory **10**(2), 103–116 (2010)
15. Cushing, J.M.: On the relationship between r and R_0 and its role in the bifurcation of equilibria of Darwinian matrix models. J. Biol. Dyn. **5**, 277–297 (2011)
16. Cushing, J.M.: A dynamic dichotomy for a system of hierarchical difference equations. J. Differ. Equ. Appl. **18**(1), 1–26 (2012)

17. Cushing, J.M.: Backward bifurcations and strong Allee effects in matrix models for the dynamics of structured populations. J. Biol. Dyn. **8**, 57–73 (2014)
18. Cushing, J.M., Henson, S.M.: Stable bifurcations in semelparous Leslie models. J. Biol. Dyn. **6**, 80–102 (2012)
19. Cushing, J.M., Li, J.: On Ebenman's model for the dynamics of a population with competing juveniles and adults. Bull. Math. Biol. **51**, 687–713 (1989)
20. Cushing, J.M., Li, J.: Juvenile versus adult competition. J. Math. Biol. **29**, 457–473 (1991)
21. Cushing, J.M., Li, J.: Intra-specific competition and density dependent juvenile growth. Bull. Math. Biol. **53**(4), 503–519 (1992)
22. Cushing, J.M., Maccracken-Stump, S.: Darwinian dynamics of a juvenile-adult model. Math. Biosci. Eng. **10**(4), 1017–1044 (2013)
23. Cushing, J.M., Zhou, Y.: The net reproductive value and stability in structured population models. Nat. Resour. Model. **8**(4), 1–37 (1994)
24. Cushing, J.M., Costantino, R.F., Dennis, B., Desharnais, R.A., Henson, S.M.: Chaos in Ecology: Experimental Nonlinear Dynamics. Theoretical Ecology Series, vol. 1. Academic Press (Elsevier Science), New York (2003). ISBN: 0-12-1988767
25. Davydova, N.V.: Old and young. Can they coexist? Ph.D. dissertation, Facultaeit der Wiskunde en Informatica, Universiteit Utrecht, Utrecht (2004)
26. Davydova, N.V., Diekmann, O., van Gils, S.A.: Year class coexistence or competitive exclusion for strict biennials? J. Math. Biol. **46**, 95–131 (2003)
27. Davydova, N.V., Diekmann, O., van Gils, S.A.: On circulant populations. I. The algebra of semelparity. Linear Algebra Appl. **398**, 185–243 (2005)
28. Dennis, B., Desharnais, R.A., Cushing, J.M., Henson, S.M., Costantino, R.F.: Estimating chaos and complex dynamics in an insect population. Ecol. Monogr. **71**(2), 277–303 (2001)
29. Diekmann, O., van Gils, S.A.: Invariance and symmetry in a year-class model. In: Buescu, J., Castro, S.D., da Silva Dias, A.P., Labouriau, I.S. (eds.) Trends in Mathematics: Bifurcations, Symmetry and Patterns, pp. 141–150. Birkhäuser Verlag, Basel (2003)
30. Diekmann, O., van Gils, S.A.: On the cyclic replicator equations and the dynamics of semelparous populations. SIAM J. Appl. Dyn. Syst. **8**(3), 1160–1189 (2009)
31. Diekmann, O., Davydova, N., van Gils, S.: On a boom and bust year class cycle. J. Differ. Equ. Appl. **11**(4), 327–335 (2005)
32. Ebenman, B.: Niche differences between age classes and intraspecific competition in age-structured populations. J. Theor. Biol. **124**, 25–33 (1987)
33. Ebenman, B.: Competition between age classes and population dynamics. J. Theor. Biol. **131**, 389–400 (1988)
34. Elaydi, S.N.: An Introduction to Difference Equations. Springer, New York (1996)
35. Henson, S.M., Cushing, J.M., Costantino, R.F., Dennis, B., Desharnais, R.A.: Phase switching in biological population. Proc. R. Soc. Lond. B **265**, 2229–2234 (1998)
36. Impagliazzo, J.: Deterministic Aspects of Mathematical Demography. Biomathematics, vol. 13. Springer, Berlin (1980)
37. Jang, S.R.-J., Cushing, J.M.: A discrete hierarchical model of intra-specific competition. J. Math. Anal. Appl. **280**, 102–122 (2003)
38. Jang, S.R.-J., Cushing, J.M. : Dynamics of hierarchical models in discrete-time. J. Differ. Equ. Appl. **11**(2), 95–115 (2005)
39. Karban, R.: Opposite density effects of nymphal and adult mortality for periodical cicadas. Ecology **65**, 1656–1661 (1984)
40. Kielhöfer, H.: Bifurcation Theory: An Introduction with Applications to Partial Differential Equations. Applied Mathematical Sciences, vol. 156. Springer, Berlin (2011)
41. King, A.A., Costantino, R.F., Cushing, J.M., Henson, S.M., Desharnais, R.A., Dennis, B.: Anatomy of a chaotic attractor: subtle model predicted patterns revealed in population data. Proc. Natl. Acad. Sci. **101**(1), 408–413 (2003)
42. Kon, R.: Nonexistence of synchronous orbits and class coexistence in matrix population models. SIAM J. Appl. Math. **66**(2), 626–636 (2005)

43. Kon, R.: Competitive exclusion between year-classes in a semelparous biennial population. In: Deutsch, A., Bravo de la Parra, R., de Boer, R., Diekmann, O., Jagers, P., Kisdi, E., Kretzschmar, M., Lansky, P., Metz, H. (eds.) Mathematical Modeling of Biological Systems. vol. II, pp. 79–90. Birkhäuser, Boston (2007)
44. Kon, R.: Permanence induced by life-cycle resonances: the periodical cicada problem. J. Biol. Dyn. **6**(2), 855–890 (2012)
45. Kon, R., Iwasa, Y.: Single-class orbits in nonlinear Leslie matrix models for semelparous populations. J. Math. Biol. **55**, 781–802 (2007)
46. Kon, R., Saito, Y., Takeuchi, Y.: Permanence of single-species stage-structured models. J. Math. Biol. **48**, 515–528 (2004)
47. Leslie, P.H.: On the use of matrices in certain population mathematics. Biometrika **33**, 183–212 (1945)
48. Leslie. P.H.: Some further notes on the use of matrices in population mathematics. Biometrika **35**, 213–245 (1948)
49. Leslie, P.H., Gower, J.C.: The properties of a stochastic model for two competing species. Biometrika **45**, 316–330 (1958)
50. Levin, S.A., Goodyear, C.P.: Analysis of an age-structured fishery model. J. Math. Biol. **9**, 245–274 (1980)
51. Li, C.-K., Schneider, H.: Applications of Perron-Frobenius theory to population dynamics. J. Math. Biol. **44**, 450–462 (2002)
52. May, R.M.: Periodical cicadas. Nature **277**, 347–349 (1979)
53. Mjϕlhus, E., Wikan, A., Solberg, T.: On synchronization in semelparous populations. J. Math. Biol. **50**, 1–21 (2005)
54. Neubert, M.G., Caswell, H.: Density-dependent vital rates and their population dynamic consequences. J. Math. Biol. **41**, 103–121 (2000)
55. Nisbet, R.M., Onyiah, L.C.: Population dynamic consequences of competition within and between age classes. J. Math. Biol. **32**, 329–344 (1994)
56. Rabinowitz, P.H.: Some global results for nonlinear eigenvalue problems. J. Funct. Anal. **7**, 487–513 (1971)
57. Ricker, W.E.: Stock and Recruitment. J. Fish. Res. Board Can. **11**(5), 559–623 (1954)
58. Robertson, S.L., Cushing, J.M.: Spatial segregation in stage-structured populations with an application to *Tribolium*. J. Biol. Dyn. **5**(5), 398–409 (2011)
59. Robertson, S.L., Cushing, J.M.: A bifurcation analysis of stage-structured density dependent integrodifference equations. J. Math. Anal. Appl. **388**, 490–499 (2012)
60. Roff, D.A.: The Evolution of Life Histories: Theory and Analysis. Chapman and Hall, New York (1992)
61. Silva, J.A.L., Hallam, T.G.: Compensation and stability in nonlinear matrix models. Math. Biosci. **110**, 67–101 (1992)
62. Smith, H.L.: Planar competitive and cooperative difference equations. J. Differ. Equ. Appl. **3**, 335–357 (1998)
63. Stearns, S.C.: The Evolution of Life Histories. Oxford University Press, Oxford, UK (1992)
64. Tschumy W.O.: Competition between juveniles and adults in age-structured populations. Theor. Popul. Biol. **21**, 255–268 (1982)
65. Wikan, A.: Dynamic consequences of reproductive delay in Leslie matrix models with nonlinear survival probabilities. Math. Biosci. **146**, 37–62 (1997)
66. Wikan, A.: Four-periodicity in Leslie matrix models with density dependent survival probabilities. Theor. Popul. Biol. **53**, 85–97 (1998)
67. Wikan, A.: Age or stage structure? Bull. Math. Biol. **74**, 1354–1378 (2012)
68. Wikan, A., Mjϕlhus, E.: Overcompensatory recruitment and generation delay in discrete age-structured population models. J. Math. Biol. **35**, 195–239 (1996)
69. Williams, K.S., Smith, K.G., Stephen F.M.: Emergence of 13-yr periodical cicadas (Cicadidae: *magicicada*): phenology, mortality, and predator satiation. Ecology **74**(4), 1143–1152 (1993)

Review on Non-Perturbative Reducibility of Quasi-Periodically Forced Linear Flows with Two Frequencies

João Lopes Dias

Abstract These are the notes of the short course "Stability of quasi-periodic dynamics" given at the Advanced School Planet Earth, Dynamics, Games and Science II held in Lisbon, Portugal, from 28 August to 6 September 2013 and organized by the International Center of Mathematics CIM - Portugal. We review some recent results concerning the stability of non-autonomous linear differential equations with a quasi-periodic forcing.

1 Non-Autonomous Linear Ode's

1.1 General Setting

Our goal is to survey some recent results on the dynamics of non-autonomous linear ordinary differential equations of the type

$$\dot{x} = A(t)\,x. \tag{1}$$

We are seeking a solution $x\colon \mathbb{R} \to \mathbb{R}^n$ for the above equation when we have a real-analytic matrix function $A\colon \mathbb{R} \to \mathrm{gl}(n, \mathbb{R})$. Here $\mathrm{gl}(n, \mathbb{R})$ stands for the Lie algebra of $n \times n$ matrices with real coefficients and $\mathrm{GL}(n, \mathbb{R})$ is the corresponding Lie group of non-singular matrices.

Any solution can be obtained from the fundamental solution

$$X\colon \mathbb{R} \to \mathrm{GL}(n, \mathbb{R})$$

J. Lopes Dias (✉)
Departamento de Matemática, ISEG and Cemapre, Universidade de Lisboa, Rua do Quelhas 6, 1200-781 Lisboa, Portugal
e-mail: jldias@iseg.ulisboa.pt

J.-P. Bourguignon et al. (eds.), *Dynamics, Games and Science*, CIM Series in Mathematical Sciences 1, DOI 10.1007/978-3-319-16118-1_13

(also called monodromy matrix, time-evolution operator, propagator, basic matrix, principal matrix, etc.) which satisfies

$$\dot{X} = A(t)\,X, \qquad X(0) = I. \tag{2}$$

In fact, the solution with initial condition x_0 is given by

$$x(t; x_0) = X(t)\,x_0.$$

Notice that the columns of $X(t)$ are linearly independent solutions.

Example 1 (Stability of Solutions of Non-Linear Ode's) Take the ordinary differential equation $\dot{x} = f(x)$ with solution $x(t; x_0)$. We can rewrite it as

$$\frac{\partial}{\partial t} x(t; x_0) = f(x(t; x_0)).$$

The stability of this solution with respect to the initial condition x_0 can be studied by looking at

$$\frac{\partial}{\partial t}\frac{\partial}{\partial x_0} x(t; x_0) = Df(x(t; x_0))\frac{\partial}{\partial x_0} x(t; x_0).$$

Its fundamental solution is therefore in the form (2).

One can focus the study on a subclass of systems (2), namely linear skew-product flows generated by vector fields in the form

$$\begin{cases} \dot{X} = A(\theta)X & \text{on } \mathrm{GL}(n, \mathbb{R}) \text{ (fiber)} \\ \dot{\theta} = \varphi(\theta) & \text{on } M \text{ (base)} \end{cases} \tag{3}$$

with $\varphi: M \to TM$ and $A: M \to \mathrm{gl}(n, \mathbb{R})$ for some compact manifold M, and initial condition (θ, X).

A fibered vector field (also called linear skew-product vector field) is the above vector field

$$v(\theta, X) = (\varphi(\theta), A(\theta)X).$$

We write it in a simpler form as $v = (\varphi, A)$. The flow generated by v is called a fibered flow (also called linear skew-product flow) and it is given by

$$\phi^t(\theta, X) = (\Theta^t(\theta), \Phi^t(\theta)X),$$

where $\Phi^t(\theta)$ is the fundamental solution of $\dot{X} = A(\Theta^t(\theta))X$ and $\Theta^t(\theta)$ is the solution of the base dynamics. The cocycle property $\phi^{t+s} = \phi^t \circ \phi^s$ implies that

$$\Phi^{t+s}(\theta) = \Phi^t(\Theta^s(\theta))\,\Phi^s(\theta) \tag{4}$$

A fibered C^r-diffeomorphism is defined as

$$f(\theta, X) = (\psi(\theta), B(\theta)X),$$

where $B: M \to \mathrm{GL}(n,\mathbb{R})$ is C^r and $\psi: M \to M$. Therefore, its action on fibered flows is

$$f \circ \phi^t \circ f^{-1}(\theta, X) = (\tilde{\Theta}^t(\theta), \tilde{\Phi}^t(\theta)X),$$

where $\tilde{\Theta}^t = \psi \circ \Theta^t \circ \psi^{-1}$ and[1]

$$\tilde{\Phi}^t = (B \circ \Theta^t\, \Phi^t\, B^{-1}) \circ \psi^{-1}.$$

On fibered vector fields the action of a fibered diffeomorphism is

$$(Df\, v) \circ f^{-1}(\theta, X) = (\tilde{\varphi}, \tilde{A}(\theta)X),$$

where $\tilde{\varphi} = (D\psi\, \varphi) \circ \psi^{-1}$ and

$$\tilde{A} = (\varphi \cdot DB\,B^{-1} + BAB^{-1}) \circ \psi^{-1}.$$

So, the vector field in the new coordinates and with a time reparametrization using $\eta \neq 0$ is

$$\tilde{v} = (\eta T\omega, \eta\tilde{A}).$$

It is easy to check that fibered diffeomorphisms and time rescalings preserve the class of fibered flows and fibered vector fields.

In this work we are going to focus in the case that the base M is the d-dimensional torus $\mathbb{T}^d = \mathbb{R}^d/\mathbb{Z}^d$, φ is a constant vector field and the fiber is restricted to the Lie group $\mathrm{SL}(2,\mathbb{R})$ of 2×2 unimodular matrices:

$$\begin{cases} \dot{X} = A(\theta)X & \text{on } \mathrm{SL}(2,\mathbb{R}) \text{ (fiber)} \\ \dot{\theta} = \omega & \text{on } \mathbb{T}^d \text{ (base)} \end{cases} \tag{5}$$

[1]We are writing $B^{-1} \circ \psi^{-1}(\theta)$ to mean $B(\psi^{-1}(\theta))^{-1}$.

with $A: \mathbb{T}^d \to \mathrm{sl}(2, \mathbb{R})$ real-analytic and $\omega \in \mathbb{R}^d$. The base dynamics is simply $\Theta^t(\theta) = \theta + \omega t \bmod \mathbb{Z}^d$.

To simplify notation whenever $\psi = \mathrm{id}$, we denote the fibered vector field A in the new fibered coordinates B as

$$B^* A = \omega \cdot DB\, B^{-1} + B\, A\, B^{-1}.$$

Example 2 (Linear One-Dimensional Schrödinger Equation) Let $V: \mathbb{T}^d \to \mathbb{R}$, $\omega \in \mathbb{R}^d$, $E \in \mathbb{R}$ and $\theta \in \mathbb{T}^d$. The second order ode with respect to $y: \mathbb{R} \to \mathbb{R}$,

$$-\ddot{y}(t) + V(\theta + \omega t) y(t) = E y(t),$$

is equivalent to

$$\begin{bmatrix} \dot{y} \\ \ddot{y} \end{bmatrix} = \begin{bmatrix} 0 & 1 \\ V(\theta + \omega t) - E & 0 \end{bmatrix} \begin{bmatrix} y \\ \dot{y} \end{bmatrix}.$$

By writing $Y = (y, \dot{y})$ one can write the above equation as a fibered vector field.

The particular simple base dynamics is preserved if we restrict the fibered diffeomorphisms to the case that ψ is an automorphism $T \in \mathrm{SL}(d, \mathbb{Z})$ of the torus. Hence, $\tilde{\Theta}^t(\theta) = \theta + T\omega t \bmod \mathbb{Z}^d$ and

$$\tilde{\Phi}^t(\theta) = B(T^{-1}\theta + \omega t)\, \Phi^t(T^{-1}\theta)\, B(T^{-1}\theta)^{-1}.$$

Furthermore, $\tilde{\varphi}(\theta) = T\omega$ and

$$\tilde{A}(\theta) = \omega \cdot DB(T^{-1}\theta)\, B(T^{-1}\theta)^{-1} + B(T^{-1}\theta)\, A(T^{-1}\theta)\, B(T^{-1}\theta)^{-1}.$$

2 Reducibility and Almost Reducibility

2.1 Reducibility

(ω, A) is **reducible** if there is a fibered coordinate change that conjugates it to a constant vector field:

$$\dot{Y} = CY.$$

That is, we want to find $C \in \mathrm{sl}(2, \mathbb{R})$ and $B: 2\mathbb{T}^d \to \mathrm{SL}(2, \mathbb{R})$ of class C^r such that

$$\Phi^t(\theta) = B(\theta + \omega t)^{-1} e^{tC} B(\theta). \tag{6}$$

Equivalently,

$$\omega \cdot DB = CB - BA.$$

Notice that we allow B to be defined in $2\mathbb{T}^d$. See Theorem 1 below to understand why we want to loose the concept of reducibility in this way.

Reducibility preserves several aspects of the dynamics. In particular, since B is a periodic and continuous map thus bounded, (6) implies that both Φ^t and e^{tC} have the same time growth (Lyapunov exponents) and boundness.

An obstruction to reducibility is non-uniform hyperbolicity. Indeed, non-uniform hyperbolic systems can not be reducible because any constant vector field which is hyperbolic is obviously uniformly hyperbolic. So,

$$\text{Non-uniformly hyperbolic} \Rightarrow \text{non-reducible.}$$

Equivalently,

$$\text{Reducible} \Rightarrow \text{uniform hyperbolic or zero Lyapunov exponents.}$$

We say that (ω, A) is **rotations reducible** if there is a conjugacy to a map $C: \mathbb{T}^d \to so(2, \mathbb{R})$.

2.2 Space of Real-Analytic Matrix-Valued Functions

Given $h > 0$, consider the set of real-analytic maps $F: \mathbb{T}^d \to gl(2, \mathbb{R})$ with Fourier expansion

$$F(\theta) = \sum_{k \in \mathbb{Z}^d} F_k e^{ik \cdot \theta}$$

and analytic extension to $\| \operatorname{Im} \theta \| < h$. We choose the norm

$$\|F\|_h = \sum_{k \in \mathbb{Z}^d} \|F_k\| e^{h\|k\|}.$$

Let $\mathscr{B}_h$ be the Banach space of such functions with finite norm $\|F\|_h < +\infty$.

2.3 Almost Reducibility

A is **almost reducible** if there exists sequences $h_n > 0$ (we might have $h_n \to 0$) and $B_n: 2\mathbb{T}^d \to SL(2, \mathbb{R})$ real-analytic on $|\operatorname{Im} \theta| < h_n$ such that B_n conjugates the system into

$$\begin{cases} \dot{X} = (A_n + F_n(\theta))X \\ \dot{\theta} = \omega \end{cases}$$

with $\|A_n\|$ bounded and

$$\lim_{n \to +\infty} \frac{\|F_n\|_{h_n}}{h_n^{\chi}} = 0$$

for any $\chi \geq 1$.

Remark 1
1. Almost reducibility is weaker than reducibility.
2. Many systems very close to constant are not reducible but are almost reducible.
3. An almost reducible system behaves like a reducible one for a long time.

3 Simple Cases

We denote by $\omega^{\perp}$ the orthogonal hyperplane to ω. It is easy to see that

$$A(\theta_0 + \omega t) = \sum_{k \in \omega^{\perp} \cap \mathbb{Z}^d} A_k e^{ik \cdot \theta_0} + \sum_{k \notin \omega^{\perp} \cap \mathbb{Z}^d} A_k e^{ik \cdot (\theta_0 + \omega t)}.$$

We can thus reduce our study to the following situations.

The first is when $\omega^{\perp} \cap \mathbb{Z}^d = \{0\}$. This is equivalent to ω being rationally independent, i.e. $\omega \cdot k \neq 0$ for any $k \in \mathbb{Z}^d \setminus \{0\}$. In two dimensions, this means that the slope of ω is an irrational number. Section 4 deals with the case $d = 2$.

If $\omega^{\perp} \cap \mathbb{Z}^d \neq \{0\}$, then either we can reduce to a lower dimensional rationally independent vector or else ω is a multiple of a rational vector in $\mathbb{Q}^d$. This last case corresponds to a periodic A. In fact, by a time rescaling we can assume that $\omega \in \mathbb{Z}^d$.

If $\omega = 0$, for a fixed initial condition θ_0,

$$\begin{cases} \dot{X} = A(\theta_0)X \\ \dot{\theta} = 0. \end{cases}$$

Hence, (2) has the solution $X(t) = e^{tA(\theta_0)}$. That is, the dynamical behavior is determined by the spectral properties of the constant matrix $A(\theta_0)$.

Example 3 $A = \begin{bmatrix} a & 0 \\ 0 & -a \end{bmatrix}$, $X(t) = \begin{bmatrix} e^{at} & 0 \\ 0 & e^{-at} \end{bmatrix}$

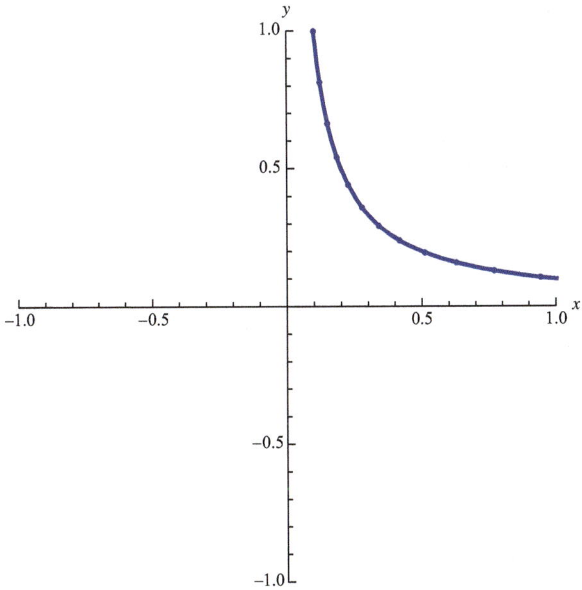

Example 4 $A = \begin{bmatrix} 0 & -b \\ b & 0 \end{bmatrix}$, $X(t) = \begin{bmatrix} \cos(bt) & -\sin(bt) \\ \sin(bt) & \cos(bt) \end{bmatrix}$

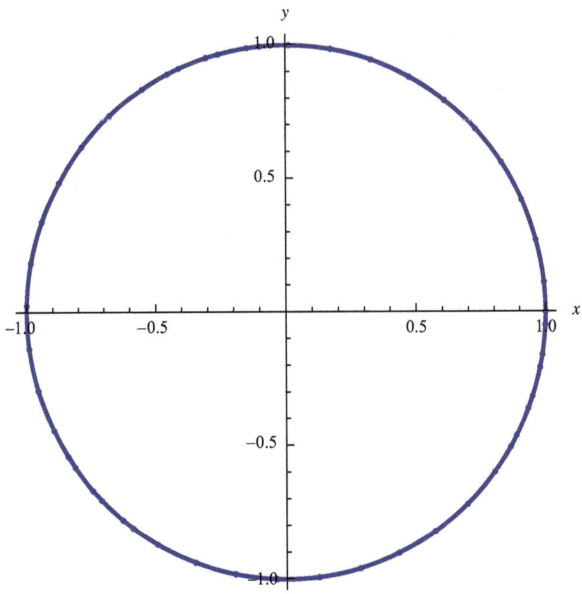

The following is a well-known result.

Theorem 1 (Floquet) *If $\omega \in \mathbb{Z}^d$ and $A \in C^r$, then (ω, A) is C^r-reducible to the constant $C = \frac{1}{2}\log \Phi^2(0) \in \mathrm{sl}(2,\mathbb{R})$.*

Proof By a torus automorphism and a time reparametrization we can reduce to the case $\omega = (1, 0, \ldots, 0)$. So, $\Phi^t(\theta) = \Phi^t(\theta_1, 0)$.

Choose

$$B(\theta) = e^{\theta_1 C}\Phi^{\theta_1}(0)^{-1}.$$

We are going to show that this is a reducibility conjugacy. Indeed, using (4),

$$\begin{aligned} B(\theta+\omega t)\,\Phi^t(\theta)\,B(\theta)^{-1} &= e^{(\theta_1+t)C}\,\Phi^{\theta_1+t}(0)^{-1}\,\Phi^t(\theta)\,\Phi^{\theta_1}(0)e^{-\theta_1 C} \\ &= e^{tC}e^{\theta_1 C}\,\Phi^{\theta_1}(0)^{-1}\Phi^t(\theta)^{-1}\,\Phi^t(\theta)\,\Phi^{\theta_1}(0)\,e^{-\theta_1 C} \\ &= e^{tC}. \end{aligned}$$

Take the canonical basis $\{e_j\}$. To address the periodicity of B let $p \in \mathbb{N}$. Now, $B(\theta + pe_j) = B(\theta)$ when $j \neq 1$ and any p. Write $C_p = \frac{1}{p}\log \Phi^p(0)$. Furthermore, using the fact that $\theta \to \Phi^t(\theta)$ is $\mathbb{Z}^d$-periodic,

$$\begin{aligned} B(\theta+pe_1) &= e^{(\theta_1+p)C_p}\Phi^{\theta_1+p}(0)^{-1} \\ &= e^{\theta_1 C_p}\Phi^p(0)\,\Phi^p(0)^{-1}\,\Phi^{\theta_1}(p,\hat\theta)^{-1} \\ &= B(\theta) \end{aligned}$$

Notice that C_1 might not be a real-valued matrix, but C_2 is. We then can take $p = 2$.

Finally, B has the regularity of Φ^t.

Example 5

$$\dot X = \begin{bmatrix} 0.5\cos t & -1 \\ 1 & -0.5\cos t \end{bmatrix} X$$

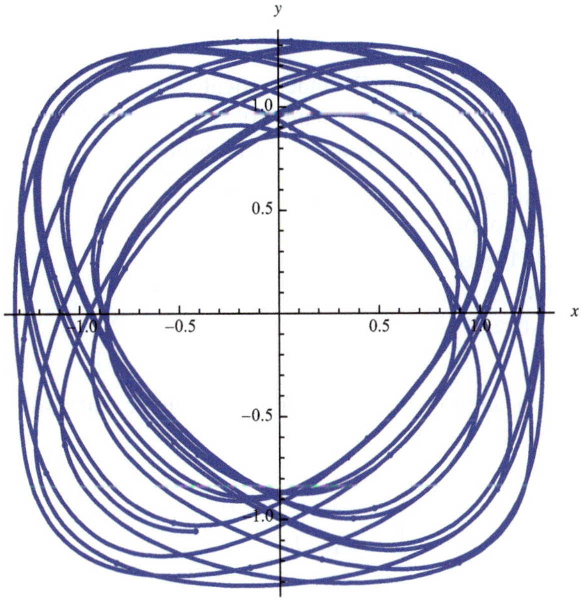

Example 6

$$\dot{X} = \begin{bmatrix} \cos t & -1 \\ 1 & -\cos t \end{bmatrix} X$$

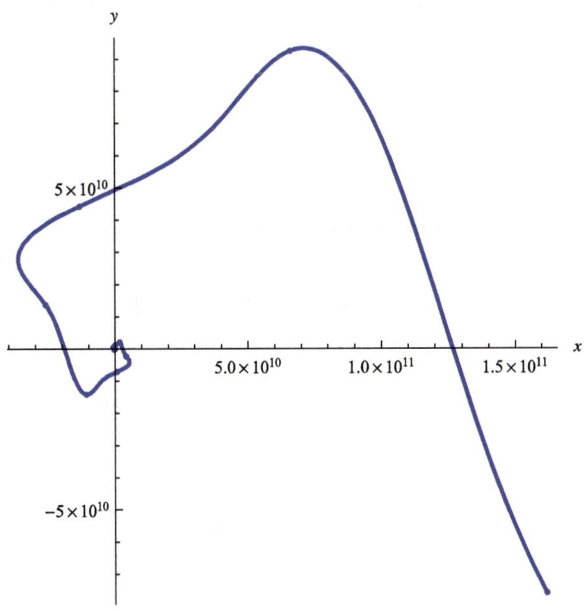

4 Results

The above problem for a rationally independent frequency vector ω has a long history. We just mention below a few names of researchers which have contributed to the subject, in order to stimulate the research for literature.

1. Bogoljubov, Mitropoliski, Samolienko (1960s)
2. Dinaburg, Sinai (1970s)
3. Moser, Pöschel (1980s)
4. Eliasson (1990s)
5. Ávila, Jitomirskaya, Krikorian, Puig, Hou, You, Zhou (2000s)
6. many others

It is our intention here to focus only on the following recent results. Consider the fibered vector field

$$\begin{cases} \dot{X} = (A + F(\theta))X \\ \dot{\theta} = (\alpha, 1) \end{cases} \tag{7}$$

where $\alpha \in (0,1) \setminus \mathbb{Q}$.

Theorem 2 (Hou-You [1]) *Let $h > 0$, $A \in \mathrm{sl}(2,\mathbb{R})$. There is $\delta = \delta(h, \|A\|) > 0$ such that for any $\|F\|_h < \delta$ and $\alpha \in (0,1) \setminus \mathbb{Q}$, (7) is almost reducible.*

Remark 2 We call the above result non-perturbative because δ does not depend on α.

By looking at the projectivized flow, i.e. the flow on $\mathbb{T}^2 \times \mathbb{P}^1$, we can compute the rotation vector of the form $(\alpha, 1, \rho)$. We call ρ the rotation number (also called internal frequency—the external frequency being $\omega = (\alpha, 1)$).

We say that ρ is ω-diophantine if there is $\gamma, \tau > 1$ such that

$$|k \cdot \omega - 2\rho| \geq \frac{\gamma^{-1}}{\|k\|^\tau}, k \in \mathbb{Z}^2 \setminus \{0\}.$$

Consider the continued fractions expansion of α

$$\alpha = \cfrac{1}{a_1 + \frac{1}{a_2 + \frac{1}{\cdots}}} = [a_1, a_2, \dots]$$

and the rational approximants $\frac{p_n}{q_n} = [a_1, \dots, a_n]$ co-prime. Thus, $\frac{p_n}{q_n} \to \alpha$. We define the number

$$\beta(\alpha) = \limsup \frac{\log q_{n+1}}{q_n}.$$

In particular $\beta(\alpha) = 0$ if α is diophantine or Brjuno.

Theorem 3 (Hou-You [1]) *Under the same conditions as in the previous theorem,*

1. *if ρ ω-diophantine, then* (7) *is rotations reducible*
2. *if ρ ω-diophantine and $\beta(\alpha) = 0$, then* (7) *is reducible*

5 Proofs

We sketch the main steps of the proofs of the above theorems. Further details, including the proofs of the lemmas, can be found in [1].

5.1 *Small Divisors*

The main difficulty is how to obtain a solution to the reducibility equation: find $C \in sl(2,\mathbb{R})$ and $B: 2\mathbb{T}^2 \to SL(2,\mathbb{R})$ close do I, i.e. $B = e^Y$, such that

$$\omega \cdot De^Y = Ce^Y - e^Y(A+F).$$

This is indeed a difficult task since problems appear already at the linearized reducibility equation.

Recall that matrices in $sl(2,\mathbb{R})$ have either real eigenvalues (hyperbolic case or parabolic for zero e-values) or pure imaginary (elliptic). Suppose $A = M\left[\begin{smallmatrix}\lambda & 0\\ 0 & -\lambda\end{smallmatrix}\right]M^{-1} = M\tilde{A}M^{-1}$ where λ is either in $\mathbb{R}$ or in $i\mathbb{R}$

Let $\tilde{F} = M^{-1}FM$ and $Z = M^{-1}YM = \left[\begin{smallmatrix}a & b\\ c & -a\end{smallmatrix}\right]$. So, the linearized reducibility equation is

$$\omega \cdot DZ = \tilde{A}Z - Z\tilde{A} - \tilde{F} = \begin{bmatrix} 0 & 2\lambda b \\ -2\lambda c & 0 \end{bmatrix} - \tilde{F}.$$

In Fourier modes ($k \in \mathbb{Z}^2 \setminus \{0\}$):

$$ik \cdot \omega Z_k = \begin{bmatrix} 0 & 2\lambda b_k \\ -2\lambda c_k & 0 \end{bmatrix} - \tilde{F}_k$$

with solution

$$Z_k = -\begin{bmatrix} \frac{\tilde{F}_{k,11}}{ik\cdot\omega} & \frac{\tilde{F}_{k,12}}{ik\cdot\omega+2\lambda} \\ \frac{\tilde{F}_{k,21}}{ik\cdot\omega-2\lambda} & \frac{\tilde{F}_{k,11}}{ik\cdot\omega} \end{bmatrix}.$$

This solution of the conjugacy equation deals with **small divisors** depending on the type of matrix A.

1. In the hyperbolic case we have $\lambda = \rho$. So we have $|ik \cdot \omega \pm 2\rho| \geq |2\rho|$. The only small divisors are then

$$\frac{1}{|k \cdot \omega|}$$

whenever $k \in \mathbb{Z}^2 \setminus \{0\}$ verifies $|k \cdot \omega| \simeq 0$.

2. The elliptic case $\lambda = i\rho$ is harder because there are more resonant modes in

$$\frac{1}{|k \cdot \omega|}, \qquad \frac{1}{|k \cdot \omega \pm 2\rho|}.$$

corresponding to all $k \in \mathbb{Z}^2 \setminus \{0\}$ for which $|k \cdot \omega|, |k \cdot \omega \pm 2\rho| \simeq 0$.

In Fig. 1 we present an example of the resonance lines, that is the points $x \in \mathbb{R}^2$ orthogonal to ω such that $x \cdot \omega = 0$ or $x \cdot \omega = \pm 2\rho$.

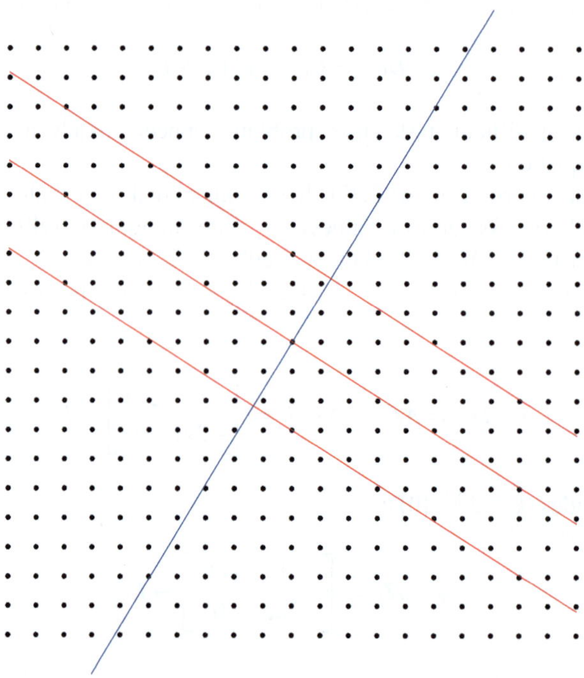

Fig. 1 Fourier modes indices in $\mathbb{Z}^2$: the *red lines* are the resonance lines, the *blue line* is the direction of ω

5.2 Truncation of Modes

Given $N \in \mathbb{N}$, consider the following truncation operator:

$$\mathscr{T}_N F(\theta) = \sum_{\|k\|<N} F_k e^{ik\cdot\theta}.$$

The remaining operator is simply defined as

$$\mathscr{R}_N F(\theta) = F(\theta) - \mathscr{T}_N F(\theta).$$

We can easily control the norm of the remaining terms. Given $h' < h$,

$$\|\mathscr{R}_N F\|_{h'} = \sum_{\|k\|\geq N} \|F_k\| e^{h\|k\|} e^{-(h-h')\|k\|} \leq e^{-N(h-h')} \|F\|_h$$

Notice also that $\|F_k\| \leq e^{-h\|k\|} \|F\|_h$ (exponential decay of Fourier coefficients for analytic functions).

5.3 Elimination of Non-Resonant Modes I

Let $\eta > 0$ and the sets of Fourier indices

$$\Lambda_1 = \{k \in \mathbb{Z}^2 \colon |k\cdot\omega| \geq \eta\}, \qquad \Lambda_2 = \{k \in \mathbb{Z}^2 \colon |k\cdot\omega \pm 2\rho| \geq \eta\}.$$

The above sets are the complement to strips around the resonance lines.

Define the projection

$$\mathbb{I}^- F(\theta) = \sum_{k\in\Lambda_1} \begin{bmatrix} 0 & -c_k \\ c_k & 0 \end{bmatrix} e^{ik\cdot\theta} + \sum_{k\in\Lambda_2} \begin{bmatrix} a_k & b_k \\ b_k & -a_k \end{bmatrix} e^{ik\cdot\theta},$$

where $F_k = \left[\begin{smallmatrix} a_k & b_k - c_k \\ b_k + c_k & -a_k \end{smallmatrix}\right]$. Notice that the elliptic part of F_k is the one related to small divisors of the type $k\cdot\omega$, and the remaining part deals with small divisors like $k\cdot\omega \pm 2\rho$.

We say that $F \in \mathscr{B}_h$ is η-resonant if $\mathbb{I}^- F = 0$. There is not much effort to construct a conjugacy of our original system to one which is non-resonant. This is because the elimination of the modes in $\mathbb{I}^- F$ does not deal with small divisors.

Lemma 1 *Let $\varepsilon < 10^{-8}$ and $\|F\|_h < \varepsilon$. Then,*

$$\begin{cases} \dot{X} = (A + F(\theta))X \\ \dot{\theta} = \omega \end{cases}$$

is conjugated to

$$\begin{cases} \dot{X} = (A + \hat{F}(\theta))X \\ \dot{\theta} = \omega \end{cases} \tag{8}$$

where $\hat{F}$ is $\varepsilon^{1/4}$-resonant and $\|\hat{F}\|_h \leq \varepsilon$.

Using the above conjugacy we obtain Fourier modes indices as in Fig. 2.

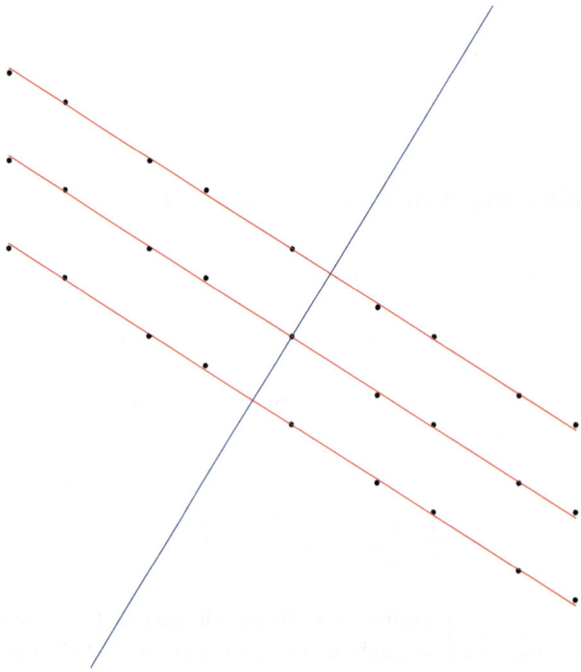

Fig. 2 Indices of the remaining Fourier modes after the elimination of the non-resonant modes

5.4 Rotation

Lemma 2 *If* $\|k^*\| = \min\{\|k\|: |k \cdot \omega \pm 2\rho| < \varepsilon^{1/4}\}$, *then* (8) *is conjugated to*

$$\begin{cases} \dot{X} = (A_1 + F_1(\theta))X \\ \dot{\theta} = \omega \end{cases} \tag{9}$$

where

$$A_1 = \begin{bmatrix} 0 & -\rho' \\ \rho' & 0 \end{bmatrix}, \qquad 2\rho' = 2\rho - k^* \cdot \omega$$

with $|2\rho'| \leq \varepsilon^{1/4}$ *and* $\|F_1\|_{h/3} \leq \varepsilon^{3/4}$.

The particular form of the resonant modes are crucial in order to have such a control on $\|F_1\|_{h/3}$. Otherwise it grows with $e^{\|k^*\|h}$.

Notice also that the analyticity width decreases, except when $k^* = 0$.

The new vector field has now modes for the indices as in Fig. 3.

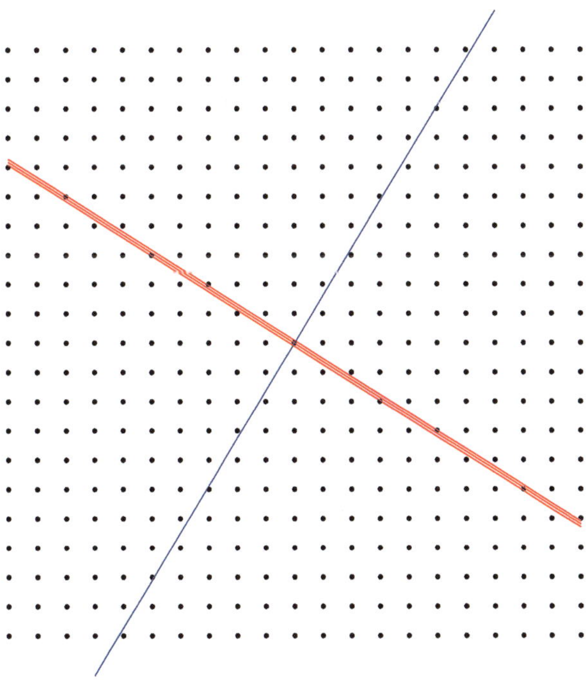

Fig. 3 The resonance lines get closer since ρ' is small, but Fourier modes are no longer restricted to resonant indices

5.5 Elimination of Non-Resonant Modes II

As before, we remove the non-resonant modes.

Lemma 3 *Equation* (9) *is conjugated to*

$$\begin{cases} \dot{X} = (A_1 + F_2(\theta))X \\ \dot{\theta} = \omega \end{cases} \tag{10}$$

where F_2 *is* $\varepsilon^{3/16}$*-resonant and* $\|F_2\|_{h/3} \leq \varepsilon^{3/4}$.

See Fig. 4.

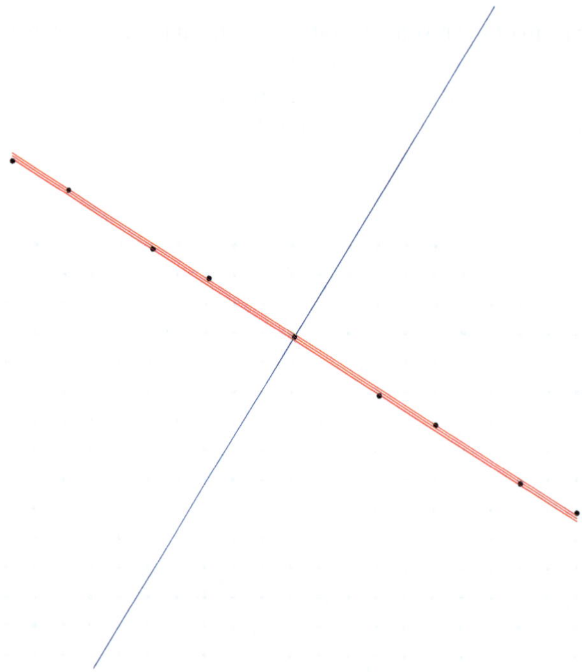

Fig. 4 Another elimination of non-resonant modes

5.6 The Structure of Resonances

Let us recall here the following well-known properties of the continued fraction expansion of α and the corresponding rational approximations p_n/q_n. We have

$$\|(p_n, q_n) - q_n(\alpha, 1)\| \leq \frac{1}{q_{n+1}},$$

and $q_n \geq \gamma^n$, where $\gamma = \frac{1+\sqrt{5}}{2}$ is the golden ratio.

For the step n we write

$$q = q_n, \quad q' = q_{n+1}, \quad p = p_n, \quad p' = p_{n+1}$$

Lemma 4 *If* $\|k\| < q'/6$ *and* $|k \cdot \omega| < 1/(7q)$*, then* $k = \ell(q, -p)$ *for some* $\ell \in \mathbb{Z}$.

From the previous lemma we know that (see also Fig. 5)

$$\mathscr{T}_{q'/6} F_2(\theta) = \sum_{k=\ell(q,-p)} F_k e^{ik\cdot\theta}.$$

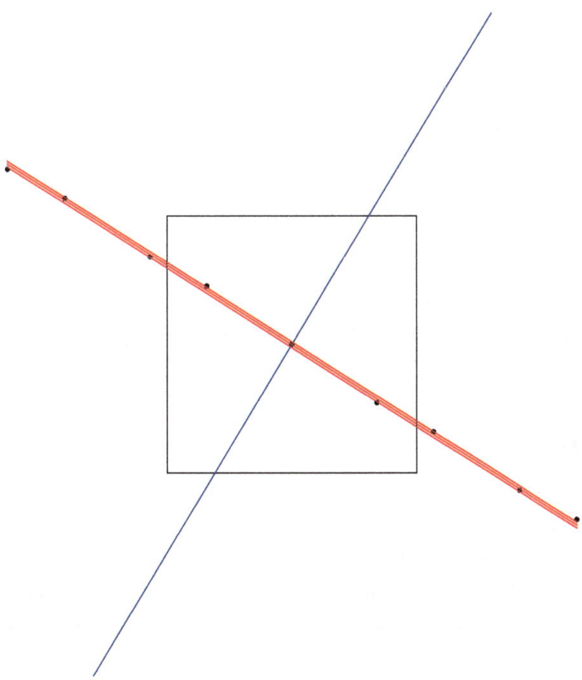

Fig. 5 In the truncation box we only find modes with indices in one line

Thus we are able to get the estimate

$$\|\mathscr{R}_{q'/6}F_2\|_{h/3} \leq \varepsilon^{3/4}e^{-q'h}.$$

5.7 Floquet Theory Revisited

By a step similar to the Floquet theorem 1, we obtain the following result.

Lemma 5 *Let* $(q, -p) \in \mathbb{Z}^2$ *fixed and*

$$F(\theta) = \sum_{\ell \in \mathbb{Z}} F_{\ell(q,-p)} e^{i\ell(q,-p)\cdot\theta}$$

Then,

$$\begin{cases} \dot{X} = F(\theta)X \\ \dot{\theta} = \omega \end{cases}$$

is reducible by a conjugacy $B(\theta)$.

Lemma 6 *Equation* (10) *is conjugated to*

$$\begin{cases} \dot{X} = (A_3 + F_3(\theta))X \\ \dot{\theta} = \omega \end{cases} \tag{11}$$

where

$$\|F_3\|_{h/4} \leq \varepsilon^{1/8}e^{-q'h}$$

$$A_3 + F_3 = B^*(A_1 + \mathscr{T}_{q'/6}F_2 + \mathscr{R}_{q'/6}F_2) = C + B^*\mathscr{R}_{q'/6}F_2$$

5.8 After One Step

In the previous sections we have constructed one step in a iterative scheme that leads to the convergence to a constant vector field. At this time we still need to perform a normalization in order to diagonalize A_3. After this we obtain a conjugated vector field whose Fourier modes indices are again as in Fig. 1 but the modes have much smaller norm.

So, one step in the scheme corresponds to

$$(A, F) \mapsto (A', F')$$

with

$$\|F'\|_{h'} \leq \min\{h^{16}, \varepsilon^{33/32}e^{-q'h}\}.$$

5.9 Convergence

Finally, we iterate the previous steps and obtain a sequence F_n with

$$\|F_n\|_{h_n} \leq \min\{h_n^{16}, e^{-q_{n+1}h_n}\}.$$

If $h_n \to 0$ the system is just almost reducible (in the real-analytic setting).

If $\lim h_n > 0$ and $\|B_n\|$ are bounded, the system is reducible. This can be achieved by imposing conditions on the rotation number that make the Rotation and Floquet Lemmas steps unnecessary after a finite number of iterations. Those steps are the ones that shorten the analyticity strip width and increase the norms of the conjugacies.

Acknowledgements I would like to express my gratitude to Alberto Pinto for the invitation to participate in the Conference and Advance School.

Reference

1. Hou, X., You, J.: Almost reducibility and non-perturbative reducibility of quasi-periodic linear systems. Invent. Math. **190**(1), 209–260 (2012)

Collateral Versus Default History

Marta Faias and Abdelkrim Seghir

Abstract This paper deals with equilibrium existence for incomplete markets economies with finitely-lived agents and infinitely-lived agents when default is allowed and borrowers have to constitute collateral in terms of durable goods. In the first model, lenders are protected by an exogenous personalized collateral. In the second model, the personalized collateral requirements are endogenously determined by a financial institution whose objective is to minimize the default rate taking into account agent's default history.

1 Introduction

The analysis of finite-horizon economies has been extended to an infinite horizon in two main ways. The first one assumes that each agent is alive for a finite number of periods and is succeeded by his offspring forming an infinite sequence of the so-called Overlapping Generations Models (OLG). The OLG models have been introduced by Samuelson [21] and used in incomplete markets without default by Florenzano–Gourdel–Páscoa [9], Schmachtenberg [23] and Seghir [24], among others. The second approach considers a finite number of infinitely-lived agents. This approach was introduced by Bewley [6] in a complete market economy and used for incomplete markets without default by Florenzano–Gourdel [8], Hernandez–Santos [14], Levine–Zame [15] and Magill–Quinzii [16, 17], among others. All these papers prove equilibrium existence in incomplete markets when assets are either numeraire or nominals, assuming that borrowers fully keep their promises. As is well known since the Hart's counterexample [13], equilibrium may fail to exist in the real case if the short sales are not bounded a priori, as the rank of the return matrix may drop.

M. Faias (✉)
Departamento de Matemática and CMA, FCT, Universidade Nova de Lisboa, Lisboa, Portugal
e-mail: mcm@fct.unl.pt

A. Seghir
Economics Department, School of Business, The American University in Cairo, Cairo, Egypt
e-mail: kseghir@aucegypt.edu

J.-P. Bourguignon et al. (eds.), *Dynamics, Games and Science*, CIM Series in Mathematical Sciences 1, DOI 10.1007/978-3-319-16118-1_14

When default is allowed, some mechanism have been imposed in the literature in order to protect lenders from total default and urge borrowers to pay, at least partially, their debt. A first mechanism requires borrowers to constitute collateral in terms of durable goods. These collateral are seized and given to the lenders in case of default. For incomplete markets of collateralized real assets, Geanakoplos–Zame [12] prove equilibrium existence for a finite-horizon model while Araujo–Páscoa–Torres-Martínez [3] show equilibrium existence, for an infinite-horizon economy with a finite number of infinitely-lived agents, without imposing any exogenous bounds on the short sales. A second mechanism assumes that borrowers suffer disutility from defaulting. For incomplete markets of nominal/numeraire assets, when borrowers suffer utility penalties proportional to their real amount of default, equilibrium existence was proved by Araujo–Monteiro–Páscoa [2] for a two-period model and by Araujo–Monteiro–Páscoa [1] for an infinite-horizon model. Moreover, in an infinite-horizon model in which these two mechanism coexist, Páscoa–Seghir [19] prove that equilibrium exists provided that utility penalties are moderate.

In this paper, we extend the model of Araujo–Páscoa–Torres-Martínez [3] to a demographic structure that includes both a finite number of infinitely-lived agents and overlapping generations. The demographic structure of this model is of interest for several reasons. First, infinitely-lived agents can be interpreted as altruistic agents who increase the preferences of their descendants, who in turn increase the preferences of their offspring... until infinity. In fact, they consider the welfare of their whole dynasty. Second, as in Muller–Woodford [18], one may suppose that certain institutions are effectively agents with infinite-horizon consumption programs. For example, Thompson [26] argues that corporations should be modeled as infinitely-lived agents while private households are finitely-lived. Third, the infinitely-lived agents (households) can be interpreted as finitely-lived individuals who inherit of their ancestors' debt while finitely-lived agents do not. It is for example the case of mortgage markets without life insurance protection. Note that such a demographic structure has been studied by Wilson [27] (for a complete market model) and Florenzano–Gourdel–Páscoa [9] (for a real incomplete market model with bounded short sales when default is not allowed). On the other hand, one may interpret the finitely-lived agents as a representation of the behavior of agents who optimize over a finite horizon, not because of their biological life span, but because of financial constraints (for example, in Seghir [24], the agents are constrained to attend the financial market at all the periods of their lifetime except at the last one). In the present, one may interpret the demographic structure as one where there are both financially constrained and unconstrained agents.

As in Araujo–Páscoa–Torres-Martínez [3], Ponzi schemes are ruled out and equilibrium existence is guaranteed without exogenous restrictions such as debt constraints or a transversality condition. It turns out that the obligation of constituting collateral in terms of durable goods guarantees that, in equilibrium, short sales are bounded node by node. These endogenous bounds rule out the possibility of Ponzi games for infinitely-lived agents as our stochastic structure is characterized by a finite number of immediate successors at each node. On the other hand, Ponzi schemes are avoided for the finitely-lived agents as they cannot attend the financial market at the last period of their lifetime. However, the presence of finitely-lived agents generates a technical problem that does not occur in the case of infinitely-lived agents. More precisely, the budget sets of finitely-lived agents may fail to be lower semicontinuous as their interior could be empty.

The objective of this paper is twofold. First, we aim at proving equilibrium existence for an infinite–horizon economy with finitely-lived agents and infinitely-lived agents when collateral requirements are personalized. Second, we study a general equilibrium model in which default taxes are restricted through some enforcement mechanism that: (1) directly affects agents' wealth, and (2) takes into account the default history of each agent. More precisely, we prove equilibrium existence in a model in which real assets are protected by *endogenous personalized collateral*. That is, in this second model, and contrary to Araujo–Páscoa–Torres-Martínez [3], collateral are personalized and are *endogenously determined* by a financial institution whose objective is to fix these personalized collateral in order to minimize the default rate. We address the equilibrium existence in such a model and study the possibility of choosing these collateral requirements in order to perfectly anticipate the equilibrium default rates and, therefore, adjusting according to traders' default history or *with the default history of their families*.

The paper is organized as follows. The model is presented in the next section. Section 3 deals with default history rate and states our main result. Section 4 presents some concluding remarks. Finally, an appendix is devoted to proofs.

2 The Model

2.1 *The Stochastic Structure*

We consider a pure-exchange economy with infinite time horizon. The stochastic structure is described by an infinite event-tree with an unique root. Formally, let $\mathcal{T} = \{0, 1, \cdots\}$ be the set of periods and let S be the set of states of nature. The revelation of information is described by a sequence of partitions of S, $(\mathbf{F}_0, \mathbf{F}_1, \ldots, \mathbf{F}_t, \ldots)$, where the number of subsets in $\mathbf{F}_t$ is finite and, for each $t \geqslant 0$, $\mathbf{F}_{t+1}$ is finer than the partition $\mathbf{F}_t$ (i.e.: $\sigma \in \mathbf{F}_{t+1}$, $\sigma' \in \mathbf{F}_t \Longrightarrow \sigma \subset \sigma'$ or $\sigma \cap \sigma' = \emptyset$).

At node 0, we assume that there is no information so that $\mathbf{F}_0 = S$. The information available at time t is assumed to be the same for all agents in the economy (symmetric information) and described by the subset σ of the partition $\mathbf{F}_t$ in which the state of nature lies.

A pair $\xi = (t, \sigma)$ where $t \in \mathcal{T}$ and $\sigma \in F_t$ is called a node and $t(\xi) = t$ is the date of node ξ. The set D consisting of all nodes is called the event-tree induced by $\mathbf{F}$.

A node $\xi' = (t', \sigma')$ is said to succeed (resp. strictly) node $\xi = (t, \sigma)$ if $t' \geqslant t$ (resp. $t' > t$) and $\sigma' \subset \sigma$. We write $\xi' \geqslant \xi$ (resp. $\xi' > \xi$). The set of nodes which succeed a node $\xi \in D$ is called the subtree $D(\xi)$ and $D^+(\xi) = \{\xi' \in D | \xi' > \xi\}$ is the set of strict successors of ξ. The subset of nodes of $D(\xi)$ at date T is denoted by $D_T(\xi)$ and the subset of nodes between $t(\xi)$ and T by $D^T(\xi)$. When ξ is the initial node, the notations are simplified to D^+, D_T, D^T.

For each $\xi \in D$, $\xi^+ = \{\xi' \in D(\xi) | t(\xi') = t(\xi) + 1\}$ is the set of immediate successors of ξ. The number of elements of ξ^+ is finite and is called the branching number $b(\xi)$ at ξ ($b(\xi) = \#\xi^+$).

If $\xi = (t, \sigma)$, $t \geqslant 1$, the unique node $\xi^- = (t-1, \sigma')$, $\sigma \subset \sigma'$ is called the predecessor of ξ.

2.2 *The Commodity Markets and Demographic Structures*

At each node $\xi \in D$, a finite number G of physical goods, indexed by $g = 1, \ldots, G$, are traded on spot markets. These goods are durable and may suffer a partial depreciation from a period to another. The structure of depreciation in the event-tree is given by a collection of functions $(Y_\xi)_{\xi \in D} : \mathbb{R}^G_+ \to \mathbb{R}^G_+$. More precisely, if one unit of a commodity g' is consumed at the node ξ^-, the consumer obtains $Y_\xi(e_{g'})_g \in \mathbb{R}_+$ units of the good g at the node ξ. So, for instance, we say that a good g is durable if, for all node ξ, $Y_\xi(e_g)_g \in \mathbb{R}_{++}$. A perishable good g is such that $Y_\xi(e_g)_g = 0$, for each node $\xi \in D$.

Let $p(\xi) = (p(\xi, g),\ g \in G)$ denotes the vector of spot prices at node ξ and $p = (p(\xi),\ \xi \in D) \in \mathbb{R}^{D \times G}_+$ be the spot price process.

Agents' incomplete participation is allowed in our model. Let $\mathcal{H}$ be the (countable) set of agents. At each node ξ, we denote by $\mathcal{H}(\xi)$ the finite set of agents who can trade on the spot markets. We denote by $\mathcal{D}^h$ the sub-tree of nodes at which agent $h \in \mathcal{H}$ can trade on the spot markets. The root of the sub-tree $\mathcal{D}^h$ is denoted by ξ^h. We say that agent h is *infinitely-lived* if for all $t \in \mathbb{N}$ there exists a node $\xi \in \mathcal{D}^h$ such that $\tilde{t}(\xi) = t$. Otherwise, agent h is said to be *finitely-lived*. We denote by $\mathcal{H}_1$ the set of finitely-lived agents and by $\mathcal{H}_2$ the set of the infinitely-lived agents.

The set $\delta\mathcal{D}^h$ denotes agent h's *terminal nodes*, that is, the set of nodes $\xi \in \mathcal{D}^h$ such that $\mathcal{D}(\xi) \cap \mathcal{D}^h = \{\xi\}$ (if such nodes exist; otherwise we suppose that $\delta\mathcal{D}^h$ is empty).

We introduce the following technical conditions:

a. For each agent $h \in \mathcal{H}$, if $\xi \in (\mathcal{D}^h - \delta\mathcal{D}^h)$ then $\xi^+ \subset \mathcal{D}^h$,
b. For each node $\xi \in \mathcal{D}$ there exists at least one agent $h \in \mathcal{H}$ such that $\xi \in (\mathcal{D}^h - \delta\mathcal{D}^h)$.

The previous conditions have also been used by Santos and Woodford [22] and Seghir and Torres-Martinez [25]. Condition b. is satisfied if there exists at least an infinitely-lived agent (i.e. if $\mathcal{H}_2 \neq \emptyset$).

We denote by $\tilde{\mathcal{H}}(\xi)$ the set of agents who have access to the financial markets at node ξ. Hence, $h \in \mathcal{H}(\xi)$ if and only if $\xi \in \mathcal{D}^h$, and $h \in \tilde{\mathcal{H}}(\xi)$ if and only if $\xi \in (\mathcal{D}^h - \delta\mathcal{D}^h)$.

At each node $\xi \in \mathcal{D}^h$, agent h can choose a *collateral-free consumption allocation* $x^h(\xi) \in \mathbb{R}_+^G$. We denote by $x^h = \left(x^h(\xi)\right)_{\xi \in \mathcal{D}^h}$ agent h's collateral-free consumption plan and by $X^h = \mathbb{R}_+^{\mathcal{D}^h \times G}$ agent h's consumption space.

Each agent $h \in \mathcal{H}$ is characterized by an endowment process $w^h \in X^h$ and a utility function $U^h : X^h \to \mathbb{R}_+$ that represents his preferences.

2.3 The Financial Structure

At each node, a finite number $J(\xi)$ of one-period real assets are available for intertemporal transaction and insurance. Given $\xi \in D$, $A^j(\xi) \in \mathbb{R}_+^G$ denotes the promises of asset j. We denote $A(\xi) = (A^j(\xi))_{j \in J(\xi-)}$, and $A := \prod_{\xi \in D} A(\xi)$. At each node $\xi \in D$, given a commodity prices $p(\xi)$, the vector $p(\xi)A^j(\xi)$ represents the financial promises value, denominated in units of account, of asset j (sold at node ξ^-). Thus, at each node $\xi \in D$, the $(b(\xi) \times J(\xi))$-matrix $V(p) := (p(\xi')A^j(\xi'))_{\substack{\xi' \in \xi^+ \\ j \in J(\xi)}}$ completely describes the default-free promises at period $t(\xi) + 1$.

Let $q(\xi) = (q(\xi, j),\ j \in J(\xi)) \in \mathbb{R}^{J(\xi)}$ be the vector of prices of the securities issued at node ξ and let $q = (q(\xi), \xi \in D)$ denote the security price process which belongs to the security price space $\prod_{\xi \in D} \mathbb{R}^{J(\xi)}$.

At each node $\xi \in (D^h - \delta D^h)$, agents can choose a portfolio $z^h(\xi) := (z_j^h(\xi),\ j \in J(\xi)) \in Z_\xi^h = \mathbb{R}^{J(\xi)}$, with $z_j^h(\xi) = \theta_j^h(\xi) - \phi_j^h(\xi)$, where:

- $\theta_j^h(\xi) \in \mathbb{R}_+$ denotes the quantity of asset j bought by agent $h \in \tilde{H}(\xi)$ at node ξ,
- $\phi_j^h(\xi) \in \mathbb{R}_+$ denotes the quantity of j sold by agent h at node ξ.

At each node $\xi \in D$, the debt of agent $h \in H(\xi)$, induced by the sale of asset $j \in J(\xi^-)$, is $p(\xi)A^j(\xi)\phi_j^h(\xi^-)$. Since default is allowed, agent $h \in H$ pays, at node ξ, an amount $\Delta_j^h(\xi)$, denominated in units of account, with $0 \leqslant \Delta_j^h(\xi) \leqslant p(\xi)A^j(\xi)\phi_j^h(\xi^-)$.

2.4 The Exogenous Personalized Collateral Requirement

In order to protect lenders against total default, we require each seller $h \in \tilde{H}(\xi)$ of one unit of an asset $j \in J(\xi)$ to constitute an exogenous personalized physical collateral $M_j^h(\xi) \in \mathbb{R}_+^G$. The lack of payment causes the seizure of the collateral by the lenders. Therefore, each agent will deliver the minimum between the value of the depreciated collateral and the original debt. That is:

$$\Delta_j^h(\xi) = \min\{p(\xi)A^j(\xi), p(\xi)Y_\xi\left(M_j^h(\xi^-)\right)\}\phi_j^h(\xi^-).$$

In view of markets' anonymity, lenders do not know how much returns they will receive on their long positions. As in Dubey et al. [7], each lender of one unit of asset j at node ξ will expect to receive the average rate of payments, denoted by $R^j(\xi)$, which will be determined endogenously in equilibrium. This average rate of payments will perfectly anticipate the market's average rate of the payments.

We define the economy $\mathcal{E}$, with exogenous personalized collateral requirements, as follows:

$$\mathcal{E} := \left[\left(X^h, Z^h, U^h, \omega^h, M^h\right)_{h\in H}, A\right].$$

Definition 1 (Budget Sets) Given vector prices (p, q) and anonymous rates of payments R, the budget set, $B^h(p, q, R)$, of agent $h \in H$ is the set of allocations $(x^h, \theta^h, \phi^h, \Delta^h)$ satisfying the following conditions:

- At node ξ^h,

$$p(\xi^h)x^h(\xi^h) + p(\xi^h)M^h(\xi^h)\phi^h(\xi^h) + q(\xi^h)z^h(\xi^h) \leqslant p(\xi^h)\omega^h(\xi^h), \tag{1}$$

- At each node $\xi \in (D^h - \delta D^h)$, $\xi \neq \xi^h$,

$$\begin{aligned} &p(\xi)x^h(\xi) + p(\xi)M^h(\xi)\phi^h(\xi) + q(\xi)z^h(\xi) + \sum_{j\in J(\xi^-)} \Delta_j^h(\xi) \\ &\leqslant \sum_{j\in J(\xi^-)} R^j(\xi)\theta_j^h(\xi^-) + p(\xi)Y_\xi\left(M^h(\xi^-)\phi^h(\xi^-) + x^h(\xi^-)\right) + p(\xi)\omega^h(\xi), \end{aligned} \tag{2}$$

- At each node $\xi \in \delta D^h$,

$$\begin{aligned} p(\xi)\left[x^h(\xi) - \omega^h(\xi)\right] \leqslant &\sum_{j\in J(\xi^-)} \left[R^j(\xi)\theta_j^h(\xi^-) - \Delta_j^h(\xi)\right] \\ &+ p(\xi)Y_\xi\left(M^h(\xi^-)\phi^h(\xi^-) + x^h(\xi^-)\right), \end{aligned}$$

where, for each $\xi \in D^h - \xi^h$, $0 \leqslant R^j(\xi) \leqslant p(\xi)A^j(\xi)$.

We define agent h's total utility at an allocation (x, θ, ϕ) as $V^h(x, \phi) := U^h\left(x(\xi) + M^h(\xi)\phi(\xi)\right)$.

We denote by W_ξ the aggregate endowment accumulated until a node ξ, that is,

$$W_\xi := \sum_{h\in H(\xi)} w^h(\xi) + Y_\xi\left(W_{\xi^-}\right), \quad \forall \xi \in D \setminus \{\xi_0\}, \tag{3}$$

where $W_{\xi_0} = \sum_{h\in H(\xi_0)} w^h(\xi_0)$.

Now, we can define the concept of equilibrium in our economy:

Definition 2 (Equilibrium) An equilibrium of $\mathcal{E}$ is a collection of vector prices $(\overline{p}, \overline{q})$, anonymous payments rates $\overline{R}$ and individual choice variables $\left(\overline{x}^h, \overline{\theta}^h, \overline{\phi}^h, \overline{\Delta}^h\right)_{h \in H}$, satisfying:

(i) For each agent $h \in H$, $(\overline{x}^h, \overline{\theta}^h, \overline{\phi}^h, \overline{\Delta}^h)$ maximizes V^h over $B^h(\overline{p}, \overline{q}, \overline{R})$.
(ii) Physical and Financial markets clear,

$$\sum_{h \in H(\xi)} [\overline{x}^h(\xi) + \overline{M}^h(\xi)\overline{\phi}^h(\xi)] = W_\xi, \tag{4}$$

$$\sum_{h \in \tilde{H}(\xi)} \overline{\theta}_j^h(\xi) = \sum_{h \in \tilde{H}(\xi)} \overline{\phi}_j^h(\xi), \quad \forall \xi \in D, \ \forall j \in J(\xi). \tag{5}$$

(iii)

$$\forall \xi \in D^T \setminus \{0\}, \ \forall j \in J(\xi^-), \ \overline{R}_j(\xi) \sum_{h \in \tilde{H}(\xi^-)} \overline{\theta}_j^h(\xi^-) = \sum_{h \in \tilde{H}(\xi^-)} \overline{\Delta}_j^h(\xi). \tag{6}$$

Condition (i) requires the optimality of agents' choices over their budget sets. Conditions (4) and (5) require the commodity and asset markets to clear. Condition (6) says that, at each node and for each asset, total effective deliveries made by the sellers are equal to total expected deliveries made by the buyers.

Proposition 1 *Let us consider an economy* $\mathcal{E} := \left[(X^h, Z^h, U^h, \omega^h, M^h)_{h \in H}, A\right]$ *satisfying the following assumptions:*

[A1]. *For each agent* $h \in H$, $U^h(x) = \sum_{\xi \in D^h} v_\xi^h(x)$. *Moreover,* $\forall \xi \in D$, *the function* $v_\xi^h : \mathbb{R}_+^G \longrightarrow \mathbb{R}$ *is continuous, strictly monotone and concave with* $v_\xi^h(0) = 0$. *In addition,* $\forall h \in H_2$, $\forall \alpha \in \mathbb{R}_+^G$, $\sum_{\xi \in D} v_\xi^h(\alpha)$ *is finite.*

[A2]. $\forall h \in H$, $\omega^h \in \operatorname{Int} X^h$,

[A3]. *There exists* $W > 0$ *such that* $W_\xi \leq W$, $\forall \xi \in D$.

[A4]. $\forall \xi \in D$, $\forall j \in J(\xi)$, $\forall h \in H(\xi)$, $M_j^h(\xi) \neq 0$.

Then, $\mathcal{E}$ *has an equilibrium* $\left(\overline{p}, \overline{q}, \overline{R}, \left(\overline{x}^h, \overline{\theta}^h, \overline{\phi}^h, \overline{\Delta}^h\right)_{h \in H}\right)$.

This model extends the model of Araujo–Páscoa–Torres-Martínez [3] of incomplete financial markets model with default and collateral requirements to the case in which borrowers have to constitute personalized collateral. This extension is very simple and the proof of equilibrium existence is close to the proof of Araujo–Páscoa–Torres-Martínez [3]. However, a technical problem arises in our model in comparison with the one of Araujo–Páscoa–Torres-Martínez [3]. More precisely, the budget correspondences of finitely-lived agents may not be lower semicontinuous, as their interior may be empty. We define modified budget sets in order to overcome this problem (see Appendix for more details).

Given personalized collateral requirements M^h, different from zero, there exists an equilibrium in finite-horizon truncated economies. It follows from the personalized non-arbitrage condition:

$$p(\xi)M_j^h(\xi) - q_j(\xi) > 0,$$

that a sequence of equilibria for truncated economies has a convergent subsequence and the cluster point is an equilibrium for the original economy.

As in Araujo–Páscoa–Torres-Martínez [3], the non-arbitrage condition guarantees that, *in equilibrium,* the commodity price vectors are bounded away node by node. Moreover, in Araujo–Páscoa–Torres-Martínez [3], this non-arbitrage condition, together with the fact that the number of alive agents is the same at all nodes, ensures that the value of the short sales is uniformly bounded from above along the event-tree. In our model, since the number of alive consumers depend on the nodes, this uniform bound no longer holds. However, we still obtain bounds on the short sales, node by node. As the stochastic structure of our model is characterized by a finite number of immediate successors at each node, we get a cluster point of the truncated economy equilibria. Finally, the short sales are protected by personalized collateral and, as in Araujo–Páscoa–Torres-Martínez [3], we do not require either debt constraints or a transversality condition to prove that this cluster point is an equilibrium for the original economy.

3 A Model with Endogenous Personalized Collateral Requirement

In this section, we use the same stochastic, financial, commodity, demographic structures of the previous model. However, to protect lenders against total default and to bound the asymptotic growth of the debts along the event-tree, we suppose that there is an institution that requires each seller $h \in \tilde{H}(\xi)$ of one unit of an asset $j \in J(\xi)$ to constitute a personalized physical collateral $M_j^h(\xi) \in \mathbb{R}_+^G \setminus \{0\}$. The main objective of this institution is to control the future default rates, without affecting too much the negotiations in the financial market. More precisely, this institution aims at maintaining the collateral at a practical level which is neither "too low" (although this allows agents to trade more, the default rate could be very high) nor "too high" (in order to have enough traders on the financial market).

Therefore, the best would be to obtain a compatibility of the personalised collateral requirement with: (1) the equilibrium default rate, and (2) the variations of commodity and asset prices.

Given prices (p, q), the default of agent $h \in \tilde{H}(\xi)$ on asset $j \in J(\xi)$, at node $\mu \in \xi^+$ is given by:

$$t^h_{\mu,j} := p(\mu)A^j(\mu)\phi^h_j(\xi) - \Delta^h_j(\mu). \tag{7}$$

The default history of agent h at node ξ is then given by the vector $\kappa^h_\xi := (t^h_{\mu,j})$, where $\mu \leqslant \xi$ and $j \in J(\mu^-)$.

The institution aims at establishing the personalized level of collateral, for agent h at node ξ, using an adjusting rule given by a function

$$F^h_{\xi,j} : \mathbb{R}^G_+ \times \mathbb{R}_+ \times \mathbb{R}^{m(\xi)}_+ \to \mathbb{R}^G_+, \tag{8}$$

where $m(\xi) = \#\{(\mu,j) : \mu \leqslant \xi,\ j \in J(\mu^-)\}$.

The main question is: *Is there any personalized collateral level M^h that allows to perfectly anticipate the default rate in equilibrium?* In other words, *is it possible to choose $M := (M^h_j(\xi),\ \xi \in D,\ j \in J(\xi),\ h \in H)$, such that there exists at least an equilibrium for which:*

$$\forall \xi \in D,\ \forall j \in J(\xi),\ \forall h \in H,\ M^h_j(\xi) = F^h_{\xi,j}\left(p(\xi), q_j(\xi), \kappa^h_\xi\right)\ ? \tag{9}$$

Agents take the requirements *as given* and they know that the lack of payment causes the seizure of the collateral by the lenders. Therefore, each borrower will choose the payments $\Delta^h_j(\xi)$ satisfying:

$$\Delta^h_j(\xi) = \min\{p(\xi)A^j(\xi), p(\xi)Y_\xi\left(M^h_j(\xi^-)\right)\}\phi^h_j(\xi^-)$$

As in the first model, each lender of one unit of asset j at node ξ will expect to receive the average rate of payments given by $R^j(\xi)$, which will be determined in equilibrium and will perfectly foresee the market average rate of payments.

Our economy $\mathcal{E}'$, with *endogenous* personalized collateral requirements, is defined as

$$\mathcal{E}' := \left[\left(X^h, Z^h, U^h, \omega^h, F^h\right)_{h\in H}, A\right].$$

Definition 3 (Budget Sets) Given vector of prices (p, q), anonymous rates of payments R, and personalized collateral requirements $(M^h)_{h\in H}$, the budget set $B^h(p, q, R, M^h)$, of agent $h \in H$ is the set of allocations $(x^h, \theta^h, \phi^h, \Delta^h)$ that satisfy the following conditions:

$$p(\xi^h)x^h(\xi^h) + p(\xi^h)M^h(\xi^h)\phi^h(\xi^h) + q(\xi^h)z^h(\xi^h) \leqslant p(\xi^h)\omega^h(\xi^h), \tag{10}$$

for each $\xi \in (D^h - \delta D^h)$, $\xi \neq \xi^h$,

$$p(\xi)x^h(\xi) + p(\xi)M^h(\xi)\phi^h(\xi) + q(\xi)z^h(\xi) + \sum_{j\in J(\xi^-)} \Delta_j^h(\xi)$$
$$\leqslant \sum_{j\in J(\xi^-)} R^j(\xi)\theta_j^h(\xi^-) + p(\xi)Y_\xi \left(M^h(\xi^-)\phi^h(\xi^-) + x^h(\xi^-)\right) + p(\xi)\omega^h(\xi), \tag{11}$$

for each $\xi \in \delta D^h$,

$$p(\xi)\left[x^h(\xi) - \omega^h(\xi)\right]$$
$$\leqslant \sum_{j\in J(\xi^-)} \left[R^j(\xi)\theta_j^h(\xi^-) - \Delta_j^h(\xi)\right] + p(\xi)Y_\xi \left(M^h(\xi^-)\phi^h(\xi^-) + x^h(\xi^-)\right),$$

where, for each $\xi \in D^h - \xi^h$, $0 \leqslant R^j(\xi) \leqslant p(\xi)A^j(\xi)$.

Now, we can define the concept of equilibrium in our economy:

Definition 4 (Equilibrium) An equilibrium of $\mathcal{E}'$ is a collection of vector prices $(\overline{p}, \overline{q})$, anonymous payments rates $\overline{R}$, personalized collateral requirements $(\overline{M}^h)_{h\in H}$ and individual choice variables $\left(\overline{x}^h, \overline{\theta}^h, \overline{\phi}^h, \overline{\Delta}^h\right)_{h\in H}$ satisfying:

(i) For each agent $h \in H$, $(\overline{x}^h, \overline{\theta}^h, \overline{\phi}^h, \overline{\Delta}^h)$ maximizes V^h over $B^h(\overline{p}, \overline{q}, \overline{R}, \overline{M}^h)$.
(ii) Physical and Financial markets clear,

$$\sum_{h\in H(\xi)} [\overline{x}^h(\xi) + \overline{M}^h(\xi)\overline{\phi}^h(\xi)] = W_\xi, \tag{12}$$

$$\sum_{h\in \tilde{H}(\xi)} \overline{\theta}_j^h(\xi) = \sum_{h\in \tilde{H}(\xi)} \overline{\phi}_j^h(\xi), \quad \forall \xi \in D,\ \forall j \in J(\xi). \tag{13}$$

(iii) The anonymous payment rates, $\overline{R}_j(\xi)$, perfectly foresee the payments $\overline{\Delta}_j^h(\xi)$,

$$\forall \xi \in D^T \setminus \{0\},\ \forall j \in J(\xi^-),\ \overline{R}_j(\xi) \sum_{h\in \tilde{H}(\xi^-)} \overline{\theta}_j^h(\xi^-) = \sum_{h\in \tilde{H}(\xi^-)} \overline{\Delta}_j^h(\xi). \tag{14}$$

Under standard assumptions on the primitives of the economy, we prove that it is possible to choose personalized collateral in order to satisfy Eq. (9). Then, our main result can be stated as follows:

Proposition 2 *Let $\mathcal{E}'$ be an economy satisfying the following assumptions:*

[A1]. *For each agent $h \in H$, $U^h(x) = \sum_{\xi \in D^h} v_\xi^h(x)$. Moreover, $\forall \xi \in D$, the function $v_\xi^h : \mathbb{R}_+^G \to \mathbb{R}$ is continuous, strictly monotone and concave with $v_\xi^h(0) = 0$. In addition, $\forall h \in H_2$, $\forall \alpha \in \mathbb{R}_+^G$, $\sum_{\xi \in D} v_\xi^h(\alpha)$ is finite.*

[A2]. $\forall h \in H$, $\omega^h \in \operatorname{Int} X^h$,

[A3]. *There exists $W > 0$ such that $W_\xi \leq W$, $\forall \xi \in D$.*

[A4]. *For each node $\xi \in D$, for each asset $j \in J(\xi)$, for each agent $h \in \tilde{H}(\xi)$, the function $F_{\xi,j}^h$ that adjusts the collateral requirements as function of the default history is increasing, continuous and uniformly bounded from below by a vector $\beta \in \mathbb{R}_+^G \setminus \{0\}$.*

Then, there exists an equilibrium $(\overline{p}, \overline{q}, \overline{R}, (\overline{M}^h)_{h\in H}, (\overline{x}^h, \overline{\theta}^h, \overline{\varphi}^h, \overline{\Delta}^h)_{h\in H})$ for which, the financial institution chooses the personalized collateral such that:

$$\overline{M}_j^h(\xi) = F_{\xi,j}^h(\overline{p}(\xi), \overline{q}_j(\xi), \overline{\kappa}_\xi^h).$$

Assumptions [A1]–[A3] are classical in the infinite horizon incomplete market models. In fact, assumption [A3] requires that for each node of the event-tree, the aggregate initial endowment accumulated until this node is uniformly bounded from above along the event-tree. This assumption is satisfied, e.g., when the depreciation structure satisfies assumption [D] of Araujo–Páscoa–Torres-Martínez [3]. Assumption [A4] requires that the collateral requirements are increasing with respect to agents' default history. Moreover, it guarantees that the personalized collateral are different from zero along the event-tree (i.e.: it does not vanish completely).

Remark 1 • The personalized collateral requirements can be interpreted as credit restrictions. In fact, given an equilibrium for which the personalized collateral perfectly foresee the default rates, we have that agents are restricted to consume at least a quantity $(p(\xi)M_j^h(\xi))_g$ of good g. Then, as the rule $F_{\xi,j}^h$ guarantees at least a positive quantity of collateral, there exists a commodity g' for which $(p(\xi)M_j^h(\xi))_{g'} > 0$. Then, if $c_\xi^h(g')$ denotes the total quantity of commodity g' consumed by agent $h \in H$, the following inequality will be satisfied in equilibrium:

$$\overline{\varphi}_j^h(\xi) \leq \frac{c_\xi^h(g')}{(p(\xi)M_j^h(\xi))_{g'}}.$$

Markovian economic models with credit restrictions which depend on the default history have been studied by Braido [5] and Sabarwal [20]. However our approach is different, since our economy does not have recursive structure and the default penalties are associated to physical requirement rather than bounds on the debt volume.

4 Concluding Remarks

In this paper, we extend the model of Araujo–Páscoa–Torres-Martínez [3] to allow for Overlapping Generations model and personalized collateral requirements. In this way, in our first model, considering the collateral to be the same for all agents, we obtain the model of Araujo–Páscoa–Torres-Martínez [3] as a particular case.

The main result of the second part of our paper is to prove the existence of an equilibrium for which the personalized collateral requirements perfectly foresees the default rate in equilibrium and is compatible with agents' default history.

The rules that can be used by the financial institution to set the optimal collateral requirement are very general, and depend on the evolution of the default rate of each agent and the price variations.

In addition, we guarantee that it is possible to obtain compatibility between personalized requirement and the default history in order to restrict agents' collateral requirement not only as a function of their personal default history but also as a function of the default history of their dynasty. This result is consistent with the practice of the mortgage markets without life insurance protection.

As we already mentioned, in the case where the adjusting rules do not depend on both the past default rate and the price evolution, we obtain the result of Araujo–Páscoa–Torres-Martínez [3] as a particular case. Moreover, Araujo–Páscoa–Torres-Martínez [4] study an extension of the model for the case of infinitely-lived assets, by analysing not just the equilibrium existence but also the existence of speculative bubbles compatible with agents' rationality. In this context, the authors allow the collateral coefficient to adjust for price variation. In particular, requirements that maintain a fixed margin value are analyzed. We could also consider in our model infinitely-lived assets and study these kind of requirements by allowing the margin to vary with the default history. Agents with high default would have to constitute higher marginal values. This is an important issue to address in future research.

Appendix

Proof of Proposition 1

The proof of this proposition is close to the proof of Araujo–Páscoa–Torres-Martínez [3]. However, because in our model the number of alive agents depend on the nodes, the uniform bound on the debt values in the former paper no longer holds. However, we obtain bounds on the short sales using similar tricks. Since the stochastic structure of our model is characterized by a finite number of immediate successors at each node, these bounds rule out the possibility of Ponzi games. Moreover, the presence of finitely-lived agents in our model leads to a new technical difficulty, namely that the interior of the budget sets of these agents could be empty, and therefore the budget correspondences may fail to be lower semicontinuous. To overcome this problem, one can define modified budget correspondences that are lower semicontinuous, and applying the Gale and Mas-Colell fixed point theorem [10, 11], we guarantee that each truncated economy has an equilibrium. These modified correspondences are close to the correspondences defined below (in the following proof of Proposition 2) but obviously they do not depend on the collateral since the personalized collateral are exogenous in our first model.

Proof of Proposition 2

Equilibria in the Truncated Economies

Let $\mathcal{E}'^T$ be the truncated economy associated with the original economy $\mathcal{E}'$, which has the same characteristics than $\mathcal{E}'$, but where we suppose that agents are constrained to stop their exchange of goods at period T and their exchange of assets at period $T-1$. Formally, for each $T > 0$, let us define the following sets:

$$\pi^{T-1} := \left\{ (p,q) \in \mathbb{R}_+^{D^T \times G} \times \prod_{\xi \in D^T} \mathbb{R}^{J(\xi)} \;\middle|\; \begin{array}{l} \forall \xi : t(\xi) < T, \ \|p(\xi)\|_1 + \|q(\xi)\|_1 = 1, \\ \forall \xi : t(\xi) = T, \ \|p(\xi)\|_1 = 1. \end{array} \right\},$$

Let us recall that for each node $\xi \in D^T$, for each asset $j \in J(\xi^-)$, one has $R^j(\xi) \leqslant \|A^j(\xi)\|_1$. Let us denote by:

$$R^T := \{R=(R^j(\xi),\ \xi \in D^T,\ j \in (\xi^-)) \mid \forall \xi \in D^T,\ \forall j \in (\xi^-))\},\ R^j(\xi) \leqslant \|A^j(\xi)\ \|_1\}.$$

Moreover, let us denote by $\mathcal{M} := \{\alpha \in \mathbb{R}_+^G \mid \beta \leqslant \alpha \leqslant W\}$.

For each $h \in H$,

$$X^{hT} := \{(x^h(\xi),\ \xi \in D) \in X^h \mid \forall \xi : t(\xi) > T,\ x^h(\xi) = 0\},$$

$$Z^{hT} := \{(z^h(\xi),\ \xi \in D) \in X^h \mid \forall \xi : t(\xi) \geqslant T,\ z^h(\xi) = 0\},$$

$$\text{and } \forall \xi : t(\xi) = T,\ \forall j \in J(\xi),\ M_j^h(\xi) = 0.$$

Moreover, the budget set of an agent $h \in H$ for the truncated economy can be defined as follows:

$$B^{hT}(p,q,R,M^h) = \left\{ (x,\theta,\phi,\Delta) \left| \begin{array}{l} p(\xi^h)x^h(\xi^h) + p(\xi^h)M^h(\xi^h)\phi^h(\xi^h) + q(\xi^h)z^h(\xi^h) \\ \leqslant p(\xi^h)\omega^h(\xi^h),\ \xi^h \in D^{T-1}, \\ p(\xi)\cdot(x^h(\xi) - \omega^h(\xi)) + p(\xi)M^h(\xi)\phi^h(\xi) + q(\xi)\cdot z^h(\xi) \\ + \sum_{j\in J(\xi^-)} \Delta_j^h(\xi) \leqslant p(\xi)[Y(\xi)x^h(\xi^-) + Y(\xi)M^h(\xi^-)\phi^h(\xi^-)] + \\ \sum_{j\in J(\xi^-)} R^j(\xi)\theta_j^h(\xi^-),\ \forall \xi \in (D^h - \delta D^h) \cap D^{T-1},\ \xi \neq \xi^h, \\ p(\xi)\cdot(x^h(\xi) - \omega^h(\xi)) + \sum_{j\in J(\xi^-)} \Delta_j^h(\xi) \leqslant \\ p(\xi)[Y(\xi)x^h(\xi^-) + Y(\xi)M^h(\xi^-)\phi^h(\xi^-)] + \\ \sum_{j\in J(\xi^-)} R^j(\xi)\theta_j^h(\xi^-), \forall \xi \in \delta D^h \cup D^T,\ \xi \neq \xi^h, \end{array} \right. \right\}$$

Moreover, for each agent $h \in H$, the utility function U^{hT} for each truncated economy $\mathcal{E}'^T$ is defined as follows:

$$U^{hT}(x^h, \theta^h, \phi^h, \Delta^h) := \sum_{\xi\in D^T} v_\xi^h(\tilde{x}^h(\xi)). \tag{15}$$

Lemma 1 *Under the assumptions stated above, an allocation* (x,θ,ϕ,D) *which satisfies the conditions (ii), (iii), (iv) and (v) of Definition 4 is bounded.*

Proof of Lemma 1 Let (x,θ,ϕ,Δ) be an allocation which satisfies the conditions (ii), (iii), (iv) and (v) of Definition 4. The bounds on x, θ and ϕ are obtained as in Araujo–Páscoa–Torres-Martínez [3]. More precisely, it follows from (ii) that:

$$\sum_{(h,g)\in H(0)\times G} [x^h(0,g) + \sum_{j\in J(0)} M_j^h(0)\phi_j^h(0)] = \sum_{(h,g)\in H\times G} \omega^h(0,g) \leqslant WH(0). \tag{16}$$

Let $\overline{Y} := \max\{(Y(\xi))_{g,g'}, (\xi,g,g') \in \times D^T \times G \times G\}$. Then, $\forall \xi \in D^T \setminus \{0\}$, one has:

$$\begin{aligned} &\sum_{(h,g)\in H(\xi)\times G} [x^h(\xi,g) + \sum_{j\in J(\xi)} M_{jg}^h(\xi)\phi_j^h(\xi)] \leqslant \\ &WH(\xi) + \overline{Y}G + \sum_{(h,g)\in H(\xi)\times G} [x^h(\xi^-,g) + \sum_{j\in J(\xi^-)} M_{jg}^h(\xi^-)\phi_j^h(\xi^-)]. \end{aligned} \tag{17}$$

It then follows from Eqs. (16) and (17) that for each node $\xi \in D^T : t(\xi) = t$ that:

$$\sum_{(h,g)\in H(\xi)\times G} [\overline{x}^h(\xi,g) + \sum_{j\in J(\xi)} M_{jg}^h(\xi)\overline{\phi}_j^h(\xi)] \leqslant WH(\xi) \sum_{n=0}^{t} (\overline{Y}G)^n. \tag{18}$$

By definition of the personalized collateral, one has that $m^h(\xi) = \min_{j \in J(\xi)} \|M_j^h(\xi)\|_1 > 0$, and hence $\forall h \in H$ one gets:

$$x^h(\xi, g) \leqslant WH(0) \sum_{n=0}^{t} (\overline{Y}G)^n := \chi(\xi) < +\infty, \tag{19}$$

$$\phi_j^h(\xi) \leqslant \frac{\chi(\xi)}{m^h(\xi)} := \alpha^h(\xi) < +\infty, \ \forall j \in J(\xi), \tag{20}$$

$$\theta_j^h(\xi) \leqslant \alpha^h(\xi) < +\infty, \ \forall j \in J(\xi), \tag{21}$$

On the other hand, since $\forall j \in J(\xi)$, $\Delta_j^h(\xi) \leqslant p(\xi)A^j(\xi)\phi_j^h(\xi^-)$ and in view of our normalization, one gets:

$$\Delta_j^h(\xi) \leqslant \|A^j(\xi)\|_1 \alpha^h(\xi^-) := \gamma^h(\xi) < +\infty. \tag{22}$$

We will denote by $\alpha(\xi) := \sup_{h \in H(\xi)} \alpha^h(\xi)$ and by $\gamma(\xi) := \sup_{h \in H(\xi)} \gamma^h(\xi)$.

For each $h \in H$, let us define:

$$B^{hT}(p, q, R, M^h, \chi, \alpha, \gamma) = \left\{ (x, \theta, \phi, \Delta) \in B^{hT}(p, q, R, M^h) \;\middle|\; \begin{array}{l} x^h(\xi, g) \leqslant 2\chi, \\ \theta_j^h(\xi) \leqslant 2\alpha(\xi), \\ \phi_j^h(\xi) \leqslant 2\alpha(\xi), \\ \Delta_j^h(\xi) \leqslant 2\gamma(\xi), \end{array} \right\}$$

Let $\mathcal{E}'^T(\chi, \alpha, \gamma)$ be an economy with the same characteristics as $\mathcal{E}'^T$ but in which the budget constraints are defined by the set $B^{hT}(p, q, R, M^h, \chi, \alpha, \gamma)$.

Lemma 2 *The truncated and compactified economy $\mathcal{E}'^T(\chi, \alpha, \gamma)$ has an equilibrium $(\overline{p}^T, \overline{q}^T, \overline{R}^T, (\overline{x}^{hT}, \overline{\theta}^{hT}, \overline{\phi}^{hT}, \overline{\Delta}^{hT})_{h \in H})$ such that $\forall h \in H$, $\forall \xi \in D^{T-1}$, $\forall j \in J(\xi^-)$, one has $\overline{M}_j^{hT}(\xi) := F_{\xi,j}^h(\overline{p}^T(\xi), \overline{q}^T(\xi), \overline{\kappa}_\xi^{hT})$.*

Proof of Lemma 2 The first new technical difficulty in our model in comparison with Araujo–Páscoa–Torres-Martínez [3] is that the budget set correspondences of the finitely-lived agents may not be lower semicontinuous (since their interior can be an empty set). Let us consider an agent $h \in H$ and define the set $B'^{hT}(p, q, R, M^h, \chi, \alpha, \gamma)$ by replacing all the inequalities in $B^{hT}(p, q, R, M^h, \chi, \alpha, \gamma)$ by strict inequalities. Moreover, let us define the correspondence

$$B''^{hT}(p, q, R, M^h, \chi, \alpha, \gamma) = \begin{cases} \{(\omega^h, 0, 0, 0)\} \text{ if } B'^{hT}(p, q, R, M^h, \chi, \alpha, \gamma) = \emptyset \\ B^{hT}(p, q, R, \chi, \alpha, \gamma) \text{ if } B'^{hT}(p, q, R, M^h, \chi, \alpha, \gamma) \neq \emptyset \end{cases}.$$

Remark 2 $\forall h \in H,\ \forall (p,q) \in \pi^T,\ \forall R \in \mathbb{R}^T,\ \forall M^h \in \mathcal{M},\ B''^{hT}(p,q,R,M^h) \neq \emptyset$ since it always contains $(\omega^i, 0)$.

Moreover, one can easily prove that B''^{hT} is lower semicontinuous. To simplify the notations, we define $v := (p,q,R,M)$, and $w := (x, \theta, \phi, \Delta)$.

For each agent $h \in H$, let us define the following correspondence:

$$\Psi^{hT}(v,w) = \begin{cases} B^{hT}(v,\chi,\alpha,\gamma) \text{ if } w \notin B^{hT}(v,\chi,\alpha,\gamma) \\ B'^{hT}(v,\chi,\alpha,\gamma) \cap P^h(w) \text{ if } w \in B^{hT}(v,\chi,\alpha,\gamma) \end{cases},$$

We also define the correspondence:

$$\Psi^{0T}(v,w) = \left\{ (p',q') \in \pi^T \;\middle|\; \begin{array}{l} \forall \xi \in D^T, \\ (p'(\xi) - p(\xi)) \cdot \sum_{h\in H} [x^h(\xi) + M^h(\xi)\phi^h(\xi) + Y(\xi)x^h(\xi^-) \\ -Y(\xi)M^h(\xi^-)\phi^h(\xi^-) - \omega^h(\xi)] + (q'(\xi) - q(\xi)) \cdot \sum_{h\in H} z^h(\xi) > 0. \end{array} \right\}$$

where $P^h(w) := \{w' \mid U^h(w') > U^h(w)\}$.

Moreover, we add the following players to this generalized game:

- Given an allocation $(x, \theta, \phi, \Delta)$, at each node $\xi \in D^{T-1}$, for each asset $j \in J(\xi)$, a financial institution chooses $M_j^h(\xi)$ in order to solve the following problem:

$$\min_{M_j^h(\xi) \in \mathcal{M}} \ [M_j^h(\xi) - F_{\xi,j}^h(p(\xi), q(\xi), \kappa_\xi^h)]^2,$$

- Given an allocation $(x, \theta, \phi, \Delta)$, at each node $\xi \in D^T \setminus \{0\}$, for each $j \in J(\xi^-)$, an auctioneer chooses $R^j(\xi) \leqslant \|A^j(\xi)\|_1$ in order to maximize:

$$[R^j(\xi) \sum_{h\in H} \theta^h(\xi^-) - \sum_{h\in H} D_j^h(\xi)]^2.$$

Since, $\forall h \in H \cup \{0\}$, Ψ^{hT} is lower semicontinuous and by definition of Ψ^{hT}, $w \notin \Psi^{hT}(v,w)$, it follows from the Gale and Mas-Colell fixed point theorem [10, 11] that there exists $(\overline{p}^T, \overline{q}^T, \overline{R}^T, (\overline{M}^{hT})_{h\in H}(\overline{x}^{hT}, \overline{\theta}^{hT}, \overline{\phi}^{hT}, \overline{\Delta}^{hT})_{h\in H}) := (\overline{v}, \overline{w})$ such that:

$$\forall h \in H \cup \{0\},\ \Psi^{hT}(\overline{v}, \overline{w}) = \emptyset.$$

That is, $\forall h \in H,\ B'^{hT}(v,\chi,\alpha,\gamma) \cap P^h(w) = \emptyset$ and

$$\begin{array}{l} (p(\xi) - \overline{p}^T(\xi)) \cdot \sum_{h\in H} [\overline{x}^{iT}(\xi) + M^h(\xi)\overline{\phi}^{hT}(\xi) - Y(\xi)\overline{x}^{hT}(\xi^-) \\ -Y(\xi)M^h(\xi^-)\overline{\phi}^{hT}(\xi^-) - \omega^h(\xi)] + (q(\xi) - \overline{q}^T(\xi)) \cdot \sum_{h\in H} \overline{z}^{hT}(\xi) \leqslant 0, \end{array} \tag{23}$$

On the other hand, the game played by the financial institution and the auctioneers yield to $M_j^h(\xi) = F_{\xi,j}^h(p(\xi), q(\xi), \kappa_\xi^h)$ and $R^j(\xi) \sum_{h \in H} \theta_j^h(\xi^-) = \sum_{h \in H} D_j^h(\xi)$. The feasibility conditions can be easily obtained using Eq. (23). □

Lemma 3 *The truncated economy $\mathcal{E}'^T$ has an equilibrium $(\overline{p}^T, \overline{q}^T, \overline{R}^T, (\overline{M}^{hT})_{h \in H}, (\overline{x}^{hT}, \overline{\theta}^{hT}, \overline{\phi}^{hT}, \overline{\Delta}^{hT})_{h \in H})$.*

Proof of Lemma 3 We have already proved that $\forall h \in H$, $B'^{hT}(v, \chi, \alpha, \gamma) \cap P^h(w) = \emptyset$. It then remains to prove that $\forall h \in H$, $B^{hT}(v, \chi, \alpha, \gamma) \cap P^h(w) = \emptyset$. This follows from a classical convexity argument. □

Asymptotic Results The techniques used in Araujo–Páscoa–Torres-Martínez [3] can be easily adapted to the case of incomplete participation and personalized collateral to show that the cluster point is an equilibrium of the original economy.

Acknowledgements M. Faias would like to thank the Fundação para a Ciência e a Tecnologia (Portuguese Foundation for Science and Technology) through PEst-OE/MAT/UI0297/2014 (CMA).

References

1. Araujo, A.P., Monteiro, P.K., Páscoa, M.R.: Infinite horizon incomplete markets with a continuum of states. Math. Financ. **6**(2), 119–132 (1996)
2. Araujo, A.P., Monteiro, P.K., Páscoa, M.R.: Incomplete markets, continuum of state and default. Econ. Theory **11**, 205–213 (1998)
3. Araujo, A.P., Páscoa, M.R., Torres-Martínez, J.P.: Collateral avoids Ponzi schemes in incomplete markets. Econometrica **70**(4), 1613–1638 (2002)
4. Araujo, A.P., Páscoa, M.R., Torres-Martínez, J.P.: Long-lived collateralized assets and bubbles. Working Paper, University of Chile, SDT 284 (2008)
5. Braido, L.H.: Trading constraints penalizing default: a recursive approach. J. Math. Econ. **44**(2), 157–166 (2008)
6. Bewley, T.: Existence of equilibria in economies with infinitely many commodities. J. Econ. Theory **4**, 514–540 (1972)
7. Dubey, P., Geanakoplos, J., Shubik, M.: Default and punishment in general equilibrium. Econometrica **73**, 1–37 (2005)
8. Florenzano, M., Gourdel, P.: Incomplete markets in infinite horizon: debt constraints versus nodes prices. Math. Finan. **6**(2), 167–196 (1996)
9. Florenzano, M., Gourdel, P., Páscoa, M.: Overlapping generations model with incomplete markets. J. Math. Econ. **18**, 357–376 (2001)
10. Gale, D., Mas-Colell, A.: An equilibrium existence theorem for a general model without ordered preferences. J. Math. Econ. **2**, 9–15 (1975)
11. Gale, D., Mas-Colell, A.: Corrections to an equilibrium existence theorem for a general model without ordered preferences. J. Math. Econ. **6**, 297–298 (1979)
12. Geanakoplos, J., Zame, W.R.: Collateral and the enforcement of intertemporal contracts. Working Paper, Yale University (2002)
13. Hart, O.: On the optimality of equilibrium when the market structure is incomplete. J. Econ. Theory **11**, 418–443 (1975)

14. Hernandez, A., Santos, M.: Competitive equilibria for infinite horizon economies with incomplete markets. J. Econ. Theory **71**, 102–130 (1996)
15. Levine, D.K., Zame, W.: Debt Constraints and equilibrium in infinite horizon economies with incomplete markets. J. Math. Econ. **26**, 103–131 (1996)
16. Magill, M., Quinzii, M.: Infinite horizon incomplete markets. Econometrica **62**(4), 853–880 (1994)
17. Magill, M., Quinzii, M.: Incomplete markets over an infinite horizon: long-lived securities and speculative bubbles. J. Math. Econ. **26**, 133–170 (1996)
18. Muller, W.J., Woodford, M.: Determinacy of equilibrium in stationary economies with both finite and infinite lived consumers. J. Econ. Theory **46**, 255–290 (1988)
19. Páscoa, M.R., Seghir, A.: Harsh default penalties lead to Ponzi schemes. Games Econ. Behav. **65**, 270–286 (2009)
20. Sabarwal, T.: Competitive equilibria with incomplete markets and endogenous bankruptcy. The B.E. J. Econ. Theory **3**(1), 1–42. De Gruyter (2003)
21. Samuelson, P.A.: An exact consumption-loan model of interest with or without the social contrivance of money. J. Polit. Econ. **66**, 467–482 (1958)
22. Santos, M., Woodford, M.: Rational asset pricing bubbles. Econometrica **65**, 19–57 (1995)
23. Schmachtenberg, R.: Stochastic overlapping generations model with incomplete markets 1: existence of equilibria. Discussion Paper No. 363–88, Department of Economics, University of Mannheim (1988)
24. Seghir, A.: An overlapping generations model with non-ordered preferences and numeraire incomplete markets. Decisions Econ. Finan. **28**(2), 95–111 (2006)
25. Seghir, A., Torres-Martinez, J.P.: Wealth transfers and the role of collateral when lifetimes are uncertain. Econ. Theory **36**(3), 471–502 (2008)
26. Thompson, E.A.: Debt instruments in both macroeconomics theory and capital theory. Am. Econ. Rev. **57**, 1196–1210 (1967)
27. Wilson, C.: Equilibrium in dynamic models with an infinity of agents. J. Econ. Theory **24**, 95–111 (1981)

Regularity for Mean-Field Games Systems with Initial-Initial Boundary Conditions: The Subquadratic Case

Diogo A. Gomes and Edgard A. Pimentel

Abstract In the present paper, we study forward-forward mean-field games with a power dependence on the measure and subquadratic Hamiltonians. These problems arise in the numerical approximation of stationary mean-field games. We prove the existence of smooth solutions under dimension and growth conditions for the Hamiltonian. To obtain the main result, we combine Sobolev regularity for solutions of the Hamilton-Jacobi equation (using Gagliardo-Nirenberg interpolation) with estimates of polynomial type for solutions of the Fokker-Planck equation.

1 Introduction

Mean-field games aim at understanding differential games with a (very) large number of rational, indistinguishable, intelligent players. Individually each player is not in position to impact the outcome of the system, however, every player is influenced by aggregate effect of the remaining agents.

This theory was introduced independently by Lasry and Lions [31–33] and Huang et al. [28, 29]. Since then it has known an intense research activity, and several authors have developed in detail a variety of problems and new directions of research. Among these we mention numerical methods [2, 3, 30], finite-state problems [14, 20, 22], obstacle problems [15], extended mean-field games [18, 24], probabilistic methods [10, 11], long-time behavior [6, 8], weak solutions [7, 38], applications to economics and environmental policy [26, 30, 34] to name only a few. For a detailed account of the recent developments and perspectives, we refer

D.A. Gomes
CEMSE Division, SRI – Uncertainty Quantification Center in Computational Science and Engineering, King Abdullah University of Science and Technology (KAUST), Thuwal 23955-6900, Saudi Arabia
e-mail: dgomes@math.ist.utl.pt

E.A. Pimentel (✉)
Department of Mathematics, Universidade Federal de São Carlos, 13.560 Sao Carlos SP, Brazil
e-mail: edgard@dm.ufscar.br

J.-P. Bourguignon et al. (eds.), *Dynamics, Games and Science*, CIM Series in Mathematical Sciences 1, DOI 10.1007/978-3-319-16118-1_15

the reader to the survey papers [1, 5, 16, 27], the lectures by Lions [35, 36] and the recent book [4].

A typical mean-field game is given by the system of partial differential equations

$$\begin{cases} -u_t + H(x, Du) = \Delta u + g(m) & \text{on } \mathbb{T}^d \times [0, T] \\ m_t - \operatorname{div}(D_p H m) = \Delta m & \text{on } \mathbb{T}^d \times [0, T], \end{cases} \tag{1}$$

where $\mathbb{T}^d$ is the d-dimensional torus, $T > 0$ is a terminal time, which is fixed. In many applications (1) is equipped with initial-terminal boundary conditions:

$$\begin{cases} u(x, T) = u_T(x) \\ m(x, 0) = m_0(x). \end{cases} \tag{2}$$

Existence of weak solutions for (1)–(2) was firstly investigated in [32]. In [38], the author addressed the existence of weak solutions to the planning problem. Smooth solutions for mean-field games in the presence of quadratic Hamiltonians were considered in [8]. However, the argument in that paper depends on a Hopf-Cole transformation which does not extend to a more general class of Hamiltonians satisfying quadratic growth conditions, except, perhaps, in very special perturbation regimes. MFG systems with nonlocal couplings were investigated in [9].

The first results concerning smooth solutions for Hamiltonians with quadratic and subquadratic growth appeared in [36]. These were substantially improved in [25]. The superquadratic case was treated in [23].

The stationary setting was first addressed in [31]. The existence of smooth solutions was discussed in [17, 21] and [24]. See also [19] for a related problem. The results in [12], which were established prior to mean-field games, ensure the existence of solutions for a class of stationary MFG systems.

In the present paper, we study a variant of (1)–(2) obtained by reversing time in the Hamilton-Jacobi equation and prescribing an initial condition u_0. This yields:

$$\begin{cases} u_t + H(x, Du) = \Delta u + g(m) & \text{on } \mathbb{T}^d \times [0, T] \\ m_t - \operatorname{div}(D_p H m) = \Delta m & \text{on } \mathbb{T}^d \times [0, T], \end{cases} \tag{3}$$

$$\begin{cases} u(x, 0) = u_0(x) \\ m(x, 0) = m_0(x). \end{cases} \tag{4}$$

The system (3)–(4) is referred to as a forward-forward mean-field game (FF-MFG). This class of systems has been considered in the realm of numerical methods to approximate the equilibrium problem, see [2]. Although these systems are well-behaved from the numerical perspective, the existence of solutions, as well as the regularity of the FF-MFG has not yet been addressed in the literature. As a main result, we obtain a set of conditions under which classical solutions for the FF-MFG systems can be shown to exist. This is done by combining a Gagliardo-Niremberg type of argument for the solution of the Hamilton-Jacobi equation with estimates for the Fokker-Planck equation, of polynomial type.

1.1 Main Assumptions

In this paper we assume that H satisfies a subquadratic growth condition and suppose further that $g(m)(x,t) = a(x)m^{\alpha}(x,t)$, where $a : \mathbb{T}^d \to \mathbb{R}_0^+$. The Hamiltonian H, $a(x)$ and the exponent α satisfy several assumptions as detailed next.

In the sequel we introduce the Assumptions under which we work.

A 1 *The Hamiltonian* $H(x,p) : \mathbb{T}^d \times \mathbb{R}^d \mapsto \mathbb{R}$

1. *is of class* C^2*;*
2. *is assumed to be, for fixed* x*, strictly convex in* p*;*
3. *is coercive with respect to* p*, i.e.,*

$$\lim_{|p|\to\infty} \frac{H(x,p)}{|p|} = +\infty,$$

and, without loss of generality, it is also assumed that $H(x,p) \geq 1$.

A 2 *The non-linearity* g *is of the form*

$$g(m)(x,t) \doteq a(x)m(x,t)^{\alpha},$$

where $a \in \mathscr{C}^1(\mathbb{T}^d)$ *is non-negative.*

A 3 *We have* $(u_0, m_0) \in \mathscr{C}^{\infty}(\mathbb{T}^d)$. *Furthermore,* $m_0 \geq \kappa_0$ *for some* $\kappa_0 > 0$.

A 4 *H satisfies the following conditions:*

$$|H(x,p)| \leq C\,|p|^{\gamma} + C$$

and

$$\left|D_pH(x,p)\right| \leq C\,|p|^{\gamma-1} + C,$$

for $1 < \gamma < 2$ *and a positive constant* $C > 0$.

Next, we state the following auxiliary result:

Lemma 1 *Let $d > 2$. Then, there exists $\alpha_{\gamma,d}$ satisfying*

$$\alpha_{\gamma,d} \geq \frac{2(-4+8\gamma-6\gamma^2+\gamma^3)}{d(4\gamma-5\gamma^2+\gamma^3)},$$

such that for $\alpha < \alpha_{\gamma,d}$, there are $0 \leq \upsilon, \zeta \leq 1 < \theta, r, p, \tilde{r}, \tilde{p}, a_\upsilon$ and b_υ satisfying (9)–(16) *as well as*

$$\frac{(\gamma-1)(4\zeta-\gamma\zeta)}{(2-\gamma)}\frac{r\upsilon\alpha}{(\theta-1)} < 1$$

and

$$\frac{(\gamma-1)(2+\gamma\zeta)}{\gamma}\frac{r\upsilon\alpha}{(\theta-1)} < 1.$$

Proof The result is established by recurring to the software Mathematica.

Notice that

$$\lim_{\gamma\to 2}\frac{2(-4+8\gamma-6\gamma^2+\gamma^3)}{d(4\gamma-5\gamma^2+\gamma^3)} = \frac{2}{d},$$

and

$$\lim_{\gamma\to 1}\frac{2(-4+8\gamma-6\gamma^2+\gamma^3)}{d(4\gamma-5\gamma^2+\gamma^3)} = +\infty.$$

A 5 *The exponent α in A2 is such that*

$$\alpha \leq \alpha_{\gamma,d}.$$

Our Assumptions include, but are not limited to, Hamiltonians given by

$$H(x,p) \doteq h(x)\left(1+|p|^2\right)^{\frac{\gamma}{2}} + V(x),$$

where $1 < \gamma < 2$, $h, V \in \mathscr{C}^2(\mathbb{T}^d)$, $h \geq 1$, and $V \geq 0$.

1.2 Main Result

The main result of this paper is the existence of classical solutions for the forward-forward mean-field games systems, as stated in the next Theorem:

Theorem 1 *Assume that the Assumptions A1–A5 are satisfied. Then,* (3) *admits a classical solution* (u, m) *under the initial-initial data* (4).

To prove Theorem 1 we introduce an approximate problem. It is done by considering the non-local operator $g_\epsilon = \eta_\epsilon * g(\eta_\epsilon * m)$, where the kernel η_ϵ is a symmetric, standard, mollifier. By doing so, one obtains:

$$\begin{cases} u_t^\epsilon + H(x, Du^\epsilon) = \Delta u^\epsilon + g(m^\epsilon) \\ m_t^\epsilon - \operatorname{div}(D_p H m^\epsilon) = \Delta m^\epsilon, \end{cases} \tag{5}$$

with $g_0 = g$, by convention. Existence of solutions to (5) can be obtained by the methods in [37] and [25]. See also [5].

1.3 Outline of the Proof

In order to establish Theorem 1 one proceeds by carefully combining estimates, of polynomial type, for the solution of the Fokker-Planck equation with bounds for u^ϵ in L^∞. A key ingredient is the following result from [25]:

Theorem 2 *Let* (u^ϵ, m^ϵ) *solve* (5)–(4). *Suppose* $m^\epsilon \in L^\infty(0, T; L^{\beta_0}(\mathbb{T}^d))$, *and assume that*

$$p > \frac{d}{2},$$

and

$$r \equiv \frac{p\,[d(\theta - 1) + 2]}{2p - d},$$

where $\theta > 1$. *Then,*

$$\int_{\mathbb{T}^d} (m^\epsilon)^{\beta_n} dx \leq C + C \left\| \left| D_p H \right|^2 \right\|^{r_n}_{L^r(0,T;L^p(\mathbb{T}^d))},$$

where

$$r_n = r \frac{\theta^n - 1}{\theta - 1},$$

$\beta_{n+1} = \theta \beta_n$, $\theta > 1$ *and* $\beta_0 \equiv 1$.

The critical bounds in L^∞ for the Hamilton-Jacobi equation are given in the following Lemma:

Lemma 2 *Let* (u^ϵ, m^ϵ) *be a solution to* (5)–(4) *and suppose that A1–A4 hold. For* $r,\ p > 1$ *satisfying*

$$p\left(\frac{r-1}{r}\right) > \frac{d}{2},$$

we have,

$$\|u^\epsilon\|_{L^\infty(0,T;L^\infty(\mathbb{T}^d))} \leq C + C\,\|g_\epsilon(m^\epsilon)\|_{L^r(0,T;L^p(\mathbb{T}^d))}\,.$$

We prove Lemma 2 in Sect. 2. Recurring to the Gagliardo-Nirenberg Theorem we have:

Theorem 3 *Let* (u^ϵ, m^ϵ) *solve* (5)–(4). *Suppose that H satisfies A1 and A4. Then, for* $1 < p, r < \infty$,

$$\|D^2 u^\epsilon\|_{L^r(0,T;L^p(\mathbb{T}^d))} \leqslant c\|g_\epsilon(m^\epsilon)\|_{L^r(0,T;L^p(\mathbb{T}^d))} + c\|u^\epsilon\|^{\frac{\gamma}{2-\gamma}}_{L^\infty(0,T;L^\infty(\mathbb{T}^d))} + C,$$

where $c,\ C > 0$ *are constants.*

We establish Theorem 3 in Sect. 3. By combining Theorems 2 and 3 with Lemma 2 one obtains bounds for $\|Du^\epsilon\|_{L^r(0,T;L^p(\mathbb{T}^d))}$, which are uniform in ϵ. Then, an additional argument based on the non-linear adjoint method, see [13], yields Lipschitz regularity for u^ϵ. See [25]. Several additional estimates for a solution (u^ϵ, m^ϵ) can be derived once Lipschitz regularity for solutions to the Hamilton-Jacobi equation is established, as outlined next. These provide further regularity for (u^ϵ, m^ϵ) and allow us to consider the limit $\epsilon \to 0$, which concludes the argument.

The remaining of this paper is organized as follows: Lax-Hopf type of estimates are established in Sect. 2. In Sects. 2 and 3 we present the proofs of Lemma 2 and Theorem 3, respectively. In Sect. 4, regularity for Du^ϵ in $L^r(0,T;L^p(\mathbb{T}^d))$ is established and, by recurring to the non-linear adjoint method, Lipschitz regularity for the solutions of the Hamiton-Jacobi equation is proved.

In Sect. 5 we prove several estimates for a solution of (3). In particular, we obtain that $\ln m^\epsilon$ is Lipschitz, and that both u^ϵ and m^ϵ are Hölder continuous. We also address the existence of solutions to the regularized problem. The proof of Theorem 1 closes the paper.

2 Lax-Hopf Estimates

When considering MFG systems equipped with initial-terminal boundary conditions, the Hamilton-Jacobi equation is inherently related to an optimal control problem. Next, we explore this idea in the setting of initial-initial boundary conditions. By doing so we manage to obtain upper bounds for u^ϵ.

Proposition 1 *Let (u^ϵ, m^ϵ) be a solution to* (5) *and let ζ solve*

$$\begin{cases} \zeta_t - \operatorname{div}(b\zeta) - \Delta\zeta = 0 \\ \zeta(x,\tau) = \zeta_\tau(x), \end{cases} \tag{6}$$

where $b : \mathbb{T}^d \times [0,T] \to \mathbb{R}^d$ is a smooth vector field and $\tau \in (0,T]$. Then

$$\int_{\mathbb{T}^d} u^\epsilon(x,\tau)\zeta(x,\tau)dx \leq \int_0^\tau \int_{\mathbb{T}^d} [L(x,b) + g_\epsilon(m^\epsilon)]\,\zeta + \int_{\mathbb{T}^d} u_0^\epsilon(x)\zeta(x,0)dx.$$

Proof For ease of notation we drop the superscript ϵ. We multiply the first equation in (5) by ζ and the first equation in (6) by u and add them. Then, integration by parts yields:

$$\frac{d}{dt}\int_{\mathbb{T}^d} u\zeta dx + \int_{\mathbb{T}^d} (H + b\cdot Du - g(m))\,\zeta dx = 0.$$

Integrating in time over $(0,\tau)$ and using the definition of $L(x,v)$ one obtains

$$\begin{aligned}\int_{\mathbb{T}^d} u(x,\tau)\zeta(x,\tau)dx &= -\int_0^\tau \int_{\mathbb{T}^d} (H + b\cdot Du - g)\,\zeta dxdt + \int_{\mathbb{T}^d} u_0(x)\zeta(x,0)dx \\ &\leq \int_0^\tau \int_{\mathbb{T}^d} [L(x,b) + g(m)]\,\zeta dxdt + \int_{\mathbb{T}^d} u_0(x)\zeta(x,0)dx,\end{aligned}$$

where the inequality follows from the definition of the Legendre transform.

We also need a lower bound for u^ϵ. This follows from the maximum principle. In what follows, Lemma 2 is proved.

Proof (Proof of Lemma 2) Set $b \equiv 0$ in Proposition 1. The result then follows by combining the Hölder inequality for convolutions and elementary properties of the heat kernel with the maximum principle.

3 Hamilton-Jacobi Equation in Sobolev Spaces

Next, we consider Sobolev estimates for solutions of the Hamilton-Jacobi equation. We recover a series of previously obtained results. See [25]. The next lemma is a simple consequence of the Gagliardo-Niremberg interpolation inequality:

Lemma 3 *Assume that $u \in W^{2,p}(\mathbb{T}^d)$. Then,*

$$\|Du\|_{L^{2p}(\mathbb{T}^d)} \leq C\|D^2u\|_{L^p(\mathbb{T}^d)}^{\frac{1}{2}}\|u\|_{L^\infty(\mathbb{T}^d)}^{\frac{1}{2}}, \tag{7}$$

for some constant $C > 0$.

Then one easily obtains:

Lemma 4 *Let* $u \in W^{1,2p}(\mathbb{T}^d)$. *Then, there exists* $C > 0$ *such that*

$$\|Du\|_{L^{\gamma p}(\mathbb{T}^d)} \leq C\|Du\|_{L^{2p}(\mathbb{T}^d)},$$

for every $1 < \gamma < 2$.

From this we get the following result

Corollary 1 *Assume that* $u \in W^{2,p}(\mathbb{T}^d)$. *Then,*

$$\|Du\|_{L^{\gamma p}(\mathbb{T}^d)} \leq C\|D^2 u\|^{\frac{1}{2}}_{L^p(\mathbb{T}^d)}\|u\|^{\frac{1}{2}}_{L^\infty(\mathbb{T}^d)}, \tag{8}$$

for some constant $C > 0$.

This Corollary implies

Lemma 5 *Assume that* (u^ϵ, m^ϵ) *is a solution of* (5)–(4) *and suppose that H satisfies A1–A4. Then, for* $1 < p, r < \infty$,

$$\|H(x, Du^\epsilon)\|_{L^r(0,T;L^p(\mathbb{T}^d))} \leq \; c\|D^2 u^\epsilon\|^{\frac{\gamma}{2}}_{L^r(0,T;L^p(\mathbb{T}^d))}\|u^\epsilon\|^{\frac{\gamma}{2}}_{L^\infty(0,T;L^\infty(\mathbb{T}^d))} + C,$$

where $c,\ C > 0$ *are constants.*

This lemma yields Theorem 3 by standard parabolic regularity.

4 Lipschitz Regularity for the Hamilton-Jacobi Equation

In what follows we combine the arguments of the Sect. 3 with polynomial estimates for the Fokker-Planck equation, as stated in Theorem 2. This yields improved regularity for the Hamilton-Jacobi equation.

Throughout this section we shall consider $1 < a < \infty$ and

$$\frac{1}{b_\upsilon} = 1 - \upsilon + \frac{\upsilon}{\theta}, \tag{9}$$

where $\theta > 1$ and $0 < \upsilon < 1$. We start by recovering some critical results from [25].

Lemma 6 *Assume that* (u^ϵ, m^ϵ) *solves* (5)–(4). *Suppose that A1–A4 hold. Then,*

$$\|m^\epsilon\|_{L^\infty(0,T;L^{b_\upsilon}(\mathbb{T}^d))} \leq C + C\left\|\left|D_p H\right|^2\right\|^{\frac{r\upsilon}{\theta}}_{L^r(0,T;L^p(\mathbb{T}^d))},$$

where

$$p > \frac{d}{2} \tag{10}$$

and

$$r = \frac{p(d(\theta - 1) + 2)}{2p - d}. \tag{11}$$

Corollary 2 *Assume that* (u^ϵ, m^ϵ) *solves* (5)–(4). *Suppose that A1–A4 hold. Assume further that* (12) *is satisfied. Then,*

$$\left\| g_\epsilon(m^\epsilon) \right\|_{L^{\frac{a}{\alpha}}(0,T;L^{\frac{b_\upsilon}{\alpha}}(\mathbb{T}^d))} \leq C + C \left\| |D_p H|^2 \right\|_{L^r(0,T;L^p(\mathbb{T}^d))}^{\frac{r\upsilon\alpha}{\theta}},$$

where $p > \frac{d}{2}$ *and* r *are given as in Lemma 6.*

Lemma 7 *Assume that* (u^ϵ, m^ϵ) *solves* (5)–(4). *Suppose that A1–A4 hold. Let* $\tilde{p}$ *and* $\tilde{r}$ *satisfy*

$$\tilde{p}\left(\frac{\tilde{r} - 1}{\tilde{r}}\right) > \frac{d}{2}, \tag{12}$$

where

$$\frac{1}{\tilde{p}} \doteq \alpha(1 - \zeta) + \frac{\alpha\zeta}{b_\upsilon} \tag{13}$$

and

$$\frac{1}{\tilde{r}} \doteq \frac{\alpha\zeta}{a}, \tag{14}$$

with $0 \leq \zeta \leq 1$. *Then*

$$\|u^\epsilon\|_{L^\infty(0,T;L^\infty(\mathbb{T}^d))} \leq C + C \, \|g_\epsilon\|^{\zeta}_{L^{\frac{a}{\alpha}}\left(0,T;L^{\frac{b_\upsilon}{\alpha}}(\mathbb{T}^d)\right)}.$$

Lemma 8 *Assume that* (u^ϵ, m^ϵ) *solves* (5)–(4). *Suppose that A1–A4 hold. Assume additionally that* (12)–(14),

$$\frac{2(\gamma - 1)r}{\gamma} = \frac{a}{\alpha} \tag{15}$$

and

$$\frac{2(\gamma-1)p}{\gamma}=\frac{b_\upsilon}{\alpha}. \tag{16}$$

Then,

$$\left\|Du^\epsilon\right\|_{L^{2(\gamma-1)r}(0,T;L^{2(\gamma-1)p}(\mathbb{T}^d))}\leq C\left\|g_\epsilon\right\|^{\frac{\zeta}{2-\gamma}+\frac{\zeta}{2}}_{L^{\frac{a}{\alpha}}(0,T;L^{\frac{b_\upsilon}{\alpha}}(\mathbb{T}^d))}+C\left\|g_\epsilon\right\|^{\frac{1}{\gamma}+\frac{\zeta}{2}}_{L^{\frac{a}{\alpha}}(0,T;L^{\frac{b_\upsilon}{\alpha}}(\mathbb{T}^d))}+C.$$

Corollary 3 *Assume that* (u^ϵ, m^ϵ) *is a solution of* (5)–(4) *and suppose that A1–A4 hold. Assume further that* (12)–(16) *are satisfied. Then,*

$$\begin{aligned}\left\|Du^\epsilon\right\|_{L^{2(\gamma-1)r}(0,T;L^{2(\gamma-1)p}(\mathbb{T}^d))}&\leq C+C\left\|Du^\epsilon\right\|^{\frac{(\gamma-1)(4\zeta-\gamma\zeta)}{(2-\gamma)}\frac{r\upsilon\alpha}{\theta}}_{L^{2(\gamma-1)r}(0,T;L^{2(\gamma-1)p}(\mathbb{T}^d))}\\&\quad+C\left\|Du^\epsilon\right\|^{\frac{(\gamma-1)(2+\gamma\zeta)}{\gamma}\frac{r\upsilon\alpha}{\theta}}_{L^{2(\gamma-1)r}(0,T;L^{2(\gamma-1)p}(\mathbb{T}^d))}.\end{aligned}$$

where $p>\frac{d}{2}$ *and* r *is given as in Lemma 6.*

Lemma 9 *Assume that* (u^ϵ, m^ϵ) *solves* (5)–(4). *Suppose that A1–A5 are satisfied. Then, there exist* $\theta>1$, r *and* p *satisfying* (10)–(11) *such that*

$$\left\|Du^\epsilon\right\|_{L^{2(\gamma-1)r}(0,T;L^{2(\gamma-1)p}(\mathbb{T}^d))}\leq C.$$

Proof Firstly, combine Lemma 1 with Corollary 3. Then, the previous computations along with an application of the weighted Young's inequality implies the result.

In what follows, the main result of this section is presented.

Theorem 4 *Assume that* (u^ϵ, m^ϵ) *solves* (5)–(4). *Suppose that A1–A5 are satisfied. Then,* $\|m^\epsilon\|_{L^\infty(0,T;L^\beta(\mathbb{T}^d))}$ *is uniformly bounded in* ϵ, *for any* $\beta>1$.

Proof For $p>\frac{d}{2}$, $\theta>1$ and $r>1$ given as in Lemma 6, Theorem 2 ensures that for any $\beta>1$ there exists r_β for which

$$\int_{\mathbb{T}^d}(m^\epsilon)^\beta(\tau,x)dx\leq C+C\left\||D_pH(x,Du^\epsilon)|^2\right\|^{r_\beta}_{L^r(0,T;L^p(\mathbb{T}^d))}.$$

By combining Lemma 9 with A4 one obtains

$$\||D_pH(x,Du^\epsilon)|^2\|_{L^r(0,T;L^p(\mathbb{T}^d))}\leq C\|Du^\epsilon\|^{2(\gamma-1)}_{L^F(0,T;L^G(\mathbb{T}^d))}+C\leq C.$$

This verifies the Theorem.

Corollary 4 *Assume that* (u^ϵ, m^ϵ) *is a solution of* (5)–(4). *Suppose that A1–A5 hold. Then,* $\|Du^\epsilon\|_{L^r(0,T;L^p(\mathbb{T}^d))}$, $\|D^2u^\epsilon\|_{L^r(0,T;L^p(\mathbb{T}^d))}$ *are uniformly bounded, in* ϵ, *for any* $p, r > 1$.

Proof It follows from Theorem 4 that $\|g_\epsilon(m^\epsilon)\|_{L^r(0,T;L^p(\mathbb{T}^d))}$ is bounded uniformly in ϵ, for any $p, r > 1$. Then $\|u\|_{L^\infty(0,T;L^\infty(\mathbb{T}^d))}$ and $\|D^2u^\epsilon\|_{L^r(0,T;L^p(\mathbb{T}^d))}$ are also bounded, because of Lemma 2 and Theorem 3. Finally, from Lemma 3

$$\|Du^\epsilon\|_{L^{2r}(0,T;L^{2p}(\mathbb{T}^d))} \leqslant C\|D^2u^\epsilon\|^{\frac{1}{2}}_{L^r(0,T;L^p(\mathbb{T}^d))}\|u^\epsilon\|^{\frac{1}{2}}_{L^\infty(0,T;L^\infty(\mathbb{T}^d))}.$$

Corollary 5 *Assume that* (u^ϵ, m^ϵ) *solves* (5)–(4). *Suppose that A1–A5 are satisfied. Then* $Du^\epsilon \in L^\infty(\mathbb{T}^d \times [0, T])$, *uniformly in* ϵ.

5 Further Regularity

Once we have established Lipschitz regularity for u^ϵ, a series of estimates can be obtained, improving the regularity of a solution (u^ϵ, m^ϵ) to (5). These are explored in [25] and recalled below.

5.1 Fokker-Planck Equation

Corollary 6 (See [25]) *Assume that* (u^ϵ, m^ϵ) *solves* (5)–(4). *Suppose that A1–A5 hold. Then*

- $D^3_{xxx}u^\epsilon, D^2_{xt}u^\epsilon \in L^2(\mathbb{T}^d \times [0, T])$, $D_{xx}u^\epsilon \in L^\infty([0, T], L^2(\mathbb{T}^d))$,
- $D^2_{xx}m^\epsilon, m^\epsilon_t \in L^2(\mathbb{T}^d \times [0, T])$, *and* $D_xm^\epsilon \in L^\infty([0, T], L^2(\mathbb{T}^d))$,
- $D^3_{xxx}m^\epsilon, D^2_{xt}m^\epsilon \in L^2(\mathbb{T}^d \times [0, T])$, $D^2_{xx}m^\epsilon \in L^\infty([0, T], L^2(\mathbb{T}^d))$ *and*
- *there is* $\mathbf{r} > d$ *such that* $D_xm^\epsilon, D^2_{xx}m^\epsilon, m^\epsilon_t \in L^{\mathbf{r}}(\mathbb{T}^d \times [0, T])$ *and then* $m^\epsilon \in C^{0,1-d/\mathbf{r}}(\mathbb{T}^d \times [0, T])$.

These bounds are uniform in ϵ.

5.2 Hopf-Cole Transformation

Consider the following logarithmic transform $w^\epsilon \doteq \ln m^\epsilon$. Hence, w^ϵ satisfies

$$w^\epsilon_t = \operatorname{div}(D_pH(x, D_xu^\epsilon)) + D_pH(x, D_xu^\epsilon)Dw^\epsilon + |Dw^\epsilon|^2 + \Delta w^\epsilon. \tag{17}$$

By investigating the regularity of solutions to (17), one obtains the following Theorem:

Theorem 5 (See [25]) *Assume that* (u^ϵ, m^ϵ) *solves* (5)–(4). *Assume also that A1–A5 hold and define* $w^\epsilon \doteq \ln m^\epsilon$. *Then, the family* $\ln m^\epsilon$ *is equi-Lipschitz.*

In particular, m is bounded by above and below.

5.3 Limit as $\epsilon \to 0$

In the sequel, we are interested in the behavior of a solution (u^ϵ, m^ϵ) to (5) as $\epsilon \to 0$. The next Lemma ensures Hölder regularity for u^ϵ.

Lemma 10 (See [25]) *Assume that* (u^ϵ, m^ϵ) *is a solution of* (5)–(4). *Assume that A1–A5 hold. Then, there exists* $\gamma \in (0,1)$ *for which*

$$\|u^\epsilon\|_{\mathscr{C}^{0,\gamma}(\mathbb{T}^d\times[0,T])} \leq C,$$

uniformly in ϵ, *for some constant* $C > 0$.

Regularity for m^ϵ in $\mathscr{C}^{0,\gamma}$, for some $\gamma \in (0,1)$, is established next.

Lemma 11 (See [25]) *Assume that* (u^ϵ, m^ϵ) *solves* (5)–(4). *Suppose that A1–A5 are satisfied. Then, there exists* $m \in \mathscr{C}^{0,\gamma}\left(\mathbb{T}^d \times [0,T]\right)$, *for some* $\gamma \in (0,1)$, *such that* $m^\epsilon \to m$ *through some subsequence, uniformly in compacts, in* $\mathscr{C}^{0,\gamma}\left(\mathbb{T}^d \times [0,T]\right)$.

Because of the non-linear nature of the Fokker-Planck equation, one must obtain convergence for the Du^ϵ. This is done in the next Lemma.

Lemma 12 (See [25]) *Assume that* (u^ϵ, m^ϵ) *is a solution of* (5)–(4). *Suppose that A1–A5 hold. Then, there is* $u \in \mathscr{C}^{0,\gamma}\left(\mathbb{T}^d \times [0,T]\right)$, *for some* $\gamma \in (0,1)$, *such that* $u^\epsilon \to u$, *through some subsequence, uniformly in compacts, in* $\mathscr{C}^{0,\gamma}\left(\mathbb{T}^d \times [0,T]\right)$. *Moreover, we also have* $Du^\epsilon \to Du$

Corollary 7 (See [25]) *Assume that* (u^ϵ, m^ϵ) *is a solution of* (5)–(4). *Suppose that A1–A5 hold. Then, the limit of* m^ϵ *as* $\epsilon \to 0$ *is a weak solution of*

$$m_t - \operatorname{div}\left(D_pH(x,Du)m\right) = \Delta m,$$

where

$$u = \lim_{\epsilon\to 0} u^\epsilon.$$

5.4 The Regularized Problem: Existence

Existence of solutions to (5) follows along the lines to those presented by the authors in [25].

Proposition 2 *There exists a solution to* (5)–(4).

5.5 Proof of Theorem 1

From Proposition 2, we know that there exists a smooth solution to (5)–(4). The key task is to pass to the limit $\epsilon \to 0$.

Lemma 10 and Corollary 6 ensure the Hölder regularity of u^ϵ and m^ϵ, uniformly in ϵ. Also, from Lemmas 11 and 12, it follows that $u^\epsilon \to u$ in $\mathscr{C}^{0,\gamma}(\mathbb{T}^d \times [0,T])$, $Du^\epsilon \to Du$ almost everywhere and $m^\epsilon \to m$ in $\mathscr{C}^{0,\gamma}(\mathbb{T}^d \times [0,T])$, as $\epsilon \to 0$.

Corollary 7 implies that m solves

$$m_t - \operatorname{div}(D_pH(x,Du)m) = \Delta m,$$

as a weak solution. Since u^ϵ is Lipschitz continuous and m^ϵ is uniformly convergent in compacts, it follows that the limit u solve

$$u_t + H(x,Du) = \Delta u + g(m),$$

in the viscosity sense. Finally, because (u,m) has the same regularity as (u^ϵ, m^ϵ), one concludes the proof.

References

1. Achdou, Y.: Finite difference methods for mean field games. In: Hamilton-Jacobi Equations: Approximations, Numerical Analysis and Applications, pp. 1–47. Springer, Heidelberg (2013)
2. Achdou, Y., Capuzzo-Dolcetta, I.: Mean field games: numerical methods. SIAM J. Numer. Anal. **48**(3), 1136–1162 (2010)
3. Achdou, Y., Camilli, F., Capuzzo-Dolcetta, I.: Mean field games: numerical methods for the planning problem. SIAM J. Control Optim. **50**(1), 77–109 (2012)
4. Bensoussan, A., Frehse, J., Yam, P.: Mean Field Games and Mean Field Type Control Theory. Springer Briefs in Mathematics. Springer, New York (2013)
5. Cardaliaguet, P.: Notes on mean-field games. Technical report. www.ceremade.dauphine.fr (2013)
6. Cardaliaguet, P.: Long time average of first order mean field games and weak KAM theory. Dyn. Games Appl. **3**(4): 473–488 (2013)
7. Cardaliaguet, P.: Weak solutions for first order mean-field games with local coupling (2013, preprint)
8. Cardaliaguet, P., Lasry, J.-M., Lions, P.-L., Porretta, A.: Long time average of mean field games. Netw. Heterogen. Media **7**(2), 279–301 (2012)
9. Cardaliaguet, P., Lasry, J.-M., Lions, P.-L., Porretta, A.: Long time average of mean field games with a nonlocal coupling. SIAM J. Control Optim. **51**(5), 3558–3591 (2013)
10. Carmona, R., Delarue, F.: Mean field forward-backward stochastic differential equations. Electron. Commun. Probab. **18**(68), 15 (2013)
11. Carmona, R., Delarue, F.: Probabilistic analysis of mean-field games. SIAM J. Control Optim. **51**(4), 2705–2734 (2013)

12. Evans, L.C.: Some new PDE methods for weak KAM theory. Calc. Var. Partial Differ. Equ. **17**(2), 159–177 (2003)
13. Evans, L.C.: Adjoint and compensated compactness methods for Hamilton-Jacobi PDE. Arch. Ration. Mech. Anal. **197**(3), 1053–1088 (2010)
14. Ferreira, R., Gomes, D.A.: On the convergence of finite state mean-field games through Γ-convergence. J. Math. Anal. Appl. **418**(1), 211–230 (2014)
15. Gomes, D., Patrizi, S.: Obstacle and weakly coupled systems problem in mean field games. Interfaces and Free Boundaries (2013, to appear)
16. Gomes, D., Saúde, J.: Mean field games models—a brief survey. Dyn. Games Appl. **4**(2), 110–154 (2014)
17. Gomes, D., Sánchez Morgado, H.: A stochastic Evans-Aronsson problem. Trans. Am. Math. Soc. **366**(2), 903–929 (2014)
18. Gomes, D., Voskanyan, V.: Extended mean-field games - formulation, existence, uniqueness and examples (2013, preprint)
19. Gomes, D., Iturriaga, R., Sánchez-Morgado, H., Yu, Y.: Mather measures selected by an approximation scheme. Proc. Am. Math. Soc. **138**(10), 3591–3601 (2010)
20. Gomes, D., Mohr, J., Souza, R.R.: Discrete time, finite state space mean field games. J. Math. Pures Appl. **93**(2), 308–328 (2010)
21. Gomes, D.A., Pires, G.E., Sánchez-Morgado, H.: A-priori estimates for stationary mean-field games. Netw. Heterogen. Media **7**(2), 303–314 (2012)
22. Gomes, D., Mohr, J., Souza, R.R.: Continuous time finite state mean-field games. Appl. Math. Opt. **68**(1), 99–143 (2013)
23. Gomes, D., Pimentel, E., Sánchez-Morgado, H.: Time dependent mean-field games in the superquadratic case. To appear in ESAIM: Control, Optimisation and Calculus of Variations (ESAIM:COVC) (2013, preprint)
24. Gomes, D., Patrizi, S., Voskanyan, V.: On the existence of classical solutions for stationary extended mean field games. Nonlinear Anal. **99**, 49–79 (2014)
25. Gomes, D.A., Pimentel, E.A., Sánchez-Morgado, H.: Time dependent mean-field games in the subquadratic case. Communications in Partial Differential Equations. **40**(1), 40–76 (2015)
26. Guéant, O.: Mean field games and applications to economics. Ph.D. Thesis, Université Paris Dauphine (2009)
27. Guéant, O., Lasry, J. M., Lions, P. L.: Mean field games and applications. In: Paris-Princeton Lectures on Mathematical Finance 2010, pp. 205–266. Springer, Berlin Heidelberg (2011)
28. Huang, M., Malhamé, R.P., Caines, P.E.: Large population stochastic dynamic games: closed-loop McKean-Vlasov systems and the Nash certainty equivalence principle. Commun. Inf. Syst. **6**(3), 221–251 (2006)
29. Huang, M., Caines, P.E., Malhamé, R.P.: Large-population cost-coupled LQG problems with nonuniform agents: individual-mass behavior and decentralized ϵ-Nash equilibria. IEEE Trans. Automat. Control **52**(9), 1560–1571 (2007)
30. Lachapelle, A., Salomon, J., Turinici, G.: Computation of mean field equilibria in economics. Math. Models Methods Appl. Sci. **20**(04), 567–588 (2010)
31. Lasry, J.-M., Lions, P.-L.: Jeux à champ moyen. I. Le cas stationnaire. C. R. Math. Acad. Sci. Paris **343**(9), 619–625 (2006)
32. Lasry, J.-M., Lions, P.-L.: Jeux à champ moyen. II. Horizon fini et contrôle optimal. C. R. Math. Acad. Sci. Paris **343**(10), 679–684 (2006)
33. Lasry, J.-M., Lions, P.-L.: Mean field games. Jpn. J. Math. **2**(1), 229–260 (2007)
34. Lasry, J.-M., Lions, P.-L., Guéant, O.: Application of mean field games to growth theory (2010, preprint)
35. Lions, P.-L.: College de France course on mean-field games (2007–2011)
36. Lions, P.-L.: Course on mean-field games. IMA, University of Minnesota. Video. http://www.ima.umn.edu/2012-2013/sw11.12-13.12/ (2012)
37. Pimentel, E.A.: Time dependent mean-field games. IST-UL. Doctoral thesis, Lisbon (2013)
38. Porretta, A.: On the planning problem for the mean-field games system. Dyn. Games Appl. **4**(2), 231–256 (2014)

A Budget Setting Problem

Orlando Gomes

Abstract Consider a typical agency relation involving a capital owner and a manager. The principal (i.e., the capital owner) has a potential budget to assign to investment projects. The effective amount of investment will be a share of the potential level, given the specific form of interaction that will be established between the principal and the agent (i.e., the manager). The budget setting problem originating from this relation is evaluated from the point of view of the manager, who wants to maximize the received budget, in an intertemporal basis. The optimal control problem is subject to a constraint, which indicates how the assigned budget evolves over time. In this constraint, a matching function takes a central role; the arguments of the function are the agent's effort to absorb new funds and the financial resources the principal has available but has not yet channeled to the manager.

1 Introduction

Agency relations and the information asymmetries they enclose are a main topic of economic analysis. Pioneer work on this subject by Akerlof [1], Spence [7], Stiglitz [8], Jensen and Meckling [5] and Fama and Jensen [3], among others, has launched a prolific literature that intends to explain how a *principal* selects an *agent* to act on her/his behalf and to pursue her/his goals. Because principal and agent have different interests and the access to relevant information probably differs among them, there are potential costs involved in this relation, for both players.

This paper proposes a simple intertemporal optimization model that deals with agency relations in a specific context, namely concerning the interaction of a capital owner with a team of managers that will undertake a series of investment projects over time. The team is selected *a priori*, and therefore we will not be concerned with adverse selection issues. The problem arises when the capital owner has to decide how much financial resources to allocate to the agent. The principal has a given potential budget to assign, but she/he will release the funds only against new

O. Gomes (✉)
Instituto Superior de Contabilidade e Administração de Lisboa (ISCAL/IPL), Av. Miguel Bombarda 20, 1069-035 Lisbon, Portugal
e-mail: omgomes@iscal.ipl.pt

J.-P. Bourguignon et al. (eds.), *Dynamics, Games and Science*, CIM Series in Mathematical Sciences 1, DOI 10.1007/978-3-319-16118-1_16

business proposals. If the managers do not make any new proposal, the principal will progressively cut the access to funds. From the point of view of the managers, they will be interested in accessing the largest possible portion of the potential budget, however collecting additional funds has a cost, related to the design of business proposals the managers will have to present to the capital owner.

Budget setting is a relevant theme of research in economics, management and management accounting. There are several approaches to this issue in the literature, for instance the ones proposed by Brekelmans [2] and Gox and Wagenhofer [4]. Our analysis differs from the mentioned studies because it takes a dynamic scenario and because it focus on the matching between the will of the capital owner in transferring funds to new projects and the effort made by managers to present new investment proposals. The matching process is essential for our analysis; we will assume a matching function that is adapted from typical labor market search and matching theory (see Pissarides [6]).

The techniques employed to solve the intertemporal optimization problem respect the conventional tools used to assess dynamic behavior in low dimensional systems (associated to our maximization problem there will be only one dynamic constraint). See, for instance, Walde [9] for a thorough analysis of the tools employed to explore this kind of modeling structure.

The remainder of the paper is organized as follows. Section 2 presents and describes the model; Sect. 3 derives optimality conditions and characterizes the steady-state; Sect. 4 approaches the stability of the derived dynamic system; Sect. 5 presents a small numerical illustration; Sect. 6 concludes.

2 The Model

Consider a continuous time modeling structure, in which a capital owner has a given potential budget to assign to investment projects. The capital owner has already chosen the team of managers that will undertake the projects and, in the specific stage we are considering, the decision under evaluation is how much financial resources should be transferred to the hands of the managers in order to achieve the best possible outcome from the point of view of the involved parties. This is an agency relation, involving a principal—agent relationship, where the capital owner will assume the role of the principal and the managers will be the agent. Principal and agent may have different or even contradictory interests and, thus, it is vital to understand how they will relate.

The maximum budget available for investment in the activities to be pursued by the hired managers is an invariant in time amount B. Not all of this budget will, presumably, be allocated to the agent and, therefore, we define share $b(t) \in (0, 1)$ as the percentage of budget B that at time t is allocated to project development. Under this simple scenario, the two involved entities will have decisions to make. The investor will be concerned with the efficiency with which the managers allocate the financial resources they receive, and wants to choose, in each period, the share of B that best serves this intent. The managers want to receive the highest possible amount of resources they can get, however they will have to incur in a cost to obtain them. Let the cost of acquisition of funds be given by variable $x(t)$. This variable may be interpreted as the costs the agent in the relation must have to support in order to convince the capital owner to release funds. Thus, variable $x(t)$ may be associated with the preparation of proposals by the agent to present projects that the principal will perceive as profitable and, thus, worth releasing additional funds. We should remark that $x(t)$ is a control variable from the point of view of the managers; they choose how much they want to expend in order to obtain the best possible outcome, from their viewpoint, which consists in maximizing, on an intertemporal basis, the difference between the received budget and the project proposal costs.

The mechanism through which funds are assigned to the managers is the result of a matching process, between the agent's project proposals effort, measured by the value of variable $x(t)$, and the budget share that at a given time period the capital owner can potentially assign to the investment projects; this corresponds to the funds, from budget B, not yet allocated to those projects, i.e., $[1 - b(t)]B$. The matching process is expressed under the form of a matching function with the following features.

Definition 1 Take a real-valued function $f : \mathbb{R}_+^2 \longrightarrow \mathbb{R}_+$. Function f is a matching function, such that $y(t) = f\{x(t), [1 - b(t)]B\}$. The output of the function, $y(t)$, represents the budget transferred from the principal to the agent, at period t, given the values of the respective inputs. Function f contemplates the following properties:

(i) f is continuous and differentiable;
(ii) f is an increasing function in both arguments: $f_x > 0, f_{(1-b)B} > 0$;
(iii) f is subject to decreasing marginal returns, relating each of its inputs: $f_{xx} < 0$, $f_{(1-b)B,(1-b)B} < 0$;
(iv) f is homogeneous of degree 1: $f\{\varepsilon x(t), \varepsilon[1 - b(t)]B\} = \varepsilon f\{x(t), [1 - b(t)]B\}$, $\forall \varepsilon > 0$;
(v) Both inputs are essential to reach a positive output, i.e., $f\{0, [1 - b(t)]B\} = f[x(t), 0] = 0$.

An explicit functional form that obeys the listed properties is, for instance, a Cobb-Douglas specification of the matching process, i.e., $f\{x(t), [1 - b(t)]B\} = Ax(t)^{\alpha}\{[1 - b(t)]B\}^{1-\alpha}$, $A > 0$, $\alpha \in (0, 1)$. The mentioned properties have, all of them, an intuitive interpretation; most importantly, they indicate that a stronger effort by the team of managers in presenting new project proposals and a larger available budget produce, evidently, a better matching result, that is translated in a more generous fund transfer. Logically, there are diminishing returns in this relation: with higher values of each of the inputs, matching will continue to occur but at a progressively lower rate. Observe, as well, that the inputs in the matching function are both indispensable to deliver a meaningful outcome: if the managers do not present any investment proposal, they will receive no funds even though these are potentially available; similarly, if the capital owner has no funds left to assign to this class of projects, the matching result will be null no matter how much effort the managers employ in presenting new proposals.

Another central assumption that we take is that the capital owner forces the agent to present innovative projects in order to continue to receive funds. To impose this behavior on the part of the agent, the principal will automatically cut a share $\lambda \in (0, 1)$ of the budget assigned in the previous period from the budget of the current period. The agent can only recover this value by presenting new projects; if it fails to do so, the budget will shrink over time and converge, in the long-run, to zero. The capital owner is not interested in maintaining an agency relation with managers who do not show a will to innovate and this mechanism functions as a way to impose the presentation of new projects in case the managers want to continue to receive funds to develop business activities.

Under the above assumptions, the following differential equation will characterize the dynamics of budget assignment,

$$\dot{b}(t) = f\{x(t), [1 - b(t)]B\} - \lambda b(t), b(0) \text{ given} \tag{1}$$

In the proposed context, the agent wants to maximize, in an intertemporal basis, the value of its available financial resources. We designate these resources by $\theta(t)$ and define them as the difference between the received budget and the costs incurred to propose new projects, i.e., $\theta(t) := b(t)B - x(t)$. Since the problem is of a dynamic nature, the objective function of the managers will be $\Theta(t) := \int_0^\infty \theta(t) \exp(-\rho t)dt$. Parameter $\rho > 0$ is the rate at which the future is discounted to the present. An infinite horizon is considered because we have not established an ending date for the agency relation; furthermore, the consideration of a positive discount rate makes far in time outcomes negligible from the current period point of view.

The above reasoning has conducted us to an optimal control problem, that the agent will want to solve, in which the value of $\Theta(t)$ is maximized, given resource constraint (1). The problem is relevant, from an economic point of view, because it contemplates a trade-off: a low effort in searching for new funds will not allow the managers to access a high budget; an excessively strong effort to collect new funds may lead to a higher budget but at an exaggerated cost; somewhere in the middle, the optimal solution will be found: by maximizing $\Theta(t)$, the agent will arrive to optimal trajectories for the two endogenous variables of this setup: the control variable $x(t)$, and the state variable $b(t)$.

Worthwhile noticing, in this environment, is the specific role of variable $b(t)$, the share of the potential budget that is delivered to the managers. This is not a control variable either for the principal or for the agent. Its value is the result of a pre-specified rule through which the capital owner attributes funds. It is the choice of the agent, which acts optimally, and the choice of the principal, concerning the values of the overall budget (B) and the rate at which it cuts managers' funds (λ), that will determine the specific path one will encounter for the assigned budget.

The next section will present the steps required to solve the optimal control problem.

3 Solution of the Optimal Control Problem

As described in the previous section, a team of managers is interested in addressing the following optimal control problem,

$$\underset{x(t)}{Max} \int_0^\infty \theta(t) \exp(-\rho t) dt$$

$$\begin{aligned} \text{subject to } &: \\ \dot{b}(t) &= f\{x(t), [1-b(t)]B\} - \lambda b(t) \\ & b(0) \text{ given} \\ \theta(t) &: = b(t)B - x(t) \end{aligned}$$

Computation of first-order conditions allows to find an equation of motion for variable $x(t)$ that is valid under the agent's optimal behavior.

Proposition 1 *Assume a Cobb-Douglas matching function. If the agent maximizes, intertemporally, the value of its financial resources, the time trajectory of its control variable, $x(t)$, will be governed by the following law of motion,*

$$\dot{x}(t) = \frac{1}{1-\alpha}\left(\rho + \lambda\frac{1-\alpha b(t)}{1-b(t)} - \alpha A\left\{\frac{[1-b(t)]B}{x(t)}\right\}^{1-\alpha}\right)x(t) \tag{2}$$

Proof To arrive to the solution of the optimization problem, we start by presenting the respective current-value Hamiltonian function,

$$H[x(t), b(t)] = b(t)B - x(t) + p(t)\left(f\{x(t), [1-b(t)]B\} - \lambda b(t)\right)$$

With the Hamiltonian function a new variable is introduced, namely the co-state or shadow-price variable $p(t)$, which can be interpreted as a kind of Lagrange multiplier for this dynamic setting. From the Hamiltonian, we draw the first-order optimality conditions of the problem,

$$H_x = 0 \Rightarrow p(t)f_x = 1$$

$$\dot{p}(t) = \rho p(t) - H_b \Rightarrow \dot{p}(t) = (\rho + \lambda)p(t) - B - f_b$$

The transversality condition $\lim_{t\to\infty} p(t)\exp(-\rho t)b(t) = 0$ must be satisfied as well. The first-order conditions correspond, under Cobb-Douglas matching, to

$$\alpha A \left\{ \frac{[1-b(t)]B}{x(t)} \right\}^{1-\alpha} p(t) = 1 \tag{3}$$

$$\dot{p}(t) = (\rho + \lambda)p(t) - B + (1-\alpha)A \left[\frac{x(t)}{1-b(t)} \right]^{\alpha} B^{1-\alpha} \tag{4}$$

The differentiation of (3) with respect to time yields

$$\frac{\dot{p}(t)}{p(t)} = (1-\alpha) \left[\frac{\dot{x}(t)}{x(t)} + \frac{\dot{b}(t)}{1-b(t)} \right] \tag{5}$$

By replacing the price motion by the corresponding expression in (4) and the motion of the budget share as given by (1), we can transform expression (5) in a dynamic equation for the control variable $x(t)$. After some computation, we arrive to differential equation (2) as presented in the proposition.

At this stage, we are in the possession of the information required to analyze the optimal dynamics of the problem under evaluation. A two-dimensional system, involving two endogenous variables is composed by Eqs. (1) and (2). The study of the dynamics requires looking at the steady-state outcome and respective stability properties. For now, in this section, we concentrate on the steady-state properties.

As it is usual in this kind of model, we define the steady-state as the long-run position for which the system eventually tends and where the respective endogenous variables have stopped growing, i.e., it will correspond to the pair of values $(x^*, b^*) = \{(x^*, b^*) : \dot{x}(t) = 0; \dot{b}(t) = 0\}$. Explicit solutions for (x^*, b^*) with respect to the parameters of the model, are not feasible to present. Nevertheless, the following relations are straightforward to obtain and will be useful when approaching the stability result,

- From (1):

$$A\left(x^*\right)^{\alpha}\left[\left(1-b^*\right)B\right]^{1-\alpha} = \lambda b^* \tag{6}$$

- From (2):

$$\alpha A \left[\frac{(1-b^*)B}{x^*} \right]^{1-\alpha} = \rho + \lambda \frac{1-\alpha b^*}{1-b^*} \tag{7}$$

The above conditions allow to state the following result

Proposition 2 *In the budget setting problem with a Cobb-Douglas matching function, a steady-state exists and it is unique.*

Proof Equation (6) may be solved in order to x^*, what delivers the outcome

$$x^* = \left\{ \frac{\lambda b^*}{A\left[(1-b^*)\,B\right]^{1-\alpha}} \right\}^{1/\alpha} \tag{8}$$

Replacing the value of x^* as presented above into (7), one obtains an equilibrium equation solely for steady-state value b^*,

$$\alpha A^{1/\alpha} \left[\frac{(1-b^*)B}{\lambda b^*} \right]^{\frac{1-\alpha}{\alpha}} = \rho + \lambda \frac{1-\alpha b^*}{1-b^*} \tag{9}$$

Although condition (9) does not allow to obtain an explicit value for b^*, it is straightforward to observe that it has a solution and that this solution is unique. The left hand side (lhs) of the condition is a decreasing function, while the right hand side (rhs) is increasing, as the first derivatives show,

$$\frac{d(\text{lhs})}{db^*} = -(1-\alpha)A^{1/\alpha} \left[\frac{(1-b^*)B}{\lambda b^*} \right]^{\frac{1-2\alpha}{\alpha}} \frac{B}{\lambda\,(b^*)^2} < 0$$

$$\frac{d(\text{rhs})}{db^*} = \lambda \frac{1-\alpha}{(1-b^*)^2} > 0$$

Thus, the lhs of (9) is a decreasing function, starting at infinity, for $b^* = 0$, and falling towards zero as b^* grows to its maximum value, $b^* = 1$. The rhs of (9) is an increasing function such that rhs$= \rho + \lambda$ for $b^* = 0$ and rhs$\rightarrow \infty$ as b^* tends to 1. This reasoning implies that the lhs and the rhs will necessarily cross and that they will cross only once in the domain defined for b^*. This proves the claim in the proposition: only one value of $b^* \in (0, 1)$ satisfies the conditions underlying the proposed analytical setup. Once in possession of the equilibrium value of the budget share, the steady-state level of x^* is straightforward to find, given (8).

In the following section, the system's stability will be addressed.

4 Stability

Having arrived to a unique steady-state point (x^*, b^*), we can now address the stability properties of such steady-state. To undertake this study, we first need to linearize the system of dynamic equations in the vicinity of (x^*, b^*). For such, one has to compute the respective Jacobian matrix, which is composed by the derivatives of each of the equations with respect to each of the endogenous variables, duly evaluated in the steady-state.

The respective computation leads to the following outcome,

$$J = \begin{bmatrix} -\lambda\frac{1-\alpha b^*}{1-b^*} & \alpha\lambda\frac{b^*}{x^*} \\ \lambda\frac{x^*}{(1-b^*)^2} + \frac{x^*}{1-b^*}\left(\rho + \lambda\frac{1-\alpha b^*}{1-b^*}\right) & \rho + \lambda\frac{1-\alpha b^*}{1-b^*} \end{bmatrix} \tag{10}$$

Matrix J in (10) is the Jacobian matrix of the linearized system,

$$\begin{bmatrix} \dot{b}(t) \\ \dot{x}(t) \end{bmatrix} = J \times \begin{bmatrix} b(t) - b^* \\ x(t) - x^* \end{bmatrix} \tag{11}$$

The stability properties of the steady-state will be dependent on the signs of the eigenvalues of matrix J. Negative eigenvalues correspond to stable dimensions and positive eigenvalues are associated with unstable dimensions. The evaluation of the matrix conducts to the following result,

Proposition 3 *The dynamic system underlying the budget setting problem, as formulated, delivers a saddle-path stable equilibrium.*

Proof One can arrive to the signs of the eigenvalues by computing the trace and the determinant of matrix (10). The value of the trace is immediately found by looking at the matrix: $Tr(J) = \rho$; the determinant will take the expression $Det(J) = -\frac{\lambda}{1-b^*}\left(\rho + \frac{\lambda}{1-b^*}\right)$.

Clearly, the trace is a positive value, while the determinant is negative, meaning that one of the eigenvalues is positive while the other is necessarily negative.[1] In this circumstance, the two-dimensional space we are dealing with involves a stable dimension and an unstable dimension, i.e., the equilibrium is saddle-path stable. The eigenvalues could also be computed directly from the matrix. In the case of this system they are relatively straightforward to derive: $\lambda_1 = -\frac{\lambda}{1-b^*} < 0$; $\lambda_2 = \rho + \frac{\lambda}{1-b^*} > 0$.

The saddle-path stable equilibrium is a convenient result in the type of optimal control problem we have just addressed. Because we have two kinds of variables, a state variable and a control variable, saddle-path stability is sufficient to guarantee convergence from any initial state (x_0, b_0) in the vicinity of the steady-state towards this second position, i.e., to point (x^*, b^*). The agent can adjust the value of x^* in order to place the system over the stable arm and, as a result, guarantee the stability of the equilibrium.

An additional step can be taken in the analysis of the stability properties. Namely, one might compute the expression of the stable trajectory. The generic expression of the stable path is given by

$$x(t) - x^* = \frac{p_2}{p_1}\left[b(t) - b^*\right]$$

[1]Recall that, for any square matrix of order 2, $Tr(J) = \lambda_1 + \lambda_2$ and $Det(J) = \lambda_1\lambda_2$, for λ_1 and λ_2 the eigenvalues of the matrix.

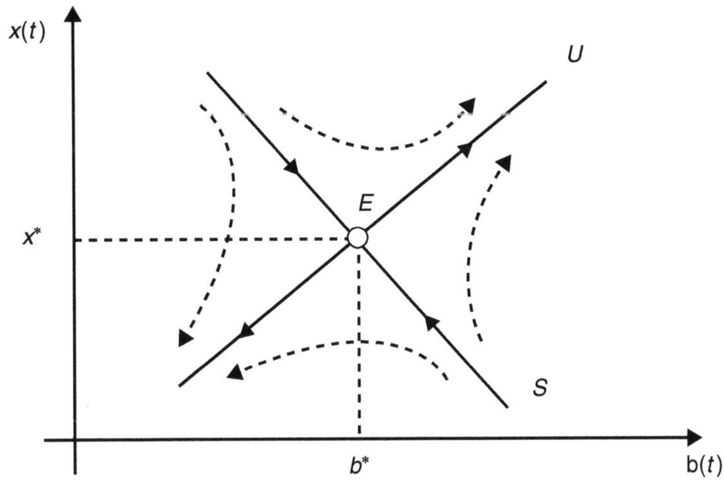

Fig. 1 Phase diagram (dynamic relation between the endogenous variables)

Elements p_1 and p_2 in the slope of the expression correspond to the elements of the eigenvector of J that relate to the negative eigenvalue (the one that represents stability). The calculus of the elements of the eigenvalue imply the following outcome:

$$x(t) - x^* = -\frac{x^*}{1 - b^*}\left[b(t) - b^*\right]$$

or, equivalently,

$$x(t) = \frac{x^*}{1 - b^*}\left[1 - b(t)\right] \tag{12}$$

The expression of the stable trajectory, (12), indicates that once the agent has adjusted her/his behavior in terms of the resources allocated to design new investment proposals, the path that will be followed and that will lead to the long-term equilibrium point is negatively sloped, i.e., as $x(t)$ increases, the budget share assigned to the managers, $b(t)$, will decline. This can be interpreted as follows: starting, for instance, at a point (x_0, b_0) for which $x_0 > x^*$ and $b_0 < b^*$, the convergence to the steady-state will be characterized by a process in which the costs incurred to gather additional budget will diminish as the budget share grows until reaching their steady-state values. Figure 1 sketches the dynamic relation between the two variables of interest.

Next section provides a small numerical example to illustrate the dynamics of the model.

Table 1 Steady-state results under different parametrizations

	$\alpha = 0.8$	$A = 0.12$	$B = 105$	$\lambda = 0.11$	$\rho = 0.075$
b^*	0.717	0.823	0.773	0.740	0.739
x^*	0.286	0.232	0.247	0.257	0.225

5 Numerical Illustration

Assume the following array of parameter values: $\alpha = 0.75$; $A = 0.1$; $B = 100$; $\lambda = 0.1$; $\rho = 0.05$. The value of α indicates that three quarters of the matching process depends on the project proposals and only one quarter is attributable to the available budget; the value of λ states that at each period 10 % of the budget previously assigned to the agent is withdrawn by the capital owner; the value chosen for ρ imposes a 5 % intertemporal discount rate. The other values are not specially relevant and just determine the scale of the analysis.

With specific numerical values, it is possible to compute the steady-state pair (x^*, b^*). The evaluation of condition (9) leads to result $b^* = 0.769$, i.e., 76.9 % of the budget the capital owner has available for the agent's activities is effectively channeled for the projects, given the matching process. The steady-state value of the cost proposal variable is obtainable from (8), $x^* = 0.247$.

If we change parameter values, the steady-state will be subject to perturbations. The direction of such perturbations should be intuitive. Table 1 indicates the impact over equilibrium of changing the value of each parameter, one at a time.

The table considers, for each case represented in a column, a different combination of parameters; the original parameterization is maintained with exception of the indicated change. This allows us to understand what is the impact over both endogenous variables, in terms of their steady-state values, when those changes occur. Results are intuitive: if the matching depends relatively more on the agent's effort in gathering additional funds, this will make b^* to decrease and x^* to increase; when the efficiency of the overall matching process increases (larger A), this implies an increase in b^* and a fall in x^* (a larger equilibrium budget is obtained with less resources allocated to attain such goal); the increase in the overall budget, relatively to the benchmark situation, does not change x^*, but makes b^* to increase; a larger automatic cut in the assigned budget will imply a new steady-state locus such that b^* falls and x^* increases; finally, a higher discount rate will signify that less attention will be given to the far future and this translates in a fall of both equilibrium values.

Let us return to the benchmark parametrization. With these values, we confirm the existence of a saddle-path stable equilibrium in this specific case, because the eigenvalues of the Jacobian matrix are $\lambda_1 = -\frac{0.1}{1-0.769} = -0.433$; $\lambda_2 = 0.05 + \frac{0.1}{1-0.769} = 0.483$. The saddle trajectory, given by (12), is $x(t) = 1.069\,[1 - b(t)]$; in this case, in the convergence towards (x^*, b^*) following the saddle path, as $b(t)$ increases one unit, $x(t)$ will fall 1.069 units.

6 Conclusion

The paper presented a dynamic optimization problem concerning the allocation of a budget from a capital owner to a manager or a team of managers. The problem is meaningful because it involves a trade-off: in this agency relation, the agent wants to maximize the value of the budget she/he can apply for, but this comes with a cost: to obtain additional funds, the manager will have to prepare new investment proposals that imply spending resources. On the principal's side, there is a maximum budget that the capital owner is willing to transfer to the manager, but the budget effectively transferred will depend on the capacity of the agent to present new projects; if these are not presented, the budget will be progressively cut over time at a constant rate.

Although simple, this theoretical structure is rich enough to deliver interesting results: an equation of motion for the control variable of the problem is derived and, evaluating such equation together with the rule that describes how the assigned budget evolves, one can address the stability of the long-run result. The respective dynamic system is saddle-path stable, what is sufficient to guarantee convergence towards the equilibrium, since one of the variables is a control variable and, thus, the corresponding value can be adjusted in order for the system to follow the saddle trajectory in the direction of the steady-state, where the values of the endogenous variables will end up by remaining constant.

In economics, as well as in other research fields, agency relations are common and subject to important discussion. The proposed model intends to contribute to this literature by furnishing a general framework of analysis. The framework is particularly suited to study the allocation of funds in situations where this allocation depends on proposals made by those who want to access the funds; for instance, the application to research grants by teams of scientists could be a relevant setting to explore further the possibilities of this setup.

References

1. Akerlof, G.A.: The market for 'Lemons': quality uncertainty and the market mechanism. Q. J. Econ. **84**, 488–500 (1970)
2. Brekelmans, R.: Stochastic models in risk theory and management accounting. Ph.D. thesis, Tilburg University (2000)
3. Fama, E.F., Jensen, M.C.: Agency problems and residual claims. J. Law Econ. **26**, 327–349 (1983)
4. Gox, R.F., Wagenhofer, A.: Economic research on management accounting. In: Hopper, T., Northcott, D., Scapens, R. (eds.) Issues in Management Accounting, Chap. 19, 3rd edn., pp. 399–424. Pearson Education Limited, Harlow (2007)
5. Jensen, M.C., Meckling, W.H.: Theory of the firm: managerial behavior, agency costs and ownership structure. J. Finan. Econ. **3**, 305–360 (1976)
6. Pissarides, C.A.: Job matching with state employment agencies and random search. Econ. J. **89**, 818–833 (1979)
7. Spence, M.: Job market signaling. Q. J. Econ. **87**, 355–374 (1973)
8. Stiglitz, J.: Incentives and risk sharing in sharecropping. Rev. Econ. Stud. **41**, 219–255 (1974)
9. Walde, K.: Applied Intertemporal Optimization. Mainz University Gutenberg Press, Mainz (2011)

Dynamic Political Effects in a Neoclassic Growth Model with Healthcare and Creative Activities

L. Guimarães, O. Afonso, and P.B. Vasconcelos

Abstract This paper extends the Ramsey-Cass-Koopmans (RCK) model by considering both a non constant number of hours worked by each individual through time and leisure, which includes healthcare and creative activities. With this extension, the seminal RCK model can be used to analyse the economic growth effects arising from governmental policies. In this context, governmental expenditures financed by lump-sum taxes and inefficient expenditures lead to a decrease in the short, medium and long-run economic growth.

1 Introduction

By considering microeconomic foundations, the Ramsey-Cass-Koopmans (RCK) model has made a great impact in the economic growth literature; however, the long-term economic growth remains unexplained (e.g., Acemoglu [1, Chaps. 2, 3 and 8]).

In the original RCK model, agents maximize their lifetime utility, dependent on the consumption level, and their labour supply is assumed to be constant. These assumptions are restrictive; for example, the number of hours worked by each individual is not constant through time and leisure, in which healthcare and creative activities are included, affects positively the utility (e.g., Fogel [4] , Ramey and Francis [7]).

The RCK is a neoclassical growth model with an endogenous saving rate. It aims at studying whether the accumulation of capital accounts for the long term growth. This is accomplished by modelling the intertemporal allocation of income, i.e, the relation between consumption and savings focusing in the dynamics. By allowing consumers to behave optimally, the analysis permit us to discuss how incentives affect the behaviour of the economy. The model deals with infinitely lived households that choose consumption and savings to maximise their dynastic utility, bearing in mind the intertemporal budget constraint.

L. Guimarães • O. Afonso (✉) • P.B. Vasconcelos
Faculdade Economia da Universidade Porto, R. Dr. Roberto Frias, 4200-464 Porto, Portugal
e-mail: 100421013@fep.up.pt; oafonso@fep.up.pt; pjv@fep.up.pt

J.-P. Bourguignon et al. (eds.), *Dynamics, Games and Science*, CIM Series in Mathematical Sciences 1, DOI 10.1007/978-3-319-16118-1_17

This paper extends the RCK model to cope with its weakness, allowing that a well-known and established model be used to analyse the economic growth effects arising from typical governmental policies (e.g., Irmen and Kuehnel [5]). In line with Fogel [4] and Ramey and Francis [7], among others, the utility function is modified to consider the fraction of time each individual devotes to healthcare and creative activities.[1] In this context, governmental expenditures financed by lump-sum taxes and inefficient expenditures lead to a decrease in the short, medium and long-run economic growth.

After these introductory remarks, the paper proceeds to characterize the set-up of the model, consumers, productive structure and laws of motion, in Sect. 2. Then, in Sect. 3 the dynamic general equilibrium is derived and, resorting to numerical computation, steady state and transitional dynamics are analyzed. Finally, in Sect. 4, the paper ends with some concluding remarks.

2 Set-up of the Model

The model will be developed considering the consumers side, Sect. 2.1, the productive side, Sect. 2.2, and deriving from these two sides the laws of motion, Sect. 2.3.

It is assumed that agents live infinitely and that the economy is populated by an invariant large set of identical households. Households divide their time between work to earn an income, and healthcare and creative activities. Additionally, they decide to spend part of their income directly on consumption and lend another part in return for future interest. The fraction of the output that is not consumed is used in investment. Also, the output of the economy is produced in perfect competition by using labour and physical capital.

2.1 Consumers

In the original RCK model, the constant-relative risk aversion instantaneous utility function, U, is represented by

$$U(C) = \begin{cases} \frac{C(t)^{1-\sigma}}{1-\sigma} & \sigma \neq 1 \\ ln(C(t)) & \sigma = 1 \end{cases} \tag{1}$$

where C is the consumption, t denotes a given time instant, σ represents the relative risk-aversion coefficient of the agents (without lost of generality, we will consider $\sigma = 1$). In Ben-Porath [2] agents decide on how to split their labour time. They may

[1]In order to isolate the effect of healthcare and creative activities on agents decisions, endogenous human-capital accumulation is not considered.

simply work or they may study. If they work, they have a higher income today while if they study, by increasing their human-capital, they might achieve a higher income in the future. Thus, the decision on how to split their labour time generates a trade-off between higher income today and higher income in the future. This division of the labor time between labor and human capital formation may also be found in one of the most important first generation of endogenous growth models (e.g., Lucas [6]). In this case, the economy has two sectors, one for the production of the final good and other for human-capital formation. In the long run, growth is explained by human-capital accumulation. Caballe and Santos [3] and Turnovsky [9] also used this approach by focusing on productive government expenditures and long run growth.

In the present model, agents devote time to healthcare and creative activities since, by doing so, they decrease the time devoted to labour yielding a higher utility. This can be interpreted as equivalent to a postponement of the entrance in the labor market. As a result, individual's utility depends not only on the consumption level but also on the fraction of labour time used for leisure. The instantaneous utility function has now an additional component of labour generating disutility:

$$U(C) = ln(C(t)) - \frac{(1-i(t))^{1+\varphi}}{1+\varphi}, \quad 0 < i(t) < 1 \tag{2}$$

where i is the fraction of time used for healthcare and creative activities, and $\varphi > 0$ is a labour coefficient as a proxy for the temporal elasticity of substitution of labour. Since $\frac{\partial U}{\partial i} > 0$ and $\frac{\partial^2 U}{\partial^2 i} < 0$ the more agents devote time to leisure the higher is utility, at decreasing rate. The intuition behind this assumption is that households like to postpone the entrance in the labour market[2]; yet, as time passes, individuals do not value this postponement and thus they prefer to work once achieved a certain age.

Agents are infinitely-lived as assumed in the original RCK model so it is their objective to maximize:

$$U(C,i) = \int_{t=0}^{+\infty} e^{-\rho t}\left[\ln C(t) - \frac{(1-i(t))^{1+\varphi}}{1+\varphi}\right] dt \tag{3}$$

where $\rho > 0$ is the discount rate.

Households accumulate assets, a, in the form of physical capital. Those assets earn returns at the interest rate $r(t)$. Households's assets stock is affected by net savings, given by the difference between income (interest and wages per unit of effective labour, w) and consumption. The flow budget constraint is

$$\dot{a}(t) = r(t)\,a(t) + w(t)\,A(t)\,[1-i(t)] - C(t) \tag{4}$$

[2]Thus, the focus is on the particular channel related to the postponement of the entrance in the labour market. This channel also accommodates the possibility of agents to switch from working to healthcare and creative activities, and vice versa.

where $\dot{a}(t)$ is the change in the assets stock, and $1 - i(t)$ is the fraction of time devoted to work. Households maximize lifetime utility subject to the budget constraint and the "no Ponzi games" condition ($\lim_{t\to\infty} a(t)\, e^{-\rho t} = 0$).

For the solution procedure we consider the Hamiltonian

$$\mathscr{H} = \ln C(t) + \lambda\,[w(t)A(t)(1 - i(t)) + a(t)r(t) - C(t)] \tag{5}$$

and compute the first order conditions.

The solution for the consumption path, which is independent of the household, is the standard Euler equation

$$\frac{\dot{C}(t)}{C(t)} = r(t) - \rho \tag{6}$$

where $\dot{C}(t)$ is the change in aggregate consumption. Moreover, the resulting expression for i is

$$i(t) = 1 - \left[\frac{w(t)A(t)}{C(t)}\right]^{\frac{1}{\varphi}} \tag{7}$$

which implies that the fraction of time devoted to healthcare and creative activities depends positively on consumption but negatively on wages. Higher wages imply higher opportunity cost connected with healthcare and creative activities and thus more time is devoted to work.

2.2 Productive Structure

Following the usual RCK approach, the production function, Y-the output, has constant returns to scale in capital, K, and labour, $L(1 - i)$, the Inada conditions are satisfied and Harrod-neutral technological-knowledge progress is considered:

$$Y(t) = K(t)^{\alpha}\,[A(t)L(t)(1 - i(t))]^{1-\alpha} \tag{8}$$

where α is the share of capital in production and A is the technological-knowledge.

Representing the capital and output per unit of effective household, respectively, $k(t) = \frac{K(t)}{A(t)L(t)} = \frac{K(t)}{A(t)}$ and $y(t) = \frac{Y(t)}{A(t)L(t)} = \frac{Y(t)}{A(t)}$, since, without loss of generality, the number of households, L, is normalised to 1. Function (8) can then by rewritten as

$$y(t) = k(t)^{\alpha}(1 - i(t))^{1-\alpha} \tag{9}$$

The change of technological-knowledge progress, $\dot{A}$, depends positively on i (healthcare and creative activities)

$$\dot{A}(t) = i(t)A(t) \tag{10}$$

What is implicit in (10) is that the increase in technological-knowledge progress is an externality from healthcare and creative activities, and since there is a very high number of agents, the impact of each of them on the technological-knowledge progress is almost null.

Under competitive markets each input earns its marginal product; thus,

$$w(t) = \frac{\partial y(t)}{\partial(1 - i(t))} = (1 - \alpha)\left[\frac{k(t)}{1 - i(t)}\right]^{\alpha} \tag{11}$$

$$r(t) = \frac{\partial y(t)}{\partial k(t)} - \delta = \alpha\left[\frac{1 - i(t)}{k(t)}\right]^{1-\alpha} - \delta \tag{12}$$

where δ is the discount rate of capital. It is important to note from (10) and (11), that the increase in technological-knowledge progress by the fraction of time i is not remunerated with wages: there is only an indirect impact on wages by the increase in A. Intuitively, households engage in healthcare and creative activities to have higher utility and not because they are increasing others productivity.

The expression for i in (7) can be rewritten considering (11).

2.3 *Laws of Motion*

Since physical capital in the economy is $K(t) = a(t)L = a(t)$, then the capital per unit of effective household is $k(t) = \frac{a(t)}{A(t)}$. Now, bearing also in mind (10), (11) and (12) in (4) yields the following path for k:

$$\dot{k}(t) = k(t)^{\alpha}\left[1 - i(t)\right]^{1-\alpha} - k(t)\left[\delta + i(t)\right] - c(t) \tag{13}$$

where $c(t) = \frac{C(t)}{A(t)L(t)} = \frac{C(t)}{A(t)}$ is the consumption per unit of effective household. This equation states that the change $k(t)$, $\dot{k}(t)$, is equal to the difference between savings, $k^{\alpha}(1 - i)^{1-\alpha} - c$, and break even investment, $k\left[\delta + i\right]$.

Considering the Euler equation (6) and (12), the path of $c(t)$ is

$$\dot{c}(t) = c(t)\left[\alpha k(t)^{\alpha-1}(1 - i(t))^{1-\alpha} - i(t)) - \delta - \rho\right] \tag{14}$$

3 General Equilibrium

Once characterised the country's economic structure, we now proceed to analyse the implications of healthcare and creative activities, which play a crucial role in the dynamic general equilibrium. We start with the steady state and then we analyse the transitional dynamics, by using, for instance, the following set of baseline parameter values: $\rho = 0.05$, $\varphi = 4$, $\delta = 0.05$ and $\alpha = 0.4$. Thus, the model is solved numerically to obtain an approximate solution. MATLAB was the software chosen, since it comprises state of the art numerical methods to solve system of ordinary differential equations.

3.1 *Steady-State and Transitional Dynamics*

The steady state can be easily computed by solving the system of nonlinear equations (13) and (14) for $\dot{k} = \dot{c} = 0$. The phase-diagram is depicted by Fig. 1 in which is also represented the stable saddle path (the eigenvalues of the Jacobian matrix evaluated at the steady state are −0.2443 and 0.1640).

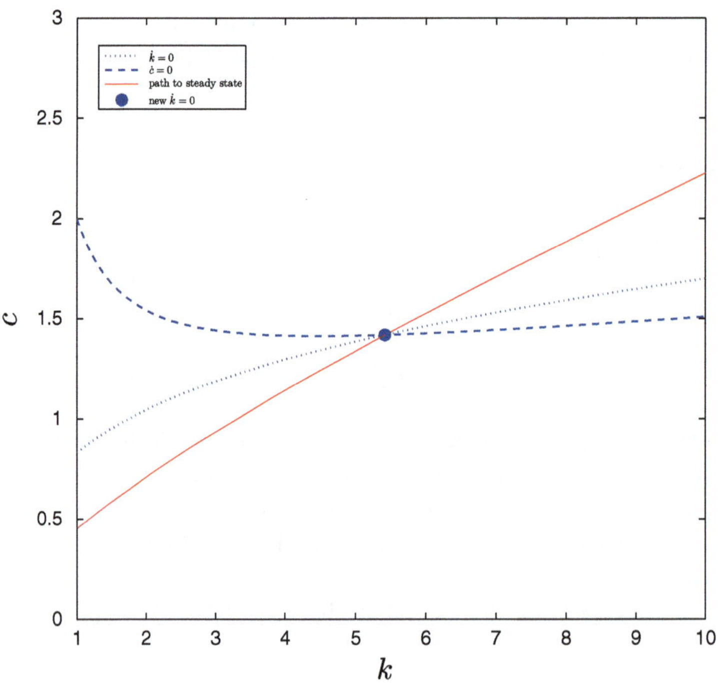

Fig. 1 Phase diagram

The $\dot{c} = 0$ curve is different from the one in the original RCK model, which is a vertical line. In this model, as the capital per unit of effective household increases, initially the consumption per unit of effective household decreases until a certain level of k. Regarding the $\dot{k} = 0$ curve, to keep k constant, the higher is c the higher has to be k. In order to exist convergence to the steady-state, both c and k must evolve in the same direction.

3.2 Government Intervention

Following Romer [8], the government buys output at rate G per unit of effective household. Additionally, it is assumed that G does not affect utility directly, is only used as public consumption, and is financed with lump-sum taxes. In this case, Eq. (13) becomes:

$$\dot{k}(t) = k(t)^{\alpha}(1 - i(t))^{1-\alpha} - k(t)\,[\delta + i(t)] - c(t) - G \tag{15}$$

A change of G from 0 to 0.1 is now considered.

The resulting effects in the phase diagram are plotted in Fig. 2. In turn, Fig. 3 depicts the immediate (short run), transitional dynamics (medium run) and the steady state levels (long run) of all relevant variables.[3]

Variable c jumps down due to the adjustment by households (immediate level effect), which does not occur in k. These two variables increase during the transitional phase towards their new steady state levels.

The immediate impact on wages results from the increase in the labour supply. As the fraction of time devoted to healthcare and creative activities jumps down, labour supply increases and thus wages decrease. Then, during the transitional phase, k increases and thus also the marginal productivity of labour; hence, the demand for labour and wages rise. In order to smooth the utility, households work less and the labour supply decreases, which also affects positively wages.

However, in the new steady-state the fraction of time househoulds devote to healthcare and creative activities falls in comparison to the previous steady state. This results from the fact that wages are higher than before while consumption is more or less the same; households face a higher opportunity cost of healthcare and creative activities.

[3] That is, consumption per unit of effective household, capital per unit of effective household, wages per unit of effective labour, interest rate, fraction of leisure time and path of the total capital and total output growth rates (which, since the number of households is fixed, are equivalent to the growth rates of output and capital per capita).

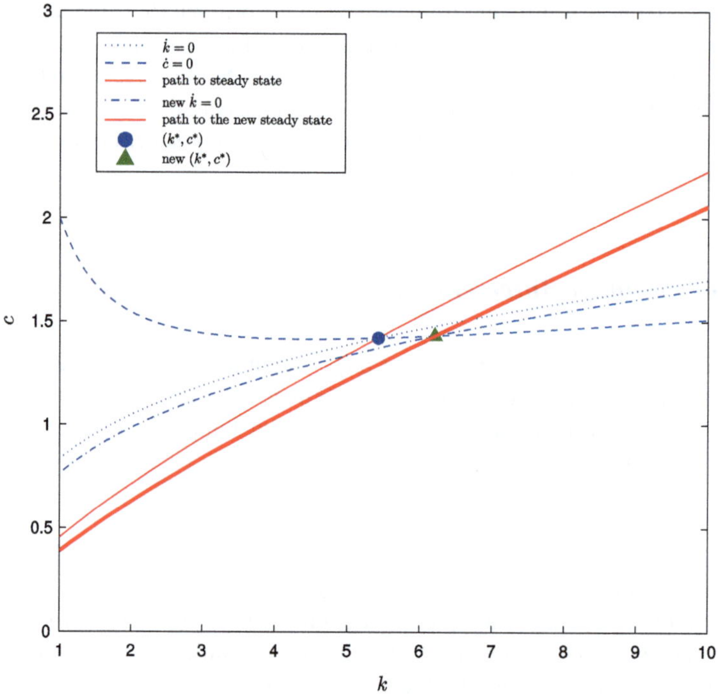

Fig. 2 Phase diagram with and without government intervention

Bearing in mind the behaviour of c and i, in the new steady state households reach lower utility. This also means, that after an increase in the government inefficiency totally financed by lump-sum taxes yield a lower growth-rate of technological knowledge arising from healthcare and creative activities.

The path of the interest rate has the opposite explanation to the one for wages per unit of effective labour. Initially, due to the jump up in the labour supply, the marginal return of investing in capital increases. On the other hand, in transition dynamics, since k is increasing and labour supply is falling, capital becomes a relatively abundant input and its marginal return falls.

Initially, k does not change. However, the growth rate of technological knowledge falls at the time of this change leading to a fall in the growth rate of capital, K. Immediately after, k starts increasing and additionally technological knowledge starts increasing at higher rates. These two facts together, lead to higher growth-rate of K than in the previous steady state. During the transition towards the new steady state, k is increasing at decreasing rates, which explains the fall in the growth rate of K. In the new steady-state, K grows at the rate of technological knowledge, which, in turn, is now lower than before.

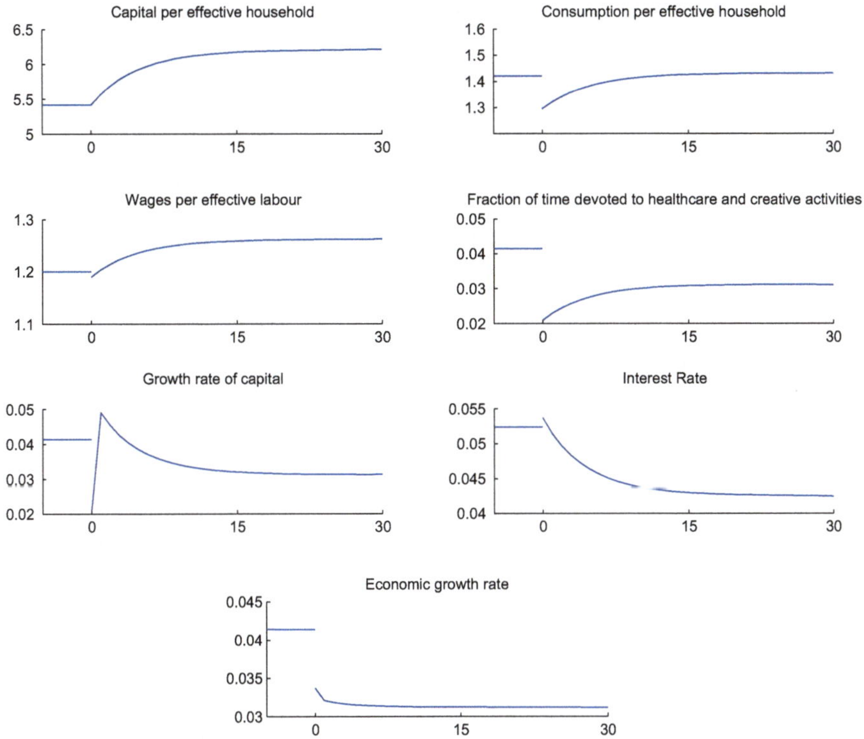

Fig. 3 Impact of government intervention on the relevant variables over time

As initially i decreases and k stays constant, there are two forces influencing the impact in the growth rate of output: the increment in the labour supply and the fall in the growth rate of technological knowledge, being, in this particular case, the latter stronger than the former. During the transitional phase, although capital and technological knowledge are increasing, the fall in the labour force lead to a fall in the growth rate of output. Hence, in the long run, an increase in the government inefficiency completely financed by lump-sum taxes lead to lower growth-rate of output.

4 Concluding Remarks

This paper presents an endogenous version of the simpler Ramsey-Cass-Koopmans model, by introducing two new features: the decision of households regarding their healthcare and creative activities responsible for the technological knowledge and an inefficient government, which expenditures are fully financed by lump-sum taxes.

In this context, a lower growth-rate of the economy can be observed. Households, concerned about the future, decrease consumption more than the increase in the government expenditures (in the original RCK model households fall consumption by the exact amount of the government expenditures). Therefore, they have a higher disposable income to invest in assets thus increasing future income. Moreover, they immediately dedicate less time to healthcare and creative activities. Then, towards the new steady state, they will increase at decreasing rates the time devoted to these activities. As a result, households expect to have a better life in the future than if they simply accommodated the increase in government expenditures with a fall in consumption.

However, to keep their level of assets in the new steady state, they have to work more even though they are consuming a bit more. Consequently, they devote less time to healthcare and creative activities and thus the growth-rate of technological knowledge falls.

We leave for future work the sensitivity analysis of the model with respect to the most relevant parameters and values of the exogeneous variables.

References

1. Acemoglu, D.: Introduction to Modern Economic Growth. Princeton University Press, Princeton (2009)
2. Ben-Porath, Y.: The production of human capital and the life cycle of earnings. J. Polit. Econ. **75**(4), 352–365 (1967)
3. Caballe, J., Santos, M.: On endogenous growth with physical and human capital. J. Polit. Econ. **101**(6), 1042–1067 (1993)
4. Fogel, R.: The Fourth Great Awakening and the Future of Egalitarianism. University of Chicago Press, Chicago/London (2000)
5. Irmen, A., Kuehnel, J.: Productive government expenditure and economic growth. J. Econ. Surv. **23**, 692–733 (2009)
6. Lucas, R.: On the mechanics of economic development. J. Monet. Econ. **22**, 3–42 (1988)
7. Ramey, V., Francis, N.: A century of work and leisure. Am. Econ. J. Macroecon. **1**(2), 189–224 (2009)
8. Romer, D.: Advanced Macroeconomics, 3rd edn. McGraw-Hill, New York (2006)
9. Turnovsky, S.J.: Fiscal policy, elastic labor supply, and endogenous growth. J. Monet. Econ. **45**, 185–210 (2000)

An Introduction to Geometric Gibbs Theory

Yunping Jiang

Abstract This is an article I wrote for Dynamics, Games, and Science. In Dynamics, Game, and Science, one of the most important equilibrium states is a Gibbs state. The deformation of a Gibbs state becomes an important subject in these areas. An appropriate metric on the space of underlying dynamical systems is going to be very helpful in the study of deformation. The Teichmüller metric becomes a natural choice. The Teichmüller metric, just like the hyperbolic metric on the open unit disk, makes the space of underlying dynamical systems a complete space. The Teichmüller metric precisely measures the change of the eigenvalues at all periodic points which are essential data needed to obtain the Gibbs state for a given dynamical system. In this article, I will introduce the Teichmüller metric and, subsequently, a generalization of Gibbs theory which we call geometric Gibbs theory.

1 Introduction

The mathematical theory of Gibbs states, an important idea originally from physics, is a beautiful mathematical theory starting from the celebrated work of Sinai [23, 24] and Ruelle [20, 21]. It leads to the study of SRB-measures in Anosov dynamical systems and, more generally, Axiom A dynamical systems due to the further work of Sinai, Ruelle, Bowen, and many other people (see [2]). A very important feature of a Gibbs state is that it is an equilibrium state. This equilibrium state plays an important role in mathematics, as well as many other areas such as physics, chemistry, biology, economy, and game theory.

An important topic in the current study of Gibbs states (in mathematics, we also call them Gibbs measures) is to study the deformation of a Gibbs state. For example, how does a Gibbs measure (or a SRB-measure) changes when the

Y. Jiang (✉)
Department of Mathematics, Queens College of the City University of New York, Flushing, NY 11367-1597, USA

Department of Mathematics, Graduate School of the City University of New York, 365 Fifth Avenue, New York, NY 10016, USA
e-mail: yunping.jiang@qc.cuny.edu

J.-P. Bourguignon et al. (eds.), *Dynamics, Games and Science*, CIM Series in Mathematical Sciences 1, DOI 10.1007/978-3-319-16118-1_18

underlying dynamical system changes? How does the density of a Gibbs measure (or SRB-measure) with respect to the Lebesgue measure changes when the underlying dynamical system changes? To study the deformation of a Gibbs measure, an appropriate metric on the space of underlying dynamical systems will be very helpful. Ruelle has proposed to use the Whitney theory (see [16, 22]). In this note I would like to introduce another metric from Teichmüller theory and, subsequently, a generalized Gibbs theory which we call geometric Gibbs theory. The Teichmüller metric closely relates to the eigenvalues at all periodic points: Given a one-dimensional (expanding) smooth dynamical system f, the essential data needed to determine the Gibbs measure is the set of eigenvalues

$$\Big\{\log \lambda_f(p) = \log |(f^n)'(p)| \;;\; p \text{ a periodic point of } f \text{ of period } n\Big\}.$$

Given two topologically conjugate dynamical systems f and g, with

$$g \circ h = h \circ f,$$

where h is the conjugacy, how can we measure the geometric difference between f and g? The answer is the set of ratios

$$\Big\{0 < \alpha(p) = \min\Big\{\frac{\log \lambda_f(p)}{\log \lambda_g(h(p))}, \frac{\log \lambda_g(h(p))}{\log \lambda_f(p)}\Big\} \leq 1\Big\}.$$

Actually, h is locally $\alpha(p)$-Hölder continuous near p but the exponent changes with p. These exponents can be measured precisely by using so-called "quasiconformal dilatation" from complex analysis (refer to [9]), that is, the Teichmüller metric. The Teichmüller metric, just like the hyperbolic metric (or Lobachevsky metric or Poincaré metric) on the open unit disk, makes the space a complete space.

This article intends to give a summary of our work in this direction. A more complete version of our work with more detailed proofs will be available in [15]. I first give a brief review of classical Gibbs theory in Sect. 2. Then, following the traditional terminology in dynamical systems, I introduce a circle g-function in Sect. 3. In Sect. 4, I give the definition of a geometric Gibbs measure associated to a circle g-function. In the same section, I show the existence of a geometric Gibbs measure for any circle g-function and the uniqueness for the constant g-function. I further show that a geometric Gibbs measure is an equilibrium state. Finally, I introduce the Teichmüller metric on the space of all circle g-functions in Sect. 5. The Teichmüller metric makes the space of all circle g-functions a complete space. I expect this new metric will play an important role in the study of deformations of geometric Gibbs measures. In particular, when a circle g-function is Hölder continuous, the corresponding geometric Gibbs measure is absolutely continuous with respect to the Lebesgue measure. Therefore, we have a density function.

We would like to study the derivative of a density function with respect to a Hölder continuous g-function and connect this derivative with the susceptibility function

$$\Psi(\lambda) = \sum_{n=0}^{\infty} \lambda^n \int_T \rho(dx) X(x) \frac{d}{dx}(A(f^n x))$$

at $\lambda = 1$, which is formally the derivative of the density with respect to a Hölder continuous g-function, as described in [16, 22].

2 Classical Gibbs Theory

Suppose $d \geq 2$ is a positive integer. Consider the symbolic dynamical system $\sigma : \Sigma \to \Sigma$, where

$$\Sigma = \{w = \cdots i_{n-1} \cdots i_1 i_0 \mid i_{n-1} \in \{0, \cdots, d-1\},\ n = 1, 2, \cdots\}$$

and

$$\sigma : w = \cdots i_{n-1} \cdots i_1 i_0 \to \sigma(w) = \cdots i_{n-1} \cdots i_1$$

is the shift map. The space $\Sigma = \prod_{\infty}^{0} \{0, 1, \cdots, d-1\}$ is a compact topological space with the product topology. *We purposely write*

$$w = \cdots i_{n-1} \cdots i_1 i_0$$

from the right to left because we will later use

$$v = j_0 j_1 \cdots j_{n-1} \cdots$$

to represent a point on the unit circle. An n-cylinder $[w]_n$ containing $w = \cdots i_{n-1} \cdots i_1 i_0$ is the subset of all elements $w' = \cdots i'_{n+m} \cdots i'_n i_{n-1} \cdots i_0$ for $i'_{n+m} \in \{0, \cdots, d-1\}$ and $m = 0, 1, \cdots$.

A real function $\phi : \Sigma \to \mathbf{R}$ is called Hölder continuous if there are two constants $C > 0$ and $0 < \tau < 1$ such that $|\phi(w) - \phi(w')| \leq C\tau^n$ as long as w and w' are in the same n-cylinder. We use $\mathscr{C}^H$ to denote the space of all Hölder continuous real functions on Σ. We call a positive Hölder continuous function ψ on Σ a Hölder potential. We also use $\mathscr{C}$ to denote the space of all continuous real functions on Σ and $\mathscr{M}$ to denote the space of all finite Borel measures on Σ, which is the dual space of $\mathscr{C}$. Then $\mathscr{M}(\sigma)$ means the space of all σ-invariant probability measures in $\mathscr{M}$, that is, the space of measures with total measure 1 and satisfying $\mu(\sigma^{-1}(A)) = \mu(A)$ for all Borel subsets A of Σ.

The classical Gibbs theory ensures that associated to each Hölder potential ψ, there is a number $P = P(\log\psi)$, called the pressure, and unique σ-invariant probability measure $\mu = \mu_\psi$, called a Gibbs measure, such that

$$C^{-1} \leq \frac{\mu([w]_n)}{\exp(-Pn + \sum_{i=0}^{n-1} \log\psi(\sigma^i(w)))} \leq C \tag{1}$$

for any n-cylinder $[w]_n$, where C is a fixed constant. A Gibbs measure depends only on a cohomologous equivalence class and is an equilibrium state in the sense that

$$P(\log\psi) = ent_\mu(\sigma) + \int_\Sigma \log\psi \, d\mu = \sup_{\nu\in\mathscr{M}(\sigma)} \Big\{ ent_\nu(\sigma) + \int_\Sigma \log\psi \, d\nu \Big\}$$

where $ent_\nu(\sigma)$ is the measure-theoretical entropy of σ with respect to ν.

In a proof of the existence and uniquess of the Gibbs measure μ for a given Hölder potential ψ, we use the Ruelle-Perron-Frobenius operator

$$\mathscr{L}_\psi\phi(w) = \sum_{\sigma(w')=w} \psi(w')\phi(w') : \mathscr{C}^H \to \mathscr{C}^H. \tag{2}$$

The Ruelle-Perron-Frobenius theorem (refer to [10] for a proof) says that there is a positive real number λ and a positive Hölder function $\rho \in \mathscr{C}^H$ such that $\mathscr{L}_\psi \rho = \lambda\rho$. Here λ is the unique, maximal, positive, simple eigenvalue of $\mathscr{L}_\psi$. Note that in this case, the pressure $P = \log\lambda$. If we consider a new Hölder potential

$$\tilde{\psi} = \frac{\psi\cdot\rho}{\lambda\cdot\rho\circ\sigma},$$

then we get a normalized Ruelle-Perron-Frobenius operator $\mathscr{L}_{\tilde{\psi}}$, that is, $\mathscr{L}_{\tilde{\psi}}1 = 1$. Let $\mathscr{L}^*_{\tilde{\psi}} : \mathscr{M} \to \mathscr{M}$ be its dual operator. Then the Gibbs measure $\mu_{\tilde{\psi}}$ is just the unique fixed point of $\mathscr{L}^*_{\tilde{\psi}}$ in this case. (The Gibbs measure $\mu_\psi = \rho\cdot\mu_{\tilde{\psi}}$.) This leads to the study of g-measure theory in Keane's paper [17] as follows.

A non-negative continuous real function g defined on Σ is called a g-function if

$$\sum_{i=0}^{d-1} g(wi) = 1. \tag{3}$$

For a g-function g, define the transfer operator

$$\mathscr{L}_g\phi(w) = \sum_{\sigma(w')=w} g(w')\phi(w') : \mathscr{C} \to \mathscr{C}.$$

It is a normalized Ruelle-Perron-Frobenius operator, that is, $\mathscr{L}_g 1 = 1$. Let $\mathscr{L}^*_g : \mathscr{M} \to \mathscr{M}$ be its dual operator. Every fixed point μ of $\mathscr{L}^*_g$ is called a g-measure associated with g. Here we always assume that μ is a probability measure.

Any $\mu \in \mathcal{M}(\sigma)$ is absolutely continuous with respect to $\tilde{\mu} = \mathcal{L}_1^* \mu$. So we have the Radon-Nikodým derivative

$$RND_{\mu,\tilde{\mu}}(w) = \frac{d\mu}{d\tilde{\mu}}(w), \quad \tilde{\mu} - a.e.\ w.$$

It is a $\tilde{\mu}$-measurable function. The following theorem is in Leddrapier's paper [18] and is used in Walters' paper [25] for the study of a generalized version of Ruelle's theorem. Let $\mathcal{B}$ be the Borel σ-algebra on Σ.

Theorem 1 *Suppose g is a g-function and $\mu \in \mathcal{M}$ is a probability measure. The following statements are equivalent:*

(i) μ *is a g-measure for g, i.e.,* $\mathcal{L}_g^* \mu = \mu$.
(ii) $\mu \in \mathcal{M}(\sigma)$ *and* $RND_{\mu,\tilde{\mu}}(w) = g(w)$ *for* $\tilde{\mu}$*-a.e.* w.
(iii) $\mu \in \mathcal{M}(\sigma)$ *and*

$$E[\phi|\sigma^{-1}(\mathcal{B})](w) = \mathcal{L}_g\phi(\sigma(w)) = \sum_{\sigma(w')=\sigma(w)} g(w')\phi(w'),\ \textit{for}\ \mu\textit{-a.e.}\ w,$$

where $E[\phi|\sigma^{-1}(\mathcal{B})]$ *is the conditional expectation of* ϕ *with respect to* $\sigma^{-1}(\mathcal{B})$.
(iv) $\mu \in \mathcal{M}(\sigma)$ *and is an equilibrium state for g in the sense that*

$$0 = ent_\mu(\sigma) + \int_\Sigma \log g\, d\mu = \sup_{\nu\in\mathcal{M}(\sigma)} \Big\{ent_\nu(\sigma) + \int_\Sigma \log g\, d\nu\Big\}.$$

Furthermore, if g is a positive Hölder continuous g-function, then $RND_{\mu,\tilde{\mu}}$ is actually a Hölder continuous function and

$$RND_{\mu,\tilde{\mu}}(w) \equiv g(w). \tag{4}$$

However, this fact may not be true in general for a merely continuous g-function. One of our goals in generalized Gibbs theory is to associate with a circle g-function (see the definition in the next section) a g-measure, which we will call a geometric Gibbs measure, such that the equality (4) holds. We will also study the uniqueness of a geometric Gibbs measure for any given circle g-function.

3 Circle g-Functions

We use

$$T = \{z \in \mathbf{C} \mid |z| = 1\}$$

to denote the unit circle in the complex plane **C**. The universal cover of T is the real line **R** with the covering map

$$z = \pi(x) = e^{2\pi ix} : \mathbf{R} \to T,$$

where $2\pi x$ is the angle of z.

Consider an orientation-preserving covering map $f : T \to T$ of degree d. We normalize it by assuming that $f(1) = 1$. We use F to denote the lift of f to **R** such that $F(0) = 0$. We have that $F(x+1) = F(x) + d$.

We use h to denote a circle homeomorphism and assume that $h(1) = 1$. Let H be the lift of h such that $H(0) = 0$. We also have $H(x+1) = H(x) + 1$.

A C^1 circle endomorphism f is called expanding if there are constants $C > 0$ and $\lambda > 1$ such that

$$(F^n)'(x) \geq C\lambda^n, \quad x \in \mathbf{R}, \quad n = 1, 2, \cdots.$$

A circle endomorphism f is $C^{1+\alpha}$ for some $0 < \alpha \leq 1$ if f' is α-Hölder continuous.

A circle homeomorphism h is called quasisymmetric if there is a constant $M \geq 1$ such that

$$M^{-1} \leq \frac{|H(x+t) - H(x)|}{|H(x) - H(x-t)|} \leq M, \quad \forall x \in \mathbf{R}, \ \forall t > 0.$$

In particular, if we can take $M = 1 + \varepsilon(t)$ for some bounded positive real function $\varepsilon(t)$ with $\varepsilon(t) \to 0^+$ as $t \to 0^+$, then h is called symmetric. For example, a C^1-diffeomorphism of T is symmetric. But, we would like to remind the reader, a symmetric homeomorphism may be totally singular, that is, it may map a positive Lebesgue measure subset to a zero Lebesgue measure subset, and vice versa.

A circle endomorphism f is called uniformly symmetric if there is a bounded real function $\varepsilon(t) > 0$ for $t > 0$ such that $\varepsilon(t) \to 0^+$ as $t \to 0^+$ and such that

$$\frac{1}{1+\varepsilon(t)} \leq \frac{|F^{-n}(x+t) - F^{-n}(x)|}{|F^{-n}(x) - F^{-n}(x-t)|} \leq 1 + \varepsilon(t), \ \forall x \in \mathbf{R}, \ \forall t > 0, \ n = 1, 2, \cdots.$$

A $C^{1+\alpha}$, for some $0 < \alpha \leq 1$, circle expanding endomorphism f is uniformly symmetric. Again we would like to remind the reader that a uniformly symmetric circle endomorphism may be totally singular.

In the terms of complex analysis, the reader can find descriptions of quasisymmetric circle homeomorphisms in [1], symmetric circle homeomorphisms in [7], and uniformly symmetric circle endomorphisms in [5].

Consider the space

$$\Sigma^+ = \{v = j_0 j_1 \cdots j_{n-1} \cdots \mid j_{n-1} \in \{0, 1, \cdots, d-1\}, \ n = 1, \cdots\}$$

and the shift map

$$\sigma^+(v) = j_1 \cdots j_{n-1} j_n \cdots : \Sigma^+ \to \Sigma^+.$$

The space $\Sigma^+ = \prod_0^\infty \{0, 1, \cdots, d-1\}$ is a compact topological space with the product topology. An n-cylinder $[v]_n^+$ containing $v = j_0 j_1 \cdots j_{n-1} \cdots$ is the subset of all points $v' = j_0 \cdots j_{n-1} j'_n j'_{n+1} \cdots$ for $j'_{n+m} \in \{0, 1, \cdots, d-1\}$ and $m = 0, 1, \cdots$. The set of all cylinders forms a topological basis of Σ^+ such that it is a compact topological space. The space Σ^+ with this topology is called the symbolic representation of the unit circle T. More precisely, for any $z = e^{2\pi i x} \in T$, we have that

$$x = x(v) = \sum_{k=0}^{\infty} \frac{j_k}{d^{k+1}}, \quad v = j_0 j_1 \cdots j_{n-1} \cdots \in \Sigma^+. \tag{5}$$

The Lebesgue metric $|v - v'| = |x(v) - x(v')|$ induces the Lebesgue measure m_0 on Σ^+.

Every uniformly symmetric circle endomorphism f is semi-conjugate to σ^+, that is, we have a projection $\pi_f : \Sigma^+ \to T$, which is 1-1 except for a countable set, such that

$$\pi_f \circ \sigma^+(v) = f \circ \pi_f(v), \quad v \in \Sigma^+.$$

This implies that any two uniformly symmetric circle endomorphisms f and g of the same degree are topologically conjugate, that is, there is a circle homeomorphism h of T such that

$$f \circ h = h \circ g.$$

Furthermore, h is a quasisymmetric circle homeomorphism (refer to [9]).

Now let us return to the space Σ. Suppose f is a uniformly symmetric circle endomorphism. For any $w = \cdots i_{n-1} \cdots i_1 i_0 \in \Sigma$, let $v_n = j_0 j_1 \cdots j_{n-2} j_{n-1}$ and $v_{n-1} = \sigma(v_n) = j_0 j_1 \cdots j_{n-2}$ where $j_0 = i_{n-1}, \cdots, j_{n-2} = i_1, j_{n-1} = i_0$. Consider two intervals on T,

$$I_{v_n} = \pi_f([v]_n^+) \subset I_{v_{n-1}} = \pi_f([v']_{n-1}^+)$$

where $v = v_n \cdots$ and $v' = v_{n-1} \cdots$ and the ratio

$$g_n(w) = \frac{|I_{v_n}|}{|I_{v_{n-1}}|}.$$

We have that

Theorem 2 (Circle g-Function [15]) *Suppose f is a uniformly symmetric circle endomorphism. Then the limiting function*

$$g(w) = \lim_{n\to\infty} g_n(w) : \Sigma \to \mathbf{R}$$

exists and is a continuous positive function. The convergence is uniform. Furthermore, if f is $C^{1+\alpha}$, then $g(w)$ is a Hölder continuous positive function and the convergence is exponential. The function $g(w)$ is called a circle g-function since it satisfies the condition (3).

Furthermore, when $d = 2$, $g(w)$ also satisfies a compatibility condition

$$\prod_{n=0}^{\infty} \frac{g(w0\overbrace{1\cdots 1}^{n})}{g(w1\underbrace{0\cdots 0}_{n})} = const, \quad \forall w \in \Sigma, \tag{6}$$

where the convergence in the formula is uniform on Σ. The conditions (3) and (6) give a complete characterization of a circle g-function as proved in [3, 4, 11]. That is, a continuous positive g-function is a circle g-function if and only if it satisfies the conditions (3) and (6). For a Hölder continuous positive g-function, it is a circle g-function if and only if it satisfies the conditions (3) and (6) and the convergence in (6) is exponential. Furthermore, the realized uniformly symmetric circle endomorphism f is $C^{1+\alpha}$. Note that, from our proof of Katok's conjecture in [12], any C^1 uniformly symmetric circle endomorphism is expanding.

We use $\mathscr{G}$ to denote the space of all circle g-functions on Σ and $\mathscr{HG}$ to denote the space of all Hölder continuous circle g-functions on Σ.

4 Geometric Gibbs Measures

For any $v = j_0 \cdots j_{n-1} \cdots \in \Sigma^+$, let $x = x(v)$ in (5). Then $z = e^{2\pi i x} \in T$. Consider the cylinder $[v]_n^+$ in Σ^+. Let $v_n = j_0 \cdots j_{n-1}$ and $w_n = i_{n-1} \cdots i_0$ where $i_0 = j_{n-1}, \cdots, i_{n-1} = j_0$. Then we have the n-cylinder $[w]_n$ in Σ where $w = \cdots w_n \in \Sigma$.

Suppose $\mu \in \mathscr{M}(\sigma)$ is non-atomic and does not take zero on any cylinders $[w]_n$. For each $n \geq 0$, take d^{n+1} intervals labeled $I_{\underbrace{00\ldots 0}_{n}}, \cdots, I_{v_n}, \cdots, I_{\underbrace{(d-1)(d-1)\ldots(d-1)}_{n}}$, each with angle length

$$|I_{v_n}| = 2\pi\mu([w]_n).$$

Arrange them counter-clockwise in numerical order of the angle of z on the unit circle T beginning at 1. Since μ is σ-invariant, we have that

$$I_{v_n} = \cup_{j=0}^{d-1} I_{v_n j}$$

and that

$$\cdots \subset I_{v_n} \subset I_{v_{n-1}} \subset \cdots \subset I_{v_1}.$$

Since μ is non-atomic and does not take zero on any cylinders $[w]_n$, $\cap_{k=1}^{\infty} I_{v_n}$ contains only one point which we denote as $h_\mu(z)$. This defines a homeomorphism $h_\mu(z) : T \to T$.

Definition 1 Suppose g is a circle g-function. A non-atomic σ-invariant probability measure μ is a geometric Gibbs measure associated with g if $f_\mu = h_\mu \circ q_d \circ h_\mu^{-1}$ is a uniformly symmetric circle endomorphism and

$$\lim_{n\to\infty} \frac{\mu([w]_n)}{\mu([\sigma(w)]_{n-1})} = g(w) \tag{7}$$

uniformly on $w \in \Sigma$.

We have proved the following theorem.

Theorem 3 (Existence [15]) *For any circle g-function g, we can find a geometric Gibbs measure $\mu = \mu(g)$ associated with it. Furthermore, if g is a Hölder continuous circle g-function, then the measure μ found in the first part of this theorem is the Gibbs measure in the classical sense.*

For the measure μ in the above theorem, we have that

$$RND_{\mu,\tilde{\mu}}(w) \equiv g(w),$$

as we expected in Sect. 2.

An important question to ask at this point is whether a geometric Gibbs measure is unique. Even when g is a Hölder continuous circle g-function, although we know it has a unique Gibbs measure in the classical sense, it may still have more than one geometric Gibbs measures. However, we have the following theorem.

Theorem 4 (Uniqueness [13, 15]) *The constant circle g-function $g(w) = 1/d$ has only one geometric Gibbs measure associated with it.*

We would like to note that a general circle g-function is very non-trivial. This makes the study of uniqueness more difficult but interesting.

Another important topic to study at this point is the ergodicity of a geometric Gibbs measure. For a Hölder continuous circle g-function g, we know that the Gibbs measure $\mu = \mu(g)$ in the classical sense is ergodic, that is, for any Borel subset A of Σ, if $\sigma^{-1}(A) = A$ and if $\mu(A) > 0$, then $\mu(A) = 1$. We would like to know the ergodicity for any geometric Gibbs measure associated with a circle g-function. We expect that the answer is affirmative.

Furthermore, we know that a geometric Gibbs measure is an equilibrium state.

Theorem 5 (Equilibrium [6, 15]) *Suppose g is a circle g-function. Every geometric Gibbs measure μ corresponding to g is an equilibrium state in the following sense,*

$$0 = ent_{\mu}(\sigma) + \int_{\Sigma} \log g \, d\mu = \sup_{\nu \in \mathcal{M}(\sigma)} \Big\{ ent_{\nu}(\sigma) + \int_{\Sigma} \log g \, d\nu \Big\}.$$

5 Teichmüller Metric

We introduce a Teichmüller metric on $\mathcal{G}$ and show that it is a complete metric. Under this metric, $\mathcal{HG}$ is dense in $\mathcal{G}$. We use [1, 7, 19] as references for the Teichmüller theory and for the quasiconformal mapping theory.

Suppose $\mathcal{US}$ is the space of all uniformly symmetric circle endomorphisms of degree d. Let $q_d(z) = z^d$ be the basepoint in $\mathcal{US}$. We first define the Teichmüller space for $\mathcal{US}$ as follows. For any $f \in \mathcal{US}$, let h_f be the conjugacy from f to q_d, i.e.,

$$f \circ h_f = h_f \circ q_d.$$

We know that h_f is quasisymmetric. Thus we can think of $\mathcal{US}$ as the space of marking pairs (f, h_f). We define an equivalence relation $\sim_T$: Two pairs $(f, h_f) \sim_T (g, h_g)$ if $h_f \circ h_g^{-1}$ is symmetric. The Teichmüller space

$$\mathcal{TUS} = \{[(f, h_f)] \mid (f, h_f) \in \mathcal{US}, \text{ with the basepoint } [(q_d, id)]\}$$

is the space of all $\sim_T$-equivalence classes $[(f, h_f)]$. We have a one-to-one correspondence between $\mathcal{G}$ and $\mathcal{TUS}$ (refer to [15]). Thus we have that

$$\mathcal{G} = \mathcal{TUS}.$$

By using $\mathcal{TUS}$, we define a Teichmüller metric on $\mathcal{G}$.

Let $\mathcal{QS}$ be the set of all quasisymmetric homeomorphisms of T. Let $\mathcal{S}$ be the subset of $\mathcal{QS}$ consisting of all symmetric homeomorphisms of T. For any $h \in \mathcal{QS}$, let $\mathcal{E}_h$ be the set of all quasiconformal extensions of h into the unit disk. For each $\tilde{h} \in \mathcal{E}_h$, let $\mu_{\tilde{h}} = \tilde{h}_{\bar{z}}/\tilde{h}_z$ be its complex dilatation. Let

$$k_{\tilde{h}} = \|\mu(z)\|_{\infty} \quad \text{and} \quad K_{\tilde{h}} = \frac{1 + k_{\tilde{h}}}{1 - k_{\tilde{h}}}.$$

Here $K_{\tilde{h}}$ is called the quasiconformal dilatation of $\tilde{h}$. Using quasiconformal dilatation, we can define a pseudo-distance in $\mathcal{QS}$ by

$$d(h_1, h_2) = \frac{1}{2} \inf\{\log K_{\tilde{h}_1 \tilde{h}_2^{-1}} \mid \tilde{h}_1 \in \mathcal{E}_{h_1}, \tilde{h}_2 \in \mathcal{E}_2\}.$$

It will induce a distance in the space $\mathcal{UT}$ of $\mathcal{QS}$ modulo the space of Möbius transformation preserving T, which is the universal Teichmüller space and which is a complete metric space and a complex manifold with complex structure compatible with the Hilbert transform. Now consider the space

$$\mathcal{AUT} = \mathcal{QS} \text{ modulo } \mathcal{S}.$$

It is called an asymptotical universal Teichmüller space. Given two cosets $\mathcal{S}h_1$ and $\mathcal{S}h_2$ in this factor space, define

$$\overline{d}(\mathcal{S}h_1, \mathcal{S}h_2) = \inf_{A,B \in \mathcal{S}} d(Ah_1, Bh_2).$$

It defines a distance on $\mathcal{AUT}$. The asymptotical Teichmüller space $(\mathcal{AUT}, \overline{d}(\cdot,\cdot))$ is a complete metric space and a complex manifold. The topology on $(\mathcal{AUT}, \overline{d}(\cdot,\cdot))$ is the finest topology which makes the projection $\pi : \mathcal{UT} \to \mathcal{AUT}$ continuous, and π is also holomorphic. Refer to [7].

An equivalent topology on the quotient space $\mathcal{AUT}$ can be defined as follows. For any $h \in \mathcal{QS}$, let $\tilde{h}$ be a quasiconformal extension of h to a small neighborhood of T in the complex plane. Suppose U is the domain of $\tilde{h}$. Let

$$\mu_{\tilde{h}}(z) = \frac{\tilde{h}_{\overline{z}}(z)}{\tilde{h}_z(z)}, \; z \in U, \quad k_{\tilde{h}} = \|\mu_{\tilde{h}}(z)\|_{\infty,U}, \quad \text{and} \quad B_{\tilde{h}} = \frac{1 + k_{\tilde{h}}}{1 - k_{\tilde{h}}}.$$

Then the boundary dilatation h is defined as

$$B_h = \inf_{\tilde{h}} B_{\tilde{h}},$$

where the infimum is taken over all quasiconformal extensions $\tilde{h}$ of h in a neighborhood of T. It is known that h is symmetric if and only if $B_h = 1$. Define

$$\tilde{d}(h_1, h_2) = \frac{1}{2} \log B_{h_2^{-1} h_1}.$$

Then it is a distance on $\mathcal{AUT}$. The two distances $\overline{d}$ and $\tilde{d}$ on $\mathcal{AUT}$ are equal.

There is a natural embedding from

$$\mathcal{G} = \mathcal{TUS} \ni g = [f, h_f] \hookrightarrow [h_f] \in \mathcal{AUT}.$$

Thus, the restriction of $\tilde{d}(\cdot,\cdot)$ on $\mathcal{G} = \mathcal{T}\mathcal{U}\mathcal{S}$ gives a distance which we denote as $d_{\mathcal{G}}(\cdot,\cdot)$. We call the space

$$\left(\mathcal{G}, d_{\mathcal{G}}(\cdot,\cdot)\right)$$

the Teichmüller space of circle g-functions.

Theorem 6 (Completeness [15]) *The Teichmüller space* $\left(\mathcal{G}, d_{\mathcal{G}}(\cdot,\cdot)\right)$ *is a complete metric space and* $\mathcal{H}\mathcal{G}$ *is dense in this space.*

Moreover, the space $\left(\mathcal{G}, d_{\mathcal{G}}(\cdot,\cdot)\right)$ has a complex Banach manifold structure (refer to [8, 15]). We would like to point out that there is a maximal norm

$$\|g\| = \max_{w\in\Sigma} |g(w)|$$

on the space $\mathcal{G}$. This also introduces a metric $d_{max}(\cdot,\cdot)$ on $\mathcal{G}$. But this metric is not complete. Moreover, this metric will not measure the change of a geometric Gibbs measure in a good sense. For example, even if $d_{max}(\cdot,\cdot)$ is small, the change of a geometric Gibbs measure could be big. So it is just like the Euclidean metric on the open unit disk. The Teichmüller metric we have introduced is precisely like the hyperbolic metric (or Lobachevsky metric or Poincaré metric) on the open unit disk. Topologies induced from both metrics are the same (refer to [14, 15]).

Acknowledgements I would like to thank my student John Adamski who read the initial version of this manuscript very carefully and found many typos and made very good suggestions to improve the exposition of this paper. This research is partially supported by the collaboration grant (#199837) from the Simons Foundation, the CUNY collaborative incentive research grants (#1861 and #2013), and awards from PSC-CUNY. This research is also partially supported by the collaboration grant (#11171121) from the NSF of China and a collaboration grant from Academy of Mathematics and Systems Science and the Morningside Center of Mathematics at the Chinese Academy of Sciences.

References

1. Ahlfors, L.V.: Lectures on Quasiconformal Mappings. Mathematical Studies, vol. 10. D. Van Nostrand Co. Inc., Toronto/New York/London (1966)
2. Bowen, R.: Equilibrium States and the Ergodic Theory of Anosov Diffeomorphisms. Springer, Berlin (1975)
3. Cui, G., Jiang, Y., Quas, A.: Scaling functions, g-measures, and Teichmüller spaces of circle endomorphisms. Discrete Contin. Dyn. Syst. **5**(3), 534–552 (1999)
4. Cui, G., Gardiner, F., Jiang, Y.: Scaling functions for circle endomorphisms. Contemp. Math. (AMS Series) **355**, 147–163 (2004)
5. Gardiner, F., Jiang, Y.: Asymptotically affine and asymptotically conformal circle endomorphisms. In: Fujikawa, E. (ed.) Infinite Dimensional Teichmüller Spaces and Moduli Spaces. RIMS Kôkyûroku Bessatsu, vol. B17, pp. 37–53 (2010)

6. Gardiner, F., Jiang, Y.: Circle endomorphisms, dual circles and Thompson's group. Contemp. Math. (AMS) **573**, 99–118 (2012)
7. Gardiner, F., Sullivan, D.: Symmetric and quasisymmetric structures on a closed curve. Am. J. Math. **114**(4), 683–736 (1992)
8. Hu, Y., Jiang, Y., Wang, Z.: Martingales for quasisymmetric systems and complex manifold structures. Ann. Acad. Sci. Fenn. Math. **38**, 1–26 (2013)
9. Jiang, Y.: Renormalization and Geometry in One-Dimensional and Complex Dynamics. Advanced Series in Nonlinear Dynamics, vol. 10, pp. xvi+309. World Scientific, River Edge (1996)
10. Jiang, Y.: A proof of existence and simplicity of a maximal eigenvalue for Ruelle-Perron-Frobenius operators. Lett. Math. Phys. **48**, 211–219 (1999)
11. Jiang, Y.: Metric invariants in dynamical systems. J. Dyn. Differ. Equ. **17**(1), 51–71 (2005)
12. Jiang, Y.: On a question of Katok in one-dimensional case. Discrete Contin. Dyn. Syst. **24**(4), 1209–1213 (2009)
13. Jiang, Y.: Symmetric invariant measures. Contemp. Math. AMS **575**, 211–218 (2012)
14. Jiang, Y.: Function model of the Teichmüller space of a closed hyperbolic Riemann surface [arXiv0810.4969v3] (2009)
15. Jiang, Y.: Geomtric Gibbs theory. Rewritten version of Teichmüller structures and dual geometric Gibbs type measure theory for continuous potentials (arXiv0804.3104v3, 2010)
16. Jiang, Y., Ruelle, D.: Analyticity of the susceptibility function for unimodal Markovian maps of the interval. Nonlinearity **18**, 2447–2453 (2005)
17. Keane, M.: Strongly mixing g-measures. Invent. Math. **16**, 309–324 (1972)
18. Ledrappier, F.: Principe variationnel et systèmes dynamiques symboliques. Z. Wahrscheinlichkeitstheorie und Verw. Gebiete **30**, 185–202 (1974)
19. Lehto, O.: Univalent Functions and Teichmüller Spaces. Springer, New York/Berlin (1987)
20. Ruelle, D.: Statistical mechanics of a one-dimensional lattice gas. Commun. Math. Phys. **9**, 267–278 (1968)
21. Ruelle, D.: A measure associated with axiom A attractors. Am. J. Math. **98**, 619–654 (1976)
22. Ruelle, D.: Differentiating the absolutely continuous invariant measure of an interval map f with respect to f. Commun. Math. Phys. **258**, 445–453 (2005)
23. Sinai, Y.G.: Markov partitions and C-diffeomorphisms. Funct. Anal. Appl. **2**(1), 64–89 (1968)
24. Sinai, Y.G.: Gibbs measures in ergodic theory. Russ. Math. Surv. **27**(4), 21–69 (1972)
25. Walters, P.: Ruelle's operator theorem and g-measures. Trans. Am. Math. Soc. **214**, 375–387 (1975)

Sphere Rolling on Sphere: Alternative Approach to Kinematics and Constructive Proof of Controllability

F. Silva Leite and F. Louro

Abstract The kinematic equations for rolling a sphere on another sphere, subject to non-holonomic constraints of non-slip and non-twist, are known and can be found in [7]. Here we present an alternative approach to derive these kinematic equations which is also suitable for describing the rolling of more general manifolds embedded in Euclidean space. This approach consists on rolling each of the manifolds separately on a common affine tangent space and then using the transitive and symmetric properties of rolling maps to derive the kinematic equations of rolling one manifold on the other. We use this approach to derive the kinematic equations for rolling an n-dimensional sphere on another one with the same dimension. It is also well known that the sphere rolling on sphere system is controllable, except when the two spheres have equal radii. This is a theoretical result that guarantees the possibility to roll one of the spheres on the other from any initial configuration to any final configuration without violating the non-holonomic constraints. However, from a practical viewpoint it is important to know how this is done. To answer this more applied question, we present a constructive proof of the controllability property, by showing how the forbidden motions can be performed by rolling without slip and twist. This is also illustrated for 2-dimensional spheres.

1 Introduction

The most classical of all non-holonomic systems is the rolling sphere, rolling without slip and without twist on the affine tangent space at a point. Another interesting example of a system subject to non-holonomic constraints is that of

---

F.S. Leite (✉)
Institute of Systems and Robotics, University of Coimbra - Pólo II, Pinhal de Marrocos, 3030-290 Coimbra, Portugal

Department of Mathematics, University of Coimbra, Largo D. Dinis, 3001-454 Coimbra, Portugal
e-mail: fleite@mat.uc.pt

F. Louro
Department of Mathematics, Rutgers University, New Brunswick, NJ, USA
e-mail: Fernando.Louro@ist.utl.pt

J.-P. Bourguignon et al. (eds.), *Dynamics, Games and Science*, CIM Series in Mathematical Sciences 1, DOI 10.1007/978-3-319-16118-1_19

a sphere rolling over another sphere of the same dimension. Rolling motions of manifolds embedded in Euclidean space $\mathbb{R}^n$ can be described by curves in the Lie group SE_n of orientation preserving isometries of the ambient space, as explained in Sharpe [13]. Other relevant studies involving rolling motions are [4, 5, 11, 14]. We take the definition in [13] and consequent properties of rolling maps to derive the kinematic equations for the rolling spheres. We also show how the forbidden motions, twists and slips, can be produced using rolling without slip/twist. This is a constructive proof of the complete controllability of the system, when the spheres have unequal radii.

The organization of the paper is as follows. The formal definition of rolling and properties of rolling maps appear in Sect. 2. The particular case of a sphere rolling on another sphere and the derivation of the corresponding kinematics are presented in Sect. 4. Finally, in Sect. 5 we include a constructive proof of controllability.

2 Rolling Maps

We refer to Sharpe [13] and Lee [9] for details concerning differential and Riemannian geometry.

Let M and N be two smooth manifolds, with the same dimension, both isometrically embedded in Euclidean space $\mathbb{R}^n$. Rolling maps describe how M rolls upon N, without slip or twist, along a curve α on M. Rolling is a rigid motion in the embedding space, subject to holonomic and non-holonomic constraints. A rolling motion is then described by the action of the isometry group on $\mathbb{R}^n$, preserving orientations. This is the special Euclidean group SE_n, the semi-direct product $\mathrm{SO}_n \ltimes \mathbb{R}^n = \{X = (R, s),\ R \in \mathrm{SO}_n,\ s \in \mathbb{R}^n\}$, with group operations

$$(R_1, s_1) \circ (R_2, s_2) = (R_1R_2, R_1s_2 + s_1),$$
$$(R, s)^{-1} = (R^{-1}, -R^{-1}s),$$

and the action on $\mathbb{R}^n$ is defined as

$$\begin{aligned} \mathrm{SE}_n \times \mathbb{R}^n &\to \mathbb{R}^n \\ (X, p) &\mapsto X(p) = Rp + s. \end{aligned}$$

We adopt the definition of a rolling map given in Sharpe [13] and write some of the constraints in terms of R and s.

Definition 1 A rolling map of M upon N, without slip or twist, along a piecewise smooth curve $\alpha : [0, t_1] \to M$ is a piecewise smooth mapping

$$\begin{aligned} X : [0, t_1] &\to \mathrm{SE}_n = \mathrm{SO}_n \ltimes \mathbb{R}^n \\ t &\mapsto X(t) = (R(t), s(t)) \end{aligned} \tag{1}$$

satisfying the following conditions:

- Rolling conditions (for all $t \in [0, t_1]$)
  - $X(t)(\alpha(t)) =: \overline{\alpha}(t) \in N$.
  - $T_{X(t)(\alpha(t))}(X(t)(M)) = T_{\overline{\alpha}(t)}N$.
- No-slip condition (for almost all $t \in [0, t_1]$):

$$\dot{\overline{\alpha}}(t) = R(t)(\dot{\alpha}(t))$$

- No-twist conditions (for almost all $t \in [0, t_1]$):
  - Tangential part: $\dot{R}(t)R^\top(t)\,(T_{\overline{\alpha}(t)}N) \subset (T_{\overline{\alpha}(t)}N)^\perp$.
  - Normal part: $\dot{R}(t)R^\top(t)\,(T_{\overline{\alpha}(t)}N)^\perp \subset T_{\overline{\alpha}(t)}N$.

The curve α on M is called the *rolling curve* and $\overline{\alpha}$ is called the *development* of α on N. The rolling conditions in the definition above are *holonomic constraints*, they correspond to admissible configurations of the two manifolds, while the non-slip and non-twist conditions are *non-holonomic constraints*. The second, normal part of the no-twist conditions is always satisfied for manifolds of co-dimension 1. For the most classical of all rolling motions: the 2-sphere rolling on the tangent space at the south pole, the admissible configurations are all positions of the sphere in which it is tangent to the plane, while the non-holonomic constraints forbid any pure translation and any rotation around an axis orthogonal to the plane.

Remark 1 It has been proven in Sharpe [13] that for each piecewise smooth curve α on M there exists a unique rolling map having α as its rolling curve. In the situation when $M = N$, the rolling map reduces to the identity map and the development curve coincides with the rolling curve. Also, if the rolling curve α belongs to the intersection of the two manifolds, then the corresponding rolling map reduces to the identity ($X(t) = (I, 0)$ satisfies all the conditions trivially) and $\alpha \equiv \overline{\alpha}$.

In what follows, if $X = X(t)$ is defined as in (1), X_* stands for the tangent map of X and X^{-1} stands for the mapping

$$\begin{aligned} X^{-1} : [0, t_1] &\to \mathrm{SE}_n = \mathrm{SO}_n \ltimes \mathbb{R}^n \\ t &\mapsto X^{-1}(t) = (R^{-1}(t), -R^{-1}(t)s(t)) \end{aligned},$$

2.1 *Properties of Rolling Motions*

The following properties can easily be proven using Definition 1 and are of particular importance for our purposes. The first two have also been derived in Sharpe [13]. Suppose that three manifolds M_1, M_2 and M_3, embedded in Euclidean space, are tangent to each other at a point $p \in M_1 \cap M_2 \cap M_3$ and that $t \mapsto \alpha_1(t)$ is a curve in M_1 satisfying $\alpha_1(0) = p$.

1. **Rolling motions are transitive**
   Suppose that M_1 rolls on M_2 with rolling map X_1, rolling curve α_1, and development curve α_2. Also suppose that M_2 rolls on M_3 with rolling map X_2, rolling curve α_2, and development curve α_3. Then M_1 rolls on M_3 with rolling map $X_2 \circ X_1$, rolling curve α_1, and development curve α_3.
2. **Rolling motions are symmetric**
   Suppose that M_1 rolls on M_2 with rolling map X_1, rolling curve α_1, and development curve α_2. Then M_2 rolls on M_1 with rolling map ${X_1}^{-1}$, rolling curve α_2, and development curve α_1.
3. **Rolling under a change of coordinates**
   If M_1 rolls on M_2 with rolling map X_1, rolling curve α_1, and development curve $\overline{\alpha}_1$ and $X_c \in \mathrm{SE}_n$ is a fixed isometry, then $X_c(M_1)$ rolls on $X_c(M_2)$ with rolling map $X_c \circ X_1 \circ {X_c}^{-1}$, rolling curve $X_c(\alpha)$ and development curve $X_c(\overline{\alpha})$.

3 Kinematic Equations of Rolling

In this section we derive the kinematic equations for the motion of a smooth manifold rolling on the affine tangent space at a point. At first glance this may seem to be very restrictive. However, due to the definition of rolling and consequent properties, the results for this particular situation are the key to study more general rolling problems, as will be illustrated later for a sphere rolling on another sphere.

Assume that M is rolling on the affine tangent space at a point, i.e. $N = T^{\mathrm{aff}}_{p_0}M$, where $p_0 = \alpha(0) = \overline{\alpha}(0)$. The kinematic equations describe the translational and the rotational velocities of the rolling motion, starting from $(R(0), s(0)) = (I, 0)$, the identity of SE_n, and so they have the form

$$\begin{cases} \dot{s}(t) = u(t) \\ \dot{R}(t) = A(t)\,R(t) \end{cases},$$

for some vector valued function u taking values in $\mathbb{R}^n$ and A taking values in $\mathfrak{so}_n$ (the Lie algebra of SO_n, consisting of the skew symmetric matrices). Conditions on these functions are determined from the holonomic and non-holonomic constraints.

When SO_n leaves M invariant, the rolling curve is always of the form $\alpha(t) = R(t)^\top p_0$, for some $R(t) \in \mathrm{SO}_n$. Under this assumption, the first rolling condition implies that $s(t) = \overline{\alpha}(t) - p_0 \in T_{p_0}M$ and, consequently, the no-slip condition becomes

$$\dot{s}(t) = -A(t)\,p_0.$$

On the other hand, the structure of $A(t) = \dot{R}(t)R^\top(t) \in \mathfrak{so}_n$ is determined from the no-twist conditions

$$
\begin{aligned}
A(t)\ T_{\overline{\alpha}(t)}N &\subset (T_{\overline{\alpha}(t)}N)^\perp, \\
A(t)\ (T_{\overline{\alpha}(t)}N)^\perp &\subset T_{\overline{\alpha}(t)}N.
\end{aligned}
$$

Consequently, for an appropriate choice of coordinates, the matrix function A has the following structure

$$
A(t) = \left[\begin{array}{c|c} 0 & b(t) \\ \hline -b^\top(t) & 0 \end{array}\right], \tag{2}
$$

where $b(t) \in \mathbb{R}^{m\times(n-m)}$. We can now write the kinematic equations for rolling the manifold M upon $N = T^{\text{aff}}_{p_0}M$:

$$
\begin{cases} \dot{s}(t) &= -A(t)\, p_0 \\ \dot{R}(t) &= A(t)\, R(t) \end{cases}, \tag{3}
$$

where $A(t)$ has the structure (2).

Remark 2 When M is the $(n-1)$-sphere S centered at the origin, with radius ρ, and p_0 is its south or north pole, then

$$
b(t) = \begin{bmatrix} u_1(t) \\ \vdots \\ u_{n-1}(t) \end{bmatrix}, \quad A(t) = \textstyle\sum_{i=1}^{n-1} u_i(t)A_{i,n},
$$

and Eqs. (3) for rolling S on its affine tangent space at p_0 reduce to the well know (see, for instance, [6]) kinematic equations

$$
\begin{cases} \dot{s}(t) = \varepsilon\rho u(t) \\ \dot{R}(t) = \left(\sum_{i=1}^{n-1} u_i(t)A_{i,n}\right) R(t) \end{cases},
$$

where $A_{i,j} = e_i e_j^\top - e_j e_i^\top$ are *the elementary skew symmetric matrices* , $\varepsilon = 1$ if p_0 is the south pole and $\varepsilon = -1$ if p_0 is the north pole. In this case, the rolling condition $X(t)(\alpha(t)) = \overline{\alpha}(t)$, where $X = (R, s)$, reduces to

$$
R(t)\alpha(t) = p_0. \tag{4}
$$

Remark 3 If the sphere is not centered at the origin, the kinematic equations can be easily derived from the above, using a convenient change of coordinates. For instance, consider the following case which will be useful later. Let M be a sphere of radius ρ, centered at the point $(0, \cdots, 0, a)^\top$. We can obtain the rolling map for the rolling motion of M on its affine tangent space at the north pole $p_0 = (0, \cdots, 0, a + \rho)^\top$ from the rolling map $X = (R, s)$ of the sphere S in previous remark and the isometry $X_\tau = (I, \tau) \in \mathrm{SE}_n$, where $\tau = (0, \cdots, 0, a)^\top$. In this case, X_τ is a pure translation and the translation vector τ sends S to $M = S + \tau$. In this situation,

$$X_\tau \circ X \circ X_\tau^{-1} = (R, -R\tau + s + \tau)$$

is the rolling map for the rolling motion of M upon its affine tangent space at the point p_0, with rolling curve $\alpha + \tau$ and development $\overline{\alpha} + \tau$.

4 A Sphere Rolling on Another Sphere

The most classical of all non-holonomic problems is that of a sphere rolling on its affine tangent space at a point. Other rolling spheres problems, including a sphere rolling on another sphere, have been studied in the literature using different approaches, but in most cases only for 2-dimensional spheres. We refer, for instance, the work of Bor and Montgomery [2], exploiting the configuration space and the connection with octonions, the work of Jurdjevic included in [6], and more recently the results of Jurdjevic and Zimmerman in [7] and of Bloch and Rojo in [1].

The kinematic equations for rolling a sphere on another sphere have been derived in [7], but we present here an alternative approach which uses the kinematic equations of a sphere rolling on its affine tangent space at a point together with the transitive and symmetric properties of rolling. This approach may be used with great success for other manifolds, as long as one knows how to roll each one on the affine tangent space at a point. And since this rolling is easier to generate, due to the simplicity of the latter space, the proposed approach provides a simpler alternative to derive the kinematic equations of rolling general manifolds embedded in Euclidean space.

We consider two spheres of the same dimension $n-1$, embedded in the Euclidean space $\mathbb{R}^n$: S_1 with radius ρ_1 and S_2 with radius ρ_2. Suppose that the sphere S_2 is centered at the origin and is stationary. Assume that S_1 is centered at the point $c = (0, \cdots, 0, -(\rho_1 + \rho_2))^\top$, so that at time $t = 0$ it is tangent to S_2 at the south pole of S_2, $p_0 = (0, \cdots, 0, -\rho_2)^\top$. Assume now that S_1 starts rolling over S_2, along a piecewise smooth curve α, satisfying $\alpha(0) = p_0$.

Our objective is to derive the kinematic equations for the rolling motion of S_1 on the outside of the stationary sphere S_2. This will be accomplished by using the kinematic equations derived in Sect. 3, for rolling a manifold on the affine tangent space at a point, together with the symmetric and transitive properties and remarks contained in Sect. 2.

Let N denote the affine tangent space to S_2 at p_0, which also coincides with the affine tangent space to S_1 at the same point. We know how to roll the spheres S_1

and S_2 on N. Consequently, we know how to roll S_1 on N and N on S_2. Thus, by transitivity, we can achieve our goal.

- **Rolling S_1 over $N = T_{p_0}^{\text{aff}} S_1$:**
  For a sphere with radius ρ_1 centered at the origin and rolling on the affine tangent space at the north pole q_0, the kinematic equations are

$$\begin{cases} \dot{s} = -A(t)\, q_0 \\ \dot{R} = A(t)\, R \end{cases}, \tag{5}$$

where $A(t) = \sum_{i=1}^{n-1} u_i(t) A_{i,n}$, for some scalar functions $u_1, \cdots, u_{n-1}$. Also, from (4), the rolling curve α satisfies

$$R(t)\alpha(t) = q_0. \tag{6}$$

So, according to Remark 3, the rolling map for S_1 over N is defined by $X_1 = (R_1, s_1) = (R, -R\tau + s + \tau)$, where τ is the translation vector $(0, \cdots, 0, -(\rho_1 + \rho_2))^\top$ with kinematic equations

$$\begin{cases} \dot{s}_1 = -A_1(t)\, (q_0 + R_1 \tau) \\ \dot{R}_1 = A_1(t)\, R_1 \end{cases}, \tag{7}$$

where $A_1 \equiv A$, having rolling curve $\alpha_1 = \alpha + \tau$ and development curve $\overline{\alpha_1} = \overline{\alpha} + \tau$. It follows from (6) that

$$R\alpha + \tau = p_0. \tag{8}$$

- **Rolling S_2 over $N = T_{p_0}^{\text{aff}} S_2$:**
  The sphere S_2 is centered at the origin and has radius ρ_2. So, ($X_2 = (R_2, s_2)$ is the rolling map for rolling S_2 over the affine tangent space at the south pole p_0, and the corresponding kinematic equations are given by:

$$\begin{cases} \dot{s}_2 = -A_2(t)\, p_0 \\ \dot{R}_2 = A_2(t)\, R_2 \end{cases}, \tag{9}$$

with $A_2(t) = \sum_{i=1}^{n-1} v_i(t) A_{i,n}$, for some scalar functions $v_1, \cdots, v_m$. Moreover, the rolling curve α_2 satisfies

$$R_2 \alpha_2 = p_0. \tag{10}$$

For our purpose, we assume that the development curve $\overline{\alpha_2}$ coincides with $\overline{\alpha_1}$. According to the symmetric property of rolling in Sect. 2, N rolls upon S_2 with rolling map $X_2 = (R_2, -R_2{}^\top s_2)$, rolling curve $\overline{\alpha_1} \in N$ and development curve $\overline{\alpha_2} \in S_2$.

- **Rolling S_1 over S_2:**
  Applying now the transitive property of rolling in Sect. 2, with $M_1 = S_1$, $M_2 = N$ and $M_3 = S_2$, we conclude the following: S_1 rolls upon S_2 with rolling map $X_3 = X_2^{-1} \circ X_1 = (R_2^\top R_1, R_2^\top(s_1 - s_2))$, rolling curve $\alpha_1 \in S_1$ and development curve $\alpha_2 \in S_2$, so that

$$X_3(\alpha_1) = \alpha_2. \tag{11}$$

We now show that, under the assumption

$$\overline{\alpha_2} = \overline{\alpha_1}, \tag{12}$$

the matrices A_1 and A_2 in (7) and (9) respectively, are related through

$$A_2 = -\frac{\rho_1}{\rho_2} A_1. \tag{13}$$

This is a consequence of the following simple calculations, where the conditions (8) and (10) are used.

$$\begin{aligned}
& X_3(\alpha_1) = \alpha_2 \\
\Leftrightarrow\ & R_2^\top R_1 \alpha_1 + R_2^\top (s_1 - s_2) = \alpha_2 \\
\Leftrightarrow\ & R_2^\top R_1 \alpha_1 + R_2^\top (s_1 - s_2) = R_2^\top p_0 \\
\Leftrightarrow\ & R_1 \alpha_1 + s_1 - s_2 = p_0 \\
\Leftrightarrow\ & R_1 \alpha + R_1 \tau + s_1 - s_2 = p_0 \\
\Leftrightarrow\ & p_0 - \tau + R_1 \tau + s_1 - s_2 = p_0 \\
\Leftrightarrow\ & s_1 - s_2 = \tau - R_1 \tau.
\end{aligned}$$

Consequently,

$$\dot{s}_2 - \dot{s}_1 = \dot{R}_1 \tau = A_1 R_1 \tau. \tag{14}$$

On the other hand, using the kinematic equations (7) and (9), we have

$$\dot{s}_2 - \dot{s}_1 = -A_2 p_0 + A_1 q_0 + A_1 R_1 \tau, \tag{15}$$

and by comparison of (14) and (15), it follows that

$$A_1 q_0 = A_2 p_0. \tag{16}$$

Finally, the relationship $A_2 = \frac{\rho_1}{\rho_2} A_1$ follows from here, taking into account the particular structure of the matrices A_1 and A_2 and the fact that $p_0 = (0 \cdots, 0, -\rho_2)^\top$ and $q_0 = (0 \cdots, 0, \rho_1)^\top$. In conclusion, we may state the following.

Theorem 1 *Suppose that S_1 starts rolling over S_2 without slip or twist along a curve α_1 satisfying $\alpha_1(0) = p_0$. Then, the corresponding rolling map is given by*

$$X_3 = ({R_2}^\top R_1, {R_2}^\top (s_1 - s_2)),$$

where s_1, s_2, R_1 and R_2 are the solutions of the following differential equations

$$\begin{cases} \dot{s}_1 &= -U(t)(p_0 - \tau + R_1\tau) \\ \dot{s}_2 &= -U(t)q_0 \\ \dot{R}_1 &= +U(t)R_1 \\ \dot{R}_2 &= -\frac{\rho_1}{\rho_2}U(t)R_2 \end{cases}, \tag{17}$$

satisfying $s_1(0) = s_2(0) = 0$ and $R_1(0)R_2(0) = I$, where

$$U(t) = \left[\begin{array}{c|c} 0 & u(t) \\ \hline -u^\top(t) & 0 \end{array}\right],$$

for some vector function u depending on the rolling curve α_1. Moreover, along the rolling motion, the point of contact p_0 traces out the curve $\alpha_2 = {R_2}^\top p_0$ on S_2.

It is straight forward to conclude from the above relations that $u(t) = -\frac{1}{\rho_1}\dot{\overline{\alpha_1}}$. Clearly u is a constant function if and only if $\overline{\alpha_1}$ is a geodesic on N. It is well known that the development of a geodesic curve is a geodesic. (This is an immediate consequence of the fact, proved in Sharpe [13, Sect. B.3], that if the development is a straight line, then the rolling curve is a geodesic.) So, the case when u is constant corresponds to the situation when the rolling curve α_1 is a geodesic on S_1 and, consequently, its development α_2 is also a geodesic on S_2.

With appropriate changes in notation, Eqs. (17) are in accordance with Proposition 2.3 in [7].

Remark 4 The kinematics for the related situation where S_1 of radius ρ_1 rolls inside S_2 of radius $\rho_2 > \rho_1$ are obtained by centering S_1 at $(0, \cdots, 0, -(\rho_2 - \rho_1))^\top$ and replacing q_0 by $(0, \cdots, 0, -\rho_1)^\top$ and τ by $(0, \cdots, 0, -(\rho_2 - \rho_1))^\top$.

5 Constructive Proof of Controllability

As before, S_1, S_2 are fixed $(n-1)$-spheres in $\mathbb{R}^n$, of radii ρ_1, ρ_2, tangent at $p_0 \in S_1 \cap S_2$, with S_1 outside S_2. Exchanging the spheres if necessary, assume $\rho_1 < \rho_2$. Rescaling, we may and will take $\rho_2 = 1$ and put $\gamma := \rho_1 < 1$.

Following Sharpe [13], we take for state space of the system where S_1 rolls on the outside of S_2 the connected component Σ of

$$\{(q, p, M) \in S_1 \times S_2 \times \mathrm{SO}_n : M\ T_qS_1 = T_pS_2\}$$

containing the trivial configuration (p_0, p_0, I). A *state space rolling motion* is any curve $\sigma : [0, t_1] \to \Sigma$ determined by a rolling motion $X = X(t) = (R, s)$ of S_1 over S_2 as follows: if X has rolling curve $\alpha_1(t)$ and development $\alpha_2(t)$, then

$$\sigma(t) = (\alpha_1(t), \alpha_2(t), R(t)).$$

In this situation, $\sigma(t_1)$ *is reachable from* $\sigma(0)$. The rolling system is *controllable* if σ_1 is reachable from σ_0, for every $\sigma_0, \sigma_1 \in \Sigma$.

A fixed Euclidean motion $X_1 \in \mathrm{SE}_n$ is *achievable* if $(p_0, X_1(p_0), X_1)$ is in Σ and is a state reachable from (p_0, p_0, I).

It is well known that the system is controllable when the spheres have unequal radii. We prove this by first showing that certain infinitesimally forbidden motions are achievable and then checking that any state transfer is achieved by a composition of such motions. This is in the spirit of [8], where a similar strategy was carried out for the n-sphere rolling on a hyperplane. The forbidden motions used are the twists and slips defined below. We will exhibit explicit piecewise geodesic rolling motions that achieve each of those motions.

When $n = 3$, the spheres are two-dimensional. The no-twist condition prevents rotations of S_1 about the axis $\overline{0p_0}$ (*twists at* p_0), while the no-slip condition forbids slipping motions which may be thought of as rotations of S_1 about an axis through the center of S_2 (the origin) and perpendicular to $\overline{0p_0}$ (these we term *slips from* p_0). In Sect. 5.1 we will show that these motions are achievable. Next, in Sect. 5.2, we will define the higher-dimensional analogues of twists and show that those are achievable as well. Finally, in the last section we establish controllability for all $n \geqslant 3$, carrying out the plan sketched above.

It will be convenient to re-parametrize system (17) so that t is the arclength of the rolling and development curves:

$$\begin{cases} s_1 - s_2 = (I - R_1)\tau \\ \dot{R}_1 \quad = +\frac{1}{\gamma} U(t) R_1 \\ \dot{R}_2 \quad = -U(t) R_2 \end{cases}. \tag{18}$$

For coordinates in which S_1 is centered at $c = (1 + \gamma)\, p_0$ and S_2 is centered at the origin, as before, but now p_0 is be the north pole of S_2, the kinematics (18) keep their form. We will use such coordinates in the following two sections.

5.1 Miming Twists and Slips for n = 3

Now $p_0 = (0, 0, 1)^\top$. With $A_{i,j}$ as in Remark 2, let $A_y := A_{1,3},\ A_x := A_{2,3},\ A_z := A_{1,2}$, and $A(\theta) := A_y \cos\theta + A_x \sin\theta$. Since $\frac{d}{dt}\big|_{t=0} \left(e^{tA(\theta)} p_0\right) = (\cos\theta, \sin\theta, 0)$, the rotation matrix $e^{tA(\theta)}$ moves p_0 in the direction with angle θ in the tangent space to S_2 at p_0, identified with the xy-plane. In this situation a twist at p_0 is the motion $\left(e^{\alpha A_z}, 0\right)$

and a slip from p_0 is $\left(e^{tA(\theta)}, 0\right)$. These correspond to the state transfers $(p_0, p_0, M) \to (p_0, p_0, e^{\alpha A_z} M)$ and $(p_0, p_0, M) \to (e^{tA(\theta)} p_0, e^{tA(\theta)} p_0, e^{tA(\theta)} M)$, respectively.

Note that if a constant control

$$U = \left[\begin{array}{cc|c} 0 & 0 & u_1 \\ 0 & 0 & u_2 \\ \hline -u_1 & -u_2 & 0 \end{array}\right]$$

has norm one ($u_1^2 + u_2^2 = 1$), then $U = A(\theta)$ for some θ and $e^{2\pi U} = I$.

A *half-tumble* is a rolling motion of S_1 over S_2 corresponding to a geodesic development with length $\pi\gamma$. If a rolling motion starting from p_0 is a sequence of two half-tumbles, then its development is $\alpha_2(t)$ with $\alpha_2(0) = p_0$, $\alpha_2(\pi\gamma) = e^{\pi\gamma V_1} p_0$, $\alpha_2(2\pi\gamma) = e^{\pi\gamma V_2} e^{\pi\gamma V_1} p_0$, for skew symmetric V_1, V_2 which relate to the control values by $U_1 = A(\theta_1) = V_1$, $U_2 = A(\theta_2) = e^{-\pi\gamma V_1} V_2 e^{\pi\gamma V_1}$. It is easy to check that if the angle between those two geodesic arcs is β (measured such that $\beta = \frac{\pi}{2}$ is a left-turn), then $\theta_2 = \pi + \theta_1 - \beta$.

By integrating (18),

$$R_1(2\pi\gamma) = e^{\pi A(\theta_2)} e^{\pi A(\theta_1)}$$

$$R_2(2\pi\gamma) = e^{\pi\gamma A(\theta_1)} e^{\pi\gamma A(\theta_2)}.$$

Therefore,

$$X_3(2\pi\gamma) = (R_3, s_3) = (R_2^\top R_1, 0). \tag{19}$$

A *tumble* is a sequence of two half-tumbles with the same control input $U = A(\theta)$, for which the rolling motion X_3 satisfies $X_3(2\pi\gamma) =, \left(e^{2\pi\gamma A(\theta)}, 0\right)$.

Proposition 1 *In dimension* 3, *any twist is achievable.*

Proof We first handle the case $0 < \gamma < \frac{1}{2}$. Given any $\alpha \in (0, \pi/2)$, consider a spherical quadrangle $[p_0, A, B, C]$ on S_2 with interior angle 2α both at p_0 and at the opposite vertex B and each of whose four arcs has the same length $T = \pi\gamma$. Note that $0 < T < \pi/2$. For definiteness, the vertices are labeled counter-clockwise, as seen from the outside of S_2 (Fig. 1).

We compute the angle β at A and C. The spherical triangle $[p_0, A, B]$ has interior angles α, β, and again α at those vertices. Let W be the arc-angle of $p_0 B$. By the law of sines of spherical trigonometry,

$$\frac{\sin\alpha}{\sin T} = \frac{\sin\beta}{\sin W} \tag{20}$$

and by the law of cosines,

$$\cos W = \cos^2 T + \sin^2 T \cos\beta. \tag{21}$$

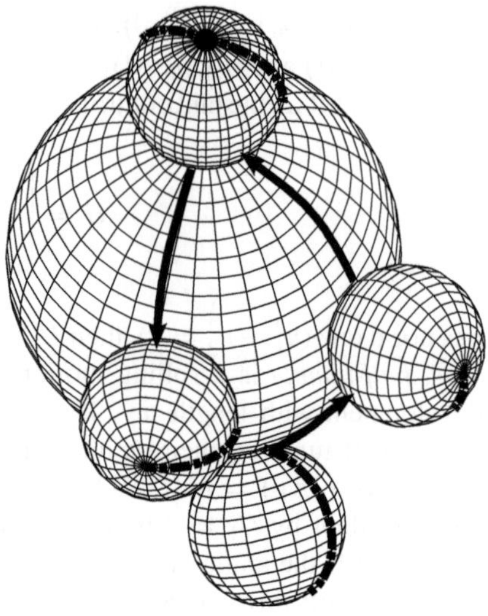

Fig. 1 A twist obtained by rolling

For convenience, put $\Gamma = \cos(T)$, so that $0 < \Gamma < 1$. Then $\cos W = \Gamma^2 + (1 - \Gamma^2)\cos\beta$. The previous relations imply

$$\cos\beta = \frac{\Gamma^2 \sin^2\alpha - \cos^2\alpha}{\Gamma^2 \sin^2\alpha + \cos^2\alpha}, \tag{22}$$

which uniquely defines $\beta \in [0, \pi]$. (Note that, for fixed Γ, $\beta(\alpha, \Gamma)$ decreases from π to 0 as α increases from 0 to $\pi/2$.)

The two left arcs form a development curve with $\theta_1 = -\alpha$, $\theta_2 = \pi - \alpha + \beta$. Using (19), the rolling motion X_{12} from p_0 along the two left arcs is $(R_{12}, 0)$ at $t = 2T$, with

$$\begin{aligned} R_{12} &= e^{TA(\theta_1)} e^{(\pi+T)A(\theta_2)} e^{\pi A(\theta_1)} \\ &= e^{TA(-\alpha)} e^{(\pi+T)A(\pi-\alpha+\beta)} e^{\pi A(-\alpha)}. \end{aligned}$$

Analogously, the two right arcs are a development with $\theta_3 = \alpha$, $\theta_4 = \pi + \alpha - \beta$ and at $t = 2T$, the rolling motion from p_0 is $(R_{34}, 0)$

$$\begin{aligned} R_{34} &= e^{TA(\theta_3)} e^{(\pi+T)A(\theta_4)} e^{\pi A(\theta_3)} \\ &= e^{TA(\alpha)} e^{(\pi+T)A(\pi+\alpha-\beta)} e^{\pi A(\alpha)}. \end{aligned}$$

Therefore, the rolling motion around the closed curve is, at $t = 4T$, $(R_{1234}, s) = (R_{34}^{-1}, 0)\,(R_{12}, 0) = (R_{34}^{-1} R_{12}, 0)$.

It is shown by elementary means in [10] that $R_{1234} = e^{(-4\alpha+2\beta)A_z}$. Alternatively, this may be seen by using the Gauss-Bonnet theorem: when a 2-sphere rolls on a plane along a closed rolling curve ψ, it undergoes a turn by an angle equal to the area enclosed by ψ. The area enclosed by the four-sided development curve is $A = 4\alpha + 2\beta - \pi$, so that S_2 is turned by that angle relative to the plane N on which both spheres roll. On the other hand, the rolling curve on S_1 is a lune of area 2β twice traversed, so that S_2 turns by 4β relative to N. The difference gives the angle of the twist. See, for example Murray and Sastry [12, p. 385].

Using (22), and fixed γ, define $f(\alpha) = 2\alpha - \beta$. One checks that $f(0) = \pi$, $f(\frac{\pi}{2}) = -\pi$. This establishes that finitely many half-tumbles suffice to achieve any twist at p_0.

If $\frac{1}{2} < \gamma < 1$, we consider a similar quadrangle as before, but now with $T = \pi(1-\gamma)$. Consider a rolling motion that traverses the same sequence of points (p_0, A, B, C, p_0), but now moving along the longer portion (of length $\tilde{T} = 2\pi - T$) of the great circle containing each successive pair of points. Since $e^{\tilde{T}A(\theta+\pi)} = e^{-TA(\theta+\pi)} = e^{TA(\theta)}$, again $X_3\left(4\tilde{T}\right) = (e^{(-4\alpha+2\beta)A_z}, 0)$.

Finally, when $\gamma = \frac{1}{2}$, it is enough to take for development a spherical lune with the poles as endpoints.

Proposition 2 *In dimension* 3, *any slip is achievable.*

Proof Suppose the slip to be constructed is the motion $(S, 0) = (e^{WA(\theta)}, 0) \in \mathrm{SE}_n$, $0 < W < \pi$. Consider a spherical triangle with vertices $p_0, A, B = Sp_0$, so that the arcs $[p_0, A]$ and $[A, B]$ have length $T = \pi\gamma$. (See Fig. 2.) It is clear that the rolling motion along that development is a sequence of two half-tumbles and that if it is preceded by a suitable twist $(e^{\xi A_z}, 0)$ at p_0, then the net effect is a slip from p_0 to B. From the construction in the previous Proposition and by symmetry, $\xi = -2\alpha + \beta$. A pure slip is achieved.

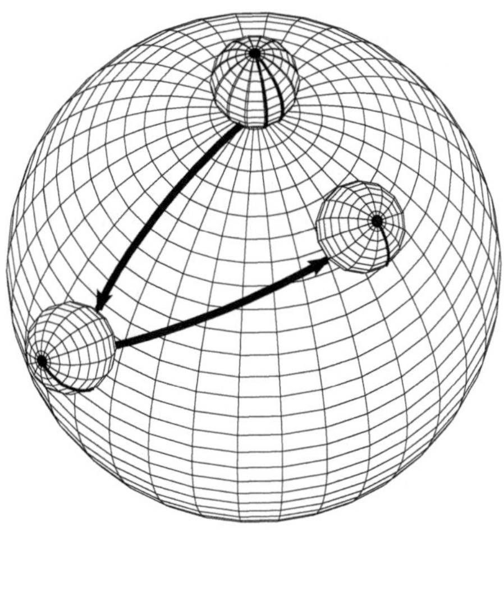

Fig. 2 A slip obtained by rolling

5.2 Miming Twists and Slips for $n \geqslant 3$

Let $n \geqslant 3$, so that S_1, S_2 are $(n-1)$-spheres in $\mathbb{R}^n$ and $S_1 \cap S_2 = \{p_0\}$, as before. We define the higher-dimension analogues of the forbidden motions of S_1 slipping from p_0 and S_1 twisting at p_0. Let $L = \text{span}\{p_0\}$. A *twist at* p_0 is $(\exp M, 0) \in \text{SE}_n$, where $M \in \mathfrak{so}_n$ and $Mp_0 = 0$. A *slip from* p_0 is $(\exp N, 0) \in \text{SE}_n$ with $N \in \mathfrak{so}_n$, $N(L) \subset L^{\perp}$, and $N\left(L^{\perp}\right) \subset L$.

Under our chosen coordinates, $p_0 = (0, \ldots, 0, 1)^{\top} \in \mathbb{R}^n$ and the requirements that $(\exp M, 0)$ be a twist at p_0 and $(\exp N, 0) \in \text{SE}_n$ be a slip from p_0 are simply that

$$M = \left[\begin{array}{c|c} \tilde{M} & 0 \\ \hline 0 & 0 \end{array}\right], \quad N = \left[\begin{array}{c|c} 0 & b \\ \hline -b^{\top} & 0 \end{array}\right], \tag{23}$$

with $\tilde{M}_{(n-1)\times(n-1)}$ skew symmetric and $b \in \mathbb{R}^{n-1}$.

Proposition 3 *Any twist* $(\exp M, 0) \in \text{SE}_n$ *is achievable.*

Proof Recall that a *Givens rotation* is a matrix of the form $\exp\left(tA_{i,j}\right)$, where $A_{i,j} = e_i e_j^{\top} - e_j e_i^{\top}$ is an elementary skew symmetric matrix, as in Remark 2, and t is scalar. Here, $1 \leqslant i,j \leqslant n-1$ for each factor of the decomposition of $\tilde{M}$. Since $\tilde{M}$ is skew symmetric, $\exp\tilde{M}$ is orthogonal of determinant one and there is a constructive procedure to decompose it as a finite product of Givens rotations [3]. In order to obtain each twist $\left(e^{tA_{i,j}}, 0\right)$, it is enough to perform the maneuvers of the previous section using only the control input entries u_i and u_j.

Proposition 4 *Any slip* $(\exp N, 0) \in \text{SE}_n$ *is achievable.*

Proof Define the vector b by the second equality in (23). We claim that there are $n-2$ twists $(\exp M_i, 0)$, all at p_0, $3 \leqslant i \leqslant n$, for which, putting

$$K_n = \exp\left(M_n\right)\cdots\exp\left(M_3\right), \tag{24}$$

and $p = (0, \ldots, 0, t)^{\top} \in \mathbb{R}^{n-1}$, one has

$$N = \left[\begin{array}{c|c} 0 & b \\ \hline -b^{\top} & 0 \end{array}\right] = K_n \left[\begin{array}{c|c} 0 & p \\ \hline -p^{\top} & 0 \end{array}\right] K_n^{-1}.$$

We check this claim by induction on the dimension $n \geqslant 3$. The base case holds with $K_3 = (e^{\psi A_z}, 0)$ and $(\cos\psi, \sin\psi)\ |b| = b$. For the step, given $N_{(n+1)\times(n+1)}$, and thus $b \in \mathbb{R}^n$, choose a twist $(\exp M_{n+1}, 0)$,

$$M_{n+1} = \left[\begin{array}{c|c} \tilde{M}_{n+1} & 0 \\ \hline 0 & 0 \end{array}\right],$$

such that $\exp\left(\tilde{M}_{n+1}\right) b = c = (0, c_2, \ldots, c_n)^\top = (0|\tilde{c})^\top$. By the induction hypothesis, there is a finite product K_n as in (24) for which

$$\left[\begin{array}{c|c} 0 & c \\ \hline -c^\top & 0 \end{array}\right] = K_n \left[\begin{array}{c|c} 0 & p \\ \hline -p^\top & 0 \end{array}\right] K_n^{-1}$$

and now we check

$$\begin{aligned} &\exp\left(M_{n+1}\right) K_n \left[\begin{array}{c|c} 0 & p \\ \hline -p^\top & 0 \end{array}\right] K_n^{-1} \exp\left(-M_{n+1}\right) \\ &= \left[\begin{array}{c|c} \exp \tilde{M}_{n+1} & 0 \\ \hline 0 & 1 \end{array}\right] \left[\begin{array}{c|c} 0 & c \\ \hline -c^\top & 0 \end{array}\right] \left[\begin{array}{c|c} \exp\left(-\tilde{M}_{n+1}\right) & 0 \\ \hline 0 & 1 \end{array}\right] \\ &= \left[\begin{array}{c|c} 0 & b \\ \hline -c^\top & 0 \end{array}\right] \left[\begin{array}{c|c} \exp\left(-\tilde{M}_{n+1}\right) & 0 \\ \hline 0 & 1 \end{array}\right] = \left[\begin{array}{c|c} 0 & b \\ \hline -b^\top & 0 \end{array}\right], \end{aligned}$$

as required.

The slip

$$\left(\exp \left[\begin{array}{c|c} 0 & p \\ \hline -p^\top & 0 \end{array}\right], 0\right)$$

corresponds to the slip $\left(e^{tA_x}, 0\right)$ with respect to the last three variables and is therefore achievable, as shown in the previous section. From Proposition 3, all of the twists $(\exp M_i, 0)$ are achievable and thus so is the slip $(\exp N, 0)$.

Slips from p_0 allow for motion from p_0 to any other point on S_2, as we now show.

Proposition 5 *For any $q \in S_2$ there is a slip S such that $Sp_0 = q$.*

Proof Put $q = (\tilde{q}, q_n)^\top$ with q_n scalar. Let b relate to N as in (23) and suppose $|b| = 1$. From

$$\exp\left(\theta N\right) = I + \left(\cos\left(\theta\right) - 1\right) \left[\begin{array}{c|c} bb^\top & 0 \\ \hline 0 & 1 \end{array}\right] + \sin\left(\theta\right) N,$$

it follows that $\exp(\theta N) p_0 = (\tilde{q}, q_n)^\top \in S_2$ when θ and b satisfy $b \sin\theta = \tilde{q}$ and $q_n = \cos\theta$, that is, $S = \exp(\theta N)$.

5.3 *Proof of Controllability*

To establish controllability, it is enough to check that $(p_0, p_0, I) \in \Sigma$ is reachable from arbitrary $\sigma \in \Sigma$, since the system is symmetric.

From $\sigma = (q, p, M)$ one reaches (p_0, p', M') by a rolling motion for which the rolling curve is a geodesic connecting q to p_0 in S_1. By a slip of S_1 from p' to p_0 one reaches (p_0, p_0, M'') and if this is in Σ, then $(M'', 0) \in \mathrm{SE}_n$ must be a twist at p_0 and therefore from (p_0, p_0, M'') one may reach (p_0, p_0, I).

Acknowledgements The work of the first author was supported by FCT project PTDC/ EEA–CRO/122812/2010.

References

1. Bloch, A., Rojo, A.: Kinematics of the rolling sphere and quantum spin. Commun. Inform. Syst **10**(4), 221–238 (2010)
2. Bor, G., Montgomery, R.: G_2 and the rolling distribution. L'Enseignement Mathématique. Revue Internationale. 2e Série **55**(1–2), 157–196 (2009)
3. Golub, G.H., Van Loan, C.F.: Matrix Computations (3rd edn.). Johns Hopkins University Press, Baltimore (1996)
4. Hüper, K., Leite, F.S.: On the geometry of rolling and interpolation curves on S^n, SO_n and Grassmann manifolds. J. Dyn. Control Syst. **13**(4), 467–502 (2007)
5. Hüper, K., Krakowski, K., Leite, F.S.: Rolling maps in a Riemannian framework. In: Cardoso, J., Hüper, K., Saraiva, P. (eds.) Textos de Matemática, vol. 43, pp. 15–30. Departamento de Matemática da Universidade de Coimbra, Coimbra (2011)
6. Jurdjevic, V.: Geometric Control Theory. Cambridge University Press, Cambridge (1997)
7. Jurdjevic, V., Zimmerman, J.A.: Rolling sphere problems on spaces of constant curvature. Math. Proc. Camb. Philos. Soc. **144**, 729–719 (2008)
8. Kleinsteuber, M., Hüper, K., Leite, F.S.: Complete controllability of the N-sphere: a constructive proof. In: Proceedings of 3rd IFAC Workshop on Lagrangian and Hamiltonian Methods for Nonlinear Control (LHMNLC'06), 19–21 July 2006, Nagoya, pp. 143–146 (2006)
9. Lee, J.M.: Riemannian Manifolds: An Introduction to Curvature. Graduate Texts in Mathematics, vol. 176. Springer, New York (1997)
10. Louro, F., Silva Leite, F.: Sphere rolling on sphere: alternative approach to kinematics and constructive proof of controllability. Departamento de Matemática da Universidade de Coimbra, pp. 12–25 (2012, preprint)
11. Montgomery, R.: A Tour of Subriemannian Geometries, their Geodesics and Applications. Mathematical Surveys and Monographs, vol. 91, p. 259. American Mathematical Society, Providence (2002)
12. Murray, R.M., Sastry, S.S., Li, Z.: A mathematical introduction to robotic manipulation. CRC Press, Boca Raton (1994)
13. Sharpe, R.W.: Differential Geometry. Springer, New York (1996)
14. Zimmerman, J.A.: Optimal control of the sphere S^n rolling on E^n. Math. Control Signals Syst. **17**, 14–37 (2005)

The Dual Potential, the Involution Kernel and Transport in Ergodic Optimization

A.O. Lopes, E.R. Oliveira, and Ph. Thieullen

Abstract Consider the shift σ acting on the Bernoulli space $\Sigma = \{1, 2, \ldots, n\}^{\mathbb{N}}$. We denote $\hat{\Sigma} = \{1, 2, \ldots, n\}^{\mathbb{Z}} = \Sigma \times \Sigma$. We analyze several properties of the maximizing probability $\mu_{\infty,A}$ of a Hölder potential $A : \Sigma \to \mathbb{R}$. Associated to $A(x)$, via the involution kernel, $W(x, y)$, $W : \hat{\Sigma} \to \mathbb{R}$, one can get the dual potential $A^*(y)$, where $(x, y) \in \hat{\Sigma}$. We denote μ_{∞,A^*} the maximizing probability for A^*. We would like to consider the transport problem from $\mu_{\infty,A}$ to μ_{∞,A^*}. In this case, it is natural to consider the cost function $c(x, y) = I(x) - W(x, y) + \gamma$, where I is the deviation function for $\mu_{\infty,A}$, as the limit of Gibbs probabilities $\mu_{\beta A}$ for the potential βA when $\beta \to \infty$. The value γ is a constant which depends on A. We could also take $c = -W$ above. We denote by $\mathscr{K} = \mathscr{K}(\mu_{\infty,A}, \mu_{\infty,A^*})$ the set of probabilities $\hat{\eta}(x, y)$ on $\hat{\Sigma}$, such that $\pi_x^*(\hat{\eta}) = \mu_{\infty,A}$, and $\pi_y^*(\hat{\eta}) = \mu_{\infty,A^*}$. We describe the minimal solution $\hat{\mu}$ (which is invariant by the shift on $\hat{\Sigma}$) of the Transport Problem, that is, the solution of

$$\inf_{\hat{\eta} \in \mathscr{K}} \int\int c(x, y)\, d\hat{\eta} = -\max_{\hat{\eta} \in \mathscr{K}} \int\int (W(x, y) - \gamma)\, d\hat{\eta}.$$

The optimal pair of functions for the Kantorovich Transport dual Problem is $(-V, -V^*)$, where we denote the two calibrated sub-actions by V and V^*, respectively, for A and A^*. We show that the involution kernel W is cyclically monotone. In other words, satisfies a twist condition in the support of $\hat{\mu}$. We analyze the question: is the support of $\hat{\mu}$ a graph? We also investigate the question of finding an explicit expression for the function $f : \Sigma \to \mathbb{R}$ whose c-subderivative determines the graph. We also analyze the same kind of problem for expanding transformations on the circle.

A.O. Lopes • E.R. Oliveira (✉)
Instituto de Matemática-UFRGS, Avenida Bento Gonçalves, 9500 Porto Alegre, RS, Brazil
e-mail: alopes@mat.ufrgs.br; oliveira.elismar@gmail.com

Ph. Thieullen
Institut de Mathématiques, Université Bordeaux, 33405 Talence, France
e-mail: philippe.thieullen@math.u-bordeaux1.fr

J.-P. Bourguignon et al. (eds.), *Dynamics, Games and Science*, CIM Series in Mathematical Sciences 1, DOI 10.1007/978-3-319-16118-1_20

1 Introduction

It seems natural to try to investigate the connections of Transport Theory with Ergodic Theory. Some results on this direction appear in [18, 32–34, 51]. Here we follow a different path.

Given a continuous function $A : \Sigma = \{1, 2, 3, .., d\}^{\mathbb{N}} \to \mathbb{R}$, we call $\mu_{\infty,A}$ a maximizing probability for A, if $\int A d\nu$ attains the maximal value in $\mu_{\infty,A}$, when the probabilities ν range among the set of invariant for the shift acting on the Bernoulli space Σ. We denote by $m(A)$ this maximal value.

Such maximizing probabilities $\mu_{\infty,A}$ can be seen as the equilibrium states at zero temperature for a system on the one dimensional lattice $\mathbb{N}$ with d spins in each site and under the influence of an interacting potential A (see [5, 8, 12, 14, 27, 35, 42, 46]).

A main conjecture on the area claims that for a generic Hölder potential A the maximizing probability has support in a unique periodic orbit for the shift (for a partial result see [12]). This conjecture was recently proved by G. Contreras (see [10]).

We address the question of finding the optimal transport plan from a certain maximizing probability to another. More precisely, we would like to consider the transport problem from $\mu_{\infty,A}$ to μ_{∞,A^*}, where $A : \Sigma = \{1, 2, 3, .., d\}^{\mathbb{N}} \to \mathbb{R}$ is a Hölder potential and A^* its dual (see [2]).

We consider here that A acts on the variable x and A^* in the variable y. A function $W(x, y)$ called the involution kernel will play an important role in the theory. The twist condition for W is a kind of convexity assumption. We will describe bellow with all details the setting we are going to consider in the present paper. We will also provide several examples to illustrate the theory.

We assume here in most (but not all) of the results that the maximizing probability $\mu_{\infty,A}$ (on Σ) for A is unique.

We denote by $\hat{\mu}$ the minimizing probability over $\hat{\Sigma} = \{1, 2, 3, .., d\}^{\mathbb{Z}} = \Sigma \times \Sigma$, for the natural Kantorovich Transport Problem associated to the $-W$, where $W(x, y)$, for $(x, y) \in \Sigma \times \Sigma$, is the involution kernel associated to A (see [2]).

We will denote by $\hat{\sigma}$ the shift on $\hat{\Sigma}$. The probability $\hat{\mu}_{max}$ denotes the natural extension of $\mu_{\infty,A}$ as described in [2].

We point out that by its very nature the Classical Transport Theory is not a Dynamical Theory (in the sense of considering invariant probabilities) [48, 53, 54]. One has to consider a cost which is obtained from dynamical properties in order to get optimal plans which are invariant for $\hat{\sigma}$.

Recent results in Ergodic Transport are [13, 22, 36, 37, 41, 44].

We will consider a cost which is the involution kernel W. First we show that:

Theorem 1 *The minimizing Kantorovich probability $\hat{\mu}$ on $\hat{\Sigma}$ associated to $-W$, where W is the involution kernel for A, is $\hat{\mu}_{max}$. Same property is true for c instead of W*

One of our main results is Theorem 5 which claims that the support of $\hat{\mu}_{max}$ is W-cyclically monotone. We do not assume the twist condition in the above result.

The calibrated subactions V play an important role in Ergodic Optimization. They can help to find the support of the maximizing probability (see [5, 27] or [12] for instance). Moreover, if we denote $R(x) = V(\sigma(x)) - V(x) - A(x) + m(A)$, then $I(x) = \sum_{n\geq 0} R(\sigma^n(x))$ defines a nonnegative lower semicontinuous function (can be infinite at several points) which is the deviation function for the family of Gibbs states associated to A when the temperature converges to zero [2] (see [3, 36] for the case of the XY model). For a class of explicit nontrivial examples of subactions V see [4].

Theorem 2 *If V is the calibrated subaction for A, and V^* is the calibrated subaction for A^*, then, the pair $(-V, -V^*)$ is the dual $(-W + I)$-Kantorovich pair of $(\mu_{\infty,A}, \mu_{\infty,A^*})$, when I is the deviation function for A.*

Finding the optimal transport measure between two probabilities is the solution of the so called relaxed problem [53]. If we want to find a measurable transformation (the Monge problem) which transfers one probability to another we need to show that the graph property is true in the support of such probability (which does not always happen if one considers a general cost function) [53].

Finally, we analyze later here the graph property for the support of the $\hat{\mu}_{max}$ (over $\hat{\Sigma} = \{1, 2, 3, .., d\}^{\mathbb{Z}}$) which is the minimizing probability for the cost function $-W$.

One can consider in the Bernoulli space $\Sigma = \{0, 1\}^{\mathbb{N}}$ the lexicographic order. In this way, $x < z$, if and only if, the first element i such that, $x_j = z_j$ for all $j < i$, and $x_i \neq z_i$, satisfies the property $x_i < z_i$. Moreover, $(0, x_1, x_2, \ldots) < (1, x_1, x_2, \ldots)$.

One can also consider the more general case $\Sigma = \{0, 1, \ldots, d-1\}^{\mathbb{N}}$, but in order to simplify the notation and to avoid technicalities, we consider only the case $\Sigma = \{0, 1\}^{\mathbb{N}}$.

Definition 1 We say a continuous $G : \hat{\Sigma} = \Sigma \times \Sigma \to \mathbb{R}$ satisfies the twist condition on $\hat{\Sigma}$, if for any $(a, b) \in \hat{\Sigma} = \Sigma \times \Sigma$ and $(a', b') \in \Sigma \times \Sigma$, with $a' > a$, $b' > b$, we have

$$G(a, b) + G(a', b') < G(a, b') + G(a', b). \quad (1)$$

The twist condition is inspired in the Aubry-Mather Theory [1, 11, 23–25]. It is a quite natural concept in Classical Optimization and Transport Theory [6, 13, 15, 40, 45, 48, 53, 54] (see [37] for dynamical examples).

The twist condition is also described by the concept of **global** cyclically monotonicity (see [53])

We point out that in Mather Theory in order to have the graph property (see [11, 43]) for the minimal action measure it is necessary to assume that Lagrangian is convex in the velocity. We need in our setting some technical assumptions to replace this important property. We believe that the twist condition is the natural one.

Definition 2 We say a continuous $A : \Sigma \to \mathbb{R}$ satisfies the twist condition, if its involution kernel W satisfies the twist condition.

The involution kernel of A is not unique (see [2]), but if the above property is true for some W, then it will also be true for any other one.

Our final result is:

Theorem 3 *Suppose the involution kernel W satisfies the twist condition on $\hat{\Sigma}$, then, the support of $\hat{\mu}_{max} = \hat{\mu}$ on $\hat{\Sigma}$ is a graph. Moreover, if $d = 2$, then there exists at most one point in the support of $\hat{\mu}$ which has two points in the support of $\hat{\mu}$ in its vertical fiber. The σ orbit of this point is a zero measure set.*

There are examples where the existence of this exceptional point occurs and this is associated to the concept of turning point (see [13, 37, 40]).

Similar results occur for the case of a general d. A similar definition can be considered for an expanding transformation on $[0, 1]$, and we are also able to get the analogous graph property result. This also includes the case of $T(x) = -2x$ (mod 1).

We present in the Appendix at the end of the paper several examples (and computations) where one can write the involution kernel W explicitly and the twist condition is satisfied. First we will explain all the preliminaries we will need later.

Consider X a compact metric space. Given a continuous transformation $f : X \to X$, we denote by $\mathcal{M}_f$ the convex set of f-invariant Borel probability measures. As usual, we consider in $\mathcal{M}_f$ the weak* topology. The standard model used in ergodic optimization is the triple $(X, f, \mathcal{M}_f)$. Given a potential $A \in C^0(X)$, we denote

$$m(A) = \max_{\nu \in \mathcal{M}_f} \int_X A(x)\, d\nu(x). \tag{2}$$

We are interested here in the characterization and main properties of A-maximizing probabilities, that is, the probabilities belonging to the set

$$\{ \mu \in \mathcal{M}_f : \int_X A(x)\, d\mu(x) = m(A) \}. \tag{3}$$

We will assume here that A is Hölder.

In the following we will also assume that the maximizing probability $\mu_{\infty,A} = \mu_\infty$ is unique.

Under reasonable hypothesis (expanding, hyperbolic, etc.) several results were obtained related to this maximizing question, among them [2, 5, 7–9, 12, 14, 23, 24, 26–28, 35, 38, 46, 50, 52]. For maximization with constraints see [20, 39]. Questions related to the dynamics on the boundary of the fat attractor appear in [37]. Naturally, if we change the maximizing notion for the minimizing one, the analogous properties will also be true.

Our focus here will be mainly on symbolic dynamics and on expanding transformations on S^1 or the interval $[0, 1]$. We recall some basic definitions (see [5] or [12] for example).

Let $\sigma : \Sigma \to \Sigma$ be a subshift of finite type defined by a matrix C of 0 and 1, where $\sigma(x_0, x_1, x_2, ..) = (x_1, x_2, x_3, ..)$. In this case we are considering $X = \Sigma =$

$\{1,2,3,..,d\}_C^{\mathbb{N}}$ and $f = \sigma$. Remind that, for a fixed $\lambda \in (0,1)$, we consider for Σ the metric $d(\mathbf{x}, \bar{\mathbf{x}}) = \lambda^k$, where $\mathbf{x} = (x_0, x_1, \ldots), \bar{\mathbf{x}} = (\bar{x}_0, \bar{x}_1, \ldots) \in \Sigma$ and $k = \min\{j : x_j \neq \bar{x}_j\}$. In this situation, given a Hölder potential $A : \{1,2,3,..,d\}^{\mathbb{N}} \to \mathbb{R}$, one should be interested in A-maximizing probabilities for the triple $(\Sigma, \sigma, \mathscr{M}_\sigma)$, where the probabilities are consider over $\mathscr{B}$, the σ-algebra of Borel of Σ. In order to simplify the notation here we will consider the full Bernoulli space (all entries of C are equal to 1).

Given an $C^{1+\alpha}$ expanding transformation T of fixed degree on S^1 and $A : S^1 \to \mathbb{R}$ we will be interested in A- maximizing probabilities on $(S^1, T, \mathscr{M}_T)$, where the probabilities are consider over $\mathscr{B}$, the σ-algebra of Borel of S^1.

One can consider the analogous setting for $C^{1+\alpha}$ expanding transformations of fixed degree over [0, 1].

Convex potentials $A : [0,1] \to \mathbb{R}$ and the transformation $T : [0,1] \to [0,1]$, given by $T(x) = 2x$ (mod 1), were considered in [29] where it was shown that the maximizing probabilities in this case are Sturmian measures. For $T(x)$ equal to $-2x$ (mod 1) however, the situation is completely different (see [31]).

Definition 3 A function $u \in C^0(\Sigma)$ is a sub-action for the potential A if, for any $\mathbf{x} \in \Sigma = \{1,2,3,\ldots,d\}_C^{\mathbb{N}}$, we have

$$u(\mathbf{x}) \leqslant u(\sigma(\mathbf{x})) - A(\mathbf{x}) + \beta_A. \tag{4}$$

Let (Σ^*, σ^*) be the dual subshift.

In the case of the full Bernoulli space (all entries of C equal 1) then $\Sigma^* = \{1,2,3,..,d\}^{\mathbb{N}}$ and $\sigma^*(y_0, y_1, y_2, \ldots) = (y_1, y_2, \ldots)$.

We consider the space of the dynamics $(\hat{\Sigma}, \hat{\sigma})$, the natural extension of (Σ, σ), as subset of $\Sigma^* \times \Sigma$. In fact, if $\mathbf{y} = (\ldots, y_1, y_0) \in \Sigma^*$ and $\mathbf{x} = (x_0, x_1, \ldots) \in \Sigma$, then $\hat{\Sigma}$ will be the set of points

$$< y, x >= (\ldots, y_1, y_0 | x_0, x_1, \ldots) \in \Sigma^* \times \Sigma,$$

such that (y_0, x_0) is an allowed word (no restrictions when we consider the full Bernoulli space). In this case

$$\hat{\sigma}\,(\ldots, y_1, y_0 | x_0, x_1, \ldots) = (\ldots, y_1, y_0, x_0 | x_1, x_2, \ldots).$$

We point out that we use here the notation $< y, x >= (x, y)$. For functions $b : \hat{\Sigma} \to \mathbb{R}$, we denote its value on $< y, x >$ by $b(x, y)$. We define the map $\tau : \hat{\Sigma} \to \Sigma$ by $\tau(x, y) = \tau_{\mathbf{y}}(\mathbf{x}) = (y_0, x_0, x_1, \ldots)$. Note that, if $\pi_x : \hat{\Sigma} \to \Sigma$ is the projection in the x coordinate, then, $\tau_y(x) = \pi_x \circ \hat{\sigma}^{-1}\,(x, y)$. We denote by $\pi_y(x, y) = y$ the projection on the second coordinate. Note that $\hat{\sigma}^{-1}(x, y) = (\tau_y(x), \sigma^*(y))$.

Definition 4 A continuous function $V : \Sigma \to \mathbb{R}$ is called calibrated subaction for A, if

$$V(x) = \max_{z : \sigma(z) = x} \big(V(z) + A(z) - m(A)\big).$$

In other terms, V is a calibrated subaction if for any $x \in \Sigma$, there exists $z \in \Sigma$, such that, $\sigma(z) = x$, and $V(z) + A(z) - m(A) = V(x)$.

Note that for all z we have $V(\sigma(z)) - V(z) - A(z) + m(A) \geq 0$. We show bellow some explicit expressions for calibrated subactions for a class of potentials A.

We point out that we will also consider here analogous results for an expanding transformation $T : S^1 \to S^1$ (or, $T : [0,1] \to [0,1]$) of class $C^{1+\alpha}$, and a Hölder potential $A : S^1 \to \mathbb{R}$ (or, $A : [0,1] \to \mathbb{R}$) as in [12]. The case $T(x) = -2x \pmod 1$ is one of the examples we have on mind.

In this case one could consider analogous problems in $S^1 \times S^1$, or, $S^1 \times \Sigma$, if one consider the symbols i which index the inverse branches τ_i of T [37, 40]. The existence of involution kernel, L.D.P. properties, etc., are also true.

The calibrated sub-action is unique (up to an additive constant) if the maximizing probability is unique (see [2, 12, 21]). We point out that we called strict in [2] what we denote here by calibrated. We will use from now on the notation of [2].

Definition 5 Given $A : \Sigma \to \mathbb{R}$ Lipchitz potential, consider $A^*(y)$ (the dual potential), where $A : \Sigma^* \to \mathbb{R}$, and $W(x,y) = W_A(x,y)$ its involution kernel.

This means, by definition that for all $< y, x >= (x,y) \in \hat{\Sigma}$

$$A^*(y) = A(\tau_y(x)) + W(\tau_y(x), \sigma^*(y)) - W(x,y). \tag{5}$$

This expression can be also written in the form

$$A^*(x,y) = A(\hat{\sigma}^{-1}(x,y)) + W(\hat{\sigma}^{-1}(x,y)) - W(x,y).$$

If A depends on just two coordinates we can take A^* as the transpose of A. Therefore, the above definition extends this concept in the case A depends on infinite coordinates on the Bernoulli space. We say A is involutive if $A = A^*$.

We address the question of regularity of the involution kernel W (is bi-Hölder) in the item (d) in the Appendix.

We denote by M the Bernoulli space or the unitary circle. Suppose T is an expanding transformation on M (T can be the shift σ or the transformation T defined above).

For a Lipchitz potential $A : M \to \mathbb{R}$ the pressure of A is the value

$$P(A) = \sup_{\mu \text{ invariant for } T} \{h(\mu) + \int A \, d\mu\},$$

where $h(\mu)$ is the Kolmogorov entropy of the invariant probability μ.

The equilibrium state for A is the probability μ which realizes the above supremum.

Given a Hölder function $A : M \to \mathbb{R}$, by definition the Ruelle operator $\mathscr{L}_A : C(M) \to C(M)$ acts on continuous functions $\phi : M \to \mathbb{R}$, in such way that, $\mathscr{L}_A(\phi) = \varphi$, where

$$\varphi(x) = \mathscr{L}_A(\phi)(x) = \sum_{T(y)=x} e^{A(y)}\,\phi(y).$$

This operator (sometimes called transfer operator) helps to understand equilibrium states in Thermodynamic Formalism. This corresponds to the analysis of the Statistical Mechanics of the one-dimensional lattice at positive temperature (see [47]). Maximizing probabilities correspond to the limit of equilibrium states when temperature goes to zero (ground states) as one can see for instance in [5].

When A is such that $\mathscr{L}_A(1) = 1$ we say that A is normalized.

The dual operator $\mathscr{L}_A^*$ acts on the space of probabilities measures on M. Given a probability μ, then, $\mathscr{L}_A^*(\mu) = \nu$ where the probability measure ν is the unique one satisfying

$$\int \phi \, d\,\mathscr{L}_A^*(\mu) = \int \phi \, d\nu = \int \mathscr{L}_A(\phi)\, d\mu$$

for any continuous function ϕ.

An important result claims that there exists a positive value λ which is simultaneous an eigenvalue for $\mathscr{L}_A$ and $\mathscr{L}_A^*$ (see [47]). This λ is the spectral radius of $\mathscr{L}_A$. This defines a main eigenfunction for $\mathscr{L}_A$ and a main eigenprobability for $\mathscr{L}_A^*$.

In [33] it is shown that the dual of the Ruelle operator $\mathscr{L}_A^*$ is a contraction for the 1-Wasserstein distance when A is normalized. The fixed point probability is the main eigenprobability for $\mathscr{L}_A^*$.

We suppose that c is a normalization constant for W in the sense that

$$\int\int e^{W(x,y)-c}\, d\nu_{A^*}(y)\, d\nu_A(x) \;=\; 1, \tag{6}$$

where ν_A and ν_{A^*} are respectively the eigen-probability for the dual Ruelle operator of A and A^* [12]. We also denote by ϕ_A and ϕ_{A^*} the corresponding eigenfunctions for $\mathscr{L}_A$. Finally, $\mu_A = \nu_A\,\phi_A =$ and $\mu_{A^*} = \nu_{A^*}\,\phi_{A^*}$ are the invariant probabilities which are the solutions of the respective pressure problems for A and A^*. For a fixed A we consider a real parameter β, and the corresponding potentials βA, and the eigenfunctions $\phi_{\beta A}$, and so on.

In Statistical Mechanics β is the inverse of temperature. In this way asymptotic results when $\beta \to \infty$ can be consider as the ones which describes the system in equilibrium at temperature zero. Note that βW is an involution kernel for βA, and its dual is βA^*.

It is known (see for instance [12]) that a sub-action V can obtained as the limit

$$V(x) = \lim_{\beta\to\infty} \frac{1}{\beta} \log \phi_{\beta A}(x). \tag{7}$$

This V is a calibrated sub-action for A (see [2, 12, 20]). We can also get a calibrated sub-action V^* for A^* using the limit

$$V^*(y) = \lim_{\beta\to\infty} \frac{1}{\beta} \log \phi_{\beta A^*}(y) \ . \tag{8}$$

From [2] (see also [42]) we have

$$\phi_{A^*}(y) = \int e^{\,W_A(x,y)-c}\, d\nu_A(x).$$

Finally, we define for each $x \in \Sigma$,

$$I(x) = \sum_{n=0}^{\infty} [\, V \circ \sigma - V - (A - m(A))\,]\, \sigma^n\,(x),$$

where V is a (any) calibrated sub-action.

The function I, where $I : \Sigma \to \mathbb{R} \cup \{\infty\}$, can have infinite values, but it is lower semi-continuous. In [2] it is shown that for any cylinder set $C \subset \Sigma$,

$$\lim_{\beta\to+\infty} \frac{1}{\beta} \log \mu_{\beta A}(C) = -\inf_{x\in C} I(x)$$

In this way we get a Large Deviation principle for $\mu_{\beta A} \to \mu_\infty$.

Remember that we denote by μ_∞^* the unique maximizing probability for A^* (it is unique because μ_∞ is unique for A, and, moreover, A and A^* are cohomologous in $\hat{\Sigma}$).

All the results described above are true for expanding transformations T of class $C^{1+\alpha}$ on the circle S^1. In this case we have to consider the natural extension $\hat{T}$ of T. This also includes the case of $T(x) = -2x$ (mod 1).

In the case $T : S^1 \to S^1$, given by $T(x) = 2x$ (mod 1), we define $\hat{T}$ in the following way: the Baker transformation associated to T, denoted by $\hat{T}(x_1, x_2)$, where $\hat{T} : [0,1]^2 \to [0,1]^2$, is such that satisfies for all $(x_1, x_2) \in [0,1]^2$, $\hat{T}(x_1, T^*(x_2)) = (T(x_1), x_2)$ (see picture bellow) . In this case $T^* : S^1 \to S^1$, with $T^*(y) = 2y$ (mod 1), $\hat{T}$ plays the role of $\hat{\sigma}$, and T^* plays the role of σ^*, on the definitions and results above.

All the above apply for an expanding transformation $T : S^1 \to S^1$, or $T : [0,1] \to [0,1]$.

The transformation $\hat{T}$ on $S^1 \times S^1$, contract vertical fibers by forward iteration and expand (and cut) vertical fibers by backward iteration.

Characterization of S

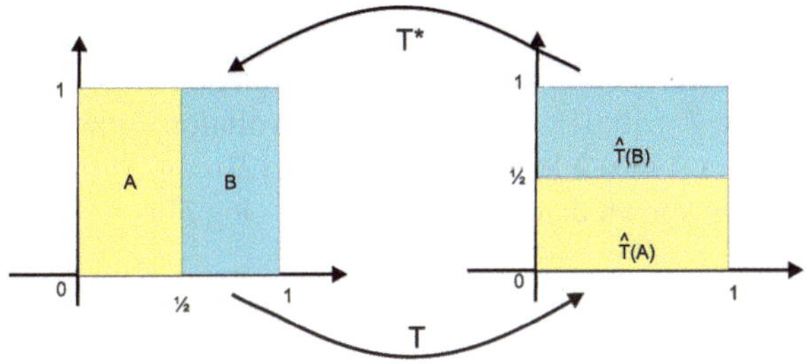

Remember that we said that $W : \hat{\Sigma} = \Sigma \times \Sigma \to \mathbb{R}$ satisfies the twist condition on $\hat{\Sigma}$, if for any $(a, b) \in \hat{\Sigma} = \Sigma \times \Sigma$ and $(a', b') \in \Sigma \times \Sigma$, with $a' > a$, $b' > b$, we have

$$W(a, b) + W(a', b') < W(a, b') + W(a', b). \tag{9}$$

We have the analogous definition for expanding transformations on the interval:

Definition 6 We say $W : [0, 1]^2 \to \mathbb{R}$ continuous satisfies the twist condition on $[0, 1]^2$, if for any $(a, b) \in [0, 1]^2$ and $(a', b') \in [0, 1]^2$, with $a' > a$, $b' > b$, we have

$$W(a, b) + W(a', b') < W(a, b') + W(a', b). \tag{10}$$

Same definition for W on $S^1 \times S^1$.

When $x, y \in [0, 1]$ (or, on S^1), the condition

$$\frac{\partial^2 W}{\partial x \, \partial y} < 0,$$

implies the twist condition for W. The twist condition can be seen as a kind of transversality condition (see [37])

Example 1 Consider the transformation $T : S^1 \to S^1$, given by $T(x) = -2x$ (mod 1) and $A(x) = a + bx + cx^2$, where a, b, c are constants and $c > 0$. In item (b) in the Appendix we show an explicit expression for the W-kernel and we prove that W satisfies the twist condition. From this, we can get an explicit expression for the calibrated subaction for a certain potential (see Remark 6 in the Appendix).

We point out that for considering the system above in S^1 we have to assume above that $A(0) = A(1)$. If we are interested in the case of $[0, 1]$ the same result can be obtained but we do not have to assume $A(0) = A(1)$.

Moreover, we also show in item (c) in the Appendix that a certain class of analytic perturbations of $A(x) = a + bx + cx^2$ produces W-kernels which are twist.

Example 2 In item (b) in the Appendix we show an example of a W-kernel for a continuous potential A, and for the action of the shift σ on the Bernoulli space $\{0, 1\}^{\mathbb{N}}$, which is twist.

Example 3 Consider the Gauss map $T(x) = \frac{1}{x} - [\frac{1}{x}]$ on $[0, 1]$.

We can define the Baker transformation associated to T, denoted by $\hat{T}(x_1, x_2)$, where $\hat{T} : [0, 1]^2 \to [0, 1]^2$. The involution kernel W for $A(x_1) = -\log T'(x_1)$ is $W(x_1, x_2) = -2 \log(1 + x_1 x_2)$ (see [2]).

It is known that the dual of $A = -\log T'$ is $A^* = -\log T'$ (see Proposition 4 in [2]).

The maximizing probability for such potential $-\log T'(x) = 2\log(x)$ is the δ-Dirac in the fixed point b, where b is the golden mean $b = \frac{\sqrt{5}-1}{2}$ (see for instance [14]). In this case $m(A) = 2\log(b)$.

Note that W is differentiable on any point $(x_1, x_2) \in [0, 1]^2$.

One can easily see that an explicit calibrated sub-action u (unique up to an additive constant because the maximizing probability is unique [20]) satisfying

$$u(x) \leqslant u(T(x)) - A(x) + m(A), \tag{11}$$

is $u(x) = W(x, b) = -2 \log(1 + x b)$.

Note that

$$\frac{\partial^2 W}{\partial x \, \partial y} < 0,$$

and, therefore, W is twist.

Example 4 Suppose $T(x) = -2x \pmod 1$, $T : [0, 1] \to [0, 1]$ and $A : [0, 1] \to \mathbb{R}$ is Hölder and monotonous. Under some assumptions on A one can get cases where the maximizing probability is unique and with support on the right fixed point p (see [31]). In the same way as in last example one can show that $V(x) = W(x, p)$ is a calibrated subaction.

If one considers on the interval $[0, 1]$ the potential $A(x) = x^2$ is under such assumptions. One can show that $A^*(y) = y^2$, and $W(x, y) = (1/3)(x^2 + y^2) - (4/3)xy$ (see Remark 6 in item (b) in the Appendix). In the same way $\frac{\partial^2 W(x,y)}{\partial x \, \partial y} < 0$.

Example 5 Consider the transformation $T : S^1 \to S^1$, given by $T(x) = -2x \pmod 1$ and $A(x) = -(x - \frac{1}{2})^2$ (a continuous potential on S^1) for which all results in [2] apply (see also [37] where it is shown in this case the graph property). The maximizing probability has support in the periodic orbit of period 2 (see [29, 30]).

One can define the continuous Baker transformation associated to T, denoted by $\hat{T}(x_1, x_2)$, where $\hat{T} : [0, 1]^2 \to [0, 1]^2$ is such that satisfies for all $(x_1, x_2) \in [0, 1]^2$, $\hat{T}(x_1, T(x_2)) = (T(x_1), x_2)$.

In this case, we show in Remark 6 in the Appendix that a smooth W-kernel is:

$$W(x, y) = -(1/3)x^2 - (1/3)y^2 + (4/3)xy - (2/3)x - (1/3)y.$$

The dual potential A^* is equal to A.

This W-kernel is **not** twist because $\frac{\partial^2 W(x,y)}{\partial x \, \partial y} > 0$.

It follows from a general result presented in [31] that any maximizing measure for this potential is $\mu_\infty = (1-t)\delta_{1/3} + t\delta_{2/3}$, where $t \in [0, 1]$, so the critical value is $m = A(1/3) = A(2/3)$.

It is easy to verify that,

$$\begin{aligned} V(x) &= (W(x,1/3) - W(1/3,1/3))\chi_{[(0,1/2)]}(x) + W(x,2/3) - W(2/3,2/3)\chi_{[1/2,1]}(x) \\ &= \max\{W(x,1/3) - W(1/3,1/3), W(x,2/3) - W(2/3,2/3)\} \end{aligned}$$

is a calibrated subaction for A.

$W(x,1/3)-W(1/3,1/3)$=*red*, $W(x,2/3) - W(2/3,2/3)$ = *blue* and ϕ=*black*—The calibrated subaction is the supremum of the two functions described in the picture

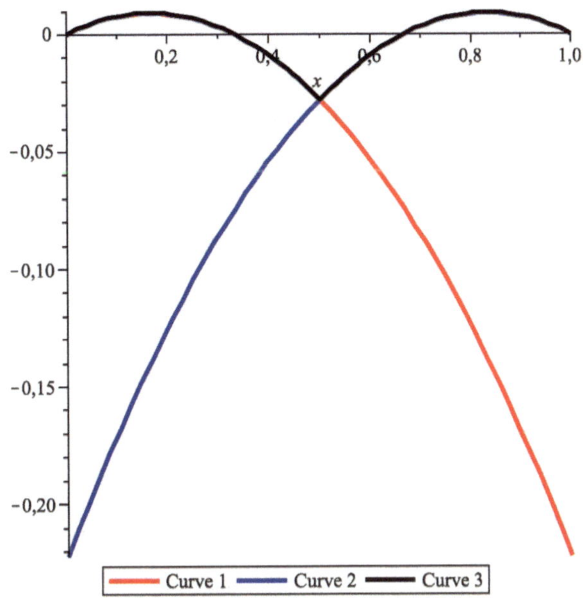

This calibrated subaction is not analytic but piecewise analytic (see [40] for more general results).

Example 6 Consider the transformation $T : S^1 \to S^1$, given by $T(x) = -2x$ (mod 1) and $A(x) = (x - \frac{1}{2})^2$ (a continuous potential on S^1) for which all results in [2] apply.

In this case we show in item (b) in the Appendix that a smooth W-kernel is:

$$W(x,y) = (1/3)x^2 + (1/3)y^2 - (4/3)xy + (2/3)x + (1/3)y,$$

the dual potential A^* is equal to A and this involution kernel W is twist.

Similar results can be obtained for $T : S^1 \to S^1$, given by $T(x) = 2x$ (mod 1) and $A(x) = -(x - \frac{1}{2})^2$ (a continuous potential on S^1)

Definition 7 Given $G : \hat{\Sigma} \to \mathbb{R}$ upper semi-continuous, and $f(x)$ continuous, where $f : \Sigma \to \mathbb{R}$, we define the G-transform of f, denoted by $f^{\#}(y)$, where $f^{\#} : \Sigma^* \to \mathbb{R}$, the function such that

$$f^{\#}(y) = \max_{x \in \Sigma} \{-f(x) + G(x,y)\}. \tag{12}$$

We can use also the notation $f^{\#}_G$, instead of $f^{\#}$, if we want to stress the dependence on G.

In this case we say that $f^{\#}$ is the G-conjugate of f [53, 54]. We use the notation of [49, p. 268]. Note that, if we add a constant to f, then new $f^{\#}$ will be obtained from the old one by subtracting the same constant. Therefore, in this case the sum $f(x) + f^{\#}(y)$ will be the same. We are interested, for example, when $G = -W$ or $G = -W + I$. A similar definition and properties can be consider for expanding transformations on $[0, 1]$.

Proposition 1 *If V is a subaction for A, then $V^{\#} = V^{\#}_W$ is a subaction for A^*.*

Proof Given y there exist z^0 such that

$$V^{\#}(\sigma^*(y)) - V^{\#}(y) = \max_{x \in \Sigma} \{-V(x) + W(x, \sigma^*(y))\} -$$

$$\max_{z \in \Sigma} \{-V(z) + W(z,y)\} =$$

$$\max_{x \in \Sigma} \{-V(x) + W(x, \sigma^*(y))\} - (-V(z_0) + W(z_0, y)) \geq$$

$$-V(\tau_y(z_0)) + W(\tau_y(z_0), \sigma^*(y)) + V(z_0) - W(z_0, y) \geq$$

$$A(\tau_y(z_0)) - m(A) + W(\tau_y(z_0), \sigma^*(y)) - W(z_0, y) =$$

$$A^*(y) - m(A) = A^*(y) - m(A^*).$$

The subaction you get by $-W$-transform is not necessarily calibrated.

Note that if we add a constant to W (the new W will be also a W-Kernel), then all of the above will be also true.

In a similar way like in the reasoning of last proposition one can get:

Proposition 2 *If V^* is a sub-action for A^*, then*

$$(V^*)^{\#}_W(x) = \max_{z \in \Sigma^*} \{-V^*(z) + W(x,z)\}$$

is a subaction for A.

Analogous definitions can be consider for an expanding transformation $T : S^1 \to S^1$. This also includes the case of $T(x) = -2x \pmod 1$.

2 The Transport Problem

We assume that the maximizing probability μ_∞ for A is unique. We denote by μ^*_∞ a fixed maximizing probability for A^*. We denote by $\mathscr{K}(\mu_\infty, \mu^*_\infty)$ the set of probabilities $\hat{\eta}(x,y)$ on $\hat{\Sigma}$, such that

$$\pi^*_x(\hat{\eta}) = \mu_\infty, \text{ and } \pi^*_y(\hat{\eta}) = \mu^*_\infty .$$

We are going to consider bellow the cost function $c(x,y) = I(x) - W(x,y) + \gamma$, which is defined for x such that $I(x) \neq \infty$.

The Kantorovich Transport Problem Given A (and all the probabilities described above) we are interested in the minimization problem

$$C(\mu_\infty, \mu^*_\infty) = \inf_{\hat{\eta}\in\mathscr{K}(\mu_\infty,\mu^*_\infty)} \int\int (I(x) - W(x,y) + \gamma)\, d\hat{\eta} =$$

$$\inf_{\hat{\eta}\in\mathscr{K}(\mu_\infty,\mu^*_\infty)} \int\int c(x,y)\, d\hat{\eta} =$$

$$\max_{\hat{\eta}\in\mathscr{K}(\mu_\infty,\mu^*_\infty)} \int\int (W(x,y) - \gamma - I(x))\, d\hat{\eta} \tag{13}$$

where, I is the deviation function for $\mu_\infty = \lim_{\beta\to\infty} \mu_{\beta A}$ (see [2]),

$$c_\beta = \int\int e^{\beta\, W(y,x)}\, d\nu_{\beta A}(x)\, d\nu_{\beta A^*}(y), \tag{14}$$

and

$$\gamma = \lim_{\beta\to\infty} \frac{1}{\beta} \log c_\beta , \tag{15}$$

as in proposition 5 in [2]. We call $c(x,y) = -W(x,y) + \gamma + I(x)$ the cost function. Therefore, c is lower semi-continuous. A probability $\hat{\eta}$ on $\hat{\Sigma}$ which attains such minimum is called an optimal transport probability. We denote it by $\hat{\mu}$. We will show later that $\hat{\mu}_{max}$, the natural extension of μ_∞, will be the optimal transport probability $\hat{\mu}$.

One of our main results is Theorem 5 which claims that: The support of $\hat{\mu}_{max}$ is c-cyclically monotone. In other words, the twist condition for c is true when restricted to the support of the maximizing probability $\hat{\mu}_{max}$.

Remark 1 Note that if we subtract the deviation function $I(x)$ of the cost function, that is, if we consider a new cost $c(x,y) = -W(x,y) + \gamma$, the problem above will not change, because I is constant zero in the support of μ_∞. In other words

$$C(\mu_\infty, \mu_\infty^*) = \inf_{\hat\eta \in \mathcal{K}(\mu_\infty, \mu_\infty^*)} \int\int (-W(x,y) + \gamma)\, d\hat\eta\,,$$

and, the optimal transport probability will be the same. In some sense this setting is nicer because the cost c is a continuous function on $\hat\Sigma$.

Definition 8 A pair of functions $f(x)$ and $f^\#(y)$ will be called c-admissible (or, just admissible for short) if

$$f^\#(y) = \min_{x\in\Sigma} \{-f(x) + c(x,y)\}. \tag{16}$$

In other words $-f^\#$ is the $-c$-conjugate of $-f$. Note that in this case, $\forall x \in \Sigma$, $y \in \Sigma^*$, we have that $f(x) + f^\#(y) \leq c(x,y)$. We denote by $\mathcal{F}$ the set of all admissible pairs $(f(x), f^\#(y))$.

The Kantorovich Dual Problem Given A and the corresponding c (W and all the probabilities described above) we are interested in the maximization problem

$$D(\mu_\infty, \mu_\infty^*) = \max_{(f,f^\#)\in\mathcal{F}} \Big(\int f d\mu_\infty + \int f^\# d\mu_\infty^* \Big). \tag{17}$$

A pair of admissible $(f, f^\#) \in \mathcal{F}$ which attains the maximum value will be called an optimal pair.

The Kantorovich duality theorem (see [53]) claims that under general conditions $D(\mu_\infty, \mu_\infty^*) = C(\mu_\infty, \mu_\infty^*)$. The main tool to prove this result is the Fenchel-Rockafellar duality Theorem.

Theorem 4 (Fenchel-Rockafellar Duality) *Suppose E is a normed vector space, Θ and Ξ two convex functions defined on E taking values in $\mathbb{R}\cup\{+\infty\}$. Denote Θ^* and Ξ^*, respectively, the Legendre-Fenchel transform of Θ and Ξ. Suppose there exists $v_0 \in E$, such that $\Theta(v_0) < +\infty$, $\Xi(v_0) < +\infty$ and that Θ is continuous on v_0.*

Then,

$$\inf_{v\in E}[\Theta(v) + \Xi(v)] = \max_{f\in E^*}[-\Theta^*(-f) - \Xi^*(f)] \tag{18}$$

We will not present the proof of this general theorem but we will present a nice geometric proof in a simple case (one-dimensional) in item (e) in the Appendix. We suppose, from now on, that the maximizing probability for A, denoted by μ_∞ is unique. We denote, as in [12] the calibrated sub-actions V and V^* by

$$V(x) = \lim_{\beta\to\infty} \frac{1}{\beta} \log \phi_{\beta A}(x) \text{ and } V^*(y) = \lim_{\beta\to\infty} \frac{1}{\beta} \log \phi_{\beta A^*}(y)\ . \tag{19}$$

The above convergence is uniform and V is (up to constant) the unique calibrated sub-action for A (see [2, 12, 20]). We will show later that $(f, f^{\#})$ such that $f(x) = -V(x)$ and $f^{\#}(y) = -V^*(y)$ is the optimal pair.

Important Property If $\hat{\mu}$ is an optimal transport probability and if $(f, f^{\#})$ is an optimal pair in $\mathscr{F}$, then the support of $\hat{\mu}$ is contained in the set

$$\{ < y, x > \in \hat{\Sigma} \mid \text{such that } (f(x) + f^{\#}(y)) = c(x,y) \}. \tag{20}$$

It follows from the prime and dual linear programming problem formulation. The condition above is the complementary slackness condition (see [17, 19, 48]).

The reciprocal of this result is also true (see [54, Remark 5.13, p. 59]).

If x and y are such that $(f(x) + f^{\#}(y)) = c(x,y)$ we say that they are realizers for the cost c. In [13] it is shown that the set of realizers for $I - W$ is an invariant set for the dynamics of $\hat{\sigma}$. In this section we are mainly concerned with the support and not with all realizers.

If one finds $\hat{\mu}$ an admissible pair $(f, f^{\#})$ satisfying the above claim (for the support), then, one solves the Kantorovich problem, that is, one finds the optimal transport probability $\hat{\mu}$.

No we will prove Theorem 1.

Proposition 3 *The minimizing Kantorovich probability $\hat{\mu}$ on $\hat{\Sigma}$ associated to $-W$ is $\hat{\mu}_{max}$.*

Proof Proposition 10 (1) in [2] claims that if $\hat{\mu}_{max}$ is the natural extension of the maximizing probability μ_∞, then for all $< p^*|p >$ in the support of $\hat{\mu}_{max}$ we have

$$-V(p) - V^*(p^*) = -W(p,p^*) + \gamma.$$

This is the same as saying that in the support of $\hat{\mu}_{max}$

$$-V(p) - V^*(p^*) = -W(p,p^*) + \gamma + I(p) = c(p,p^*),$$

because I is zero in the support of μ_∞. Then if $-V(x)$ and $-V^*(y)$ is an admissible pair, then $\hat{\mu}_{max}$ is the optimal transport probability for such $c(x,y)$. This will be shown in the next proposition. We will show bellow that the $-c$-transform of V is V^*.

Note that if W is a W-Kernel for A, for all β, we have that βW is a W-Kernel for βA. We denote by c_β the normalizing constant for βW, as in [2]. It is known that $\frac{1}{\beta} \log c_\beta = \gamma$.

Now we will show Theorem 2.

Proposition 4 *The pair $(-V, -V^*)$ is admissible.*

Proof For a fixed y we have to show that

$$-V^*(y) = (-V)^{\#}_c = \inf_{x \in \Sigma} \{-(-V(x)) + c(x,y)\}.$$

This is the same as

$$V^*(y) = \sup_{x\in\Sigma} \{ (-V(x)) - c(x,y) \} = \sup_{x\in\Sigma} \{ -V(x) - (\gamma - W(x,y) + I(x))\} ,$$

or, for all x

$$-V^*(y) \le V(x) + c(x,y) . \tag{21}$$

From Proposition 3 in [2] (we just write here $W(x,y)$, instead of $W(y,x)$ there) we have

$$\phi_{\beta A^*}(y) = \int e^{\beta\, W_A(x,y) - c_\beta} \frac{1}{\phi_{\beta A}(x)} \, d\mu_{\beta A}(x) = \int e^{\beta\, W_A(x,y) - c_\beta - \log\phi_{\beta A}(x)} \, d\mu_{\beta A}(x).$$

Consider now the limit

$$V^*(y) = \lim_{\beta\to\infty} \frac{1}{\beta} \log(\phi_{\beta A^*}(y)) =$$

$$\lim_{\beta\to\infty} \frac{1}{\beta} \log \int e^{\beta\, W_A(x,y) - c_\beta - \log\phi_{\beta A}(x)} \, d\mu_{\beta A}(x).$$

From [12] the function $\frac{1}{\beta} \log(\phi_{\beta A}(x))$ converges uniformly with β to $V(x)$. Therefore, one can write

$$\lim_{\beta\to\infty} \frac{1}{\beta} \log \int e^{\beta\, W_A(x,y) - c_\beta - \log\phi_{\beta A}(x)} \, d\mu_{\beta A}(x) =$$

$$\lim_{\beta\to\infty} \frac{1}{\beta} \log \int e^{\beta\, (W_A(x,y) - \gamma - V(x))} \, d\mu_{\beta A}(x)$$

Now, by Varadhan's Integral Lemma [16] we obtain

$$V^*(y) = \sup_x \{ W_A(x,y) - \gamma - V(x) - I(x)\} = \sup_x \{ -V(x) + W(x,y) - \gamma - I(x)\},$$

where I is the deviation function.

Finally, we get that $\hat{\mu}_{max}$ is the optimal transport probability for such $c(x,y)$. From now on we will use either the notation $\hat{\mu}$ or $\hat{\mu}_{max}$ for the optimal transport probability. In [40] Transport Theory is used as a tool to show that in some cases the calibrated subaction is piecewise analytic. In [13] some generic properties of the potential A is considered and special results about the realizers of the $W - I$ are obtained.

The last theorem says: for any $y \in \Sigma^*$ we have

$$V^*(y) = \sup_{x\in\Sigma} \{ -V(x) - c(x,y)\}. \tag{22}$$

Note that when $y = p^*$, for p^* in the support of μ^*_∞, the supremum

$$V^*(p^*) = \sup_x\{-V(x) + W(x,p^*) - \gamma - I(x)\} = \sup_x\{-V(x) - c(x,p^*)\},$$

is realized at $x = p$, for p in the support of μ_∞ (with $< p^*, p >$ in the support of $\hat{\mu}$).

Remark 2 Remember that, if the maximizing probability for A^* is unique, then there is a unique calibrated sub-action for A^* (up to additive constant) [2, 20].

Analogous definitions and properties can be obtained for $T : S^1 \to S^1$. This also includes the case of $T(x) = -2x \pmod 1$. We could likewise consider the analogous problem for A^*: given A^* (obtained from A) fixed, denote $I^* : \Sigma^* \to \mathbb{R}$, the non-negative deviation function for $\mu_{\beta A^*} \to \mu^*_\infty$. Denote $c^*(x,y) = (I^*(y) - W(x,y) + \gamma)$.

Then, consider the problem

$$C(\mu_\infty, \mu^*_\infty) = \inf_{\hat{\eta}\in\mathscr{K}(\mu_\infty,\mu^*_\infty)} \int\int (I^*(y) - W(x,y) + \gamma)\, d\,\hat{\eta} =$$

$$\inf_{\hat{\eta}\in\mathscr{K}(\mu_\infty,\mu^*_\infty)} c^*(x,y)\, d\,\hat{\eta} = \inf_{\hat{\eta}\in\mathscr{K}(\mu_\infty,\mu^*_\infty)} \int\int (-W(x,y) + \gamma)\, d\,\hat{\eta},$$

which have the same minimizing measures, as for the minimization for $c(x,y) = (I(x) - W(x,y) + \gamma)$ among probabilities on $\mathscr{K}(\mu_\infty, \mu^*_\infty)$.

Note also that from Proposition 3 in [2] we have

$$\phi_{\beta A}(x) = \int e^{\beta\, W_A(x,y) - c_\beta} \frac{1}{\phi_{\beta A^*}(y)}\, d\mu_{\beta A^*}(y) =$$

$$\int e^{\beta^*\, W_A(x,y) - c_\beta - \log \phi_{\beta A^*}(y)}\, d\mu_{\beta A^*}(y).$$

In the same way as before one can show that for any $x \in \Sigma$, we have

$$V(x) = (-V^*)^{\#}_{c^*} = \sup_{y\in\Sigma^*}\{-V^*(y) - c^*(x,y)\}. \tag{23}$$

Note that $c(x,y) = c^*(x,y)$ in the support of the minimizing $\hat{\mu}_{max}$ for c (or for c^*).

Remark 3 It is not necessarily true that $(\,(-V^*)^{\#}_{c^*}\,)^{\#}_{c^*} = -V^*$. However, the expression is true when restricted to the support of the optimal transport probability $\hat{\mu}_{max}$. In the same way $(\,(-V)^{\#}_{c}\,)^{\#}_{c} = -V$ in the support of $\hat{\mu}_{max}$.

3 Graph Properties and the Twist Condition

Consider a lower semi-continuous cost function $c(x,y)$ on $\hat{\Sigma}$ (or, a continuous cost function $-W(x,y)$ on $\hat{\Sigma}$). We refer the reader to [48, 53, 54] and [19] for general references on optimal mass transportation problems.

Definition 9 A set $S \subset \hat{\Sigma}$ is called c-cyclically monotone, if for any finite number of points (x_j, y_j) in S, $j \in \{1, 2, \ldots, n\}$, and any permutation σ of the n letters, we have

$$\sum_{j=1}^{n} c(x_j, y_j) \leq \sum_{j=1}^{n} c(x_{\sigma(j)}, y_j). \tag{24}$$

Proposition 5 (See Theorem 2.3 [19]) *For a continuous function $c(x,y) \geq 0$, where $\hat{\Sigma}$, if $\rho \in \mathcal{K}(\mu_\infty, \mu_\infty^*)$ is optimal for c, then, ρ has a c-cyclically monotone support.*

Corollary 1 *The support of $\hat{\mu}_{max}$, the natural extension of μ_∞ is c-cyclically monotone.*

We will present bellow in the next theorem a direct proof of this fact.

Definition 10 A function $f : \Sigma \to \mathbb{R} \cup \{\infty\}$ is c-concave, if there exist a set $A \subset \Sigma \times \mathbb{R}$ such that

$$f(y) = \sup_{(x,\lambda)\in A} \{c(x,y) + \lambda\}$$

Definition 11 A function $f : X \to \mathbb{R} \cup \{\infty\}$ is c-convex, if $(-f)$ is c-concave.

Definition 12 Given $x \in \Sigma$, the set $\hat{\partial}_c f(x)$ is the set of $y \in \hat{\Sigma}$ such that, for all $z \in \Sigma$ we have

$$f(z) - f(x) \leq c(z,y) - c(x,y)$$

In this case we say y is a c-sub-derivative for f in x.

An important problem is to know, for a certain given x, if the $\hat{\partial}_c f(x)$ has cardinality 1.

Proposition 6 (See Theorem 2.7 in [19], Lemma 2.1 in [49] and Section 4 in [48]) *For $S \subset \hat{\Sigma}$ to be c-cyclically monotone, it is necessary and sufficient that $S \subset \hat{\partial}_c(f)(x) = \{(x,y)\,|\, f(z) - f(x) \leq c(z,y) - c(x,y)\,,\ \forall z \in X\}$, for some c concave f, where $f : \Sigma \to \mathbb{R} \cup \{\infty\}$.*

Moreover: f is defined in the following way: choose $(x_0, y_0) \in S$, then

$$f(x) = \inf_{n\in\mathbb{N},\,(x_j,y_j)\in S,\,1\le j\le n} [\,(\,c(x,y_n) - c(x_n,y_n)\,) +$$

$$(\,c(x_n,y_{n-1}) - c(x_{n-1},y_{n-1})\,) + \dots$$

$$+(\,c(x_2,y_1) - c(x_1,y_1)\,) + (\,c(x_1,y_0) - c(x_0,y_0)\,)\,].$$

Note that if $S \subset \hat{\Sigma}$ is a graph, then for each $x \in \Sigma$ in the x-projection of S, we have that $\hat{\partial}_c(f)(x)$ has cardinality 1. Consider fixed $(x_0, y_0), (x_1, y_1)$ in the support of $\hat{\mu}_{max}$ and $(x_0, y_1), (x_1, y_0) \in \hat{\Sigma}$. Given a function $f(x, y)$ we denote

$$\Delta_f\,((x_0,y_1),(x_1,y_0)) = (f(x_0,y_0) + f(x_1,y_1)) - (f(x_0,y_1) + f(x_1,y_0)), \qquad (25)$$

and

$$b(x,y) = I(x) + \gamma - W(x,y) + V(x) + V^*(y). \qquad (26)$$

The c-cyclically monotone condition for the support of $\hat{\mu}_{max}$ will follow from the claim

$$\Delta_c\,((x_0,y_1),(x_1,y_0)) = (\,c(x_0,y_0) + c(x_1,y_1)) - (\,c(x_0,y_1) + c(x_1,y_0)\,) \le 0. \qquad (27)$$

This is so because any permutation of letters can be obtained by a series of composition of transformations that exchange just two letters. It will follow from the proof bellow that $\Delta_c \circ \sigma = \Delta_c$.

The next result does not assume a global assumption on twist condition for c.

Theorem 5 *Given $A : \Sigma \to \mathbb{R}$ Hölder, then $c(x,y) = I(x) - W(x,y) + \gamma \ge 0$, for all $(x,y) \in \Sigma$. Moreover, for $(x_0,y_0), (x_1,y_1)$ in the support of $\hat{\mu}_{max}$, we have $\Delta_c \le 0$. Therefore, the support of $\hat{\mu}_{max}$ is c-cyclically monotone. In other words, the twist condition for c (or, for W) is true when restricted to the support of the maximizing probability $\hat{\mu}_{max}$.*

Proof First we point out that $\Delta_c = \Delta_b$. We will show that under our hypothesis is true that $\Delta_b \le 0$. First note that

$$[\,V^* \circ \hat{\sigma}^{-1} - V^* - A^*\,]\,\hat{\sigma}\,(x,y) = [V^* - V^* \circ \hat{\sigma} - A - W + W \circ \hat{\sigma}]\,(x,y) =$$

$$[\gamma + V(x) + V^*(y) - W(x,y)] + [V \circ \hat{\sigma} - V - A](x,y) -$$

$$[\gamma + V \circ \hat{\sigma} + V^* \circ \hat{\sigma} - W \circ \hat{\sigma}]\,(x,y).$$

Remember (see [2]) that

$$I(x) = \sum_{n=0}^{\infty} [V \circ \sigma - V - A]\,\hat{\sigma}^n\,(x,y)$$

We denote

$$I_n(x,y) = \sum_{k=0}^{n-1} [V \circ \sigma - V - A] \circ \hat{\sigma}^k (x,y) = I_n(x),$$

and

$$R_n(x,y) = I_n(x,y) + [\gamma + V(x) + V^*(y) - W(x,y)] - [\gamma + V + V^* - W]\hat{\sigma}^n (x,y).$$

We claim that if (x,y) is in the support of $\hat{\mu}_{max}$, then $b(x,y) = 0$. Moreover, for all $(x,y) \in \Sigma$, we have $b(x,y) \geq 0$. One can prove this result by means of Varadhan's Integral Lemma [16] with the same reasoning as in the last proposition of the previous section. We will give bellow a direct proof of the claim.

Either $I(x) = \infty$, and the claim is trivially true or $I(x)$ is finite. In this case, any accumulation point of $\hat{\sigma}^n(x,y)$ will be in the support of $\hat{\mu}_{max}$.

Moreover, $b(x,y) = R(x,y) = \lim_{n\to\infty} R_n(x,y) \geq 0$. As in the support of $\hat{\mu}_{max}$, we have that $R(x,y) = 0$, then, $b(x,y) = 0$. In any case $R(x,y) \geq 0$. This shows the claim. We point out that $\Delta_c = \Delta_b = \Delta_W$ in the case $I(x)$ is finite.

We also remark that if (x_0, y_0) is in support of $\hat{\mu}_{max}$, then as $R(x_0, y_0)$ is zero, it follows that $R(x_0, y)$ is finite. This is so because (x_0, y) is in the stable manifold of (x_0, y_0) and

$$R_n(x_0, y) - R_n(x_0, y_0) = \sum_{k=1}^{n} \{ [V^* \circ \hat{\sigma}^{-1} - V^* - A^*]\hat{\sigma}^k(x_0, y) - [V^* \circ \hat{\sigma}^{-1} - V^* - A^*]\hat{\sigma}^k(x_0, y_0) \}.$$

Finally, if (x_0, y_0) and (x_1, y_1) are both in the support of $\hat{\mu}_{max}$, then $R(x_0, y_1) < \infty$, $R(x_1, y_0) < \infty$ and $I(x_0) = 0 = I(x_1)$. In this case, for any (x,y) of the form (x_0, y_0), (x_1, y_1), (x_1, y_0), or (x_0, y_1)

$$R(x,y) = I(x,y) + [\gamma + V + V^* - W](x,y) = b(x,y).$$

As we know that R is non-negative, then

$$[b(x_0, y_0) + b(x_1, y_1)] - [b(x_1, y_0) + b(x_0, y_1)] = 0 - [b(x_1, y_0) + b(x_0, y_1)] \leq 0.$$

This shows that $\Delta_b \leq 0$.

We did not use the twist condition above. Note that we could alternatively consider the function $g : \Sigma \to \mathbb{R}$ defined in the following way: choose $(x_0, y_0) \in S$, then

$$g(x) = \inf_{n\in\mathbb{N},\,(x_j,y_j)\in S,\,1\leq j\leq n} [\,(\,W(x, y_n) - W(x_n, y_n)\,)+$$

$$(\,W(x_n, y_{n-1}) - W(x_{n-1}, y_{n-1})\,) + \ldots$$

$$+(\,W(x_2, y_1) - W(x_1, y_1)\,) + (\,W(x_1, y_0) - W(x_0, y_0)\,)\,],$$

which has the advantage of just taking into account a continuous function W. The graph property for $S =$ support of $\hat{\mu}$, and all kinds of different considerations can be obtained from such g. We want to show now that if W satisfies the twist condition and the maximizing probability for A is unique, then the support of $\hat{\mu}$ on $\hat{\Sigma}$ is a graph. Our proof works for the Venously space $\{0, 1, 2, .., d\}^{\mathbb{N}}$ as well for the interval $[0, 1]$ [considering T either conjugated to $2x$ (mod 1) or to $-2x$ (mod 1)].

Consider the cost $c(x, y) = I(x) - W(x, y) - \gamma$, and a subset $S \subset X \times Y$ c-cyclically monotone.

Lemma 1 *Suppose the c satisfies the twist condition and let S be a c-cyclically monotone subset, if $(a, b), (a', b') \in S$ and $a \neq a'$ and $b \neq b'$, then $a < a'$ and $b > b'$, or $a > a'$ and $b < b'$.*

Proof Indeed, suppose $a < a'$ then, if $b < b'$, the twist condition on W implies that

$$c(a, b) + c(a', b') > c(a, b') + c(a', b).$$

On the other hand, S is c-cyclically monotone subset, so

$$c(a, b) + c(a', b') \leq c(a, b') + c(a', b),$$

that is an absurd.

A similar property is true for W. This Lemma means that the correct figure associated to a pair of points in S is given by:

We point out that, in principle, could exist points z of S in the vertical fiber passing by a or in the horizontal fiber passing by b.

Now we will show Theorem 3.

Theorem 6 (Graph Theorem) *Suppose the involution kernel W satisfies the twist condition and let $\hat{\mu}$ be the c-minimizing measure of probability to the transport problem, then $S =$ supp $\hat{\mu}$ is a graph in x (up to an orbit of measure zero), moreover this graph is monotone not increasing.*

Proof In fact we will just use the twist condition for W on the support of the optimal transport probability. In order to get advantage of the geometrical and combinatorial arguments we will present pictures for the case of a transformation $T : [0, 1] \to$

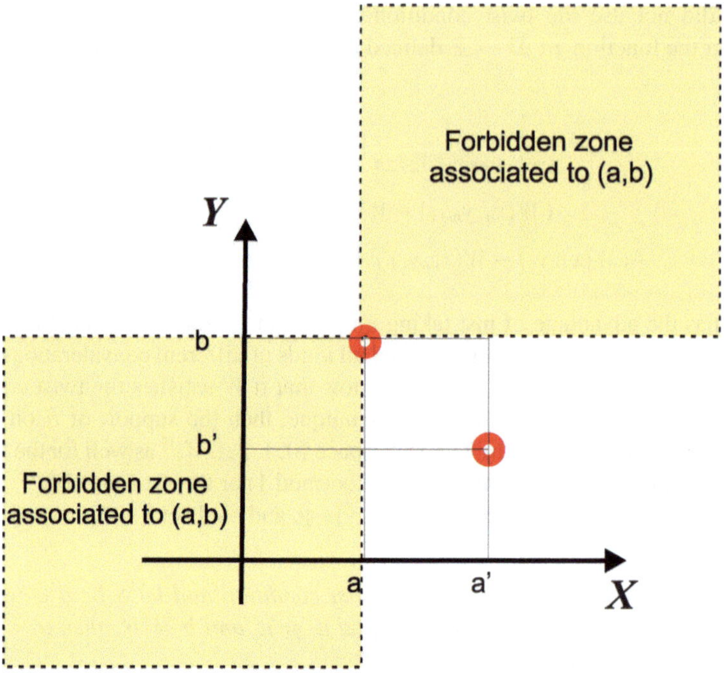

Characterization of S

Characterization of S

$[0, 1]$, given by $T(x) = 2x \pmod 1$. Define $v^+(x) = \max\{y|(x, y) \in S\}$ and $v^-(x) = \min\{y|(x, y) \in S\}$. In order to prove that supp $\hat{\mu}$ is a graph we need to prove that $v^-(x) = v^+(x)$ for any x in the support of μ_∞. We say that a point (x, y) in the support of $\hat{\mu}$ is non-graph, if there exist another point of the form (x, z), in the support of $\hat{\mu}$, and such that $z \neq y$. Note that the image of two points in the support of $\hat{\mu}$ on the fiber over x will go on two different points in the support of $\hat{\mu}$ on the fiber over $\sigma(x)$. That is, the forward image by $\hat{\sigma}^n$ of non-graph points will go on non-graph points. This maybe can not be true for backward images by $\hat{\sigma}^n$.

Suppose the support of the maximizing probability μ_∞ (unique) is a periodic orbit. If S is not a graph, then $v^-(x) < v^+(x)$ for some x. As the transformation $\hat{\sigma}$ contracts each fiber by forward iteration, we have that, the image of the interval fiber from $(x, v^-(x))$ to $(x, v^+(x))$, by a finite iterate of $\hat{\sigma}$, goes inside the fiber $(x, v^-(x))$ to $(x, v^+(x))$. Therefore, σ^* has a periodic point in the support of μ^*_∞. If the maximizing probability μ_∞ is unique for A, then μ^*_∞ is unique for the maximization problem for A^*. In this case the support of μ^*_∞ is this periodic orbit. Therefore, there is a minimal distance (in vertical fiber) between non-graph points and this is in contradiction with the contraction on vertical fibers. The conclusion is that S is a graph if the support of the maximizing probability μ_∞ is a periodic orbit.

Remark 4 In the case of the shift, if suppμ_∞ is a periodic orbit, one can easily show that if suppμ_∞ = the orbit by σ of $(a_0, a_1, \ldots, a_{(n-1)}, a_0, \ldots)$ then suppμ^*_∞ = orbit by σ^* of $(a_{(n-1)}, \ldots, a_2, a_1, a_0, a_{(n-1)}, \ldots)$.

Support of $\hat{\mu}$ in the periodic case

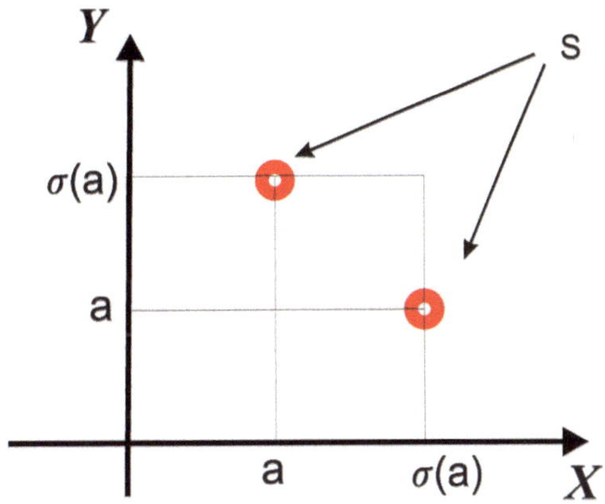

Support of $\hat{\mu}$ in the periodic case.

We suppose from now on that the support of the maximizing probability μ_∞ is not a periodic orbit.

Characterization of S

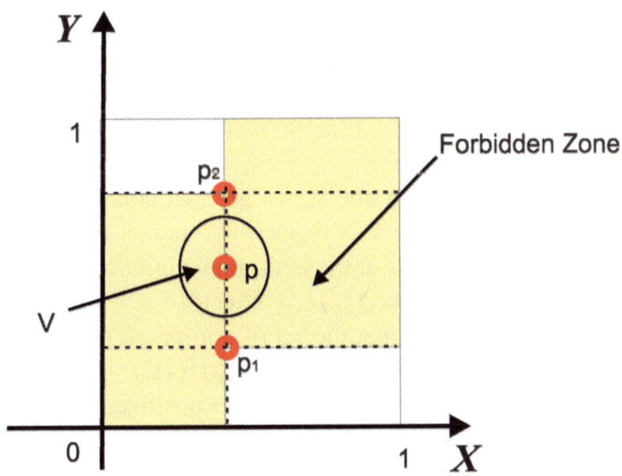

Characterization of S

Suppose, that $v^-(x) < v^+(x)$ for some x, then we claim that there is no other point in support of $\hat{\mu}$ in the fiber by x between $p_1 = v^-(x)$ and $p_2 = v^+(x)$. Indeed, from the above picture we see that if there exists a point (x, p) in the support of $\hat{\mu}$ such that $v^-(x) = p_1 < p < p_2 = v^+(x)$, then, as $\hat{\mu}$ is ergodic, should exist a point

(q_1, q_2) in a small neighborhood V of (x, p) such that returns by a forward n-iterate by $\hat{\sigma}$ to V.

This iterate has to return to the fiber, and this contradicts the fact that the support of the maximizing probability μ_∞ is not a periodic orbit.

If the support of μ_∞ is not a periodic orbit, then we claim that does not exist two pairs $(x_1, y_1), (x_1, z_1)$ and $(x_2, y_2), (x_2, z_2)$, in the support of $\hat{\mu}$, such that, the orbits by σ of x_1 and x_2 are different.

In order to simplify the argument and the notation we consider bellow $T^*(x) = 2x \pmod 1$, but we point out the reasoning apply to any expanding transformation of degree d. Given y_n and z_n, $n = 1, 2$, there exists a rational point of the form $s_n = \frac{q}{2^k}$, with $0 < q < 2^k$, $q, k \in \mathbb{N}$, such that $y_n < s_n < z_n$, $n = 1, 2$. Consider the s_n determined by the smallest possible value k.

The pair of points $\hat{T}^{-r}(x_n, y_n)$ and $\hat{T}^{-r}(x_n, z_n)$, $r \geq 0$, determine non-graph points in the same fiber, for any $r > 0$, until time $r = k$. In time $r = k-1$, it happens for the first time that the horizontal fiber through $1/2$ cuts the vertical segment connecting $\hat{T}^{-(k-1)}(x_n, y_n)$ and $\hat{T}^{-(k-1)}(x_n, z_n)$.

In this way, for each n, we get a horizontal forbidden region A_n (a horizontal strip from one vertical side to the other vertical side of $[0, 1] \times [0, 1]$) determined by such pair $\hat{T}^{k-1}(x_n, y_n)$ and $\hat{T}^{-(k-1)}(x_n, z_n)$, $n = 1, 2$, which contains the horizontal fiber through $1/2$.

If we apply the argument for $n = 1$, then the next forbidden region A_2 for $n = 2$ will contain the previous one A_1. Moreover, considering the full forbidden region determined by these two pair of points we reach a contradiction.

In the picture bellow we show the final pair of points q_1 and q_2 in a $\hat{\sigma}$-orbit (in the same vertical fiber) which has the property that its images p_1 and p_2 are on different sides of the upper and down rectangles. The images of p_1 and p_2 by $\hat{\sigma}$ are not anymore in the same vertical fiber (neither their future iterates). There is no room for getting a different pair of p_1 and p_2 like this (because of the forbidden region).

In this way, from above, we get that could exist just one orbit of x by σ such that over the fiber over x there is two points in the support. That is, the projection $K \subset \Sigma$ on the x-axis of the non-graph points have to be the orbit of a single point x. Therefore, $\mu_\infty(K) = \sum_k \mu_\infty(\{\sigma^k(x)\})$.

We assume first that the set of non-graph points have probability 1 and we will reach a contradiction. Indeed, $\mu_\infty(\{\sigma^k(x)\}) \geq \mu_\infty(\{\sigma^j(x)\})$, for $k \geq j$, and the μ_∞ probability of the set $\{x\}$ is zero or is positive.

Remember that the support of $\hat{\mu}$ is invariant by $\hat{\sigma}$. Now we will show that, indeed, if there exists non-graph points, this set has probability 1.

Note that if the vertical fiber by $x \in \Sigma$ is such that $v^-(x) < v^+(x)$, then $\sigma(x)$ also has this property. If the transformation $\hat{\sigma}$ we consider preserves orientation in the vertical fiber then the iterates are in the same order. Otherwise they exchange order. That is, the set of points (x, y) which are not graph point are invariant by forward iteration by $\hat{\sigma}$. Moreover, $\hat{\sigma}$ is a forward contraction in vertical fibers. Denote by $B = \{ (x, v^+(x))\}$ in the support of $\hat{\mu}$ such that $\{v^-(x) < v^+(x) \}$. The set B is the upper part of the non-graph part of the set S.

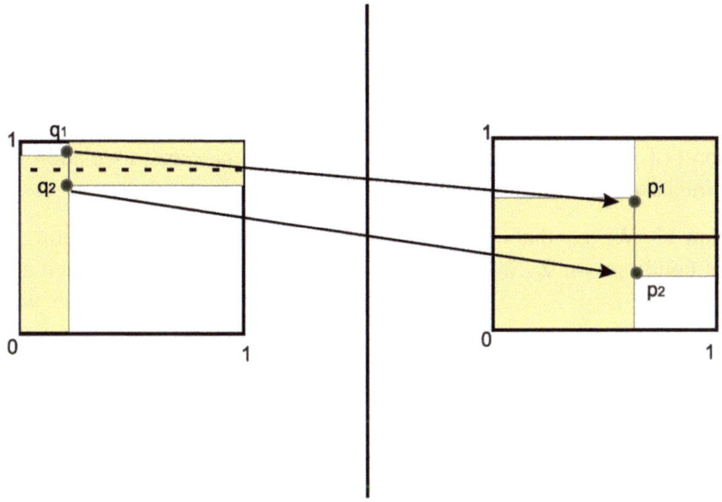

The dynamics on the support

The dynamics on the support

We will show that $\hat{\mu}(B) = 0$ or $\hat{\mu}(B) = 1$. We suppose first that $\hat{\sigma}$ preserves order in the fiber by forward iteration. Consider $\tilde{B}$ the set $\{ (x, y)\}$ in the support of $\hat{\mu}$ such that for some $n \geq 0$ we have $\{\hat{\sigma}^n (x, y) \in B\}$. Note that as B is forward invariant, once $\hat{\sigma}^n (x, y) \in B$, for some fixed n, then $\hat{\sigma}^m (x, y) \in B$, for any $m \geq n$.

We will show that $\hat{\sigma}^{-1}\tilde{B} = \tilde{B}$. The fact that $\hat{\sigma}^{-1}\tilde{B} \subset \tilde{B}$ follows easily from the definition of $\tilde{B}$. Given $x \in \tilde{B}$, there exists $n \geq 0$ such that $\hat{\sigma}^n (x, y) \in B$. If $n \geq 1$, then $\hat{\sigma}^{n-1} (\hat{\sigma}(x, y)) \in B$ and, therefore, $(x, y) \in \hat{\sigma}^{-1}\tilde{B}$. In the other case $(x, y) \in B$, but then $(\hat{\sigma}(x, y)) \in B$, because $\hat{\sigma}$ preserves order in the fiber, and does not exist more than two points in the vertical fiber over $\sigma(x)$ which are in S. Therefore, $(x, y) \in \hat{\sigma}^{-1}\tilde{B}$.

As $\hat{\mu}$ is ergodic, then $\hat{\mu}(\tilde{B}) = 0$ or $\hat{\mu}(\tilde{B}) = 1$.

If $\hat{\mu}(\tilde{B}) = 1$, then take a Birkhoff point $z \in \tilde{B}$ for the ergodic probability $\hat{\mu}$. Therefore, we get that the asymptotic frequency of visit to the set $C = \{ (x, v^-(x))\}$ in the support of $\hat{\mu}$ such that $\{v^-(x) < v^+(x) \}$ (the bellow part of the non-graph part of set S) is zero. Finally, we get that $\hat{\mu}(C) = 0$. In the same way $\hat{\mu}(B) = 1$.

If $\hat{\mu}(\tilde{B}) = 0$, we get that $\hat{\mu}(B) = 0$. Now, using a similar argument for the lower part of the non-graph part we get that $\hat{\mu}(C) = 1$.

This shows that the π_1 projection of the non-graph points has probability one and this proves the theorem.

The above reasoning also applies to $T(x) = -2x$ (mod 1) and to the shift in the Bernoulli space.

4 Selection of Minimizing Sequences

In this section we want to exhibit a nice expression for the function f (defined before) such that, the set $\{(x, \hat{\partial}_c f(x)) \,|\, x \in \}$ support $\{\mu_\infty\}$ = support of $\hat{\mu}_{max}$, in the case the support of $\hat{\mu}_{max}$ is a periodic orbit. In the end of the section we address briefly the general case.

Definition 13 We say that $c : \hat{\Sigma} = \Sigma \times \Sigma \to \mathbb{R}$, upper semicontinuous, satisfies the twist condition on $\hat{\Sigma}$, if (bellow we just consider values of c which are finite) for any $(a, b) \in \hat{\Sigma} = \Sigma \times \Sigma$ and $(a', b') \in \Sigma \times \Sigma$, with $a' > a$, $b' > b$, we have

$$c(a, b) + c(a', b') > c(a, b') + c(a', b). \tag{28}$$

If W is twist and $c(x, y) = I(x) - W(x, y) + \gamma$, then c is twist. We assume from now on this property.

Theorem 7 *Suppose the support of $\hat{\mu}_{max}$ is a periodic orbit. Choose (x_0, y_0) in such way that $x_0 \in \Sigma$ is the smaller point in the projection and $y_0 \in \hat{\Sigma}$ the smaller on the fiber over x_0. From the above, in this case for any given $z \in \Sigma$, the f defined before is such that*

$$\begin{aligned} f(z) = [\,(\,c(z, y_n) - c(x_n, y_n)\,)+ \\ (\,c(x_n, y_{n-1}) - c(x_{n-1}, y_{n-1})\,)\ + \ldots \\ + \ldots + \{(\,c(x_3, y_2) - c(x_2, y_2)\,)\} + \\ (\,c(x_2, y_1) - c(x_1, y_1)\,) + (\,c(x_1, y_0) - c(x_0, y_0)\,)\ ], \end{aligned}$$

where we use all the possible x_i which are in the support of the maximizing probability for A on the left of z, and for each x_i we choose the corresponding y_i. In the notation of f above, the last one $(x_n, y_n) = (x_n(z), y_n(z))$ is such that $(x_n(z), y_n(z)) = (x_{k-1}, y_{k-1})$. Which means $n = k - 1$.

Moreover, $x_0 < x_1 < x_2 < \ldots < x_n$.

If $z = x_k$ for some element x_k in the support of μ_A, then, in the notation of f above, if $x_{k-1} < z < x_k$, then $(x_n, y_n) = (x_n(z), y_n(z))$ is such that $(x_n(x_k), y_n(x_k)) = (x_{k-1}, y_{k-1})$. The case $z = x_k$ is include in the expression above for f. In this case $x_k = x_{n+1}$ following the above notation. The index of the x_i has no dynamical meaning.

Proof Consider the cost $c(x, y) = I(x) - W(x, y) - \gamma$, and a subset $S \subset X \times Y$ c-cyclically monotone. Also, assume that c verifies the twist condition: If $a < a'$ and $b < b'$ then $c(a, b) + c(a', b') > c(a, b') + c(a', b)$.

In this way, the definition of c implies that: $W(a, b) + W(a', b') < W(a, b') + W(a', b)$.

Define $\Delta(x, x', y) = W(x, y) - W(x', y)$, so the twist condition can be restated as: if $a < a'$, and $b < b'$, then, $\Delta(a, a', b) < \Delta(a, a', b')$.

Therefore, if we define the map $y \to \Delta(a, a', y)$ we get a increasing map.
Observe that:

(i) $\Delta(x, x', y) = -\Delta(x', x, y)$
(ii) $\Delta(x, x, y) = 0$
(iii) $\Delta(x, x', y) + \Delta(x', x'', y) = \Delta(x, x'', y)$

In particular the map, $y \to \Delta(a', a, y)$ is decreasing if $a' > a$.
Using the fact that $c(x, y) = I(x) - W(x, y) - \gamma$ we get,

$$\partial_c f(x) = \{y \in Y | f(z) - f(x) \leq I(z) - I(x) - [W(z, y) - W(x, y)], \forall z \in X\}.$$

We know that S is c-cyclically monotone, if and only if, $S \subset \hat{\partial}_c f(x_0)$ where f is a c-convex function given by:

$$f(z) = \min_{(x_i, y_i) \subset S, i=1..n} \sum_{i=0}^{n} c(x_{i+1}, y_i) - c(x_i, y_i),$$

where $(x_0, y_0) \in S$ is as fixed point and $x_{n+1} = z$. Using $c(x, y) = I(x) - W(x, y) - \gamma$ we get,

$$f(z) = min_{(x_i, y_i) \subset S, i=1..n} \sum_{i=0}^{n} I(x_{i+1}) - I(x_i) - [W(x_{i+1}, y_i) - W(x_i, y_i)] =$$

$$= min_{(x_i, y_i) \subset S, i=1..n} \sum_{i=0}^{n} I(x_{i+1}) - I(x_i) + [\Delta(x_i, x_{i+1}, y_i)] =$$

$$= min_{(x_i, y_i) \subset S, i=1..n} I(z) - I(x_0) + \sum_{i=0}^{n} \Delta(x_i, x_{i+1}, y_i).$$

Lemma 2 *If, $(x_i, y_i) \subset S, i = 0, 1, 2$ is such that $x_0 < x_1 < x_2 < z$ and $y_2 < y_1 < y_0$ then, $\Delta(x_0, x_1, y_0) + \Delta(x_1, z, y_1) > \Delta(x_0, x_1, y_0) + \Delta(x_1, x_2, y_1) + \Delta(x_2, z, y_2)$ (Figs. 1 and 2).*

Proof Observe that, $\Delta(x_1, z, y_1) = \Delta(x_1, x_2, y_1) + \Delta(x_2, z, y_1) > \Delta(x_1, x_2, y_1) + \Delta(x_2, z, y_2)$, because $\Delta(x_2, z, \cdot)$ is increasing and $y_1 > y_2$.

Lemma 3 *If, $(x_i, y_i) \subset S, i = 0, 1, 2$ is such that $x_0 < x_1 < z < x_2$ and $y_2 < y_1 < y_0$ then, $\Delta(x_0, x_1, y_0) + \Delta(x_1, z, y_1) < \Delta(x_0, x_1, y_0) + \Delta(x_1, x_2, y_1) + \Delta(x_2, z, y_2)$.*

In particular, $\Delta(x_0, x_1, y_0) + \Delta(x_1, z, y_1) < \Delta(x_0, x_2, y_0) + \Delta(x_2, z, y_2)$ (Figs. 3 and 4).

Proof Observe that, $\Delta(x_1, z, y_1) = \Delta(x_1, x_2, y_1) + \Delta(x_2, z, y_1) < \Delta(x_1, x_2, y_1) + \Delta(x_2, z, y_2)$, because $\Delta(x_2, z, \cdot)$ is decreasing and $y_1 > y_2$.

Fig. 1 Bad

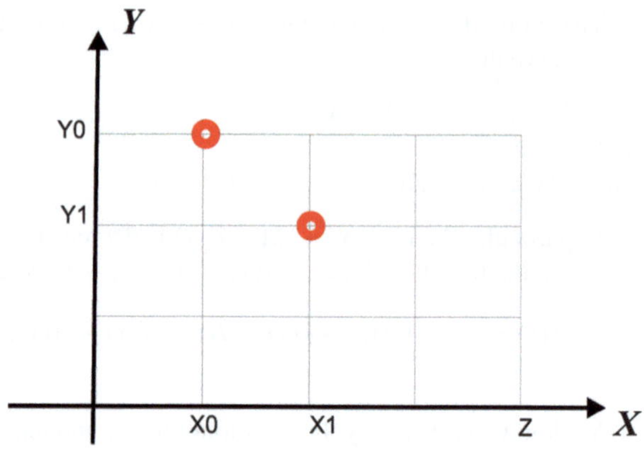

Fig. 2 Good

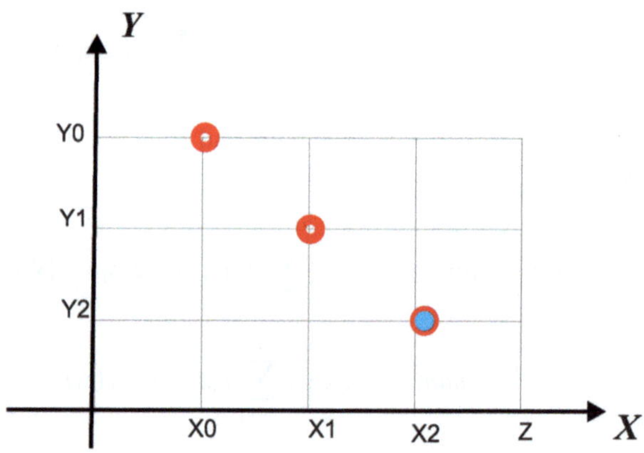

Fig. 3 Bad

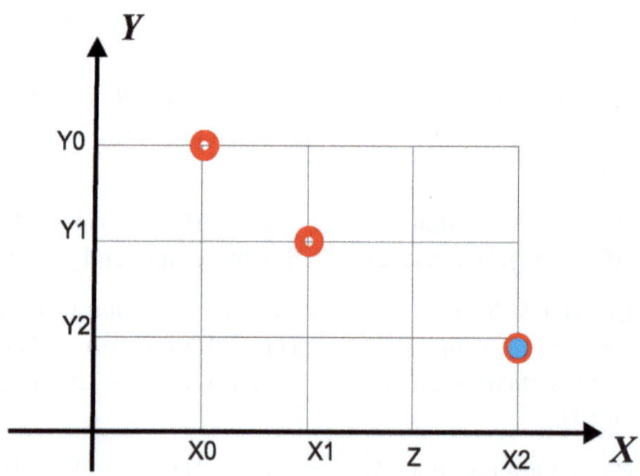

Fig. 4 Good

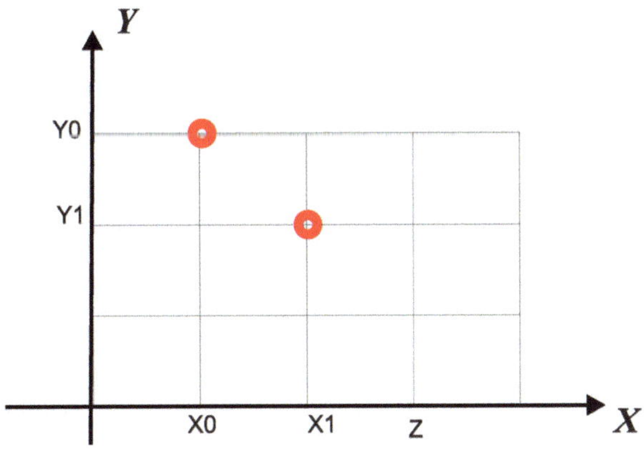

Now observe that,
$\Delta(x_0, x_2, y_0) + \Delta(x_2, z, y_2) = \Delta(x_0, x_1, y_0) + \Delta(x_1, x_2, y_0) + \Delta(x_2, z, y_2) >$
$\Delta(x_0, x_1, y_0) + \Delta(x_1, x_2, y_1) + \Delta(x_2, z, y_2) > \Delta(x_0, x_1, y_0) + \Delta(x_1, z, y_1)$.

Now one can generalize the idea above: Suppose that, $(x_i, y_i) \subset S, i = 0, 1, 2, \ldots, n$ is such that $x_0 < x_1 < \ldots < x_k < z < x_{k+1} < \ldots < x_n$ and $y_n < \ldots < y_2 < y_1 < y_0$, then, $\Delta(x_0, x_1, y_0) + \Delta(x_1, x_2, y_1) + \ldots + \Delta(x_k, z, y_k) < \Delta(x_0, x_1, y_0) + \Delta(x_1, x_2, y_1) + \ldots + \Delta(x_n, z, y_n)$.

In order to see this, we proceed by induction in the right side of the inequality above:
$\Delta(x_{n-1}, x_n, y_{n-1}) + \Delta(x_n, z, y_n) > \Delta(x_{n-1}, x_n, y_{n-1}) + \Delta(x_n, z, y_{n-1}) = \Delta(x_{n-1}, z, y_{n-1})$.

In this step we discard the pair (x_n, y_n). We must to repeat this process while $n - j > k$, discarding all points in the right side of z. So the conclusion is, that we can discard all points in the right side of z decreasing the sum, and we can introduce a point between the last point in the left size of z, and z, decreasing the sum.

We discard $(x_2, y_2), (x_3, y_3), (x_4, y_4)$, from right size and insert (A, B) between (x_1, y_1) and z (Figs. 5 and 6).

The case in which $z < x_0$ must be analyzed now:

Observe that:
$\Delta(x_0, x_1, y_0) + \Delta(x_1, x_2, y_1) + \Delta(x_2, x_3, y_2) + \Delta(x_3, x_4, y_3) + \Delta(x_4, x_5, y_4) + \Delta(x_5, z, y_5) >$
$\Delta(x_0, x_1, y_0) + \Delta(x_1, x_2, y_1) + \Delta(x_2, x_3, y_2) + \Delta(x_3, x_4, y_3) + [\Delta(x_4, x_5, y_4) + \Delta(x_5, z, y_4)] =$
$\Delta(x_0, x_1, y_0) + \Delta(x_1, x_2, y_1) + \Delta(x_2, x_3, y_2) + \Delta(x_3, x_4, y_3) + \Delta(x_4, z, y_4)$,

and successively to eliminate 4 and 3.

Now we check what happen with permutations of the order in the projected points.

Note that the sum $\sum_{i=0}^{n} c(x_{i+1}, y_i) - c(x_i, y_i)$ can change by sorting the sequence of points $(x_i, y_i) \subset S, i = 1..n$. So we need to consider the natural question about the better way to rename this points.

Fig. 5 Bad

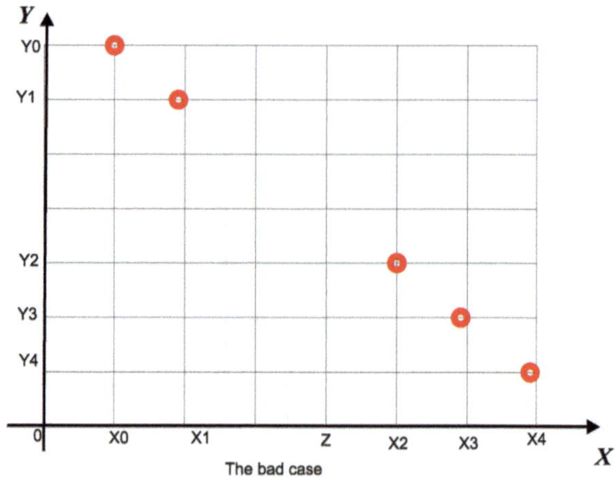

Fig. 6 Good

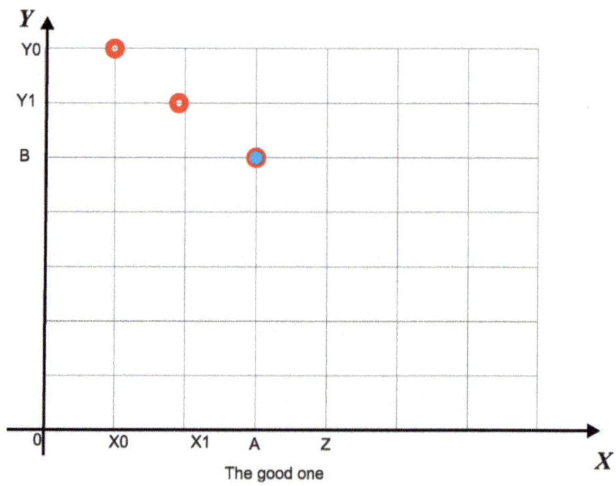

Please, check the below figure (Figs. 7 and 8):

We claim that it is possible discard all the points at the right side of z and also all the points between x_0 and z that are no ordered in order to minimize the sum above.

In fact:

$$\Delta(x_0, x_1, y_0) + \Delta(x_1, x_2, y_1) +$$

$$\Delta(x_2, x_3, y_2) + \Delta(x_3, x_4, y_3) + [\Delta(x_4, x_5, y_4) + \Delta(x_5, z, y_5)] >$$

$$\Delta(x_0, x_1, y_0) + \Delta(x_1, x_2, y_1) + \Delta(x_2, x_3, y_2) + [\Delta(x_3, x_4, y_3) + \Delta(x_4, z, y_4)] >$$

$$\Delta(x_0, x_1, y_0) + [\Delta(x_1, x_2, y_1) + \Delta(x_2, x_3, y_2)] + [\Delta(x_3, z, y_3)] >$$

$$\Delta(x_0, x_1, y_0) + [\Delta(x_1, x_3, y_1) + \Delta(x_3, z, y_3)] >$$

$$\Delta(x_0, x_1, y_0) + \Delta(x_1, z, y_1).$$

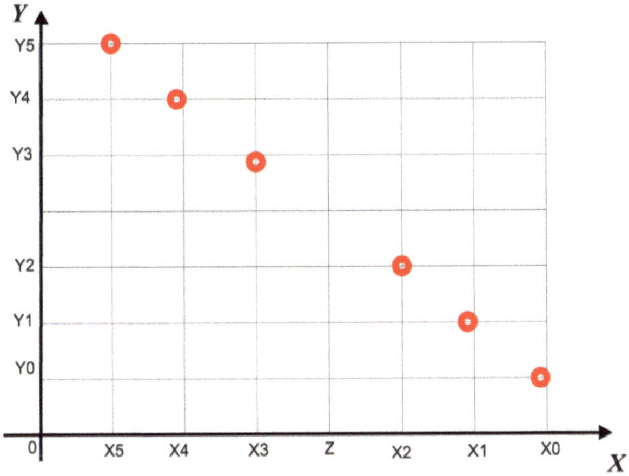

Fig. 7 Bad

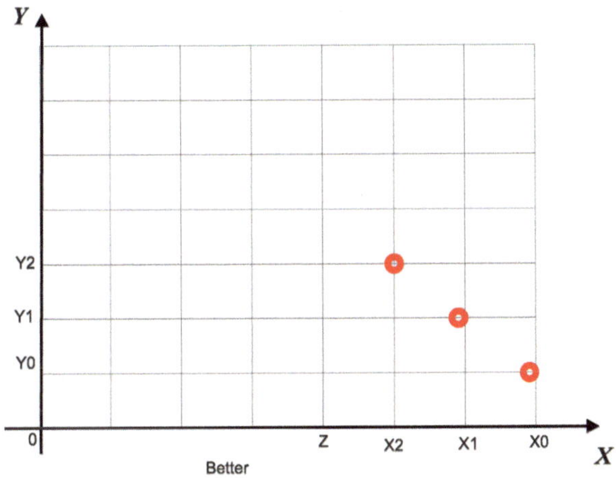

Fig. 8 Good

So the sequence $(x_0, y_0), (x_1, y_1)$ in this order minimize this sum. We know that the graph property is true. But suppose we have a more general case where $\Delta(x, z, y)$ can be consider and we do not have the graph property.

Consider the sequence $(x_0, y_0), (x_1, y_1)$ and suppose $z > x_1 > x_0$. Additionally suppose that $(x_1, .) \cap S \neq \{y_1\}$, so we can compares the sum $\Delta(x_0, x_1, y_0) + \Delta(x_1, z, y_1)$ with $\Delta(x_0, x_1, y_0) + \Delta(x_1, z, y)$ for any $y \in (x_1, .) \cap S \neq \{y_1\}$.

We claim that this function is monotone increasing in y.

In fact suppose that $y' < y_1 < y'' < y_0$, as in Figs. 9 and 10. Observe that, $\Delta(x_1, z, y_1) < \Delta(x_1, z, y'')$ and $\Delta(x_1, z, y_1) > \Delta(x_1, z, y')$ because $x_1 < z$. The conclusion is that if the support of $\hat{\mu}_{max}$ is a periodic orbit, then, we choose

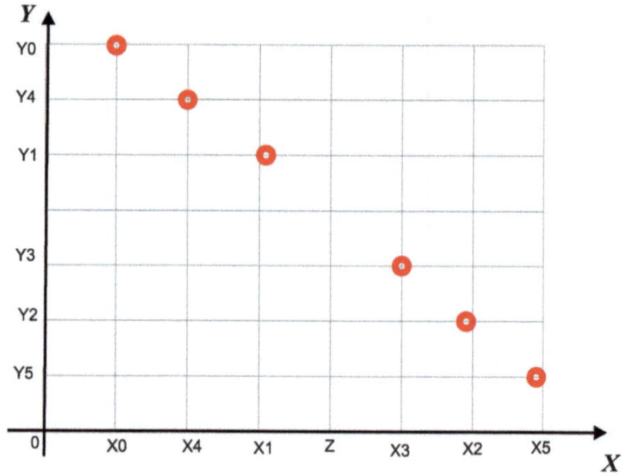

Fig. 9 Too bad

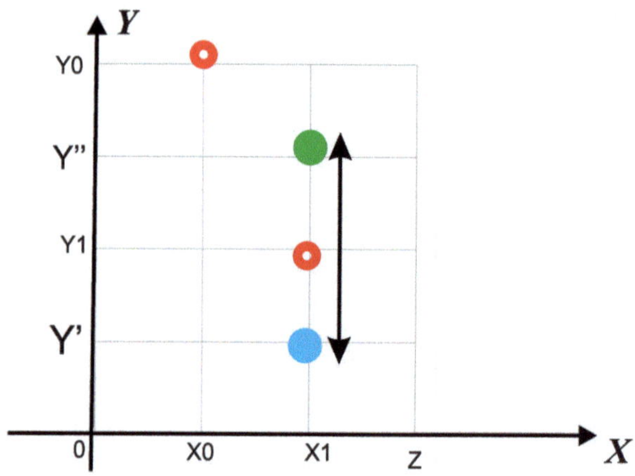

Fig. 10 Going down is better

(x_0, y_0) in the support of $\hat{\mu}_{max}$. From the above, in this case given $z \in \Sigma$, then

$$f(z) = [\,(\,c(z, y_n) - c(x_n, y_n)\,)\,]+$$
$$(\,c(x_n, y_{n-1}) - c(x_{n-1}, y_{n-1})\,)\,\,+\ldots$$
$$+\ldots+\{(\,c(x_3, y_2) - c(x_2, y_2)\,)\,\}+$$
$$(\,c(x_2, y_1) - c(x_1, y_1)\,) + [(\,c(x_1, y_0) - c(x_0, y_0)\,)\,\,].,$$

where we use all the possible x_i, $i = 1, 2, .., n$, on the left of z, and for each x_i we choose the corresponding y_i such that (x_i, y_i) is in the support of $\hat{\mu}_{max}$. Moreover, $x_0 < x_1 < x_2 < \ldots < x_n$.

Finally, we can say that $\hat{\partial}_c f(x_k) = y_k$, for any k.

One can get similar results for the function g (obtained just from the kernel W) defined before.

From the reasoning above (for the case of W satisfying the twist condition), in the case μ_∞ is not a periodic orbit, then in definition of f, the infimum is not attained in a finite sequence of x_n in the support of μ_∞.

Acknowledgements The first author is partially supported by CNPq, CAPES and FAPERGS.

Appendix

Here we consider first the shift $\Sigma = \{0,1\}^{\mathbb{N}}$, and Σ as a metric space with the usual distance:

$$d(x,y) = \begin{cases} 0, \text{ if} & x = y \\ (1/2)^n, \text{ if} & n = \min\{i \mid x_i \neq y_i\}. \end{cases}$$

Additionally, we suppose that Σ is ordered by $x < y$, if $x_i = y_i$ for $i = 1..n-1$, and $x_n = 0$ and $y_n = 1$.

As the usual, we consider the dynamical system (Σ, σ) where $\sigma : \Sigma \to \Sigma$ is given by $\sigma(x) = \sigma(x_1, x_2, x_3, \ldots) = (x_2, x_3, x_4, \ldots)$.

(a) Potentials and the Involution Kernel

As usual we denote

$$\tau_x^*(y) = (x_1, y_1, y_2, y_3, \ldots) \text{ and } \tau_y(x) = (y_1, x_1, x_2, x_3, \ldots),$$

and

$$\hat{\sigma}(x,y) = (\sigma(x), \tau_x^*(y)) \text{ and } \hat{\sigma}^{-1}(x,y) = (\tau_y x, \sigma^*(y)),$$

the skew product map, where $\sigma^*(y = (y_1, y_2, y_3, \ldots)) = (y_2, y_3, y_4, \ldots)$.

We also define $\tau_{k,y} x = (y_k, y_{k-1}, \ldots y_2, y_1, x_0, x_1, x_2, \ldots)$, where $x = (x_0, x_1, x_2, \ldots)$, $y = (y_1, y_2, y_3, \ldots)$. In a similar way we define $\tau_{k,y}^* x$.

Given a continuous function $A : \Sigma \to \mathbb{R}$, remember that a continuous function $W : \Sigma \times \Sigma \to \mathbb{R}$ is an involution kernel for A if $(W \circ \hat{\sigma}^{-1} - W + A \circ \hat{\sigma}^{-1})(x,y)$ does not depends on x; In this case the continuous function $A^*(y) = (W \circ \hat{\sigma}^{-1} - W + A \circ \hat{\sigma}^{-1})(x,y)$ is called the W-dual potential of A.

As in [2] we define the cocycle $\Delta_A(x, x', y)$, where

$$\Delta_A(x,x',y) = \sum_{n \geq 1} A \circ \hat{\sigma}^{-n}(x,y) - A \circ \hat{\sigma}^{-n}(x',y) = \sum_{n \geq 1} A \circ \tau_{n,y}(x) - A \circ \tau_{n,y}(x'),$$

and its dual version $\Delta_{A^*}(x, y, y')$, where

$$\Delta_{A^*}(x,y,y') = \sum_{n \geq 1} A^* \circ \hat{\sigma}^{n}(x,y) - A^* \circ \hat{\sigma}^{n}(x,y') = \sum_{n \geq 1} A^* \circ \tau_{n,x}^*(y) - A^* \circ \tau_{n,x}^*(y').$$

Note that:

(i) $\Delta_A(x,x',y) = -\Delta_A(x',x,y)$, in particular $\Delta_A(x,x,y) = 0$,
(ii) $\Delta_A(x,x',y) + \Delta_A(x',x'',y) = \Delta_A(x,x'',y)$,
(iii) $\Delta_A(x,x',y) = \Delta_A(\tau_y x, \tau_y x', \sigma^*(y)) + [A \circ \tau_y x - A \circ \tau_y x']$,

and the same relations are true for $\Delta_{A^*}(x,y,y')$.

Using this properties one can prove that, for any involution kernel we have $W(x,y) - W(x',y) = \Delta_A(x,x',y)$ and $W(x,y) - W(x,y') = \Delta_{A^*}(x,y,y')$.

From this fact, we get that the difference between two involution kernels for A is a continuous function of y: $\{\text{Involution kernels for } A\}/C^0(\Sigma) = W^0$, where $W^0(x,y) = \Delta_A(x,x',y)$ for a fix $x' \in \Sigma$ is called a fundamental involution kernel of A. Indeed, the property (iii) shows that W^0 is an involution kernel for A.

On the other hand, given another involution kernel, W we have $W(x,y) - W(x',y) = \Delta_A(x,x',y)$, thus

$$W(x,y) = W(x',y) + \Delta_A(x,x',y) = W(x',y) + W^0(x,y) = g(y) + W^0(x,y),$$

where $g(y) = W(x',y) \in C^0(\Sigma)$.

As an example we compute the general dual potential. First for $W^0(x,y) = \Delta_A(x,x',y)$ we get:

$$\begin{aligned} A_0^*(y) &= \quad (W^0(\tau_y x, \sigma^*(y))) - W^0(x,y) + A(\tau_y x) \\ &= \Delta_A(\tau_y x, x', \sigma^*(y)) - \Delta_A(x,x',y) + A(\tau_y x) \\ &= \qquad A(\tau_y x') + \Delta_A(\tau_y x', x', \sigma^*(y)). \end{aligned}$$

Given another involution kernel, W we have $W(x,y) = W(x',y) + W^0(x,y)$ thus

$$A^*(y) = (W \circ \hat{\sigma}^{-1} - W + A \circ \hat{\sigma}^{-1})(x,y) = W(x', \sigma^*(y)) - W(x',y) + A_0^*(y).$$

(b) The Twist Property of an Involution Kernel

If $A : \Sigma \to \mathbb{R}$ is a potential and W an arbitrary involution kernel for A, as we said before, W has the twist property, if for any, $a, b, a', b' \in \Sigma$

$$W(a,b) + W(a',b') < W(a,b') + W(a',b),$$

provided that $a < a'$ and $b < b'$.

If we rewrite this inequality as,

$$\begin{aligned} W(a,b) + W(a',b') &< W(a,b') + W(a',b) \\ W(a,b) - W(a',b) &< W(a,b') - W(a',b') \\ \Delta_A(a,a',b) \quad &< \quad \Delta_A(a,a',b'), \end{aligned}$$

we get an alternative criteria for the twist property, that is, W has the twist property, if for any, $a, a' \in \Sigma$ the function $y \to \Delta_A(a,a',y)$, is strictly increasing, provided that $a < a'$.

Remark 5 This characterization shows a very important fact. The twist property is a property of A, so we can said that A is a twist potential or equivalently A has a twist involution kernel (as, obviously other involution kernel is also twist).

Remark 6 As an initial approximation we can consider a different setting of dynamics. Let $T(x) = -2x \,(\bmod 1)$, and

$$\tau_0 x = -\frac{1}{2}x + \frac{1}{2}, \text{ and } \tau_1 x = -\frac{1}{2}x + 1,$$

the inverse branches that defines the skew maps (that are not the actual natural extension of T):

$$\hat{T}(x, y) = (T(x), \tau_x^*(y)) \text{ and } \hat{T}^{-1}(x, y) = (\tau_y x, T^*(y)).$$

So, one can compute an involutive (that is, $A^*(y) = A(y)$) smooth kernel for $A_1(x) = x$ and $A_2(x) = x^2$ given by

$$W_1(x, y) = -\frac{1}{3}(x + y) \text{ and } W_2(x, y) = \frac{1}{3}(x^2 + y^2) - \frac{4}{3}xy.$$

As a corollary we get that any potential $A(x) = a + bx + cx^2$ has a smooth involution kernel given by $W(x, y) = a + bW_1(x, y) + cW_2(x, y)$.

Here and in the next paragraphs, we will denote

$$W_A(x, y) := a + bW_1(x, y) + cW_2(x, y),$$

where $A(x) = a + bx + cx^2$ is a polynomial of degree 2.

We observe that the twist property can be derived from the positivity of the second mix derivative of the involution kernel when it is smooth. Note that,

$$\frac{\partial^2 W_1}{\partial x \partial y} = 0, \text{ and } \frac{\partial^2 W_2}{\partial x \partial y} = -\frac{4}{3},$$

thus W_1 is not twist and W_2 is. Actually any potential $A(x) = a + bx + cx^2$ where $c > 0$ is twist.

Remark 7 In this remark we are going to consider the case of $A(x) = a + bx + cx^2$ where $c < 0$ (not twist). In this case we will be able to compute the calibrated subaction explicitly, which, we believe, it is interesting in itself.

As a first example consider $A(x) = -(x - 1)^2$ which is a convex potential.

From [30, 31] we get that the unique maximizing measure for this potential is $\mu_\infty = \delta_{2/3}$, so the critical value is $m = A(2/3)$. Using the fact that $m = A(2/3)$ one can show that there is a unique (up to constants) calibrated subaction ϕ given by:

$$\phi(x) = W(x, 2/3) - W(2/3, 2/3) = -\frac{1}{3}x^2 + \frac{2}{9}x$$

where the kernel is given by

$$W(x,y) = -(1/3)x^2 - (1/3)y^2 + (4/3)xy - (2/3)x - (2/3)y.$$

As a second example consider $A(x) = -(x-\frac{1}{2})^2$ which it is also a concave potential.

The general arguments in [31] shown that any maximizing measure for this potential is $\mu_\infty = (1-t)\delta_{1/3} + t\delta_{2/3}$, where $t \in [0,1]$, so the critical value is $m = A(1/3) = A(2/3)$. In this case the involutive smooth involution kernel is:

$$W(x,y) = -(1/3)x^2 - (1/3)y^2 + (4/3)xy - (2/3)x - (1/3)y.$$

It is easy to verify that,

$$\phi(x) = V_1(x)\chi_{[(0,1/2)]}(x) + V_2(x)\chi_{[1/2,1]}(x) = \max\{V_1(x), V_2(x)\},$$

is indeed a calibrated subaction for A, where
$V_1(x) = W(x,1/3) - W(1/3,1/3) = \Delta(x,1/3,1/3) = -(1/3)x^2 + (1/9)x$,
$V_2(x) = W(x,2/3) - W(2/3,2/3) = \Delta(x,2/3,2/3) = -(1/3)x^2 + (5/9)x - 2/9$,
Note that,

$$\begin{aligned}
\phi(\tau_0 x) &= V_1(\tau_0 x)\chi_{[(0,1/2)]}(\tau_0 x) + V_2(\tau_0 x)\chi_{[1/2,1]}(\tau_0 x) \\
&= V_1(\tau_0 x) = \Delta(\tau_0 x, 1/3, 1/3) \\
&= \Delta(\tau_{1/3}x, \tau_{1/3}1/3, T^*1/3) \\
&= \Delta(x,1/3,1/3) - [A(\tau_{1/3}x) - A(\tau_{1/3}1/3)] \\
&= V_1(x) - [A(\tau_0 x) - m].
\end{aligned}$$

Thus $\phi(\tau_0 x) + A(\tau_0 x) - m = V_1(x)$. Analogously, $\phi(\tau_1 x) + A(\tau_1 x) - m = V_2(x)$ so

$$\begin{aligned}
\phi(x) &= \max\{V_1(x), V_2(x)\} \\
&= \max\{\phi(\tau_0 x) + A(\tau_0 x) - m, \phi(\tau_1 x) + A(\tau_1 x) - m\} \\
&= \max_{y\in\Sigma}\{\phi(\tau_y x) + A(\tau_y x) - m\}.
\end{aligned}$$

(c) Twist Criteria

Is natural to consider a criteria for the twist property for a class of functions that has a small dependence on the cubic (or higher order) terms. Let $P_2^+ = \{p(x) = a + bx + cx^2 \mid c > 0\}$ be the set of strictly convex polynomial. Consider $p \in P_2^+$, and define

$$\mathscr{C}_\varepsilon(p) = \{A \in C^3([0,1]) | A(x) = p(x) + \varepsilon R(x), \text{ where } \frac{\partial R}{\partial x} \in C^3([0,1])\}$$

Theorem 8 *For any $p \in P_2^+$, there exists $\varepsilon > 0$ such that all $A \in \mathscr{C}_\varepsilon(p)$ is twist.*

Proof Consider $p \in P_2^+$ fixed. So, p has a smooth and involutive involution kernel given by

$$W_p(x,y) - (a + bW_1 + cW_2)(x,y),$$

that is, $p^*(y) = p(y)$, where $W_1(x,y) = -\frac{1}{3}(x+y)$ and $W_2(x,y) = \frac{1}{3}(x^2+y^2) - \frac{4}{3}xy$, are the involution kernel associated to x and x^2 respectively. Let, $A = p + \varepsilon R \in \mathscr{C}_\varepsilon(p)$, and W_R be the involution kernel for R. Since R is C^3 we get that, its corresponding involution kernel W_R is C^2 in the variable x. Using the linearity of the cohomological equation, we get $W_A(x,y) = p(W)(x,y) + \varepsilon W_R(x,y)$, and differentiating with respect to x, we have

$$\begin{aligned}\tfrac{\partial}{\partial x}W_A(x,y) = (b\tfrac{\partial}{\partial x}W_1 + c\tfrac{\partial}{\partial x}W_2)(x,y) + \varepsilon\tfrac{\partial}{\partial x}W_R(x,y) = \\ -\tfrac{1}{3}b + \tfrac{2}{3}cx - \tfrac{4}{3}cy + \varepsilon\tfrac{\partial}{\partial x}W_R(x,y)\end{aligned}$$

Since $-\frac{4}{3}c < 0$, and $\frac{\partial}{\partial x}W_R(x,y) \in C^0([0,1]^2)$ the compactness of $[0,1]^2$ implies that $\frac{\partial}{\partial x}W_A(x,\cdot)$ is a strictly decreasing function for any ε small enough, which is sufficient to ensure the twist property.

Remark 8 If, $A \in C^\infty([0,1])$ is strongly convex, we can consider a perturbation of A of order 2 given by

$$B_\varepsilon(x) = A(0) - A'(0)x + \frac{A''(0)}{2}x^2 + \varepsilon\sum_{n\geq 3}\frac{A^{(n)}(0)}{n!}x^n \in \mathscr{C}_\varepsilon(p_A),$$

where $p_A = A(0) - A'(0)x + \frac{A''(0)}{2}x^2 \in P_2^+$. Thus, we can find $\varepsilon_0 > 0$ such that B_ε is twist for any $0 < \varepsilon < \varepsilon_0$.

(d) The Involution Kernel is Bi-Hölder

We consider now $T(x) = 2x$ (mod 1) on the interval $[0,1]$ and the shift σ on $\Omega = \{0,1\}^{\mathbb{N}}$. A natural question is the regularity of the involution kernel W. We denote τ_j, $j = 0,1$ the two inverse branches of T. Given $w = (w_1, w_2, \ldots) \in \{0,1\}^{\mathbb{N}}$ we denote by $\tau_{k,w}$ the transformation in $[0,1]$ given by $\tau_{k,w}(x) = (\tau_{w_k} \circ \tau_{w_{k-1}} \circ \ldots \circ \tau_{w_1})(x)$. We have that, for a fixed x_0

$$\Delta(x, x_0, w) = \sum_{k=1}^{\infty} A(\tau_{k,w}(x)) - A(\tau_{k,w}(x_0))$$

and, the involution kernel W can be described as: for any (x,w) we have $W(x,w) = \Delta(x, x_0, w)$. It is easy to see that W is Hölder on the variable x. Consider $a, b \in \Omega$ and suppose that $d(a,b) = 2^{-n}$. In this way $a_j = b_j, j = 1, 2\ldots, n-1, n$. We denote $\bar{a} = \sigma^n(a)$ and $\bar{b} = \sigma^n(b)$.

Proposition 7 *Suppose A is $\alpha-$Hölder. Consider $a, b \in \Omega$ such that $d(a,b) = 2^{-n}$. For a fixed $x \in [0,1]$ we have $\mid W(x,a) - W(x,b) \mid \leq C\,(2^{-n})^{\alpha}$.*

Proof Note that for $z = \tau_{n,a}(x) = \tau_{n,b}(x)$ and $z_0 = \tau_{n,a}(x_0) = \tau_{n,b}(x_0)$ we have

$$W(x,a) - W(x,b) = \sum_{k=1}^{\infty} A(\tau_{k,a}(x)) - A(\tau_{k,a}(x_0)) - A(\tau_{k,b}(x)) + A(\tau_{k,b}(x_0)) =$$

$$\sum_{k=1}^{\infty} [A(\tau_{k,a}(x)) - A(\tau_{k,b}(x))] - [A(\tau_{k,a}(x_0)) - A(\tau_{k,b}(x_0))] =$$

$$\sum_{k=1}^{\infty} [A(\tau_{k,\bar{a}}(z)) - A(\tau_{k,\bar{b}}(z))] - [A(\tau_{k,\bar{a}}(z_0)) - A(\tau_{k,\bar{b}}(z_0))].$$

Note also that $|z - z_0| \leq d(a,b) = 2^{-n}$. Consider $z = z_0 + h$, then

$$A(\tau_{k,\bar{a}}(z_0 + h)) - A(\tau_{k,\bar{a}}(z_0)) \leq C_A\, d(\tau_{k,\bar{a}}(z_0 + h), \tau_{k,\bar{a}}(z_0))^{\alpha} \leq$$
$$C_A\,(2^{-k}\,h)^{\alpha} = C_A(2^{-k})^{\alpha}\,h^{\alpha}.$$

Then,

$$\sum_{k=1}^{\infty} [A(\tau_{k,\bar{a}}(z)) - A(\tau_{k,\bar{a}}(z_0))] - [A(\tau_{k,\bar{b}}(z)) - A(\tau_{k,\bar{b}}(z_0))]$$

$$\leq C_A \sum_{k=1}^{\infty} 2\,(2^{-k})^{\alpha}\,h^{\alpha} \leq C_A \sum_{k=1}^{\infty} 2\,(2^{\alpha})^{-k}\,h^{\alpha} \leq C\,d(a,b)^{\alpha}.$$

From the above we get:

Theorem 9 *If $A : S^1 \to \mathbb{R}$ is Hölder then $W : S^1 \times \{0,1\}^{\mathbb{N}} \to \mathbb{R}$ is bi-Hölder.*

(e) **The Fenchel-Rockafellar Theorem** Given $f : \mathbb{R} \to \mathbb{R}$ defined on the variable x, the Legendre transform of f, denoted by f^*, is the function on the variable p defined by

$$f^*(p) = \sup_{x \in \mathbb{R}} \{p\,x - f(x)\}.$$

Theorem 10 (Fenchel-Rockafellar) *Suppose $f(x)$ is smooth strictly convex, $f : \mathbb{R} \to \mathbb{R}$, and, $g(x)$ is smooth strictly concave, $g : \mathbb{R} \to \mathbb{R}$. Denote by f^* and g^* the corresponding Legendre transforms on the variable p. Then,*

$$\inf_{x \in \mathbb{R}} \{f(x) \; - \; g(x)\} \; = \; \sup_{p \in \mathbb{R}} \{g^*(p) \; - \; f^*(p)\}$$

Fig. 11 The infimum

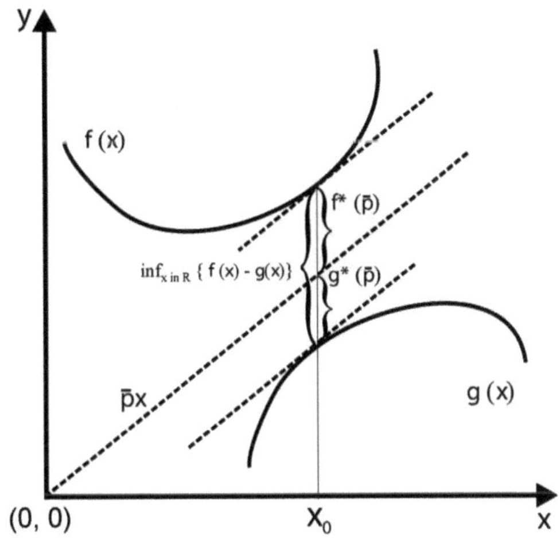

Fig. 12 The supremum

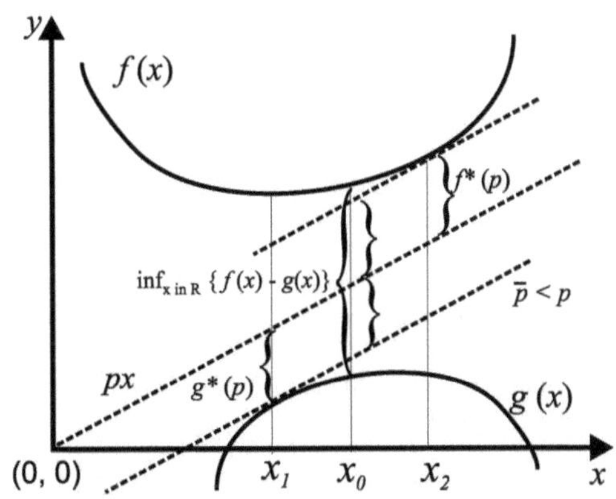

Proof By convexity and concavity properties we have that there exists x_0 such that

$$\inf_{x \in \mathbb{R}} \{f(x) - g(x)\} = f(x_0) - g(x_0).$$

It is also true that $f'(x_0) - g'(x_0) = 0$. Denote by $\overline{p}$ that value $\overline{p} = f'(x_0)$. We illustrate the proof via two pictures in a certain particular case. Figure 11 shows a geometric picture of the position and values of $f(x_0) - g(x_0)$, $g^*(\overline{p})$ and $f^*(\overline{p})$. Note that in this picture we have that $f(x_0) - g(x_0) > 0$. This picture also shows the graph of $\overline{p}\,x$ as a function of x. We observe that the Legendre transform is not linear on the function. Let's consider different values of p and estimate $f^*(p)$ and $g^*(p)$. Suppose first $p > \overline{p}$. In Fig. 12 we show the graph of $p\,x$, and the values of $f^*(p)$ and $g^*(p)$. We denote by x_2 the value such that

$$f^*(p) = \sup_{x \in \mathbb{R}}\{p\,x - f(x)\} = p\,x_2 - f(x_2).$$

Note that $x_2 > x_0$. We denote by x_1 the value such that

$$0 < g^*(p) = \sup_{x\in\mathbb{R}}\{px - g(x)\} = px_1 - g(x_1).$$

Note that $x_1 < x_0$.
Note also that $f^*(p)$ and $g^*(p)$ have different signs. From this picture one can see that $g^*(p) - f^*(p) < f(x_0) - g(x_0)$. In the case $p < \overline{p}$ a similar reasoning can be done.

References

1. Bangert, V.: Mather sets for twist maps and geodesics on tori. In: Dynamics Reported, vol. 1, pp. 1–56. Wiley, Chichester (1988)
2. Baraviera, A., Lopes, A.O., Thieullen, Ph.: A large deviation principle for equilibrium states of holder potentials: the zero temperature case. Stoch. Dyn. **6**, 77–96 (2006)
3. Baraviera, A.T., Cioletti, L.M., Lopes, A.O., Mohr, J., Souza, R.R.: On the general *XY* Model: positive and zero temperature, selection and non-selection. Rev. Math. Phys. **23**(10), 1063–1113 (2011)
4. Baraviera, A., Lopes, A.O., Mengue, J.: On the selection of subaction and measure for a subclass of potentials defined by P. Walters. Ergodic Theory Dyn. Syst. **33**(5), 1338–1362 (2013)
5. Baraviera, A., Leplaideur, R., Lopes, A.O.: Ergodic optimization, zero temperature limits and the max-plus algebra. In: Mini-course in XXIX Coloquio Brasileiro de Matemática, Rio de Janeiro (2013)
6. Bhattacharya, P., Majumdar, M.: Random Dynamical Systems. Cambridge University Press, Cambridge (2007)
7. Bissacot, R., Garibaldi, E.: Weak KAM methods and ergodic optimal problems for countable Markov shifts. Bull. Braz. Math. Soc. **41**(3), 321–338 (2010)
8. Bousch, T.: Le poisson n'a pas d'arêtes. Ann. Inst. Henri Poincaré Probab. Stat. **36**, 489–508 (2000)
9. Bousch, T.: La condition de Walters. Ann. Sci. l'École Normale Supérieure **34**, 287–311 (2001)
10. Contreras, G.: Ground states are generically a periodic orbit, Arxiv (2013)
11. Contreras, G., Iturriaga, R.: Global minimizers of autonomous Lagrangians. 22° Colóquio Brasileiro de Matemática, IMPA (1999)
12. Contreras,G., Lopes, A.O., Thieullen, Ph.: Lyapunov minimizing measures for expanding maps of the circle. Ergodic Theory Dyn. Syst. **21**, 1379–1409 (2001)
13. Contreras, G., Lopes, A.O., Oliveira, E.: Ergodic Transport Theory, periodic maximizing probabilities and the twist condition. In: D. Zilberman, A. Pinto (eds.) Modeling, Optimization, Dynamics and Bioeconomy. Springer Proceedings in Mathematics, pp. 183–219. Springer, Cham (2014)
14. Conze, J.P., Guivarc'h, Y.: Croissance des sommes ergodiques et principe variationnel, manuscript circa (1993)
15. Delon, J., Salomon, J., Sobolevski, A.: Fast transport optimization for Monge costs on the circle. SIAM J. Appl. Math. **7**, 2239–2258 (2010)
16. Dembo, A., Zeitouni, O.: Large Deviations Techniques and Applications. Springer, New York (1998)
17. Evans, L., Gomes, D.: Linear programming interpretation of Mather's variational principle. ESAIM Control Optim. Cal. Var. **8**, 693–702 (2002)

18. Galatolo, S., Pacifico, M.: Lorenz-like flows: exponential decay of correlations for the Poincaré map, logarithm law, quantitative recurrence. Ergodic Theory Dyn. Syst. **30**(6), 1703–1737 (2010)
19. Gangbo, W., McCann, R.J.: The geometry of optimal transportation. Acta Math. **177**, 113–161 (1996)
20. Garibaldi, E., Lopes, A.O.: Functions for relative maximization. Dyn. Syst. **22**, 511–528 (2007)
21. Garibaldi, E., Lopes, A.O., Thieullen, Ph.: On calibrated and separating sub-actions. Bull. Braz. Math. Soc. **40**(4), 577–602 (2009)
22. Garibaldi, E., Lopes, A.O.: The effective potential and transshipment in thermodynamic formalism at temperature zero. Stoch. Dyn. **13**(1), 1250009 (13 p) (2013)
23. Garibaldi, E., Thieullen, Ph.: Minimizing orbits in the discrete Aubry-Mather model. Nonlinearity **24**(2), 563–611 (2011)
24. Garibaldi, E., Thieullen, Ph.: Description of some ground states by Puiseux technics. J. Stat. **146**(1), 125–180 (2012)
25. Gole, C.: Sympletic Super-Twist Maps. World Scientific, Singapore (1998)
26. Hunt, B.R., Yuan, G.C.: Optimal orbits of hyperbolic systems. Nonlinearity **12**, 1207–1224 (1999)
27. Jenkinson, O.: Ergodic optimization. Discrete Continuous Dyn. Syst. Ser. A **15**, 197–224 (2006)
28. Jenkinson, O.: Every ergodic measure is uniquely maximizing. Discrete Continuous Dyn. Syst. Ser. A **16**, 383–392 (2006)
29. Jenkinson, O.: A partial order on x2 -invariant measures. Math. Res. Lett. **15**(5), 893–900 (2008)
30. Jenkinson, O.: Optimization and majorization of invariant measures. Electron. Res. Announc. Am. Math. Soc. **13**, 1–12 (2007)
31. Jenkinson, O., Steel, J.: Majorization of invariant measures for orientation-reversing maps. Ergodic Theory Dyn. Syst. **30**(5), 1471–1483 (2010)
32. Kloeckner, B.: Optimal Transport and dynamics of circle expanding maps acting on measures. Ergodic Theory Dyn. Syst. **33**(2), 529–548 (2013)
33. Kloeckner, B., Lopes, A.O., Stadlbauer, M.: Contraction in the Wasserstein metric for some Markov chains, and applications to the dynamics of expanding maps, preprint (2014)
34. Kloeckner, B., Giulietti, P., Lopes, A.O., Marcon, D.: On the Geometry of Thermodynamical Formalism, preprint (2014)
35. Leplaideur, R.: A dynamical proof for the convergence of Gibbs measures at temperature zero. Nonlinearity **18**(6), 2847–2880 (2005)
36. Lopes, A., Mengue, J.: Duality theorems in ergodic transport. J. Stat. Phys. **149**(5), 921–942 (2012)
37. Lopes, A.O., Oliveira, E.R.: On the thin boundary of the fat attractor, preprint UFRGS (2011)
38. Lopes, A.O., Thieullen, Ph.: Sub-actions for Anosov diffeomorphisms. Astérisque **287**, 135–146 (2003)
39. Lopes, A.O., Thieullen, P.: Mather measures and the Bowen-Series transformation. Ann. Inst. Henri Poincaré Analyse non Linéaire **23**, 663–682 (2006)
40. Lopes, A.O., Oliveira, E.R., Smania, D.: Ergodic transport theory and piecewise analytic subactions for analytic dynamics. Bull. Braz. Math. Soc. **43**(3), 467–512 (2012)
41. Lopes, A.O., Mengue, J.K., Mohr, J., Souza, R.R.: Entropy, pressure and duality for Gibbs plans in ergodic transport. Bull. Braz. Math. Soc. (to appear)
42. Lopes, A., Mengue, J.K., Mohr, J., Souza, R.R.: Entropy and variational principle for one-dimensional lattice systems with a general a-priori probability: positive and zero temperature. Ergodic Theory Dyn. Syst. (to appear)
43. Mather, J.: Action minimizing invariant measures for positive definite Lagrangian Systems. Math. Z. **207**(2), 169–207 (1991)
44. Mengue, J.K., Oliveira, E.R.: Duality results for Iterated Function Systems with a general family of branches, preprint Arxiv (2014)

45. Mitra, T.: Introduction to dynamic optimization theory. In: Majumdar, M., Mitra, T., Nishimura, K. (eds.) Optimization and Chaos. Studies in Economic Theory. Springer, Heidelberg (2000)
46. Morris, I.D.: A sufficient condition for the subordination principle in ergodic optimization. Bull. Lond. Math. Soc. **39**(2), 214–220 (2007)
47. Parry, W., Pollicott, M.: Zeta functions and the periodic orbit structure of hyperbolic dynamics. In: Astérisque vol. 187–188. Société mathématique de France, Paris (1990)
48. Rachev, S., Ruschendorf, L.: Mass Transportation Problems, vol. 1 and 2. Springer, New York (1998)
49. Ruscheendorf, L.: On c-optimal random variables. Stat. Probab. Lett. **27**, 267–270 (1996)
50. Savchenko, S.V.: Cohomological inequalities for finite Markov chains. Funct. Anal. Appl. **33**, 236–238 (1999)
51. Souza, R.R.: Ergodic and Thermodynamic Games, preprint (to appear in Stochastics and Dynamics, 2014)
52. Tal, F.A., Zanata, S.A.: Maximizing measures for endomorphisms of the circle. Nonlinearity **21**, 2347–2359 (2008)
53. Villani, C.: Topics in Optimal Transportation. AMS, Providence (2003)
54. Villani, C.: Optimal Transport: Old and New. Springer, Berlin (2009)

Rolling Maps for the Essential Manifold

L. Machado, F. Pina, and F. Silva Leite

Abstract Computer vision problems typically have geometric constraints. When two cameras view a 3D scene from two distinct positions, or a single camera views a 3D scene from two different locations, there are a number of geometric relations between the 3D points and their projections onto the 2D images. These relations lead to constraints between the image points. In particular, the epipolar constraint encodes the relation between correspondences across two images of the same scene. In a calibrated setting, the epipolar constraint is parameterized by essential matrices, which form the Essential Manifold. The reconstruction of a video from several images of a scene can be formulated as an interpolation problem on this manifold. An approach that simplifies the generation of an interpolating curve consists in projecting the problem to a linear manifold where it can be solved easily, and then projecting back the solution on the nonlinear manifold. The projection is realized by rolling the Essential Manifold, without slip and twist, over an affine tangent space. This gives particular relevance to rolling motions in the context of certain computer vision problems. Having this in mind, we derive the kinematic equations for the rolling motions of the Essential Manifold and present explicit solutions when it rolls along geodesics.

L. Machado
Institute of Systems and Robotics, University of Coimbra - Pólo II, Pinhal de Marrocos, 3030-290 Coimbra, Portugal

Department of Mathematics, University of Trás-os-Montes and Alto Douro, Quinta de Prados, 5001-801 Vila Real, Portugal
e-mail: lmiguel@utad.pt

F. Pina
Department of Mathematics, University of Coimbra, Largo D. Dinis, 3001-454 Coimbra, Portugal
e-mail: fpina@mat.uc.pt

F.S. Leite (✉)
Institute of Systems and Robotics, University of Coimbra - Pólo II, Pinhal de Marrocos, 3030-290 Coimbra, Portugal

Department of Mathematics, University of Coimbra, Largo D. Dinis, 3001-454 Coimbra, Portugal
e-mail: fleite@mat.uc.pt

J.-P. Bourguignon et al. (eds.), *Dynamics, Games and Science*, CIM Series in Mathematical Sciences 1, DOI 10.1007/978-3-319-16118-1_21

1 Introduction

Computer vision is a challenging topic which is being used in a wide variety of real world applications, such as earth observation, optical character recognition, 3D model building, medical imaging, machine inspection, automotive safety, match move, motion capture, surveillance, fingerprint recognition and biometrics. We refer to Szeliski [13] and references therein for details concerning multiple applications in this area.

The problem of recovering structure and motion from a sequence of images, also known as stereo matching, is a crucial problem in computer vision and continues to be one of the most active research areas with remarkable progress in imaging and computing hardware (see also Ma et al. [10]). The Essential Manifold plays an important role in this area since it encodes the epipolar constraint. The classical problem of reconstructing a scene, or a video, from several images of the scene can be formulated as an interpolation problem on the Essential Manifold. Typically, it is given an ordered set of time-labeled essential matrices, $E_1, \ldots, E_n$ relating n different consecutive camera views (*snapshots*), and the objective is to calculate a continuum of additional virtual views by computing a smooth interpolating curve through the E_i's. According to Hüper and Silva Leite [6], interpolation problems on manifolds can be efficiently solved via rolling techniques. This approach enables to transform a difficult problem on a curved space into an easy problem on a flat space. Therefore, in order to implement an interpolation algorithm on the Essential Manifold it is particularly important to study rolling motions of this manifold over an affine tangent space where classical interpolation methods may then be applied. There are other problems in the area of computer vision where rolling methods have been used successfully. We refer to Caseiro et al. [1] for a novel application of rolling to solve multi-class classification problems on manifolds.

The classical definition of rolling, without slip and without twist, is presented in Sharpe [12] for manifolds embedded in Euclidean spaces, namely $\mathbb{R}^n$ equipped with the Euclidean metric. These rolling motions result from the action of the group of orientation preserving isometries of the ambient space, which is the special Euclidean group SE_n. Although, according to Nash Theorem [11], every finite dimensional Riemannian manifold can be smoothly isometrically embedded in a sufficiently high-dimensional Euclidean space, finding an appropriate embedding is not necessarily an easy task. For that reason, the concept of rolling has been extended to manifolds embedded in a general Riemannian manifold in Hüper et al. [7]. In the present work though, we consider the Essential Manifold embedded in an appropriate Euclidean space, but since elements in this manifold have a matrix representation, we follow the approach in Hüper and Silva Leite [6] and adjust Sharpe's definition so that the matrix structure is not destroyed.

The organization of the paper is the following. In Sect. 2, we introduce the notions of essential matrix, epipolar constraint and Essential Manifold, and describe the Riemannian structure of the Essential Manifold, following the approach given in Helmke et al. [4] and Ma et al. [10]. Section 3 starts with the notion of a rolling map, which describes the rolling motion of a submanifold of a general Riemannian manifold over another submanifold of equal dimension. This is based on the work of Hüper et al. [7]. The main results, dedicated to the rolling motions of the Essential Manifold over the affine tangent space at a point, appear after the general definition of rolling. More specifically, we adjust the general conditions to our particular case, derive the kinematic equations of rolling, and solve those equations explicitly when the rolling curves are geodesics on the manifold. The paper ends with some remarks and directions for further research.

2 The Essential Manifold

2.1 *Geometric Formulation*

It is well known from computer vision literature that the intrinsic projective geometry between two views of the same scene is independent of the scene structure and only depends on the cameras internal parameters and relative pose (see, for instance, Hartley and Zisserman [3]). In this paper we deal with calibrated cameras, that is, we assume that the camera parameters are known. We also assume that the scene is static and, for simplicity, we admit that the images are taken by two identical pinhole cameras, with focal length equal to one. The two cameras are denoted by C_1 and C_2 and the corresponding images of the scene structure p are denoted by X_1 and X_2, respectively, as shown in Fig. 1.

Each camera is represented by an orthonormal reference frame and can therefore be described as a change of coordinates relatively to an inertial reference frame. Without loss of generality, we can assume that the inertial frame corresponds to one of the two cameras, say C_1, while the other is positioned and oriented according to

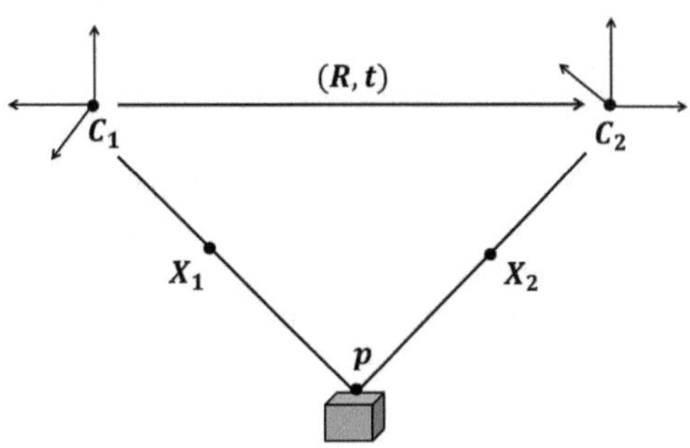

Fig. 1 Geometry between two views of the same scene structure

an element (R, s) of the special Euclidean group $SE_3 = SO_3 \ltimes \mathbb{R}^3$, where R denotes a rotation and s represents a translation vector of the displacement of the first camera C_1 into the second one C_2. Let s_1, s_2 and s_3 be the coordinates of s with respect to the first camera basis ($s = [s_1\ s_2\ s_3]^\top$) and $x_1, x_2 \in \mathbb{R}^3$ be the homogeneous coordinates of the projection of the same point p onto the two image planes of the cameras. If we call $X_1 \in \mathbb{R}^3$ and $X_2 \in \mathbb{R}^3$ the 3D coordinates of the point p relative to the two camera frames, they are related by a rigid body motion:

$$X_2 = RX_1 + s,$$

where $X_i = \lambda_i x_i$, $i = 1, 2$, can be written in terms of the image points x_i, $i = 1, 2$ and the depths $\lambda_i, i = 1, 2$, $(\lambda_i > 0)$. So, the last equation can be written as

$$\lambda_2 x_2 = R\lambda_1 x_1 + s. \tag{1}$$

Consider the isomorphism

$$\widehat{(\cdot)}\colon \mathbb{R}^3 \longrightarrow \mathfrak{so}_3$$

$$s = [s_1\ s_2\ s_3]^\top \longmapsto \hat{s} := \begin{bmatrix} 0 & -s_3 & s_2 \\ s_3 & 0 & -s_1 \\ -s_2 & s_1 & 0 \end{bmatrix},$$

between $\mathbb{R}^3$ and the Lie algebra of SO_3, which is the set of all 3×3 skew-symmetric matrices, here denoted by $\mathfrak{so}_3$. It is well known and trivial to prove that for any vector $x \in \mathbb{R}^3$, $\hat{s}x = s \times x$ ($\times$ denotes the cross product). Multiplying (on the left) both sides of the Eq. (1) by $\hat{s}$ we then obtain

$$\lambda_2 \hat{s} x_2 = \lambda_1 \hat{s} R x_1.$$

Now, by taking the inner product of both sides of the previous equation with x_2, it follows

$$x_2^\top \hat{s} R x_1 = 0, \tag{2}$$

which is called the *epipolar constraint* (Longuet-Higgins [9]). This intrinsic constraint is independent of depth information and decouples the problem of motion recovery from 3D structure. This problem consists in finding $(R, s) \in SE_3$ using the known image points x_1 and x_2 and the epipolar constraint. The matrix $E = \hat{s}R$ in (2), which captures the relative orientation between the two cameras, is called the *essential matrix* and the set of all such matrices is the so-called Essential Manifold.

2.2 Riemannian Structure of the Normalized Essential Manifold

For many of the applications concerning essential matrices, it is enough to work with a subset of normalized matrices, those of the form $\hat{s}R$, where the translation vector s has norm 1. This set, referred in the literature as the Normalized Essential Manifold, is defined as

$$\mathscr{E} = \left\{ \hat{s}R : \hat{s} \in \mathfrak{so}_3,\ R \in \mathrm{SO}_3,\ \frac{1}{2}\mathrm{tr}(\hat{s}^\top \hat{s}) = 1 \right\}.$$

As a consequence of a result in Huang and Faugeras [5], concerning a characterization of essential matrices in terms of their singular values, we can say that all normalized essential matrices have singular values $\{1, 1, 0\}$. Therefore, using the singular value decomposition, any matrix E in $\mathscr{E}$ can be written as

$$E = UE_0V^\top, \text{ for some } U,\ V \in \mathrm{SO}_3 \text{ and } E_0 = \begin{bmatrix} I_2 & 0 \\ 0 & 0 \end{bmatrix}.$$

Also, as pointed out in Helmke et al. [4], the Normalized Essential Manifold can also be represented by pairs $(UE_0U^\top, UV^\top)$, where U, V and E_0 are as above. That is, $\mathscr{E} = \mathscr{G}(2,3) \times \mathrm{SO}_3$, where $\mathscr{G}(2,3)$ is the isospectral manifold consisting of the 3×3 real symmetric projection matrices of rank 2 (a Grassmann manifold). From now on we use this parametrization so that $\mathscr{E} = \{(UE_0U^\top, UV^\top) : \quad U,\ V \in \mathrm{SO}_3\}$. We may replace $UV^\top$ by an arbitrary rotation matrix R to obtain the following definition of the Normalized Essential Manifold that will be used throughout the rest of the paper. Also, for the sake of brevity we omit the word normalized and call it simply Essential Manifold.

Definition 1 The *Essential Manifold* is the 5-dimensional smooth manifold defined as

$$\mathscr{E} := \left\{ \left(UE_0U^\top, R\right) : \quad U,\ R \in \mathrm{SO}_3 \right\}, \tag{3}$$

where

$$E_0 = \begin{bmatrix} I_2 & 0 \\ 0 & 0 \end{bmatrix}. \tag{4}$$

The Essential Manifold can be considered embedded in the Euclidean space $\mathfrak{s}_3 \times \mathbb{R}^{3\times 3}$, where $\mathfrak{s}_3$ denotes the set of all 3×3 real symmetric matrices. The natural metric in this embedding space is defined as

$$\langle (J,K), (L,M) \rangle_{\mathfrak{s}_3 \times \mathbb{R}^{3\times 3}} = \langle J, L \rangle_{\mathfrak{s}_3} + \langle K, M \rangle_{\mathbb{R}^{3\times 3}}, \tag{5}$$

where the metric $\langle .,. \rangle$ in the right hand side is related to the Frobenius norm for matrices, that is, $\langle A, B \rangle = \mathrm{tr}(A^\top B)$. With the above parametrization of the Essential Manifold, the tangent space at a point $P_0 = (\Theta_0 E_0 \Theta_0^\top, R_0) \in \mathscr{E}$ and the corresponding orthogonal space are, respectively, given by:

$$T_{P_0}\mathscr{E} = \left\{ \left(\Theta_0 \left[\Omega, E_0 \right] \Theta_0^\top,\, R_0 C \right) : \Omega, C \in \mathfrak{so}_3 \right\} \tag{6}$$

or, equivalently,

$$T_{P_0}\mathscr{E} = \left\{ \left(\Theta_0 \begin{bmatrix} 0 & \Lambda \\ \Lambda^\top & 0 \end{bmatrix} \Theta_0^\top,\, R_0 C \right) : \Lambda \in \mathbb{R}^{2\times 1},\, C \in \mathfrak{so}_3 \right\} \tag{7}$$

and

$$(T_{P_0}\mathscr{E})^\perp = \left\{ \left(\Theta_0 \begin{bmatrix} B & 0 \\ 0 & b \end{bmatrix} \Theta_0^\top,\, R_0 S \right) : B \in \mathfrak{s}_2,\, b \in \mathbb{R},\, S \in \mathfrak{s}_3 \right\}. \tag{8}$$

Note that, as expected, the dimensions of the above spaces match with the dimension of the embedding space, which is 15. Indeed, $\dim(T_{P_0}\mathscr{E}) = 5$ and $\dim\left((T_{P_0}\mathscr{E})^\perp\right) = 10$.

3 Rolling the Essential Manifold

We start this section with the important definition of a rolling map for general manifolds and then specialize to the situation when the Essential Manifold rolls, without slip and twist, over the affine tangent space at a point. The kinematic equations for this rolling motion are therefore derived.

3.1 Rolling Maps

We gather the necessary information about rolling maps, so that we can describe the rolling motion of the Essential Manifold later. As already mentioned, the classical definition of a rolling map, for manifolds embedded in Euclidean spaces, appeared first in Sharpe [12]. In the meanwhile, it has been refined and generalized in order to accommodate manifolds embedded in a general Riemannian manifold. We follow closely the notations in Hüper and Silva Leite [6], but include a more general definition contained in Hüper et al. [7]. In this context, a rolling map describes how two connected manifolds M_0 and M_1 of the same dimension n, both isometrically embedded in the same Riemannian complete m-dimensional manifold $\overline{M}$ ($1 \leq n < m$), roll on each other without slipping and twisting. These motions are described by the action of the group of isometries on the embedding manifold $\overline{M}$, which preserve orientations. Let us recall that, if $\overline{M}$ is equipped with the tensor metric g, an isometry on $\overline{M}$ is a diffeomorphism $l : \overline{M} \to \overline{M}$ which preserves g, that is, $l^* g = g$, where l^* denotes the pullback of l. Furthermore, the group of isometries on $\overline{M}$, denoted by $\mathsf{Isom}(\overline{\mathsf{M}})$ is a Lie group, whose dimension is never greater than $m(m+1)/2$ (Kobayashi [8]). The rolling map will be defined as a curve in the connected component of $\mathsf{Isom}(\overline{\mathsf{M}})$ that contains the identity, satisfying several conditions to be presented in Definition 2. This subgroup will be denoted by $\overline{G}$. So, a rolling map on a closed interval $I = [0, \tau] \subset \mathbb{R}$ ($\tau > 0$) can be described using the following pair of mappings:

$$\begin{aligned} h : I &\to \overline{G} \\ t &\mapsto h(t) \end{aligned} \quad \text{and} \quad \begin{aligned} h(t) : \overline{M} &\to \overline{M} \\ p &\mapsto q = h(t)(p) \end{aligned}.$$

Let $x \in \overline{M}$ be a point and $\eta \in T_x\overline{M}$ be a tangent vector. This means that there exists a smooth curve $y :]-\varepsilon, \varepsilon[\to \overline{M}$ such that $y(0) = x$ and $\dot{y}(0) = \eta$. We denote by $h(t)_*$ the pushforward (differential) of $h(t)$. Also, from the action of $\overline{G}$ on $\overline{M}$, we may define the following actions, which will be used in the definition of rolling map (Definition 2).

$$\dot{h}(t)(x) := \frac{\mathrm{d}}{\mathrm{d}\sigma} \left[h(\sigma)(x) \right] \Big|_{\sigma=t}, \tag{9}$$

$$\left(\dot{h}(t) \circ h(t)^{-1} \right)(x) := \frac{\mathrm{d}}{\mathrm{d}\sigma} \left[\left(h(\sigma)\, h(t)^{-1} \right)(x) \right] \Big|_{\sigma=t}, \tag{10}$$

$$\left(\dot{h}(t) \circ h(t)^{-1} \right)_* (\eta) := \frac{\mathrm{d}}{\mathrm{d}\sigma} \left[\left(\dot{h}(t) \circ h(t)^{-1} \right)(y(\sigma)) \right] \Big|_{\sigma=0}. \tag{11}$$

Definition 2 Let M_0 and M_1 be two n-dimensional connected manifolds isometrically embedded in an m-dimensional complete Riemannian manifold $\overline{M}$ and let $\overline{G}$ be the connected component of the group of isometries of $\overline{M}$ that contains the identity. A *rolling map of* M_1 *over* M_0, *without slipping and twisting*, is a smooth curve $h : I \to \overline{G}$, satisfying, for all $t \in I$, the following three properties:

1. *Rolling conditions*: There exists a smooth curve $\alpha_1 : I \to M_1$, such that

   a. $h(t)(\alpha_1(t)) \in M_0$;
   b. $T_{h(t)(\alpha_1(t))}(h(t)(M_1)) = T_{h(t)(\alpha_1(t))}M_0$.

   The curve α_1 is called the *rolling curve* and the curve $\alpha_0 : I \to M_0$, defined by

   $$\alpha_0(t) = h(t)(\alpha_1(t)), \tag{12}$$

   is called the *development of* α_1 *on* M_0.
2. *No-slip condition*:

   $$\left(\dot{h}(t) \circ h(t)^{-1} \right)(\alpha_0(t)) = 0. \tag{13}$$

3. *No-twist conditions*:

   a. (Tangential part)

$$\left(\dot{h}(t) \circ h(t)^{-1}\right)_* \left(T_{\alpha_0(t)}M_0\right) \subset (T_{\alpha_0(t)}M_0)^{\perp}, \tag{14}$$

   b. (Normal part)

$$\left(\dot{h}(t) \circ h(t)^{-1}\right)_* (T_{\alpha_0(t)}M_0)^{\perp} \subset T_{\alpha_0(t)}M_0. \tag{15}$$

Remark 1

1. Rolling along piecewise-smooth curves only requires a minor adjustment in the conditions, involving derivatives, of the previous definition, replacing "for all t" by "for almost all t".
2. The first rolling condition means that, during the motion, the development curve α_0 is being drawn on M_0 by the point of contact of the moving manifold $h(t)(M_1)$ and the static manifold M_0. The second rolling condition means that, at each time t, both manifolds $h(t)(M_1)$ and M_0 have the same tangent space.
3. The no-slip condition is equivalent to $\dot{\alpha_0}(t) = h(t)_*(\dot{\alpha_1}(t))$. So, this condition has the interpretation that the velocities of the rolling curve and of its development at the point of contact are the same.
4. An interpretation for the no-twist conditions is not so easy to obtain. But Godoy et al. in [2] proved that these conditions can be given an interesting geometric interpretation as follows:

   a. Tangential part: A vector field $Y(t)$ is tangent parallel along the curve $\alpha_1(t)$ if, and only if, $V(t) = h(t)_*Y(t)$ is tangent parallel along $\alpha_0(t)$.
   b. Normal part: A vector field $Z(t)$ is normal parallel along the curve $\alpha_1(t)$ if, and only if, $V(t) = h(t)_*Z(t)$ is normal parallel along $\alpha_0(t)$.

5. In Sharpe [12], it has been proven that given any smooth curve on M_0, there exists a unique rolling map along that curve. This property of existence and uniqueness has been generalized to any Riemannian submanifolds in Hüper et al. [7].

3.2 Rolling Maps for the Essential Manifold

In this section we specialize the rolling maps to the particular situation when M_1 is the Essential Manifold $\mathscr{E}$, and M_0 is the affine tangent space to $\mathscr{E}$ at a particular point P_0, $T^{\mathrm{aff}}_{P_0}\mathscr{E}$. Notice that M_0 and M_1 are assumed to be embedded submanifolds of $\overline{M} = \mathfrak{s}_3 \times \mathbb{R}^{3\times 3}$, endowed with the Riemannian metric defined in (5). The approach we take here follows that of Hüper and Silva Leite [6], where the rolling of Grassmann manifolds and of rotation groups has been studied. We recall that, according to our definition of the Essential Manifold given in (3), elements in $\mathscr{E}$

are represented by pairs. We must define the group of isometries $\overline{G}$ of $\overline{M}$. For that, let us start with the Lie group $G = \mathrm{SO}_3 \times \mathrm{SO}_3 \times \mathrm{SO}_3$. It is an easy task to show that it acts transitively on $\mathscr{E}$ via equivalence:

$$\begin{array}{rcl}\overline{\sigma}: & G \times \mathscr{E} & \longrightarrow \mathscr{E} \\ & \Big((U,V,W),(\Theta E_0\Theta^\top, R)\Big) & \longmapsto \Big(U\Theta E_0 \Theta^\top U^\top, VRW^\top\Big)\end{array}. \tag{16}$$

Consider now the group $\overline{G} = G \ltimes \left(\mathfrak{s}_3 \times \mathbb{R}^{3\times 3}\right)$, with the product rule

$$\begin{aligned}&\Big(U_1, V_1, W_1, X_1, Y_1\Big)\Big(U_2, V_2, W_2, X_2, Y_2\Big) \\ &\qquad = \Big(U_1U_2, V_1V_2, W_1W_2, U_1X_2U_1^\top + X_1, V_1Y_2W_1^\top + Y_1\Big),\end{aligned}$$

and inverse

$$\Big(U,V,W,X,Y\Big)^{-1} = \Big(U^\top, V^\top, W^\top, -U^\top X U, -V^\top Y W\Big). \tag{17}$$

The group $\overline{G}$ is connected and acts on $\mathfrak{s}_3 \times \mathbb{R}^{3\times 3}$ via

$$\begin{array}{rcl}\overline{G} \times \left(\mathfrak{s}_3 \times \mathbb{R}^{3\times 3}\right) & \longrightarrow & \mathfrak{s}_3 \times \mathbb{R}^{3\times 3} \\ \Big((U,V,W,X,Y),(A,B)\Big) & \longmapsto & \left(UAU^\top + X, VBW^\top + Y\right)\end{array}. \tag{18}$$

We can conclude that $\overline{G}$ is the isometry group of $\overline{M} = \mathfrak{s}_3 \times \mathbb{R}^{3\times 3}$.

Now, if $P_0 = (\Theta_0 E_0 \Theta_0^\top, R_0)$ is an arbitrary point in $\mathscr{E}$, $\alpha_1 : [0,\tau] \to \mathscr{E}$, defined by $\alpha_1(t) = \left(U(t)\Theta_0 E_0 \Theta_0^\top U(t)^\top, V(t) R_0 W(t)^\top\right)$ is a curve on $\mathscr{E}$ starting from P_0 at $t = 0$. The transitive action of G on $\mathscr{E}$ defined by (16), ensures that any curve on $\mathscr{E}$ has this form. Our goal is to find conditions under which the map

$$\begin{aligned}h : \left[0,\tau\right] &\longrightarrow \overline{G} \\ t &\longmapsto h(t) = (U(t)^\top, V(t)^\top, W(t)^\top, X(t), Y(t))\end{aligned}, \tag{19}$$

is a rolling map of the Essential Manifold $\mathscr{E}$ over its affine tangent space $T_{P_0}^{\mathrm{aff}}\mathscr{E}$, along

$$\alpha_1(t) = \left(U(t)\Theta_0 E_0 \Theta_0^\top U(t)^\top, V(t) R_0 W(t)^\top\right),$$

with development curve

$$\alpha_0(t) = h(t)(\alpha_1(t)) = \left(\Theta_0 E_0 \Theta_0^\top + X(t), R_0 + Y(t)\right) = P_0 + Z(t) \in M_0, \tag{20}$$

where $Z(t) = (X(t), Y(t)) \in \mathfrak{s}_3 \times \mathbb{R}^{3\times 3}$.

First, we must rewrite (9)–(11) for our particular situation.

Let (A, B) be a point in $\mathfrak{s}_3 \times \mathbb{R}^{3\times 3}$ and $(\xi, \eta) \in \mathfrak{s}_3 \times \mathbb{R}^{3\times 3}$ be a tangent vector to a smooth curve $t \in]-\varepsilon, \varepsilon[\longmapsto y(t) = \big(A(t), B(t)\big) \in \mathfrak{s}_3 \times \mathbb{R}^{3\times 3}$ that satisfies $y(0) = \big(A(0), B(0)\big) = (A, B)$ and $\dot{y}(0) = (\xi, \eta)$. Then, since

$$h(t)(A, B) = \big(U(t)^\top A U(t) + X(t), V(t)^\top B W(t) + Y(t)\big),$$

one gets

$$\begin{aligned}
&\dot{h}(t)((A, B)) \\
&\quad = \frac{\mathrm{d}}{\mathrm{d}\sigma}\left[h(\sigma)((A, B))\right]\bigg|_{\sigma=t} \\
&\quad = \frac{\mathrm{d}}{\mathrm{d}\sigma}\left[\big(U(\sigma)^\top A U(\sigma) + X(\sigma), V(\sigma)^\top B W(\sigma) + Y(\sigma)\big)\right]\bigg|_{\sigma=t} \\
&\quad = \big(\dot{U}(t)^\top A U(t) + U(t)^\top A \dot{U}(t) + \dot{X}(t), \dot{V}(t)^\top B W(t) + V(t)^\top B \dot{W}(t) + \dot{Y}(t)\big).
\end{aligned} \tag{21}$$

This is the counterpart of (9). Now,

$$\begin{aligned}
&h(\sigma)\, h(t)^{-1} \\
&\quad = \big(U(\sigma)^\top U(t), V(\sigma)^\top V(t), W(\sigma)^\top W(t), -U(\sigma)^\top U(t) X(t) U(t)^\top U(\sigma) + X(\sigma), \\
&\qquad - V(\sigma)^\top V(t) Y(t) W(t)^\top W(\sigma) + Y(\sigma)\big),
\end{aligned} \tag{22}$$

so that, the counterpart of (10) is

$$\begin{aligned}
&\big(\dot{h}(t) \circ h(t)^{-1}\big)((A, B)) \\
&\quad = \frac{\mathrm{d}}{\mathrm{d}\sigma}\left[\big(h(\sigma)\, h(t)^{-1}\big)((A, B))\right]\bigg|_{\sigma=t} \\
&\quad = \frac{\mathrm{d}}{\mathrm{d}\sigma}\bigg|_{\sigma=t}\Big[\big(U^\top(\sigma) U(t) A U(t)^\top U(\sigma) - U^\top(\sigma) U(t) X(t) U^\top(t) U(\sigma) + X(\sigma), \\
&\qquad V^\top(\sigma) V(t) B W(t)^\top W(\sigma) - V(\sigma)^\top V(t) Y(t) W(t)^\top W(\sigma) + Y(\sigma)\big)\Big] \\
&\quad = \big(\dot{U}(t)^\top U(t) A + A U(t)^\top \dot{U}(t) - \dot{U}(t)^\top U(t) X(t) - X(t) U(t)^\top \dot{U}(t) + \dot{X}(t), \\
&\qquad \dot{V}(t)^\top V(t) B + B W(t)^\top \dot{W}(t) - \dot{V}(t)^\top V(t) Y(t) - Y(t) W(t)^\top \dot{W}(t) + \dot{Y}(t)\big).
\end{aligned} \tag{23}$$

Finally, the counterpart of (11) is written as

$$\begin{aligned}
&\left(\dot h(t)\circ h(t)^{-1}\right)_*((\xi,\eta))\\
&=\frac{\mathrm{d}}{\mathrm{d}\sigma}\left[\left(\dot h(t)\circ h(t)^{-1}\right)(A(\sigma),B(\sigma))\right]\Big|_{\sigma=0}\\
&=\left(\dot U(t)^\top U(t)\xi+\xi U(t)^\top \dot U(t),\dot V(t)^\top V(t)\eta+\eta W(t)^\top \dot W(t)\right).
\end{aligned}\tag{24}$$

3.3 *The Kinematic Equations of Rolling*

In this section we derive the kinematic equations for the rolling motion by imposing the no-slip and no-twist conditions on $h(t)$ given by (19). Taking into account (23) and the expression for α_0 given by (20), the no-slip condition (13) can be rewritten as

$$\begin{cases}\dot U(t)^\top U(t)\Theta_0E_0\Theta_0^\top+\Theta_0E_0\Theta_0^\top U(t)^\top \dot U(t)+\dot X(t)=0\\ \dot V(t)^\top V(t)R_0+R_0W(t)^\top \dot W(t)+\dot Y(t)=0\end{cases}.\tag{25}$$

If we define the skew-symmetric matrices Ω_U, Ω_V and Ω_W by

$$\Omega_U:=\Theta_0^\top \dot U^\top U\Theta_0,\quad \Omega_V:=R_0^\top \dot V^\top VR_0,\quad \Omega_W:=R_0\dot W^\top WR_0^\top,\tag{26}$$

the no-slip condition takes the form

$$\begin{cases}\dot X(t)=-\Theta_0\left[\Omega_U(t),E_0\right]\Theta_0^\top\\ \dot Y(t)=\Omega_W(t)R_0-R_0\Omega_V(t)\end{cases}.\tag{27}$$

Now, using (24), the tangential part of the no-twist conditions is equivalent to showing that, for all $(\xi,\eta)\in T_{\alpha_0(t)}M_0$,

$$\left(\dot U^\top U\xi+\xi U^\top \dot U,\dot V^\top V\eta+\eta W^\top \dot W\right)\in(T_{\alpha_0(t)}M_0)^\perp.\tag{28}$$

But, $T_{\alpha_0(t)}M_0=T_{P_0}\mathscr{E}$ (and similarly for the normal space). So, taking into account the notations (26), the tangential part of the no-twist conditions (28) is equivalent to

$$\left(\left[\Theta_0\Omega_U\Theta_0^\top,\xi\right],R_0\Omega_VR_0^\top\eta-\eta R_0^\top\Omega_WR_0\right)\in(T_{P_0}\mathscr{E})^\perp,\tag{29}$$

for all $(\xi,\eta)\in T_{P_0}\mathscr{E}$. But, according to (6), for $(\xi,\eta)\in T_{P_0}\mathscr{E}$, we have

$$\xi=\Theta_0\begin{bmatrix}0&\Lambda\\ \Lambda^\top&0\end{bmatrix}\Theta_0^\top,\quad \Lambda\in\mathbb{R}^{2\times1}\quad\text{and}\quad \eta=R_0C,\quad C\in\mathfrak{so}_3.\tag{30}$$

Hence, writing the skew-symmetric matrix Ω_U as

$$\Omega_U = \begin{bmatrix} \Omega_1 & \Omega_2 \\ -\Omega_2^\top & 0 \end{bmatrix},$$

where $\Omega_1 \in \mathfrak{so}_2$, $\Omega_2 \in \mathbb{R}^{2\times 1}$, and taking into account that

$$\left[\Theta_0 \Omega_U \Theta_0^\top, \xi\right] = \Theta_0 \begin{bmatrix} \Omega_2 \Lambda^\top + \Lambda \Omega_2^\top & \Omega_1 \Lambda \\ -\Lambda^\top \Omega_1 & -2\Omega_2^\top \Lambda \end{bmatrix} \Theta_0^\top,$$

the characterization of the orthogonal space (8) enables us to conclude that

$$\Omega_1 \Lambda = 0, \quad \text{for all} \quad \Lambda \in \mathbb{R}^{2\times 1}.$$

This implies that $\Omega_1 = 0$ and, therefore, Ω_U must have the constrained structure

$$\Omega_U = \begin{bmatrix} 0 & \Omega_2 \\ -\Omega_2^\top & 0 \end{bmatrix}. \tag{31}$$

Additionally, when $\eta = R_0 C$, $C \in \mathfrak{so}_3$, the second component in (29) should be of the form $R_0 S$, with $S \in \mathfrak{s}_3$. This requires that the matrix $\left(\Omega_V C - C R_0^\top \Omega_W R_0\right)$ is symmetric, for all $C \in \mathfrak{so}_3$. Using this requirement, and after some simple calculations, one concludes that this is equivalent to

$$\left[\Omega_V + R_0^\top \Omega_W R_0, C\right] = 0, \quad \text{for all} \quad C \in \mathfrak{so}_3.$$

Hence, $\Omega_V + R_0^\top \Omega_W R_0 = 0$, that is

$$\Omega_V = -R_0^\top \Omega_W R_0. \tag{32}$$

Therefore, the *tangential part of the no-twist conditions* for the Essential Manifold is equivalent to requiring that

$$\Omega_U = \begin{bmatrix} 0 & \Omega_2 \\ -\Omega_2^\top & 0 \end{bmatrix} \qquad \text{and} \qquad \Omega_V = -R_0^\top \Omega_W R_0. \tag{33}$$

Finally, we must impose the normal part of the no-twist conditions, which is equivalent to showing that, for all $(\xi, \eta) \in (T_{\alpha_0(t)} M_0)^\perp$,

$$\left(\dot{U}^\top U \xi + \xi U^\top \dot{U}, \dot{V}^\top V \eta + \eta W^\top \dot{W}\right) \in T_{\alpha_0(t)} M_0. \tag{34}$$

But it turns out that if conditions (33) hold, the normal part of the no-twist conditions holds as well. Indeed, the previous condition is equivalent to

$$\left(\left[\Theta_0 \Omega_U \Theta_0^\top, \xi\right], R_0 \Omega_V R_0^\top \eta - \eta R_0^\top \Omega_W R_0\right) \in T_{\alpha_0(t)} M_0. \tag{35}$$

So, since $(\xi, \eta) \in (T_{\alpha_0(t)}M_0)^{\perp} = (T_{P_0}\mathscr{E})^{\perp}$, we must have

$$\xi = \Theta_0 \begin{bmatrix} B & 0 \\ 0 & b \end{bmatrix} \Theta_0^{\top}, \quad B \in \mathfrak{s}_2, \; b \in \mathbb{R} \qquad \text{and} \qquad \eta = R_0 S, \quad S \in \mathfrak{s}_3. \tag{36}$$

Hence, using (33), after some calculations we obtain that

$$\left[\Theta_0 \Omega_U \Theta_0^{\top}, \xi\right] = \Theta_0 \begin{bmatrix} 0 & \Omega_2 b - B\Omega_2 \\ -\Omega_2^{\top} B + b\Omega_2^{\top} & 0 \end{bmatrix} \Theta_0^{\top},$$

which is in accordance with the characterization of the tangent space (7). Moreover, the second component presented in relation (35) should be of the form $R_0 C$, with $C \in \mathfrak{so}_3$ and, taking into account (33), this requires that the matrix $(\Omega_V S + S\Omega_V)$ must be skew-symmetric, for all $S \in \mathfrak{s}_3$. A few computations show that this requirement is verified. Thus, *the no-twist conditions reduce to Eq. (33).*

Now, if the second condition in (33) is used in (27), one obtains

$$\begin{cases} \dot{X}(t) = -\Theta_0\left[\Omega_U(t), E_0\right]\Theta_0^{\top} \\ \dot{Y}(t) = -2R_0\Omega_V(t) \end{cases}. \tag{37}$$

The no-slip condition reduces to Eq. (37).

We can now state the main theorem.

Theorem 1 *Let $\Omega_U(t), \Omega_V(t) \in \mathfrak{so}_3$ with $\Omega_U = \begin{bmatrix} 0 & \Omega_2 \\ -\Omega_2^{\top} & 0 \end{bmatrix}$, $\Omega_2 \in \mathbb{R}^{2\times 1}$. If (U, V, W, X, Y) is the solution of the following system of differential equations, evolving on $\overline{G}$,*

$$\begin{cases} \dot{U}(t) = -U(t)\Theta_0\Omega_U(t)\Theta_0^{\top} \\ \dot{V}(t) = -V(t)R_0\Omega_V(t)R_0^{\top} \\ \dot{W}(t) = W(t)\Omega_V(t) \\ \dot{X}(t) = -\Theta_0\left[\Omega_U(t), E_0\right]\Theta_0^{\top} \\ \dot{Y}(t) = -2R_0\Omega_V(t) \end{cases}, \tag{38}$$

with initial condition at the identity element of $\overline{G}$, that is, $(U(0), V(0), W(0), X(0), Y(0)) = (I, I, I, 0, 0)$, then

$$t \mapsto h(t) = (U(t)^{\top}, V(t)^{\top}, W(t)^{\top}, X(t), Y(t)) \in \overline{G}$$

is a rolling map (in the sense of Definition 2) of the Essential Manifold $\mathscr{E}$ over the affine tangent space at the point $P_0 = (\Theta_0 E_0 \Theta_0{}^\top, R_0)$, along the rolling curve

$$t \mapsto \alpha_1(t) = \left(U(t)\Theta_0 E_0 \Theta_0^\top U(t)^\top, V(t) R_0 W(t)^\top\right),$$

with development curve

$$t \mapsto \alpha_0(t) = \left(\Theta_0 E_0 \Theta_0^\top + X(t), R_0 + Y(t)\right).$$

Proof We have already proved, before the statement of the theorem, that Eq. (38) encode the no-slip and the no-twist conditions. Since the curve α_1 clearly lives in the manifold $\mathscr{E}$ and $\alpha_0(t) = h(t)(\alpha_1(t)) = P_0 + Z(t)$, with $Z(t) = (X(t), Y(t))$, to complete the proof it is enough to show that $Z(t) \in T_{P_0}\mathscr{E}$. But since $\Omega_U(t)$ and $\Omega_V(t)$ are skew-symmetric, it follows from the last two equations of (38) that $\dot{Z}(t) \in T_{P_0}\mathscr{E}$. This, together with the initial condition $Z(0) = 0$, implies that $Z(t) \in T_{P_0}\mathscr{E}$, that is, $\alpha_0(t) \in T_{P_0}^{\mathrm{aff}}\mathscr{E}$.

Remark 2 Equations (38), which encode the non-holonomic constraints of no-slip and no-twist are called the **kinematic equations** for rolling the Essential Manifold over the affine tangent space at the point P_0.
The choice of Ω_U and Ω_V completely determine the solutions of the kinematic equations and, consequently, the rolling curve (and its development). For that reason, we say that these two functions are the "control functions" of the motion.

3.4 Rolling Along Geodesics

For the special situation where the control functions are constant, say $\Omega_U(t) = \Omega_U$ and $\Omega_V(t) = \Omega_V$, the solution of the kinematic equations (38), with initial condition $(U(0), V(0), W(0), X(0), Y(0)) = (I, I, I, 0, 0)$, can be solved explicitly and

$$\begin{cases} U(t) = \Theta_0 e^{-t\Omega_U} \Theta_0^\top \\ V(t) = R_0 e^{-t\Omega_V} R_0^\top \\ W(t) = e^{t\Omega_V} \\ X(t) = -t\Theta_0 \left[\Omega_U, E_0\right] \Theta_0^\top \\ Y(t) = -2tR_0 \Omega_V \end{cases} . \tag{39}$$

In this case, the rolling curve

$$t \mapsto \alpha_1(t) = \left(\Theta_0 e^{-t\Omega_U} E_0 e^{t\Omega_U} \Theta_0^\top, R_0 e^{-2t\Omega_V}\right) \tag{40}$$

is a geodesic on $\mathscr{E}$, passing through P_0 (at $t = 0$) and, consequently,

$$t \mapsto \alpha_0(t) = P_0 + (X(t), Y(t)) = P_0 + t\left(\Theta_0\left[E_0, \Omega_U\right]\Theta_0^\top, 2R_0\Omega_V^\top\right) \tag{41}$$

is also a geodesic in the affine tangent space $T_{P_0}^{\text{aff}}\mathscr{E}$, satisfying $\alpha_0(0) = P_0$. The second statement is obvious since a geodesic in the affine space is a straight line. The first statement can be checked differentiating α_1 twice and noticing that $\ddot{\alpha}_1(t)$ belongs to $(T_{\alpha_1(t)}\mathscr{E})^\perp$. Indeed, using the fact that $e^{t\Omega_U}\Omega_U e^{-t\Omega_U} = \Omega_U$ and the anticommutativity of the matrix commutator, we can write

$$\begin{aligned}\dot{\alpha}_1(t) &= \left(\Theta_0 e^{-t\Omega_U}\left[E_0, e^{t\Omega_U}\Omega_U e^{-t\Omega_U}\right]e^{t\Omega_U}\Theta_0^\top, R_0 e^{-2t\Omega_V}(-2\Omega_V)\right)\\ &= \left(\Theta_0 e^{-t\Omega_U}\left[E_0, \Omega_U\right]e^{t\Omega_U}\Theta_0^\top, R_0 e^{-2t\Omega_V}(-2\Omega_V)\right).\end{aligned} \tag{42}$$

Differentiating again and simplifying, we obtain

$$\ddot{\alpha}_1(t) = \left(\Theta_0 e^{-t\Omega_U}\left[\left[E_0, \Omega_U\right], \Omega_U\right]e^{t\Omega_U}\Theta_0^\top, R_0 e^{-2t\Omega_V}(4\Omega_V^2)\right). \tag{43}$$

Hence, taking into account that

$$\left[\left[E_0, \Omega_U\right], \Omega_U\right] = \begin{bmatrix} -2\Omega_2\Omega_2^\top & 0 \\ 0 & 2\Omega_2^\top\Omega_2 \end{bmatrix}, \tag{44}$$

with $-2\Omega_2\Omega_2^\top$ a symmetric matrix and $2\Omega_2^\top\Omega_2$ a real number, we have

$$\ddot{\alpha}_1(t) = \left(\Theta_0 e^{-t\Omega_U}\begin{bmatrix} -2\Omega_2\Omega_2^\top & 0 \\ 0 & 2\Omega_2^\top\Omega_2 \end{bmatrix}e^{t\Omega_U}\Theta_0^\top, R_0 e^{-2t\Omega_V}(4\Omega_V^2)\right), \tag{45}$$

which is in accordance with (8). So, $\ddot{\alpha}_1(t)$ belongs to $(T_{\alpha_1(t)}\mathscr{E})^\perp$, that is, the covariant derivative of $\dot{\alpha}_1$ is identically zero and, thus, (40) is a geodesic on $\mathscr{E}$. We summarize the previous in the following corollary of Theorem 1.

Corollary 1 *If the control functions Ω_U and Ω_V are constant skew-symmetric matrices, then*

$$h(t) = \left(\Theta_0 e^{t\Omega_U}\Theta_0^\top, R_0 e^{t\Omega_V}R_0^\top, e^{-t\Omega_V}, -t\Theta_0\left[\Omega_U, E_0\right]\Theta_0^\top, -2tR_0\Omega_V\right)$$

is the rolling map of the Essential Manifold $\mathscr{E}$ over $T_{P_0}^{\text{aff}}\mathscr{E}$, without slipping and twisting, along the geodesic

$$t \mapsto \alpha_1(t) = \left(\Theta_0 e^{-t\Omega_U}E_0 e^{t\Omega_U}\Theta_0^\top, R_0 e^{-2t\Omega_V}\right) \in \mathscr{E}$$

with development curve

$$t \mapsto \alpha_0(t) = P_0 + Z(t) = P_0 + (X(t), Y(t)) = P_0 + t\left(\Theta_0 \left[E_0, \Omega_U\right] \Theta_0^\top, 2R_0\Omega_V^\top\right),$$

also a geodesic in the affine tangent space $T_{P_0}^{aff}\mathscr{E}$.

4 Final Remarks

We have derived the kinematic equations of rolling the Essential Manifold over the affine tangent space at a point. The Essential Manifold plays a crucial role in 3D computer vision and these rolling motions may be used to efficiently solve interpolation problems on this manifold, by reducing them to simpler interpolation problems on a flat space, the affine space. The kinematic equations of rolling can be seen as control systems evolving on the group of isometries of the embedding space, whose controls are the functions Ω_U and Ω_V. That is, choosing the controls is equivalent to defining the rolling curve. This motivates several questions concerning controllability and optimal control of rolling motions, issues that will be under investigation in the near future.

Acknowledgements The work of the first and third authors was supported by FCT project PTDC/EEA-CRO/122812/2010.

References

1. Caseiro, R., Martins, P., Henriques, J.F., Silva Leite, F., Batista, J.: Rolling Riemannian manifolds to solve the multi-class classification problem. In: Proc. IEEE Conference on Computer Vision and Pattern Recognition (CVPR 2013), pp. 41–48, Oregon, 25–27 June 2013
2. Godoy, M., Grong, E., Markina, I., Silva Leite, F.: An intrinsic formulation of the problem on rolling manifolds. Int. J. Dyn. Control Syst. **18**(2), 181–214 (2012)
3. Hartley, R., Zisserman, A.: Multiple View Geometry in Computer Vision, 2nd edn. Cambridge University Press, Cambridge (2004)
4. Helmke, U., Hüper, K., Lee, P.Y., Moore, J.: Essential Matrix Estimation Using Gauss-Newton Iterations on a Manifold. Int. J. Comput. Vis. **74**(2), 117–136 (2007)
5. Huang, T.S., Faugeras, O.D.: Some properties of the E Matrix in two-view motion estimation. IEEE Trans. Pattern Anal. Mach. Intell. **11**(12), 1310–1312 (1989)
6. Hüper, K., Silva Leite, F.: On the geometry of rolling and interpolation curves on S^n, SO_n and Grassmann manifolds. J. Dyn. Control Syst. **13**(4), 467–502 (2007)
7. Hüper, K., Krakowski, K., Silva Leite, F.: Rolling maps in a Riemannian framework. In: Cardoso, J., Knut, K., Saraiva, P. (eds.) Textos de Matemática, vol. 43, pp. 15–30. Departamento de Matemática da Universidade de Coimbra, Portugal (2011)
8. Kobayashi, S.: Transformation Groups in Differential Geometry. Springer, Berlin (1972)
9. Longuet-Higgins, H.: A computer algorithm for reconstructing a scene from two projections. Nature **293**, 133–135 (1981)

10. Ma, Y., Soatto, S., Kosecka, J., Sastry, S.: An Invitation to 3D Vision: From Images to Geometric Models. Springer, New York (2004)
11. Nash, J.: The imbedding problem for Riemannian manifolds. Ann. Math. **63**(1), 20–63 (1956)
12. Sharpe, R.W.: Differential Geometry. Springer, New York (1997)
13. Szeliski, R.: Computer Vision: Algorithms and Applications. Texts in Computer Science. Springer, New York (2011)

Singleton Free Set Partitions Avoiding a 3-Element Set

Ricardo Mamede

Abstract The definition and study of pattern avoidance for set partitions, which is an analogue of pattern avoidance for permutations, begun with Klazar. Sagan continued his work by considering set partitions which avoid a single partition of three elements, and Goyt generalized these results by considering partitions which avoid any family of partitions of a 3-element set. In this paper we enumerate and describe set partitions, even set partitions and odd set partitions without singletons which avoid any family of partitions of a 3-element set. The characterizations of these families allow us to conclude that the corresponding sequences are P-recursive. We also construct Gray codes for the sets of singletons free partitions that avoid a single partition of three elements.

1 Introduction

Enumeration of pattern-avoiding objects such as permutations, words or compositions, is a very active area of research, with connections to several areas of mathematics. In 1996, Klazar [5] extended the notion of pattern avoidance for permutations, words and compositions to set partitions by analyzing set partitions that avoid the patterns *abab* and *aabb*. Sagan [10] continued this work by considering set partitions which avoid a single partition of a 3-element set. Since then this notion has been studied by many authors (see [7] and the references therein for a comprehensive survey). In particular, Goyt [3] generalized Sagan's results by considering partitions, even partitions and odd partitions that avoid any family of partitions of a 3-element set. In his book [7], T. Mansour proposed the study of pattern avoidance in set partitions without singletons, that is set partitions whose blocks have at least two elements, as a research direction. Following this suggestion, we continue the work of Sagan and Goyt on pattern avoidance in set partitions, considering set partitions without singletons that avoid any family of partitions of a 3-element set. To this end, we need some definitions.

R. Mamede (✉)
CMUC, Department of Mathematics, University of Coimbra, 3001-501 Coimbra, Portugal
e-mail: mamede@mat.uc.pt

J.-P. Bourguignon et al. (eds.), *Dynamics, Games and Science*, CIM Series in Mathematical Sciences 1, DOI 10.1007/978-3-319-16118-1_22

For integers $m \leq n$ define the interval $[m, n] = \{m, m+1, \ldots, n\}$ with the special case $[1, n] = [n]$. A *partition* π of a set $S \subseteq [n]$, $n \geq 1$, is a collection of nonempty disjoint subsets $B_1, \ldots, B_t$ of S, called *blocks*, whose union is S. We will write $\pi \vdash S$ and $b(\pi) = t$ to denote the number of blocks of π. A block with only one element is said to be a *singleton*. A partition is said to be in *standard form* if it is written as $\pi = B_1/B_2/\cdots/B_t$, where the blocks are listed in ascending order according to their smallest element. Generally, we will not use braces and commas in the blocks unless they are needed for clarity. For example, if $\pi = 13/245/6/7$ then $\pi \vdash [7]$ with $b(\pi) = 4$.

The set of all set partitions of $[n]$, $n \geq 1$, will be denoted by

$$\Pi_n = \{\pi : \pi \vdash [n]\}.$$

If S is a subset of the integers with cardinality $\#S = n$, then the standardization map corresponding to S is the unique order-preserving bijection $st_S : S \to [n]$. When S is clear from the context we drop the subscript. For example, if $S = \{2, 5, 7\}$ then $st(2) = 1, st(5) = 2$ and $st(7) = 3$. Thus, if $\pi = 27/5$ its standardization is $st(\pi) = 13/2$.

A set *subpartition* of a set partition $\pi = B_1/B_2/\cdots/B_t$ of S is a set partition π' of $S' \subseteq S$ such that each block of π' is contained in a different block of π. For example, $27/5$ is a subpartition of $1356/27/4$ but not of $1357/26/4$. Let $\pi \in \Pi_k$ be a given set partition called the *pattern*. We say that a partition $\sigma \in \Pi_n$ *contains the pattern* π if there exists a set subpartition σ' of σ such that $st(\sigma') = \pi$. In this case, σ' is called an occurrence of the pattern π in σ. If σ has no occurrences of π, then we say that σ *avoids* the pattern π. For example, $\sigma = 16/23/45$ avoids the pattern 123 but contains the pattern $13/2$ since the standardization of the subpartition $\sigma' = 16/2$ is $13/2$. In this context, for $R \subseteq \Pi_k$ we use the notation

$$\Pi_n(R) = \{\sigma \in \Pi_n : \sigma \text{ avoids every pattern } \pi \in R\}.$$

The set $\Pi(R)$, with $R \subseteq \Pi_3$, was studied by Sagan [10] when $\#R = 1$ and by Goyt [3] for $\#R \geq 2$. Denote by Π_n' the set of all singleton free partitions of $[n]$, and given $R \subseteq \Pi_k$ a subset of patterns, let

$$\Pi_n'(R) = \{\sigma \in \Pi_n' : \sigma \text{ avoids every pattern } \pi \in R\}$$

be the set of all singleton free partitions of $[n]$ that avoid all partitions of R. When $R = \{\pi\}$, we simplify the notation and write $\Pi_n'(\pi)$.

In the next section we characterize the set $\Pi_n'(\pi)$, and give exact formulas and generating functions for its cardinal, for various patterns π, including all $\pi \vdash [3]$. We then use these results to characterize and enumerate $\Pi_n'(R)$, for any $R \subset \Pi_3$. In Sect. 3 we present the notion of sign of a partition, defined in [3], and enumerate the set of singleton free signed partitions of $[n]$ which avoid any family of patterns of

Π_3. The study of P-recursiveness associated with permutation patterns begun with Gessel [2] and Noonan-Zeilberger [9], and was applied to set partitions by Sagan [10]. In Sect. 4 we show that although Π'_n is not P-recursive, the sets of singleton free partitions and singleton free sign partitions that avoid any pattern $\pi \vdash [3]$ are P-recursive.

The last section is devoted to the combinatorial generation of the elements of $\Pi'_n(\pi)$, with $\pi \vdash [3]$. The search for combinatorial algorithms that list all elements of a given combinatorial class of objects is common to several scientific topics and has many applications (see [1] for an exhaustive bibliography). Among all lists of a combinatorial class, there is a special kind of list called Gray code, where two successive objects in the list are encoded in such a way that their codes differ as little as possible. In such a list the generation of its elements is usually faster, and the computational cost to produce the list is, in general, smaller. Moreover, the usual recursive structure of a Gray code may throw new light on the combinatorial class. To this end we construct Gray codes for the sets $\Pi'_n(\pi)$, for all $\pi \vdash [3]$ for which the set is not trivial, where each partition in the list is obtained from its immediate predecessor by changing the block of at most two elements.

2 Singleton Free Set Partitions

We start by considering the case $\Pi'_n(\pi)$, with π a pattern in Π_3, namely 123, 1/23, 12/3, 1/2/3 and 13/2. Following the notation of [10] for exponential generating functions, we let

$$F_I(x) = \sum_{i \in I} \frac{x^i}{i!}, \tag{1}$$

for a set I of nonnegative integers. In particular, when $I = [0, m]$, we write

$$\exp_m(x) = \sum_{i=0}^{m} \frac{x^i}{i!}.$$

Let $a^I_{n,\ell}$ denote the number of partitions of $[n]$ with ℓ blocks with cardinalities in the set $I \subseteq \mathbb{N}$. As $F_I(x)$ is the exponential generating function for the number of ways an n-set can form a block with size in the set I, it follows that (see, for example, [8] or [15])

$$\sum_{n \geq 0} a^I_{n,\ell} \frac{x^n}{n!} = \frac{F_I(x)^\ell}{\ell!} \tag{2}$$

is the exponential generating function for the number of partitions of $[n]$ with ℓ blocks, each of them having sizes in the set I. Finally, given a pattern π, we write

$$F_\pi(x) = \sum_{n\geq 0} \#\Pi'_n(\pi)\frac{x^n}{n!}. \tag{3}$$

The distinction between (1) and (3) will be clear, since we denote patterns by Greek letters and sets of integers by capital Latin letters.

For example, with $I = \mathbb{N} \setminus \{1\}$, it follows that $\#\Pi'_n$ is the sum over all $\ell \geq 0$ of the numbers $a^I_{n,\ell}$, and thus the exponential generating function for the number of singleton free set partitions of $[n]$ is

$$F(x) = \sum_{n\geq 0} \#\Pi'_n\frac{x^n}{n!} = \sum_{n,\ell\geq 0} a^I_{n,\ell}\frac{x^n}{n!} = \sum_{\ell\geq 0}\frac{(e^x-1-x)^\ell}{\ell!} = \exp(e^x-1-x). \tag{4}$$

A partition $\sigma \vdash [n]$ is *layered* if it is of the form $[1,i]/[i+1,j]/[j+1,k]/\cdots/[\ell+i,n]$. A partition σ is said to be a *matching* if $\#B \leq 2$, for all block B of σ. When the cardinality of each block is exactly 2 the partition is called a *perfect matching*. The characterization of the set partitions in $\Pi_n(\pi)$, for $\pi \in \Pi_3$, obtained by Sagan [10], will be used repeatedly, so we state it below.

Theorem 1 (Sagan) *For $n \geq 1$,*

$$\begin{aligned}\Pi_n(1/2/3) &= \{\sigma \in \Pi_n : b(\sigma) \leq 2\},\\ \Pi_n(123) &= \{\sigma \in \Pi_n : \sigma \text{ is a matching}\},\\ \Pi_n(13/2) &= \{\sigma \in \Pi_n : \sigma \text{ is layered}\}.\end{aligned}$$

Given positive integers $i < m$, let π^i_m be the layered pattern

$$1/2/\cdots/i-1/i(i+1)/i+2/\cdots/m$$

in Π_m, where all blocks are singletons with the exception of $B_i = \{i, i+1\}$.

Theorem 2 *For $n \geq 2$,*

$$\begin{aligned}\Pi'_n(\pi^i_m) &= \{\sigma \in \Pi'_n : b(\sigma) \leq m-2\},\\ F_{\pi^i_m}(x) &= \exp_{m-2}(\exp(x)-1-x).\end{aligned}$$

Proof Since π^i_m has $m-1$ blocks, it is clear that if $b(\sigma) \leq m-2$ then σ avoids the pattern π^i_m. Reciprocally, let $\sigma \in \Pi'_n(\pi^i_m)$ and assume that $b(\sigma) \geq m-1$. Let $B_1, \ldots, B_{m-1}$ be $m-1$ blocks of σ, each of them with at least two elements, ordered by their least element: $B_j = \{a_j, \ldots\}$, with

$$a_1 < a_2 < \cdots < a_{m-1}.$$

Next, let $B'_i, \ldots, B'_{m-1}$ be the blocks $B_i, \ldots, B_{m-1}$ ordered by their largest element: $B'_j = \{\ldots, b_j\}$, with

$$b_i < b_{i+1} < \cdots < b_{m-1}.$$

Then,

$$a_1 < a_2 < \cdots < a_i < b_i < b_{i+1} < \cdots < b_{m-1}$$

and $a_1/a_2/\cdots/a_i\, b_i/b_{i+1}/\cdots/b_{m-1}$ is a copy of π_m^i in σ, a contradiction.

It follows that the number of partitions in $\Pi'_n(\pi_m^i)$ is the sum over all $0 \leq \ell \leq m-2$ of the number $a^I_{n,\ell}$ of partitions of $[n]$ with ℓ blocks, each of them with at most two elements:

$$\#\Pi'_n(\pi_m^i) = \sum_{\ell=0}^{m-2} a^I_{n,\ell},$$

with $I = \{2, 3, \ldots\}$. Thus, we can use (2) to write

$$\begin{aligned} F_{\pi_m^i}(x) &= \textstyle\sum_{n\geq 0} \#\Pi'_n(\pi_m^i)\frac{x^n}{n!} = \sum_{n\geq 0}\sum_{\ell=0}^{m-2} a^I_{n,\ell}\frac{x^n}{n!} = \sum_{\ell=0}^{m-2}\left(\sum_{n\geq 0} a^I_{n,\ell}\frac{x^n}{n!}\right) \\ &= \textstyle\sum_{\ell=0}^{m-2}\frac{F_I(x)^\ell}{\ell!} = \exp_{m-2}(\exp(x) - 1 - x). \end{aligned}$$

Two patterns σ and π are said to be *Wilf-equivalent* [7], denoted by $\sigma \sim \pi$, if the number of elements of the sets $\Pi'_n(R)$ and $\Pi'_n(T)$ are the same for all $n \geq 1$. The last result shows that $\pi_m^i \sim \pi_m^j$, for $i, j < m$.

Corollary 1 *The patterns π_m^i, for $1 \leq i \leq m-1$, are Wilf-equivalent.*

Corollary 2 *For $n \geq 2$,*

$$\Pi'_n(12/3) = \Pi'_n(1/23) = \{12\cdots n\},$$
$$F_{1/23}(x) = F_{12/3}(x) = e^x - x.$$

Proof It follows from the last results since $12\cdots n$ is the only partition in Π'_n with a single block.

Theorem 3 *For $n \geq 2$,*

$$\Pi'_n(12\cdots m) = \{\sigma \in \Pi_n : 2 \leq \#B \leq m-1, \textit{ for all block } B \in \sigma\},$$
$$F_{12\cdots m}(x) = \exp(\exp_{m-1}(x) - 1 - x).$$

Proof The characterization of the elements of $\Pi'_n(12\cdots m)$ is clear, since a partition contains a copy of $12\cdots m$ if and only if it has a block with at least m elements. It follows that $\#\Pi'_n(12\cdots m)$ is the sum of the numbers $a^I_{n,\ell}$ of partitions of $[n]$ with $\ell \geq 0$ blocks with cardinalities in the set $I = [2, m-1]$. Again, we use (2) to write:

$$F_{12\cdots m}(x) = \sum_{n\geq 0} \#\Pi'_n(12\cdots m)\tfrac{x^n}{n!} = \sum_{n,\ell\geq 0} a^I_{n,\ell}\tfrac{x^n}{n!} = \sum_{\ell\geq 0}\tfrac{F_I(x)^\ell}{\ell!}$$
$$= \exp(\exp_{m-1}(x) - 1 - x).$$

The *double factorial* of an odd positive integer $2i-1$ is defined as the product of all positive odd integers up to $2i-1$:

$$(2i-1)!! = (2i-1)(2i-3)\cdots 5\cdot 3\cdot 1.$$

Corollary 3 *For $n \geq 2$,*

$$\Pi'_n(123) = \{\sigma \in \Pi_n : \sigma \text{ is a perfect matching}\},$$
$$\#\Pi'_n(123) = \begin{cases}(2k-1)!! & \text{if } n = 2k\\ 0 & \text{otherwise}\end{cases}.$$

Proof The characterization of $\Pi'_n(123)$ is a consequence of the previous theorem, and it can also be deduced from Theorem 1. Moreover, we can write

$$F_{123}(x) = \sum_{n\geq 0}\frac{\left(\frac{x^2}{2!}\right)^n}{n!} = \sum_{n\geq 0}\frac{(2n)!}{2^n n!}\frac{x^{2n}}{(2n)!},$$

and since $\frac{(2n)!}{2^n n!} = (2n-1)!!$ the result follows.

Theorem 4 *For $n \geq 2$,*

$$\Pi'_n(1/2/\cdots/m) = \{\sigma \in \Pi'_n : b(\sigma) \leq m-1\},$$
$$F_{1/2/\cdots/m}(x) = \exp_{m-1}(\exp(x) - 1 - x).$$

Proof The characterization of the partitions of $[n]$ that avoid the pattern $1/2/\cdots/m$ is clear from the definitions, and the generating function follows from (1), since:

$$F_{1/2/\cdots/m}(x) = \sum_{n\geq 0}\#\Pi'_n(1/2/\cdots/m)\frac{x^n}{n!} = \sum_{n\geq 0}\sum_{\ell=0}^{m-1} a^I_{n,\ell}\frac{x^n}{n!} =$$
$$= \sum_{\ell=0}^{m-1}\left(\sum_{n\geq 0} a^I_{n,\ell}\frac{x^n}{n!}\right) = \sum_{\ell=0}^{m-1}\frac{F_I(x)^\ell}{\ell!} = \exp_{m-1}(\exp(x) - 1 - x),$$

where $a^I_{n,\ell}$ is the number of partitions of $[n]$ with $\ell \leq m-1$ blocks, and I is the set of all integers greater than, or equal to 2.

When $m = 3$ the function $F_{1/2/3}(x)$ generates the sequence A000295 in Sloane's Encyclopedia [14].

Corollary 4 *We have*

$$\Pi'_n(1/2/3) = \{\sigma \in \Pi'_n : b(\sigma) \leq 2\},$$
$$\#\Pi'_n(1/2/3) = 2^{n-1} - n, \text{ for } n \geq 3,$$

with $\#\Pi'_0(1/2/3) = \#\Pi'_2(1/2/3) = 1$ *and* $\#\Pi'_1(1/2/3) = 0$.

Proof From the generating function given in the last result, we have

$$\begin{aligned} F_{1/2/3}(x) &= 1 + (e^x - 1 - x) + \tfrac{(e^x-1-x)^2}{2} = \tfrac{1}{2} + \tfrac{x^2}{2} + \tfrac{e^{2x}}{2} - xe^x \\ &= 1 + \tfrac{x^2}{2} + \textstyle\sum_{n\geq 1}(2^{n-1} - n)\tfrac{x^n}{n!}, \end{aligned}$$

and the result follows.

The *Eulerian number* $e(n,m)$ is the number of permutations $p_1p_2\cdots p_n$ of $[n]$ with exactly m descents, that is, m places in which $p_j > p_{j+1}$, for $1 \leq j \leq n-1$. Let $E(n,m)$ be the set of all permutations of $[n]$ with exactly m descents.

Theorem 5 *There is a bijection between* $\Pi'_n(1/2/3)$ *and* $E(n-1,1)$, *for* $n \geq 1$.

Proof Using the description of $\Pi'_n(1/2/3)$ as the partitions of Π'_n having one or two blocks, its cardinality $2^{n-1} - n$ for $n \geq 3$ can be obtained directly as follows. If $\sigma \in \Pi'_n$ has only one block then $\sigma = 12\cdots n$. Otherwise, $\sigma = B_1/B_2$, with

$$B_1 = \{1\} \cup S,$$

where $S \subset [2,n]$ has i elements, for some $1 \leq i \leq n-3$. Thus,

$$\#\Pi'_n(1/2/3) = 1 + \sum_{i=1}^{n-3}\binom{n-1}{i} = \sum_{i=0}^{n-1}\binom{n-1}{i} - n = 2^{n-1} - n.$$

On the other hand, a permutation $p = p_1p_2\cdots p_{n-1}$ of $[n-1]$ with exactly one descent must satisfy

$$p_1 < \cdots < p_k, \quad p_k > p_{k+1}, \quad p_{k+1} < \cdots < p_{n-1},$$

for some $1 \leq k \leq n-2$. Thus, to give such a permutation is to give a set $S = \{p_1,\ldots,p_k\}$ with k elements of $[n-1]$ such that $p_1 < \cdots < p_k$ and $p_{k+1} < \cdots < p_n$. There will be a descent at position k if and only if $S \neq \{1,\ldots,k\}$. We identify permutations in $E(n-1,1)$ with sets $S \subset [n-1]$ such that $S \neq [k]$. Therefore,

$$e(n-1,1) = \sum_{k=1}^{n-1}\left(\binom{n-1}{k}-1\right) = \left(\sum_{k=0}^{n-1}\binom{n-1}{k}\right) - n = 2^{n-1} - n.$$

We can now give an explicit bijection $\psi : E(n-1,1) \to \Pi'_n(1/2/3)$, for $n \geq 3$. Note that for $n = 1$ or 2 the result is trivial.

Let $S = \{p_1, \ldots, p_k\} \subset [n-1]$, $S \neq [k]$, with $p_1 < \cdots < p_k$. If $\#S \neq n-2$, we set

$$\psi(S) = \{1, p_1+1, \ldots, p_k+1\}/B,$$

where B is the complement of $\{1, p_1+1, \ldots, p_k+1\}$ in $[n]$, having $\#B \geq 2$. If $\#S = n-2$, then we must have $S = \{1, \ldots, \hat{i}, \ldots, n-1\}$, for some $i \in [n-2]$, where $\hat{i}$ means that the integer i is not in S. In this case, we put

$$\psi(S) = \begin{cases} \{1,2,\ldots,i\}/\{i+1,\ldots,n\}, & \text{if } i \neq 1 \\ \{1,2,\ldots,n\}, & \text{if } i = 1 \end{cases}.$$

From its construction, the partition $\psi(S)$ has one or two blocks, each with at least two elements. Moreover, note that the partition $\{1,2,\ldots,i\}/\{i+1,\ldots,n\}$ must be obtained via the map ψ from a uniquely determined set $S \subset [n-1]$ with $\#S = n-2$, for otherwise we would have $S = \{1, \ldots, i-1\}$, a contradiction. Henceforth, we can easily conclude that ψ is a bijection.

Denote by F_n the n-th Fibonacci number which is defined by the recurrence relation

$$F_n = F_{n-1} + F_{n-2}, \quad n \geq 2,$$

with the initial conditions $F_0 = 0$ and $F_1 = 1$ (sequence A000045 in [14]).

Theorem 6 *For $n \geq 1$,*

$$\Pi'_n(13/2) = \{\sigma \in \Pi'_n : \sigma \text{ is layered}\},$$
$$\#\Pi'_n(13/2) = F_{n-1}.$$

Proof It is clear that if σ is layered then σ avoids the pattern $13/2$. Reciprocally, let B_1 be the block of $\sigma \in \Pi'_n(13/2)$ having the integer 1, and let $i > 1$ be the largest integer of B_1. Note that if there is an integer $1 < j < i$ such that j is not in B_1, then $st(1i/j) = 13/2$. Thus, we must have $B_1 = [1, i]$. Iterating this process we find that σ is layered.

For the enumeration part, note that $\#\Pi'_1(13/2) = F_0 = 0$ and $\#\Pi'_2(13/2) = F_1 = 1$. We claim that the number of elements of $\Pi'_n(13/2)$ is equal to the sums of

Table 1 Singleton free partitions avoiding a 3-letter pattern

π	$\Pi'_n(\pi)$	$\#\Pi'_n(\pi)$
12/3	$12\cdots n$	1
1/23	$12\cdots n$	1
1/2/3	Partitions with at most two blocks	$2^{n-1}-n$
13/2	Layered partitions	F_{n-1}
123	Perfect matchings	$(2k-1)!!$ if $n=2k$ 0 otherwise

the cardinals of $\Pi'_{n-2}(13/2)$ and $\Pi'_{n-1}(13/2)$, for $n \geq 3$. Consider the map

$$\phi : \Pi'_{n-2}(13/2) \cup \Pi'_{n-1}(13/2) \longrightarrow \Pi'_n(13/2),$$

where the image of the singleton free layered partition σ of, respectively, $[n-2]$ or $[n-1]$ is obtained by adding, respectively, the block $\{n-1, n\}$ to σ, or by adding the integer n to the block containing the letter $n-1$. The map ϕ is a bijection, since if τ is a layered partition of $[n]$, then the block B containing n must also contain the integer $n-1$. Therefore, if $B = \{n-1, n\}$, then τ is the image of the layered partition of $[n-2]$ obtained by removing B from τ, and if $\#B \geq 3$, then it is the image of the layered partition of $[n-1]$ obtained from τ by removing the letter n. Thus, we find that $\#\Pi'_n(13/2) = \#\Pi'_{n-2}(13/2) + \#\Pi'_{n-1}(13/2)$ and the result follows (Table 1).

Corollary 5 *The number of layered set partitions of* $[n]$ *with at least one singleton is given by* $2^{n-1} - F_{n-1}$.

Proof It follows from the previous result and the number 2^{n-1} of layered partitions of $[n]$ obtained by Sagan [10].

We consider now the classification and enumeration of the set of singleton free partitions that avoid a set R of patterns of Π_3, with $\#R \geq 2$. Note that since $12/3 \sim 1/23$, if both patterns $12/3$ and $1/23$ are in R, then $\Pi'_n(R) = \Pi'_n(R \setminus \{1/23\})$. Therefore, without loss of generality we may consider only the patterns $12/3, 1/2/3, 13/2$ and 123. The following proposition is a consequence of Corollaries 2–4 and Theorem 6.

Proposition 1 *Let* $R = \{12/3, \pi\} \subset \Pi_3$. *Then, for* $n \geq 3$

$$\Pi'_n(R) = \begin{cases} \emptyset, & \text{if } \pi = 123 \\ \{12\cdots n\}, & \text{otherwise} \end{cases}.$$

It follows that $\Pi'_n(\Pi_3) = \emptyset$. The results for $\Pi'_n(R)$, with $\#R = 2$ or 3, are easy to prove, so we omit the proofs. Table 2 describes these sets and gives their enumeration for $n \geq 3$.

Table 2 Singleton free partitions with more than one restriction

R	$\Pi_n'(R)$	$\#\Pi_n'(R)$
$\{12/3, \pi\}$	$\emptyset$ if $\pi = 123$	0 if $\pi = 123$
	$\{12\cdots n\}$ if $\pi \neq 123$	1 if $\pi \neq 123$
$\{123,13/2\}$	$\{12/34/\cdots/(n-1)n\}$ if n even	1 if n even
	$\emptyset$ if n odd	0 if n odd
$\{123,1/2/3\}$	$\emptyset$ if $n \neq 4$	0 if $n \neq 4$
	$\{12/34, 13/24, 14/23\}$ if $n = 4$	3 if $n = 4$
$\{13/2,1/2/3\}$	$\{1\cdots i/(i+1)\cdots n : i \in [2, n-2]\} \cup \{12\cdots n\}$	$n-2$
$\{12/3, 13/2, 1/2/3\}$	$\{12\cdots n\}$	1
$\{12/3, 123, \pi\}$	$\emptyset$ for $\pi = 1/2/3$ or $\pi = 13/2$	0
$\{13/2, 123, 1/2/3\}$	$\{12/34\}$ if $n = 4$	1 if $n = 4$
	$\emptyset$ if $n \neq 4$	0 if $n \neq 4$

3 Even and Odd Singleton Free Set Partitions

In this section we consider the number of even and odd singleton free set partitions that avoid a set R of patterns of Π_3. A partition $\sigma \vdash [n]$ with $b(\sigma) = k$ has *sign*

$$sgn(\sigma) = (-1)^{n-k}.$$

Definition 1 A set partition σ of $[n]$ is *even* if $sgn(n) = 1$, and is *odd* if $sgn(n) = -1$ [3]. We denote by $E\Pi_n'$ (resp. $O\Pi_n'$) the set of all singleton free even (resp. odd) set partitions of $[n]$. Given $R \subset \Pi_3$, let $E\Pi_n'(R)$ (resp. $O\Pi_n'(R)$) be the set of all singleton free even (resp. odd) set partitions of $[n]$ that avoid the patterns in R.

The *complement* σ^c of a set partition $\sigma = B_1/B_2/\cdots/B_k \vdash [n]$, is the set partition $\sigma = B_1^c/B_2^c/\cdots/B_k^c$ where

$$B_i^c = \{n - a + 1 : a \in B_i\}.$$

As mentioned in [3], the sign of σ is the same as the sign of σ^c. Therefore, since $12/3 \sim 1/23$, we obtain the following lemma.

Lemma 1 *For* $n \geq 1$,

$$\#E\Pi_n'(12/3) = \#E\Pi_n'(1/23),$$
$$\#O\Pi_n'(12/3) = \#O\Pi_n'(1/23).$$

We start by considering single restrictions.

Theorem 7 *For $n \geq 1$,*

$$E\Pi'_n(12/3) = \begin{cases} \emptyset, & \text{if } n \text{ is even} \\ \{12\cdots n\}, & \text{if } n \text{ is odd} \end{cases},$$

and

$$O\Pi'_n(12/3) = \begin{cases} \emptyset, & \text{if } n \text{ is odd} \\ \{12\cdots n\}, & \text{if } n \text{ is even} \end{cases}.$$

Proof By Corollary 2, the set $\Pi'_n(12/3)$ has only the one block partition $12\cdots n$, which will be even if n is odd, and will be odd otherwise.

Theorem 8 *For $n \geq 1$,*

$$E\Pi'_n(1/2/3) = \begin{cases} \{\sigma \in \Pi'_n : b(\sigma) = 2\}, & \text{if } n \text{ is even} \\ \{12\cdots n\}, & \text{if } n \text{ is odd} \end{cases},$$

$$\#E\Pi'_n(1/2/3) = \begin{cases} 2^{n-1} - n - 1, & \text{if } n \text{ is even} \\ 1, & \text{if } n \text{ is odd} \end{cases};$$

and

$$O\Pi'_n(1/2/3) = \begin{cases} \{\sigma \in \Pi'_n : b(\sigma) = 2\}, & \text{if } n \text{ is odd} \\ \{12\cdots n\}, & \text{if } n \text{ is even} \end{cases},$$

$$\#O\Pi'_n(1/2/3) = \begin{cases} 2^{n-1} - n - 1, & \text{if } n \text{ is odd} \\ 1, & \text{if } n \text{ is even} \end{cases}.$$

Proof By Corollary 4, the $2^{n-1} - n$ partitions of $[n]$ that avoid the pattern $1/2/3$ are the ones having one or two blocks. As in the previous result, the only partition $12\cdots n$ with one block is even if n is odd, and is odd otherwise. On the other hand, if σ is one of the $2^{n-1} - n - 1$ partitions of $[n]$ with two blocks, then it will have the same parity as n. Thus, the result holds.

Theorem 9 *If n is an odd integer then $E\Pi'_n(123) = O\Pi'_n(123) = \emptyset$.*
If $n = 2k \geq 1$, then

$$E\Pi'_n(123) = \Pi'_n(123) \text{ and } O\Pi'_n(123) = \emptyset, \text{ if } k \text{ is even}$$

and

$$O\Pi'_n(123) = \Pi'_n(123) \text{ and } E\Pi'_n(123) = \emptyset, \text{ if } k \text{ is odd.}$$

Proof It follows from Corollary 3, since when $n = 2k$, all perfect matchings of $[n]$ have k blocks, and thus its parity is the same of that of k.

Theorem 10 *For* $n \geq 1$,

$$E\Pi'_n(13/2) = \{\sigma \in \Pi'_n : \sigma \textit{ is layered and } b(\sigma) \textit{ has the parity of } n\},$$
$$\#E\Pi'_n(13/2) = \tfrac{1}{2}\left(\tfrac{\alpha^n-\beta^n}{\alpha-\beta}\right) - \tfrac{1}{2}\left(\tfrac{\gamma^n-\delta^n}{\gamma-\delta}\right),$$

where

$$\alpha = \frac{1+\sqrt{5}}{2}, \quad \beta = \frac{1-\sqrt{5}}{2}, \quad \gamma = -\frac{1}{2} + \frac{\sqrt{3}}{2}i, \quad \delta = -\frac{1}{2} - \frac{\sqrt{3}}{2}i$$

are the roots of the equation $x^4 + 2x^3 + x^2 - 1 = 0$.

Proof The description of the set $E\Pi'_n(13/2)$ follows from Theorem 6 and the definitions. For the enumeration part, we start by noticing that

$$\#E\Pi'_n(13/2) = \#O\Pi'_{n-2}(13/2) + \#O\Pi'_{n-1}(13/2)$$

since, as in the proof of Theorem 6, any partition $\sigma \in \#E\Pi'_n(13/2)$ is uniquely obtained from a partition in $\Pi'_{n-2}(13/2)$, with parity different from n, by adding the block $\{n-1, n\}$, or from a partition from $\Pi'_{n-1}(13/2)$, with parity different from n, by adding the integer n to the block having the letter $n-1$. Therefore, using Theorem 6 we can write

$$\begin{aligned}
\#E\Pi'_n(13/2) &= \#O\Pi'_{n-2}(13/2) + \#O\Pi'_{n-1}(13/2)\\
&= \#\Pi'_{n-2}(13/2) - \#E\Pi'_{n-2}(13/2) + \#\Pi'_{n-1}(13/2) - \#E\Pi'_{n-1}(13/2)\\
&= F_{n-3} + F_{n-2} - \#E\Pi'_{n-2}(13/2) - \#E\Pi'_{n-1}(13/2).
\end{aligned}$$

Thus, the sequence formed by the cardinalities $a_n := \#E\Pi'_n(13/2)$, for $n \geq 0$, satisfies the recurrence relation

$$a_n = F_{n-3} + F_{n-2} - a_{n-2} - a_{n-1}, \text{ for } n \geq 3 \tag{5}$$

with the initial conditions $a_0 = a_1 = a_2 = 0$.

Recalling that $F(x) = \dfrac{x}{1-x-x^2}$ is the generating functions for the Fibonacci numbers (see [6]), and setting $G(x) = \sum_{n\geq 0} a_n x^n$, from the recurrence (5) we obtain

$$\begin{aligned}
G(x) &= \sum_{n\geq 3} (F_{n-3} + F_{n-2} - a_{n-2} - a_{n-1})\, x^n\\
&= x^3 \sum_{n\geq 0} F_n x^n + x^2 \sum_{n\geq 1} F_n x^n - x^2 G(x) - xG(x)\\
&= x^2(x+1)F(x) - (x^2+x)G(x),
\end{aligned}$$

that is, the generating function for the number of partitions in $E\Pi'_n(13/2)$ is

$$G(x) = \frac{x^2(x+1)}{(1-x-x^2)(1+x+x^2)}.$$

Let $\alpha = \frac{1+\sqrt{5}}{2}, \beta = \frac{1-\sqrt{5}}{2}, \gamma = -\frac{1}{2} + \frac{\sqrt{3}}{2}i, \delta = -\frac{1}{2} - \frac{\sqrt{3}}{2}i$ be the roots of the equation $(1-x-x^2)(1+x+x^2) = 0$. By the Binet formula [6], we have

$$\frac{x}{1-x-x^2} = \sum_{n\geq 0} \frac{\alpha^n - \beta^n}{\alpha - \beta} x^n.$$

In a similar way, we can write $1 + x + x^2 = (1-\gamma x)(1-\delta x)$, and thus

$$\frac{x}{1+x+x^2} = \frac{x}{(1-\gamma x)(1-\delta x)} = \frac{1}{\gamma - \delta}\left(\frac{1}{1-\gamma x} - \frac{1}{1-\delta x}\right) = \sum_{n\geq 0} \frac{\gamma^n - \delta^n}{\gamma - \delta} x^n.$$

Finally, noticing that

$$G(x) = \frac{1}{2}\left(\frac{x}{1-x-x^2}\right) - \frac{1}{2}\left(\frac{x}{1+x+x^2}\right),$$

we get the desired result.

The sequence generated by $\Pi'_n(13/2)$, $n \geq 2$, is sequence A093040 in Sloane’ s Encyclopedia [14]. Since the set $\Pi'_n(13/2)$ is the union of the disjoint sets $E\Pi'_n(13/2)$ and $O\Pi'_n(13/2)$, from the last theorem we get the analogous result for singleton free odd set partitions that avoid the pattern 13/2, which corresponds to sequence A094686 in [14].

Corollary 6 *For $n \geq 1$,*

$$O\Pi'_n(13/2) = \{\sigma \in \Pi'_n : \sigma \textit{ is layered and } b(\sigma) \textit{ has not the parity of } n\},$$

$$\#O\Pi'_n(13/2) = \frac{1}{2}\left(\frac{\alpha^n - \beta^n}{\alpha - \beta}\right) + \frac{1}{2}\left(\frac{\gamma^n - \delta^n}{\gamma - \delta}\right),$$

where

$$\alpha = \frac{1+\sqrt{5}}{2}, \quad \beta = \frac{1-\sqrt{5}}{2}, \quad \gamma = -\frac{1}{2} + \frac{\sqrt{3}}{2}i, \quad \delta = -\frac{1}{2} - \frac{\sqrt{3}}{2}i$$

are the roots of the equation $x^4 + 2x^3 + x^2 - 1 = 0$.

Table 3 Singleton free even partitions with more than one restriction

R	$E\Pi'_n(R)$	$\#E\Pi'_n(R)$
$\{12/3, 1/2/3\}$	$\emptyset$ if n is even	0
	$\{12\cdots n\}$ if n is odd	1
$\{12/3,123\}$	$\emptyset$	0
$\{12/3,13/2\}$	$\emptyset$ if n is even	0
	$\{12\cdots n\}$ if n is odd	1
$\{1/2/3,123\}$	$\emptyset$ if $n \neq 4$	0
	$\{12/34, 13/24, 14/23\}$ if $n = 4$	3
$\{1/2/3,13/2\}$	$\{1\cdots i/(i+1)\cdots n : 2 \leq i \leq n-2\}$ if n is even	$n-3$
	$\{12\cdots n\}$ if n is odd	1
$\{123,13/2\}$	$\{12/34/\cdots/(n-1)n\}$ if $n = 2k$ with k even	1
	$\emptyset$ otherwise	0
$\{12/3, 123, \pi\}$	$\emptyset$ for $\pi = 1/2/3$ or $\pi = 13/2$	0
$\{12/3, 1/2/3, 13/2\}$	$\emptyset$ if n is even	0
	$\{12\cdots n\}$ if n is odd	1
$\{123,1/2/3,13/2\}$	$\{12/34\}$ if $n = 4$	1
	$\emptyset$ if $n \neq 4$	0
$\#R \geq 4$	$\emptyset$	0

Proof If $H(x)$ is the generating function for the numbers $\#O\Pi'_n(13/2)$, then by the previous theorem,

$$H(x) = F(x) - G(x) = \frac{1}{2}\sum_{n\geq 0}\left(\frac{\alpha^n - \beta^n}{\alpha - \beta} + \frac{\gamma^n - \delta^n}{\gamma - \delta}\right)x^n,$$

and the result follows.

We consider next the description and enumeration of the sets $E\Pi'_n(R)$ and $O\Pi'_n(R)$ where $\#R \geq 2$ and $n \geq 2$. As before, by Lemma 1, we have $E\Pi'_n(R) = E\Pi'_n(R \setminus \{12/3\})$ and $O\Pi'_n(R) = O\Pi'_n(R \setminus \{12/3\})$, so we need to consider only the patterns $12/3, 1/2/3, 123$ and $13/2$. Tables 3 and 4 give the results for $E\Pi'_n(R)$ and $O\Pi'_n(R)$, $n \geq 2$. The proofs are direct consequences of the theorems above.

4 P-Recursion

A sequence $(a_n)_{n\geq 0}$ is said to be P-recursive (short for *polynomial recursive*) if there exist polynomials $p_0(x), p_1(x) \ldots, p_d(x)$ with $p_d(x) \neq 0$, such that

$$p_0(n)a_n + p_1(n)a_{n+1} + \cdots + p_d(n)a_{n+d} = 0,$$

Table 4 Singleton free odd partitions with more than one restriction

R	$O\Pi'_n(R)$	$\#O\Pi'_n(R)$
$\{12/3, 1/2/3\}$	$\emptyset$ if n is odd	0
	$\{12\cdots n\}$ if n is even	1
$\{12/3, 123\}$	$\emptyset$	0
$\{12/3, 13/2\}$	$\emptyset$ if n is odd	0
	$\{12\cdots n\}$ if n is even	1
$\{1/2/3, 123\}$	$\emptyset$	0
$\{1/2/3, 13/2\}$	$\{1\cdots i/(i+1)\cdots n : 2 \leq i \leq n-2\}$ if n is odd	$n-3$
	$\{12\cdots n\}$ if n is even	1
$\{123, 13/2\}$	$\{12/34/\cdots/(n-1)n\}$ if $n = 2k$ with k odd	1
	$\emptyset$ otherwise	0
$T=\{12/3, 1/2/3, 13/2\}$	$\emptyset$ if n is odd	0
	$\{12\cdots n\}$ if n is even	1
$\#R \geq 3,\ R \neq T$	$\emptyset$	0

for all $n \geq 0$. That is, $(a_n)_{n\geq 0}$ satisfies a homogeneous linear recurrence of finite degree with polynomial coefficients [12]. The above relation defines a_{n+d} in terms of the values of $a_n, a_{n+1}, \ldots, a_{n+d-1}$, provided $p_d(n) \neq 0$, and can be used to compute the sequence of values a_{n+d} with relatively low computational cost, for n large enough. Our objective in this section is to identify the sequences $\#\Pi'_n(\pi)$, $\#E\Pi'_n(\pi)$ and $\#O\Pi'_n(\pi)$, $n \geq 1$, for $\pi \vdash [3]$, which are P-recursive.

Closely related with P-recursive sequences is the notion of D-finite (short for *differentiably finite*) formal power series [11]. A power series $f(x)$ is D-finite if there exist finitely many polynomials $p_0(x), p_1(x), \ldots, p_m(x)$ with $p_m(x) \neq 0$ such that

$$p_0(x)f(x) + p_1(x)f^{(1)}(x) + \cdots + p_m(x)f^{(m)}(x) = 0, \tag{6}$$

where $f^{(i)}(x) = d^if/dx^i$.

An example of a D-finite function is $f(x) = e^x$, since $f(x) - f'(x) = 0$. Similarly, any linear combination of series of the form $x^m e^{ax}$ ($m \in \mathbb{N}, a \in \mathbb{R}$) is D-finite, since such series satisfy a linear homogeneous differential equation with constant coefficients.

The following result, proved by Stanley in [11], was also mentioned in Jungen [4].

Theorem 11 *A sequence $(a_n)_{n\geq 0}$ is P-recursive if and only if its ordinary generating function $f(x) = \sum_{n\geq 0} a_n x^n$ is D-finite.*

Sagan [10] proved the following analogous result for exponential generating functions.

Theorem 12 *A sequence $(a_n)_{n\geq 0}$ is P-recursive if and only if its exponential generating function $f(x) = \sum_{n\geq 0} a_n x^n/n!$ is D-finite.*

A formal power series is said to be *algebraic* if there exist polynomials $p_0(x)$, $p_1(x), \ldots, p_d(x)$, not all zero, such that

$$p_0(x) + p_1(x)f(x) + \cdots + p_d(x)f(x)^d = 0. \tag{7}$$

The smallest positive integer d for which (7) hold is called the *degree* of $f(x)$. It is simple to see that an algebraic power series $f(x)$ has degree 1 if and only if $f(x)$ is rational. The following result asserts that all algebraic power series are D-finite (see [12]).

Theorem 13 *If $f(x)$ is an algebraic power series then $f(x)$ is D-finite*

The converse of this result is false, since, for instance, the power series $f(x) = e^x$ is D-finite but not algebraic.

We will also need the following result of Stanley [12].

Theorem 14 *If $f(x)$ and $g(x)$ are D-finite, then any linear combination $af(x) + bg(x)$ is also D-finite.*

If $f(x)$ is D-finite and $g(x)$ is algebraic with $g(0) = 0$, then the composition $f(g(x))$ is D-finite.

We start our analysis by showing that $\#\Pi'_n$, $n \geq 1$, does not form a P-recursive sequence.

Proposition 2 *The sequence $\#\Pi'_n$, $n \geq 1$, is not P-recursive.*

Proof The proof follows essentially the same argument used by Sagan in [10] to show that $\#\Pi_n$ is not P-recursive. By contradiction, assume that the sequence $\#\Pi'_n$ is P-recursive. Then, its generating function

$$F(x) = e^{e^x - 1 - x},$$

determined in (4), must be D-finite by Theorem 12, and so it must satisfy Eq. (6) for some polynomials $p_0(x), p_1(x), \ldots, p_d(x)$. A simple induction shows that the i-th derivative of $F(x)$ can be written as

$$\frac{d^i}{dx^i}F(x) = F(x)\left(a_0^i + a_1^i e^x + a_2^i e^{2x} + \cdots + a_{i-1}^i e^{(i-1)x} + e^{ix}\right),$$

for some constants a_j^i, $j = 0, 1, \ldots, i-1$. Thus, taking the derivatives in Eq. (6) and dividing by $F(x)$, which is never zero, we get

$$q_0(x) + q_1(x)e^x + \cdots + q_d(x)e^{dx} = 0,$$

where

$$q_i(x) = p_i(x) + \sum_{k=i+1}^{d} a_i^k p_k(x).$$

Moreover, since the $p_i(x)$ are not all zero, the same is true for the $q_i(x)$. But this implies that e^x is algebraic, a contradiction.

Theorem 15 *For any $m \geq 1$, the following sequences are P-recursive, for $n \geq 1$:*

$$\#\Pi'_n(12\cdots m), \quad \#\Pi'_n(\pi^i_m), \quad \#\Pi'_n(1/2/\cdots/m).$$

Furthermore, for any $\pi \vdash [3]$, the sequences $\#\Pi'_n(\pi)$, $\#E\Pi'_n(\pi)$ and $\#O\Pi'_n(\pi)$, $n \geq 1$, are P-recursive.

Proof The exponential generating function for the numbers $\#\Pi'_n(12\cdots m)$, $n \geq 1$, is given by $F_{12\cdots m}(x) = \exp(\exp_{m-1}(x) - 1 - x)$. We have already seen that $f(x) = e^x$ is D-finite, and $g(x) = \exp_{m-1}(x) - 1 - x$ is algebraic since it is a polynomial. Thus, by Theorem 14 the composition $f(g(x)) = F_{12\cdots m}(x)$ is D-finite.

The exponential generating functions $\exp_{m-2}(e^x - 1 - x)$ and $\exp_{m-1}(e^x - 1 - x)$, respectively, for the numbers $\#\Pi'_n(\pi^i_m)$ and $\#\Pi'_n(1/2/\cdots/m)$, $n \geq 1$, are D-finite since these functions are linear combinations of series of the form $x^m e^{ax}$, with $m \in \mathbb{N}$ and $a \in \mathbb{R}$, and thus satisfy a linear homogeneous differential equation with constant coefficients.

Finally, note that by the results of Sects. 2 and 3, the generating functions for $\#\Pi'_n(\pi)$, $\#E\Pi'_n(\pi)$ and $\#O\Pi'_n(\pi)$, for each $\pi \vdash [3]$, are either specifications of the functions above, or rational functions, and thus are D-finite.

Since the generating functions of all sequences considered are D-finite, we can use Theorems 11 and 12 to conclude that all these sequences are P-recursive.

5 Gray Codes

A Gray code for a class of combinatorial objects is a list of these objects so that the transition from one object in the list to its successor takes only a "small change" (see [13] for a comprehensive survey). The definition of "small change" depends on the particular class of objects. In our case, we define the distance $d(\pi, \omega)$ between two partitions π, ω of $[n]$ as the minimum number of letters that must be moved between blocks of π, possibly creating a new block, so that the resulting partition is ω.

If the maximum distance between any two consecutive elements of a Gray code is k, then we say that the Gray code has distance k.

In this section, we describe Gray codes with distance 2 for the sets $\Pi'_n(\pi)$, for $\pi = 1/2/3, 123, 13/2$. The remaining cases $\pi = 12/3$ and $1/23$ are trivial. We point out that 2 is the minimum possible distance for a Gray code for these sets. Except for $\pi = 123$, the partition $12\cdots n$ belongs to $\Pi'_n(\pi)$, and therefore, the distance between $12\cdots n$ and any other partition must be at least equal to 2. The set $\Pi'_{2n}(123)$ is formed by perfect matchings, and again in this case, 2 is the minimum distance between two elements of this set.

We start with the case $\Pi'_n(13/2)$, for which we need the following definitions.

Definition 2 Given a singleton free partition $\sigma = B_1/\cdots/B_t$ of $[n-j]$, $j = 1, 2$, define the partition σ^n of $[n]$ as

$$\sigma^n = \begin{cases} B_1/\cdots/B_t \cup \{n\}, & \text{if } j = 1 \\ B_1/\cdots/B_t/\{n-1, n\}, & \text{if } j = 2 \end{cases}.$$

Definition 3 Let $\sigma = B_1/\cdots/B_{t-1}/B_t$ and π be layered singleton free partitions of $[n]$. We say that σ and π form a *good pair* if whenever $\#B_{t-1} \geq 3$ and $B_t = \{n-1, n\}$, then $B_{t-1} \cup \{n-1, n\}$ is not a block of π.

Lemma 2 *If σ, π is a good pair of $\Pi'_{n-j}(13/2)$ and $d(\sigma, \pi) \leq 2$ then σ^n, π^n is also a good pair of $\Pi'_n(13/2)$ and $d(\sigma^n, \pi^n) \leq 2$, for $j = 1, 2$.*

Proof If σ, π is a good pair, it follows from the definitions of good pair and σ^n that σ^n, π^n is also a good pair. Assume that $d(\sigma, \pi) \leq 2$. This means that one or two integers moved between blocks of σ to get ω, and the same is true for the partitions σ^n and π^n. Since σ^n and π^n are obtained from σ and π by inserting n is the last block, or by inserting the block $\{n-1, n\}$, the only non trivial situation to analyze is when $j = 1$ and the last block, say $B_t = \{n-2, n-1\}$, of σ vanishes in π. That is, $\sigma = B_1/\cdots/B_{t-1}/B_t$ and $\tau = B_1/\cdots/B_{t-1} \cup B_t$. In this case, we have $\sigma^n = B_1/\cdots/B_{t-1}/B_t \cup \{n\}$ and $\tau^n = B_1/\cdots/B_{t-1} \cup B_t \cup \{n\}$. But since σ and π form a good pair, we must have $B_{t-1} = \{n-4, n-3\}$, and therefore π^n is obtained from σ^n by moving the integers $n-4$ and $n-3$ to the last block. It follows that $d(\sigma^n, \pi^n) = 2$.

Note that if we drop the good pair condition in the last lemma, we may have layered singleton free partitions σ and π of $[n-1]$ with distance 2 such that the distance of σ^n and π^n is greater than 2. For instance, $d(123/45, 12345) = 2$ but $d(123/456, 123456) = 3$.

Theorem 16 *For each $n \geq 4$ there is a Gray code sequence with distance 2,*

$$\pi_1, \pi_2 \ldots, \pi_s,$$

for $\Pi'_n(13/2)$ such that any two consecutive elements are good pairs, $\pi_1 = 12\cdots n$ and $\pi_s = 12\cdots(n-2)/(n-1)n$.

Proof The list $1234, 12/34$ is a good pair and forms a Gray code with distance 2 for $\Pi'_4(13/2)$. Assume the result for integers less than n, with $n > 4$, and let

$$\alpha_1, \ldots, \alpha_s \text{ and } \beta_1, \ldots, \beta_t$$

be Gray codes with distance 2 for $\Pi'_{n-2}(13/2)$ and $\Pi'_{n-1}(13/2)$, respectively, in the conditions of the theorem. Then

$$\beta_1^n = 12\cdots(n-1)n,$$

$$\beta_t^n = 12\cdots(n-3)/(n-2)(n-1)n,$$

Table 5 Gray codes for $\Pi'_n(13/2)$, $n = 2, \ldots, 8$

$\Pi'_2(13/2)$	12
$\Pi'_3(13/2)$	123
$\Pi'_4(13/2)$	1234, 12/34
$\Pi'_5(13/2)$	12345, 12/345, 123/45
$\Pi'_6(13/2)$	123456, 12/3456, 123/456, 12/34/56, 1234/56
$\Pi'_7(13/2)$	1234567, 12/34567, 123/4567, 12/34/567, 1234/567, 123/45/67, 12/345/67, 12345/67
$\Pi'_8(13/2)$	12345678, 12/345678, 123/45678, 12/34/5678, 1234/5678, 123/45/678, 12/345/678, 12345/678, 1234/56/78, 12/34/56/78, 123/456/78, 12/3456/78, 123456/78

$$\alpha_1^n = 12\cdots(n-2)/(n-1)n, \text{ and}$$
$$\alpha_s^n = 12\cdots(n-4)/(n-3)(n-2)/(n-1)n.$$

Thus, β_t^n and α_s^n is a good pair with $d(\beta_t^n, \alpha_s^n) = 2$ and we may use Lemma 2 to conclude that any other two consecutive partitions of the sequence

$$\beta_1^n, \ldots, \beta_t^n, \alpha_s^n, \ldots, \alpha_1^n. \tag{8}$$

form a good pair and have distance less than, or equal to 2. Moreover, from the construction used in the proof of Theorem 6, we find that this sequence is an exhaustive list of the elements of $\Pi'_n(13/2)$. This means that the list (8) is a Gray code with distance 2 for $\Pi'_n(13/2)$ in the conditions of the theorem (Table 5).

Theorem 17 *For each $n \geq 4$ there is a Gray code sequence with distance 2 for $\Pi'_n(1/2/3)$ which starts with $12\cdots n$ and is followed by $1n/2\cdots(n-1)$.*

Proof For $n = 4$, the list $1234,\ 14/23,\ 13/24,\ 12/34$ is a Gray code with distance 2. By induction, assume that

$$\alpha_0, \alpha_1, \ldots, \alpha_t$$

is a Gray code sequence with distance 2 for $\Pi'_{n-1}(1/2/3)$, for some $n-1 \geq 4$, with $\alpha_0 = 12\cdots(n-1)$ and $\alpha_1 = 1(n-1)/23\cdots(n-2)$. Recalling that each partition in $\Pi'_{n-1}(1/2/3)$ has one or two blocks, given $\alpha = B_1/B_2 \in \Pi'_{n-1}(1/2/3)$ define

$${}^n\alpha = B_1 \cup \{n\}/B_2 \quad \text{and} \quad \alpha^n = B_1/B_2 \cup \{n\}.$$

For each $i = 1, \ldots, n-1$, let $\beta_i = in/1\cdots\hat{i}\cdots(n-1)$, where $\hat{i}$ means that the integer i is not in the block, and let L be the sequence of partitions in $\Pi'_n(1/2/3)$ defined by:

$$L = 12\cdots n,\ \beta_1,\ \beta_2,\ \ldots, \beta_{n-1}, \alpha_1^n,\ \alpha_2^n, \ldots, \alpha_t^n,\ {}^n\alpha_t, \ldots,\ {}^n\alpha_2,\ {}^n\alpha_1.$$

Table 6 Gray codes for $\Pi'_n(1/2/3)$, $n = 2, 3, 4, 5, 6$

$\Pi'_2(1/2/3)$	12
$\Pi'_3(1/2/3)$	123
$\Pi'_4(1/2/3)$	1234, 14/23, 24/13, 12/34
$\Pi'_5(1/2/3)$	12345, 15/234, 25/134, 35/124, 45/123, 14/235, 24/135, 12/345, 125/34, 245/13, 145/23
$\Pi'_6(1/2/3)$	123456, 16/2345, 26/1345, 36/1245, 46/1235, 56/1234, 15/2346, 25/1346, 35/1246, 45/1236, 14/2356, 24/1356, 12/3456, 125/346, 245/136, 145/236, 1456/23, 2456/13, 1256/34, 126/345, 246/135, 146/235, 456/123, 356/124, 256/134, 156/234

It is clear from the definitions that each consecutive partitions in L have distance 2. Moreover, note that by Corollary 4, the number of elements in L is

$$\begin{aligned}\#L &= 2\left(\#\Pi'_{n-1}(1/2/3) - 1\right) + n \\ &= 2\left(2^{n-2} - (n-1) - 1\right) + n \\ &= 2^{n-1} - n.\end{aligned}$$

That is, L is an exhaustive list of the elements in $\Pi'_n(1/2/3)$, and therefore is a Gray code sequence with distance 2 for $\Pi'_n(1/2/3)$ (Table 6).

In the next theorem we construct a Gray code with distance 2 for the perfect matchings of $[2k]$, that is, for the set $\Pi'_{2k}(123)$, $k \geq 2$. The next lemma, whose proof is clear from the definitions, characterizes perfect matchings with distance 2.

Lemma 3 *Two perfect matchings of* $[2k]$ *have distance 2 if and only if all but two of their blocks are equal.*

Let $\alpha = B_1/\cdots/B_{k-1}$ be a perfect matching of $[n]$ with $n = 2(k-1)$, written in standard form. For each $j = 1, \ldots, k-1$ let $B_j = \{a, b\}$ with $a < b$, and define

$$\begin{aligned}\alpha^0 &= B_1/\cdots/B_j/\cdots/B_{k-1}/\{n-1, n\}, \\ \alpha^{j1} &= B_1/\cdots/B_{j-1}/\{a, n\}/B_{j+1}/\cdots/B_{k-1}/\{b, n-1\}, \text{ and} \\ \alpha^{j2} &= B_1/\cdots/B_{j-1}/\{b, n\}/B_{j+1}/\cdots/B_{k-1}/\{a, n-1\}.\end{aligned}$$

Lemma 4 *Let* α *and* α_1 *be two perfect matchings of* $[2(k-1)]$ *with distance 2, and* $j \in [k-1]$. *Then,*

1. $d(\alpha^0, \alpha_1^0) = 2$;
2. $d(\alpha^0, \alpha^{j\ell}) = 2$, *for* $\ell = 1, 2$;
3. $d(\alpha^{j1}, \alpha^{j2}) = 2$;
4. $d(\alpha^{j\ell}, \alpha_1^{j\ell}) = 2$ *for* $\ell = 1, 2$.

Proof The first three conditions are clear since all but two of the blocks of each of the pairs of partitions α^0, α_1^0, $\alpha^0, \alpha^{j\ell}$ and α^{j1}, α^{j2} are equal.

Let $\alpha = B_1/\cdots/B_{k-1}$ and $\alpha_1 = B_1'/\cdots/B_{k-1}'$ be perfect matchings of $[2(k-1)]$, written in standard form and such that $d(\alpha, \alpha_1) = 2$. Let $n = 2(k-1), j \in [k-1]$ and assume that $B_j = \{a, b\}$, with $a < b$, so that

$$\alpha^{j1} = B/\{a, n\}/\{c, d\}/\{b, n-1\} \quad \text{and} \quad \alpha^{j2} = B/\{b, n\}/\{c, d\}/\{a, n-1\}, \tag{9}$$

where $B = B_1/\cdots/\hat{B}_j/\cdots/\hat{B}_q/\cdots/B_{k-1}$. Since the distance between α and α_1 is 2, there must be a block $B_q = \{c, d\}$ of α, with $c < d$, and two integers $j', q' \in [k-1]$ such that $B_\ell' = B_\ell$ for $\ell \neq j', q'$, and either

$$B_{j'}' = \{a, c\} \text{ and } B_{q'}' = \{b, d\} \quad \text{or} \quad B_{j'}' = \{a, d\} \text{ and } B_{q'}' = \{b, c\}.$$

Now, if $B_j = B_j'$, then it is clear that $d(\alpha^{j\ell}, \alpha_1^{j\ell}) = 2$ since all but two of the blocks of these partitions are equal, for $\ell = 1, 2$. So, assume that $B_j \neq B_j'$. We have two cases to consider: $a < c$ or $c < a$. We consider only the case $a < c$, the other case being analogous. Then, we have $j < q$ and $j' < q'$, and this implies that

$$B_j' = B_{j'}' = \{a, c\} \text{ and } B_{q'}' = \{b, d\} \quad \text{or} \quad B_j' = B_{j'}' = \{a, d\} \text{ and } B_{q'}' = \{b, c\}.$$

In the first case we have

$$\alpha_1^{j1} = B/\{a, n\}/\{b, d\}/\{c, n-1\} \text{ and } \alpha_1^{j2} = B/\{c, n\}/\{b, d\}/\{a, n-1\},$$

and in the second

$$\alpha_1^{j1} = B/\{a, n\}/\{b, c\}/\{d, n-1\} \text{ and } \alpha_1^{j2} = B/\{d, n\}/\{b, c\}/\{a, n-1\}.$$

In both cases, comparing the expressions of $\alpha^{j\ell}$ given in (9) with that of $\alpha_1^{j\ell}$, for $\ell = 1, 2$, we conclude that their distance is 2.

Theorem 18 *For each integer $k \geq 1$, there is a Gray code sequence for $\Pi_{2k}'(123)$ with distance 2.*

Proof The proof is by induction on $k \geq 1$. For $k = 1$ and $k = 2$, the lists 12 and 12/34, 13/24, 14/23 are Gray codes with distance 2. Assume the result for $k - 1 \geq 2$, and let

$$L_{k-1} = \alpha_1, \alpha_2, \ldots, \alpha_s,$$

be a Gray code sequence for $\Pi_{2(k-1)}'(123)$ with distance 2, where $s = (2k-3)!!$ by Corollary 3.

Table 7 Gray codes for $\Pi'_n(123)$, $n = 2, 4, 6$

$\Pi'_2(123)$	12
$\Pi'_4(123)$	12/34, 13/24, 14/23
$\Pi'_6(123)$	12/34/56, 16/34/25, 26/34/15, 36/24/15, 46/23/15, 16/23/45, 16/24/35, 13/24/56, 13/26/45, 12/36/45, 12/46/35, 13/46/25, 14/36/25, 14/26/35, 14/23/45

For each $i = 1, \ldots, k-1$, let R_i be the list of all $2s$ partitions $\alpha_j^{i\ell}, j = 1, \ldots, s$ and $\ell = 1, 2$, starting with α_i^{i1} and ending in α_{i+1}^{i1}, defined by:

$$R_i = \alpha_i^{i1},\ \alpha_{i-1}^{i1}, \ldots,\ \alpha_1^{i1},\ \alpha_1^{i2},\ \alpha_2^{i2}, \ldots,\ \alpha_s^{i2},\ \alpha_s^{i1},\ \alpha_{s-1}^{i1}, \ldots, \alpha_{i+1}^{i1}.$$

Finally, let

$$L_k = \alpha_1^0,\ R_1,\ \alpha_2^0\ R_2, \ldots, \alpha_{k-1}^0,\ R_{k-1},\ \alpha_k^0,\ \alpha_{k+1}^0, \ldots, \alpha_s^0.$$

By construction, all partitions in L_k are perfect matchings and, by Lemma 4, any two consecutive partitions in L_k are distinct and have distance 2. Moreover, the list L_k exhausts all elements of $\pi_{2k}(123)$, since its cardinal is given by

$$\begin{aligned}\#L_k &= s + (k-1)2s \\ &= (2k-3)!! + (k-1)2\,((2k-3)!!) \\ &= (1 + 2k - 2)((2k-3)!!) \\ &= (2k-1)!!\end{aligned}$$

Therefore, L_k is a Gray code with distance 2 for $\Pi'_{2k}(123)$ (Table 7).

Acknowledgements This work was partially supported by the Centro de Matemática da Universidade de Coimbra (CMUC), funded by the European Regional Development Fund through the program COMPETE and by the Portuguese Government through the FCT - Fundação para a Ciência e a Tecnologia under the project PEst-C/MAT/UI0324/2011.

References

1. Bacchelli, S., Barcucci, E., Grazzini, E., Pergola, E.: Exhaustive generation of combinatorial objects by ECO. Acta Inform. **40** (8), 585–602 (2004)
2. Gessel, I.M.: Symmetric functions and P-recursiveness. J. Combin. Theory Ser. A **53**(2), 257–285 (1990)
3. Goyt, A.M.: Avoidance of partitions of a three-element set. Adv. Appl. Math. **41**(1), 95–114 (2008)

4. Jungen, R.: Sur les séries de Taylor n'ayant que des singularités algébrico-logarithmiques sur leur cercle de convergence. Comment. Math. Helv. **3**(1), 266–306 (1931)
5. Klazar, M.: On abab-free and abba-free set partitions. Eur. J. Combin. **17**(1), 53–68 (1996)
6. Koshy, T.: Fibonacci and Lucas Numbers with Applications. Wiley-Interscience, New York (2001)
7. Mansour, T.: Combinatorics of Set Partitions. CRC Press [Taylor and Francis Group], Boca Raton (2013)
8. Bóna, M.: A Walk Throught Combinatorics: An Introduction to Enumeration and Graph Theory. World Scientific, Singapore (2006)
9. Noonan, J., Zeilberger, D.: The enumeration of permutations with a prescribed number of "forbidden" patterns. Adv. Appl. Math. **17**4, 381–407 (1996)
10. Sagan, B.E.: Pattern avoidance in set partitions. Ars Comb. **94**, 79–96 (2010)
11. Stanley, R.P.: Differentiably finite power series. Eur. J. Comb. **1**2, 175–188 (1980)
12. Stanley, R.P.: Enumerative Combinatorics, vol. 2. Cambridge University Press, Cambridge (1999)
13. Savage, C.: A survey of combinatorial gray codes. SIAM Rev. **39**4, 605–629 (1997)
14. Sloane, N.: The encyclopedia of integer sequences (2013). http://oeis.org
15. Wilf, H.S.: Generatingfunctionology, 2nd edn. Academic Press, Boston, MA (1994)

Some Results on the Krein Parameters of an Association Scheme

Vasco Moço Mano, Enide Andrade Martins, and Luís Almeida Vieira

Abstract We consider association schemes with d classes and the underlying Bose-Mesner algebra, $\mathscr{A}$. Then, by taking into account the relationship between the Hadamard and the Kronecker products of matrices and making use of some matrix techniques over the idempotents of the unique basis of minimal orthogonal idempotents of $\mathscr{A}$, we prove some results over the Krein parameters of an association scheme.

1 Introduction

The concept of association scheme was defined by Bose and Shimamoto in 1952, [4], and constitutes a powerful algebra and combinatorics tool that has a wide range of applications from statistics, [2, 4], combinatorial designs, [2–4], coding theory, [6], group theory, [8, 9], or character theory, [7]. One can observe an association scheme with d classes as a general and more complex combinatorial structure. In fact, each relation of an association scheme corresponds to an undirected graph and, as a particular example, an association scheme with just two classes is equivalent to a strongly regular graph and its complement.

In this work we consider association schemes with d classes and the corresponding Bose-Mesner algebra, $\mathscr{A}$, that is the algebra spanned by the matrices of the association scheme, as well as the unique basis of minimal orthogonal idempotents $\{E_0, \ldots, E_d\}$ associated to $\mathscr{A}$. We consider some special sums and products of these idempotents to prove some results over the Krein parameters of the association scheme.

This paper is organized as follows. In Sect. 2 the theory of association schemes is surveyed, while in Sect. 3 we present some important notation and matrix theory

V.M. Mano (✉) • E.A. Martins
Department of Mathematics, CIDMA-Center for Research and Development in Mathematics and Applications, University of Aveiro, 3810-193 Aveiro, Portugal
e-mail: vascomocomano@gmail.com; enide@ua.pt

L.A. Vieira
Department of Mathematics, Faculty of Sciences, CMUP-Center of Research of Mathematics, University of Porto, Rua do Campo Alegre, 687, 4169-007 Porto, Portugal
e-mail: lvieira@fe.up.pt

J.-P. Bourguignon et al. (eds.), *Dynamics, Games and Science*, CIM Series in Mathematical Sciences 1, DOI 10.1007/978-3-319-16118-1_23

results. Then, in Sect. 4, we prove some results over the Krein parameters of an association scheme, namely a new upper bound for some of the Krein parameters. We finish the paper with two examples of association schemes which proves the optimality of our bound (Sect. 5).

2 Association Schemes and the Bose-Mesner Algebra

In this section we present relevant concepts for our work which can be seen, for instance, in [1].

An *association scheme* with *d associate classes* on a finite set X is a partition of $X \times X$ into sets $R_0, R_1, \ldots, R_d$, that are relations on X such that

(*i*) $R_0 = \{(x,x) : x \in X\}$;
(*ii*) if $(x,y) \in R_i$, then $(y,x) \in R_i$, for all x, y in X and i in $\{0, 1, \ldots, d\}$;
(*iii*) for all i, j, l in $\{0, 1, \ldots, d\}$ there is an integer p^l_{ij} such that, for all (x,y) in R_l

$$\left|\{z \in X : (x,z) \in R_i \text{ and } (z,y) \in R_j\}\right| = p^l_{ij}.$$

The numbers p^l_{ij} are called the *intersection numbers* of the association scheme. In the case we have $(x,y) \in R_i$, the elements x and y of X are called *i-th associates*. It is usual to observe the intersection numbers as the entries of the so called *intersection matrices* $L_0, L_1, \ldots, L_d$, with $(L_i)_{lj} = p^l_{ij}$, where $L_0 = I_n$.

The definition presented above is due to Bose and Shimamoto, [4], and by axiom (*ii*) the relations R_i are all symmetric. This is why an association scheme defined in this way is normally called *symmetric*. A more general definition can be seen in [6]. Along this text we will only consider symmetric association schemes.

The associate classes $R_0, R_1, \ldots, R_d$ of a symmetric association scheme can be described by their adjacency matrices $A_0, A_1, \ldots, A_d$, where each A_i is a matrix of order n defined by $(A_i)_{xy} = 1$, if $(x,y) \in R_i$, and $(A_i)_{xy} = 0$, otherwise. We also have

(*a*) $A_0 = I_n$;
(*b*) $\sum_{i=0}^d A_i = J_n$;
(*c*) $A_i = A_i^\top$, $\forall i \in \{0, 1, \ldots, d\}$;
(*d*) $A_iA_j = \sum_{l=0}^d p^l_{ij} A_l$, $\forall i, j \in \{0, 1, \ldots, d\}$;

where I_n and J_n are the identity and the all ones matrices of order n, respectively, and $A^\top$ denotes the transpose of A. Note that equality (*b*) implies that the matrices A_i, $i \in 0, 1, \ldots, d$, are linearly independent. It is also well known (see [1, Lemma 1.3]) that the symmetry of the scheme asserts that $p^l_{ij} = p^l_{ji}$ and thus $A_iA_j = A_jA_i$, for all $i, j \in \{0, 1, \ldots, d\}$.

We can acknowledge $A_1, A_2, \ldots, A_d$ as adjacency matrices of undirected simple graphs $G_1, G_2, \ldots, G_d$, with common vertex set V. Two vertices u and v of V are *i-related* if uv is an edge in G_i, for $i \in \{1, 2, \ldots, d\}$.

The simpler association schemes are those with only one class. It corresponds to $A_0 = I_n$ and $A_1 = J_n - I_n$. Since G_1 is the complete graph this situation is out of interest. The next simpler case regards symmetric association schemes with two classes which is equivalent to strongly regular graphs. In fact, we have $A_0 = I_n$, A_1, $A_2 = J_n - A_1 - I_n$, where A_1 and A_2 correspond to the adjacency matrices of a strongly regular graph and it's complement, respectively. Conversely, if A is the adjacency matrix of a strongly regular graph, then $I_n, A, J_n - A - I_n$ form an association scheme with two classes.

The matrices $A_0, A_1, \ldots, A_d$ of a symmetric association scheme generate a commutative algebra, $\mathscr{A}$, with dimension $d+1$, of symmetric matrices with constant diagonal. This algebra is called the *Bose-Mesner algebra* of the scheme because it was firstly studied by these two mathematicians in [3]. Note that $\mathscr{A}$ is an algebra with respect to the usual matrix product as well as to the *Hadamard* (or *Schur*) *product*, defined for two matrices A, B of order n as the componentwise product: $(A \circ B)_{ij} = A_{ij}B_{ij}$. The algebra $\mathscr{A}$ is commutative and associative relatively to this product with unit J_n.

An element E in $\mathscr{A}$ is an *idempotent* if $E^2 = E$. Two idempotents E and F in $\mathscr{A}$ are orthogonal if $EF = 0$. The Bose-Mesner algebra $\mathscr{A}$ has a unique basis of minimal orthogonal idempotents $\{E_0, \ldots, E_d\}$ such that

$$E_iE_j = \delta_{ij}E_i,$$

$$\sum_{i=0}^{d} E_i = I_n,$$

where $\delta_{ij} = 1$, if $i = j$ and $\delta_{ij} = 0$, otherwise, for any i, j natural numbers. Let $\mathscr{A}$ be an association scheme with d classes. If $A_j \in \mathscr{A}, j \in \{0, 1, \ldots, d\}$ has $d + 1$ distinct eigenvalues, namely $\lambda_0, \lambda_1, \ldots, \lambda_d$, the idempotents E_i can be obtained as the projectors associated to the matrix A_j through the equality:

$$E_i = \prod_{l=0, l\neq i}^{d} \frac{A_j - \lambda_l I_n}{\lambda_i - \lambda_l}. \tag{1}$$

Besides the intersection numbers already introduced in the beginning of the section each association scheme contains three more families of parameters: the eigenvalues, the dual eigenvalues and the Krein parameters. In fact, there are scalars $p_i(j)$ and $q_i(j)$ such that, for all $i \in 0, 1, \ldots, d$, we have

$$A_i = \sum_{j=0}^{d} p_i(j)E_j \text{ and} \tag{2}$$

$$E_i = \sum_{j=0}^{d} q_i(j)A_j, \tag{3}$$

where the numbers $p_i(j)$ and $q_i(j)$ are the *eigenvalues* and the *dual eigenvalues* of the scheme, respectively. We also define the *eigenmatrix*, $P = (P_{ij})$, and the *dual eigenmatrix*, $Q = (Q_{ij})$, each with dimension $(d+1) \times (d+1)$, as $P_{ij} = p_j(i)$ and $Q_{ij} = q_j(i)$, respectively. From (2) and (3) one can deduce that $PQ = I_n$. As a consequence, the dual eigenvalues are determined by the eigenvalues of $\mathscr{A}$.

Finally, the *Krein parameters* discovered by Scott [13], of an association scheme with d classes are the numbers q_{ij}^l, with $i, j, l \in \{0, 1, \ldots, d\}$, such that

$$E_i \circ E_j = \sum_{l=0}^{d} q_{ij}^l E_l.$$

These parameters can be seen as dual parameters of the intersection numbers and they are determined by the eigenvalues of the scheme. The Krein parameters of an association scheme with d classes can also be considered as the entries of the matrices $L_0^*, L_1^*, \ldots, L_d^*$, such that $(L_i^*)_{lj} = q_{ij}^l$, which are called the *dual intersection matrices* of the scheme.

3 Matrix Tools

In this section we introduce some notation and some Matrix Theory results that are used in our work in Sect. 4.

We denote by $\mathscr{M}_n(\mathbb{R})$ the space of n dimensional square matrices with real entries and by $\mathscr{M}_{m,n}(\mathbb{R})$ the space of $m \times n$ matrices with real entries. The space of hermitian matrices with complex entries and dimension n is denoted by $Herm_n(\mathbb{C})$ and $Sym_n(\mathbb{R})$ denotes the space of n dimensional real symmetric matrices. Besides the Hadamard product already introduced in Sect. 2, we denote by $\otimes$ the *Kronecker product*, for matrices $C = [c_{ij}] \in \mathscr{M}_{m,n}(\mathbb{R})$ and $D = [d_{ij}] \in \mathscr{M}_{p,q}(\mathbb{R})$, defined by

$$C \otimes D = \begin{pmatrix} c_{11}D & \cdots & c_{1n}D \\ \vdots & \ddots & \vdots \\ c_{m1}D & \cdots & c_{mn}D \end{pmatrix}.$$

The next result is of central importance in the proof of our results. For $B \in Sym_n(\mathbb{R})$, we denote the eigenvalues of B in increasing order by $\lambda_1(B) \leq \lambda_2(B) \leq \cdots \leq \lambda_n(B)$.

Theorem 1 ([10, Eigenvalues Interlacing Theorem]) *Let $A \in Sym_n(\mathbb{R})$ and A_r denote any principal submatrix of A. Then, the eigenvalues of A_r interlace those of A in the sense that:*

$$\lambda_i(A) \leq \lambda_i(A_r) \leq \lambda_{n-r+i}(A),$$

for each $1 \leq i \leq r$.

Note that A_r is obtained by deleting $n - r$ rows and the corresponding columns from A.

The next result shows that $A \circ B$ is a principal submatrix of $A \otimes B$. Note that $A(\alpha, \beta)$ denotes a submatrix of $A \in \mathcal{M}_{m,n}(\mathbb{R})$ determined by some index sets α and β.

Lemma 1 ([11, Lemma 5.1.1]) *If* $A, B \in \mathcal{M}_{m,n}(\mathbb{R})$, *then*

$$A \circ B = (A \otimes B)(\alpha, \beta)$$

in which $\alpha = \{1, m+2, 2m+3, \ldots, m^2\}$ *and* $\beta = \{1, n+2, 2n+3, \ldots, n^2\}$. *In particular, if* $m = n$, $A \circ B$ *is a principal submatrix of* $A \otimes B$.

By Lemma 1 and since the eigenvalues of $A \otimes B$ are the product between the eigenvalues of A with the eigenvalues of B, we have the following corollary of Theorem 1, (see [11]).

Corollary 1 *If* $A, B \in Herm_n(\mathbb{C})$ $(Sym(n, \mathbb{R}))$, *then:*

(*i*) $\lambda_{\min}(A \circ B) \geq \lambda_{\min}(A)\lambda_{\min}(B)$;
(*ii*) $\lambda_{\max}(A \circ B) \leq \lambda_{\max}(A)\lambda_{\max}(B)$;

where $\lambda_{\min}(X)$ *and* $\lambda_{\max}(X)$ *denote the least eigenvalue and the greatest eigenvalue of the matrix* X, *respectively.*

4 Some Results on the Krein Parameters of an Association Scheme

In this section we make use of the tools presented in Sect. 3 to prove some results over the Krein parameters of an association scheme.

The following result establishes a formula for the calculation of the Krein parameters of an association scheme.

Proposition 1 *Consider an association scheme with* d *classes and let* $j, k, l \in \{0, 1, \ldots, d\}$. *Then*

$$q_{jk}^{l} = \sum_{m=0}^{d} Q_{mj} Q_{mk} P_{lm}, \tag{4}$$

with P *and* Q *the eigenmatrix and the dual eigenmatrix of the association scheme, respectively.*

Proof Let $\{A_0, A_1, \ldots, A_d\}$ be an association scheme with d classes, P and Q the eigenmatrix and the dual eigenmatrix of the association scheme, respectively, and $\{E_0, E_1, \ldots, E_d\}$ the unique basis of minimal orthogonal idempotents of the underlying Bose-Mesner algebra $\mathcal{A}$.

Let $j, k, l \in \{0, 1, \ldots, d\}$. We have $E_j = \sum_{i=0}^{d} Q_{ij}A_i$ and $E_k = \sum_{i=0}^{d} Q_{ik}A_i$. Therefore,

$$E_j \circ E_k = \sum_{i=0}^{d} Q_{ij}Q_{ik}A_i.$$

Also, we have the equality

$$E_j \circ E_k E_l = \sum_{i=0}^{d} Q_{ij}Q_{ik}A_i E_l.$$

From (2), we conclude that $A_i E_l = P_{li}E_l$. Thus

$$E_j \circ E_k E_l = \sum_{i=0}^{d} Q_{ij}Q_{ik}P_{li}E_l. \tag{5}$$

Since, $E_j \circ E_k = \sum_{i=0}^{d} q_{jk}^{i} E_i$, we have $q_{jk}^{l} E_l = E_j \circ E_k E_l$. Therefore, from (5), we have

$$q_{jk}^{l} = \sum_{i=0}^{d} Q_{ij}Q_{ik}P_{li}.$$

□

Making use of the entries of the matrices P and Q, the formula given by equality (4) allow us to easily calculate the Krein parameters of an association scheme. Furthermore, the Krein parameters of an association scheme satisfy the following results.

Theorem 2 *The Krein parameters of an association scheme with d classes satisfy the following properties.*

1. *For $l \in \{0, 1, \ldots, d\}$ the following equality holds:*

$$\sum_{0 \leq i,j \leq d} q_{ij}^{l} = 1. \tag{6}$$

2. *For $l, r \in \{0, 1, \ldots, d\}$, we have*

$$\sum_{\substack{0 \leq i \leq r-1 \\ r \leq j \leq d}} q_{ij}^{l} \leq \frac{1}{2}. \tag{7}$$

Proof Consider an association scheme with d classes, the underlying Bose-Mesner algebra, $\mathscr{A}$, and $\{E_0, E_1, \ldots, E_d\}$ the unique basis of minimal orthogonal idempotents of $\mathscr{A}$.

1. From equality

$$\left(\sum_{i=0}^{d} E_i\right) \circ \left(\sum_{j=0}^{d} E_j\right) = I_n,$$

we conclude that, for $l \in \{0, 1, \ldots, d\}$,

$$\sum_{j=0}^{d} \left(\sum_{i=0}^{d} E_i \circ E_j\right) E_l = E_l,$$

from which (6) naturally arises.

2. Let $r \in \{1, 2, \ldots, d\}$ and B be the following matrix

$$B = (E_0 + E_1 + \cdots + E_{r-1}) \otimes (E_r + E_{r+1} + \cdots + E_d) \\ + (E_r + E_{r+1} + \cdots + E_d) \otimes (E_0 + E_1 + \cdots + E_{r-1}).$$

Since B is an idempotent matrix its eigenvalues belong to the set $\{0, 1\}$. By Lemma 1, we observe that matrix B has a principal submatrix, C, given by

$$C = (E_0 + E_1 + \cdots + E_{r-1}) \circ (E_r + E_{r+1} + \cdots + E_d) \\ + (E_r + E_{r+1} + \cdots + E_d) \circ (E_0 + E_1 + \cdots + E_{r-1}),$$

and since the Hadamard product is commutative, C is given simply by

$$C = 2(E_0 + E_1 + \cdots + E_{r-1}) \circ (E_r + E_{r+1} + \cdots + E_d).$$

Now, applying Theorem 1, we conclude that, for $l \in \{0, 1, \ldots, d\}$,

$$0 \leq 2 \sum_{\substack{0 \leq i \leq r-1 \\ r \leq j \leq d}} q_{ij}^l \leq 1$$

and inequality (7) follows immediately.

□

The following result is a consequence of Theorem 2.

Corollary 2 *For each $l \in \{0, 1, \ldots, d\}$, the Krein parameters of an association scheme with d classes satisfy the following properties:*

1. $\sum_{i=0}^{d} q_{ii}^{l} \leq 1$;
2. $\min_{i \in \{0,\ldots,d\}} \{q_{ii}^{l}\} \leq \frac{1}{d+1}$.

Our last result establishes a new upper bound for some of the Krein parameters of an association scheme.

Theorem 3 *If $l, i, j \in \{0, 1, \ldots, d\}$, $i \neq j$, then*

$$q_{ij}^{l} \leq \frac{1}{2}.$$

Furthermore, if there exists an $i \in \{0, 1, \ldots, d\}$, $i \neq j$, such that $q_{ii}^{l} \neq 0$, then the inequality presented is strict.

Proof Consider an association scheme with d classes, the underlying Bose-Mesner algebra, $\mathscr{A}$, and $\{E_0, E_1, \ldots, E_d\}$ the unique basis of minimal orthogonal idempotents of $\mathscr{A}$.

Let $i, j \in \{0, 1, \ldots, d\}$. The matrix

$$B = \sum_{\substack{0 \leq r \leq d \\ r \neq j}} (E_r \otimes E_r) + E_i \otimes E_j + E_j \otimes E_i$$

is an idempotent matrix which has a principal submatrix, C, given by

$$C = \sum_{\substack{0 \leq r \leq d \\ r \neq j}} (E_r \circ E_r) + E_i \circ E_j + E_j \circ E_i,$$

(see Lemma 1). For each $l \in \{0, 1, \ldots, d\}$ we also have that

$$\left[\sum_{\substack{0 \leq r \leq d \\ r \neq j}} (E_r \circ E_r) + E_i \circ E_j + E_j \circ E_i \right] E_l$$

$$= \sum_{\substack{0 \leq r \leq d \\ r \neq j}} [(E_r \circ E_r) E_l] + \left(E_i \circ E_j\right) E_l + \left(E_j \circ E_i\right) E_l$$

$$= \sum_{\substack{0 \leq r \leq d \\ r \neq j}} (q_{rr}^{l} E_l) + q_{ij}^{l} E_l + q_{ij}^{l} E_l.$$

Since the eigenvalues of an idempotent matrix are either 0 or 1, by Theorem 1, the eigenvalues of C are bounded by 0 and 1 and therefore, for each $l \in \{0, 1, \ldots, d\}$ we have

$$0 \leq \sum_{\substack{0 \leq r \leq d \\ r \neq j}} (q^l_{rr}) + 2q^l_{ij} \leq 1. \tag{8}$$

By property (1.) of Corollary 2, from (8), we conclude the statements of Theorem 3. □

5 Some Examples

In this section we present two examples for our upper bound of the Krein parameters of an association scheme. The first example is based on the notation presented in the paper [5].

Example 1 Let n be an even natural number and $U_{i,j} \in \mathscr{M}_n(\mathbb{R})$ be the matrices defined by $\left(U_{i,j}\right)_{pq} = \delta_{ip}\delta_{jq}$, for $i, j, p, q \in \{1, 2, \ldots, n\}$. Let $m = \frac{n}{2} + 1$. Now we consider the family of matrices $\mathscr{F} = \{B_i\}_{i \in \{1,\ldots,m\}}$ such that:

- $B_1 = I_n$;
- $B_r = \sum_{l=r}^{n} U_{l,l-r+1} + \sum_{l=r}^{n} U_{l-r+1,l} + \sum_{l=1}^{r-1} U_{n-r+1+l,l} + \sum_{l=1}^{r-1} U_{l,n-r+1+l}$, $r = 2, \ldots, m$;
- $B_m = \sum_{l=1}^{m-1} U_{n-m+1+l,l} + \sum_{l=1}^{m-1} U_{l,n-m+1+l}$.

From the definition, the matrices A_j, $j \in \{2, \ldots, m\}$ are symmetric matrices and have null diagonal elements.

For $i = 0, 1, \ldots, n-1$, let the matrices C_i be defined by the formula

$$(C_i)_{pq} = \begin{cases} 1 \text{ if } q = p \oplus_n i, \\ 0 \text{ if } q \neq p \oplus_n i \end{cases},$$

where $\oplus_n$ denotes the sum modulo n. Then we have that the matrices B_j, for $j \in \{2, \ldots, m\}$, are given by:

$$B_1 = C_0;$$
$$B_j = C_{j-1} + C_{n-j+1},\ j \in \{2, \ldots, m-1\};$$
$$B_m = C_{m-1}.$$

Since the matrices C_i are commutative, then the family $\mathscr{F} = \{B_i\}_{i \in \{1,\ldots,m\}}$ is also commutative.

Now we construct the following association scheme with two classes $\mathscr{A} = \{A_0, A_1, A_2\}$ where:

$$A_0 = I_n;$$
$$A_1 = \sum_{i=2}^{m-1} B_i;$$
$$A_2 = J_n - A_1 - I_n;$$

where J_n is the all ones matrix. The minimal polynomial of A_1 is given by

$$p(\lambda) = \lambda(\lambda + 2)(\lambda - n + 2),$$

since A_1 is the adjacency matrix of a strongly regular graph with parameters $(n, n-2, n-4, n-2)$ and eigenvalues 0, -2 and $n-2$ (for detailed information on the parameters and the eigenvalues of a strongly regular graph see [12]).

From equality (1) with $j = 1$ and since $A_1^2 = (n-2)I_n + (n-2)A_1 + (n-4)(J_n - A_1 - I_n)$, then the unique basis of minimal idempotents of $\mathscr{A}$ is the set $\{E_0, E_1, E_2\}$ such that

$$E_0 = \frac{1}{n}J_n;$$
$$E_1 = \frac{1}{2}I_n - \frac{1}{2}\left(J_n - A_1 - I_n\right);$$
$$E_2 = \frac{n-2}{2n}I_n - \frac{1}{n}A_1 + \frac{n-2}{2n}\left(J_n - A_1 - I_n\right).$$

Then the Krein parameter q_{12}^1 can be written as

$$q_{12}^1 = \frac{n-2}{2n},$$

which converges to $1/2$ when n tends to infinity.

Since the association schemes of two classes are particular cases of association schemes, we may conclude, from Example 1 that the upper bound $1/2$ for the Krein parameters q_{ij}^l, for $i \neq j$, in Theorem 3, is optimal for an association scheme with any number of classes.

In our final example we present a family of association schemes with three classes constructed from symmetric designs. This family has an infinite number of elements and it is presented and studied in [14], where the following definition can be seen.

Let $\mathscr{P}$ be a set of points and $\mathscr{B}$ be a set of blocks, where a *block* is a subset of $\mathscr{P}$. Then, the ordered pair $(\mathscr{P}, \mathscr{B})$ is a *symmetric design* with parameters (n, k, c), with $c < k$, if it satisfies the following properties:

(i) $\mathscr{B}$ is a subset of the power set of $\mathscr{P}$;
(ii) $|\mathscr{P}| = |\mathscr{B}| = n$;
(iii) $\forall b \in \mathscr{B}$, $|b| = k$;
(iv) $\forall p \in \mathscr{P}$, $|\{b \in \mathscr{B} : p \in b\}| = k$;
(v) $\forall p_1, p_2 \in \mathscr{P}, p_1 \neq p_2$, $|\{b \in \mathscr{B} : p_1, p_2 \in b\}| = c$;
(vi) $\forall b_1, b_2 \in \mathscr{B}$, $b_1 \neq b_2$, $|\{p \in \mathscr{P} : p \in b_1 \wedge p \in b_2\}| = c$.

Example 2 Given a symmetric design with parameters (n, k, c), we build a three class association scheme, as in [14], in the following manner. Let $X = \mathscr{P} \cup \mathscr{B}$. We define the following relations in $X \times X$:

$$\begin{aligned}
R_0 &= \{(x,x) : x \in X\};\\
R_1 &= \{(x,y) \in \mathscr{P} \times \mathscr{B} : x \in y\} \cup \{(y,x) \in \mathscr{B} \times \mathscr{P} : x \in y\};\\
R_2 &= \{(x,y) \in \mathscr{P} \times \mathscr{P} : x \neq y\} \cup \{(x,y) \in \mathscr{B} \times \mathscr{B} : x \neq y\};\\
R_3 &= \{(x,y) \in \mathscr{P} \times \mathscr{B} : x \notin y\} \cup \{(y,x) \in \mathscr{B} \times \mathscr{P} : x \notin y\}.
\end{aligned}$$

Through the axioms $(i) - (vi)$ of a symmetric design it is proved that R_0, R_1, R_2, R_3 constitute an association scheme with three classes over X. From the relations above we compute the intersection matrices of the association scheme, given by $L_0 = I_4$,

$$L_1 = \begin{pmatrix} 0 & k & 0 & 0 \\ 1 & 0 & k-1 & 0 \\ 0 & c & 0 & k-c \\ 0 & 0 & k & 0 \end{pmatrix}, \quad L_2 = \begin{pmatrix} 0 & 0 & n-1 & 0 \\ 0 & k-1 & 0 & n-k \\ 1 & 0 & n-2 & 0 \\ 0 & k & 0 & n-k-1 \end{pmatrix},$$

$$L_3 = \begin{pmatrix} 0 & 0 & 0 & n-k \\ 0 & 0 & n-k & 0 \\ 0 & k-c & 0 & n-2k+c \\ 1 & 0 & n-k-1 & 0 \end{pmatrix}.$$

Now, using axioms $(a)-(d)$ of the matrices of the Bose-Mesner algebra, $\mathscr{A} = \{A_0, A_1, A_2, A_3\}$, we obtain their multiplication table.

$\times$	A_0	A_1	A_2	A_3
A_0	A_0	A_1	A_2	A_3
A_1	A_1	$kA_0 + cA_2$	$(k-1)A_1 + kA_3$	$(k-c)A_2$
A_2	A_2	$(k-1)A_1 + kA_3$	$(n-1)A_0 + (n-2)A_2$	$(n-k)A_1 + (n-k-1)A_3$
A_3	A_3	$(k-c)A_2$	$(n-k)A_1 + (n-k-1)A_3$	$(n-k)A_0 + (n-2k+c)A_2$

Making use of the multiplication table of the matrices of $\mathscr{A}$, we can calculate the powers of A_1 to obtain the following polynomial:

$$p_{A_1}(\lambda) = \lambda^4 + (-k^2 - k + c)\lambda^2 + k^2(k-c), \tag{9}$$

such that $p_{A_1}(A_1) = \mathscr{O}_n$, where $\mathscr{O}_n$ denotes the n dimensional null matrix. Then A_1 has four distinct eigenvalues and therefore the least natural number such that the set $\{I_n, A_1, A_1^2, \ldots, A_1^k\}$ is linear dependent is 4. Then, we conclude that the polynomial (9) is the minimal polynomial of A_1.

Applying formula (1) in order to matrix A_1, considering the eigenvalues of the polynomial (9), $\lambda_0 = k$, $\lambda_1 = -k$, $\lambda_2 = \sqrt{k-c}$ and $\lambda_3 = -\sqrt{k-c}$, and taking into account the equality

$$(n-1)c = k(k-1), \tag{10}$$

satisfied by these symmetric designs with parameters (n,k,c), see [12], we obtain the elements of the unique basis of minimal orthogonal idempotents of $\mathscr{A}$:

$$E_0 = \frac{A_0 + A_1 + A_2 + A_3}{2n} = \frac{J_n}{2n};$$

$$E_1 = \frac{A_0 - A_1 + A_2 - A_3}{2n};$$

$$E_2 = \frac{(n-1)\sqrt{k-c}A_0 + (n-k)A_1 - \sqrt{k-c}A_2 - kA_3}{2n\sqrt{k-c}};$$

$$E_3 = \frac{(n-1)\sqrt{k-c}A_0 - (n-k)A_1 - \sqrt{k-c}A_2 + kA_3}{2n\sqrt{k-c}}.$$

Now we apply equalities (2) and (3) to compute the matrices P and Q, respectively:

$$P = \begin{pmatrix} 1 & k & n-1 & n-k \\ 1 & -k & n-1 & k-n \\ 1 & \sqrt{k-c} & -1 & -\sqrt{k-c} \\ 1 & -\sqrt{k-c} & -1 & \sqrt{k-c} \end{pmatrix}, \quad Q = \frac{1}{2n}\begin{pmatrix} 1 & 1 & n-1 & n-1 \\ 1 & -1 & -\frac{k-n}{\sqrt{k-c}} & \frac{k-n}{\sqrt{k-c}} \\ 1 & 1 & -1 & -1 \\ 1 & -1 & -\frac{k}{\sqrt{k-c}} & \frac{k}{\sqrt{k-c}} \end{pmatrix}.$$

Finally, we obtain the dual intersection matrices of this association scheme by applying formula (4) from Proposition 1 and taking into account equality (10): $L_0^* = I_4/2n$,

$$L_1^* = \frac{1}{2n}\begin{pmatrix} 0 & 1 & 0 & 0 \\ 1 & 0 & 0 & 0 \\ 0 & 0 & 0 & 1 \\ 0 & 0 & 1 & 0 \end{pmatrix},$$

$$L_2^* = \frac{1}{2n}\begin{pmatrix} 0 & 0 & n-1 & 0 \\ 0 & 0 & 0 & n-1 \\ 1 & 0 & \frac{n-2}{2} + \frac{n-2k}{2\sqrt{k-c}} & \frac{n-2}{2} - \frac{n-2k}{2\sqrt{k-c}} \\ 0 & 1 & \frac{n-2}{2} - \frac{n-2k}{2\sqrt{k-c}} & \frac{n-2}{2} + \frac{n-2k}{2\sqrt{k-c}} \end{pmatrix},$$

$$L_3^* = \frac{1}{2n}\begin{pmatrix} 0 & 0 & 0 & n-1 \\ 0 & 0 & n-1 & 0 \\ 0 & 1 & \frac{n-2}{2} - \frac{n-2k}{2\sqrt{k-c}} & \frac{n-2}{2} + \frac{n-2k}{2\sqrt{k-c}} \\ 1 & 0 & \frac{n-2}{2} + \frac{n-2k}{2\sqrt{k-c}} & \frac{n-2}{2} - \frac{n-2k}{2\sqrt{k-c}} \end{pmatrix}.$$

From the dual intersection matrices presented above, it is possible to extract some evidence of the optimality of the upper bound $1/2$, for the Krein parameters q_{ij}^l, with $i \neq j$, presented in Theorem 3. In fact, we can observe that

$$q_{23}^0 = (L_2^*)_{03} = \frac{n-1}{2n}$$

and this value converges to $1/2$, when n tends to infinity.

With these two examples we show that the upper bound presented in Theorem 3, for the Krein parameters q_{ij}^l, with $i \neq j$ of any association scheme, is optimal and cannot be improved in the general case.

Acknowledgements

1. Enide Andrade Martins and Vasco Moço Mano were partially supported by Portuguese funds through CIDMA—Center for Research and Development in Mathematics and Applications, and the Portuguese Foundation for Science and Technology ("FCT—Fundação para a Ciência e Tecnologia"), within project PEst-OE/MAT/UI4106/2014.
2. Luís Vieira research partially funded by the European Regional Development Fund through the program COMPETE and by the Portuguese Government through the FCT—Fundação para a Ciência e a Tecnologia under the project PEest-C/MAT/UI0144/2013.
3. The authors would like to thank the anonymous referees for their careful revision and their relevant comments and suggestions that improved the paper.

References

1. Bailey, R.A.: Association Schemes, Designed Experiments, Algebra and Combinatorics. Cambridge University Press, Cambridge (2004)
2. Bose, R.C.: Strongly regular graphs, partial geometries and partially balanced designs. Pac. J. Math. **13**, 389–419 (1963)
3. Bose, R.C., Mesner, D.M.: On linear associative algebras corresponding to association schemes of partially balanced designs. Ann. Math. Stat. **30**, 21–38 (1959)
4. Bose, R.C., Shimamoto, T.: Classification and analysis of partially balanced incomplete block designs with two associate classes. J. Am. Stat. Assoc. **47**, 151–184 (1952)
5. Cardoso, D.M., Vieira, L.A.: Euclidean Jordan algebras with strongly regular graphs. J. Math. Sci. **120**(1), 881–894 (2004)
6. Delsarte, P.: An algebraic approach to the association schemes of coding theory. Philips Res. Rep. Suppl. **10**, 97 (1973)
7. Hanaki, A.: Character of association schemes and normal closed subsets. Graphs Combin. **19** (3), 363–369 (2003)
8. Higman, D.G.: Coherent configurations part I: ordinary representation theory. Geometriae **4**, 1–32 (1975)
9. Higman, D.G.: Coherent configurations part II: weights. Geometriae **5**, 413–424 (1976)
10. Horn, R., Johnson, C.R.: Matrix Analysis. Cambridge University Press, Cambridge (1985)
11. Horn, R., Johnson, C.R.: Topics in Matrix Analysis. Cambridge University Press, Cambidge (1991)
12. Lint, J.H.V., Wilson, R.M.: A Course in Combinatorics. Cambridge University Press, Cambridge (2006)
13. Scott, L.L. Jr.: A condition on Higman's parameters. Notices Am. Math. Soc. **20** (A-97), 721-20-45 (1973)
14. Shakan, G., Xin, Y.: Q-Polynomial Association Schemes with Irrational Eigenvalues. A Major Qualifying Project submitted to the Faculty of the Worcester Polytechnic Institute (2012)

A Periodic Bivariate Integer-Valued Autoregressive Model

Magda Monteiro, Manuel G. Scotto, and Isabel Pereira

Abstract In this paper, a bivariate integer-valued autoregressive model with periodic structure is introduced and studied in some detail. The model can be view as a generalization of the one considered in Pedeli and Karlis (Stat. Model. 11:325–349, 2011). Emphasis is placed on models with periodic bivariate Poisson innovations. Basic probabilistic and statistical properties of the model are discussed as well as parameter estimation and forecasting. The proposed model is applied to a bivariate data series concerning the monthly number of fires in neighbor counties, Aveiro and Coimbra, in Portugal.

1 Introduction

Periodically correlated processes play an important role in the analysis of a variety of data sets drawn from different areas such as economy [5, 6, 8], hydrology [19–22], and signal processing [7] just to mention a few. Further examples can be viewed in [10, 13] and the references therein. It is worth to mention that a large part of the literature on this topic is devoted to the continuous-valued Periodic AutoRegressive Moving Average (PARMA) models which are extensions of the commonly used ARMA models, having parameters which vary periodically in time. In contrast, however, the analysis of (univariate) periodically correlated time series of counts has not received much attention in the literature. The work in [13] introduced the periodic integer-valued autoregressive model of order one driven by a periodic sequence of independent Poisson-distributed random variables. The authors analyzed basic probabilistic and statistical properties of this models, namely the existence and uniqueness of a periodically stationary and causal process, its second-order structure, and issues related with parameter estimation. An application of the

M. Monteiro (✉)
Escola Superior de Tecnologia e Gestão de Águeda and CIDMA, Universidade de Aveiro, Aveiro, Portugal
e-mail: msvm@ua.pt

M.G. Scotto • I. Pereira
Departamento de Matemática and CIDMA, Universidade de Aveiro, Aveiro, Portugal
e-mail: mscotto@ua.pt; isabel.pereira@ua.pt

J.-P. Bourguignon et al. (eds.), *Dynamics, Games and Science*, CIM Series in Mathematical Sciences 1, DOI 10.1007/978-3-319-16118-1_24

model proposed by [13] for the analysis of the number of hospital admissions per week caused by influenza can be found in [14]. In [9] was introduced a general class of periodic non-negative integer-valued moving average processes driven by a sequence of periodic integer-valued random variables with regularly varying tails. The authors analyzed some extremal properties related with this class of processes.

A related important problem which has not been addressed yet is the development of the bivariate integer-valued autoregressive model with periodic structure. This work aims at giving a contribution towards this direction. Many phenomena have in their essence a periodic structure and there are several potential applications for this class of models. For instance, they can be applied in the environmental area, to model the monthly number of fires in neighbor counties (see Fig. 6); in epidemiological area, in the analysis of monthly (or daily) number of infections of different diseases related to each other, or in economy through the analysis of the monthly number of short term unemployed and long term unemployed or the monthly number of arrival flights and departure flights from an airport.

The literature on bivariate (and also multivariate) time series models for counts based on thinning operators is still in its infancy. An important contribution was made by Franke and Subba Rao in [4] who introduced the multivariate integer-valued autoregressive (MINAR) model of order one based on the binomial thinning operator, while a multivariate generalized INAR of order p was proposed by Latour in [12] in which matrices operate on vectors using the generalized thinning operator. More recently, Pedeli and Karlis introduced, in [15], the bivariate INAR (BINAR) model of order one with bivariate Poisson and bivariate negative binomial innovations. Pedeli and Karlis's model is defined as

$$\boldsymbol{X}_t = \boldsymbol{A} \circ \boldsymbol{X}_{t-1} + \boldsymbol{Z}_t \equiv \begin{bmatrix} \alpha_1 & 0 \\ 0 & \alpha_2 \end{bmatrix} \circ \begin{bmatrix} X_{1,t-1} \\ X_{2,t-1} \end{bmatrix} + \begin{bmatrix} Z_{1,t} \\ Z_{2,t} \end{bmatrix}, \; t \in \mathbb{Z}, \tag{1}$$

where the (binomial) thinning operator "$\circ$" is defined as $\alpha_j \circ X_j \overset{d}{=} \sum_{i=1}^{X_j} U_i(\alpha_j)$, being $U_i(\alpha_j)$, for $i = 1, \ldots, X_j$, i.i.d. Bernoulli random variables with success probability $\alpha_j \in [0, 1]$, for $j = 1, 2$. Furthermore, the authors assumed that all thinning operations are performed independently of each other and of $(\boldsymbol{Z}_t) \equiv (\boldsymbol{Z}_t : t \in \mathbb{N})$ and that the thinning operations at each time t and $\boldsymbol{Z}_t$ are independent of $(\boldsymbol{X}_s)$ for $s < t$. Moreover, $(Z_{1,t}, Z_{2,t})$ are assumed to be independent $\mathbb{N}^2$-valued random pairs. The authors illustrated the performance of the BINAR(1) model through an empirical application to the joint modeling of the number of daytime and nighttime road accidents in the Netherlands for the year 2001. It is important to refer that in Pedeli and Karlis' model the autoregression matrix $\boldsymbol{A}$ is diagonal which means that there is no cross-autocorrelation in the counts; see also [16] for further details. A bivariate INAR model that accounts for cross-autocorrelation in the counts has been recently proposed by Boudreault and Charpentier in [2]. In order to also account for negative correlation between the time series, Karlis and Pedeli introduced in [11] a family of bivariate INAR(1) processes where negative cross-correlation is introduced through the innovations in terms of appropriate bivariate copulas.

Extensions for bivariate INAR(1) models with positively correlated geometric marginals can be found in [18]. Bivariate INMA models based on the binomial thinning operator and non cross-autocorrelation in the count were proposed by Quoreshi, in [17] and by Brännäs and Quoreshi [3] who report an application to the number of transactions in intra-day data of stock.

In this work, the model proposed by Pedeli and Karlis, in [15], is generalized by assuming periodic time-varying parameters and periodic bivariate sequences of innovations, i.e., expression (1) takes the form

$$\boldsymbol{X}_t = \boldsymbol{A}_t \circ \boldsymbol{X}_{t-1} + \boldsymbol{Z}_t \equiv \begin{bmatrix} \phi_{1,t} & 0 \\ 0 & \phi_{2,t} \end{bmatrix} \circ \begin{bmatrix} X_{1,t-1} \\ X_{2,t-1} \end{bmatrix} + \begin{bmatrix} Z_{1,t} \\ Z_{2,t} \end{bmatrix}, \; t \in \mathbb{Z} \tag{2}$$

with $\phi_{j,t} = \alpha_{j,i}$, for $t = i + kT\,(i = 1, \ldots, T)$, $j = 1, 2$, and $k \in \mathbb{N}_0$. In this framework, the thinning operator is defined as

$$\phi_{j,t} \circ X_{j,t-1} \stackrel{d}{=} \sum_{i=1}^{X_{j,t-1}} U_{i,t}(\phi_{j,t}),$$

where $(U_{m,t}(\phi_{j,t}))$ is a periodic sequence of independent Bernoulli random variables with success probability $P(U_{m,t}(\phi_{j,t}) = 1) = \phi_{j,t}$. Note that by the properties of the binomial thinning operator

$$X_{j,t} = \phi_{j,t} \circ X_{j,t-1} + Z_{j,t}, \; j = 1, 2. \tag{3}$$

It is assumed that $(\boldsymbol{Z}_t)$ forms a periodic sequence of independent random vectors with mean $\boldsymbol{\delta}_t := [\delta_{1,t}\ \delta_{2,t}]'$ being $\delta_{j,t} = \lambda_{j,i}$ and covariance matrix $\boldsymbol{\Sigma}_t$ where $\sigma^2_{j,t} = \upsilon_{j,i}\lambda_{j,i}$, with $\upsilon_{j,i} > 0, \sigma_{12,t} := \varphi_i$, for $j = 1, 2$, and $t = i + kT\,(i = 1, \ldots, T,\ k \in \mathbb{N}_0)$. Furthermore, for each t, $Z_{j,t}$ is assumed to be independent of $X_{j,t-1}$ and $\phi_{j,t} \circ X_{j,t-1}$. To avoid ambiguity T is taken as the smallest positive integer satisfying (2).

Throughout the rest of the work the model in (2) will be referred to as periodic bivariate integer-valued autoregressive model of order one (PBINAR(1), in short) with period $T \in \mathbb{N}$. Basic probabilistic and statistical properties of the PBINAR(1) model will be studied in some detail in the subsequent sections. Moreover, parameter estimation and forecasting will be also discussed.

The rest of the paper is organized as follows: in Sect. 2, we demonstrate the existence and uniqueness of a periodically stationary and causal PBINAR(1) process satisfying (2). Furthermore, expressions for the mean, variance, and autocovariance function are also derived. Parameter estimation is covered in Sect. 3. Forecasting is addressed in Sect. 4. In Sect. 5.2 we present a simulation study with a comparison between the different predictors referred in the previous section. An application to real data concerning to the monthly number of fires in Aveiro and Coimbra is presented in Sect. 6.Finally, some concluding remarks are given in Sect. 7.

2 Basic Properties of the PBINAR(1) Model

The analysis of the existence and uniqueness of a periodically stationary and causal PBINAR(1) process follows easily by the arguments given by Pedeli and Karlis in [15], Sect. 2, since $(\boldsymbol{X}_t)$ with $t = i + kT$ $(i = 1, \ldots, T)$ is a strictly stationary process. By iterating the equation in (3) and after rearranging some terms, it follows by Proposition 2.1 in [13] that for $i = 1, \ldots, T$ the stationary distribution of $(X_{j,i+kT})$ is given by that of

$$V_{j,i} = \sum_{m=1}^{\infty}\sum_{a=0}^{T-1}(\beta_{i,i}^{(j)}\beta_{T,a}^{(j)}(\beta_{T,T}^{(j)})^{m-1}) \circ Z_{j,T(m+1)-a} + \sum_{m=0}^{i-1}\beta_{i,m}^{(j)} \circ Z_{j,i-m}, \; j = 1, 2, \tag{4}$$

with

$$\beta_{t,m}^{(j)} := \begin{cases} \prod_{l=0}^{m-1} \phi_{j,t-l} & l > 0 \\ 1 & l = 0 \end{cases},$$

for $m \leq t$. Note that $(\beta_{t,m}^{(j)})$ is T-periodic and that $\beta_{t+kT,m}^{(j)} = \beta_{t,m}^{(j)}$ for $t > m$, satisfying that for $i = 1, 2, \ldots, T$, and $k \in \mathbb{N}_0$, $\beta_{t,i+kT}^{(j)} = \beta_{t,i}^{(j)}\left(\beta_{T,T}^{(j)}\right)^k$, $\beta_{i+T,i+a}^{(j)} = \beta_{i,i}^{(j)}\beta_{T,a}^{(j)}$ and $\beta_{T,T}^{(j)} = \prod_{m=1}^{T}\alpha_{j,m}$. The series on the right-hand side of (4) converges almost surely and also in L_2.

From the representation in (4) the mean and autocovariance function of $(\boldsymbol{X}_t)$ can be obtained.

Lemma 1 *The mean value, variance and autocovariance structure of* $(X_{j,t})$, *for* $j = 1, 2$ *and* $t = i + kT$ *with* $i = 1, 2, \ldots, T$, $(T \in \mathbb{N})$, *and* $k \in \mathbb{N}_0$ *are given by*

1. *Mean value:*

$$\mu_{j,i} \equiv E[X_{j,t}] = \frac{\sum_{m=0}^{i-1}\beta_{i,m}^{(j)}\lambda_{j,i-m} + \beta_{i,i}^{(j)}\sum_{m=0}^{T-i-1}\beta_{T,m}^{(j)}\lambda_{j,T-m}}{1 - \beta_{T,T}^{(j)}};$$

2. *Variance:*

$$V[X_{j,t}] =$$

$$K_j(\beta)\sum_{m=0}^{i-1}\left\{\beta_{T,T}^{(j)}\beta_{i,m}^{(j)}\lambda_{j,i-m} + \beta_{i,m}^{(j)}(1-\beta_{i,m}^{(j)})\lambda_{j,i-m} + \left(\beta_{i,m}^{(j)}\right)^2 \upsilon_{j,i-m}\lambda_{j,i-m}\right\} +$$

$$+ K_j(\beta) \sum_{m=0}^{T-i-1} \Big\{ \beta_{T,T}^{(j)} \beta_{i,i}^{(j)} \beta_{T,m}^{(j)} \lambda_{j,T-m} +$$

$$+ \beta_{i,i}^{(j)} \beta_{T,m}^{(j)} \left(1 - \beta_{i,i}^{(j)} \beta_{T,m}^{(j)}\right) \lambda_{j,T-m} + \left(\beta_{i,i}^{(j)} \beta_{T,m}^{(j)}\right)^2 \upsilon_{j,T-m} \lambda_{j,T-m} \Big\}$$

with $K_j(\beta) := 1/[1 - (\beta_{T,T}^{(j)})^2]$. *The convention* $\sum_{m=0}^{-1} = 0$ *is adopted.*

3. *Autocovariance structure:*

$$\omega_i := Cov(X_{1,i}, X_{2,i}) = Cov(X_{1,i+kT}, X_{2,i+kT})$$

$$= \frac{1}{1 - \beta_{T,T}^{(1)} \beta_{T,T}^{(2)}} \beta_{i,i}^{(1)} \beta_{i,i}^{(2)} \sum_{m=0}^{T-i-1} \beta_{T,m}^{(1)} \beta_{T,m}^{(2)} \varphi_{T-m} +$$

$$+ \frac{1}{1 - \beta_{T,T}^{(1)} \beta_{T,T}^{(2)}} \sum_{m=0}^{i-1} \beta_{i,m}^{(1)} \beta_{i,m}^{(2)} \varphi_{i-m}$$

and

$$Cov(X_{1,t+h}, X_{2,t}) = \frac{\beta_{t+h,h}^{(1)}}{1 - \beta_{T,T}^{(1)} \beta_{T,T}^{(2)}} \beta_{i,i}^{(1)} \beta_{i,i}^{(2)} \sum_{m=0}^{T-i-1} \beta_{T,m}^{(1)} \beta_{T,m}^{(2)} \varphi_{T-m} +$$

$$+ \frac{\beta_{t+h,h}^{(1)}}{1 - \beta_{T,T}^{(1)} \beta_{T,T}^{(2)}} \sum_{m=0}^{i-1} \beta_{i,m}^{(1)} \beta_{i,m}^{(2)} \varphi_{i-m}$$

$$Cov(X_{1,t}, X_{2,t+h}) = \frac{\beta_{t+h,h}^{(2)}}{1 - \beta_{T,T}^{(1)} \beta_{T,T}^{(2)}} \beta_{i,i}^{(1)} \beta_{i,i}^{(2)} \sum_{m=0}^{T-i-1} \beta_{T,m}^{(1)} \beta_{T,m}^{(2)} \varphi_{T-m} +$$

$$+ \frac{\beta_{t+h,h}^{(2)}}{1 - \beta_{T,T}^{(1)} \beta_{T,T}^{(2)}} \sum_{m=0}^{i-1} \beta_{i,m}^{(1)} \beta_{i,m}^{(2)} \varphi_{i-m}$$

with $\varphi_i := Cov(Z_{1,i+T}, Z_{2,i+T})$.

Proof The results follows by straightforward, although tedious, calculations. We skip the details.

Remark The mean $\mu_{j,i}$ and ω_i can be calculated through the expressions

$$\mu_{j,i} = \beta_{i,i}^{(j)} \left(\mu_{j,T} + \frac{1}{\beta_{i,i}^{(j)}} \sum_{k=0}^{i-1} \beta_{i,k}^{(j)} \lambda_{j,i-k} \right), \; i = 1, \ldots, T, \; j = 1, 2.$$

and

$$\omega_i = \sum_{k=0}^{i-1} \beta_{i,k}^{(1)} \beta_{i,k}^{(2)} \varphi_{i-k} + \beta_{i,i}^{(1)} \beta_{i,i}^{(2)} \omega_T, \ i = 1, \ldots, T,$$

respectively.

Note that the probability generating function (pgf) of $\boldsymbol{X}_t$, for $t = i + kT$, takes the form

$$\begin{aligned} &G_{X_{i+kT}}(s_1, s_2) \equiv G_{X_{1,i+kT}, X_{2,i+kT}}(s_1, s_2) = \\ &= G_{X_{1,0}}(1 - \beta_{i,i}^{(1)} (\beta_{T,T}^{(1)})^k + \beta_{i,i}^{(1)} (\beta_{T,T}^{(1)})^k s_1) G_{X_{2,0}}(1 - \beta_{i,i}^{(2)} (\beta_{T,T}^{(2)})^k + \beta_{i,i}^{(2)} (\beta_{T,T}^{(2)})^k s_2) \times \\ &\times \prod_{m=1}^{k} \prod_{a=0}^{T-1} G_{T-a;Z_1,Z_2} \left(1 - \beta_{i,i}^{(1)} \beta_{T,a}^{(1)} (\beta_{T,T}^{(1)})^{m-1} (1-s_1), 1 - \beta_{i,i}^{(2)} \beta_{T,a}^{(2)} (\beta_{T,T}^{(2)})^{m-1} (1 - s_2)\right) \\ &\times \prod_{m=0}^{i-1} G_{i-m;Z_1,Z_2} \left(1 - \beta_{i,m}^{(1)} + \beta_{i,m}^{(1)} s_1, 1 - \beta_{i,m}^{(2)} + \beta_{i,m}^{(2)} s_2\right), \end{aligned} \tag{5}$$

where $G_{i;Z_1,Z_2}$ represents the pgf of $\mathbf{Z}_{i+kT}$. The expression in (5) reduces to

$$\begin{aligned} &G_{X_{i+kT}}(s_1, s_2) = \\ &\prod_{m=1}^{+\infty} \prod_{a=0}^{T-1} G_{T-a;Z_1,Z_2} \left(1 - \beta_{i,i}^{(1)} \beta_{T,a}^{(1)} (\beta_{T,T}^{(1)})^{m-1} (1 - s_1), 1 - \beta_{i,i}^{(2)} \beta_{T,a}^{(2)} (\beta_{T,T}^{(2)})^{m-1} (1-s_2)\right) \\ &\times \prod_{m=0}^{i-1} G_{i-m;Z_1,Z_2} \left(1 - \beta_{i,m}^{(1)} + \beta_{i,m}^{(1)} s_1, 1 - \beta_{i,m}^{(2)} + \beta_{i,m}^{(2)} s_2\right), \end{aligned} \tag{6}$$

as k tends to infinity.

Remark For the particular case in which $\mathbf{Z}_{i+kT}$ ($i = 1, \ldots, T$) follows the bivariate Poisson distribution (Johnson et al., 1997, p. 125)

$$\begin{aligned} &P(Z_{1,i+kT} = z_1, Z_{2,i+kT} = z_2) = \\ &= e^{-(\lambda_{1,i}+\lambda_{2,i}-\varphi_i)} \sum_{m=0}^{\min(z_1,z_2)} \frac{(\lambda_{1,i} - \varphi_i)^{z_1-m}}{(z_1 - m)!} \frac{(\lambda_{2,i} - \varphi_i)^{z_2-m}}{(z_2 - m)!} \frac{\varphi_i^m}{m!}, \end{aligned} \tag{7}$$

where $\lambda_{1,i}, \lambda_{2,i} > 0$ and $\varphi_i \in [0, \min(\lambda_{1,i}, \lambda_{2,i}))$, then

$$G_{X_{i+kT}}(s_1, s_2) = \exp\{\mu_{1,i}(s_1 - 1) + \mu_{2,i}(s_2 - 1) + \omega_i(s_1 - 1)(s_2 - 1)\}.$$

Parameters φ_i represent the covariance between the two time series within the ith period, for $i = 1, \ldots, T$. The previous remark lead us to the following result.

Theorem 1 *The marginal distribution of* $(\boldsymbol{X}_{i+kT})$ *for* $i = 1, \ldots, T$ *and* $k \in \mathbb{N}_0$ *is bivariate Poisson with parameters* $(\mu_{1,i}, \mu_{2,i}, \omega_i)$ *if and only if* $(\mathbf{Z}_{i+kT})$ *is bivariate Poisson with parameters* $(\lambda_{1,i}, \lambda_{2,i}, \varphi_i)$.

3 Parameter Estimation

Consider a finite time series $(\boldsymbol{X}_1, \ldots, \boldsymbol{X}_{NT})$ from the PBINAR(1) model in (2), where N represents the number of complete cycles. Let $\boldsymbol{\theta} := (\boldsymbol{\alpha}_1, \boldsymbol{\alpha}_2, \boldsymbol{\lambda}_1, \boldsymbol{\lambda}_2, \boldsymbol{\varphi})$ with $\boldsymbol{\alpha}_j = (\alpha_{j,1}, \ldots, \alpha_{j,T})$, $\boldsymbol{\lambda}_j = (\lambda_{1,T}, \ldots, \lambda_{j,T})$, for $j = 1, 2$ and $\boldsymbol{\varphi} = (\varphi_1, \ldots, \varphi_T)$ be the vector of unknown parameters. Without loss of generality it is assumed that $X_0 = x_0$. Note that the transition probabilities in this case take the form

$$
\begin{aligned}
p_i(\boldsymbol{y}|\boldsymbol{x}) &:= P(\boldsymbol{X}_{i+kT} = \boldsymbol{y}|\boldsymbol{X}_{i-1+kT} = \boldsymbol{x}) \\
&= \sum_{m_1=0}^{M_1} \sum_{m_2=0}^{M_2} P(\alpha_{1,i} \circ X_{1,i-1+kT} = m_1, \alpha_{2,i} \circ X_{2,i-1+kT} = m_2 | \boldsymbol{X}_{i-1+kT} = \boldsymbol{x}) \times \\
&\quad \times P(Z_{1,i+kT} = y_1 - m_1, Z_{2,i+kT} = y_2 - m_2)
\end{aligned}
$$

with $\boldsymbol{y} := [y_1 \ y_2]'$, $\boldsymbol{x} := [x_1 \ x_2]'$, $M_1 := \min(x_1, y_1)$ and $M_2 := \min(x_2, y_2)$. The CML-estimator $\hat{\boldsymbol{\theta}}$ of $\boldsymbol{\theta}$ is obtained by maximizing the conditional log-likelihood function

$$
l(\boldsymbol{\theta}) := \ln(L(\boldsymbol{\theta})) = \sum_{n=0}^{N-1} \sum_{i=1}^{T} \ln\left(p_i(\boldsymbol{x}_{i+Tn}|\boldsymbol{x}_{i-1+Tn})\right).
$$

Numerical maximization is straightforward with standard statistical packages.

Note that from Theorem 2.2. in [1], since $(\boldsymbol{X}_t)$ is a Markov chain, under standard assumptions, we can obtain asymptotically normality of the CML-estimators.

Theorem 2 *The CML-estimator* $\hat{\boldsymbol{\theta}}$ *of* $\boldsymbol{\theta}$ *is asymptotically normal, i.e,*

$$
\sqrt{N}(\hat{\boldsymbol{\theta}} - \boldsymbol{\theta}) \xrightarrow{d} N(\boldsymbol{0}, I^{-1}(\boldsymbol{\theta})),
$$

where $I(\boldsymbol{\theta})$ *is the Fisher information matrix.*

The choice of the joint distribution for the innovation bivariate process determines the properties of the underlying bivariate process. In the univariate case, the most frequently distributions assumed to the innovation process are the Poisson distribution and the negative binomial distribution. The first is appropriate for modeling equidispersed data and has the advantage that the stationary distribution has a closed form, also a Poisson distribution, while the second is adequate to model overdispersed count data. In the standard bivariate INAR model presented in [15], the bivariate distributions assumed for the innovation process were the bivariate Poisson and the bivariate negative binomial. In the periodic case we will also give emphasis to these two distributions.

3.1 Innovations with Periodic Bivariate Poisson Distribution

In the case of the PBINAR with periodic bivariate Poisson distribution for the innovation process, the transition probabilities are given by

$$\begin{aligned} p_i(\boldsymbol{y}|\boldsymbol{x}) &:= P(\boldsymbol{X}_{i+kT} = \boldsymbol{y}|\boldsymbol{X}_{i-1+kT} = \boldsymbol{x}) \\ &= e^{-(\lambda_{1,i}+\lambda_{2,i}-\varphi_i)} \sum_{m_1=0}^{M_1} \sum_{m_2=0}^{M_2} \sum_{l=0}^{L} \frac{\varphi_i^l}{l!} \prod_{j=1}^{2} C_{m_j}^{x_j} \alpha_{j,i}^{m_j} (1-\alpha_{j,i})^{x_j-m_j} \frac{(\lambda_{j,i}-\varphi_i)^{y_j-m_j-l}}{(y_j-m_j-l)!}, \end{aligned}$$

with $L := \min(y_1 - m_1, y_2 - m_2)$. In this case, from the partial derivatives of first order the following system is obtained

$$\begin{cases} \displaystyle\sum_{n=0}^{N-1} \frac{x_{1,i-1+nT}}{1-\alpha_{1,i}} \left(\frac{p_i(\boldsymbol{x}_{i+Tn} - (1,0)|\boldsymbol{x}_{i-1+Tn} - (1,0))}{p_i(\boldsymbol{x}_{i+Tn}|\boldsymbol{x}_{i-1+Tn})} - 1 \right) = 0 \\ \displaystyle\sum_{n=0}^{N-1} \frac{x_{2,i-1+nT}}{1-\alpha_{2,i}} \left(\frac{p_i(\boldsymbol{x}_{i+Tn} - (0,1)|\boldsymbol{x}_{i-1+Tn} - (0,1))]}{p_i(\boldsymbol{x}_{i+Tn}|\boldsymbol{x}_{i-1+Tn})} - 1 \right) = 0 \\ \displaystyle\sum_{n=0}^{N-1} \frac{p_i(\boldsymbol{x}_{i+Tn} - (1,0)|\boldsymbol{x}_{i-1+Tn})}{p_i(\boldsymbol{x}_{i+Tn}|\boldsymbol{x}_{i-1+Tn})} = N \\ \displaystyle\sum_{n=0}^{N-1} \frac{p_i(\boldsymbol{x}_{i+Tn} - (0,1)|\boldsymbol{x}_{i-1+Tn})}{p_i(\boldsymbol{x}_{i+Tn}|\boldsymbol{x}_{i-1+Tn})} = N \\ \displaystyle\sum_{n=0}^{N-1} \frac{p_i(\boldsymbol{x}_{i+Tn} - (1,1)|\boldsymbol{x}_{i-1+Tn})}{p_i(\boldsymbol{x}_{i+Tn}|\boldsymbol{x}_{i-1+Tn})} = N \end{cases}$$

for $i = 1, \ldots, T$. Analytical estimates for the above system cannot be found. Thus, to solve this system numerical procedures have to be employed. In order to find standard errors for the parameter estimates associated to the Theorem 2, the diagonal entries of the Hessian matrix are related to the expressions below and all other entries are calculated in a very straightforward manner:

$$\frac{\partial^2 l(\theta)}{\partial \alpha_{1,i}^2} = \frac{1}{(1-\alpha_{1,i})^2} \times$$

$$\times \sum_{n=0}^{N-1} \left\{ -x_{1,i-1+nT} + x_{1,i-1+nT}(x_{1,i-1+nT} - 1) \frac{p_i(\boldsymbol{x}_{i+Tn} - (2,0)|\boldsymbol{x}_{i-1+Tn} - (2,0))}{p_i(\boldsymbol{x}_{i+Tn}|\boldsymbol{x}_{i-1+Tn})} + \right.$$

$$+ 2x_{1,i-1+nT} \frac{p_i(\boldsymbol{x}_{i+Tn} - (1,0)|\boldsymbol{x}_{i-1+Tn} - (1,0))}{p_i(\boldsymbol{x}_{i+Tn}|\boldsymbol{x}_{i-1+Tn})} -$$

$$\left. - \left(x_{1,i-1+nT} \frac{p_i(\boldsymbol{x}_{i+Tn} - (1,0)|\boldsymbol{x}_{i-1+Tn} - (1,0))}{p_i(\boldsymbol{x}_{i+Tn}|\boldsymbol{x}_{i-1+Tn})} \right)^2 \right\}; \tag{8}$$

$$\frac{\partial^2 l(\theta)}{\partial \lambda_{1,i}^2} = \sum_{n=0}^{N-1} \left\{ \frac{p_i(\boldsymbol{x}_{i+Tn} - (2,0)|\boldsymbol{x}_{i-1+Tn})}{p_i(\boldsymbol{x}_{i+Tn}|\boldsymbol{x}_{i-1+Tn})} - \left(\frac{p_i(\boldsymbol{x}_{i+Tn} - (1,0)|\boldsymbol{x}_{i-1+Tn})}{p_i(\boldsymbol{x}_{i+Tn}|\boldsymbol{x}_{i-1+Tn})} \right)^2 \right\}; \tag{9}$$

$$\frac{\partial^2 l(\theta)}{\partial \varphi_i^2} = \sum_{n=0}^{N-1} \left(\frac{p_i(\boldsymbol{x}_{i+Tn} - (1,1)|\boldsymbol{x}_{i-1+Tn})}{p_i(\boldsymbol{x}_{i+Tn}|\boldsymbol{x}_{i-1+Tn})} \right) \varphi_i' - \frac{\partial^2 l(\theta)}{\partial \varphi_i \partial \lambda_{1,i}} - \frac{\partial^2 l(\theta)}{\partial \varphi_i \partial \lambda_{2,i}},$$

for $i = 1, \ldots, T$. The second derivates in order to $\alpha_{2,i}^2$ and $\lambda_{2,i}^2$ are as in (8) and (9) with $(2,0)$ and $(1,0)$ replaced by $(0,2)$ and $(0,1)$. In the case of $\alpha_{2,i}^2$, $x_{1,i-1+nT}$ has to be replaced by $x_{2,i-1+nT}$ in (8).

3.2 *Innovations with Periodic Bivariate Negative Binomial Distribution*

The case of the PBINAR with periodic bivariate negative binomial innovations is more flexible than the periodic Poisson BINAR(1). The transition probabilities are given by

$$p_i(y|\boldsymbol{x}) = \left(\frac{\beta_i^{-1}}{\lambda_{1,i} + \lambda_{2,i} + \beta_i^{-1}} \right)^{\beta_i^{-1}} \sum_{m_1=0}^{M_1} \sum_{m_2=0}^{M_2} \frac{1}{\Gamma(\beta_i^{-1} + x_1 - m_1 + x_2 - m2)\Gamma(\beta_i^{-1})} \times$$

$$\times \prod_{j=1}^{2} \frac{C_{m_j}^{x_j}}{\Gamma(x_j - m_j + 1)} \alpha_{j,i}^{m_j} (1-\alpha_{j,i})^{x_j - m_j} \left(\frac{\lambda_{j,i}}{\lambda_{1,i} + \lambda_{2,i} + \beta_i^{-1}} \right)^{y_j - m_j},$$

where $\lambda_{1,i}$, $\lambda_{2,i}$, $\beta_i > 0$, for all $i \in \{1, 2, \cdots, T\}$, are the parameters associated with the periodic negative binomial bivariate distribution. Furthermore, the $\lambda_{1,i}$ and $\lambda_{2,i}$ are the mean of each component in season i and β_i is, for each season, the parameter associated with the overdispersion. In fact, the variance, $\sigma^2_{j,i}$ is equal to $\lambda_{j,i}(1 + \beta_i\lambda_{j,i})$. The covariance between the two components, in each season i, is $\varphi_i = \lambda_{1,i}\lambda_{2,i}\beta_i$, $i = 1, \cdots, T$, which only allows for positive correlation.

In this case the expressions of the partial derivatives of the log likelihood do not have a simple form as in the previous subsection.

4 Forecasting

In this section we consider the forecasting of future values $\boldsymbol{X}_{i+NT+h}$ of the periodic Poisson BINAR(1) process given past observations up through time $i + NT$, for $i = 1, \ldots, T$. Throughout the rest of the section it shall be assumed that $h = j + kT$, for $j \in \{1, \ldots, T\}$. First note that by iterating equation (3) it follows that $X_{m,i+NT+h}$ can be expressed as

$$X_{m,i+NT+h} \overset{d}{=} \beta^{(m)}_{j+i,j}(\beta^{(m)}_{T,T})^k \circ X_{m,i+NT} + V_{m,j+i+kT},$$

where

$$V_{m,j+i+kT} = \sum_{r=0}^{j-1} \beta^{(m)}_{j+i,r} \circ Z_{m,j+i-r+NT} + \sum_{w=0}^{k-1}\sum_{r=0}^{T-1} \beta^{(m)}_{j+i+T(N+k),r+j+Tw} \circ Z_{i+T(N+k-w)-r}.$$

As in the univariate case,

$$\beta^{(m)}_{j+i,j}(\beta^{(m)}_{T,T})^k \circ X_{m,i+NT} | X_{m,i+NT} \sim Bi(X_{m,i+NT}, \beta^{(m)}_{j+i,j}(\beta^{(m)}_{T,T})^k),\ m = 1, 2.$$

Moreover, $V_{1,j+i+kT}$ and $V_{2,j+i+kT}$ are independent of $X_{1,i+NT}$ and $X_{2,i+NT}$, respectively with joint pgf

$$G_{V_{1,j+i+kT},V_{2,j+i+kT}}(s_1, s_2) = \prod_{r=0}^{j-1} G_{j+i-r;Z_1,Z_2}\left(1 - \beta^{(1)}_{j+i,r}(1 - s_1), 1-\beta^{(2)}_{j+i,r}(1-s_2)\right) \times$$

$$\times \prod_{w=0}^{k-1}\prod_{r=0}^{T-1} G_{i-r+T;Z_1,Z_2}\left(1 - \beta^{(1)}_{j+i+TN+k),r+j+Tw}(1-s_1), 1-\beta^{(2)}_{j+i+T(N+k),r+j+Tw}(1-s_2)\right).$$

Furthermore, it is assumed that $\boldsymbol{Z}_{i+kT}$ follows the bivariate Poisson distribution in (7) with parameters $(\delta_{1,t}, \delta_{2,t}, \psi_t)$ such that for $t = i + kT$, $\delta_{1,t} = \lambda_{1,i}$, $\delta_{2,t} = \lambda_{2,i}$

and $\psi_t = \varphi_i$. In this case, the joint pgf above takes the form

$$G_{V_{1,j+i+kT},V_{2,j+i+kT}}(s_1, s_2) = \exp\left\{\left(\mu_{1,<i+j>} - (\beta_{T,T}^{(1)})^k \beta_{j+i,j}^{(1)}\mu_{1,i}\right)(s_1 - 1)\right\} \times \tag{10}$$

$$\times \exp\left\{\left(\mu_{2,<i+j>} - (\beta_{T,T}^{(2)})^k \beta_{j+i,j}^{(2)}\mu_{2,i}\right)(s_2 - 1)\right\} \times$$

$$\times \exp\left\{\left(\omega_{<i+j>} - (\beta_{T,T}^{(1)}\beta_{T,T}^{(2)})^k \beta_{j+i,j}^{(1)}\beta_{j+i,j}^{(2)}\omega_i\right)\right.$$

$$\left.(s_1 - 1)(s_2 - 1)\right\}$$

with

$$< i+j >:= \begin{cases} i+j, & i+j \leq T \\ i+j-T, & i+j > T \end{cases}.$$

Note that the expression in (10) is the joint pgf of the bivariate Poisson distribution in (7) with parameters $(\nu_{1,i}, \nu_{2,i}, \nu_{3,i})$, being

$$\nu_{1,i} := \mu_{1,<i+j>} - (\beta_{T,T}^{(1)})^k \beta_{j+i,j}^{(1)}\mu_{1,i};$$

$$\nu_{2,i} := \mu_{2,<i+j>} - (\beta_{T,T}^{(2)})^k \beta_{j+i,j}^{(2)}\mu_{2,i};$$

$$\nu_{3,i} := \omega_{<i+j>} - (\beta_{T,T}^{(1)}\beta_{T,T}^{(2)})^k \beta_{j+i,j}^{(1)}\beta_{j+i,j}^{(2)}\omega_i.$$

Thus, the distribution of $(X_{1,i+NT+h}, X_{2,i+NT+h})$ given $(X_{1,i+NT}, X_{2,i+NT})$ is the convolution of two binomial distributions, with parameters $(X_{1,i+NT}, \beta_{j+i,j}^{(1)}(\beta_{T,T}^{(1)})^k)$ and $(X_{2,i+NT}, \beta_{j+i,j}^{(2)}(\beta_{T,T}^{(2)})^k)$ respectively, with the bivariate distribution which has the joint pgf given in (10).

The discussion above leads to the following result.

Theorem 3 *For the bivariate Poisson periodic model* $\mathbf{X}_{i+NT+h}|\mathbf{X}_{i+NT}$, $h = j + kT$, *for* $j \in \{1, 2, \ldots, T\}$ *and* $k \in \mathbb{N}_0$, *the following properties hold:*

(a) The pgf of $\mathbf{X}_{i+NT+h}|\mathbf{X}_{i+NT}$ *is given by*

$$G_{\mathbf{X}_{i+NT+h}|\mathbf{X}_{i+NT}=(x_{1,i+NT},x_{2,i+NT})}(s_1, s_2) =$$

$$= (1 - \beta_{j+i,j}^{(1)}(\beta_{T,T}^{(1)})^k(1 - s_1))^{x_{1,i+NT}}(1 - \beta_{j+i,j}^{(2)}(\beta_{T,T}^{(2)})^k(1 - s_2))^{x_{2,i+NT}}$$

$$\times \exp\left\{\left(\mu_{1,<i+j>} - (\beta_{T,T}^{(1)})^k \beta_{j+i,j}^{(1)}\mu_{1,i}\right)(s_1 - 1)\right\} \times$$

$$\times \exp\left\{\left(\mu_{2,<i+j>} - (\beta_{T,T}^{(2)})^k \beta_{j+i,j}^{(2)}\mu_{2,i}\right)(s_2 - 1)\right\} \times$$

$$\times \exp\left\{\left(\omega_{<i+j>} - (\beta_{T,T}^{(1)}\beta_{T,T}^{(2)})^k \beta_{j+i,j}^{(1)}\beta_{j+i,j}^{(2)}\omega_i\right)(s_1 - 1)(s_2 - 1)\right\};$$

(b) $E[X_{m,i+NT+j+kT}|X_{m,i+NT}] = \left(\beta^{(m)}_{j+i,j}(\beta^{(m)}_{T,T})^k\right) X_{m,i+NT} + \mu_{m,<i+j>} - (\beta^{(m)}_{T,T})^k \beta^{(m)}_{j+i,j}$ $\mu_{m,i}$, $m = 1,2$;

(c) $V[X_{m,i+NT+j+kT}|X_{m,i+NT}] = \left(\beta^{(m)}_{j+i,j}(\beta^{(m)}_{T,T})^k\right)\left(1 - \beta^{(m)}_{j+i,j}(\beta^{(m)}_{T,T})^k\right) X_{m,i+NT} + \mu_{m,<i+j>} - (\beta^{(m)}_{T,T})^k \beta^{(m)}_{j+i,j}\mu_{m,i}$, $m = 1,2$;

(d) $Cov(X_{1,i+NT+j+kT}, X_{2,i+NT+j+kT}|X_{1,i+NT}, X_{2,i+NT}) = \omega_{<i+j>} - (\beta^{(1)}_{T,T}\beta^{(2)}_{T,T})^k \beta^{(1)}_{j+i,j}$ $\beta^{(2)}_{j+i,j}\omega_i$.

(e) *As* $k \to +\infty$, $\boldsymbol{X}_{i+NT+h}|\boldsymbol{X}_{i+NT}$ *has a bivariate Poisson distribution with parameters* $(\mu_{1,<i+j>}, \mu_{2,<i+j>}, \omega_{<i+j>})$.

To make a h-step ahead prediction we use the mode of the distribution of $\boldsymbol{X}_{i+NT+h}|\boldsymbol{X}_{i+NT}$ or the mean, median and mode of the marginal distributions of each component of $\boldsymbol{X}_{i+NT+h}|\boldsymbol{X}_{i+NT}$. The median and mode are considered estimates coherent with the model whereas the mean is considered an incoherent estimate since may not produce an integer value.

5 Simulation Study for a Particular Periodic Poisson BINAR(1) Model

In this section a simulation study is conducted to illustrate the theoretical findings given in the previous section for the periodic Poisson BINAR(1) model and to assess the small, moderate and large sample behavior of the CML estimators. A comparison between the different predictors is also made in this section.

Throughout the analysis it shall be assumed that $T = 4$. The simulation study contemplates the following combination of α's, λ's and φ's: $\boldsymbol{\alpha}_1 = (0.5, 0.9, 0.3, 0.8)$, $\boldsymbol{\alpha}_2 = (0.85, 0.4, 0.7, 0.2)$, $\boldsymbol{\lambda}_1 = (4, 2, 8, 5)$, $\boldsymbol{\lambda}_2 = (1.5, 5, 3, 10)$ and $\boldsymbol{\varphi} = (1, 1.5, 2.3, 3.8)$. We simulated times series of length $n = NT = 500, 1{,}200, 2{,}000$ with 200 independent replicates.

5.1 Estimation

For all simulated model as well as all replicates, the CML estimates of the parameters were calculated and the results are summarized in Table 1. The estimates were calculated through numerical routines in R software which need initial values to start the optimization procedure. In this case we used the CML estimates for the parameters α's and λ's from each marginal PINAR model obtained by the use of the bisection method which does not require initial values. The initial values for the covariance parameters of the innovation bivariate process were found through the use of sample covariance of each season combined with the first equation of the third point of Lemma 1. The results obtained with these initial values were similar

Table 1 Maximum likelihood estimates for $\boldsymbol{\theta}$

n	500	1,200	2,000
$\hat{\alpha}_{1,1}$	0.500 (0.056)	0.499 (0.027)	0.502 (0.019)
$\hat{\alpha}_{1,2}$	0.893 (0.070)	0.901 (0.008)	0.901 (0.005)
$\hat{\alpha}_{1,3}$	0.289 (0.069)	0.296 (0.011)	0.300 (0.031)
$\hat{\alpha}_{1,4}$	0.790 (0.083)	0.799 (0.019)	0.800 (0.013)
$\hat{\alpha}_{2,1}$	0.845 (0.065)	0.848 (0.009)	0.850 (0.007)
$\hat{\alpha}_{2,2}$	0.398 (0.066)	0.401 (0.035)	0.401 (0.027)
$\hat{\alpha}_{2,3}$	0.689 (0.063)	0.700 (0.021)	0.698 (0.015)
$\hat{\alpha}_{2,4}$	0.198 (0.085)	0.199 (0.053)	0.193 (0.041)
$\hat{\lambda}_{1,1}$	3.947 (0.635)	4.027 (0.366)	3.980 (0.305)
$\hat{\lambda}_{1,2}$	2.010 (0.293)	1.997 (0.111)	2.008 (0.090)
$\hat{\lambda}_{1,3}$	8.048 (0.998)	8.051 (0.531)	7.97 (0.440)
$\hat{\lambda}_{1,4}$	5.014 (0.747)	5.025 (0.243)	4.989 (0.174)
$\hat{\lambda}_{2,1}$	1.508 (0.199)	1.525 (0.134)	1.495 (0.099)
$\hat{\lambda}_{2,2}$	4.956 (0.758)	4.980 (0.429)	5.011 (0.310)
$\hat{\lambda}_{2,3}$	3.073 (0.416)	2.998 (0.215)	3.003 (0.149)
$\hat{\lambda}_{2,4}$	9.933 (1.116)	10.031 (0.537)	10.053 (0.416)
$\hat{\varphi}_1$	0.889 (0.299)	0.976 (0.234)	0.952 (0.193)
$\hat{\varphi}_2$	1.403 (0.422)	1.478 (0.236)	1.499 (0.182)
$\hat{\varphi}_3$	2.216 (0.473)	2.241 (0.303)	2.234 (0.258)
$\hat{\varphi}_4$	3.813 (0.957)	3.830 (0.385)	3.838 (0.300)

Standard errors in parentheses

to those obtained by the use of the true values of the parameters as initial values in the optimization procedure.

Figures 1, 2, and 3 display boxplots of the biases of the estimates for $\boldsymbol{\alpha}_j$ and $\boldsymbol{\lambda}_j$ for $j = 1, 2$, and $\boldsymbol{\varphi}$.

From Table 1, it can be observed that the standard errors of the estimators rapidly decrease to zero as n increases with special emphasis to the parameters related to the binomial thinning and the parameters associated with the average of the innovations. Furthermore, Figs. 1, 2, and 3 reveal that the estimates of $\boldsymbol{\alpha}_1$ and $\boldsymbol{\alpha}_2$, componentwise, tend to be biased to the left and negatively skewed which implies that the CML estimation has a tendency of underestimating the $\boldsymbol{\alpha}$'s mainly in the case of small sample sizes. It also can be seen that CML estimation has a better performance componentwise, regardless the component of the bivariate model, for both α and λ, when thinning parameter is superior to 0.5. For $\boldsymbol{\varphi}$'s this tendency is less obvious and is related with the magnitude of both thinning parameters. As expected, however, both the bias and skewness approach zero as the sample size increases. This is in agreement with the asymptotic properties of the CML estimators, namely unbiasedness and consistency.

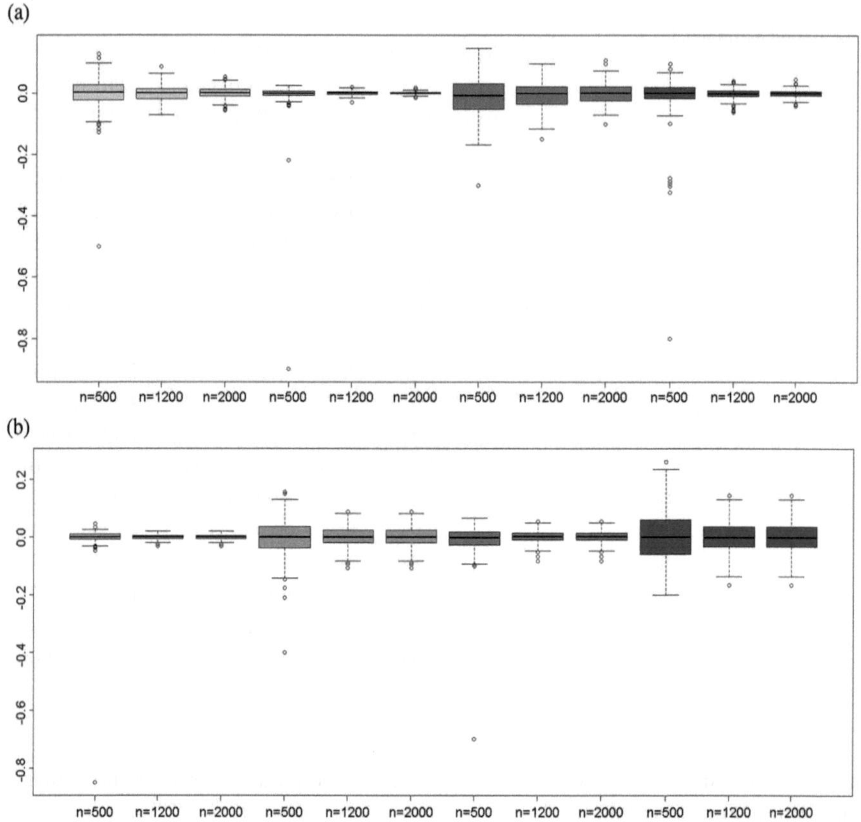

Fig. 1 Boxplots for (**a**) the biases of the CML estimators $\hat{\boldsymbol{\alpha}}_1$ and (**b**) the biases of the CML estimators $\hat{\boldsymbol{\alpha}}_2$. In (**a**), *from left to right* the first three boxplots display the biases of $\hat{\alpha}_{1,1}$ for $n = 500, 1{,}200, 2{,}000$. The subsequent three boxplots show the same information for $\hat{\alpha}_{1,2}$, the next three for $\hat{\alpha}_{1,3}$, and the last three boxplots for $\hat{\alpha}_{1,4}$. The boxplots in (**b**) show the same information for the four components of $\hat{\boldsymbol{\alpha}}_2$

5.2 *Prediction*

To compare and analyze the different predictors previously mentioned in Sect. 4 the realizations of the PBINAR model were used to make h-step ahead predictions, from one to twenty. Consider $\hat{X}_{i,t+h}$, $\hat{m}_{i,t+h}$ and $\hat{mo}_{i,t+h}$ respectively the estimators of the mean, median and mode of the marginal conditional distribution $X_{i,n+h}|X_{i,n}$. In addition to these estimators was also used the mode of the joint conditional distribution $\boldsymbol{X}_{n+h}|\boldsymbol{X}_n$, $\hat{\boldsymbol{mo}}_{t+h}$.

In the different predictors the CML estimates were plugged-in in the prediction probability functions. To assess the performance of each estimator with the increase of dimension different measures were used. For the conditional mean is was considered the square root of the mean squared error (RMSE) while the mean

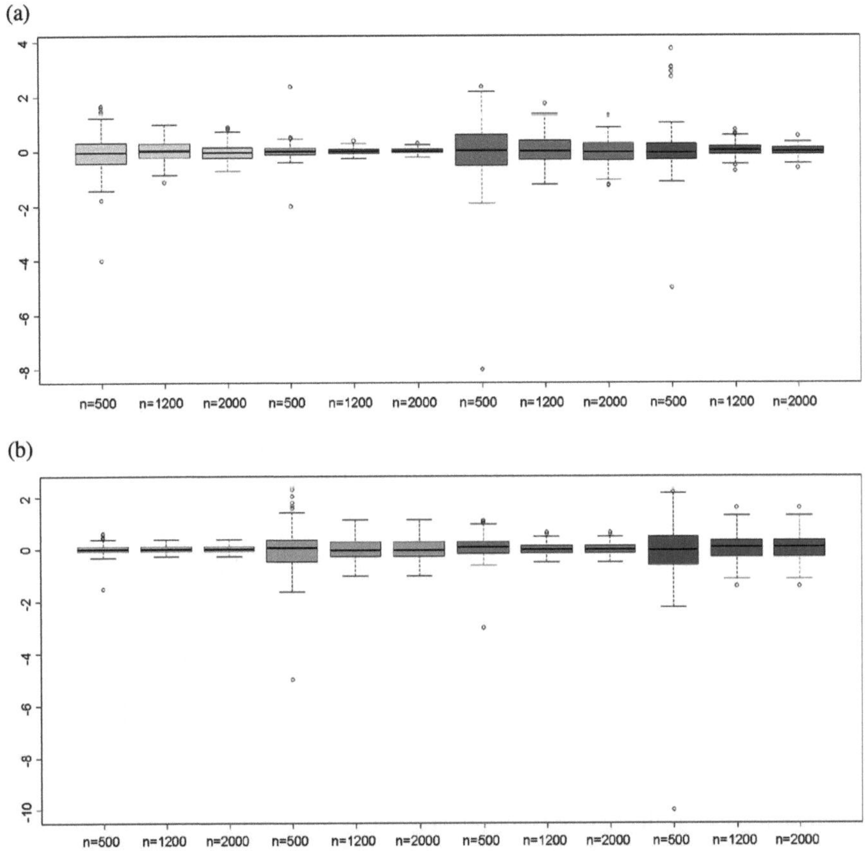

Fig. 2 Boxplots for (**a**) the biases of the CML estimators $\hat{\boldsymbol{\lambda}}_1$ and (**b**) the biases of the CML estimators $\hat{\boldsymbol{\lambda}}_2$. In (**a**), *from left to right* the first three boxplots display the biases of $\hat{\lambda}_{1,1}$ for $n = 500, 1{,}200, 2{,}000$. The subsequent three boxplots show the same information for $\hat{\lambda}_{1,2}$, the next three for $\hat{\lambda}_{1,3}$, and the last three boxplots for $\hat{\lambda}_{1,4}$. The boxplots in (**b**) show the same information for the four components of $\hat{\boldsymbol{\lambda}}_2$

absolute error (MAE) was used to evaluate the performance of the conditional median. For the conditional marginal mode and the mode of the joint distribution the loss function everything or nothing (LFEN) was used to evaluate their performance. This last function is defined by

$$LFEN = \frac{1}{2mh} \sum_{k=1}^{m} \sum_{h=1}^{20} \sum_{i=1}^{2} I\left(x_{i,t+h}^{(k)}\right),$$

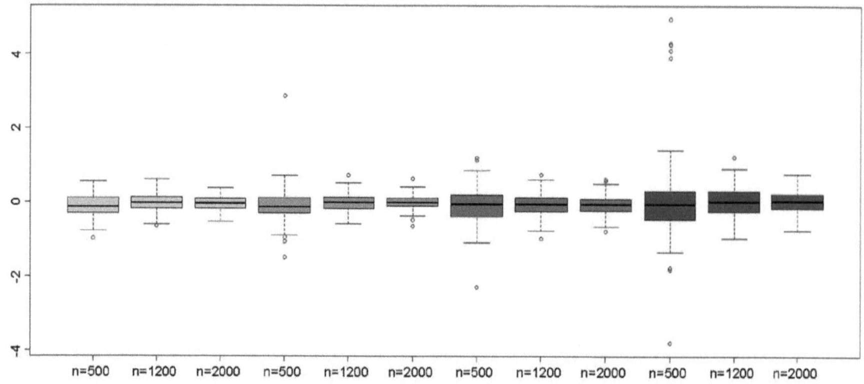

Fig. 3 Boxplots for the biases of the CML estimators $\hat{\boldsymbol{\varphi}}$. *From left to right* the first three boxplots display the biases of $\hat{\varphi}_1$ for $n = 500, 1{,}200, 2{,}000$. The subsequent three boxplots show the same information for $\hat{\varphi}_2$, the next three for $\hat{\varphi}_3$, and the last three boxplots for $\hat{\varphi}_4$

Table 2 RMSE, MAE, LFEN e MPAE of 20 predictions h-step ahead

	$\hat{X}_{t+h}$		$\hat{m}_{t+h}$		$\hat{mo}_{t+h}$		$\hat{\boldsymbol{mo}}_{t+h}$	
	RMSE	MPAE	MAE	MPAE	LFEN	MPAE	LFEN	MPAE
$n = 500$	3.94	0.317	3.13	0.313	0.696	0.305	0.700	**0.300**
$n = 1{,}200$	3.93	0.325	3.12	0.321	0.704	0.312	0.694	**0.294**
$n = 2{,}000$	3.88	0.310	3.07	0.308	0.695	0.300	0.691	**0.295**

where m represents the number of replicates and

$$I\left(x_{i,t+h}^{(k)}\right) = \begin{cases} 1 \text{ if } |\hat{mo}_{i,t+h}^{(k)} - x_{i,t+h}^{(k)}| > 1 \\ 0 \text{ if } |\hat{mo}_{i,t+h}^{(k)} - x_{i,t+h}^{(k)}| \leq 1 \end{cases}.$$

In order to compare the performance of the different predictors it was used the measure mean percentage absolute error (MPAE) given by

$$MPAE = \frac{1}{2Hm} \sum_{h=1}^{H} \sum_{k=1}^{m} \sum_{i=1}^{2} |X_{i,t+h}^{*(k)} - X_{i,t+h}^{(k)}| / X_{i,t}^{(k)},$$

where m represents the number of replicates and H the number of predictions and $X_{i,t+h}^{*}$ represents one of the predictors used according to the methodology.

Table 2 presents a summary of the measures used to compare the predictors in the considered scenario.With the increase of n the measures RMSE, MAE and LFEN tend to decrease and the comparison of MPAE between predictors allows us to conclude that the mode (see bold values in Table 2) of the joint conditional distribution of $\boldsymbol{X}_{n+h}|\boldsymbol{X}_n$ is the one that has a better performance for all dimensions

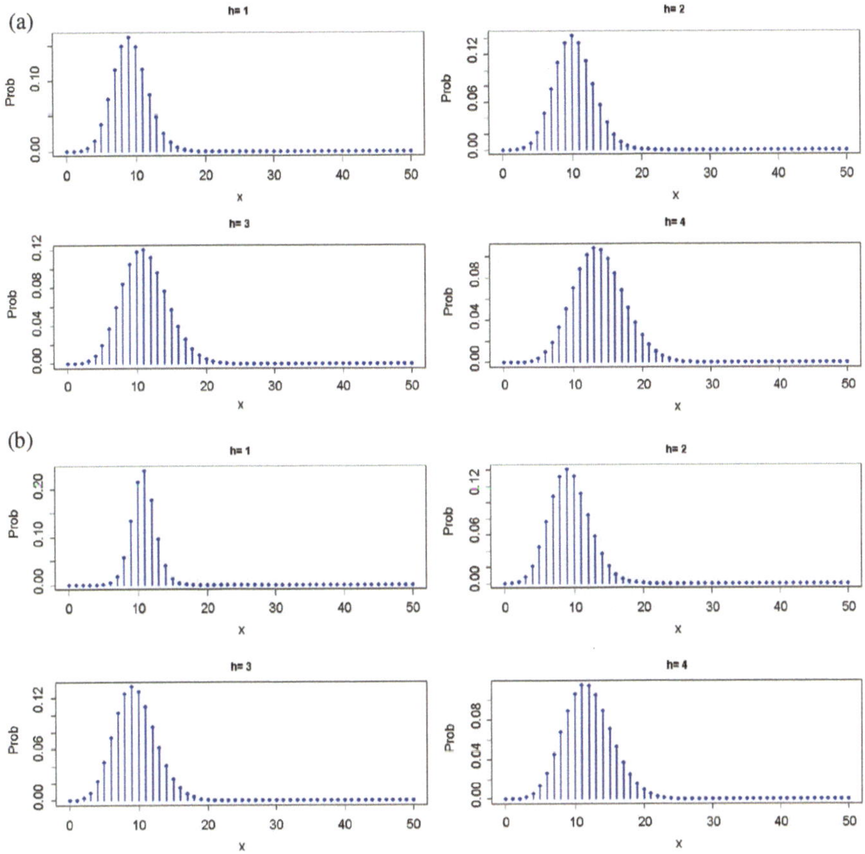

Fig. 4 Plots for the h-step-ahead predictive marginal distributions for each component, $P(x_{k,T+h}|x_{k,T})$, $k = 1, 2$ and $h = 1, 2, 3, 4$; (**a**) component 1 (**b**) component 2

that were used. Figure 4 presents the h-step-ahead predictive marginal distributions for each component, $P(x_{k,T+h}|x_{k,T})$, $k = 1, 2$ and $h = 1, 2, 3, 4$ for a particular realization of PBINAR. For each component, 1 and 2, the mode for the first season is respectively 9 and 11, for the second season is 10 and 9, for the third season is 11 and 9 and for the fourth season is 13 and 11. Figure 5 shows the h-step-ahead joint predictive distribution $P((x_{1,T+h}, x_{2,T+h})|(x_{1,T}, x_{2,T}))$, $h = 1, 2, 3, 4$. For the first season the mode is (10,11), for the second season this pair is (11,9), for the third season is (10,9) and in the last season is (11,9).

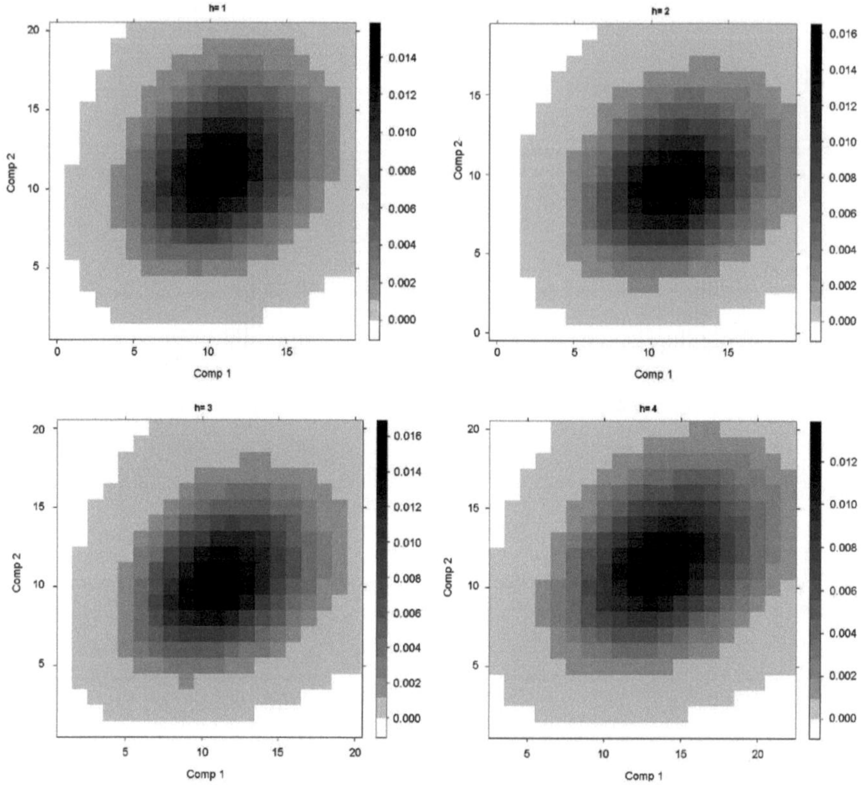

Fig. 5 Plots for the h-step-ahead joint predictive distribution $P(\boldsymbol{x}_{T+h}|\boldsymbol{x}_T)$, $h = 1, 2, 3, 4$

6 Application

The data used in this application refer to the monthly number of fires in the neighbor counties Aveiro and Coimbra (Portugal) during the period 1980–2010 (Fig. 6). A visual inspection of the sample ACF functions (Fig. 7) reveal a non-decaying structure in the autocorrelation of the time series with a periodic pattern of 12 months. Figure 8 presents monthly sample means, variances and cross correlations, where it can be seen the existence of overdispersion for both series in almost every months.

Since the bivariate distributions for the innovations discussed in Sect. 3 only allow for non-negative correlations and in August the sample correlation between the two series is negative (−0.2), we tested the significance of this correlation which, for the usual significance levels, not rejected the null hypothesis. Hence, in order to model the data we considered both the bivariate Poisson INAR(1) model and the bivariate INAR(1) model with bivariate negative binomial innovations. The results are shown in Table 3. Comparing the log-likelihood one can see that negative

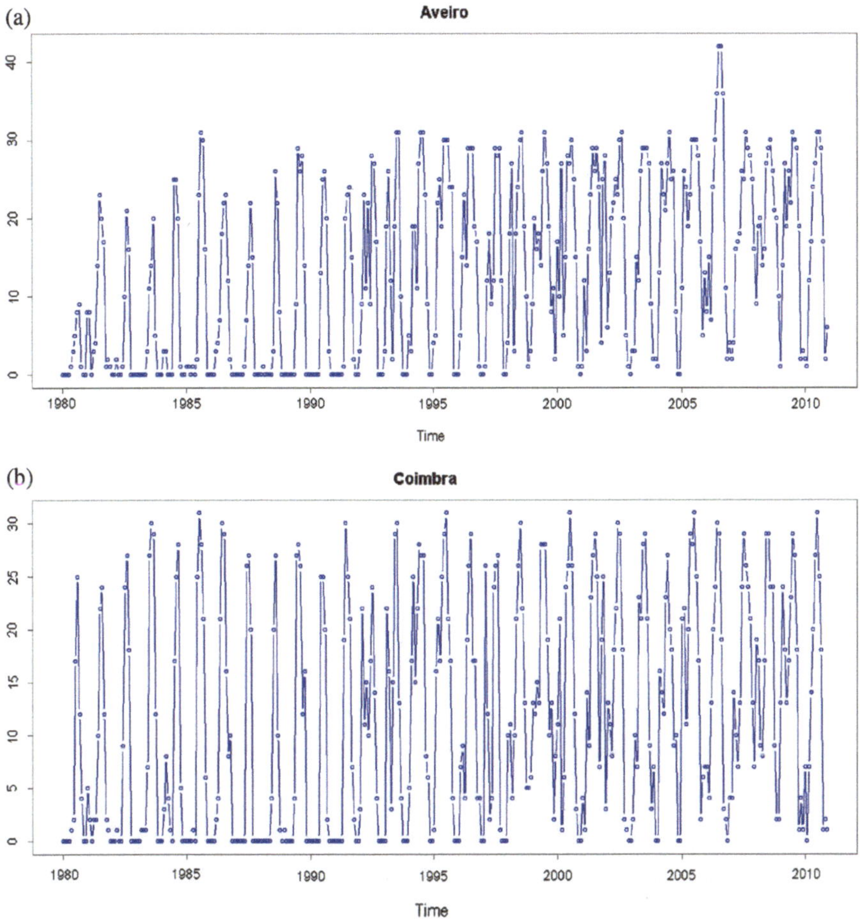

Fig. 6 Monthly number of fires in Aveiro and Coimbra Counties

binomial BINAR(1) model provides a better fit and can suits more properly the overdispersion. The standard errors of the estimates obtained by the two models (derived numerically from the Hessian) show that fitting a BINAR(1) model with negative binomial innovations generally improves the precision of the estimates. On the other hand it is apparent that ignoring the overdispersion might lead to incorrect standard errors and hence incorrect inferences.

It can be noticed that according to Coimbra data values the selected model presents several thinning parameter estimates that are not significant, which means that in the correspondent months the number of fires is being modeled only through the innovation process.

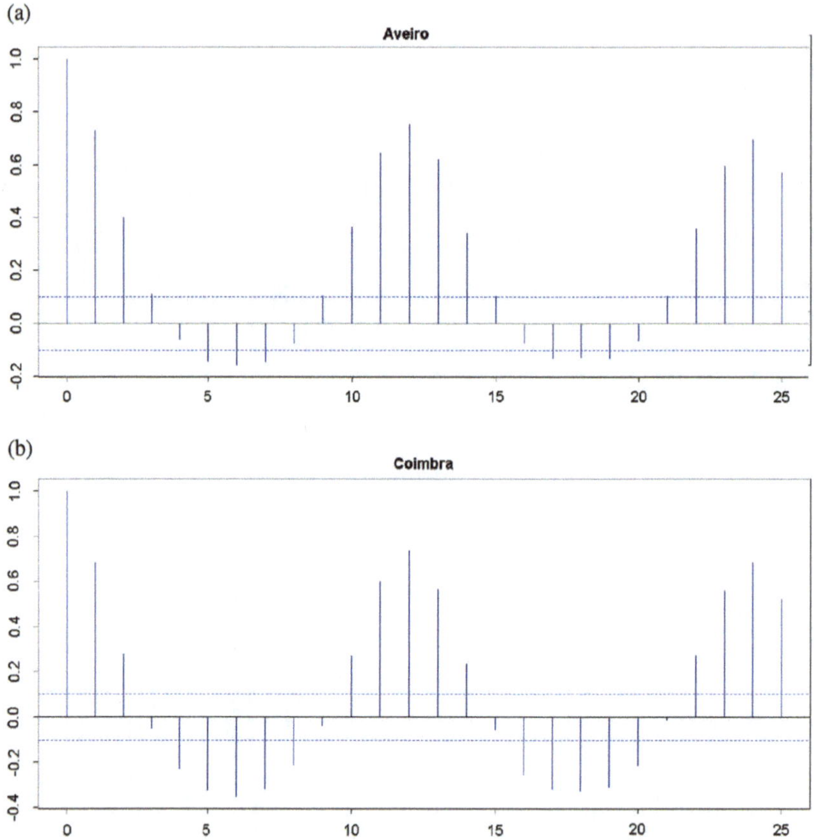

Fig. 7 Sample ACF for the Monthly number of fires in Aveiro and Coimbra Counties

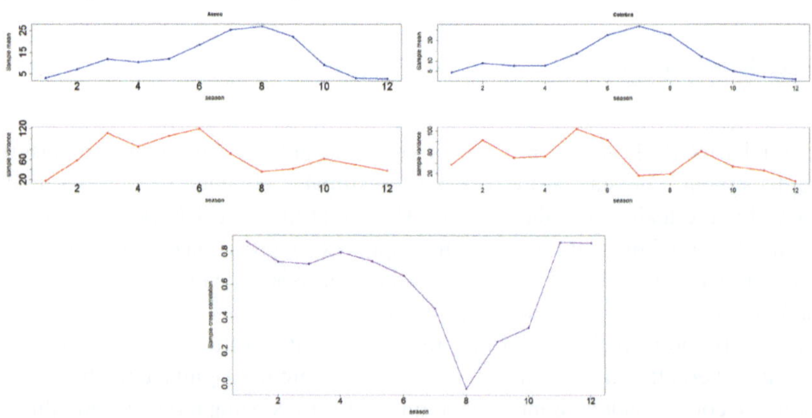

Fig. 8 Plots of sample monthly means, sample monthly variances and sample monthly correlation

Table 3 Maximum likelihood estimates from fitting a PBINAR(1) (standard errors in brackets)

	Bivariate Poisson innovations					Bivariate negative binomial innovations				
	Aveiro		Coimbra			Aveiro		Coimbra		
	α_1	λ_1	α_2	λ_2	φ	α_1	λ_1	α_2	λ_2	β
January	0.1060	2.9770	0.6681	3.7315	1.6701	0.0181	3.2176	0.2926	4.2206	2.6571
	(0.0689)	(0.3550)	(0.1388)	(0.3838)	(0.3348)	(0.0148)	(0.1861)	(0.0599)	(0.2475)	(0.1522)
February	0.6607	5.1049	0.3266	7.5138	2.5066	0.4566	5.9164	7.48×10^{-8}	8.9086	2.1722
	(0.0815)	(0.4562)	(0.0860)	(0.5979)	(0.4919)	(0.0260)	(0.3176)	(4.321×10^{-8})	(0.4491)	(0.1072)
March	0.5866	7.5864	0.3150	4.8207	2.6052	0.0758	11.2336	1.37×10^{-6}	7.6248	1.9480
	(0.0571)	(0.5978)	(0.0628)	(0.6408)	(0.6471)	(0.0281)	(0.5481)	(6.329×10^{-7})	(0.4209)	(0.0995)
April	0.4329	5.3729	0.2778	4.8860	2.7946	0.1066	9.2260	0.1393	6.7097	1.8519
	(0.0467)	(0.6200)	(0.0653)	(0.5909)	(0.5873)	(0.0187)	(0.4704)	(0.0180)	(0.3343)	(0.09766)
May	0.6323	5.3062	0.6375	8.7860	3.0208	0.3906	7.8411	0.1026	12.9453	1.4148
	(0.0473)	(0.5819)	(0.0798)	(0.7810)	(0.7396)	(0.0165)	(0.3552)	(0.0287)	(0.5560)	(0.0703)
June	0.7273	9.7064	0.4102	16.9757	6.2075	0.6552	10.2336	0.3188	17.5072	0.2958
	(0.0488)	(0.7594)	(0.1002)	(1.5285)	(1.5600)	(0.0138)	(0.2486)	(0.0307)	(0.5006)	(0.0167)
July	0.6017	14.1938	0.4099	17.6025	0.0000	0.6235	13.8509	0.3823	18.4261	0.0099
	(0.0660)	(1.3366)	(0.0984)	(2.3167)	(3.0087)	(0.0141)	(0.2752)	(0.0226)	(0.5595)	(0.0455)
August	0.6181	11.4198	0.0000	22.6772	0.0174	0.6179	11.3858	0.0095	22.4222	0.0090
	(0.0935)	(2.4159)	(0.2511)	(6.7811)	(3.9038)	(0.0164)	(0.4410)	(0.00345)	(0.9851)	(0.0056)
September	0.5154	8.1003	0.1691	8.2935	2.2187	0.6209	5.2111	1.20×10^{-7}	12.3697	0.3678
	(0.0790)	(2.1458)	(0.0605)	(1.4313)	(1.0357)	(0.0162)	(0.5053)	(7.342×10^{-8})	(0.8204)	(0.03477)
October	0.2524	3.6976	0.1238	3.7248	0.8342	0.0783	7.5334	0.0805	4.2500	1.1266
	(0.0412)	(0.8998)	(0.0430)	(0.5911)	(0.4089)	(0.0117)	(0.3724)	(0.0071)	(0.1793)	(0.0708)
November	0.1226	2.0905	0.0334	2.1160	1.4792	0.0107	3.9292	1.5×10^{-6}	2.8756	6.1111
	(0.0291)	(0.3283)	(0.0392)	(0.3236)	(0.2487)	(0.0028)	(0.4236)	(8.321×10^{-7})	(0.367)	(0.4074)
December	0.6191	0.8417	0.1865	0.8631	0.4477	0.4296	1.5329	1.66×10^{-7}	1.3540	6.9367
	(0.0559)	(0.1874)	(0.0685)	(0.2030)	(0.1656)	(0.0189)	(0.1563)	(7.781×10^{-8})	(0.1473)	(0.5059)
Log-Lik.	−2881.097					−2068.415				
AIC	5882.194					4256.830				

7 Conclusions

In this article, a family of bivariate integer-valued autoregressive model of order one with periodic structure was proposed. This family is a generalization of the BINAR model of Pedeli and Karlis (2011)[15]. Likelihood-based estimators for model parameters were derived and their asymptotic properties obtained and prediction was also addressed.

As referred throughout, an important limitation of Pedeli and Karlis' model is that the autoregression matrix is diagonal which means that it causes no cross-autocorrelation in the counts. This is also true for the PBINAR model. Therefore, extensions for PBINAR models accounting for cross-autocorrelation is also an impeding problem. Moreover, similar to what happens with conventional PARMA models, PBINAR models can have an inordinately large number of parameters. Therefore, the development of procedures for dimensionality reduction remains an important topic for future work.

Acknowledgements This work was supported by Portuguese funds through the CIDMA—Center for Research and Development in Mathematics and Applications, and the Portuguese Foundation for Science and Technology ("FCT Fundação para a Ciência e a Tecnologia"), within project PEst-OE/MAT/UI4106/2014.

References

1. Billingsley, P.: Statistical Inference for Markov Processes. University of Chicago Press, Chicago (1961)
2. Boudreault, M., Charpentier, A.: Multivariate integer-valued autoregressive models applied to earthquake counts. Technical report (2011)
3. Brännäs, K., Quoreshi, A.M.M.S.: Integer-valued moving average modelling of the number of transactions in stocks. Appl. Financ. Econ. **20**, 1429–1440 (2010)
4. Franke, J., Subba Rao, T.: Multivariate first-order integer-valued autoregressions. Technical Report. Forschung Universitat Kaiserslautern (1993)
5. Franses, P.H.: A multivariate approach to modeling univariate seasonal time series. J. Econ. **63**, 133–151 (1994)
6. Franses, P.H., Paap, R.: Periodic Time Series. Oxford University Press, Oxford (2004)
7. Gardner, W.A., Napolitano, A., Paura, L.: Cyclostationary: half a century of research. Signal Process. **86**, 639–697 (2006)
8. Haldrup, N., Hyllerberg, S., Pons, G., Sansó, A.: Common periodic correlation features and the interaction of stocks and flows in daily airport data. J. Bus. Econ. Stat. **25**, 21–32 (2007)
9. Hall, A., Scotto, M.G., Cruz, J.: Extremes of integer-valued moving average sequences. Test **19**, 359–374 (2010)
10. Hurd, H.L., Miamee, A.: Periodically Correlated Random Sequences: Spectral Theory and Practice. Wiley, New Jersey (2007)
11. Karlis, D., Pedeli, X.: Flexible bivariate INAR(1) processes using copulas. Commun. Stat. Theory Meth. **42**, 723–740 (2013)
12. Latour, A.: The multivariate GINAR(p) process. Adv. Appl. Probab. **29**, 228–248 (1997)
13. Monteiro, M., Scotto, M.G., Pereira, I.: Integer-valued autoregressive processes with periodic structure. J. Stat. Plann. Inference **140**, 1529–1541 (2010)

14. Moriña, D., Puig, P., Ríos, J., Vilella, A., Trilla, A.: A statistical model for hospital admissions caused by seasonal diseases. Stat. Med. **30**, 3125–3136 (2011)
15. Pedeli, X., Karlis, D.: A bivariate INAR(1) process with application. Stat. Model. **11**, 325–349 (2011)
16. Pedeli, X., Karlis, D.: On composite likelihood estimation of a multivariate INAR(1) model. J. Time Ser. Anal. **34**, 206–220 (2013)
17. Quoreshi, A.M.M.S.: Bivariate time series modelling of financial count data. Commun. Stat. Theory Meth. **35**, 1343–1358 (2006)
18. Ristic, M.M., Nastic, A.S., Jayakumar, K., Bakouch, H.S.: A bivariate INAR(1) time series model with geometric marginals. Appl. Math. Lett. **25**, 481–485 (2012)
19. Salas, J.D.: Analysis and modeling of hydrologic time series. In: Maidment, D.R. (ed.) Handbook of Hydrology, Chap. 19. McGraw-Hill, New York (1993)
20. Tesfaye, Y.G., Meerschaert, M.M., Anderson, P.L.: Identification of PARMA models and their application to the modeling of riverflows. Water Resour. Res. **42**(W01419), 11 (2006)
21. Ursu, E., Turkman, K.F.: Periodic autoregressive model identification using genetic algorithms. J. Time Ser. Anal. **33**, 398–405 (2012)
22. Vecchia, A.V.: Periodic autoregressive-moving average (PARMA) modelling with applications to water resources. Water Res. Bull. **21**, 721–730 (1985)

4. [illegible] D., [illegible] R., [illegible] Wells, M., [illegible] thermal transitions [illegible]
5. [illegible] (NARCI) [illegible]
6. [illegible] model. J. [illegible]
7. [illegible] matrices and [illegible] with [illegible] Appl. Mat.
8. [illegible]
9. [illegible]
10. [illegible]
11. [illegible]
12. [illegible] ARMA [illegible] approximations [illegible] 27, 71–79 (19[illegible])

The Macrodynamics of Employment Under Uncertainty

Paulo R. Mota and P. B. Vasconcelos

Abstract Inthe context of the current Eurozone crisis, the study of the effects of uncertainty in the macrodynamics of employment is a topic of major importance. This paper tackles this challenging question. At a first step a non-ideal relay hysteresis type microeconomic model of employment adjustment with uncertainty is presented. Then, an aggregation mechanism is explicitly considered in order to analyse the aggregate level of employment. Finally, as a new feature, uncertainty is considered endogenously determined by the actual state of the economy. Aggregate time-series built from micro monthly data on a representative sample of Portuguese manufacturing firms is used on a computational implementation of the linear play model of hysteresis. Results illustrate that uncertainty enhances the hysteretic behaviour of employment in small firms, but this effect is not significant for large ones.

1 Introduction

Firms, from almost all sectors of the economy, do not permanently adjust the number of employees to accommodate demand shocks. This has been confirmed by early empirical studies.[1] Their reaction, on the contrary, is often discontinuous

[1] See, e.g., [16, 26], and for the Portuguese case [23, 30, 38].

P.R. Mota (✉)
University of Porto - School of Economics and Business and NIFIP (Núcleo de Investigação em Finanças Públicas e Política Monetária da Universidade do Porto), Rua Dr. Roberto Frias, 4200-464 Porto, Portugal
e-mail: mpaulo@fep.up.pt

P.B. Vasconcelos
University of Porto - School of Economics and Business, Rua Dr. Roberto Frias, 4200-464 Porto, Portugal

CMUP (Centro de Matemática da Universidade do Porto), Rua do Campo Alegre 687, 4169-007 Porto, Portugal
e-mail: pjv@fep.up.pt

J.-P. Bourguignon et al. (eds.), *Dynamics, Games and Science*, CIM Series in Mathematical Sciences 1, DOI 10.1007/978-3-319-16118-1_25

and lumpy, with periods of inaction punctuated by episodes of large adjustment, a clear sign of non-convex adjustment costs.

It's well established in the literature, Theory of Optimal Inertia, that when a firm decision involves non-convex adjustment costs, as for the case of employment, it is rational to postpone decisions until uncertainty has lower down to conformable levels.[2] Moreover, in the presence of uncertainty, non-convex employment adjustment costs at the firm level may be relevant to the dynamics of aggregate employment [30, 31].

In Portugal, as well as in other Eurozone countries, at the present, there is a high level of economic, financial and regulatory uncertainty. This uncertainty may reduce the employment reaction to macroeconomic policy, leading to a problem of *uncertainty trap*.[3] The sources of uncertainty are: (1) the size of the government debt (and deficit) to GDP ratio and the doubts about the success of the adjustment program; (2) the long term economic and social effects of the austerity measures imposed by the IMF, the European Commission and the ECB; (3) doubts about the permanent or temporary character of public wage cuts; (4) doubts about the announced intention of a tax reform, and about the permanent or temporary nature of tax increases; (5) constant application of piecemeal reforms in the labor market; (6) great uncertainty surrounding the projections of the main macroeconomic variables, such the GDP growth rate and the government deficit to GDP ratio that are constantly being revised.

If a firm is uncertainty about whether an aggregate demand shock is transitory or permanent it may delay its investment and employment decisions.[4] Thus, as the size of the employment inaction band increases with the level of uncertainty, hysteresis can be quite strong even if the government applies measures towards the deregulation of the labour market.

The main contribution of this paper is to analyse the effect of uncertainty in the macrodynamics of employment. It also aims to show that uncertainty could result from attempts to reduce non-convex adjustment costs, or from austerity measures that depress the economic in the short-run. This is of high importance, particularly in the context of the current Eurozone crisis.

We begin by introducing a simple microeconomic non-ideal relay-type model of employment adjustment with uncertainty. Then, an aggregation mechanism is explicitly considered in order to analyse the macrodynamics of employment. Finally, uncertainty is considered endogenously determined by the actual state of the economy, which encompasses the novelty of the work.

Portuguese firm-level monthly data spanning from 1995 until 2008 is used. Notably, Portugal provides a good case to study labour demand driven hysteresis since it has one of the strictest employment protection legislation in Europe, which is a source of non-convexities in the adjustment technology.

---

[2] See, e.g., [7, 8, 12, 15, 21].

[3] See [4, 11].

[4] See [17].

A switching employment equation is estimated from a computational implementation of the linear play model of hysteresis over aggregate time-series built from firm-level data. This equation mimics the behaviour upon which for small changes of labour demand there is a weak reaction of employment level, whereas for large changes there is a strong reaction. It is assumed that the splitting factor is a positive function of the magnitude of non-convex employment adjustment costs and the level of uncertainty in the economy.[5]

This paper is structured as follows. Section 2 describes the model along with some implementation details. Section 3 presents our empirical results, and Sect. 4 concludes.

2 The Model and Its Implementation

2.1 *Micro Foundations of Discontinuous Adjustment at the Firm Level*

Let us assume a competitive market where each price taker active firm, j, ($j = 1, \ldots J$), employs one unit of employment, $n_{j,t}$, at the unit wage cost w_j, and one unit of firm specific capital, $k_{j,t}$, that costs $r_j \times k_{j,t}$ and produces $y_{j,t} = n_{j,t} \times k_{j,t} = 1$ units of output, which it sells at a unit price P_j (the revenue is simply the output price).[6] If inactive, the firm produces no output and employs zero units of employment. Furthermore, every individual plant must pay a fixed and constant cost in time to enter (hire a worker and to acquire firm specific physical assets), H_j, or to leave the market (fire its single worker), F_j.[7] In this model, switching the state of activity leads to a complete depreciation of firing and hiring costs, the reason for which these are regarded as sunk costs. Assuming a discount factor, $\delta = \frac{1}{1+i}$, where i is the risk free interest rate, and considering a profit maximising problem of the individual firm, with discrete time and an infinite plan horizon, a previously inactive firm will only enter the market if hiring costs are recovered. Hence the entry (expanding) trigger

---

[5]We follow [8].

[6]We are assuming that all firms face a common demand schedule ($P_{j,t} = P_t$) and that the wage rate and the real cost of capital is constant over time, but not necessarily across firms.

[7]On the hiring side, examples of adjustment costs are the costs of advertising, recruiting and training the new workers, including the costs resulting from disruption in production when the new workers are hired. On the firing side, adjustment costs include mandatory advance notice requirements, severance pay and other procedural inconveniences to dismissal caused by employment protection legislation. A significant part of these adjustment costs are related to personnel and legal departments to deal with hires and fires and thus fixed, i.e., independent of the number of workers that are hired or dismissed (see, e.g., [27, 32]). A firm also incurs in irreversible cost to buy physical assets like firm specific equipment or intangible assets such as reputation, acquired by investments in marketing and advertising, or technical knowledge (see, e.g., [24, 35, 36]). Other non-firm specific investments like office equipment, cars, trucks and computers can have a resale value well below their purchase cost due to the "lemons" problem [36, , p. 1111].

price, $P^C_{entry,j}$, exceeds the wage and the interest cost of firm non-specific physical capital by $\frac{1}{1+i}H_j$. Conversely, a previously active firm will exit the market if losses exceed firing costs. Hence exit (contracting) trigger price, $P^C_{exit,j}$, is below the wage plus the interest cost of firm non-specific physical capital by $\frac{1}{1+i}F_j$.[8]

The employment demand function of the plant j, may be represented by the non-ideal relay hysteresis operator, $R_{P^C_{exit,j},P^C_{entry,j}}(P_t) = n_{j,t}$[9]:

$$n_{j,t} = \begin{cases} 1, \; n_{j,t-1} = 0 \wedge P_t \geq w_j + r_j \times k_j + \frac{i}{i+1}H_j \vee \\ \quad n_{j,t-1} = 1 \wedge P_t > w_j + r_j \times k_j - \frac{i}{i+1}F_j \\ 0, \; n_{j,t-1} = 0 \wedge P_t < w_j + r_j \times k_j + \frac{i}{i+1}H_j \vee \\ \quad n_{j,t-1} = 1 \wedge P_t \leqslant w_j + r_j \times k_j - \frac{i}{i+1}F_j \end{cases} \tag{1}$$

The first expression of Eq. (1) refers to a situation where firm j enters or stays active while the second specifies the situation where firm j stays inactive or exits.

Since $P^C_{entry,j} = w_j + r_jk_j + \frac{i}{i+1}H_j$[10] is greater than $P^C_{exit,j} = w_j + r_jk_j - \frac{i}{i+1}F_j$[11] the difference between these two trigger points, $\frac{i}{i+1}(H_j + F_j)$, creates an employment band of inaction or hysteresis band (see [15, 21, 36]). The employment band of inaction depends positively on the fixed adjustment costs and negatively on the interest rate. When $i \longrightarrow 0$ the band of inaction tends towards zero, and when $i \longrightarrow \infty$ the band of inaction tends to $H_j + F_j$. Thus, the higher the interest rate the higher the importance of the fixed adjustment costs.

The model implies discontinuous employment adjustment. Each plant requires an aggregate positive demand shock $P_t > P^C_{entry,j}$ to hire its workforce and an aggregate negative demand shock $P_t < P^C_{exit,j}$ to dismiss it. Demand shocks within the range $P^C_{exit,j} < P_t < P^C_{entry,j}$ do not cause any action in employment (see [12, 14, 15, 19, 21], for a Theory of Optimal Inaction in the presence of non-convex employment adjustment costs). Moreover, P_t, is not sufficient to determine the plant's state of employment. The whole history of the system, summarised in n_{jt} must be taken into account. Thus, the system is characterised by path dependence and non-linearity.

In order to illustrate the effect of uncertainty on entry/job creation decision and on exit/job destruction decision, we assume that output prices are random, rather than known with certainty. The firm must choose the level of output and employment before the output price is observed. Instead of considering permanent

[8] In this setting, the decision to enter is akin to the hiring decision, and the decision to exit is akin to the firing decision. This simplification does not change the conclusions of the model as we can consider a firm divided into single production units, with every unit represented individually [8].

[9] See [25] for a complete description of the model.

[10] The value function for a firm that enters the market is $V_{j,t} = \frac{P_t - w_t - r_jk_j}{1-\delta} - H_j$, while the value function for remaining outside the market is 0. Therefore, the entry condition is $\frac{P_t - w_t - r_jk_j}{1-\delta} - H_j > 0$.

[11] The value function for a firm to remain active is $V_{j,t} = \frac{P_t - w_t - r_jk_j}{1-\delta}$, while the value function for exiting the market is $-F_j$. Therefore, the exit condition is $\frac{P_t - w_t - r_jk_j}{1-\delta} < -F_j$.

uncertainty, which requires dynamic programming tools as in [19], we introduce uncertainty by considering an expected future stochastic one-time shock in the price level that generates revenue uncertainty, in line with [7, 10]. As our objective is the aggregation up to the macro level, we model uncertainty in a simple way, assuming a nonrecurring single stochastic change in the output price, which can be either positive, $+\mu$, or negative,$-\mu$, in a discrete time model. We consider that both realizations of the shock have the same probability of $1/2$. In this case, $P_{t+1} = P_t \pm \mu \Rightarrow E(P_{t+1}) = P_t$ and from period $t + 1$ on the firm will decide under certainty again.[12]

Admitting that the future path of the price is uncertain, waiting can have a positive value since it brings more information about the evolution of the price level. With uncertainty, a previously inactive/active firm has three possible strategies: (1) stay inactive/active; (2) enter/exit the market; (3) wait and make a decision after the realization of the stochastic shock. If the firm has the possibility of delaying its entry decision, it faces a trade-off: waiting has the benefits mentioned above, but it also has the cost of foregoing the profits earned, if entry had occurred.

Thus, uncertainty introduces an additional cost of entering (opportunity cost) that is the value of the option to wait (see [5, 13, 19, 20, 22, 36]). With uncertainty, the opportunity to make an investment and enter in the market is akin to a financial American call option, while the decision to exit the market is akin to a financial put option (see [1, 36], for the dynamics of capital stock).

In the case of uncertainty, the labour demand function of the individual firm (which corresponds to the supply function) can be described as a non-linear hysteretic transformation of a stochastic input, P_t[13]:

$$n_{j,t} = \begin{cases} 1, \; n_{j,t-1} = 0 \wedge P_t \geq w_j + r_j \times k_j + \frac{i}{i+1} H_j + \frac{1}{1+2i}\mu \quad [entry] \vee \\ \quad n_{j,t-1} = 1 \wedge P_t > w_j + r_j \times k_j - \frac{i}{i+1} F_j \quad [stay\ active] \vee \\ \quad n_{j,t-1} = 1 \wedge w_j + r_j \times k_j - \frac{i}{i+1} F_j - \frac{1}{1+2i}\mu < P_t \leq \\ \qquad \leq w_j + r_j \times k_j - \frac{i}{i+1} F_j \quad [wait\ in\ activity] \\ 0, \; n_{j,t-1} = 0 \wedge P_t < w_j + r_j \times k_j + \frac{i}{i+1} H_j \quad [stay\ inactive] \vee \\ \quad n_{j,t-1} = 0 \wedge w_j + r_j \times k_j + \frac{i}{i+1} H_j \leq P_t < \\ \qquad < w_j + r_j \times k_j + \frac{i}{i+1} H_j + \frac{1}{1+2i}\mu \quad [wait\ in\ activity] \vee \\ \quad n_{j,t-1} = 1 \wedge P_t \leq w_j + r_j \times k_j - \frac{i}{i+1} F_j - \frac{1}{1+2i}\mu \quad [exit] \end{cases} \tag{2}$$

Combining both triggers under uncertainty, the width of the band of inaction is[14]:

$$P^U_{entry,j} - P^U_{exit,j} = P^C_{entry,j} - P^C_{exit,j} + \frac{2\delta}{2-\delta}\mu = \frac{i}{i+1}\left(H_j + F_j\right) + \frac{2\mu}{1+2i} \tag{3}$$

[12]Although, we consider only revenue uncertainty, there could be also uncertainty in input costs (like in the interest rates), exchange rate uncertainty, and tax and regulatory policies uncertainty (see [22, p. 14]).

[13]See [7, 10] for more detail.

[14]See [7] for more detail.

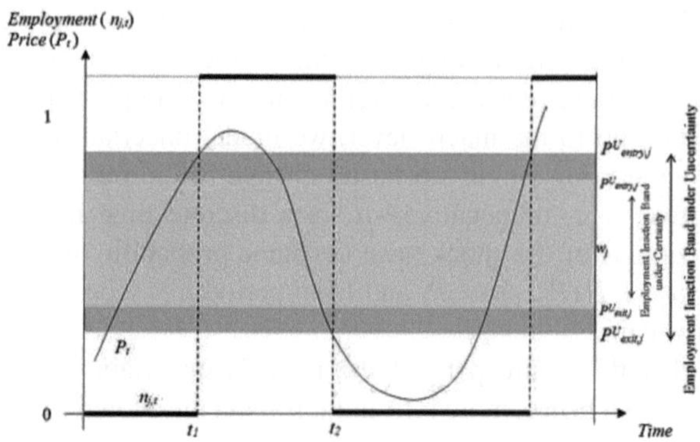

Fig. 1 Micro hysteresis loop—employment demand for plant *j* according to the non-ideal relay model

where $P^U_{entry,j}$ and $P^U_{exit,j}$ are the entry and the exit triggers under uncertainty respectively. Thus, uncertainty in the future behaviour of prices widens the employment band of inaction. The option value of waiting effect raises the optimal entry threshold, increasing the probability of a firm to stay inactive even if current demand increases; similarly on the opposite sense, the waiting effect lowers the optimal exit threshold, increasing the probability of a firm staying active in face of a decrease in demand (see also [21, p. 121], and [7, p. 275]). Thus hysteresis effects are amplified.

Figure 1 illustrates a hypothetical dynamics of the price level, P_t, and the correspondent path of the employment level of firm j, $n_{j,t}$, in the non-ideal relay. Considering some price dynamics, a previous inactive firm will enter the market when the price is greater than $P^U_{entry,j}$, which occurs at time t_1, and will exit the market when the price is lower than $P^U_{exit,j}$, which occurs at time t_2. Between t_1 and t_2 the price level dynamics does not induce any action of the level of employment.

We conclude, also, from Eq. (2) that the lower the interest rate the higher the importance of uncertainty for the width of the band of inaction. This has indeed important policy implications. First, the effectiveness of expansionary monetary policy via cutting interest rates is lower when uncertainty is large [9, 11, 22].[15] Second, frequent interest rate changes by the central bank induce additional uncertainty, which reduces the sensibility of investment to the interest rates [9]. Third, the lower the interest rate the lower the effect of the monetary policy on the output as uncertainty increases [see Eq. (3)], result that entails an *uncertainty trap* (see [4, 11]). Fourth, when uncertainty is large, if monetary policy is to be effective, large variations of the Central Bank's key interest rates are needed—possibly in the range of the 50 basis points [7]. In the Eurozone with the ECB key interest rate for

[15]Belke [6] extends this argument to the effect of fiscal stimulus package to deal with the recent crisis.

the main refinancing operations set at 0.05 %, a lower point was reached where no range of cuts has significant impact in the economy.

This result highlights the importance of hysteresis caused by uncertainty in a context of low interest rate, as it is the present situation in many advanced economies. In fact, in a low interest rate environment, fixed employment adjustment costs are not as relevant as uncertainty in generating hysterical effects [see Eq. (3)]. Indeed, hysteresis can be quite strong even for small values of the fixed hiring and firing costs (see [19]).

2.2 The Problem of Aggregation

The aggregate economy is represented as a set of the potential number of active heterogeneous firms, J, in a limiting triangle with area T, each one acting according to Eq. (1):

$$T = \left\{ \left(P^U_{exit,j},\ P^U_{entry,j} \right) : P^U_{entry,j} \geqslant P^U_{exit,j}\ ,\ P^U_{exit,j} \geqslant P^U_{exit,min}\ ,\ P^U_{entry,j} \leqslant P^U_{entry,max} \right\}$$

where $P^U_{exit,min}$ is the exiting threshold for the less demanding firm or unit of labour, and $P^U_{entry,max}$ is the entering threshold of the most demanding firm or unit of labour. In this setting, the dynamics of aggregate employment is fully described by the Preisach operator, $\Phi\,[P\,(t)]$ (for a more complete explanation of the Preisach model of hysteresis see [29][16]:

$$\Phi\,[P\,(t)] = N\,(t) = \int\int_T u\left(P_{exit,j}, P_{entry,j}\right) R_{P_{exit,j},P_{entry,j}}\, dP_{exit,j}\, dP_{entry,j} \tag{4}$$

where $N\,(t)$ is the aggregate employment at time t, and $u\left(P_{exit,j},\ P_{entry,j}\right)$ is the density function of the individual firms in T.

The distance of the relays from the origin is determined by the variable cost, $w_j + r_j \times k_j$, and the orthogonal distance of the relays from the 45°-line is a positive function of the non-convex employment, capital adjustment costs, and uncertainty. The more important the widening effect of the employment band of inaction at the micro level is, due to the uncertainty, the weaker the reaction of aggregate employment to its forcing variables. Comparing with the case where firms, distributed uniformly in the Preisach triangle, T, decide under certainty, each relay is displaced to the northwest, and zones of inaction emerge at the macro level (see [7, 8]). We are considering that the price level (the input variable) in our

[16]Preisach-type models of hysteresis have been used as a vehicle to describe the macrodynamics of economic systems—see [2, 3, 18] for an early application to economic problems.

hysteresis model is exogenous, and we concentrate in the endogenous[17] character of uncertainty.

Furthermore, we specify a time-dependent Preisach Model. Let us assume that the unemployment rate in period t is given by the area of inactive firms in the Preisach Triangle, as a proportion of the area of the Preisach Triangle, T, that represents the potential number of active firms each one employing one worker[18]:

$$U(t) = \frac{T - \Phi\,[\mathrm{P}\,(\mathrm{t})]}{T} \tag{5}$$

We admit further that the level of uncertainty, μ, depends on the variation of the unemployment rate. Increasing unemployment is the result of the malfunctioning of the economy and an indicator of a lack of aggregate demand and thus affects business confidence, consequently:

$$\mu = f\,[\Delta U\,(t)] \tag{6}$$

Equations (2), (5) and (6) imply that a decrease of the aggregate demand, captured by the price level, originates an increase of the unemployment rate leading to an increase in the distance between the entry and exit thresholds for every firm:

$$(P^U_{entry,j} - P^U_{exit,j})(U(t)) = \frac{i}{i+1}(H_j + F_j) + 2\frac{f(\Delta U(t))}{1+2i} \tag{7}$$

Consequently this causes a displacement of the Preisach triangle. Thus, the process with uncertainty reinforces the hysteresis at the macro level.

To exemplify how the model works consider the hypothetical dynamics of the price level displayed in Fig. 2 and the assumption that $P(t_0) < P_{exit,min}$.

In this situation, all the relays are switched off, i.e., all the firms are outside the market, employing zero workers ($n_{j,0} = 0, \forall j$), and aggregate employment is zero (see Fig. 3a). Subsequently, the price increases monotonically, reaching a local maximum at t_1, $P(t_1)$. All relays with $P_{entry,j} \leqslant P(t_1)$ are switched on and all firms hire one worker. The relays are now divided into two sets T_0 and T_1, the set of the relays that are, respectively, switched off and on (Fig. 3b). Next, at t_2, if the

---

[17]In fact, the market price can be influenced by the internal market dynamics as in [37], but this is not essential here.

[18]The determinantes the unemployment rate are complex. Unemployment can be caused by: (1) the time people take to move between jobs (frictional unemployment); (2) a mismatch of skills in the labour market due to a lack of occupational and geographical mobility, and by technological change (structural unemployment); (3) a lack of aggregate demand (cyclical unemployment). Here we emphasise the third cause. We consider the firms as potential units of labour, with the set of all potential units of labour representing all the jobs that can potentially be created in the economy [28], and unemployment occurs when the economy is bellow full capacity. Moreover, cyclical unemployment may be transformed into structural unemployment by hysteresis mechanisms blurring the distinction between these two types of unemployment.

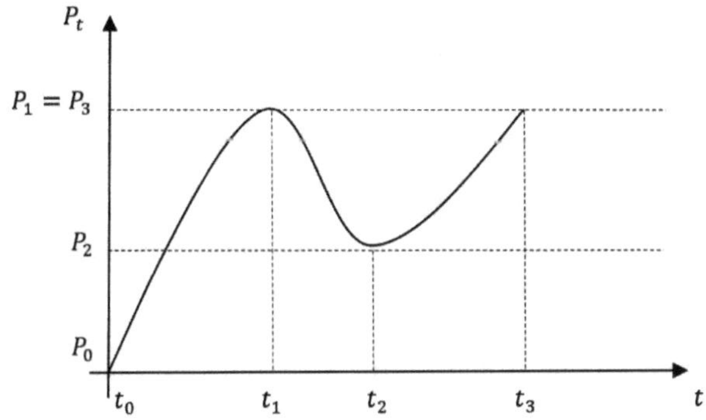

Fig. 2 Hypothetical price level dynamics

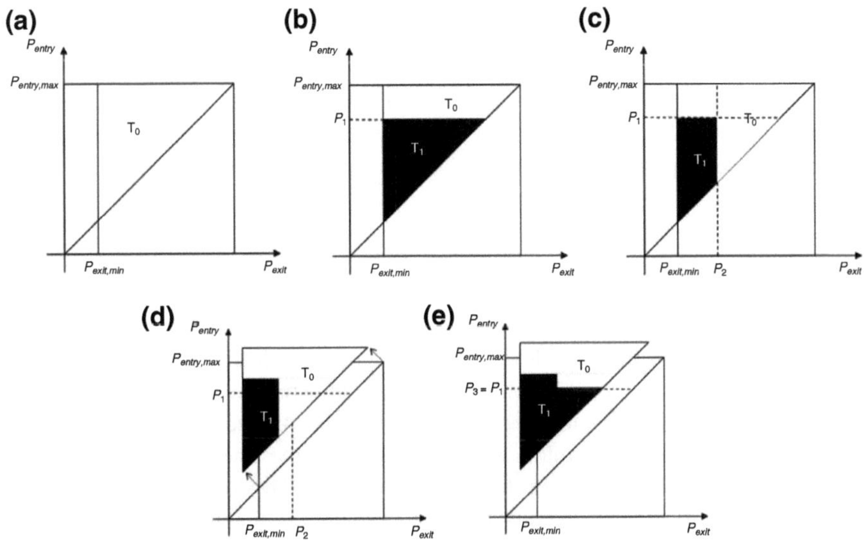

Fig. 3 Aggregate employment dynamics accordingly to the Preisach model (**a**) Initial State (**b**) Price Increase to P_1 (**c**) Price Decrease to P_2 (**d**) Preisach Triangle Displacement due to Uncertainty (**e**) Price Increase to P_1

aggregate price level decreases to a local minimum $P(t_2)$, those relays for which $P_{exit,j} \geq P(t_2)$ are switched off and the firms dismiss the worker (Fig. 3c).

This dynamics traces a staircase line dividing in two the area of the Preisach triangle, T,—one part where the relays are on, representing the aggregate level of employment, and the other where they are off, representing the level of unemployment in the economy.[19] The vertex coordinates of the staircase line correspond to

[19]Given the potential number of units of labour that can be created in the economy (the area of the Preisach Triangle—T), the unemployment rate results from a lack of demand represented by the price level, and thus it should be considered as involuntary.

the sequence of the past non-dominated extrema of the input variable. At the macro level, the system retains a selective memory of past shocks, represented by the staircase line, since the level of employment is defined by the sequence of non-dominated maxima and minima at the aggregate price level. Depending on this sequence, the relationship between aggregate employment and the price level is represented by different branches. Whenever the direction of the price path changes, a continuous branch-to-branch transition occurs, causing multi-branch non-linearity. A non-dominated demand shock, as the one that occurs at t_5, clears the effect of the previous dominated extrema from the memory bank. Thus, the coordinates of the staircase partition between T_0 and T_1 are removed from the memory.

Let us consider now the effect of the time varying uncertainty. When the price level decreases from t_1 to t_2, the unemployment increases from the area T_0 in Fig. 3b to the area T_0 in Fig. 3c originating an increase of the level of uncertainty for every firm, accordingly to Eq. (5). The consequence is the displacement of the Preisach Triangle to Norwest, as in Fig. 3d. Assuming that the price level increases again to a level $P(t_3) = P(t_1)$, all the relays with $P_{entry,j} \leqslant P(t_3)$ switch on leading to an increase of the aggregate level of employment (decrease of the level of unemployment). However, comparing Fig. 3b, e the transitory shock in the aggregate demand caused a permanent increase of the unemployment rate.

2.3 Model Implementation

The transition from the firm level to the macro level leads to a change in the hysteresis properties (see [29]). While micro adjustment is discontinuous, for macrodynamics the adjustment occurs continuously. A piecewise-linear approximation of the Preisach operator, where the slope of the linear functions changes every time the price reaches an extremum, is well-suited to describe a case like this. The linear play hysteresis operator is able to capture the feature that the aggregate demand can produce permanent effects on employment.

The play operator, P_r, is characterised by horizontal reversible inner branches of the same length (the play segment) and upward sloping linear limiting branches (the spurt segments),[20] giving rise to counter-clockwise oriented loops. In this model the memory effect is captured by the difference between two adjacent lines (the play and the spurt). If β_1 denotes the slope of the flatter line (the play), then $\beta_1 + \beta_2$ is the slope of the steeper one (the spurt), and β_2 is the memory or remanence parameter [8]:

$$\frac{dN_t}{dP_t} = \beta_1 + d \times \beta_2, \quad \text{with } d = \begin{cases} 0, & \text{on the play lines} \\ 1, & \text{on the spurt lines} \end{cases} \tag{8}$$

[20] See [22, p. 15].

As the slope of limiting branches is fixed, the operator is characterised by a single constant—its input threshold value or the magnitude of the play segment. The initial value of the operator state, the pair $[P_r(t_0), P(t_0,)]$, together with the future values of thc input, $P(t)$, determine the value of the employment, $N(t)$ (see, for more detail, [39]).

The linear play hysteresis operator is implemented empirically via a linear switching employment equation with an unknown splitting factor—the play—capturing the non-linear play hysteresis effects.

In the empirical work, we use aggregate sales in manufacturing, S_t, as a proxy of the state of aggregate demand represented in Eq. (4) by P_t. Following [8], we consider that the change in S_t (the variable that causes hysteresis) may occur along the play segment, $PLAY_t$, in which case it is referred to as Δa_t, or on the spurt line, in which case it is referred to as $\Delta SPURT_t$:

$\Delta S_t = \Delta a_t + \Delta SPURT_t$, with

$$\Delta SPURT_t = \begin{cases} sign(\Delta S_t) \times (|\Delta S_t - PLAY_t,|), & (|\Delta S_t - PLAY_t,|) > 0 \\ 0, \text{ otherwise} \end{cases} \tag{9}$$

The change in aggregate employment, N_t, induced by a change in real sales, is divided between a weak reaction in the play area and a strong reaction described by the spurt line when S_t changes sufficiently:

$$\Delta N_t = \beta_1 \Delta a_t + (\beta_1 + \beta_2)\,\Delta SPURT_t, \quad \text{with} \quad |\beta_1| < |\beta_1 + \beta_2| \tag{10}$$

The location of the play line is shifted vertically by movements on the spurt line in the direction of the change in employment. The cumulative vertical displacement of the play line, induced by all previous movements on both spurt lines, is expressed as:

$$V_{t-1} = \sum_{i=0}^{t-1} \Delta SPURT_i \tag{11}$$

Thus, the realization of N_t can be expressed as a shift in V_t induced by past spurts and the current change in the independent variable, ΔS_t:

$$N_t = C + V_{t-1} + \Delta N_t \tag{12}$$

and using (10) and (11) we obtain:

$$N_t = C + \beta_2 \sum_{i=0}^{t-1} \Delta SPURT_i + \beta_1 \Delta a_t + (\beta_1 + \beta_2)\,\Delta SPURT_t \tag{13}$$

Using (10) and bearing in mind that $\sum_{i=0}^{t-1} \Delta SPURT_i + \Delta SPURT_i = \sum_{i=0}^{t} \Delta SPURT_i$, it results:

$$N_t = C + \beta_2 \sum_{i=0}^{t} \Delta SPURT_i + \beta_1 S_t \tag{14}$$

Summing and subtracting $\beta_1 \sum_{i=0}^{t-1} \Delta S_i$, making $\beta_0 = C - \beta_1 \sum_{i=0}^{t-1} \Delta S_i$, and considering $\sum_{i=0}^{t-1} \Delta S_i + \Delta S_i = \sum_{i=0}^{t} \Delta S_i = S_t$ and $\sum_{i=0}^{t} \Delta SPURT_i = SPURT_t$ we have:

$$N_t = C + \beta_0 + \beta_1 S_t + \beta_2 \Delta SPURT_t \tag{15}$$

where $SPURT_t$ is a filtered aggregate sales series (dependent on the play value), which summarises all preceding and present movements on the spurt lines causing a structural shift of the current relationship between employment and sales.

Based on Eq. (15), we estimate the following model (we add a time trend, T).

$$\begin{cases} N_t = \beta_0 + \beta_1 S_t + \beta_2 \Delta SPURT_t + \beta_3 T + \mu_t \\ SPURT_t = f\left(PLAY_t\right) \\ PLAY_t = \gamma + \delta \sigma_{S_t}, \quad \gamma, \delta \geqslant 0 \\ \sigma_{S_t} = \frac{1}{n-1} \sum_{i=t-k}^{t-1} \left[\left(S_i - \overline{S} \right)^2 \right] \end{cases} \tag{16}$$

where the artificial variable $SPURT_t$ is computed from Eqs. (9) to (15), assuming a variable play value; $PLAY_t = \gamma + \delta \sigma_{S_t}$. The splitting factor, $PLAY_t$, is modelled as a positive function of the fixed employment adjustment costs, captured by parameter γ, and also a positive function of the degree of uncertainty captured by parameter δ. We consider, following [21, p. 116], that the level of uncertainty is determined by the variance of aggregate sales taken as a proxy of the level of the aggregate demand. Accordingly, we use the moving standard deviation of the logarithm of aggregate sales, σ_{S_t}.[21] Higher values of σ_{S_t} correspond to more uncertainty and therefore to wider play.[22]

In this framework, β_1 gives the reaction along the play, while $\beta_1 + \beta_2$ the reaction along the spurt segment. In the presence of hysteresis we expect $\beta_2 > 0$.

[21] In the estimation we set $K = 3$.

[22] There is a certain lack of consensus in literature concerning the best way to construct a proxy for uncertainty. Nonetheless, typically uncertainty is captured by moving variances of the relevant variables like output, inflation, real wages, interest rates, exchanges rates, etc. (see [17], for a survey).

Following the algorithm described in [8], a MATLAB program to generate the spurt variable was developed and implemented, which in turn requires estimation of the play width. The algorithm for the variable play model works as follows:

1. *Load data:*
   - *n, number of elements taken for analysis;*
   - S_t*, real sales;*
   - N_t*, employment;*
   - UI_t*, the proxy of uncertainty;*
   - *T, the time trend.*
2. *Build a grid for the variable play:*
   Given:
   - h_{min}*, the minimum value to be considered on the grid for the variable play;*
   - h_{max}*, the maximum value to considered on the grid for the variable play;*
   - h_{prec}*, the required precision;*

   Compute:
   - $h = \frac{h_{max}-h_{min}}{h_{prec}}$*,number of points on the grid.*
3. *Execute a grid search over a set of admissible values of the play and spurt:*
   (a) *for each pair* (γ, δ)*:*
   - *for* $i = 1, \ldots, g + 1$*;*
   - *for* $j = 1, \ldots, h + 1$*;*

   (b) *recognise the switches and compute the values of the play and spurt:*
   - *define* $\delta = h_{min} + (j + 1) \times h_{prec}$*;*
   - *compute* $play = \gamma + \delta \times UI$*;*
   - *build* $spurt = f(play)$*; and*

   (c) *estimate the employment equation by OLS and compute the corresponding* R^2*:*
   - $X = [ones\,(n, 1)\,,\ spurt\,(1:n)\,,\ (1:n)']\,;$
   - $[R^2\,(i,j)\,,\ \beta\,(i,j)\,,\ :] = R^2\_function(x, y).$
4. *Select the pair* (γ, δ) *that maximises the goodness of fit of the employment equation (as measure by* R^2*):*
   - $\left[R^2_{\max\_vec}, i_{\max\_vec}\right] = \max\left(R^2\right)\,;$
   - $\left[R^2_{\max}, j_{\max}\right] = \max\left(R^2_{\max\_vec}\right)\,; i_{\max\_vec} = i_{\max\_vec}(j_{\max});$
   - $play = h_{min} + (j_{\max} - 1) \times h_{prec}.$

It is worth mentioning that: (1) the *spurt* $= f(play)$ at step 3.2. provides the necessary computations to evaluate changes in the spurt variable due to changes in the input variable—see, for more detail, [8, p.191]; (2) the β coefficients for the estimation of the R^2 are computed using a numerically stable and computationally efficient procedure based on QR factorisation; and (3) the non-linearity inherent to hysteresis is captured by the hysteresis variable, while the rest of the model is kept linear.

The test for the presence of hysteresis consists of checking the ability of the hysteretic transformed input variable, $SPURT_t$, to explain the observed aggregate employment dynamics. The strategy is to test whether the non-linear model, which includes hysteresis, provides better results than the linear one.

We start by studying the stationarity of the series by applying the augment Dickey-Fuller units root test to the series in levels and to their first differences. This step is necessary in order to check if the series are integrated of the same order. To rule out the possibility of a spurious regression and to verify the existence of a true equilibrium relationship between the variables, we test for the existence of cointegration using the Johansen Test Procedure.[23]

However, because the series are non-stationary, we re-estimate the cointegrating regression by Fully Modified least Squares (FM-OLS) proposed by Phillips and Hansen [34] and developed by Phillips [33], which is an asymptotically efficient estimator of long-run economic relationships.

Finally, we test for the significance of the transformed sales variable, hysteresis implies $\beta_2 > 0$ in Eq. (16), using new statistics called fully-modified Wald tests, which are asymptotically distributed chi-squared criteria, and facilitate inference in integrated series of order one, $I(1)$, regression models. To verify the comparative explanatory value of the models, we also perform a test on the increase in the goodness of fit of the regression with only the original sales variable when we add the hysteresis transformed variable.

3 Data

For the simulations, we use firm level data from the "Inquérito Mensal à Indústria—Volume de Negócios e Emprego", a monthly mandatory mail survey of manufacturing firms with at least ten employees, run by Statistics Portugal.

Data include the number of employees in the firm, $n_{j,t}$, and total sales, $s_{j,t}$. We use 168 waves of the survey, from January 1995 to December 2008. On average, 2,616 firms answered each month, totalling 439,488 records (*firms* × *months*) over the entire 14-year period. The distributions of firms by number of employees and industry in the starting period are reported in Tables 1 and 2.

This data set was used to build the aggregate time series of employment, N_t, and real sales, S_t (defined as the nominal value of sales obtained by aggregation over our

[23]We apply the Trace Test performed with four lags in the VAR representation and with an intercept and time trend in the cointegration equation. We report the results of testing the null hypothesis of no cointegration ($r = 0$) against the existence of at least one cointegrated vector (r).

Table 1 Distribution of firms by size (1995:01)

	Number of firms	Proportion of firms (%)
$10 \leqslant n < 19$	299	13.45
$20 \leqslant n < 49$	528	23.75
$50 \leqslant n < 99$	477	21.46
$100 \leqslant n < 199$	410	18.44
$200 \leqslant n < 500$	374	16.82
$n \geqslant 500$	135	6.07
Total	2,223	100.00

Table 2 Distribution of firms by activity sector (1995:01)

	Number of firms	Proportion of firms (%)
Mining	91	4.09
Food, tobacco and beverages	290	13.05
Textile, leather and shoes	447	20.11
Furniture and wood	310	13.95
Paper and printing	151	6.79
Chemicals, petroleum and rubber and plastic products	182	8.19
Non metallic mineral products	184	8.28
Primary metals	50	2.25
Machinery, fabricated metals, motors and cars and other transport material	498	22.40
Electricity and gas	20	0.90
Total	2,223	100.00

micro data, and deflated using CPI from OECD—Main Economic Indicators). The variables were seasonally adjusted.

4 Estimation Results

We start by applying the augment Dickey-Fuller unit root test to find the order of integration of the series. Table 3 shows the augmented Dickey-Fuller test statistic for the levels and for the first difference of the variables. For all the variables in levels the augmented Dickey-Fuller test statistic is larger than the 5 % critical value (−3.445) indicating that we do not reject the hypothesis the existence of a unit root. We do not reject the hypothesis of stationary of the first difference of the series. In this case, the augmented Dickey-Fuller test statistic is smaller than the 5 % critical value for all the variables. Thus, all the variables used in the regressions are non-stationary and are $I(1)$. Moreover, the transformed series (the spurt variable) tends to reflect the stationary properties of the original sales series variable.

Table 3 Augmented Dickey-Fuller test statistics (5 % critical value: −3.445)

Variable	Whole sample		Small firms		Large firms	
	Level	First difference	Level	First difference	Level	First difference
N_t	−0.48	−14.97	−1.31	−19.84	−1.39	−11.46
S_t	−2.15	−15.78	−2.70	−9.28	−2.71	−13.99
$SPURT_t$	−1.39	−11.02	−0.94	−12.34	−1.87	−11.06

Table 4 Estimated play parameters

	Whole sample		Small firms		Large firms	
	Constant play	Variable play	Constant play	Variable play	Constant play	Variable play
γ	0.106	0.102	0.170	0.176	0.074	0.05
δ	–	0.200	–	0.200	–	0.194
Average play width		0.108		0.187		0.056

To obtain asymptotically unbiased estimates of the parameters, we estimate Eq. (16) using FM-OLS.[24]

Through the process of grid search over parameters γ and δ described in Sect. 3, the estimated average values of $PLAY_t$ are 0.108, 0.187 and 0.056 for the whole sample, for the subsample of small firms, and for the subsample of large firms respectively (see Table 4).[25] These results are consistent with the presence of an employment band of inaction, which is found to be wider for small firms.[26]

The estimation results (see Table 5) show that the coefficient that captures the reaction along the play, β_1, is not significantly different from zero (for a 5 % significance level), while the coefficient that captures the additional reaction along the spurt, β_2, is. The estimated β_2 are 0.366, 0.632, and 0.246 for the whole sample and for the subsamples of the small and large firms respectively (t statistics are 6.952, 8.628 and 2.728 respectively). This is evidence of the reaction along the play being weaker than the reaction along the spurt, which is true in all of the cases considered.

In order to distinguish the impact of uncertainty from the non-convex employment adjustment costs on the presence of hysteresis, we test the hypothesis H_0: $\delta = 0$ against H_0: $\delta > 0$. The F-Statistic (for $K = 6$ parameter and $m = 1$

[24]By applying Johansen cointegrating test to the three samples, we do not reject the hypothesis of a single cointegrating vector relating the variables. The trace test statistic, 52.405, 58.81, and 48.166 for the whole sample, for the subsample of small firms and for the subsample of large firms respectively, is greater than the 5 % critical value (42.91).

[25]We also report the estimation results for the case of a constant splitting factor (play) in the employment equation.

[26]As the estimated play width is greater for the sub sample of the small firms, the linear play algorithm originates a transformed series, which us smoother than in the case of the large firms.

Table 5 Estimation results

	Whole sample		Small firms		Large firms	
	Constant play	Variable play	Constant play	Variable play	Constant play	Variable play
Cons	11.349[a] (11.64)	11.405[a] (11.825)	6.124[a] (7.488)	6.274[a] (8.853)	9.484[a] (6.284)	11.186[a] (6.431)
S_t	0.038 (0.829)	0.035 (0.779)	0.058 (1.109)	0.049 (1.081)	0.080 (1.116)	−0.001 (−0.012)
$SPURT_t$		0.366[a] (6.952)		0.632[a] (8.628)		0.246[a] (2.728)
T	−0.002[a] (−26.56)	−0.002[a] (−26.64)	−0.001[a] (−15.01)	−0.001[a] (−17.38)		−0.002[a] (−16.61)
R^2	0.891	0.892	0.786	0.815		0.848
DW	0.135	0.130	0.470	0.494		0.100

[a]Significant at 5 %. t-statistics are in parentheses

restriction) for a comparison of the unrestricted ($\delta > 0$) and the restricted case with $\delta = 0$ is 1.17 for the whole sample, 19.75 for small firms and 3.32 for large firms.[27] Consequently, uncertainty contributes to explaining the dynamics of aggregate employment through hysteresis mechanisms, mainly, in the case of small firms.

5 Conclusion

Our results highlight the importance of hysteresis caused by uncertainty in a context of low interest rate, as it is the present situation in many advanced economies. In fact, in a low interest rate environment, fixed employment adjustment costs are less important to generate hysteresis effects, but on the contrary uncertainty is more relevant. Therefore, hysteresis can be quite strong even for small values of the fixed hiring and firing costs.

Although hysteresis is pervasive across different firm's class size, due to non-convex adjustment costs, uncertainty is especially important for the employment dynamics of the small firms. A possible reason is that due to the structure of management they are less able to develop mechanisms to deal with uncertainty.

This result is highly relevant as micro and small firms represent 97.8 % of the total number of Portuguese firms, and they have a share of 53 % of the number of employees.[28]

[27] $F(\delta = 0) = \frac{\left(R^2_{unrestricted} - R^2_{restricted}\right)/m}{\left(1 - R^2_{unrestricted}\right)/(N-K)}$.

[28] We are relying on 2008 data from Statistics Portugal.

This implies that measures toward the deregulation of the labour market to increase flexibility in working time; to make wages and labour costs more responsive to market pressures; to weaken employment security provisions and unemployment benefit systems, could be of no use to increase the level of employment due to hysteresis effects caused by uncertainty in the labour demand. Besides, even if government reduces the level of fixed employment adjustment costs to a minimum, a substantial source of irreversibility remains due to non-convex costs of physical capital adjustment.

References

1. Abel, A.B., Dixit, A.K., Eberly J.C., Pindyck, R.S.: Options, the value of capital, and investment. Q. J. Econ. **111**(3), 753–777 (1996)
2. Amable, B., Henry, J., Lordon, F., Topol, R.: Unit-root in the wage-price spiral is not hysteresis in unemployment. J. Econ. Stud. **20**(1–2), 123–135 (1993)
3. Amable, B., Henry, J., Lordon, F., Topol, R.: Hysteresis revisited: a methodological approach. In: Cross, R. (ed.) The Natural Rate of Unemployment: Reflections on 25 Years of the Hypothesis, pp. 181–200. Cambridge University Press, Cambridge (1995)
4. Aoki, M., Yoshikawa, H.: Uncertainty, policy ineffectiveness, and long stagnation for the macroeconomy. Jpn. World Econ. **18**(3), 261–272 (2006)
5. Baldwin, R., Krugman, P.: Persistent trade effects of large exchange rate shocks. Q. J. Econ. **104**(4), 635–654 (1989)
6. Belke, A.: Fiscal stimulus packages and uncertainty in times of crisis. Econ. Policy Open Econ. **39**(1), 25–45 (2009)
7. Belke, A., Göcke, M.: A simple model of hysteresis in employment under exchange rate uncertainty. Scott. J. Polit. Econ. **46**(3), 260–286 (1999)
8. Belke, A., Göcke, M.: Exchange rate uncertainty and employment: an algorithm describing the play. Appl. Stoch. Model. Bus. Ind. **17**, 181–204 (2001)
9. Belke, A., Göcke, M.: Monetary Policy (In)-Effectiveness Under Uncertainty: Some Normative Implications for European Monetary Policy. Diskussionsbeitge Aus Dem Insitut Fr Volkswirtschaftslehere, Universitt Hohebheim, Stuttgart (2003)
10. Belke, A., Göcke, M.: Real options effects of employment: does exchange rate uncertainty matters for aggregation. Ger. Econ. Rev. **6**(2), 185–203 (2005)
11. Belke, A., Göcke, M.: Monetary policy and investment decision a stilized treatment of the uncertainty trap. Mimeo, Universities of Stuttgart-Hohenheim and Giessen (2006)
12. Bentolila, S., Bertola, G.: Firing costs and labour demand: how bad is eurosclerosis? Rev. Econ. Stud. **57**(3), 381–402 (1990)
13. Bernanke, B.S.: Irreversibility, uncertainty, and cyclical investment. Q. J. Econ. **98**(1), 85–106 (1983)
14. Bertola, G.: Job security, employment and wages. Eur. Econ. Rev. **34**(4), 851–879 (1990)
15. Bertola, G.: Labor turnover costs and average labor demand. J. Labor Econ. **10**(4), 389–411 (1992)
16. Caballero, R.J., Engel, E., Haltiwanger, J.: Aggregate employment dynamics: building from microeconomic evidence. Am. Econ. Rev. **87**(1), 115–137 (1997)
17. Carruth, A., Dickerson, A., Henly, A.: What do we know about investment under uncertainty? J. Econ. Surv. **14**(2), 119–153 (2000)
18. Cross, R.B.: The macroeconomic consequences of discontinuous adjustment: selective memory of non-dominated extrema. Scott. J. Polit. Econ. **41**(2), 212–221 (1994)
19. Dixit, A.: Entry and exit decisions under uncertainty. J. Polit. Econ. **97**(3), 620–638 (1989)

20. Dixit, A.: Analytical approximations in models of hysteresis. Rev. Econ. Stud. **58**, 141–151 (1991)
21. Dixit, A.: Investment and hysteresis. J. Econ. Perspect. **6**(1), 107–132 (1992)
22. Dixit, A., Pindyck, R.: Investment Under Uncertainty. Princeton University Press, Princeton (1994)
23. Ejarque, J.M., Portugal, P.: Labor adjustment costs in a panel of establishments: a structural approach. IZA Discussion Papers Series, vol. 3091 (2007)
24. Folta, T.B., Johnson, D.R., O'Brien J.: Uncertainty, irreversibility, and the likelihood of entry: an empirical assessment of the option to defer. J. Econ. Organ. **61**(3), 432–452 (2006)
25. Göcke, M.: Various concepts of hysteresis applied in economics. J. Econ. Surv. **16**, 167–188 (2002)
26. Hamermesh, D.S.: Labor demand and the structure of adjustment costs. Am. Econ. Rev. **79**(4), 674–689 (1989)
27. Hamermesh, D.S., Pfann, G.A.: Adjustment costs in factor demand. J. Econ. Lit. **34**(3), 1264–1292 (1996)
28. Lang, D.: Involuntary unemployment in a path-dependent system: the case of strong hysteresis. In: Arestis, P., Sawyer, M. (eds.) Path Dependence and Macroeconomics, pp. 80–118. Palgrave Macmillan, Basingstoke (2009)
29. Mayergoyz, I.D.: Mathematical models of hysteresis and their applications. Elsevier, Amsterdam (2003)
30. Mota, P.R., Vasconcelos, P.B.: Non-convex adjustment costs, hysteresis, and the macrodynamics of employment. J. Post Keynesian Econ. **35**(1), 93–112 (2012)
31. Mota, P.R., Varejão, J., Vasconcelos, P.B.: Hysteresis in the dynamics of employment. Metroeconomica **63**(4), 661–692 (2012)
32. Nickel, S.J.: Employment and labour demand. Economica **45**(180), 329–345 (1978)
33. Phillips, P.C.B.: Fully modified least squares and vector autoregression. Econometrica **63**(5), 1023–1078 (1995)
34. Phillips, P.C.B., Hansen, B.E.: Statistical inference in instrumental variables regression with I(1) processes. Rev. Econ. Stud. **57**, 99–125 (1990)
35. Pindyck, R.S.: Irreversible investment, capacity choice and the value of the firm. Am. Econ. Rev. **78**(5), 969–985 (1988)
36. Pindyck, R.: Irreversibility, uncertainty, and investment. J. Econ. Lit. **29**(3), 1110–1148 (1991)
37. Piscitelli, L., Grinfeld, M., Lamba, H., Cross, R.: On entry and exit in response to aggregate shocks. Appl. Econ. Lett. **6**(9), 569–572 (1999)
38. Varejão, J., Portugal, P.: Employment dynamics and the structure of labor adjustment costs. J. Labor Econ. **25**(1), 137–165 (2007)
39. Visitin, A.: Differential Models of Hysteresis. Springer, New York (1994)

A State Space Model Approach for Modelling the Population Dynamics of Black Scabbardfish in Portuguese Mainland Waters

Isabel Natário, Ivone Figueiredo, and M. Lucília Carvalho

Abstract Black scabbardfish (*Aphanopus carbo* Lowe, 1839) is a widely distributed species across the Atlantic ocean. In Portuguese mainland waters the existing specimens are immature (not able to reproduce). It is admitted that they have migrated from the West of the British Isles and that they remain in the area for some years, until they attain an adequate size or physiological conditions which allow them to migrate and reproduce elsewhere.

The present study aims to model the dynamics of the population of black scabbardfish living in the International Council for the Exploration of the Seas Division IXa, for which disaggregated data are available, although within the context of a larger population. With this purpose, a state-space model is used, which enables the estimation of the unknown abundance (latent process) by exploring its dependency relationship with the observational data on the species fishing landings in that area. The population is partitioned into length groups and the population evolution process is subdivided into biological related subprocesses.The estimation is achieved within a Bayesian paradigm, where all the available biological information is incorporated in the prior distributions of the parameters of the subprocesses. Later, short-term trajectories of the population living in IXa are studied, via simulations that are constructed based on different management scenarios.

I. Natário (✉)
CMA-FCT-UNL, Departamento de Matemática, Faculdade de Ciências e Tecnologia, Universidade Nova de Lisboa, 2829-516 Caparica, Portugal
e-mail: icn@fct.unl.pt

I. Figueiredo
Instituto Português do Mar e da Atmosfera (IPMA), Lisboa, Portugal
e-mail: ifigueiredo@ipma.pt

M. Lucília Carvalho
CEAUL, Faculdade de Ciências da Universidade de Lisboa, Lisboa, Portugal
e-mail: mlucilia.carvalho@gmail.com

J.-P. Bourguignon et al. (eds.), *Dynamics, Games and Science*, CIM Series in Mathematical Sciences 1, DOI 10.1007/978-3-319-16118-1_26

1 Introduction

Black scabbardfish (BSF), *Aphanopus carbo* Lowe, 1839, is a deep water species widely distributed across the Atlantic Ocean. The species population dynamics in the NE Atlantic is currently considered to display a migratory behaviour essentially driven by feeding and reproduction considerations [1, 4, 14]. Spawning seems to occur in southern areas like Madeira and Canary Islands [2, 4, 9]. The recruits migrate to most northern areas such as the West of the British Isles (BI) and the Faroes Islands, where they grow for a few years. Afterwards juveniles seem to move south again towards off mainland Portugal (P), International Council for the Exploration of the Seas (ICES) Division IXa, growing to an adult size before leaving for maturing and spawning in southern areas.

The main objective of this work is to come up with a model that accommodates the previous assumptions about the BSF population dynamics and that is able to estimate the BSF population abundance in P, which is an unknown time series.

In the NE Atlantic the BSF is an important commercial resource, being mainly fished in three different spots [2, 6], in BI—deep-water trawl fishery—in P and off the Madeira Archipelago—artisanal longlines fishery. BSF catches represent 0.01 % of the Portuguese gross domestic product and 5 % of the total landed fish in value in Portugal (average values for the decade 2002–2012, source Instituto Nacional de Estatística). The kind of data that are available are landings data, observed as time series, running in parallel with the related unknown BSF population abundance time series. However, at the moment, almost only data from P are available with enough detail.

So, the idea is to use a state space model for specifying the relation between the unobserved state process (population abundance) and the observational process (landings), which also allows the incorporation of the available knowledge on the biology and the spatial dynamics (including migrations) of the species, following the extensions in [12] of the population projection matrix models in [3]. Like this we are able to estimate the latent fish abundance, along the time, as well as its credible intervals, and also to produce estimates of the species vital parameters and fishing mortality. Short-term projections of population abundances can also be obtained by simulation, considering different scenarios regarding abundance and fishing effort, enabling comparisons between different fishing policies and associated uncertainties.

The estimation is done within the Bayesian paradigm, which is better suited to incorporate the a prior biological information on the species dynamics. A sequential importance sampling scheme [7] is implemented for estimation, a quite computer intensive one as not only the model parameters but as well as the latent fish abundance over time are to be estimated.

In Sect. 2 the data and model are detailed and carefully explained, including the estimation and the projections, in Sect. 3 the corresponding results are presented and finally in Sect. 4 they are discussed.

2 Model and Data

The model for the BSF population dynamics that is proposed accommodates what is known about the species, which is not much. It is observed that in P there are not very small (young) specimens, as are observed in BI. It is then postulated that in P the BSF population is composed by immigrants from BI that remain in the area for some years. However, it is observed that in P the specimens are immature, that is, they are not able to reproduce. So, in P, the BSF specimens that attain an adequate size or physiological condition leave the area, emigrate, to reproduce somewhere else (Madeira, Canárias,...).

Any other existing biological information on species should be included, such as survival to natural mortality and life stage transition, here defined through length classes. Also survival to fishing must be considered.

The BSF population is better modelled if partitioned into length groups, corresponding to different age groups, including in class C1 those specimens with length inferior to 50 cm (recruits), nonexistent in P, in class C2 those with length between 50 and 103 cm (juveniles) and in class C3 those larger than 103 cm (adults). In P, the number of specimens in C2 group is annually augmented due to the entrance of specimens arriving from BI, observed to occur in the first semester of the year. The number of specimens in C3 group is annually reduced due to the exit of a fraction of specimens to elsewhere, observed to occur in the second semester of the year.

Hereupon, in order to estimate the unknown population abundance, in number, by exploring its dependency relationship with the observational data on landings, the population dynamics is modelled through a state space model, constituted by two component processes that run in parallel, a latent representing the unknown population abundance and an observational being the observed landings (both vectorial). Like this it is possible to model the evolution in time of the state process, the unknown population, decomposing it into subprocesses, which describe the main biological aspects of the species life cycle. The chosen time unit is the semester, denoted by s, due to the distinct semestrial migration pattern mentioned before.

2.1 Data

The landings data in P come from a small size commercial vessel fleet (n = 16) operating there (Sesimbra landing port), targeting the BSF species, being the fishing grounds the hard bottoms along the slopes of canyons (depth: 800–1200 m), and employing deep-water longline as fishing gear. The landings occur, typically, three times a week, and are divided per length group. The data, provided by the

Portuguese General Directorate of Fisheries and Aquaculture, are in weight, being converted to number by means of the mean weight by length group, estimated from the DCF/EU landings sampling program.

2.2 Model Structure

As stated before the P population state vector in each semester s is constituted by the number of C2 and C3 group members, $n_{C2,s}$ and $n_{C3,s}$, respectively. It is further convenient to subdivide these into the ones that were fished (F) and those that were not ($\overline{F}$). Furthermore, for modelling purposes, it is advisable to also include in this population state vector the number of fish in BI, four semesters before s, $n_{BI,s}$, although its estimation is not a goal, but its consideration allows a better evaluation of the fish flux entering P. So, shortening, we have the following state vector:

$$\boldsymbol{n}_s' = \left(n_{BI,s}\ n_{C2,s}(\overline{F})\ n_{C3,s}(\overline{F})\ n_{C2,s}(F)\ n_{C3,s}(F)\right)$$

The stochastic state process is based on the deterministic general process, $\boldsymbol{n}_s = P\boldsymbol{n}_{s-1}$, where P is a Lefkovitch projection matrix [3]. The complexity of this dynamics is better captured and modeled by further subdividing the state process into subprocesses [12], each of which only depends on the subprocess that occurred immediately before:

- Semestrial evolution of BI population, which is supposed to be fairly unknown and consequently not divided into subprocesses:

$$U_s - \textit{Not detailed evolution in BI}$$

- Semestrial evolution of the population in P, the main concern of this model, is divided into four subprocesses:

$$\begin{aligned} M_s &- \textit{Survival to natural mortality} \\ T_s &- \textit{Class transition} \\ D_s &- \textit{Displacement by migration -} \\ &\quad \textit{entrance and departures of immature adults} \\ F_s &- \textit{Survival to fishing} \end{aligned}$$

The deterministic formulation, $\boldsymbol{n}_s = P\boldsymbol{n}_{s-1} = FDTMU\,\boldsymbol{n}_{s-1}$ evolved to the stochastic formulation via conditional expectations of the state process, which is assumed to be a first order Markov process. It is further assumed that all the individuals in the population act identically and independently. Denoting by $\mathbf{u}_s^X$

the state vector after subprocesses X happens, $X = U, M, T, D, F$, the population evolution is done accordingly to the subprocesses described below:

Not detailed evolution in the BI, only allowing some variation in the BI population abundance:

$$\boldsymbol{u}_s^U \sim \boldsymbol{H}_s^U(\boldsymbol{n}_{s-1}) : \begin{pmatrix} u_{BI,s}^U & \sim \mathrm{N}(n_{BI,s-1}, 0.1 \times n_{BI,s-1}) \\ u_{C2,s}^U & = n_{C2,s-1}(\overline{F}) \\ u_{C3,s}^U & = n_{C3,s-1}(\overline{F}) \end{pmatrix}$$

Survival to natural mortality, representing p_M the probability of surviving to natural mortality:

$$\boldsymbol{u}_s^M \sim \boldsymbol{H}_s^M(\boldsymbol{u}_s^U) : \begin{pmatrix} u_{BI,s}^M & = u_{BI,s}^U \\ u_{C2,s}^M & \sim \mathrm{Bi}\left(u_{C2,s}^U, p_M\right) \\ u_{C3,s}^M & \sim \mathrm{Bi}\left(u_{C3,s}^U, p_M\right) \end{pmatrix}$$

Class transition, representing p_{23} the probability of a specimen in group $C2$ grows into group $C3$:

$$\boldsymbol{u}_s^T \sim \boldsymbol{H}_s^T(\boldsymbol{u}_s^M) : \begin{pmatrix} u_{BI,s}^T & = u_{BI,s}^M \\ u_{C2,s}^T & = u_{C2,s}^M - X[u_{C2,s}^M], \text{ with } X[u_{C2,s}^M] \sim \mathrm{Bi}\left(u_{C2,s}^M, p_{23}\right) \\ u_{C3,s}^T & = u_{C3,s}^M + X[u_{C2,s}^M] \end{pmatrix}$$

Displacement by migration: this differs according to whether s is odd (immigration to P) or even (emigration from P); let λ represents the probability that a specimen from BI, which has immigrated four semesters earlier, arrives alive in P and p_E represent the probability that a specimen exits P by migration:

$$\boldsymbol{u}_s^D \sim \boldsymbol{H}_s^D(\boldsymbol{u}_s^T) : \begin{pmatrix} u_{BI,s}^D & = u_{BI,s}^T \\ u_{C2,s}^D & = u_{C2,s}^T + I_s, \text{ with } I_s \sim \mathrm{Bi}\left(u_{BI,s}^T, \lambda\right) \\ u_{C3,s}^D & = u_{C3,s}^T \end{pmatrix}, \text{ s odd}$$

$$\boldsymbol{u}_s^D \sim \boldsymbol{H}_s^D(\boldsymbol{u}_s^T) : \begin{pmatrix} u_{BI,s}^D & = u_{BI,s}^T \\ u_{C2,s}^D & = u_{C2,s}^T \\ u_{C3,s}^D & = u_{C3,s}^T - E_s \text{ with } E_s \sim \mathrm{Bi}\left(u_{C3,s}^T, p_E\right) \end{pmatrix}, \text{ s even}$$

Survival to fishing: survival to fishing probabilities relate to the mortality rates $F_{Ci,s}$ through $1 - \phi_{Ci,s} = exp(-F_{Ci,s})$. Fishing mortality rates are estimated by linking them to the fishing effort on the basis of the catchability coefficient. A full recruitment model with log-normal error term is considered, $F_{Ci,s} = q_{Ci}E_s$ [10], where E_s is a standardized fishing effort in semester s derived from estimates of the quantity catch-per-unit-of-effort (CPUE), obtained through the adjustment of a GLM model to them, where the covariates involved, besides the temporal variations, reflect the specificities of the different elements of the fishing fleet [8]. Further note that an adjustment is made in group $C2$ (through probability $p_{LargeC2}$, fixed) to account for those smaller specimens in this group that are never caught (smaller than 70 cm).

$$\boldsymbol{n}_s = \boldsymbol{u}_s^F \sim \boldsymbol{H}_s^F(\boldsymbol{u}_s^D) : \begin{pmatrix} n_{BI,s} & = u_{BI,s}^D \\ n_{C2,s}(\overline{F}) & \sim \mathrm{Bi}(u_{C2,s}^D \times p_{LargeC2}, 1-\phi_{C2,s}) \\ n_{C3,s}(\overline{F}) & \sim \mathrm{Bi}(u_{C3,s}^D, 1-\phi_{C3,s}) \\ n_{C2,s}(F) & = u_{C2,s}^D - n_{C2,s}(\overline{F}) \\ n_{C3,s}(F) & = u_{C3,s}^D - n_{C3,s}(\overline{F}) \end{pmatrix}$$

The observational process $\boldsymbol{y}_s' = (y_{BI,s}\ y_{C2,s}\ y_{C3,s})$, is a stochastic function of the unknown states $\boldsymbol{n}_s$, representing $y_{BI,s}$ the estimated number of BI fish in semester $s-4$ and representing $y_{C2,s}$ and $y_{C3,s}$ the BSF catches in $C2$ and $C3$ length groups in semester s. Normal measurement errors with constant coefficients of variation are considered for these:

$$y_{BI,s} \sim \mathrm{N}\left(n_{BI,s}, \psi_{BI}^2.n_{BI,s}^2\right)$$

$$y_{Ci,s} \sim \mathrm{N}\left(n_{C_{i,s}}(F), \psi_{Ci}^2.n_{C_{i,s}}^2(F)\right), \quad i = 2,\ 3$$

To summarize, the state space model can then be described by

$$g_0(\boldsymbol{n}_0; \Theta), \qquad g_s(\boldsymbol{n}_s|\boldsymbol{n}_{s-1}; \Theta); \qquad f_s(\boldsymbol{y}_s|\boldsymbol{n}_s; \Theta),$$

with parameters $\Theta = (p_M, p_{23}, \lambda, p_E, q_{C2}, q_{C3}, \psi_{C2}, \psi_{C3}, \psi_{BI})$, and

$$g_s(\boldsymbol{n}_s|\boldsymbol{n}_{s-1}; \Theta) = \int_{\boldsymbol{u}_s^F}\int_{\boldsymbol{u}_s^D}\int_{\boldsymbol{u}_s^T}\int_{\boldsymbol{u}_s^M}\int_{\boldsymbol{u}_s^U} g^U(\boldsymbol{u}_s^U|\boldsymbol{n}_{s-1}; \Theta) g^M(\boldsymbol{u}_s^M|\boldsymbol{u}_s^U; \Theta) \times$$

$$\times\ g^T(\boldsymbol{u}_s^T|\boldsymbol{u}_s^M; \Theta) g^D(\boldsymbol{u}_s^D|\boldsymbol{u}_s^T; \Theta) g^F(\boldsymbol{n}_s|\boldsymbol{u}_s^D; \Theta) \mathrm{d}\boldsymbol{u}_s^U \mathrm{d}\boldsymbol{u}_s^M \mathrm{d}\boldsymbol{u}_s^T \mathrm{d}\boldsymbol{u}_s^D \mathrm{d}\boldsymbol{u}_s^F.$$

The initialization of this Markov process is done considering the specimens caught in each group in the first semester and simulating the probability of fishing, for each group, using the prior distribution of the corresponding q parameters. With these we can initialize the number of fish caught in each group as the ones observed in the first semester of the observed time series, estimate the total number of specimens in each group by dividing the number of fish caught in each group by the corresponding fishing probability, and then initialize the number of non-caught fishes in each group as the difference between the total and the fished ones.

2.3 Estimation

The estimation is done within a Bayesian paradigm, implying non-trivial integration of the several probability density functions, which is accomplished through sequential Monte Carlo. The algorithm proposed by Liu and West [7] is implemented to do sequential importance sampling with resampling, specially intended for space state models. It is though necessary to specify prior distributions for the parameters, where all the biological knowledge available about them is included.

Importance sampling is a technique that is used when direct sampling from a target probability distribution is not feasible, but we can generate samples from an alternative and easier trial distribution, and then weight them properly to be used as samples from the target distribution. Sequential importance sampling with resampling is a technique where, for space state models, the generation of the unknown states is carried out using as trial density function the state equation $g_s(\boldsymbol{n}_s|\boldsymbol{n}_{s-1};\Theta)$ and weights which are proportional to the observation density (filtering). The algorithm yields estimates of $\mathbf{n}_s|\mathbf{y}^s$, ($\mathbf{y}^s = (y_1,\ldots,y_s)$) and parameter densities at each time point s. At the last time point S we get an estimate of the posterior density of Θ.

To overcome a problem of "particle deplection" (particles with relative large sizes tend to be chosen many times and dominate) kernel smoothing of parameter vectors has been implemented at each time step, adding a small perturbation to parameter values, increasing the diversity of parameters values in vicinity of the parameter space. Also, auxiliary particle filter was implemented, where an initial "auxiliary" resample is taken from the population at time s, with weights calculated according to the expected likelihood of the states at time $s+1$, given the data as time $s+1$. This resampled set of particles is then projected forward from time s to time $s+1$, and "corrected" using likelihood weights just as with filter, except that the likelihood weights must take account of the auxiliary resampling stage.

Model adequacy is based mainly on the inspection of the estimated credible intervals for the latent abundances, specially those relating to the number of catch fishes, for which we have data to compare. The expected deviance is also calculated.

The prior distributions for the model parameters were selected so that all the available biological information was incorporated there, see Table 1. The common non-informative gamma priors were chosen for the dispersions of the observation errors.

2.4 Simulation

For management purposes it is interesting to be able to simulate several scenarios regarding the BSF abundance, according to several levels of exploitation, for example. Taking the estimates of the parameters and states for the last year for

Table 1 Description of the subprocesses parameters

Parameter	Subprocess	Prior distribution	Description
p_M	Surviving	Beta	*Probability of surviving to natural mortality.* $E[p_M] = e^{-M}$, assuming an exponential model for the individual lifetime. M is estimated [5] as $(e^{-M/2})^{\text{age}_{\max}} = p_{unf}$, where p_{unf} is the proportion of the unfished population that attends the maximum age, $\text{age}_{\max}$, considered to be respectively equal to 0.05 and 46 semesters [13]
p_{23}	Class transition	Beta	*Probability of a specimen to transit from C2 to C3.* $E[p_{2,3}]$ is determined using: (1) the interval of total length for specimens that are likely to transit to length group $C3$ in one semester; (2) the probability of a specimen living in P belongs to this length interval (determined assuming the Von Bertallanffy model for growth with parameter estimates given in [13])
λ	Immigration	Beta	*Probability that a BI specimen that has immigrated four semesters before arrives alive in P.* $E[\lambda]$ is calculated as the mean of the relative decrease of the BI CPUE between the odd and even semesters
p_E	Emigration	Beta	*Probability that a specimen emmigrates from P.* $E[p_E]$ is calculated as the mean of the relative decrease of Portuguese CPUE between the odd and even semesters
ϕ_{C2}	Survival to	Lognormal	*Probability that a C_2 or C_3 specimen die due to fishing.* The information available to estimate catchability is insufficient, so vague log-normal distributions are adopted as prior of distributions of q_{C2} and q_{C3}

which data were available as inputs of a simulation study, and assuming that the population dynamics remains almost identically except for some little differences, three different scenarios were considered: no fishing, an increase of 5 % in the fishing effort in P and a decrease of 20 % in the BI abundance, with no change in the fishing effort in P.

3 Results

The estimation algorithm as well as the predictions were programmed in R [11]. It was tuned according to [7] and, in order to implement it, 1,000,000 initial particles were used. The Monte Carlo error, evaluated by averaging the coefficient of variation for all the parameter mean estimates, based on five runs, has a value of 5.8 %. This section describes the main findings.

3.1 Estimation

The obtained estimates were the posterior distributions of all the model parameters and the posterior estimates of all state vector components, summarized in their posterior medians and 95 % credible intervals. Figure 1 depicts the estimates of the catches and the observed data, showing the good adjustment between them, and relatively narrow credible intervals.

Figure 2 displays the estimated abundances for C2 and C3 groups, both depicting slightly increasing trends. Note however the high variability of the estimates in the beginning of the time series, possibly due to a not so good initialization or to not so much informative priors.

Figure 3 displays the prior and posterior distributions of parameters p_M, p_{23}, λ, p_E, q_{C2} and q_{C3}. From here it can be seen that the parameters who had more modified the corresponding parameter distribution by the data were λ and q_{C2}, q_{C3}, for which less prior information was available.

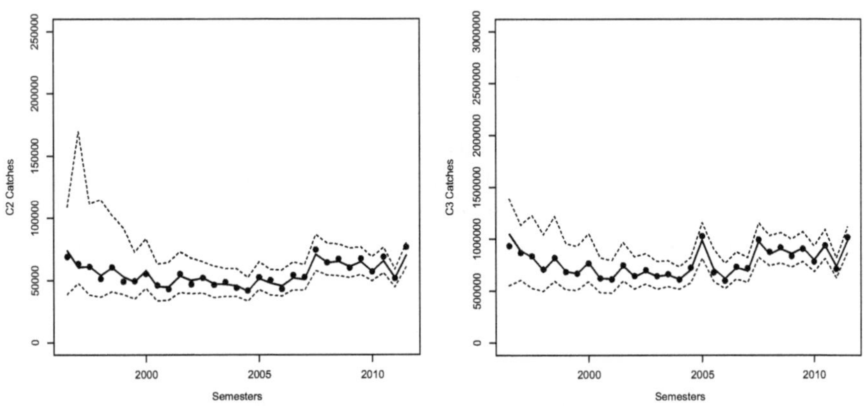

Fig. 1 C2 (*left*) and C3 (*right*) group estimates (*lines*) and observed (*dots*) catches

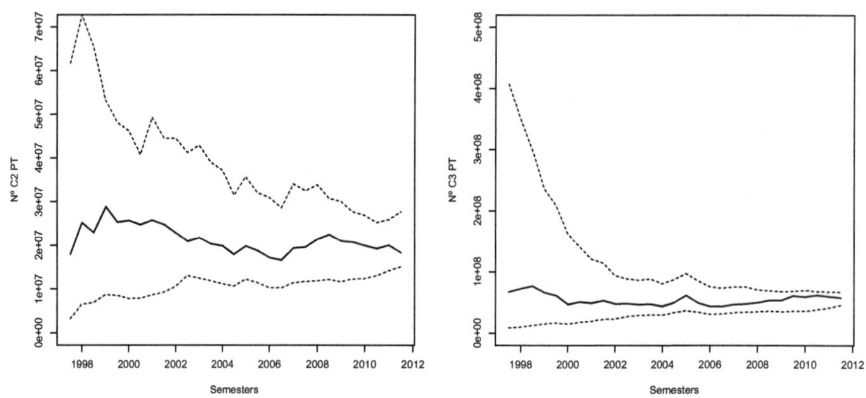

Fig. 2 C2 (*left*) and C3 (*right*) group abundance estimates

3.2 Simulation

Figure 4 depicts the simulation results for the three scenarios considered: scenario with no fishing, scenario with an increase of 5 % in fishing and the scenario where a decrease of 20 % in the BI abundance of the resource occurs. From here it can be seen that both scenarios involving changes produce the same result of decreasing *C*3 predicted abundance. Interestingly, *C*2 predicted abundance do not change significantly between the three scenarios, possibly because this group corresponds to a much smaller part of the population.

4 Discussion

The model applied here to BSF dynamics has some important advantages over other deterministic alternatives commonly used for the same purpose. It allows simultaneous incorporation in the model of fishery data and existing prior information on the species life cycle, presents a flexible way to incorporate the different biological aspects of the life cycle in a modular way, allows unprecise data by including observational errors and weak information on the population process parameters by using non-informative priors. This model not only provides abundance estimates as it allows short-term predictions with just a little more effort.

As to the estimation results it is worth noting that data essentially altered the distributions of the catchabilities q_{C2} and q_{C3}, increasing their mean values that were most probably taken to be very small, and of the probability of a BI specimen that has immigrated four semesters before arrives alive in P, λ, for which a prior ignorance state was the starting point.

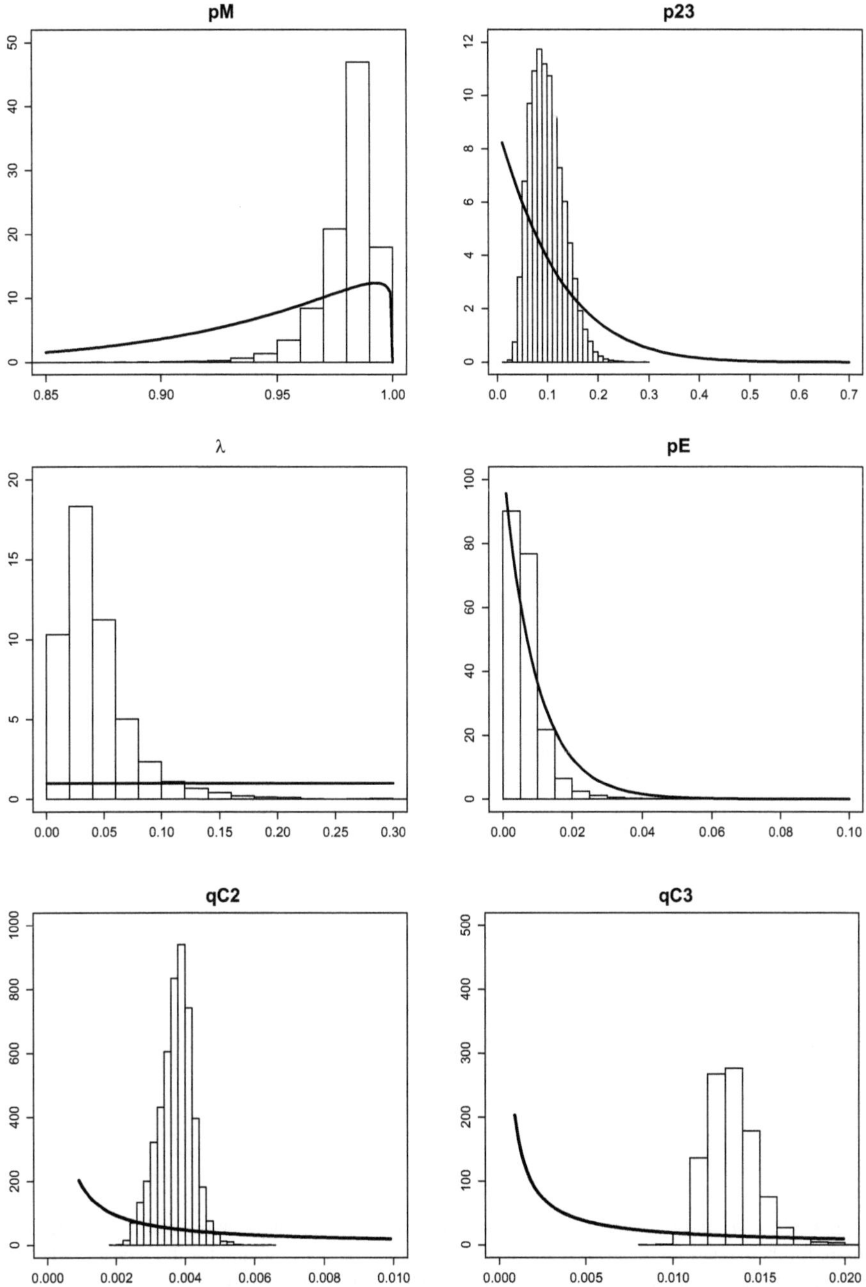

Fig. 3 Prior (*line*) and estimated posterior (*histogram*) densities for parameters p_M and p_{23} (*upper row*), λ and p_E (*middle row*) and q_{C2} and q_{C3} (*bottom row*)

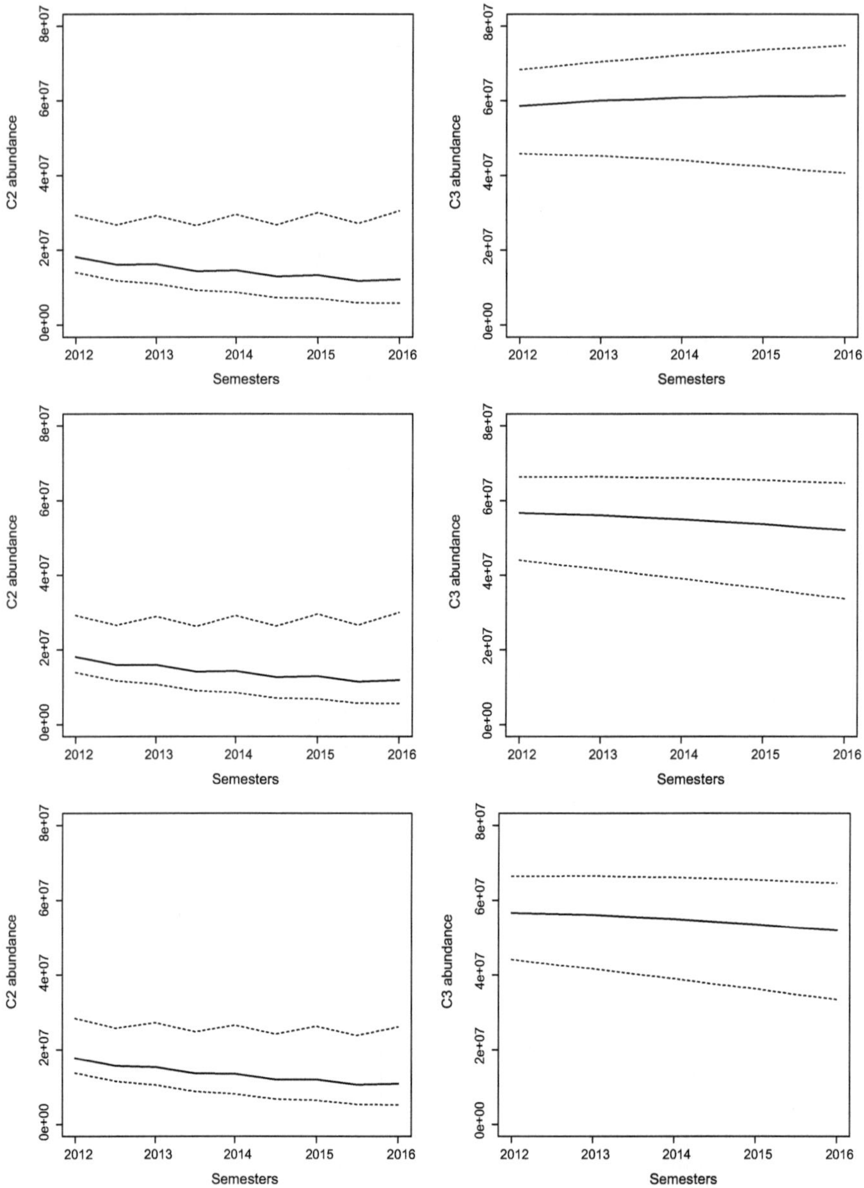

Fig. 4 C2 (*left*) and C3 (*right*) group simulated abundance for the scenario with no fishing (*top row*), for the scenario with an increase of 5 % in fishing (*middle row*) and for the scenario where a decrease of 20 % in the BI abundance of the resource occurs (*bottom row*)

Finally, as far as predictions are concerned, the slightly increasing estimated abundance trend that was observed in the case where no fishing was assumed was inverted when an increase in the fishing regime or a decrease in immigrants was imposed.

As argued before this model is very modularized, being quite easy to incorporate the population dynamics of the BSF in the other places assumed to belong to its life spatial cycle. That should be the next step of this work, for which data are being gathered. Naturally that when that happens it will be possible to obtain more accurate predictions about the resource abundance in the different sea areas where it exists and where it is being fished, allowing a better understanding of its evolution for different fishing scenarios in the different places and helping to review the actual fishing quotas if necessary.

Acknowledgements This work is financed by National Funds through FCT—Fundação para a Ciência e a Tecnologia—in the scope of projects UID/MAT/00297/2013 and PEst-OE/MAT/UI0006/2014 and by EU Funds in the scope FP7-DEEPFISHMAN project, management and monitoring of deep-sea fisheries and stocks, Grant agreement no.: 227390.

References

1. Anonymous: final report of the EU study project CT97/0084 - environment and biology of deep-water species *Aphanopus carbo* in the NE Atlantic: basis for its management (BASBLACK). DGXIV European Commission (2000)
2. Bordalo-Machado, P., Fernandes, A.C., Figueiredo, I., Moura, O., Reis, S., Pestana, G., Gordo, L.S.: The black scabbardfish (*Aphanopus carbo* Lowe, 1839) fisheries from the Portuguese mainland and Madeira Island. Sci. Mar. **73S2**, 63–76 (2009)
3. Caswell, H.: Matrix population Models: Construction, Analysis and Interpretation, 2nd edn. Sinauer Associates, Sunderland (2001)
4. Figueiredo, I., Bordalo-Machado, P., Reis, S., Sena-Carvalho, D., Blasdale, T., Newton, A., Gordo, L.S.: Observations on the reproductive cycle of the black scabbardfish (*Aphanopus carbo* Lowe, 1839) in the NE Atlantic. IVES J. Mar. Sci. **60**, 774–779 (2003)
5. Hoenig, J.M.: Empirical use of longevity data to estimate mortality rates. Fish. Bull. US **81**(4), 898–903 (1983)
6. ICES: Report of the Working Group on the Biology and Assessment of Deep-Sea Fisheries Resources (WGDEEP). ICES CM 2012/ACOM:17 (2012)
7. Liu, J., West, M.: Combining parameter and state estimation in simulation-based filtering. In: Doucet, A., Freitas, N., Gordon, N. (eds.) Sequential Monte Carlo Methods in Practice, pp 197–224. Springer, Berlin (2001)
8. Maunder, M.N., Punt, A.E.: Standardizing catch and effort data: a review of recent approaches. Fish. Res. **70**(2), 141–159 (2004)
9. Pajuelo, J.G., González, J.A., Santana, J.I., Lorenzo, J.M., García-Mederos, A., Tuset,V.: Biological parameters of the bathyal fish black scabbardfish (*Aphanopus carbo* Lowe, 1839) off the Canary Islands, Central-east Atlantic. Fish. Res. **92**, 140–147 (2008)
10. Quinn, T.J., Deriso, R. B.: Quantitative Fish Dynamics. Oxford University Press, Oxford (1999)
11. R Core Team: R: a language and environment for statistical computing. R Foundation for Statistical Computing, Vienna, Austria (2014). ISBN:3-900051-07-0. http://www.R-project.org/http://www.R-project.org/

12. Thomas, L., Buckland, S.T., Newman, K.B., Harwood, J.: A unified framework for modelling wildlife population dynamics. Aust. N. Z. J. Stat. **47**, 19–34 (2005)
13. Vieira, A.R., Farias, I., Figueiredo, I., Neves, A., Morales-Nin, B., Sequeira, V., Martins, M.R., Gordo, L.S.: Age and growth of black scabbardfish (*Aphanopus carbo* Lowe, 1839) in the southern NE Atlantic. Sci. Mar. **73S2**, 33–46 (2009)
14. Zilanov, V.K., Shepel, L.I.: A contribution to the ecology of black scabbardfish *Aphanopus carbo* Lowe in the North Atlantic (In Russian). Vopr. Ikhtiol. **93**, 737–740 (1975)

Entropy and Negentropy: Applications in Game Theory

Eduardo Oliva

Abstract The concept of entropy has been applied to such different fields as thermodynamics, cosmology, biology, chemistry, information theory and economics. An interesting application of entropy in the latter field is the existence of a complete ordering of information structures represented by the decrease in entropy, computed à la Shannon, of the agent's beliefs. In this paper we will apply this entropy ordering to information structures used in experiments assessing the role of communication in coordination games.

1 Introduction

Since the early works of Carnot, Clausius and Boltzmann [8, 10], entropy has proven to be one of the most fruitful concepts in science. Although entropy is best known for the role that it plays in classical thermodynamics and the so called *second law*, applications of entropy can be found in such different fields as cosmology (i.e. the arrow of time, Beckenstein-Hawking entropy of a black hole [9]), chemistry and biology (Schrödinger's and Brillouin's concept of *negentropy* [3, 11]) and, surprisingly, in language, cryptography and information theory with the seminal work of Shannon on a mathematical theory of communication [12]. This connection between information (the information theoretic *Shannon entropy*) and the mecano-statistical properties of physical systems (the thermodynamic *Boltzmann entropy*) allows one to apply physical techniques (as the Boltzmann-Gibbs distribution [7]) to economic problems (i.e. the statistical mechanics of money distribution [13]).

In this paper we will use another application of entropy in economics: deciding which piece of information is better for an agent. A decision maker values information that reduces the uncertainty about the true state of nature. Nevertheless,

E. Oliva (✉)
Universidad Nacional de Educación a Distancia (UNED), Madrid, Spain

Laboratoire d'Optique Appliquée, ENSTA ParisTech, École Polytechnique ParisTech, CNRS, Palaiseau, France

Laboratoire de Physique des Gaz et des Plasmas, Université Paris Sud, CNRS, Orsay, France
e-mail: eduardo.oliva@u-psud.fr

J.-P. Bourguignon et al. (eds.), *Dynamics, Games and Science*, CIM Series in Mathematical Sciences 1, DOI 10.1007/978-3-319-16118-1_27

gathering information usually has a cost. Thus, the decision maker must choose the most informative piece of information (from now *information structure*) at a given cost. This is a hard choice, since the ranking of information depends upon at least:

- The agent's priors about the true state of nature.
- The preferences and wealth of the agent.
- The decision problem itself.

It turns out that, under certain conditions, a complete ordering of the informativeness of information structures exists and it is represented by the decrease in Shannon's entropy (i.e. increase in negentropy) of the agent's beliefs. We will apply these tools to the study of communication in coordination games. In [6] the effect of nonbinding, preplay communication in bilateral coordination games was studied. Two different information structures, one-way communication (only one agent sends a signal) and two-way communication (both agents send a signal at the same time) were used. The experiment implied the counterintuitive result that the dominated strategy plays a fundamental role. Thus, a model where the agents have private values (altruists and egoists) was proposed to explain the results. We will apply the information ordering to discern which information structure (one-way or two-way) is better in the signaling problem faced by an agent that knows its private value (altruist, egoist) but does not know its opponent's private value and wants to maximize its expected utility.

The structure of the rest of the paper is as follows. In Sect. 2 we will explain the experiment reported in [6] and the altruist-egoist model. Section 3 will be devoted to describe the information structures used in the experiment. In Sect. 4 the entropy ordering will be explained and applied to the experiment's information structures. Finally, the paper will close with a few remarks and conclusions.

2 A Cooperative Coordination Game: Altruist-Egoist Model

In [6] the effect of "cheap talk" (i.e. nonbinding, preplay communication at no cost) on the equilibrium selection in coordination games was studied. Two different information structures (one-way and two-way communication) and different coordination games (with or without a dominated cooperative strategy) were considered. Table 1 shows a coordination game with two Pareto-ranked Nash equilibra, (1,1) and (2,2), and a dominated cooperative strategy (3) that leads to a Pareto-dominating solution (3,3). The structure of this coordination game is intended to shed some light on the processes of equilibrium selection and thus it does not try to model any *real world* situation. Indeed, this coordination game is a mixture of two well known games: the *stag hunt* (entries for strategies (1) and (2) in Table 1) and the *prisonner's dilemma* (entries for strategies (1) and (3) in the same table). Nevertheless, it is possible to devise some *ad hoc* situation where this game applies, for example, in the adoption of technologies. Let's suppose that both players are enterprises that make business between them via some software. They can: (1) continue working with the same software; (2) make an upgrade to a new version (3) buy a better (but expensive) software that is compatible with the old version of the actual software but not with the upgraded version (because it is too new). This new software is expensive to maintain but it is really user friendly for the clients of the enterprise (i.e. the other player). Upgrading the software (2) or buying new software (3) has a cost. Thus, if the other enterprise does not upgrade it (1), the payoff will be less. Worse, if the enterprise plays (3) and the oponent (1), the costs will strongly reduce the payoff while the oponent will be glad to use a user-friendly software at a cost zero (i.e. its payoff is increased). When one enterprise upgrades (2) and the other buys (3), the systems are incompatible and thus no business can be done, dropping the payoffs.

In [6] it was shown that one-way communication increases the play of the Pareto-dominant equilibrium (2,2) while two-way communication does not always decrease the frequency of coordination failures, i.e. a result different from (2,2) and (3,3).

More surprisingly, the dominated strategy (3) was announced and played a non negligible number of times. Furthermore, in [5] it was found that variation on the payoffs of dominated strategies influence the selection of a Nash equilibrium, which cannot be explained assuming self-interested, rational players.

A model that can explain this behavior supposes that not all players are self-interested. A percentage ρ of the players are altruists that receive, in addition of the payoffs shown in Table 1, a *warm glow* δ when playing the cooperative strategy (3). In the previous example, this *warm glow* corresponds to the possibility of attracting new clients due to the improved interface. When $c - f \geq \delta \geq a - b$ strategy (3) is neither dominant nor dominated for altruists players and (3) is the best response to (3) and (1). Assuming that ρ is common knowledge, the game becomes a game of imperfect information as being altruist or egoist is a private value. Now, preplay communication will help to signal (or conceal) types of players and influence the selection of an equilibrium. Depending on the type of communication and the proportion of altruist players, different kinds of equilibria will appear and disappear, as demonstrated in the appendix.

Table 1 Cooperative coordination game ($a > b > c > d > e$) with multiple Nash equilibria (1,1) and (2,2), and a dominated strategy (3)

	1	2	3
1	d,d (350,350)	d,e (350,250)	a,f (1000,0)
2	e,d (250,350)	c,c (550,550)	f,f (0,0)
3	f,a (0,1000)	f,f (0,0)	b,b (600,600)

The cooperative solution (3,3) Pareto-dominates both equilibria. In each cell, the first number is row player's payoff and the second is column player's payoff. The values used in [6] are shown in parentheses

2.1 Equilibria in One-Way Communication

- When $\rho > \max\left(\frac{c-d}{a-d}, \frac{c-f-\delta}{b-f}\right)$ there exists a nonrevealing equilibrium in which all players announce (3), egoists play (1) and altruists play (3). All other announcements lead to the play of (2).
- When $\frac{c-d}{a-d} > \rho \geq \frac{c-(f+\delta)}{b-f}$ there is a totally revealing equilibrium in which altruists announce and play (3) and egoists announce and play (2). Egoists will play (2) in response to an announce of (2) and (1) when (3) is announced. Altruists will play (2) when (2) is announced and (3) when (3) is announced.
- When $\frac{c-(f+\delta)}{b-f} > \rho$ Egoists and altruists will announce (2) and play (2).

2.2 Equilibria in Two-Way Communication

- When $\rho > \max\left(\frac{c-d}{a-d}, \frac{c-f-\delta}{b-f}\right)$ there exists a nonrevealing equilibrium in which all players announce strategy (3). When (3,3) is announced, an altruist player will play (3) and an egoist (1). Other announcements lead to the play of (2).
- When $\max\left(\frac{c-d}{a-d}, \frac{c-f-\delta}{b-f}\right) > \rho$, there exists a sequential Nash equilibrium in which all altruists announce (3), a proportion $\epsilon = \frac{\rho}{1-\rho}\frac{c-a}{d-c}$ of egoists announce (3) and the reminder announce (2). When announced (3,3) altruists will play (3) and egoists (1). If other pair of announcements are made, the strategy played by both players is (2).

As explained in Sect. 1, we will apply the entropy ordering to discern which of the above communication structures (one-way or two-way) is preferred for a player that knows her own private value (altruist or egoist) but does not know her opponent's private value.

3 Information Structures

Consider a decision maker facing a problem under uncertainty. This situation can be characterized by

- $X = (x_1, \ldots, x_K)$ A vector composed by the K different states of nature.
- $\Phi = (\phi_1, \ldots, \phi_K)$ The vector of prior probabilities, ϕ_j, of x_j being the true state of nature.
- $S = (s_1, \ldots, s_M)$ A vector composed by the M possible signals that the decision maker can observe.
- $Q = (q_1, \ldots, q_M)$ The vector of unconditional probabilities, q_j, of observing signal s_j.

An information structure consists of:

- The vector of unconditional probabilities of each signal, Q.
- The matrix $\Pi^{M\times K}$ of posterior probabilities. The element $\pi_{mk} = p(x_k|s_m)$ is the probability that the state of the world is x_k when the signal s_m has been observed.

The product of the unconditional probabilities of each signal and the matrix of posterior probabilities must be equal to the vector of prior probabilities $\Phi = Q\Pi$. Another useful relationship is obtained applying Bayes' theorem to the elements of Π

$$\pi_{mk} = p(x_k|s_m) = \frac{\alpha_{km}\phi_k}{q_m} \tag{1}$$

where $\alpha_{km} = p(s_m|x_k)$ is the probability of signal s_m being observed when the state of nature is x_k.

Once defined, we can express the information structures corresponding to both communication structures studied in [6]. The states of nature are $X = (altruist, egoist)$ with prior probabilities $\Phi = (\rho, 1-\rho)$. The possible signals that the opponent can announce are $S = (1, 2, 3)$. The probabilities Q of each signal and the matrix Π

$$\Pi = \begin{pmatrix} p(altruist|1)\ p(egoist|1) \\ p(altruist|2)\ p(egoist|2) \\ p(altruist|3)\ p(egoist|3) \end{pmatrix}$$

depend on the communication structure that we are studying.

3.1 One-Way Communication

- When $\rho > \max\left(\frac{c-d}{a-d}, \frac{c-f-\delta}{b-f}\right)$, all players announce (3), so $Q_{\rho>\max\left(\frac{c-d}{a-d}, \frac{c-f-\delta}{b-f}\right)} = (0, 0, 1)$. The announcement of (3) carries no information. The probability of the opponent being an altruist player after having announced (3) is ρ, whereas $1-\rho$ is the probability of being egoist. The components of interest of Π are thus

$$\Pi_{\rho>\max\left(\frac{c-d}{a-d}, \frac{c-f-\delta}{b-f}\right)} = \begin{pmatrix} - & - \\ - & - \\ \rho & 1-\rho \end{pmatrix}$$

- When $\frac{c-d}{a-d} \geq \rho \geq \frac{c-(f+\delta)}{b-f}$, altruist players announce (3) and egoist players announce (2). Then, $Q_{\frac{c-d}{a-d}\geq\rho\geq\frac{c-(f+\delta)}{b-f}} = (0, 1-\rho, \rho)$. This is a totally revealing equilibrium, as the components of interest of Π manifest

$$\Pi_{\frac{c-d}{a-d}\geq\rho\geq\frac{c-(f+\delta)}{b-f}} = \begin{pmatrix} - & - \\ 0 & 1 \\ 1 & 0 \end{pmatrix}$$

- When $\frac{c-(f+\delta)}{b-f} > \rho$, all players announce (2). Thus, $Q_{\frac{c-(f+\delta)}{b-f}>\rho} = (0, 1, 0)$ and the components of interest of Π are

$$\Pi_{\frac{c-(f+\delta)}{b-f}>\rho} = \begin{pmatrix} - & - \\ \rho & 1-\rho \\ - & - \end{pmatrix}$$

3.2 Two-Way Communication

- When $\rho > \max\left(\frac{c-d}{a-d}, \frac{c-f-\delta}{b-f}\right)$, both the player and opponent will announce (3). So, the probability of receiving each signal is $Q_{\rho>\max\left(\frac{c-d}{a-d}, \frac{c-f-\delta}{b-f}\right)} = (0, 0, 1)$. The announcement of the oponent (and that of the player) carries no information. As expected, the matrix Π is

$$\Pi_{\rho>\max\left(\frac{c-d}{a-d}, \frac{c-f-\delta}{b-f}\right)} = \begin{pmatrix} - & - \\ - & - \\ \rho & 1-\rho \end{pmatrix}$$

- When $\max\left(\frac{c-d}{a-d}, \frac{c-f-\delta}{b-f}\right) > \rho$, the most interesting case appears. Now, egoist players will mix signals, announcing signal (2) and (3) with probability $1-\epsilon$ and ϵ respectively. Altruist players will always announce (3). Thus, $q_2 = (1-\rho)(1-\epsilon)$, $q_3 = \rho + (1-\rho)\epsilon$ and $Q_{\max\left(\frac{c-d}{a-d}, \frac{c-f-\delta}{b-f}\right)>\rho} = \left(0, 1-\frac{a-d}{c-d}\rho, \rho\frac{a-d}{c-d}\right)$ The components of interest of Π are computed taking advantage of Bayes' theorem, as expressed by Eq. (1). The probability of signal (3) being announced conditioned to being the oponent an altruist is $p(3|altruist) = \alpha_{altruist,3} = 1$. The probability of signal (3) being announced conditioned to being the oponent an egoist is $p(3|egoist) = \alpha_{egoist,3} = \epsilon$. Similarly, $p(2|egoist) = \alpha_{egoist,2} = 1-\epsilon$. Since $\Phi = (\rho, 1-\rho)$, the matrix Π is

$$\Pi_{\max\left(\frac{c-d}{a-d}, \frac{c-f-\delta}{b-f}\right)>\rho} = \begin{pmatrix} - & - \\ 0 & 1 \\ \frac{c-d}{a-d} & \frac{a-c}{a-d} \end{pmatrix}$$

For all cases described above, it is straightforward to demonstrate that $\Phi = Q\Pi$.

4 Entropy Ordering of Information Structures

A decision-maker with some initial prior about the probabilities of each state of nature and with the possibility to buy some information faces a dilemma: which piece of information buy at some cost or, stated in a different way, when one information structure is more informative than another.

The first answer to this question appeared in the seminal work of Blackwell [1, 2]. According to Blackwell's ordering, an information structure is more informative than another if the latter is equivalent to receiving the informative signal of the former structure with noise (i.e. the less informative is a garbled version of the most informative). Usually, it is not possible to compare information structures with this method (i.e. finding a garbling matrix that relates both information structures), thus Blackwell ordering is incomplete.

In a recent paper [4], Cabrales, Gossner and Serrano present an informativeness ordering that is complete and it is represented by the decrease of entropy (or increase of negentropy) of the agent's beliefs. This ordering depends only on the agent's prior but it is independent of his preferences, initial wealth and decision problem. Thus, when some general properties on the agent's utility function and payoffs related to the states of nature [4] hold, ordering the information reduces to compute the negentropy of the information process.

The negentropy of this process is

$$I = H(\Phi) - \sum_{m=1}^{M} q_m H(\pi_{m\cdot}) \tag{2}$$

where $\pi_{m\cdot}$ is the m row of Π and H(p) is the entropy computed *à la Shannon*

$$H(p) = -\sum_{j} p_j \ln p_j \tag{3}$$

By continuity $0 \cdot \ln 0 = 0$. With these definitions and the information structures computed in Sect. 3 we can compute the negentropy of one-way and two-way communication.

4.1 One-Way Communication

For all cases, we have

$$H(\Phi) = -\left[\rho \ln \rho + (1-\rho)\ln(1-\rho)\right] \tag{4}$$

- When $\rho > \max\left(\frac{c-d}{a-d}, \frac{c-f-\delta}{b-f}\right)$ the information structure is

$$Q_{\rho > \max\left(\frac{c-d}{a-d}, \frac{c-f-\delta}{b-f}\right)} = (0, 0, 1)$$

$$\Pi_{\rho > \max\left(\frac{c-d}{a-d}, \frac{c-f-\delta}{b-f}\right)} = \begin{pmatrix} - & - \\ - & - \\ \rho & 1-\rho \end{pmatrix}$$

The entropy of the third row is

$$H(\pi_{3\cdot}) = -\left[\rho \ln \rho + (1-\rho)\ln(1-\rho)\right]$$

and the negentropy associated is identically zero $I_{\rho > \frac{c-d}{a-d}} = 0$. In this case, one-way communication gives no information about the private value of the opponent.

- When $\frac{c-d}{a-d} \geq \rho \geq \frac{c-(f+\delta)}{b-f}$, the information structure is

$$Q_{\frac{c-d}{a-d} \geq \rho \geq \frac{c-(f+\delta)}{b-f}} = (0, 1-\rho, \rho)$$

$$\Pi_{\frac{c-d}{a-d} \geq \rho \geq \frac{c-(f+\delta)}{b-f}} = \begin{pmatrix} - & - \\ 0 & 1 \\ 1 & 0 \end{pmatrix}$$

The entropy of each row (i.e. of the probability distribution of the state of nature when a signal is emitted) is $H(\pi_{2\cdot}) = H(\pi_{3\cdot}) = -\left[0 \cdot \ln 0 + 1 \cdot \ln 1\right] = 0$. The incertitude has disappeared and the real state of nature (the type of the oponent) has been revealed. The negentropy is thus $I = H(\Phi)$, the initial entropy, and it is maximal.

- When $\frac{c-(f+\delta)}{b-f} > \rho$ the information structure is

$$Q_{\frac{c-(f+\delta)}{b-f} > \rho} = (0, 1, 0)$$

$$\Pi_{\frac{c-(f+\delta)}{b-f} > \rho} = \begin{pmatrix} - & - \\ \rho & 1-\rho \\ - & - \end{pmatrix}$$

The entropy of the second row is

$$H(\pi_{2\cdot}) = -\left[\rho \ln \rho + (1-\rho)\ln(1-\rho)\right]$$

and the negentropy associated is identically zero $I_{\frac{c-(f+\delta)}{b-f} > \rho} = 0$. In this case, one-way communication gives no information about the private value of the opponent.

4.2 Two-Way Communication

As in the previous information structure, Eq. (4) gives the entropy of the prior probability distribution.

- When $\rho > \max\left(\frac{c-d}{a-d}, \frac{c-f-\delta}{b-f}\right)$, there is no difference with the one-way case and thus, the negentropy is identically zero. No information is obtained when the signal is observed.
- When $\max\left(\frac{c-d}{a-d}, \frac{c-f-\delta}{b-f}\right) \geq \rho$ the messages convey some information. In this case, the information structure is

$$Q_{\max\left(\frac{c-d}{a-d},\frac{c-f-\delta}{b-f}\right)\geq\rho} = \left(0, 1 - \frac{a-d}{c-d}\rho, \rho\frac{a-d}{c-d}\right)$$

$$\Pi_{\max\left(\frac{c-d}{a-d},\frac{c-f-\delta}{b-f}\right)\geq\rho} = \begin{pmatrix} - & - \\ 0 & 1 \\ \frac{c-d}{a-d} & \frac{a-c}{a-d} \end{pmatrix}$$

The entropy of each row is

$$H(\pi_{2\cdot}) = 0$$

$$H(\pi_{3\cdot}) = -\left[\frac{c-d}{a-d}\ln\frac{c-d}{a-d} + \frac{a-c}{a-d}\ln\frac{a-c}{a-d}\right]$$

and the corresponding negentropy is $I = H(\Phi) - \frac{a-d}{c-d}\rho H(\pi_{3\cdot}) < H(\Phi)$

4.3 Comparing Both Information Structures

With these values, it is easy to conclude which information structure is preferred.

- When $\rho > \max\left(\frac{c-d}{a-d}, \frac{c-f-\delta}{b-f}\right)$, the signals observed do not convey any information and thus both information structures are equivalent.
- When $\frac{c-d}{a-d} \geq \rho \geq \frac{c-(f+\delta)}{b-f}$, the negentropy of one-way communication is $I_{one-way} = H(\Phi)$, a totally revealing equilibrium, whereas the negentropy of two-way communication is $I_{two-way} = H(\Phi) - \frac{a-d}{c-d}\rho H(\pi_{3\cdot}) < I_{one-way}$.
- When $\frac{c-f-\delta}{b-f} > \rho$, the negentropy of one-way communication is zero and the negentropy of two-way communication is $I_{two-way} = H(\Phi) - \frac{a-d}{c-d}\rho H(\pi_{3\cdot}) > 0$

Thus, there is an interval where one-way communication is preferred and another interval where the preferred structure is two-way communication. Nevertheless, it is worth to note that this last interval can be made arbitrarily small just by making the free parameter δ (the *warm-glow*) approach its limiting value $\delta \to c - f$, increasing at the same time the interval where one-way communication is preferred.

5 Conclusions

In this paper we have applied the entropy ordering of Cabrales, Gossner and Serrano [4] to a generalized egoist-altruist model of a cooperative coordination game [6]. We have obtained the result that, when the *warm glow* altruists receive when playing the cooperative strategy is high enough, one-way communication is more informative than two-way and thus preferred for any agent. Although this result is not surprising, given the existence of a totally revealing equilibrium in one-way communication, this method would be useful in other situations: when announcements are binding or when gathering or emitting information has a cost.

Appendix

In this section we will find some of the equilibria of the general game shown in Table 1.

Warm Glow Payoff

The *warm glow* δ that altruists players add to the payoffs shown in Table 1 when playing the cooperative strategy (3) makes this strategy neither dominated nor dominant for altruist players and strategy (3) is the best response to both (3) and (1) for this type of players. The latter condition is met when, respectively

$$b + \delta \geq a \tag{5}$$

$$f + \delta \geq d \tag{6}$$

while the former is met whenever some of these conditions (7), (8) or (9) hold but not all at the same time (if it were the case, (3) would be a dominant strategy).

$$f + \delta \geq d \tag{7}$$

$$f + \delta \geq c \tag{8}$$

$$b + \delta \geq a \tag{9}$$

Since if inequality (8) holds, then inequality (7) will also hold and since inequality (9) is required for (3) to be the best response to (3) and inequality (7) is required for (3) to be the best response to (1), we choose inequality (8) to be false. Thus, we have the condition that $c - f \geq \delta \geq a - b$.

One-Way Communication

Proposition 1 *If $\rho > \max\left(\frac{c-d}{a-d}, \frac{c-f-\delta}{b-f}\right)$, there is a nonrevealing equilibrium in which all players announce (3), egoists play (1) and altruists play (3). Any other announcement leads to the play of (2) for both types of player.*

Proof For egoist players, the strategy of announcing (3) and playing (1) should result in a higher payoff than announcing and thus playing (2). Then, $a\rho+d(1-\rho) > c$ or equivalently $\rho > \frac{c-d}{a-d}$. For altruist players, announcing and playing (3) should give a higher payoff than announcing and playing (2). Thus, $\rho(b+\delta)+(1-\rho)(f+\delta) > c$ or equivalently $\rho > \frac{c-f-\delta}{b-f}$.

Proposition 2 *If $\frac{c-d}{a-d} > \rho > \frac{c-f-\delta}{b-f}$, there exists a totally revealing equilibrium in which altruists announce and play (3) and egoists announce and play (2). Egoists and altruists play (2) in response to an announcement of (2). When (3) is announced, egoists play (1) and altruists play (3).*

Proof As shown above, when $\frac{c-d}{a-d} \geq \rho$ egoists prefer to announce and play (2). When (3) is announced, egoists will play (1) as it is the best response to an altruist announcing and playing (3). Furthermore, when (2) is announced, playing (2) is the best response. As $\rho > \frac{c-f-\delta}{b-f}$ announcing and playing (3) for altruists dominates announcing and playing (2). For altruists, the best response to an altruist announcing (3) is playing (3) and to an egoists announcing (2) is playing (2).

Proposition 3 *If $\frac{c-f-\delta}{b-f} \geq \rho$, there is a nonrevealing equilibrium in which all players announce and play (2).*

Proof As shown above, when $\frac{c-f-\delta}{b-f} \geq \rho$ altruists will prefer to announce and play (2) rather than announcing and playing (3). As egoists cannot induce altruists to play (3) due to the low proportion of altruist players, they also announce and play (2). This is true even if $\rho > \frac{c-d}{a-d}$, as altruists will never play (3).

Two-Way Communication

Proposition 4 *If $\rho > \max\left(\frac{c-d}{a-d}, \frac{c-f-\delta}{b-f}\right)$, there is a nonrevealing equilibrium in which all players announce (3), egoists play (1) and altruists play (3). Any other pair of announcements will lead to the play of (2).*

Proof As in one-way communication, the payoff for egoists players when announcing (3) is $a\rho + d(1-\rho)$ which is greater than c if $\rho > \frac{c-d}{a-d}$. The same reasoning applies for altruists, being their payoff when they announce (3) $\rho(b+\delta)+(1-\rho)(f+\delta)$ greater than the payoff of announcing (2) when $\rho > \frac{c-f-\delta}{b-f}$.

Proposition 5 *If* $\frac{c-d}{a-d} > \rho$*, there exists an equilibrium in which all altruists announce (3) and egoists announce (3) with probability* $\epsilon = \frac{\rho}{1-\rho}\frac{a-c}{c-d}$ *and (2) with probability* $1-\epsilon$*. When both announcements are (3) altruists will play (3) and egoists will play (1). All other pair of announcements lead to the play of (2).*

Proof For the egoists to be indifferent with respect to the announcement of (2) and (3) we have

$$\rho a + (1-\rho)\left[\epsilon d + (1-\epsilon)c\right] = c \tag{10}$$

Right side of Eq. (10) is the payoff of announcing (2). If the egoist player announces (3), receives an announcement of (3) and plays (1), he will win a with probability ρ (he was confronted to an altruist) or d with a probability $(1-\rho)\epsilon$ (the opponent announcing (3) was an egoist). If he receives an announcement of (2), he is confronted with an egoist, plays (2) and gains c. The probability of this event is $(1-\rho)(1-\epsilon)$. In order to be indifferent, we have that $\epsilon = \frac{\rho}{1-\rho}\frac{a-c}{c-d}$.

For altruists, announcing (3) is better than announcing (2) if, following the same reasoning as above, we have

$$\rho(b+\delta) + (1-\rho)\left[\epsilon(f+\delta) + (1-\epsilon)c\right] \geq c \tag{11}$$

Substituting the value of ϵ in Eq. (11) we obtain, after some algebra

$$\rho\left[(b+\delta-c)(c-d) + (f+\delta-c)(a-c)\right] \geq 0 \tag{12}$$

The first term inside brackets is positive while the second is negative. Indeed, $(f+\delta-c) < 0$. Thus, we can rewrite inequality (12) as $(b+\delta-c)(c-d) \geq (c-\delta-f)(a-c)$ where all of its terms are positive. Since $b+\delta-c \geq a-c$ we just have that $c-d \geq c-(f+\delta)$. As $d \leq f+\delta$, which is required for the *warm glow* conditions, we have that inequality (12) holds for any value of its parameters.

Proposition 6 *If* $\frac{c-f-\delta}{b-f} > \rho$*, there exists an equilibrium in which all altruists announce (3) and egoists announce (3) with probability* $\epsilon = \frac{\rho}{1-\rho}\frac{a-c}{c-d}$ *and (2) with probability* $1-\epsilon$*. When both announcements are (3) altruists will play (3) and egoists will play (1). All other pair of announcements lead to the play of (2). In addition to this, there exists another equilibrium in which altruists and egoists both announce and play (2).*

Proof Since $\frac{c-f-\delta}{b-f} > \rho$, altruists have no incentive to announce and play (3) unless egoists reveal themselves announcing (2) with probability $1-\epsilon$. For egoists being indifferent between announcing (2) and (3) we have

$$\rho\left[\epsilon a + (1-\epsilon)c\right] + (1-\rho)\left[\epsilon^2 d + 2\epsilon(1-\epsilon)c + (1-\epsilon)^2 c\right] = c \tag{13}$$

The first term in Eq. (13) is the payoff when facing an altruist that announces (3). With probability ϵ the egoist announces (3), receives (3) and plays (1) and with probability $1-\epsilon$ announces (2), receives (3) and plays (2). The second term is the payoff when facing an egoist. With probability ϵ^2 both announce (3) and thus play (1). The other possible announcements lead to the play of (2). After some algebra it can be shown that either $\epsilon = 0$ or $\epsilon = \frac{\rho}{1-\rho}\frac{a-c}{c-d}$. For $\epsilon = 0$, egoists always announce and play (2) and thus altruists always announce and play (2).

When $\epsilon = \frac{\rho}{1-\rho}\frac{a-c}{c-d}$, altruist's payoff should be greater or equal than c

$$\rho(b+\delta) + (1-\rho)\left[\epsilon(f+\delta) + (1-\epsilon)c\right] \geq c \tag{14}$$

Since Eqs. (11) and (14) are similar, we can apply the same analysis obtaining the conclusion that inequality (14) identically holds.

References

1. Blackwell, D.: Comparison of experiments. In: Proceedings of the Second Berkeley Symposium on Mathematical Statistics and Probability, pp. 93–102 (1951)
2. Blackwell, D.: Equivalent comparison of experiments. Ann. Math. Stat. **24**, 265–272 (1953)
3. Brillouin, L.: The negentropy principle of information. J. Appl. Phys. **24**(9), 1152–1163 (1953). doi:http://dx.doi.org/10.1063/1.1721463. http://www.scitation.aip.org/content/aip/journal/jap/24/9/10.1063/1.1721463
4. Cabrales, A., Gossner, O., Serrano, R.: Entropy and the value of information for investors. Am. Econ. Rev. **103**(1), 360–377 (2013)
5. Cooper, R.W., DeJong, D.V., Forsythe, R., Ross, T.W.: Selection criteria in coordination games: some experimental results. Am. Econ. Rev. **80**(1), 218–233 (1990)
6. Cooper, R.W., DeJong, D.V., Forsythe, R., Ross, T.W.: Comunication in coordination games. Q. J. Econ. **107**(2), 739–771 (1992)
7. Feynman, R.P.: Statistical Mechanics. Westview Press, Boulder (1998)
8. Haar, D.T.: Foundations of statistical mechanics. Rev. Mod. Phys. **27**, 289–338 (1955). doi:10.1103/RevModPhys.27.289. http://www.link.aps.org/doi/10.1103/RevModPhys.27.289
9. Hawking, S.W.: Black hole explosions? Nature **248**, 30–31 (1974)
10. Maruyama, K., Nori, F., Vedral, V.: *Colloquium*: the physics of Maxwell's demon and information. Rev. Mod. Phys. **81**, 1–23 (2009). doi:10.1103/RevModPhys.81.1. http://www.link.aps.org/doi/10.1103/RevModPhys.81.1
11. Schrödinger, E.: What is Life? Cambridge University Press, Cambridge (1992)
12. Shannon, C.E.: A mathematical theory of communication. Bell Syst. Tech. J. **27**, 379–423; 623–656 (1948)
13. Yakovenko, V.M., Rosser, J.B.: *Colloquium* : Statistical mechanics of money, wealth, and income. Rev. Mod. Phys. **81**, 1703–1725 (2009). doi:10.1103/RevModPhys.81.1703. http://www.link.aps.org/doi/10.1103/RevModPhys.81.1703

Micro-Econometric Analysis of New Household Formation in Spain

Orlando Montoro Peinado

Abstract This paper begins a new line in the estimate and classification of New household formation in Spain. It starts with the study of the emancipation of young people dependent of their parents and proposes a micro-econometric analysis to find and measure socioeconomic factors that affect the decisions youngsters make when they leave their parent's home. In the first place a discrete choice model three level nested multinomial logit based in population characteristics is proposed. In order to improve the results avoiding systematic biases and making use of all the information in the data source, the model is replaced by a sequence of three binary logits.

The period of study extends from 2008 to 2011 so it will be useful to find evidence of how the economic crisis has affected the current trends of Spanish growing New household formation levels and increasing emigration of young dependents. The gap between Spain and the rest of European countries concerning Emancipation and New household formation levels is reducing since the last nineties but the high level of unemployment in the current crisis has supposed a brake in that trend.

1 Introduction

Even when in 2007 there was already a gap between demand and supply for new housing making bigger the housing bubble in Spain, it is clear that a main factor of long term new housing demand is the New household[1] formation. On the other hand, the family structure and household composition is essential in the study of savings pattern, consumption and in general people's economic behaviour. This can be seen, for example, in the Spanish current economic crisis where consumption in

[1]Household stands for the whole group of people living together in the same house.

O.M. Peinado (✉)
National University of Distance Education of Spain, C/ Paseo Senda del Rey S/N, 28040 Madrid, Spain
e-mail: orlando.montoro@meyss.es

J.-P. Bourguignon et al. (eds.), *Dynamics, Games and Science*, CIM Series in Mathematical Sciences 1, DOI 10.1007/978-3-319-16118-1_28

Table 1 Estimated parameters

X	β_1	β_2	β_3
Intercept	−3.7102	2.6964	−2.4504
Family income	−0.00000234	−0.00002	0.000038
Personal income	0.000027	0.000058	−0.00000615
Age	0.0647	−0.00429	0.0457
Year-2008	−0.6618	1.8073	−1.436
Year-2009	0.054	0.7941	−0.989
Genre-men	−0.2519	−0.0918	0.2845
Employment-no	−0.9287	−2.0216	0.573

one family nucleus[2] has been held by another family nucleus of the same household or another member of what it is known as the *extended family*.[3] This support has made consumption to decrease less than expected while family debts have run higher than expected.

The family structure and household composition are fundamental also for the comprehension of social behaviour. Spain, like the rest of Mediterranean countries, tends to have a delayed emancipation calendar. Spanish young people usually leave their parents' home later and have fewer experiences working abroad or living alone before forming their own family than the rest of European people. While the economic crisis has stimulated emigration inside Europe, it has slowed down the tendency to a growing Emancipation and New household formation levels which were reducing the gap with the rest of Europe. These tendencies are marked, for instance, in the estimated model by the negative sign of the unemploymentś parameters $\beta_1 = -0.9$ and $\beta_2 = -2$ detailed in Table 1.

The National Statistical Institutes have answered this demand for information and have begun to improve their household projections integrating them into the rest of statistics as Labour Force Survey or National Accounts. The household projections methods can be divided into static and dynamic ones. Among the static methods are the *Household Head* method used in [3] and the *Propensity Method* used currently in the National Statistical Institute of Spain, INE. The static methods need less information but they are more limited in results and less flexible to juncture changes.

The dynamic projection methods are based on hypothesis about future fluxes of New household formation which are applied to current household and population stocks in order to estimate future number of households and their typification. Not only they offer more information than static ones but dynamic methods can include juncture information in their models so they are more flexible to economy changes.

[2]Family nucleus stands for a couple living together, a couple and their children living together or just one parent living with some of her/his children.

[3]Extended family stands for a family nucleus and other relatives.

In [1] a probabilistic prediction model is developed. That model defines a specific household's typification and the states of household's members depending mainly on their relationships. Then a matrix of transitional probabilities between states is estimated. Applying to this matrix a projected population conditioned to each type of household it is set a new row of households by type which is considered a short term estimate of the number of households and its typification. In [4] it is developed the macro-simulation LIPRO program that processes all the information.

Evidence found in this paper could improve probabilities of changed between states in the transition matrix of the mentioned methods so not only demographic but juncture economic information could be included in the household projections to better them.

Another category of studies, as in [2], highlights how in the last two decades the One-person households have grown spectacularly in Spain, have diversified its composition and are no longer exclusively from rural areas becoming significant in big cities also. Between 1970 and 2001 the share of population living alone grew from 2 to 7 %. This growth was caused partly by the fact that more young people decided to emancipate and live alone.

On the other hand, in [5] it is noticed that the new ways to live in Spain had begun in the 90s but it was in 2000 on when they rocketed. For instance, young people from 25 to 34 years old living alone increase from 112.173 in 1991 to 346.290 in 2001 according to the population censuses information.

This tendency was confirmed in 2011 where the share of One-person household grew to 23 % in 2011. The micro-econometric analysis of New household formation is coherent with all this results and can improve them providing ways to estimate this information not only in census years but between and beyond that.

2 Econometric Model

To distill Emancipation from the process of leaving home, the target population Young Dependent People is defined as people between 16 and 39 years old living with their parents, at least one of them, not living with any partner and not being parents-or if they are parents, not living with any of their children-. These people are included in their parents' family nucleus supposedly since they were born so they are considered never to have left home and the act of leaving home for first time is considered as the act of Emancipation.

The New household formation among Young Dependent People as defined is almost 15 times higher than New household formation among non Young Dependent People in 2010 which means that 77.5 % of new households formed in 2010 is owed to Young Dependent People.

For this study the considered typification of New household formation is: One-person household, Two Partners, Partition and Joining an Existing household. Two Partners stands for two people who begin to live together coming from different households of origin. Partitions are divisions of a family nucleus or divisions of a household with more than one family nucleus where at least two members of the household leave.

Joining an Existing household stands for people who change their own household to be part of another existing one or to found a new one with more than one person from other household of origin.

The share of New household formation based on Partitions and Joining and Existing household in 2010 represents less than 10 % of the total New household formation. Moreover, these two types of New household formation among Young Dependent People is completely residual.

Restricting the research to Young Dependent People as defined and to the first two types of New household formation, One-person household and Two Partners, it gets explained more than 70 % of the phenomena of New household formation and, joining the young people's emigration it is explained almost total Emancipation phenomena. The emancipation of young dependents naturally relates to social, demographic and economic reasons so it has been considered an appropriate dimension to start working.

Therefore for the target population the process consists of deciding emancipate and then choosing between going abroad or starting a New household and in this case whether it is a One-person or Two Partners household. The factors to take into account for this purpose are genre, age, household income, personal income, year and employment. Only main effects are considered so at this stage dependency between states is left out. This means, among other things, that this model does not make differences between young dependents that never have left parents' home from those who had already gone but at the period of study they are living at their parents' home again. The economic reasons for living home in these two groups are considered to be the same. It is not possible either to differentiate between new Two Partners households where both people have emancipated at the same time from the ones where one of its members were already emancipated from her/his parents before beginning this new relationship.

The data source for the study is the longitudinal EU-SILC 2008–2011 related to Spain so another two advantages of this study are that it can be replicated and compared at European level and it can be continued in time. This survey tracks households for 4 years, including new households funded by people previously living in households included in the initial sample. The sample design of the EU-SILC is a rotating panel with partial replacement of the sample of 25 % annually. Information is available for both households and individuals that constitute them. For households there is an annual cross-elevation[4] factor and for people aged 16 or older we use a longitudinal elevation factor.

[4]The elevation factor is the amount of total population represented by each case in the study. The cross- elevation factor is the representation of one case over the total population 1 year. The longitudinal elevation factor is the representation of the same case over the population who is present two consecutive years.

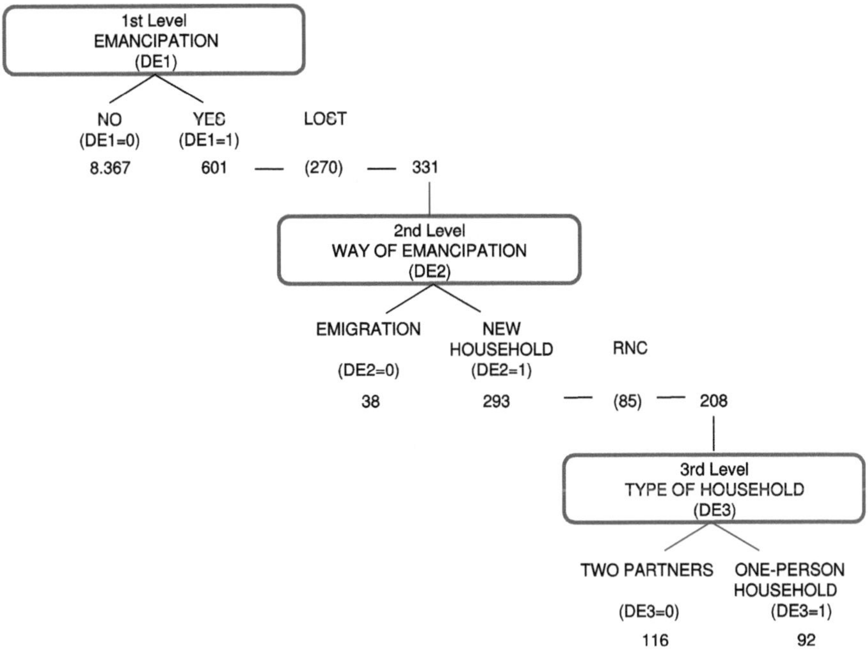

Fig. 1 Decision tree

The proposed model to analyze the decision process of emancipation is a three-level nested logit as shown in Fig. 1.

The original data includes 270 cases, named LOST next to 45 % of total cases of emancipation- of young dependents for whom it is known they have emancipated but the type of emancipation they have chosen is not known. There are also 85 cases, named RNC next to 30 % of New household formation- of young dependents for whom it is known they have started a New household but it is not known the type of household they have formed.

In order to include these cases to make maximum use of the survey results and avoid systematic biases, the three-level nested logit is replaced with a sequence of three binary logits[5] and conditional probabilities are estimated at each level.[6] This way all available cases at each level are included. In first level, wether to leave the

[5]In fact, in a decision process as the one described in Fig. 1 these two models are mathematically the same except for the missing data -LOST and RNC cases-.

[6]For instance, according to the data in Fig. 1, the next probabilities are considered:

$$P(DE2 = 1|DE1 = 1) = \frac{Population represented by the 293 cases where DE2 = 1}{Population represented by the 331 cases where DE1 = 1 different from LOST}$$

$$P(DE3 = 1|DE2 = 1) = \frac{Population represented by the 92 cases where DE3 = 1}{Population represented by the 208 cases where DE2 = 1 different from RNC}$$

parental home is decided, LOST cases are included. In second level, the conditional probability of whether to form a new home or leave the country is decided, LOST cases are not considered but RNC cases are included. In third level conditional between living alone or starting a domestic partnership it is chosen so neither LOST nor RNC cases can be included.

The proportion of young people who take each emancipation decision is estimated as marginal probabilities obtained from the conditional ones estimated at each level. For instance, marginal probability of New household formation (DE2=11) is the compound probability of marginal Emancipation (DE1=1) and the conditional New household formation (DE2=1 restricted to DE1=1) and marginal probability of One-person household P(DE3=111) is the compound of marginal probability of New household formation (DE2=11) and conditional One-person household (DE3=1 restricted to DE2=11).

$$\text{Probability of New household formation}$$
$$P(DE2 = 11) = P(DE1 = 1)P(DE2 = 1|DE1 = 1)$$
$$\text{Probability of One-person Household formation}$$
$$P(DE3 = 111) = P(DE2 = 11)P(DE3 = 1|DE2 = 11)$$

Estimated probabilities by binary logit based in characteristics take the general form $\frac{1}{1+e^{X\beta}}$ for decisions DE $= 0$ and $\frac{e^{X\beta}}{1+e^{X\beta}}$ for decisions DE $= 1$ where X are the characteristics taken in account and β is the parameter which measures the effect of the correspondent characteristic in the decision process. Characteristics X are known for every individual in the study while β must be estimated. As a result, these are the expressions to be adjusted:

$$\text{Living with parents } P(DE1 = 0) = \frac{1}{1 + e^{X\beta_1}}$$

$$\text{Emancipation } P(DE1 = 1) = \frac{e^{X\beta_1}}{1 + e^{X\beta_1}}$$

$$\text{Emigration } P(DE2 = 10) = \frac{e^{X\beta_1}}{1 + e^{X\beta_1}} \frac{1}{1 + e^{X\beta_2}}$$

$$\text{New household formation } P(DE2 = 11) = \frac{e^{X\beta_1}}{1 + e^{X\beta_1}} \frac{e^{X\beta_2}}{1 + e^{X\beta_2}}$$

$$\text{Two Partner household } P(DE3 = 110) = \frac{e^{X\beta_1}}{1 + e^{X\beta_1}} \frac{e^{X\beta_2}}{1 + e^{X\beta_2}} \frac{1}{1 + e^{X\beta_3}}$$

$$\text{One-person household } P(DE3 = 111) = \frac{e^{X\beta_1}}{1 + e^{X\beta_1}} \frac{e^{X\beta_2}}{1 + e^{X\beta_2}} \frac{e^{X\beta_3}}{1 + e^{X\beta_3}}$$

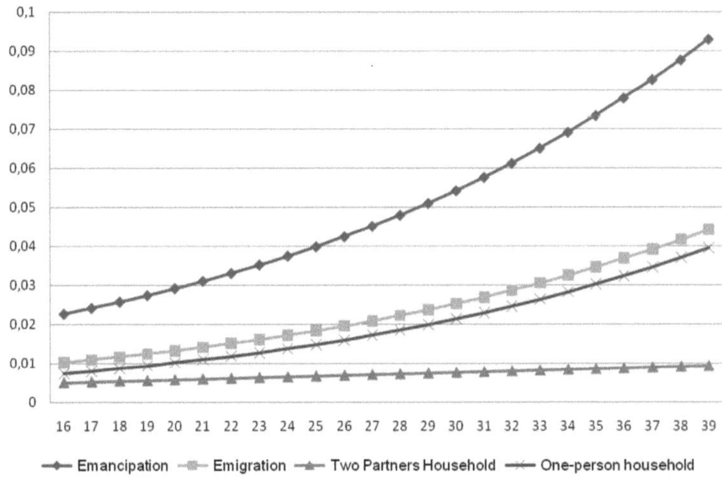

Fig. 2 Probability of emancipation for unemployed dependents in 2010 per age, average personal and family incomes

3 Results and Discussion

Related to the goodness of fit every parameter at every level has a p-value lower than 0,0001 what is quite good to consider them different to 0 and ROC curve[7] for each parameter is around 75 % what is considered good test also.The research is completed with an influence analysis of LOST and RNC cases and dependency between states analysis. It is suggested also to build a ROC curve for the three decision levels integrated as one and to make considerations about the independence of irrelevant alternatives clause that underlies the model.

As a trial, this paper focuses on unemployment effects over Emancipation and New household formation. To understand the effect that each factor has over the emancipation decision process, it is easy to see how the direction of the effect that each factor has at each decision's level depends on the sign of the estimated parameter. Table 1 shows how unemployment negatively affects the decision of emancipation at level 1.At the second stage, in case of deciding to emancipate, unemployment encourages young dependents to go abroad rather than to form a New household and, in the event of starting a household unemployment encourages to be a One-person rather than a Two Partners household.

It is worth also building tendency charts as in Fig. 2 that shows how forming a Two Partners household is quite unlikely at every age for unemployed dependents

[7]In a binary prediction model, a Receiver Operating Characteristic,ROC curve, is a representation for different threshold settings of the fraction of true positive rate vs. false positive rate. For this analysis three ROC curves are calculated −i=1, 2, 3-. True positive rate stands for the rate of well-predicted DEi=1 survey cases and false positive rate stands for the rate of wrong-predicted DEi=0 -survey cases where DEi=1 is predicted by the model though real decision has been DEi=0-.

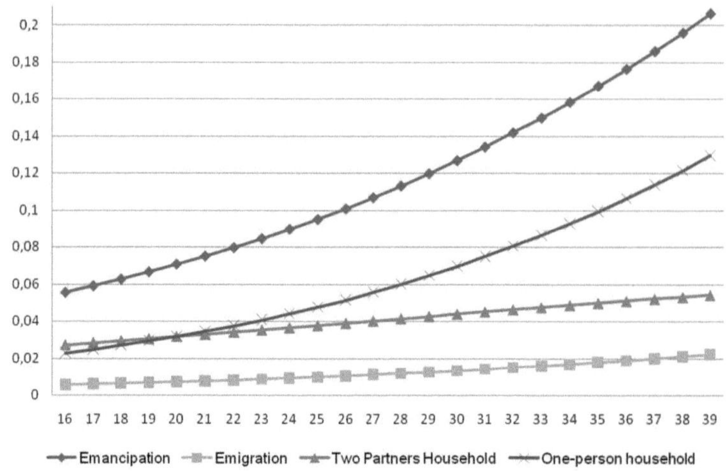

Fig. 3 Probability of emancipation for employed dependents in 2010 per age, average personal and family incomes

with average personal and household incomes. As forming a Two Partners household stands mainly for new couples where at least one of them is previously living with her/his parents, unemployment reveals here as an important brake to form new family nucleus when it means to leave the parent's home. This does not exactly mean that probabilities in Table 3 should be bigger for One-person than for Two Partners in case of unemployment because there are other factors to take into account as the ones included in the study and the probability's levels in each case before the crisis. In Fig. 3 it is shown the same figure for employed people. Shares of emigration are higher for unemployed dependents while New household formation is higher for employed dependents.

Another kind of results proposed to analyze the data are tables of shares of dependents who emancipated depending on the different states they can represent. Table 2 shows the emancipation shares per year in the period 2008–2010 for each type of emancipation. It seems 2009 to be the big year for new Two Partners households while Emigration and One-person households keep growing; One-person households grew to be the most important form of emancipation in 2010 while Emigration became a significant way to emancipate in the same year.

In Table 3 it is shown how the employment factor affects to the emancipation decisions. There are clearly higher levels of emancipation founding a New household for employed people and higher levels of going abroad for unemployed ones. It is interesting also to note how One-person households for employed people grew intensively in the period while it stagnated for the unemployed ones. Out of this trial, it is interesting also to note how Spanish young women have begun to cut distances with men concerning the age of emancipation beginning to leave their

Table 2 Shares of emancipation per year and type

Young dependents	DE1=1 emancipation	DE2=10 emigration	DE3=110 two partners	DE3=111 on-person
2008	4.7	0.2	3.3	1.2
2009	8.1	0.9	4.5	2.7
2010	7.5	1.4	2.4	3.6
Total	7.2	1.0	3.3	2.9

Table 3 Shares of emancipation per year and type, employ and unemployed people

Young unemployed dependents	DE1=1 emancipation	DE2=10 emigration	DE3=110 two partners	DE3=111 on-person
2008	1.9	0.3	1.2	0.5
2009	3.8	1.1	1.6	1.1
2010	3.5	1.7	0.7	1.1
Total	3.4	1.3	1.1	1.0
Young employed dependents	DE1=1 emancipation	DE2=10 emigration	DE3=110 two partners	DE3=111 on-person
2008	7.5	0.1	5.4	2.0
2009	13.9	0.5	8.4	5.0
2010	13.5	1.1	5.0	7.4
Total	12.5	0.7	6.2	5.6

parent's home earlier and how the family and personal incomes affect the decisions of emancipation. Other interesting results concern the different effects of family and personal incomes over level, age and ways of emancipation.

4 Conclusions

This analysis measures the effects of economic factors for Young Dependent People in the New household formation in Spain within the period 2008–2011. By comparison between Young Unemployed and Employed dependents in Table 3 it is easy to see how unemployment has slowed down the increasing One-person New household formation and the increasing general emancipation rates in Spain. By comparison between Figs. 2 and 3 it is also easy to conclude that unemployment has delayed the age of emancipation. Finally emigration of Young Dependent People has increased from 0.2 to 1.4 making it a significant new way of emancipation. This way it shows how except for the current economic situation Spain tends to narrow the gap with the rest of European countries regarding Emancipation and New household formation. These results can be applied to a projected population and households' distribution to estimate Emigration and New household formation under certain circumstances that regular population projections do not consider. Usually population projections are conservative regarding juncture changes and do not anticipate some economic effects as higher youth emigration rates that this kind of models do. A final word on making estimates this way and taking households' distribution from population surveys is to bear in mind the systematic bias that underestimates the number of One-person and Two Partners households that some of these surveys suffer.

Acknowledgements This is a paper of a complete research presented as a final work of the Master in Economy which conducts to the Ph.d. in Economy at the National University of Distance Education of Spain. I would like to express my most sincere gratitude to Mr. Jose María Labeaga Azcona, for his direction as my academic tutor. I also deeply thanks Mr. Alberto A. Alvarez and Mr. Alberto A. Pinto for the opportunity to present my work at the MPE 2013 congress.

References

1. Alho, J., Keilman, N.: On future household structure. J. R. Stat. Soc. A. Stat. Soc. **173**(1), 117–143 (2010)
2. López-Villanueva, C., Pujadas, I.: Transformaciones sociodemográficas y territoriales de los hogares unipersonales en españa. Boletín De La AGE **55**, 153–182 (2011)
3. Rodríguez, J., Fellinger, E., Domínguez, J.: Hogares en españa (2008)
4. Van Imhoff, E., Keilman, N.: LIPRO 2.0: An application of a dynamic demographic projection model to household structure in the Netherlands Swets and Zeitlinger Amsterdam (1991)
5. Vilá, A.A., Villanueva, C.L.: Familias, hogares y viviendas en las regiones metropolitanas. el caso de barcelona. Cadernos Metrópole.ISSN (Impresso) 1517–2422; (Eletrónico) 2236–9996, (17) (2007)

An Adaptive Approach for Skin Lesion Segmentation in Dermoscopy Images Using a Multiscale Local Normalization

Jorge Pereira, Ana Mendes, Conceição Nogueira, Diogo Baptista, and Rui Fonseca-Pinto

Abstract Skin cancer is one of the most common malignancies in humans. Early detection of suspicious skin signs is critical to prevent this kind of malignancy, and various disciplines can play a crucial role in its detection. The lesion border is especially relevant for diagnosis, and provides information on the shape of the lesion, growth path, and growth rate. Digital image processing methods can be used to perform automatic lesion border detection; nonetheless, the presence of artifacts may induce artificial borders, thereby jeopardizing the efficiency of automatic detection algorithms. Artifact removal is a necessary pre-processing step to improve the accuracy quality of the border identification.

In this work, we present a method to identify and remove artifacts in dermoscopic images. This pre-processing step enhances the output of the segmentation of the lesion. This process is based on several applications of the Local Normalization, which is a method that increases the local contrast between local pixels, improving the overall quality of the image, especially with non-uniform illumination. The

J. Pereira (✉)
IT - Instituto de Telecomunicações, Leiria, Portugal
e-mail: jpereira@co.it.pt

A. Mendes
School of Technology and Management, Polytechnic Institute of Leiria, Leiria, Portugal
e-mail: aimendes@ipleiria.pt

C. Nogueira
School of Technology and Management, Polytechnic Institute of Leiria, Leiria, Portugal

CMAT, University of Minho, Braga, Portugal
e-mail: conceicao.veloso@ipleiria.pt

D. Baptista
School of Technology and Management, Polytechnic Institute of Leiria, Leiria, Portugal

CIMA, University of Évora, Évora, Portugal
e-mail: diogo.baptista@ipleiria.pt

R. Fonseca-Pinto
IT - Instituto de Telecomunicações, Leiria; School of Technology and Management, Polytechnic Institute of Leiria, Leiria, Portugal
e-mail: rui.pinto@ipleiria.pt

J.-P. Bourguignon et al. (eds.), *Dynamics, Games and Science*, CIM Series in Mathematical Sciences 1, DOI 10.1007/978-3-319-16118-1_29

process is scale sensitive and uses a multi-scale approach adaptable to every shape and size of skin lesions.

1 Introduction

Amongst all diseases and related research, cancer is still a major challenge for science. It is seen as a senseless part of the natural life cycle of humans, and constitutes a high cost issue for government budget.

Skin cancer is classified as a function of the cells from which it expands. Basal Cell Carcinoma (BCC) emerges from the lower layer of the epidermis, Squamous Cell Cancer (SCC) emerges from the middle layer of the epidermis and melanoma is derived from melanocytes, which are pigment producing cells. BCC is the most common form of skin cancer and the least dangerous. SCC is the second most common form of skin cancer, but most likely to spread. Approximately 65 % of all SCC and 36 % of all BCC arise in lesions that previously were diagnosed as actinic keratosis, also known as solar keratosis [5], are the most common pre-cancer, and the majority is derived from UV rays and consequent insulation.

Although melanoma type of cancer is the least common, it is also the most aggressive, most likely to spread, and quickly becomes fatal [13].

In the United States, skin cancer is the most common form of cancer, and over the past three decades more people have had skin cancer than all other cancers combined [17, 20]. Melanoma is also the most common form of cancer for young adults (25–29 years old) and the second most common form of cancer for young people (15–29 years old) [2].

Several studies in Europe have documented increases of melanoma incidence in the last few decades [1, 6, 11, 18, 21]. In the particular case of Portugal the estimated incidence for 2012 was 7.5 per 100,000, mortality 1.6 per 100,000 and prevalence at 1, 3 and 5 years 12.08, 33.99 and 53.93 % respectively [8].

Melanocytic lesion is a term used to describe a region of the skin that differs in color from the surrounding area. This difference in color (discoloration) is often a benign nevus found in great number over the entire body and regularly called age-spot. A relation between common acquired nevi and dysplastic nevi as precursors of cutaneous melanoma has been found [19, 23], thus a change in the melanocytic lesion characteristics constitutes a marker of warmness and should be investigated. Early detection and monitoring of suspicions lesions is crucial for the disease prognosis. Dermatologists use epiluminescence microscopy, dermatoscopy, or dermoscopy as is usually referred, to perform early diagnosis of melanocytic lesions and to track the progression thereof.

Dermoscopy uses a polarized light source and a magnifying lens allowing the identification of dozens of morphological features such as pigment networks, dots/globules, streaks, blue-white areas, and blotches [15]. A fluid is usually spread on the skin surface to minimize light scattering, and therefore increases the performance of this technique. The use of this fluid together with the presence of hairs in the skin surface, conducts to conspicuous artifacts in dermoscopic images. The classification of some melanocytic lesions is sometimes difficult, even for experienced specialists. The lesion border is especially relevant for diagnosis since it allows gathering information on the shape of the lesion, growth path, and growth rate. Lesion border detection algorithms applied to dermoscopic images have been widely used in recent works with dermoscopic images [3, 7, 9, 12, 14]. Early detection of suspected skin lesions requires periodic monitoring. Currently dermatologists often resort to digital dermoscopes and computer storage of the information. Computers can also be used to perform automatic lesion border detection. The presence of artifacts may induce artificial borders, thereby jeopardizing the efficiency of automatic detection algorithms. Artifact removal is a required pre-processing step to improve the quality of detection.

2 Artifact Removal and Border Detection

Dermoscopic images involve some artifacts directly related to this kind of images, i.e. hairs and air bubbles. The correct outline of lesion borders is critical for diagnosis, and the efficiency of automatic lesion border detection is hampered by artifacts.

2.1 Color Transformation and Rescaling

Dermoscopic images acquired by dermoscopes are true-color images with a typical resolution of 768 × 512 pixels. To implement the proposed methodology, for the segmentation, a transformation of the original RGB color space images into a gray-scale color space is performed. This color transformation is appropriate in this case since the reduction of data is important for the functioning of the algorithm and improves their accuracy. Upon the conventional ways of performing this transform, a process based on the weighted average of all three RGB channels is implemented. Each RBG channel, $P_{i,j,m}$ in the color space transform is defined in (1).

$$P_{i,j,m} = \frac{\left(p_{i,j,m}\right)^2}{\sqrt{\left(p_{i,j,R}\right)^2 + \left(p_{i,j,G}\right)^2 + \left(p_{i,j,B}\right)^2}} ; for \begin{array}{c} i = 1,2,\ldots,L \\ j = 1,2,\ldots,C \\ m = R,G,B \end{array} \tag{1}$$

where $p_{i,j,m}$ represents the pixel (i, j) of the m channel in the image, C is the number of columns in the RGB matrix and L is the number of rows. Once performed the weighted average above defined (1), a non-normalized image is obtained. This results in an irregular distribution of the pixel intensities which leads to the creation of a blind region. As it can be observed in Fig. 1b, there may be some

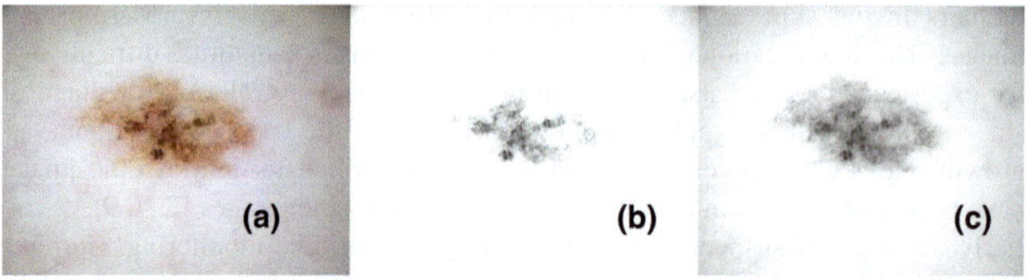

Fig. 1 Color transformation and rescaling. (**a**) Original true color *RGB* image. (**b**) Weighted average grayscale. (**c**) Grayscale final image

important details for the borders veiled in that blind region. In order to recover that information, while preserving the conversion, a rescaling process is applied to normalize the image. The final result of the transformation methodology, allows to compute a gray scale image of the skin lesion and to begin disposing of undesirable data, as shown in Fig. 1c.

2.2 Removing Brighter Artifacts

To increase the local contrast between neighbor pixels, a Local Normalization (*LN*) method is used to improve the overall quality of the image, especially with non-uniform illumination and shading artifacts. A similar method was used in [16] with X-ray lung images.

The *LN* is defined as

$$LN = \frac{I - \bar{I}}{\sqrt{\overline{(I^2)} - (\bar{I})^2}} \tag{2}$$

where I is the original image and $\bar{I}$ is the result between I and a Gaussian kernel. This kernel affords *LN* to be a scale sensitive method, as their results will depend on the chosen Gaussian sigma coefficient. Larger sigma allows to larger objects on the image to be enhanced over the smaller objects. When applying lower sigma in *LN* process, we can observe the opposite, with preferential enhancement of smaller objects. As the *LN* methodology is a sensitive method it can be used to solve the most common problems in segmentation of dermoscopic images, like specific air bubbles and hair artifacts. Air bubbles are usually smaller and brighter than other elements on the image, therefore it is possible to identify them with a low sigma value. With the contrast increment of these bubbles allied with some high thresholding values (the brighter spots are symptomatic of high intensity values), the removal of these areas is possible from the original data. In this process some few details may be lost, as shown in Fig. 2b. This absence occurs out of the

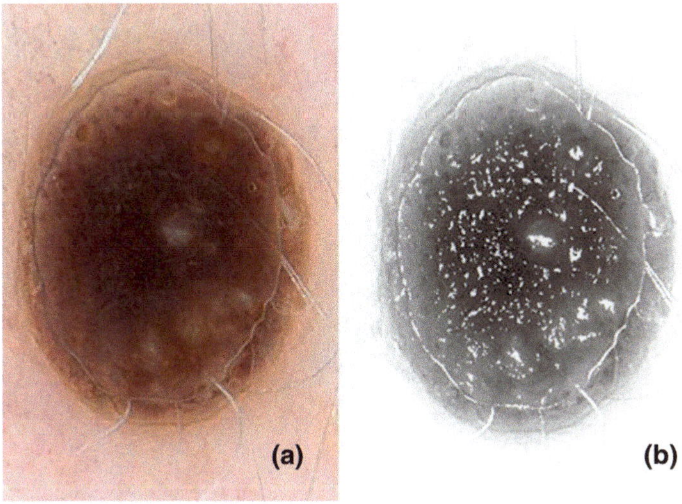

Fig. 2 Local Normalization transformation. (**a**) Real RGB image. (**b**) Image after *LN* and thresholding

border boundaries of the image, and they can be recovered later, after the proper segmentation.

2.3 *Region of Interest*

The notion of Region Of Interest (ROI) is very useful in several fields of image processing [4, 10, 22] as it allows to work on a very specific area of the input image. This process reduces the computational time, a large number of artifacts and non-productive areas. At the same time, it preserves all the characteristics of the target object, which in this case is the skin lesion.

When one has an image dataset with the same resolution, it is possible, by using *LN*, to compute the ROI for dermoscopy images. Visually, it creates valleys around the image elements, which allows a simple threshold to compute a region for each one of them. By assumption, it is assumed that the skin lesions are located somewhere on the central region of the image data. After thresholding every valley, only the central one is kept, representing our target object. The final process is given by morphological operations refinement in order to achieve a more suitable shape and distinct region. The result can be seen in Fig. 3.

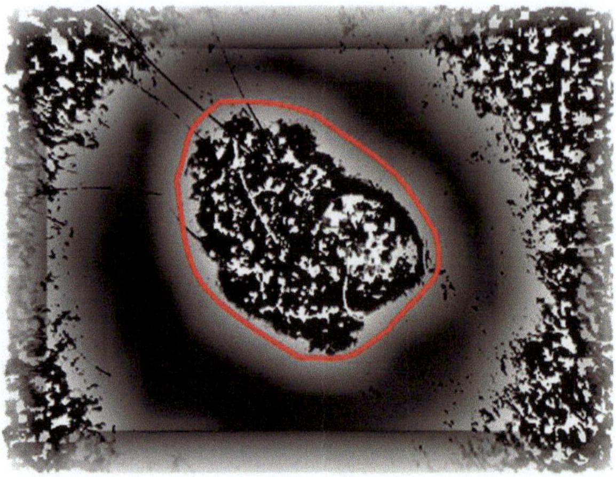

Fig. 3 Segmentation step using ROI

3 Multiscale Local Normalization (MLN)

3.1 Borders Candidates Detection and Local Normalization (LN)

As is known, skin lesions may have several shapes and sizes, so a single scale method cannot achieve high quality borders definition. If the lesion size is compatible with the *LN* scale, it should present good results, but if the lesion has a different size order, it may induce errors to the final detection. In order to avoid these issues and improve the accuracy and universality of this method, the detection of the borders candidates is performed with the application of a multi-scale approach of the *LN*. Based on the previous assumptions, it is assumed that the scale is directly proportional to the lesion. The larger the target, the higher the scale applied to define it. In order to acknowledge their size order, the area of the ROI is computed. It is expected that this area differs from image to image, depending directly on the lesion size and shape. In the end, the scale used will vary with the area of the ROI; applying small scales to small ROI, and larger scales to larger regions.

After performing the *LN*, in order to enhance the target main features, it is also applied a threshold depending on the maximum intensity of the resulting image, so it can adapts to every lesions and their image characteristics. This is a very useful approach, since not every lesion has the same intensity values (color and illumination features dependence), and consequently the same response level to the previous *LN* application. At this stage, the border definition is not completed yet, as it may contain some image artifacts corrupting the real borders (Fig. 4).

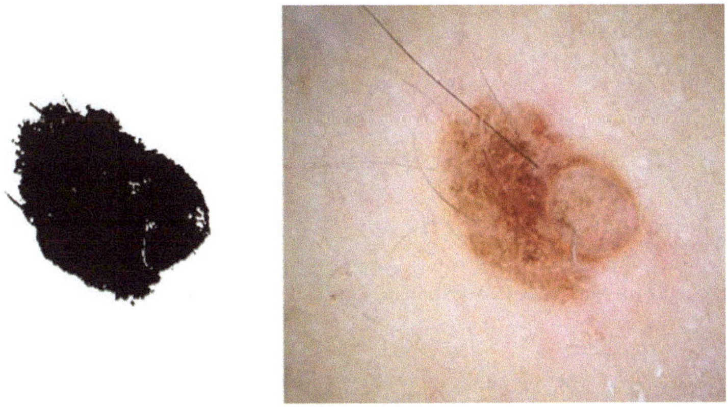

Fig. 4 Multiscale Local Normalization (MLN) before border smoothing

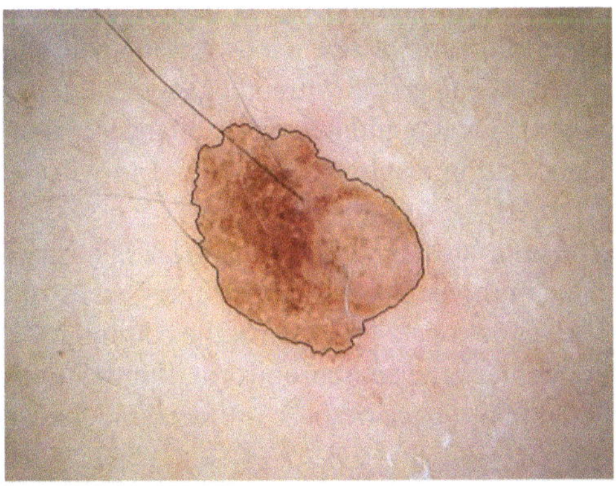

Fig. 5 Multiscale Local Normalization (MLN) final output

3.2 *Borders Morphological Refining*

The final step of the algorithm consists in refining the border candidates previously detected. It begins by applying an average filtering, to the border pixels, in order to smooth their lines and define a more natural and suitable border shape. This process also eliminates some noisy artifacts of these areas. This method is finalized, using morphological operations. Firstly to thin and find the skeleton of the smoothed lines, which will be the main structure of the borders, and secondly by deleting exterior branches, that should correspond mainly to cross-border hairs and other possible artifacts. In Fig. 5 the final segmentation is presented joint with the original image.

In Fig. 6, three more examples of the MLN algorithm performance are presented.

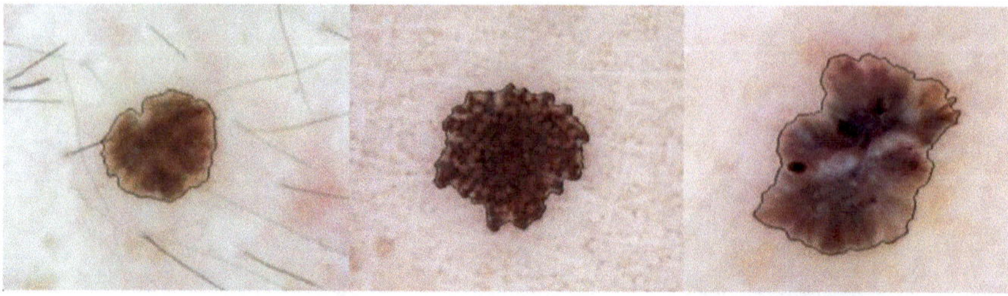

Fig. 6 Examples of Multiscale Local Normalization (MLN) from images with different morphologies

4 Conclusions and Further Work

The present work provides a very promising methodology to apply in the context of the segmentation as the follow-up of suspicious melanocytic skin lesions. Although it lacks some ground truth comparison with other segmentation techniques, the segmentation output presents great qualitative performance as shown on the presented examples. These segmentation results are very close to what is assumed to be the real lesion borders.

These qualitative results substantiate that the *MLN* is a very suitable method for this area, beyond their multitasking on almost every steps of the algorithm. The *MLN* also showed very good robustness against the common artifacts found in these images i.e. hairs, air bubbles or even non-uniform illumination, showing promising results with or without the presence of these artifacts in dermoscopic images.

Acknowledgements This work was partly supported by the CENTRO-07-ST24-FEDER-002022 / QREN.

The third author's research was supported by the Research Centre of Mathematics of the University of Minho with the Portuguese Funds from the "Fundação para a Ciência e a Tecnologia", through the Project PEstOE/MAT/UI0013/2014.

References

1. Baumert, J., Schmidt, M., Giehl, K.A., et al.: Time trends in tumour thickness vary in subgroups: analysis of 6475 patients by age, tumour site and melanoma subtype. Melanoma Res. **19**, 24–30 (2009)
2. Bleyer, A., O'Leary, M., Barr, R., Ries, L.A.G.: Cancer epidemiology in older adolescents and young adults 15 to 29 years of age, including SEER incidence and survival: 1975–2000. National Cancer Institute, Bethesda, MD (2006)
3. Celebi, M.E., Aslandogan, Y.A., Stoecker, W.V., Iyatomi, H., Oka, H., Chen, X.: Unsupervised border detection in dermoscopy images. Skin Res. Technol. **13**(4), 454–462 (2007)
4. Chen, M., Han, J., Hu, X., Jiang, X., Guo, L., Liu, T.: Survey of encoding and decoding of visual stimulus via FMRI: an image analysis perspective. Brain Imaging Behav. **8**(1), 7–23 (2014)

5. Criscione, V.D., Weinstock, M.A., Naylor, M.F., Luque, C., Eide, M.J., Bingham, S.F.: Actinic keratoses natural history and risk of malignant transformation in the veterans affairs tropical tretinoin chemoprevention trial. Cancer **115**, 2523–2530 (2009)
6. Downing, A., Newton-Bishop, J.A., Forman, D.: Recent trends in cutaneous malignant melanoma in the Yorkshire region of England; incidence, mortality and survival in relation to stage of disease, 1993–2003. Br. J. Cancer **95**, 91–95 (2006)
7. Erkol, B., Moss, R.H., Stanley, R.J., Stoecker, W.V., Hvatum, E.: Automatic lesion boundary detection in dermoscopy images using gradient vector flow snakes. Skin Res. Technol. **11**(1), 17–26 (2005)
8. Ferlay, J., Steliarova-Foucher, E., Lortet-Tieulent, J., Rosso, S., Coebergh, J.W.W., Comber, H., Forman, D., Bray, F.: Cancer incidence and mortality patterns in Europe: estimates for 40 countries in 2012. Eur. J. Cancer **49**(6), 1374–1403 (2013)
9. Fonseca-Pinto, R., Caseiro, P., Andrade, A.: Image empirical mode decomposition (IEMD) in dermoscopic images: artefact removal and lesion border detection. In: Proceedings of Signal Processing Pattern Recognition and Applications - IASTED International Conference, pp. 341–345 (2009)
10. Garcia-Alvarez, J.C., Diaz-Solarte, P.C.: Correlation analysis for quality assessment of Region-of-Interest-coded images. In: 8th Computing Colombian Conference (8CCC), 8th, pp. 1–6 (2013)
11. Holterhues, C., Vries, E., Louwman, M.W., et al.: Incidence and trends of cutaneous malignancies in the Netherlands, 1989–2005. J. Invest. Dermatol. **130**, 1807–1812 (2010)
12. Iyatomi, H., Oka, H., Saito, M., et al.: Quantitative assessment of tumor extraction from dermoscopy images and evaluation of computer-based extraction methods for automatic melanoma diagnostic system. Melanoma Res. **16**(2), 183–190 (2006)
13. Kasper, D.L., Braunwald, E., Fauci, A., et al.: Harrison's Principles of Internal Medicine, 16th edn. McGraw-Hill, New York (2005)
14. Melli, R., Grana, C., Cucchiara, R.: Comparison of color clustering algorithms for segmentation of dermatological images. In: Proceedings of the SPIE Medical Imaging Conference, vol. 6144, pp. 3S1–3S9 (2006)
15. Menzies, S.W., Crotty, K.A., Ingwar, C., McCarthy, W.H.: An Atlas of surface microscopy of pigmented skin lesions: dermoscopy. McGraw-Hill, Sydney, Australia (2003)
16. Ribeiro, R.: Lung nodule detection in chest radiographs. MSc Thesis, University of Porto (2008)
17. Rogers, H.W., Weinstock, M.A., Harris, A.R., et al.: Incidence estimate of nonmelanoma skin cancer in the United States 2006. Arch. Dermatol. **146**(3), 283–287 (2010)
18. Sant, M., Allemani, C., Santaquilani, M., et al.: EUROCARE-4 (2209), Survival of cancer patients diagnosed in 1995–1999. Results and commentary. Eur. J. Cancer **45**, 931–991 (2009)
19. Skender-Kalnenas, T., English, R., Heenan, P.: Benign melanocytic lesions: risk markers or precursors of cutaneous melanoma? J. Am. Acad. Dermatol. **33**(6), 1000–1007 (1995)
20. Stern, R.S.: Prevalence of a history of skin cancer in 2007: results of an incidence-based model. Arch. Dermatol. **146**(3), 279–282 (2010)
21. Tryggvadottir, L., Gislum, M., Hakulinen, T., et al.: Trends in the survival of patients diagnosed with malignant melanoma of the skin in the Nordic countries 1964–2003 followed up to the end of 2006. Acta. Oncol. **49**, 665–672 (2010)
22. Yang, J., Shi, Y.: Finger-vein ROI localization and vein ridge enhancement. Pattern Recogn. Lett. **33**(12), 1569–1579 (2012)
23. Whiteman, D.C., Pavan, W.J., Bastian, B.C.: The melanomas: a synthesis of epidemiologival clinical, histopathological, genetic, and biological aspects, supporting distinct subtypes, causal pathways, and cells of origin. Pigment Cell Melanoma Res. **24**(5), 879–897 (2011)

Chaotic Dynamics and Synchronization of von Bertalanffy's Growth Models

J. Leonel Rocha, Sandra M. Aleixo, and Acilina Caneco

Abstract This chapter concerns dynamics, bifurcations and synchronization properties of von Bertalanffy's functions, a new class of continuous one-dimensional maps, which was first studied in [22]. This family of unimodal maps is proportional to the right hand side of von Bertalanffy's growth equation. We provide sufficient conditions for the occurrence of stability, period doubling, chaos and non admissibility of von Bertalanffy's dynamics. These dynamics are dependent on the variation of the intrinsic growth rate of the individual weight, which is given by $r = r(K, W_\infty)$, where K is von Bertalanffy's growth rate constant and W_∞ is the asymptotic weight. A central point of our investigation is the study of bifurcations structure for this class of functions, on the two-dimensional parameter space (K, W_∞). Another important approach in this work is the study of synchronization phenomena of von Bertalanffy's models in some types of networks: paths, grids and lattices. We study the synchronization level when the local dynamics vary and the topology of the network is fixed. This variation is expressed by the Lyapunov exponents, as a function of the intrinsic growth rate r. Moreover, we present some results about the evolution of the network synchronizability, as the number of nodes increases, keeping fixed the local dynamics, in some types of networks: paths, grids and lattices. We also discuss the evolution of the network synchronizability as the number of edges increases. To support our results, we present numerical simulations for these types of networks.

1 Introduction and Motivation

A variety of growth curves have been developed to model general growth processes, since the study of population dynamics is one of the major research topics of the present time. Classical growth models such as logistic, Gompertz, Richards,

J.L. Rocha (✉) • S.M. Aleixo
Instituto Superior de Engenharia de Lisboa - ISEL, ADM and CEAUL,
Rua Conselheiro Emídio Navarro, 1, 1959-007 Lisboa, Portugal
e-mail: jrocha@adm.isel.pt; sandra.aleixo@adm.isel.pt

A. Caneco
CIMA-UE, Rua Romão Ramalho, 59, 7000-671 Évora, Portugal
e-mail: acilina@adm.isel.pt

J.-P. Bourguignon et al. (eds.), *Dynamics, Games and Science*, CIM Series in Mathematical Sciences 1, DOI 10.1007/978-3-319-16118-1_30

Blumberg and von Bertalanffy equations continue to be widely and frequently used to describe several demographic, economic, ecological, biological and medical processes. In particular, one of the most familiar growth equation used to describe the growth of marine populations, namely fishes, seabirds, marine mammals, invertebrates, reptiles and sea turtles is von Bertalanffy's equation, see for example [11] and references therein. This growth equation remains one of the most popular flexible growth equations to model fish weight growth, since it was presented by von Bertalanffy for this aim in 1938, see [28] and [29]. For a certain population, the growth of an individual, regarded as an increase in its length or weight with increasing age, is commonly modeled by a mathematical equation that represents the growth of an "average" individual in the population. One of the most important functions that have been used to analyze the increase in average length or weight of fish is von Bertalanffy's model, see for example [4] and [7].

On the other hand, an important feature of our world is the tendency of different systems to achieve common rhythms, namely, the tendency for synchronization. Synchronization is a fundamental nonlinear phenomenon, which can be observed in many real systems, in physics, chemistry, mechanics, engineering, secure communications or biology, see for example [1]. It can be observed in living beings, on the level of single cells, physiological subsystems, organisms and even on the level of large populations. Sometimes, this phenomenon is essential for a normal functioning of a system, e.g. for the performance of a pacemaker, where the synchronization of many cells produce a macroscopic rhythm that governs respiration and heart contraction. In other cases, the synchrony leads to a severe pathology, e.g. in case of the Parkinson's disease, when locking of many neurons leads to the tremor activity. Biological systems use internal circadian clocks to efficiently organize physiological and behavioral activity within the 24-h time domain. For some species, social cues can serve to synchronize biological rhythms. Social influences on circadian timing might function to tightly organize the social group, thereby decreasing the chances of predation and increasing the likelihood of mating, see for example [6]. Almost all seabirds breed in colonies; colonial and synchronized breeding is hypothesized to reduce predation risk and increases social interactions, thereby reducing the costs of breeding. Moreover, it is believed that synchronization may promote extinctions of some species. Full synchronism may have a deleterious effect on population survival because it may lead to the impossibility of a recolonization in case of a large global disturbance, see [26]. Understand the aggregate motions in the natural world, such as bird flocks, fish schools, animal herds, or bee swarms, for instance, would greatly help in achieving desired collective behaviors of artificial multi-agent systems, such as vehicles with distributed cooperative control rules.

Motivated by the interest and relevance of the study of growth models and the synchronization phenomenon, we propose to study in this work the chaotic dynamics and synchronization of von Bertalanffy's growth models. The layout of this paper is as follows. In Sect. 2, we present a new dynamical approach to von Bertalanffy's growth equation: a new class of one-dimensional discrete dynamical systems, a family of unimodal maps which was first studied in [22], designated by von Bertalanffy's functions. In Sect. 2.1 at Lemma 1, we provide sufficient conditions for the occurrence of stability, period doubling, chaos and non admissibility of von Bertalanffy's dynamics. These dynamics are dependent on the variation of the intrinsic growth rate of the individual weight r, for which the dynamics remains inside the $[0, 1]$ interval. The intrinsic growth rate is given by $r = r(K, W_\infty) = \frac{K}{3} \times W_\infty^{\frac{2}{3}} > 0$, where K is von Bertalanffy's growth rate constant and W_∞ is the asymptotic weight. Section 2.2 is devoted to the study of bifurcations structure for the von Bertalanffy functions, on the two-dimensional parameter space (K, W_∞). To support our results, we present fold and flip bifurcations curves and numerical simulations of the bifurcation diagram.

In Sect. 3, we study the synchronization and desynchronization phenomena of von Bertalanffy's models in some types of networks: paths, grids and lattices. In Sect. 3.1 are given preliminaries notions and results on graph and synchronization theories. The synchronization interval is presented in terms of the network connection topology, expressed by its Laplacian matrix and of the Lyapunov exponent of the network's nodes. In Sect. 3.2 at Proposition 1 we provide and discuss sufficient conditions for the decreasing of the amplitude of the network synchronization interval, for each type of networks considered. In Sect. 4, we give numerical simulations on some kinds of networks, evaluating its synchronization interval and amplitude of this interval, for several values of the intrinsic growth rates r. The networks considered have in each node the same dynamical system, defined by von Bertalanffy's functions. Finally, we discuss our results and provide some relevant conclusions: how the synchronization interval changes with increasing of the number of vertices in each type of networks and with increasing of Lyapunov exponent, when fixing the network topology. We also observe and discuss some desynchronization phenomena.

2 Chaotic Dynamics and Bifurcations in Von Bertalanffy's Growth Models

An usual form of von Bertalanffy's growth function, one of the most frequently used to describe chick growth in marine birds and in general marine growths, is given by

$$W_t = W_\infty \left(1 - e^{-\frac{K}{3}(t-t_0)}\right)^3, \tag{1}$$

where W_t is the weight at age t, W_∞ is the asymptotic weight, K is von Bertalanffy's growth rate constant and t_0 is the theoretical age the chick would have at weight zero, see [4] and [7]. On the other hand, the special case of the Bernoulli differential equation

$$g\left(W_t\right) = \frac{dW_t}{dt} = \frac{K}{3}\, W_t^{\frac{2}{3}} \left(1 - \left(\frac{W_t}{W_\infty}\right)^{\frac{1}{3}}\right), \tag{2}$$

it was introduced by von Bertalanffy to model fish weight growth, see [28] and [29]. The *per capita* growth rate, associated to this growth model, is given by

$$h(W_t) = \frac{g(W_t)}{W_t} = \frac{K}{3} W_t^{-\frac{1}{3}} \left(1 - \left(\frac{W_t}{W_\infty}\right)^{\frac{1}{3}}\right). \tag{3}$$

The following subsections are devoted to the detailed study of dynamics and bifurcations approaches of von Bertalanffy's growth models.

2.1 Chaotic Dynamics of Von Bertalanffy's Functions

In this contribution we consider a new class of one-dimensional discrete dynamical systems, a family of unimodal maps which was first studied in [22], designated by von Bertalanffy's functions, $f_r : [0, 1] \to [0, 1]$, defined by

$$f_r(x) = r\, x^{\frac{2}{3}} \left(1 - x^{\frac{1}{3}}\right), \tag{4}$$

with $x = \frac{W_t}{W_\infty} \in [0, 1]$ the normalized weight and $r = r(K, W_\infty) = \frac{K}{3} \times W_\infty^{\frac{2}{3}} > 0$ an intrinsic growth rate of the individual weight, see some examples at Fig. 1. This family of functions is proportional to the right hand side of von Bertalanffy's equation, Eq. (2). Remark that, the study which we present depends on two biological parameters: von Bertalanffy's growth rate constant K and the asymptotic weight W_∞. The following conditions are satisfied:

(A1) f_r is continuous on $[0, 1]$;
(A2) f_r has an unique critical point $c = (2/3)^3 \in]0, 1[$;
(A3) $f_r'(x) \neq 0, \forall x \in]0, 1[\setminus \{c\}, f_r'(c) = 0$ and $f_r''(c) < 0$;
(A4) $f_r \in C^3 (]0, 1[)$ and the Schwarzian derivative of f_r, given by

$$S(f_r(x)) = \frac{f_r'''(x)}{f_r'(x)} - \frac{3}{2}\left(\frac{f_r''(x)}{f_r'(x)}\right)^2,$$

verifies $S(f_r(x)) < 0, \forall x \in]0, 1[\setminus \{c\}$ and $S(f_r(c)) = -\infty$.

Conditions $(A1)$–$(A4)$ are essential to prove the stability of the only positive fixed point, [25]. The negative Schwarzian derivative ensures a "good" dynamic behavior of the models: continuity and monotonicity of topological entropy, order in the succession of bifurcations, the existence of an upper limit to the number of stable orbits and the non-existence of wandering intervals, [13] and [27]. See [24] for a topological dynamics approach of unimodal maps. The unimodal maps theory has proved to be useful in many branches of science. In population dynamics, aiming

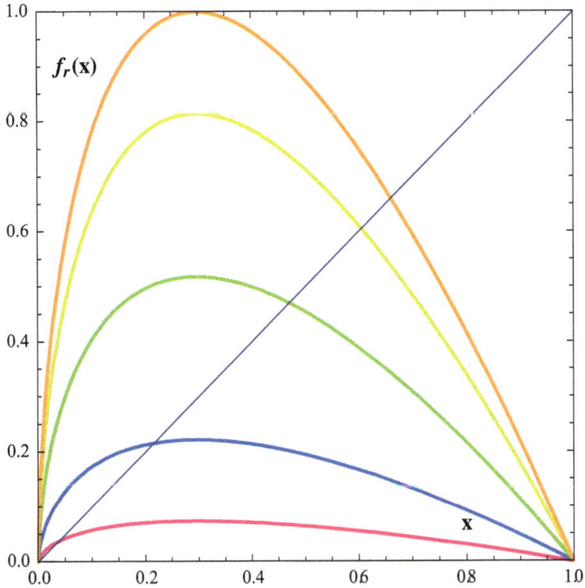

Fig. 1 Graphics of von Bertalanffy's functions $f_r(x)$, Eq. (4), for several values of intrinsic growth rate r (0.5 (*magenta*), 1.5, 3.5, 5.5 and 6.75 (*orange*))

to model the growth of a certain species, the use of these families has been frequent. A similar approach is used for example in [19, 20] and [21].

Von Bertalanffy's functions have two fixed points, given by

$$A_0 = 0 \text{ and } A_{K,W_\infty} \equiv A_r = \left(\frac{r}{r+1}\right)^3.$$

For these models, the extinction region and the semistability curve have no expressive meaning. We verify that, $\lim_{x\to 0^+} f_r'(x) > 1$. So, the fixed point $A_0 = 0$ is unstable and the origin's basin of attraction is empty, except at most a set of measure zero. For this reason it is difficult to identify for this models *per capita* growth rates, Eq. (3), less than one for all densities, to the extinction case, and *per capita* growth rates strictly less than one for all densities, except at one population density, to the semistability case, except at most a set of measure zero.

In the next result we provide sufficient conditions dependent on the variation of the intrinsic growth rate r, for which the dynamics remains inside the unit interval.

Lemma 1 *Let $f_r(x)$ be von Bertalanffy's functions, given by Eq. (4), with $r \in \mathbb{R}^+$ and satisfying* $(A1) - (A4)$ *conditions.*

(*i*) *(Stability region of the fixed point A_r) If $0 < r < \frac{5}{3}$, then there is a linearly stable fixed point $A_r \in]0,1[$ whose basin of attraction is $]0,1[$;*

(ii) *(Period doubling and chaotic regions) If* $\frac{5}{3} < r < \frac{3^3}{2^2}$, *then the interval* $[f_r^2(c), f_r(c)]$ *is forward invariant with basin of attraction* $]0, 1[$;

(iii) *(Chaotic semistability curve) If* $r = \frac{3^3}{2^2}$, *then* $[0, 1]$ *is invariant and verifies that*

$$\bigcup_{n \geq 0} f_r^n(x) = [0, 1] \text{ and } \lim_{n \to \infty} \frac{1}{n} |Df_r^n(x)| > 0,$$

for Lebesgue almost every $x \in [0, 1]$.

Proof Consider that the fixed point A_r is given by $A_r = \left(\frac{r}{r+1}\right)^3$. If $0 < r < \frac{5}{3}$, then $|f_r'(A_r)| < 1$. Therefore, the fixed point A_r is linearly stable. By Modified Singer's Theorem, see [27], the fixed point A_r is the only linearly stable fixed point in $]0, 1[$ and the immediate basin of A_r includes the orbit of the critical point c. So, the interval $[c, f_r(c)]$ is contained in the immediate basin of A_r. As the point A_r is the only fixed point in $]0, 1[$, this implies that $f_r(x) > x$ on $]0, A_r[$. Thus, the interval $]0, f_r(c)]$ is also contained in the basin of attraction of the fixed point A_r. Considering that the von Bertalanffy functions f_r map the interval $[f_r(c), 1[$ into $]0, f_r^2(c)]$ and $]0, f_r^2(c)] \subset]0, f_r(c)[$, then the interval $]0, 1[$ is the basin of attraction of the fixed point A_r.

In the second case, if $\frac{5}{3} < r < \frac{3^3}{2^2}$, then the fixed point A_r is not linearly stable. In this case, it is verified that $f_r(x) > x$ for $x \in]0, c]$ and f_r has no fixed point at $]0, c]$. This implies that all the orbits of every points $x \in]0, 1[$ enters on the interval $[f_r^2(c), f_r(c)]$, after a finite time of iterations. As $f_r'(x) < 0$ for $x \in]c, 1]$, then f_r maps the interval $[c, f_r(c)]$ into $[f_r^2(c), f_r(c)]$. On the other hand, considering that $f_r(x) > x$ for $x \in]0, c]$, then f_r maps the interval $[f_r^2(c), c]$ into $[f_r^2(c), f_r(c)]$. Therefore,

$$f_r\left([f_r^2(c), f_r(c)]\right) \subseteq [f_r^2(c), f_r(c)],$$

i.e., the interval $[f_r^2(c), f_r(c)]$ is forward invariant with basin of attraction $]0, 1[$.

Finally, if $r = \frac{3^3}{2^2}$, or an equivalent way $f_r(c) = 1$, then it appears that the maximum size growth of the population is equal to the critical density at $r = \frac{3^3}{2^2}$. Clearly, the fixed point A_r is linearly unstable. Since it is verified that $f_r'(x) > 0$ for $x \in]0, c[$, then the von Bertalanffy functions f_r map $]0, c]$ into $]0, f_r(c)]$. Also, since $f_r'(x) < 0$ for $x \in]c, f_r(c)]$ and $f_r(c) = 1$, then f_r maps the interval $[c, f_r(c)]$ into $[0, 1]$. So, $[0, 1]$ is invariant, which is called invariant absorbing segment of level one, see [15]. To show that this interval admits complex dynamics it suffices to check the conditions for which the von Bertalanffy functions f_r on $[0, 1]$ admit an ergodic absolutely continuous invariant measure, see the results presented in [16]. In fact, the von Bertalanffy functions satisfy $(A1) - (A4)$ conditions. Also, it is verified that $f_r^2(c) = 0$ and $f_r(0) = 0$, then it follows that $f_r^n(c) \neq c$, $\forall n > 2$. Considering that,

$$\lim_{x \to 0+} f_r'(x) > 1 \text{ and } f_r^2(c) = 0,$$

the Modified Singer Theorem implies that the von Bertalanffy functions f_r on $[0, 1]$ have no attracting periodic points. Therefore, from the theorem presented by Misiurewicz in [16] and Birkhoff's Ergodic Theorem follow the properties of *(iii)*. Thus, the results are proved. □

2.2 Bifurcations of Von Bertalanffy's Functions

In this section we investigate the dynamical complexity of the proposed models at (K, W_∞) parameter plane. The analysis of their bifurcations structure is done based on the bifurcation diagram, see Fig. 2. We will make use of the fold and flip bifurcations, related with some cycles of order $n \in \mathbb{N}$. We recall that an order n cycle $(x_1, x_2, \ldots, x_n)$ is stable (or attractive) iff $\left|\frac{\partial f_r^n}{\partial x}(x_j)\right| < 1,\ \forall j = 1, 2, \ldots, n$.

The fold bifurcation corresponds to the appearance of two order n cycles, one stable and the other unstable, when it is verified $\frac{\partial f_r^n}{\partial x}(x_j) = 1,\ \forall j = 1, 2, \ldots, n$. On the other hand, the flip bifurcation corresponds to the change of stability of an order n cycle and the appearance of an order $2n$ cycle. Before the bifurcation, the order n cycle is stable, after the bifurcation, the order n cycle is unstable and the $2n$ cycle is stable. At the bifurcation it is verified that, $\frac{\partial f_r^n}{\partial x}(x_j) = -1,\ \forall j = 1, 2, \ldots, n$. For more details about bifurcation theory see for example [14] and [15].

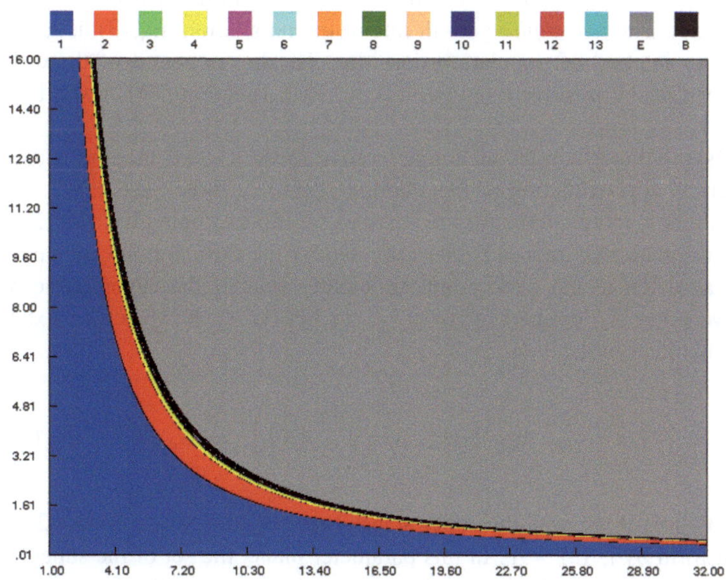

Fig. 2 Bifurcation diagram of von Bertalanffy's functions $f_r(x)$ in the (K, W_∞) parameter plane. The *blue* region is the stability region. The period doubling and chaotic regions correspond to the cycles shown on *top of figure*. The *gray* region is the non admissible region

In a general way, to von Bertalanffy's functions $f_r(x)$, defined by Eq. (4), with $r = \frac{K}{3} \times W_\infty^{\frac{2}{3}} \in \mathbb{R}^+$, the fold and flip bifurcation curves relative to a cycle of order n are determined as follows. If $x \in [0, 1[$ is a point of an order n cycle that satisfies the equations

$$f_r^n(x) = x \text{ and } \frac{\partial f_r^n}{\partial x}(x) = 1 \tag{5}$$

then there exists a solution φ_n, such that the fold bifurcation curves relative to a cycle of order $n \in \mathbb{N}$ are given by $W_\infty = \varphi_n(x; K)$, and are denoted by $\Lambda_{(n)_0}$. On the other hand, if $x \in [0, 1[$ is such that,

$$f_r^n(x) = x \text{ and } \frac{\partial f_r^n}{\partial x}(x) = -1 \tag{6}$$

then exists a solution ψ_n, such that the flip bifurcation curves relative to a cycle of order $n \in \mathbb{N}$ are given by $W_\infty = \psi_n(x; K)$, and are denoted by Λ_n.

In particular, to von Bertalanffy's functions $f_r(x)$, defined by Eq. (4), the fold bifurcation curve of the fixed points A_0 and A_r, corresponding to Eq. (5) for $n = 1$, has no meaning at (K, W_∞) parameter plane. Because for von Bertalanffy's growth models does not exist an extinction region. Note that the fold bifurcation curve $\Lambda_{(1)_0}$ is the bifurcation curve which defines the transition between the extinction region and the stability region, see for example [19] and [23]. A behavior of stability is defined when a population persists for intermediate initial densities and otherwise goes extinct. The *per capita* growth rate of the population, Eq. (3), is greater than one for an interval of population densities. The lower bound of these densities correspond to the positive fixed point A_r of each function $f_r(x)$, given by Eq. (4), see Fig. 1.

The symbolic dynamics techniques prove to be a good method to determine a numerical approximation to the stability region (in blue), see Fig. 2. For more details about symbolic dynamics techniques see for example [20]. In the (K, W_∞) parameter plane, this region is characterized by the critical point iterates that are always attracted to the fixed point sufficiently near of the super stable or super attractive point $\tilde{A}_r$, defined by $f_r(c) = c$. Let $\bar{A}_r \in]0, 1[$ be the fixed points sufficiently near of $\tilde{A}_r$, then

$$\lim_{n\to\infty} f_r^n(c) = \bar{A}_r, \text{ for } \left(3K^{-1}A_r^{\frac{1}{3}}\left(1 - A_r^{\frac{1}{3}}\right)\right)^{\frac{3}{2}} < W_\infty(K) < \hat{W}_\infty(K)$$

where $\hat{W}_\infty(K)$ represents the super stable curve of the cycle of order 2, given in implicit form by $f_r^2(c) = c$. In this parameter plane, the set of the super stable or super attractive points $\tilde{A}_r$ defines the super stable curve of the fixed point. In the region before reaching the super stable curve, the symbolic sequences associated to the critical points orbits are of the type CL^{∞}. After this super stable curve, the symbolic sequences are of the type CR^{∞}.

On the other hand, the flip bifurcation curve Λ_1 correspondent to Eq. (6), with $r = \frac{K}{3} \times W_{\infty}^{\frac{2}{3}} > 0$, for $n = 1$, i.e., the flip bifurcation curve of the nonzero stable fixed point A_r, is given by

$$x = \left(\frac{r}{r+1}\right)^3 \quad \text{and} \quad \psi_1(x;K) = \left(\frac{5}{K}\right)^{\frac{3}{2}}. \tag{7}$$

Note that the flip bifurcation curve Λ_1 is the bifurcation curve which defines the transition between the stability region and the period doubling region, such as established in Lemma 1 for $r = \frac{5}{3}$.

The period doubling region corresponds to the parameters values, to which the population weight oscillates asymptotically between 2^n states, with $n \in \mathbb{N}$. In period-doubling cascade, the symbolic sequences correspondent to the iterates of the critical points are determined by the iterations $f_r^{2^n}(c) = c$. Analytically, these equations define the super-stability curves of the cycle of order 2^n. The period doubling region is bounded below by the curve of the intrinsic growth rate values where the period doubling starts, $\hat{W}_{\infty}(K)$, correspondent to the 2-period symbolic sequences $(CR)^{\infty}$. Usually, the upper bound of this region is determined using values of intrinsic growth rate r, corresponding to the first symbolic sequence with non null topological entropy. Commonly, the symbolic sequence that identifies the beginning of chaos is $\left(CRLR^3\right)^{\infty}$, a 6-periodic orbit, see for example [20] and [21].

On the (K, W_{∞}) parameter plane, at Fig. 2, the region between the blue and the gray regions corresponds to period doubling region and chaotic region, also stated in Lemma 1 (*ii*). The period doubling region is bounded below by the flip bifurcation curve of the stable fixed point nonzero A_r, Λ_1. The upper limit of this region is defined by the accumulation value of the flip bifurcation curves of the cycle of order 2^n, of the stable fixed points nonzero, see [14] and [15]. This bifurcation curve is denoted by Λ_{∞} and from Eq. (7) we have,

$$\Lambda_{\infty} = \lim_{n\to\infty} \psi_{2^n}(x;K)$$

with $x \in [0,1[$ a fixed point. In Fig. 2 the period doubling regions are well evidenced, highlighting in particular the cycles of order 2 and 4.

In the chaotic region of the (K, W_{∞}) parameter plane, the evolution of the population size is *a priori* unpredictable. The maps are continuous on the interval with positive topological entropy whence they are chaotic and the Sharkovsky ordering is verified, see [24]. At this case, the populations can persist at a semistable chaotic interval. The symbolic dynamics are characterized by iterates of the functions f_r that originate orbits of several types, which already present chaotic patterns of behavior. The topological entropy is a non-decreasing function in order to the parameter r, until reaches the maximum value ln 2 (consequence of the negative Schwartzian derivative). In [20] and [21] can be seen a topological order with several symbolic sequences and their topological entropies, which confirm this result to others growth models.

The chaotic region is upper bounded by the chaotic semistability curve, as stated in Lemma 1 (*iii*). This bifurcation curve is denoted by Λ_{NA} and is given by

$$\Lambda_{NA} = \left\{(K, W_\infty) \in \mathbb{R}^2 : f_r(c) = 1\right\}$$

$$= \left\{(K, W_\infty) \in \mathbb{R}^2 : W_\infty = \zeta(K), \text{ with } \zeta(K) = \left(\tfrac{3^4}{4K}\right)^{\frac{3}{2}}\right\}.$$

The chaotic semistability curve corresponds to the transition between chaotic region and no admissible region. In Fig. 2, the gray region is the no admissible region. At this region the graphic of any von Bertalanffy's function is no longer totally in the invariant set $[0, 1]$. Almost all trajectories of f_r (besides a hyperbolic set of zero measure) leave the interval $[0, 1]$ and escape to infinity. The maps under these conditions are not good models for populations dynamics. For more details about the bifurcation structure on this type of growth models see for example [19] and [23].

3 Synchronization and Desynchronization of Von Bertalanffy's Models

The synchronization of coupled chaotic systems depends on several factors, including the strength of the coupling, reflected in the value of the coupling parameter, the network topology and the dynamic characteristics of the system that exists at each vertex. Given that a network can be mathematically represented by a graph, the theory of dynamic networks is a combination of graph theory and nonlinear dynamics. One might think on the behavior of the network according to the local dynamics at each node, assuming that the network structure is fixed, or to admit that the network has a dynamic topology that evolves according to certain rules, but the state of the nodes are fixed. Thus, the emphasis of the study may be placed on the local dynamic, on the global dynamics or, which is most interesting, on a combination of both.

The dynamics in the nodes is determined by the intrinsic growth rate parameter r of von Bertalanffy's model, which influences the associated Lyapunov exponent. So, in this work we study the synchronizability when the local Lyapunov exponent vary and the topology of the network is fixed. Moreover, are present some results about the evolution of the network synchronizability, when the number of nodes or the number of edges increase, keeping fixed the local dynamic.

3.1 Preliminaries

Mathematically, networks are described by graphs, directed and undirected, and the theory of dynamical networks is a combination of graph theory and nonlinear dynamics. A graph G is a set $G = (V, E)$ where $V = V(G)$ is a nonempty set of N vertices or nodes (N is called the order of the graph) and $E = E(G) \subseteq V(G) \times V(G)$ is the set of m pairs of vertices that are called edges or links e_{ij} that connect two vertices v_i and v_j. The matrix $A = A(G) = \left[a_{ij}\right]$, is called the adjacency matrix. For a non weighted graph, it carries an entry 1 at the intersection of the *ith* row and the *jth* column if there is an edge from v_i to v_j, where v_i, $v_j \in V(G)$. When there is no edge, the entry will be 0. If the graph is not directed, $a_{ij} = a_{ji}$ and the matrix $A(G)$ is symmetric. The degree of a node v_i, represented by k_i, is the number of edges incident on it, i.e., $k_i = \sum_{j=1, j\neq i}^{N} a_{ij}$. Considering the diagonal matrix $D = D(G) = \left[d_{ij}\right]$, where $d_{ii} = k_i$, then $L = D - A$ is called the Laplacian matrix. The eigenvalues of L are all non negative reals and are contained in the interval $[0, \min\{N, 2\Delta\}]$, where Δ is the maximum degree of the vertices. The spectrum of L may be ordered, $\lambda_1 = 0 \leqslant \lambda_2 \leqslant \cdots \leqslant \lambda_N = \lambda_{max}$. The second eigenvalue λ_2 is know as the algebraic connectivity or Fiedler value and plays a special role in the graph theory. We will denote by $\lambda_2(G)$ the lower non zero eigenvalue of the Laplacian of graph G. The larger $\lambda_2(G)$ is, the more difficult it is to separate the graph G in disconnected parts. The graph is connected if and only if $\lambda_2 \neq 0$. In fact, the multiplicity of the null eigenvalue λ_1 is equal to the number of connected components of the graph. As we will see later, the bigger λ_2, the easier the network synchronizes.

Consider a network of N identical chaotic dynamical oscillators, described by a connected graph, with no loops and no multiple edges. In each node the dynamics of the oscillators is defined by $\dot{x}_i = f(x_i)$, with $f : \mathbb{R}^n \to \mathbb{R}^n$ and $x_i \in \mathbb{R}^n$ the state variables of the node i. The state equations of this network, in the discretized form, are given by

$$x_i(k+1) = f(x_i(k)) + c \sum_{j=1}^{N} l_{ij}\, x_j(k), \tag{8}$$

with $i = 1, 2, \ldots, N$, where $c > 0$ is the coupling parameter, $A = \left[a_{ij}\right]$ is the adjacency matrix and $L = \left(l_{ij}\right) = D - A$ is the Laplacian matrix or coupling configuration of the network. The network given by Eq. (8) achieves asymptotic synchronization if $x_1(t) = x_2(t) = \ldots = x_N(t) \to e(t)$ as $t \to \infty$, where $e(t)$ is a solution of an isolate node (equilibrium point, periodic orbit or chaotic attractor), satisfying $\dot{e}(t) = f(e(t))$.

In a chaotic system it is important to measure the sensitivity with respect to initial conditions. One way to do that is to compute the Lyapunov exponents that measure the average rate at which nearby trajectories diverge from each other. Consider the trajectories x_k and y_k, starting from x_0 and y_0, respectively. If both trajectories are, until time k, always in the same linear region, we can write

$$|x_k - y_k| = e^{\gamma k}|x_0 - y_0|, \text{ where } \gamma = \frac{1}{k}\sum_{j=0}^{k-1} \ln |f_r'(x_j)|.$$

The Lyapunov exponent of a trajectory x_k is defined by

$$\mu = \lim_{k\to+\infty} \frac{1}{k}\sum_{j=0}^{k-1} \ln |f_r'(x_j)|, \tag{9}$$

whenever it exists. The computation of the Lyapunov exponent μ gives the average rate of divergence (if $\mu > 0$), or convergence (if $\mu < 0$) of the two trajectories from each other, during the time interval $[0, k]$, see for example [9]. In particular, for von Bertalanffy's functions, the Lyapunov exponents depend on one biological parameters: the intrinsic growth rate r. In Fig. 3 one can observe the behaviour of the Lyapunov exponent estimate when the intrinsic growth rate increases.

It is known that the network given by Eq. (8), with identical chaotic nodes, is synchronized if the coupling parameter c belongs to the synchronization interval

$$\frac{1-e^{-\mu}}{\lambda_2} < c < \frac{1+e^{-\mu}}{\lambda_{max}} \tag{10}$$

where λ_2 and λ_{max} are, respectively, the smaller non zero and the larger eigenvalues of the Laplacian matrix L and μ is the Lyapunov exponent of each individual n-dimensional node, see [12].

Note that the synchronization occurs for values of the coupling parameter c such that $\frac{1+e^{-\mu}}{\lambda_{max}} > \frac{1-e^{-\mu}}{\lambda_2}$, which implies that it is a necessary condition for

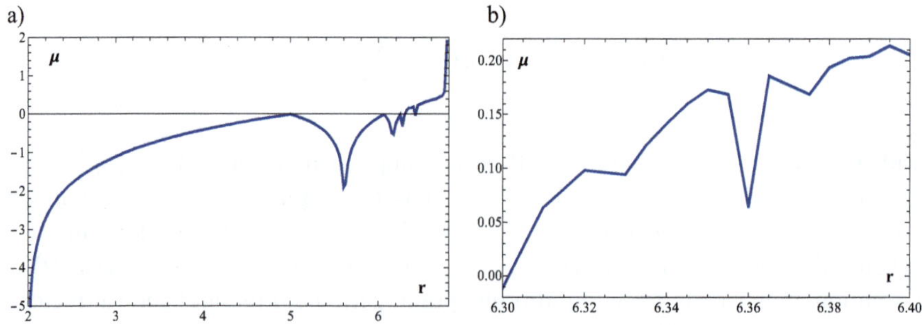

Fig. 3 Lyapunov exponents for von Bertalanffy's functions Eq. (4), as a function of the intrinsic growth rate r. This figure has been obtained by numerical simulations using 5,000 iterations. The zoom in (**b**) show the values of the Lyapunov exponents used in Tables 1 and 2 . (**a**) Lyapunov exponents for von Bertalanffy's functions. (**b**) Zoom of (**a**) with $r \in [6.3, 6.4]$

synchronization that

$$\mu < \ln(2K+1), \text{ where } K = \frac{\lambda_1 - \lambda_2}{\lambda_2 - \lambda_{max}}, \tag{11}$$

see [12] for more details.

Considering Eq. (10), the amplitude of the network synchronization interval is given by

$$\alpha = \frac{(\lambda_2 - \lambda_{max}) + (\lambda_2 + \lambda_{max})e^{-\mu}}{\lambda_2\, \lambda_{max}},$$

where λ_2 and λ_{max} are, respectively, the smaller non zero and the larger eigenvalues of the Laplacian matrix. From Eq. (10) we conclude that fixing the dynamics f_r in the nodes, the amplitude of the synchronization interval will be as larger as much bigger is the eigenratio R, which is defined by

$$R = \frac{\lambda_2}{\lambda_{max}}. \tag{12}$$

3.2 Synchronization of Paths, Grids and Lattices

Since we are studying the phenomenon of synchronization in a population model, it makes sense to consider the influence in each individual of its closest neighbours. So, in our work we will pay special attention to some particular types of networks: paths, grids and lattices.

A path graph P_N is a sequence of N vertices such that from each of its vertices there is an edge to the next vertex in the sequence, see [2]. The eigenvalues of the Laplacian matrix of the path P_N are given by

$$\lambda_i\,(P_N) = 2 - 2\cos\frac{\pi\,(i-1)}{N} = 4\sin^2\frac{\pi\,(i-1)}{2N}, \tag{13}$$

with $i = 1, \ldots, N$, see for example [3] and [17].

Let G be a graph with $|V(G)| = N$ vertices and H a graph with $|V(H)| = M$ vertices. The cartesian product of graphs G and H, denoted by $G\Box H$, is the graph with vertex set $V(G\Box H) = V(G) \times V(H)$ where there is an edge between two vertices (u_1, u_2) and (v_1, v_2) of the cartesian product if and only if $u_1 = v_1$ and $u_2v_2 \in E(H)$ or $u_2 = v_2$ and $u_1v_1 \in E(G)$. The cartesian product of two paths P_N and P_M is a $N \times M$ grid graph, which is denoted by $G_{N\times M} = P_N\Box P_M$. It is known that the Laplacian eigenvalues of cartesian product $G\Box H$, see [18], are

$$\lambda_i(G) + \lambda_j(H), \text{ with } i = 1, 2, \ldots, |V(G)| \text{ and } j = 1, 2 \ldots, |V(H)|.$$

As a consequence, one has

$$\lambda_2 (G\Box H) = min \{\lambda_2 (G), \lambda_2 (H)\},$$

and

$$\lambda_{max} (G\Box H) = \lambda_{max} (G) + \lambda_{max} (H).$$

This result allows to determine the Laplacian spectrum of a $N \times M$ grid graph. In the particular case of $N = M$, one has

$$\lambda_2 (G_{N\times N}) = min \{\lambda_2 (P_N), \lambda_2 (P_N)\} = \lambda_2 (P_N) = 4\sin^2 \frac{\pi}{2N},$$

and

$$\lambda_{max} (G_{N\times N}) = \lambda_{max} (P_N) + \lambda_{max} (P_N) = 2\lambda_{max} (P_N) = 8\sin^2 \frac{\pi(N-1)}{2N}. \tag{14}$$

In the case of $N < M$, considering that $\lambda_2 (P_N)$ given by Eq. (13) is a decreasing function with N, it follows

$$\lambda_2 (G_{N\times M}) = min \{\lambda_2 (P_N), \lambda_2 (P_M)\} = \lambda_2 (P_M) = 4\sin^2 \frac{\pi}{2M},$$

and

$$\begin{aligned} \lambda_{max} (G_{N\times M}) &= \lambda_{max} (P_N) + \lambda_{max} (P_M) \\ &= 4\ \sin^2 \frac{\pi(N-1)}{2N} + 4\sin^2 \frac{\pi(M-1)}{2M}. \end{aligned}$$

The case $N > M$ is similar.

A (N, k)-lattice graph, denoted by $L_{(N,k)}$, is a $2k$-regular graph in which the N vertices are put in a circle and each vertex is connected to its $2k$ nearest neighbours. The eigenvalues of the Laplacian matrix of a (N, k)-lattice are $\upsilon_1 = 0$ and

$$\upsilon_{i+1} = 2k - 2\sum_{n=1}^{K} \cos \frac{2\pi in}{N} = 4\sum_{n=1}^{K} \sin^2 \frac{\pi in}{N} = 2k + 1 - \frac{\sin \frac{\pi i(2k+1)}{N}}{\sin \frac{\pi i}{N}}, \tag{15}$$

with $i = 1, 2, \ldots N-1$. For more details see for example [5] and [10]. Note that these eigenvalues are not sorted in an ascending order. In fact, $\upsilon_1 = \lambda_1 = 0$, $\upsilon_2 = \lambda_2$, but $\lambda_{max} = \max_{1\leq i\leq N} \upsilon_i$. The larger eigenvalue of the Laplacian matrix L is obtained from Eq. (15), with $i = \frac{N+b}{4}$, where b is the smaller integer such that $N + b$ is a multiple of 4, with $b = 0, 1, 2, 3$. Then, denoting the constant $2k + 1 = q$, we can write

$$\lambda_{max} = \upsilon_{\frac{N+b}{4}+1} = 2k + 1 - \frac{\sin \frac{(2k+1)(N+b)\pi}{4N}}{\sin \frac{(N+b)\pi}{4N}} = q - \frac{\sin \frac{q(N+b)\pi}{4N}}{\sin \frac{(N+b)\pi}{4N}}. \tag{16}$$

On the other hand, from Eq. (15), for $i = 1$, we also have

$$\lambda_2 = \upsilon_2 = 2k + 1 - \frac{\sin \frac{\pi(2k+1)}{N}}{\sin \left(\frac{\pi}{N}\right)} = q - \frac{\sin \frac{q\pi}{N}}{\sin \frac{\pi}{N}}. \tag{17}$$

With the above expressions for the eigenvalues of the three kinds of graphs, we can obtain results about the eigenratio, R, and therefore about the amplitude, α, of the synchronization interval.

Proposition 1 *Let G_1 and G_2 be two graphs of the same type and α_i be the amplitude of synchronization interval of the network associated with G_i, with $i = 1, 2$. Then $\alpha_2 \leq \alpha_1$ in the following conditions:*

(i) If the two networks are path graphs P_{N_i}, with N_i vertices, $i = 1, 2$ and $N_2 \geq N_1$;
(ii) If the two networks are grid graphs $G_{N_i \times M_i} = P_{N_i} \square P_{M_i}$, with N_i and M_i vertices of each path P_{N_i} and P_{M_i}, respectively, $i = 1, 2$ and

  (1) $N_i = M_i$ for $i = 1, 2$, or
  (2) $N_i \neq M_i$ for $i = 1, 2$, $M_2 \geq M_1$ and $N_1 = N_2$, or
  (3) $N_i \neq M_i$ for $i = 1, 2$, $N_2 \geq N_1$ and $M_1 = M_2$;

(iii) If the two networks are lattice graphs $L_{(N_i,k)}$, with N_i vertices, $i = 1, 2$ and $N_2 \geq N_1$.

Proof

(i) Considering the expressions given by Eq. (13) of the eigenvalues of a path P_N, with N vertices, the eigenratio R in Eq. (12) becomes

$$R_{P_N}(N) = \frac{\lambda_2}{\lambda_{max}} = \frac{\sin^2 \frac{\pi}{2N}}{\sin^2 \frac{\pi}{2N}\frac{(N-1)}{}} = \tan^2 \frac{\pi}{2N},$$

which is a decreasing function with the number of vertices N, since

$$R'_{P_N}(N) = -\frac{\pi \tan \frac{\pi}{2N}}{N^2 \cos^2 \frac{\pi}{2N}} < 0, \text{ for } N > 1.$$

So, if $N_2 \geq N_1$ then $R_{P_{N_2}}(N_2) \leq R_{P_{N_1}}(N_1)$, and consequently $\alpha_2 \leq \alpha_1$.

(ii)

(1) Considering the expressions given by Eq. (14) of the eigenvalues of a grid $G_{N\times N}$, with $N\times N$ vertices, the eigenratio R comes

$$R_{G_{N\times N}}(N) = \frac{1}{2}R_{P_N}(N) = \frac{1}{2}\tan^2\frac{\pi}{2N}.$$

So $R_{G_{N\times N}}(N)$ is also a decreasing function with the number of vertices N.

(2) If $N \neq M$ one has $N < M$ or $N > M$. Since the proof is similar, we only considerer $N < M$. In this case, the eigenratio is

$$R_{G_{N\times M}}(N,M) = \frac{\lambda_2(P_M)}{\lambda_{max}(P_N) + \lambda_{max}(P_M)} = \frac{\sin^2\frac{\pi}{2M}}{\sin^2\frac{\pi(N-1)}{2N} + \sin^2\frac{\pi(M-1)}{2M}}. \tag{18}$$

Note that $\lambda_{max}(P_N)$ given by Eq. (13) is an increasing function with N. If N is fixed and M increases, then $\lambda_{max}(P_N)$ is fixed, $\lambda_2(P_M)$ decreases and $\lambda_{max}(P_M)$ increases, so $R_{G_{N\times M}}(N,M)$ decreases.

(3) If $N \neq M$, $N < M$, N increases and M is fixed, then $\lambda_2(P_M)$ and $\lambda_{max}(P_M)$ are fixed and $\lambda_{max}(P_N)$ increases, so $R_{G_{N\times M}}(N,M)$, given by Eq. (18) decreases.

As mentioned before, the decreasing of R implies the decreasing of α.

(iii) Using the expressions given by Eqs. (16) and (17) of the eigenvalues of a $(N,k)-$lattice $L_{(N,k)}$, with N vertices, the eigenratio is

$$R_{L_{(N,k)}}(N) = \frac{\lambda_2}{\lambda_{max}} = \frac{q - \frac{\sin\frac{q\pi}{N}}{\sin\frac{\pi}{N}}}{q - \frac{\sin\frac{q(N+b)\pi}{4N}}{\sin\frac{(N+b)\pi}{4N}}} = \frac{q - \csc\frac{\pi}{N}\sin\frac{q\pi}{N}}{q - \csc\frac{(N+b)\pi}{4N}\sin\frac{q(N+b)\pi}{4N}}.$$

$R_{L_{(N,k)}}(N)$ is a continuous function, such that $R'_{L_{(N,k)}}(q) < 0$ and $R'_{L_{(N,k)}}(N)$ has no zeros for $N \geq q$. So $R'_{L_{(N,k)}}(N) < 0$ for $\forall N \geq q$. This is sufficient to prove that $R_{L_{(N,k)}}(N)$, with $N \in \mathbb{N}$, decreases with the number of vertices N. So, the amplitude α is also decreasing with N.

□

In the previous proposition it is stated that when the number of vertices of a graph decreases, the synchronization improves. On the other hand, one has more results about the evolution of the network synchronizability. When the number of edges decreases, maintaining the number of vertices, the synchronization worsens. In fact, one has the following result.

Proposition 2 *Considering two networks associated with graphs G_1 and G_2, such that G_1 and G_2 have the same number N of vertices and $G_1 \subseteq G_2$, then $c_2 \leq c_1$, where c_i denotes the lower bound of the synchronization interval of the graph G_i, with $i = 1, 2$.*

This result, see [23], is a consequence of Corollary 3.2. of [8], since under the conditions stated in Proposition 2, it is verified that $\lambda_2(G_1) \leq \lambda_2(G_2)$. The previous results concern the evolution of the synchronizability when the network topology evolves and the local dynamics is fixed. If the network topology is fixed and the local dynamics varies, one has the following result, see also [23].

Proposition 3 *Consider a network given by Eq. (8) with N nodes, having in each node the same chaotic dynamical system, with Lyapunov exponent μ. If the network topology is fixed, then the amplitude of the network synchronization interval decreases as the local Lyapunov exponent μ increases.*

4 Numerical Simulations

To support our approaches, we consider some examples of paths, grids and lattices. In each case we evaluate the eigenvalues of the Laplacian matrix and the synchronization interval, for a set of values of the local Lyapunov exponent.

4.1 Paths

First are considered the paths P_N, with $N = 4$ and $N = 6$ nodes, having in each node the same model, the von Bertalanffy function f_r given by Eq. (4). See Fig. 4.

If, for instance, $N = 4$, see Fig. 4a, the adjacency matrix A, the diagonal matrix D and the Laplacian matrix L are given by

$$A = \begin{bmatrix} 0&1&0&0 \\ 1&0&1&0 \\ 0&1&0&1 \\ 0&0&1&0 \end{bmatrix}, \quad D = \begin{bmatrix} 1&0&0&0 \\ 0&2&0&0 \\ 0&0&2&0 \\ 0&0&0&1 \end{bmatrix} \quad \text{and} \quad L = D - A = \begin{bmatrix} 1&-1&0&0 \\ -1&2&-1&0 \\ 0&-1&2&-1 \\ 0&0&-1&1 \end{bmatrix}.$$

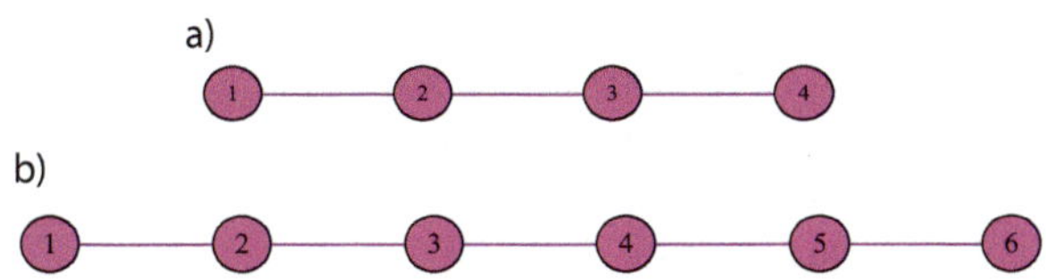

Fig. 4 Graphs of paths P_N with $N = 4$ vertices in (**a**) and with $N = 6$ vertices in (**b**)

So, the network correspondent to the graph in Fig. 4a, according to Eq. (8), is defined by the system

$$\begin{cases} \dot{x}_1 = f_r(x_1) + c(x_1 - x_2) \\ \dot{x}_2 = f_r(x_2) + c(-x_1 + 2x_2 - x_3) \\ \dot{x}_3 = f_r(x_3) + c(-x_2 + 2x_3 - x_4) \\ \dot{x}_4 = f_r(x_4) + c(-x_3 + x_4) \end{cases}.$$

For this path graph the eigenvalues of the Laplacian matrix are $\lambda_1 = 0, \lambda_2 = 2-\sqrt{2}$, $\lambda_3 = 2$ and $\lambda_4 = 2 + \sqrt{2}$. Considering Eq. (11), there is desynchronization if $\mu < 0.347$. In Table 1 we present a list of some Lyapunov exponents obtained by numerical simulation using 5,000 iterations, for several values of the intrinsic growth rate r. For all the values of r in this Table, the Lyapunov exponent μ are such that, there is a synchronization interval for P_4. We evaluate the synchronization interval and its amplitude for each of these values of μ. If, for instance, we consider $r = 6.320$, the Lyapunov exponent of $f_r(x)$ is $\mu \approx 0.098$, Eq. (9). Then, concerning Eq. (10), this graph synchronizes if $\frac{1-e^{-0.098}}{2-\sqrt{2}} < c < \frac{1+e^{-0.098}}{2+\sqrt{2}} \Leftrightarrow 0.160 < c < 0.558$ and the amplitude of the synchronization interval is 0.398. In Table 1 are presented the results obtained for the other values of r.

For the path P_6 of Fig. 4b, the eigenvalues of the Laplacian matrix are $\lambda_1 = 0$, $\lambda_2 = 2 - \sqrt{3}$, $\lambda_3 = 1$, $\lambda_4 = 2$, $\lambda_5 = 3$ and $\lambda_6 = 2 + \sqrt{3}$. With these values, similar calculations were made, which are presented in Table 1. For P_6 there are several values of μ, such that the lower bound of the synchronization interval is larger then the upper bound, so the desynchronization phenomena occurs. These

Table 1 Lyapunov exponent, μ, synchronization interval, $\left]\frac{1-e^{-\mu}}{\lambda_2}, \frac{1+e^{-\mu}}{\lambda_{max}}\right[$, and amplitude of this interval, for several intrinsic growth rates r, for the paths (a) and (b) of Fig. 4

r	μ	Synchronization interval of paths		Amplitude α	
		P_4	P_6	P_4	P_6
6.305	0.043	$]0.072, 0.573[$	$]0.158, 0.525[$	0.501	0.367
6.310	0.064	$]0.106, 0.568[$	$]0.231, 0.519[$	0.462	0.288
6.320	0.098	$]0.160, 0.558[$	$]0.349, 0.511[$	0.398	0.162
6.330	0.095	$]0.154, 0.559[$	$]0.337, 0.512[$	0.405	0.175
6.335	0.122	$]0.196, 0.552[$	$]0.428, 0.505[$	0.356	0.077
6.340	0.141	$]0.225, 0.547[$	$]0.492, 0.501[$	0.322	0.009
6.350	0.173	$]0.272, 0.539[$	(∗)	0.267	(∗)
6.355	0.169	$]0.266, 0.540[$	(∗)	0.275	(∗)
6.360	0.095	$]0.155, 0.559[$	$]0.338, 0.512[$	0.405	0.174
6.365	0.186	$]0.290, 0.536[$	(∗)	0.246	(∗)
6.370	0.177	$]0.277, 0.538[$	(∗)	0.261	(∗)
6.375	0.169	$]0.266, 0.540[$	(∗)	0.274	(∗)

In the cases denoted by (∗) the desynchronization phenomenon occurs, see Eq. (11)

cases are denoted by (∗) in Table 1 there is no synchronization interval in these cases.

4.2 Grids

In this subsection are considered the grid graphs $G_{N\times M}$, with $N\times M = 2\times 3,\ 2\times 4,\ 3\times 3,\ 4\times 4$ presented in Fig. 5, having in each node the same von Bertalanffy function.

For the grid graph $G_{2\times 3}$, Fig. 5a, the eigenvalues of the Laplacian matrix are $\lambda_1 = 0$, $\lambda_2 = 1$, $\lambda_3 = 2$, $\lambda_4 = 3$, $\lambda_5 = 3$ and $\lambda_6 = 5$. Then, concerning Eq. (11), there is synchronization if $\mu < 0.405$. For the grid graph $G_{2\times 4}$, Fig. 5b, the eigenvalues of the Laplacian matrix are $\lambda_1 = 0$, $\lambda_2 = 2-\sqrt{2}$, $\lambda_3 = 2$, $\lambda_4 = 2$, $\lambda_5 = 4-\sqrt{2}$, $\lambda_6 = 2+\sqrt{2}$, $\lambda_7 = 4$ and $\lambda_8 = 4+\sqrt{2}$. Then, considering Eq. (11), there is synchronization if $\mu < 0.217$. For the grid graph $G_{3\times 3}$, Fig. 5c, the eigenvalues of the Laplacian matrix are $\lambda_1 = 0$, $\lambda_2 = 1$, $\lambda_3 = 1$, $\lambda_4 = 2$, $\lambda_5 = 3$, $\lambda_6 = 3$, $\lambda_7 = 4$, $\lambda_8 = 4$ and $\lambda_9 = 6$. Then, concerning Eq. (11), there is synchronization if $\mu < 0.336$. For these three grid graphs, the desynchronization phenomena do not

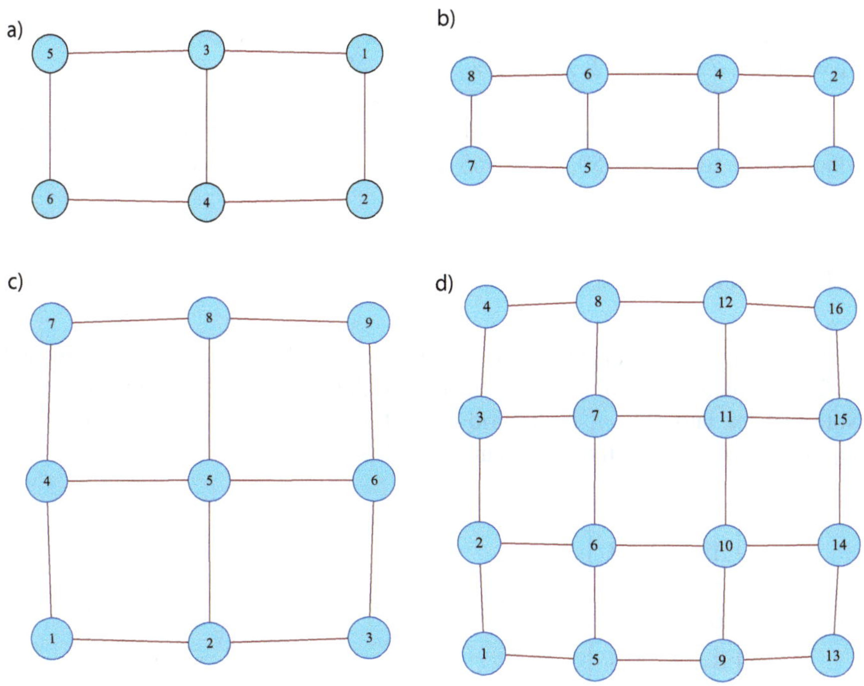

Fig. 5 Graphs of grids $G_{N\times M}$, with $N\times M = 2\times 3$ in (**a**), $N\times M = 2\times 4$ in (**b**), $N\times M = 3\times 3$ in (**c**) and $N\times M = 4\times 4$ in (**d**)

occurs for any value of r considered in Table 2. For the grid graphs $G_{4\times4}$, Fig. 5d, the Laplacian matrix is

$$L = D - A = \begin{bmatrix} 2 & -1 & 0 & 0 & -1 & 0 & 0 & 0 & 0 & 0 & 0 & 0 & 0 & 0 & 0 & 0 \\ -1 & 3 & -1 & 0 & 0 & -1 & 0 & 0 & 0 & 0 & 0 & 0 & 0 & 0 & 0 & 0 \\ 0 & -1 & 3 & -1 & 0 & 0 & -1 & 0 & 0 & 0 & 0 & 0 & 0 & 0 & 0 & 0 \\ 0 & 0 & -1 & 2 & 0 & 0 & 0 & -1 & 0 & 0 & 0 & 0 & 0 & 0 & 0 & 0 \\ -1 & 0 & 0 & 0 & 3 & -1 & 0 & 0 & -1 & 0 & 0 & 0 & 0 & 0 & 0 & 0 \\ 0 & -1 & 0 & 0 & -1 & 4 & -1 & 0 & 0 & -1 & 0 & 0 & 0 & 0 & 0 & 0 \\ 0 & 0 & -1 & 0 & 0 & -1 & 4 & -1 & 0 & 0 & -1 & 0 & 0 & 0 & 0 & 0 \\ 0 & 0 & 0 & -1 & 0 & 0 & -1 & 3 & 0 & 0 & 0 & -1 & 0 & 0 & 0 & 0 \\ 0 & 0 & 0 & 0 & -1 & 0 & 0 & 0 & 3 & -1 & 0 & 0 & -1 & 0 & 0 & 0 \\ 0 & 0 & 0 & 0 & 0 & -1 & 0 & 0 & -1 & 4 & -1 & 0 & 0 & -1 & 0 & 0 \\ 0 & 0 & 0 & 0 & 0 & 0 & -1 & 0 & 0 & -1 & 4 & -1 & 0 & 0 & -1 & 0 \\ 0 & 0 & 0 & 0 & 0 & 0 & 0 & -1 & 0 & 0 & -1 & 3 & 0 & 0 & 0 & -1 \\ 0 & 0 & 0 & 0 & 0 & 0 & 0 & 0 & -1 & 0 & 0 & 0 & 2 & -1 & 0 & 0 \\ 0 & 0 & 0 & 0 & 0 & 0 & 0 & 0 & 0 & -1 & 0 & 0 & -1 & 3 & -1 & 0 \\ 0 & 0 & 0 & 0 & 0 & 0 & 0 & 0 & 0 & 0 & -1 & 0 & 0 & -1 & 3 & -1 \\ 0 & 0 & 0 & 0 & 0 & 0 & 0 & 0 & 0 & 0 & 0 & -1 & 0 & 0 & -1 & 2 \end{bmatrix}.$$

The eigenvalues of the Laplacian matrix are $\lambda_1 = 0$, $\lambda_2 = 2-\sqrt{2}$, $\lambda_3 = 2-\sqrt{2}$, $\lambda_4 = 2(2-\sqrt{2})$, $\lambda_5 = 2$, $\lambda_6 = 2$, $\lambda_7 = 4-\sqrt{2}$, $\lambda_8 = 4-\sqrt{2}$, $\lambda_9 = 2+\sqrt{2}$, $\lambda_{10} = 2+\sqrt{2}$, $\lambda_{11} = 4$, $\lambda_{12} = 4$, $\lambda_{13} = 4$, $\lambda_{14} = 4+\sqrt{2}$, $\lambda_{15} = 4+\sqrt{2}$ and $\lambda_{16} = 2(2+\sqrt{2})$. Then, considering Eq. (11), there is synchronization if $\mu < 0.172$, which occurs for $r = 6.350$, $r = 6.365$ and $r = 6.370$ considered in Table 2.

4.3 Lattices

In this subsection are considered the lattices $L_{(N,k)}$, with $(N,k) = (4,1), (6,1)$ and $(6,2)$ presented in Fig. 6, having in each node the same von Bertalanffy function. Note that the lattice $L_{(4,1)}$ is the grid $G_{2\times2}$. For the lattice $L_{(6,2)}$, see Fig. 6c, the adjacency matrix A and the Laplacian matrix L are

$$A = \begin{bmatrix} 0&1&1&0&1&1 \\ 1&0&1&1&0&1 \\ 1&1&0&1&1&0 \\ 0&1&1&0&1&1 \\ 1&0&1&1&0&1 \\ 1&1&0&1&1&0 \end{bmatrix} \quad \text{and} \quad L = D - A = \begin{bmatrix} 4&-1&-1&0&-1&-1 \\ -1&4&-1&-1&0&-1 \\ -1&-1&4&-1&-1&0 \\ 0&-1&-1&4&-1&-1 \\ -1&0&-1&-1&4&-1 \\ -1&-1&0&-1&-1&4 \end{bmatrix}.$$

Table 2 Lyapunov exponent, μ, synchronization interval, $\left]\frac{1-e^{-\mu}}{\lambda_2}, \frac{1+e^{-\mu}}{\lambda_{max}}\right[$, and amplitude of this interval, $\frac{1+e^{-\mu}}{\lambda_{max}} - \frac{1-e^{-\mu}}{\lambda_2}$, for several intrinsic growth rates r, for the grids (a), (b), (c) and (d) of Fig. 5

		Synchronization interval of grids				Amplitude α			
r	μ	$G_{2\times3}$	$G_{2\times4}$	$G_{3\times3}$	$G_{4\times4}$	$G_{2\times3}$	$G_{2\times4}$	$G_{3\times3}$	$G_{4\times4}$
6.305	0.043	]0.042, 0.392[	]0.072, 0.362[	]0.042, 0.326[	]0.072, 0.287[	0.350	0.290	0.284	0.215
6.310	0.064	]0.062, 0.388[	]0.106, 0.358[	]0.062, 0.323[	]0.106, 0.284[	0.326	0.252	0.261	0.178
6.320	0.098	]0.094, 0.381[	]0.160, 0.352[	]0.094, 0.318[	]0.160, 0.279[	0.287	0.192	0.224	0.119
6.330	0.095	]0.090, 0.382[	]0.154, 0.353[	]0.090, 0.318[	]0.154, 0.280[	0.292	0.199	0.228	0.126
6.335	0.122	]0.115, 0.377[	]0.196, 0.348[	]0.115, 0.314[	]0.196, 0.276[	0.262	0.152	0.199	0.080
6.340	0.141	]0.132, 0.374[	]0.225, 0.345[	]0.132, 0.311[	]0.225, 0.274[	0.242	0.120	0.179	0.049
6.350	0.173	]0.159, 0.368[	]0.272, 0.340[	]0.159, 0.307[	(*)	0.209	0.068	0.148	(*)
6.355	0.169	]0.156, 0.369[	]0.266, 0.341[	]0.156, 0.307[	]0.266, 0.270[	0.213	0.075	0.151	0.005
6.360	0.095	]0.091, 0.382[	]0.155, 0.353[	]0.091, 0.318[	]0.155, 0.280[	0.291	0.198	0.227	0.125
6.365	0.186	]0.170, 0.366[	]0.290, 0.338[	]0.170; 0.305[	(*)	0.196	0.048	0.135	(*)
6.370	0.177	]0.162, 0.368[	]0.277, 0.339[	]0.162, 0.306[	(*)	0.206	0.062	0.144	(*)
6.375	0.169	]0.156, 0.369[	]0.266, 0.341[	]0.156, 0.307[	]0.266, 0.270[	0.213	0.075	0.151	0.004

In the cases denoted by (*) the desynchronization phenomenon occurs, see Eq. (11)

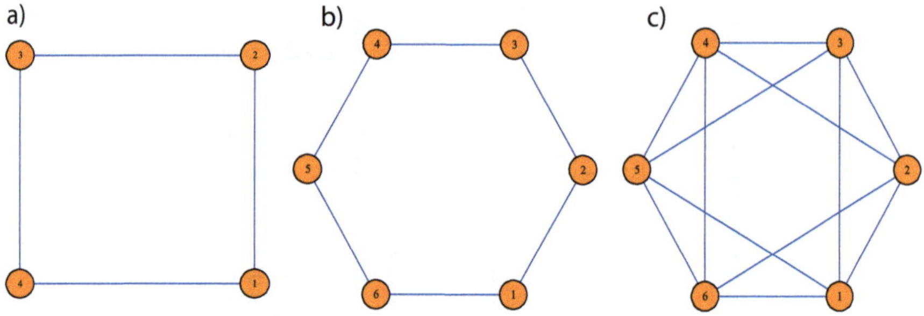

Fig. 6 Lattices $L_{(N,k)}$, with $(N,k) = (4,1)$ in (**a**), $(N,k) = (6,1)$ in (**b**) and $(N,k) = (6,2)$ in (**c**). From (**a**) to (**b**) the total number of vertices of the network increases maintaining the number of neighbors of each node, and from (**b**) to (**c**) increases the number of neighbors of each node, but the total number of vertices of the network remains the same

Table 3 Lyapunov exponent, μ, synchronization interval, $\left]\frac{1-e^{-\mu}}{\lambda_2}, \frac{1+e^{-\mu}}{\lambda_{max}}\right[$, and amplitude of this interval, $\frac{1+e^{-\mu}}{\lambda_{max}} - \frac{1-e^{-\mu}}{\lambda_2}$, for several intrinsic growth rates r, for the lattices (a), (b) and (c) of Fig. 6

		Synchronization interval of lattices			Amplitude α		
r	μ	$L_{(4,1)}$	$L_{(6,1)}$	$L_{(6,2)}$	$L_{(4,1)}$	$L_{(6,1)}$	$L_{(6,2)}$
6.330	0.095	$]0.045, 0.477[$	$]0.090, 0.477[$	$]0.0225668, 0.318[$	0.432	0.387	0.295
6.350	0.173	$]0.080, 0.460[$	$]0.159, 0.460[$	$]0.040, 0.307[$	0.380	0.301	0.267
6.400	0.206	$]0.093, 0.453[$	$]0.186, 0.453[$	$]0.047, 0.302[$	0.360	0.267	0.255
6.500	0.297	$]0.128, 0.436[$	$]0.257, 0.436[$	$]0.064, 0.291[$	0.308	0.179	0.226
6.550	0.347	$]0.147, 0.427[$	$]0.293, 0.427[$	$]0.073, 0.285[$	0.280	0.134	0.211
6.600	0.377	$]0.157, 0.421[$	$]0.314, 0.421[$	$]0.079, 0.281[$	0.264	0.107	0.202
6.650	0.406	$]0.167, 0.417[$	$]0.334, 0.417[$	$]0.083, 0.278[$	0.250	0.083	0.194
6.700	0.463	$]0.185, 0.407[$	$]0.371, 0.407[$	$]0.093, 0.272[$	0.222	0.037	0.179
6.730	0.506	$]0.199, 0.401[$	$]0.397, 0.401[$	$]0.099, 0.267[$	0.202	0.003	0.168
6.740	0.533	$]0.207, 0.397[$	$(*)$	$]0.103, 0.265[$	0.190	$(*)$	0.161
6.750	0.598	$]0.225, 0.388[$	$(*)$	$]0.112, 0.258[$	0.163	$(*)$	0.146

In the cases denoted by $(*)$ the desynchronization phenomenon occurs, see Eq. (11)

For the lattice $L_{(6,2)}$ the eigenvalues of the Laplacian matrix are $\lambda_1 = 0$, $\lambda_2 = \lambda_3 = \lambda_4 = 4$ and $\lambda_5 = \lambda_6 = 6$. If we consider, for instance, $r = 6.60$, the Lyapunov exponent of $f_r(x)$ is 0.377, Eq. (9). Then, taking into account Eq. (10), this lattice synchronizes if $\frac{1-e^{-0.377}}{4} < c < \frac{1+e^{-0.377}}{6} \Leftrightarrow 0.079 < c < 0.281$ and the amplitude of the synchronization interval is 0.202. For more examples see Table 3. The lattice $L_{(6,1)}$ correspondent to the Fig. 6b has eigenvalues of the Laplacian matrix $\lambda_1 = 0$, $\lambda_2 = \lambda_3 = 1$, $\lambda_4 = \lambda_5 = 3$ and $\lambda_6 = 4$. Thus, for the same $r = 6.600$, the lattice synchronizes if $0.314 < c < 0.421$ and the amplitude of this interval is 0.107. Moreover, to the lattice $L_{(4,1)}$ in Fig. 6a, the eigenvalues of the Laplacian matrix are $\lambda_1 = 0$, $\lambda_2 = \lambda_3 = 2$ and $\lambda_4 = 4$. For the same $r = 6.600$, the lattice synchronizes if $0.157 < c < 0.421$ and the amplitude of this interval is 0.264. In Table 3 are presented more examples, where we computed the synchronization interval for several values of the intrinsic growth rate r, for all these lattices (a), (b) and (c) of Fig. 6.

5 Discussions and Conclusions

In the first part of this work we study the dynamical behaviour and the bifurcations structure of von Bertalanffy's functions. For this class of functions, we prove sufficient conditions, in Lemma 1, for the initial population densities to which the stability, period doubling, chaos and non admissibility occur. We provide the bifurcation analysis of von Bertalanffy's functions, in the two-dimensional parameter space (K, W_∞). Fold and flip bifurcations curves and numerical simulations of the bifurcation diagram for von Bertalanffy's functions are presented. As von Bertalanffy's functions are used to model population dynamics, it makes sense to consider the influence in an individual of its neighbors. Therefore, in this study, we consider some particular types of networks such as paths, grids and lattices. So, it was studied the synchronizability of these networks, having in each node a von Bertalanffy's function, in terms of the r parameter and also in terms of the network topology. It was concluded that:

- The amplitude of the synchronization interval decreases if one consider two networks of the same type, in the following cases: the two networks are paths P_N, with increasing number N of vertices; the two networks are grids $G_{N\times M}$, with the increasing of one of the indexes N or M, maintaining fixed the other; or the two networks are lattices $L_{(N,k)}$, with increasing number N of vertices;
- The synchronizability improves if an edge is added to the network graph;
- The amplitude of the network synchronization interval decreases if the local Lyapunov exponent increases, when fixing the network topology.

Note that in the case of grids, it is false to assume that the increasing of the number of vertices implies the decreasing of the amplitude of the synchronization interval. See for example grids $G_{2\times 4}$ and $G_{3\times 3}$ in Table 2.

Considering values of the parameter r in the chaotic region some numerical simulations were performed. Observing Tables 1, 2 and 3, all previous results can be confirmed.

In future works we will study growth models of von Bertalanffy's type, which incorporate Allee effect. In fact, species extinction is currently a major focus of ecological research. It is believed that synchronization may promote extinctions of some species. Full synchronism may have a deleterious effect on population survival because it may lead to the impossibility of a recolonization in case of a large global disturbance. So, it is our aim to investigate the relation between synchronization and Allee effect in new types of von Bertalanffy's models.

Acknowledgements Research partially sponsored by national funds through the Fundação Nacional para a Ciência e Tecnologia, Portugal – FCT, under the project PEst-OE/MAT/UI0006/2014, CEAUL, CIMA and ISEL. The authors are grateful to Prof. Danièle Fournier-Prunaret for having made the image of Fig. 2. The authors are grateful to the anonymous referees for a careful checking of the details and for helpful comments that improved this work.

References

1. Balanov, A., Janson, N., Postnov, D., Sosnovtseva, O.: Synchronization: From Simple to Complex. Springer, NewYork (2009)
2. Bondy, J.A., Murty, U.S.R.: Graph Theory with Applications. North-Holland, New York (1976)
3. Brouwer, A.E., Haemers, W.H.: Spectra of Graphs. Springer, New York (2012)
4. Cailliet, G.M., Smith, W.D., Mollet, H.F., Goldman, K.J.: Age and growth studies of chondrichthyan fishes: the need for consistency in terminology, verification, validation, and growth function fitting. Environ. Biol. Fish. **77**, 211–228 (2006)
5. Chen, J., Lu, J., Zhan, C., Chen, G.: Laplacian Spectra and Synchronization Processes on Complex Networks, Handbook of Optimization in Complex Networks, Springer Optimization and Its Applications, 81–113 (2012)
6. Davidson, A.J., Menaker, M.: Birds of a feather clock together sometimes: social synchronization of circadian rhythms. Curr. Opin. Neurol. **13**(6), 765–769 (2003)
7. Essington, T.E., Kitchell, J.F., Walters, C.J.: The von Bertalanffy growth function, bioenergetics, and the consumption rates of fish. Can. J. Fish. Aquat. Sci. **58**, 2129–2138 (2001)
8. Fiedler, M.: Algebraic connectivity of graphs. Czechoslov. Math. J. **2**, 298–305 (1973)
9. Hasler, M., Maistrenko, Y.L.: An introduction to the synchronization of chaotic systems: coupled skew tent maps. IEEE Trans. Circuits Syst. – I, **44**(10), 856–866 (1987)
10. Jamakovic, A. Van Mieghem, P.: The Laplacian Spectrum of Complex Networks, European Conference on System, Oxford, pp. 25–26 (2006)
11. Karpouzi, V.S., Pauly, D.: In: Palomares, M.L.D., Pauly, D. (eds.) Life-History Patterns in Marine Birds. Fisheries Center Research Reports 16 (10), Von Bertalanffy Growth Parameters of Non-Fish Marine Organisms, 27–43, Canada, 2008. The Fisheries Center, University of British Columbia
12. Li, X., Chen, G.: Synchronization and desynchronization of complex dynamical networks: an engineering viewpoint. IEEE Trans. on Circ. Syst. –I, **50** (11), 1381–1390 (2003)
13. Melo, W., van Strien, S.: One-Dimensional Dynamics, Springer, New York (1993)
14. Mira, C.: Chaotic Dynamics. From the One-Dimensional Endomorphism to the Two-Dimensional Diffeomorphism. World Scientific, Singapore (1987)
15. Mira, C., Gardini, L., Barugola, A., Cathala, J-C.: Chaotic Dynamics in Two-Dimensional Noninvertible Maps. World Scientific, Singapore (1996)
16. Misiurewicz, M., Absolutely continuous measures for certain maps of an interval. Inst. Hautes Études Sci. Publ. Math. **53**, 17–51 (1981)
17. Mohar, B.: The Laplacian Spectrum of Graphs. In: Alavi, Y., Chartrand, G., Oellermann, O.R., Schwenk, A.J. (eds.) Graph Theory, Combinatorics, and Applications. Wiley, New York (1991)

18. Mohar, B.: Some Applications of Laplace Eigenvalues of Graphs, In: Hahn, G., Sabidussi, G. (eds.) Graph Symmetry: Algebraic Methods and Applications. Kluver Academic Publishers, Dordrecht (1997)
19. Rocha, J.L., Fournier-Prunaret, D., Taha, A-K.: Strong and weak Allee effects and chaotic dynamics in Richards' growths. Discrete Contin. Dyn. Syst.-Ser.B **18, 3**, 2397–2425 (2013)
20. Rocha, J.L., Aleixo, S.M.: An extension of gompertzian growth dynamics: Weibull and Fréchet models. Math. Biosci. Eng. **10**, 379–398 (2013)
21. Rocha, J.L., Aleixo, S.M.: Dynamical analysis in growth models: Blumberg's equation. Discrete Contin. Dyn. Syst.-Ser.B **18**, 783–795 (2013)
22. Rocha, J.L., Aleixo, S.M., Caneco, A.: Synchronization in von Bertalanffy's models. Chaotic Model. Simul. **4**, 519–528 (2013)
23. Rocha, J.L., Aleixo, S.M., Caneco, A.: Synchronization in Richards' chaotic systems. J. Appl. Nonlinear Dyn. **3**(2), 115–130 (2014)
24. Sharkovsky, A.N., Kolyada, S.F., Sivak, A.G., Fedorenko, V.V.: Dynamics of One-Dimensional Maps. Kluwer Academic Publishers, Netherlands (1997)
25. Schreiber, S.J.: Chaos and population disappearances in simple ecological models. J. Math. Biol. **42**, 239–260 (2001)
26. Silva, J.A.L., Giordani, F.T.: Density-dependent migration and synchronism in metapopulations. Bull. Math. Biol. **68**, 451–465 (2006)
27. Singer, D.: Stable orbits and bifurcations of maps of the interval. SIAM J. Appl. Math. **35**, 260–267 (1978)
28. Von Bertalanffy, L.: A quantitative theory of organic growth. Hum. Biol. **10**, 181–213 (1938)
29. Von Bertalanffy, L.: Quantitative laws in metabolism and growth. Q. Rev. Biol. **32**, 217–231 (1957)

Three Dimensional Flows: From Hyperbolicity to Quasi-Stochasticity

Alexandre A.P. Rodrigues

Abstract In the present survey, we give an overview of some recent developments on examples of differential equations whose flows have heteroclinic cycles and networks; we fit some properties of their nonwandering sets into the classic theory of hyperbolic and pseudo-hyperbolic sets.

1 Introduction

The existence of heteroclinic cycles in systems with symmetry is no longer a surprising feature. There are several examples of cycles arising in differential equations symmetric under the action of a specific compact Lie group [27]. Similarities among them and the identification of some strange attractors in their flows are the purpose of the present survey.

Several definitions of heteroclinic cycles and networks have been given in the literature. Throughout the present survey, we use the following definition valid for a finite dimensional system of ordinary equations (ODE):

Definition 1 A *heteroclinic cycle* is a finite collection of invariant saddles $\{\xi_1, \ldots, \xi_n\}$ of the ODE together with a set of heteroclinic connections $\{\gamma_1, \ldots, \gamma_n\}$ where γ_j is a solution of the ODE such that:

$$\lim_{t \to -\infty} \gamma_j = \xi_j \qquad \text{and} \qquad \lim_{t \to +\infty} \gamma_j = \xi_{j+1}$$

and $\xi_{n+1} \equiv \xi_1$. When $n = 1$, we say that the set $\{\xi_1, \gamma_1\}$ is a *homoclinic cycle*. A *heteroclinic network* is a connected union of heteroclinic cycles.

We start with a chronological perspective on the subject.

A.A.P. Rodrigues (✉)
Centro de Matemática da Universidade do Porto and Faculdade de Ciências da Universidade do Porto, Rua do Campo Alegre 687, 4169–007 Porto, Portugal
e-mail: alexandre.rodrigues@fc.up.pt

J.-P. Bourguignon et al. (eds.), *Dynamics, Games and Science*, CIM Series in Mathematical Sciences 1, DOI 10.1007/978-3-319-16118-1_31

1.1 A Chronological Perspective

One of the goals of *Dynamical Systems* is to describe the asymptotic behavior of systems for which an evolution rule is known. For continuous-time systems, the evolution rule is given generically by a differential equation. Since mathematical models given by differential equations are simplifications of the real world, one aims to understand whether the asymptotic behavior remains the same if the differential equation is slightly perturbed. The first attempt in tackling the problem is naturally to solve the differential equations, which turns out to be impossible in most cases. In the nineteenth century, Poincaré proposed to combine methods from other subjects to find qualitative information on the dynamics without finding explicitly the solutions. This qualitative analysis attained full maturity in his remarkable contribution to the Celestial Mechanics [50] which is considered to be the birth of Dynamical Systems as a mathematical discipline.

In the 1930s, in the context of diffeomorphisms, the Poincaré's direction has been followed by Birkhoff [15, 16] in the phenomenon of transverse homoclinic points. Nowadays, we recognize the existence of transverse homoclinic points as a paradigm of chaos. This phenomenon has been completely explained by Smale [66] in the sixties with the geometric concept of *horseshoe*, a simple two-dimensional model containing infinitely many periodic orbits in a compact manifold. The horseshoe as well as the hyperbolic toral automorphism were unified by the notion of hyperbolicity [66, 67]: an invariant subset of the phase space such that the tangent space at each point splits into two transverse directions that are uniformly contracted under forward and backward iterations, respectively.

The notion of structural stability of systems introduced by Andronov and Pontryagin [7] (in the thirties) is connected with the uniform hyperbolicity together with a transversality condition of the invariant manifolds of the critical points—more details in Sect. 2. The theory of hyperbolic systems was developed from the sixties to the eighties and gave a mathematical foundation that deterministic systems may present chaos in a robust fashion. Nevertheless, uniform hyperbolicity is "less universal" than one might think: in fact strict uniform hyperbolicity rarely occurs in applications. This leads to the study of different classes of systems that are "robust" and non-hyperbolic. The study of differential equations whose flows have homo and heteroclinic cycles provide remarkable challenges for this view. The complete taxonomy of these sets is an open problem; an informal overview of the corresponding chaotic systems is being done in the present paper.

1.2 The Chaos

One of the most important findings in science in the twentieth century was the discovery of dynamical chaos, characterized by the high sensitivity to initial conditions.

Many modern problems associated with systems involving high energies, velocities, game theory, geomagnetic field are modeled by multidimensional nonlinear differential equations—see for instance [2, 11, 17, 43, 52] and references therein. The study of such systems has revealed numerous new concepts in nonlinear dynamics. Dynamical chaos has also been characterized and studied by statistical and ergodic methods, including experimental analysis of correlation functions.

The simplest attractors are the ones consisting of a simple equilibrium or a periodic solution. In the hyperbolic case, the other extreme corresponds to attractors consisting of the whole ambient manifold, like the one induced by hyperbolic toral automorphisms [24] and also the geodesic flows on surfaces of negative curvature [8]. At this point, it would be interesting to find examples of differential equations whose flows contain (hyperbolic or not) attractors that do not cover the whole manifold. The existence of heteroclinic structures in the flow is a crucial step towards this subject.

1.3 The Role of Examples

The majority of mathematicians were not led to their results by a process of deduction from general features of the vector field, but rather by a scrupulous examination of properly chosen particular examples and observations of concrete numerical simulations. The generalizations come later since it is easier to generalize an established result than to discover new arguments.

Since the work by E. Noether and H. Weyl in the first half of twentieth century, symmetries play a major theoretical role in mathematics. In some examples, the complex nature of the geometry can be described analytically because they are close to a highly symmetric differential equation which, by construction, exhibits special features.

There is a vast catalog of exotic phenomena associated with heteroclinic connections. Possibilities include connections among chaotic saddles, connections between chaos and non-chaos, cycling chaos and complex networks of connections with “random switching” [5, 43].

Along this survey, we briefly outline some valuable facts about hyperbolic and pseudo-hyperbolic sets, following the classification of Shilnikov [58, 62]. The main purpose of the author is to fit some examples of regular and chaotic dynamics involving heteroclinic cycles into the folklore theory of hyperbolic and pseudo-hyperbolic sets in three-dimensional manifolds. We do not aim to define rigorously all the concepts and terminology. On the way, we will suggest some references to the reader.

2 Global Perspective of Hyperbolicity

In an attempt to identify which features are common among stable systems, Smale introduced the notion of *hyperbolicity*. Remarkably it turned out that structurally stable systems are essentially the hyperbolic ones plus a transversality condition. In the sixties and in the seventies, a complete theory of hyperbolic dynamical systems has been developed, culminating with the proof of the C^1-Stability Conjecture in the nineties [31, 42]. Based on [18], in what follows we present some classic results on this theory together with some terminology and notation.

2.1 Uniform Hyperbolicity

Let M be a n-dimensional compact riemannian smooth manifold (possibly without boundary) and $\Lambda \subset M$ be a compact and invariant set.

According to the reference [49], a diffeomorphism $f : M \to M$ is called *uniformly hyperbolic* on Λ if there is a decomposition of the tangent bundle of M at Λ,

$$T_\Lambda M = E^s \oplus E^u,$$

such that $df|_{E^s}$ and $df^{-1}|_{E^u}$ are uniform contractions.

For a smooth vector field X, define the associated global flow as the family of diffeomorphisms $(X^t)_{t\in\mathbf{R}}$ satisfying:

(1) X^0 is the identity map;
(2) $X^{t+s} = X^t \circ X^s$ for all $t, s \in \mathbf{R}$ and
(3) $\frac{d}{dt}X^t(q)|_{t=t_0} = X(X^{t_0}(q)))$ for all $q \in M$ and $t_0 \in \mathbf{R}$.

Conversely, a given flow $(X^t)_{t\in\mathbf{R}}$ determines a unique vector field X whose associated flow is precisely $(X^t)_{t\in\mathbf{R}}$.

A flow $\{X^t\}_{t\in\mathbf{R}}$ generated by a vector field X is *hyperbolic* if there exists a decomposition

$$T_\Lambda M = E^s \oplus E^c \oplus E^u$$

where E^c is one-dimensional and tangent to the flow X (outside the equilibria, which we assume to be finite) for which the following conditions hold:

$$||dX^t|_{E^s}|| < Ce^{\lambda t} \qquad \text{and} \qquad ||dX^{-t}|_{E^u}|| < Ce^{\lambda t}, \tag{1}$$

for $t \in \mathbf{R}$, $C > 0$ and $\lambda \in (0, 1)$. We may adapt the Riemannian metric in order to get $C = 1$—see [64]. In the cases where such decompositions occur in the whole manifold, we refer to such *globally hyperbolic* diffeomorphisms or flows as *Anosov* ones.

Associated to the solutions of a hyperbolic set there are *stable* and *unstable* manifolds, corresponding to the contracting and expanding sub-bundles by the action of *df* (*dX* in the case of a flow). In what follows *dist* denotes the distance on M induced by the Riemannian norm.

If $\Lambda \subset M$ is a hyperbolic set, by the Invariant Manifold Theory [33], it follows that for every $p \in \Lambda$ the sets:

$$W^s(p) = \left\{ q \in M : \lim_{t \to +\infty} dist(X^t(q), X^t(p)) = 0 \right\}$$

and

$$W^u(p) = \left\{ q \in M : \lim_{t \to -\infty} dist(X^t(q), X^t(p)) = 0 \right\}$$

are invariant manifolds tangent to E^s_p and E^u_p respectively, at p.

A hyperbolic set $\Lambda \subset M$ is a *basic set* if it is topologically transitive and isolated (i.e. $\Lambda = \bigcap_{t \in \mathbf{R}} X^t(U)$ for some neighbourhood U of Λ). We refer to a *hyperbolic* system as satisfying *Axiom A* when its non-wandering set is hyperbolic and the closure of the non-wandering points of f coincides with the set of its periodic points.

A diffeomorphism f is C^1-structurally stable (or simply *robust*), if for any C^1 arbitrarily small perturbation g of f, there is a homeomorphism h of the phase space such that:

$$\forall x \in M, \qquad h \circ f(x) = g \circ h(x).$$

For flows, we require the existence of a homeomorphism h (close to the Identity map) sending trajectories of the initial flow to the trajectories of any small C^1-perturbation.

2.2 Examples

Hyperbolic attractors are the sets for which the *Axiom A* of Smale is valid, and hence, they are structurally stable. In the context of diffeomorphisms, for hyperbolic attractors, periodic orbits as well as homo/heteroclinic cycles are everywhere dense. An important property is that "inside" hyperbolic connected attractors all trajectories have the same Morse index, i.e. the stable (respectively, unstable) manifold of all periodic orbits have the same dimension.

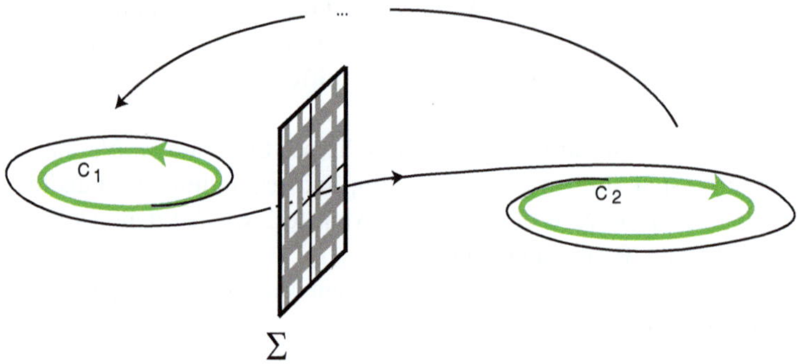

Fig. 1 In a three-dimensional topological sphere, the authors of [56] exhibit an explicit example of a heteroclinic network associated to three hyperbolic (non-trivial) periodic solutions, whose neighborhood contains a uniformly hyperbolic suspended horseshoe. If Σ is a section transverse to the cycle, the shape of the suspended horseshoe is depicted and is consistent with the geometrical structure given in [61]

Classic examples of chaotic hyperbolic attractors include Anosov's systems [24], Smale-Williams' solenoid [21, Ch. 2.5], Plykin's attractors, etc. In the context of continuous-time dynamics, the first examples of a non-trivial hyperbolic set (different from an equilibrium point or a periodic solution) was the geodesic flow on a Riemannian manifold with negative curvature studied by Anosov [8].

Based on the works [6, 61], restricted to a three-dimensional topological sphere, Rodrigues et al. [56] presented an example of a heteroclinic network associated to three hyperbolic (non-trivial) periodic solutions, whose neighbourhood contains a robustly transitive hyperbolic set where the first return map (to a section transverse to the cycle) is topologically conjugated to a Bernoulli shift with infinitely many symbols—see Fig. 1.

The results of [56] are consistent with the work of Doering [22] which says that on a three-dimensional compact manifold, an invariant compact set without equilibria is robustly transitive if it is Anosov. Furthermore, the theory developed in [56] allows us to conclude that the dynamics near the heteroclinic network exhibits:

- *heteroclinic switching*: there are trajectories that visit neighbourhoods of the saddles following all the heteroclinic connections of the network in any given order;
- *chaotic double cycling*: there are trajectories that follow each cycle on the network making any prescribed number of turns near the saddles, for any given bi-infinite sequence of turns.

Other type of dynamical behaviour might occur but they are probably far from the network. More details in [5, 56].

Realizations of hyperbolic attractors in the form of maps or differential equations are difficult to find in specific applications. The attractors reported in the next sections fit beyond (strict) uniform hyperbolicity.

3 Strange Attractors: Pseudo-Hyperbolicity

The hyperbolic attractors are robust and, historically, were the only ones known to be robust until the appearance of the Lorenz attractor. In this section, we start by describing this classic example, emphasizing the type of pseudo-hyperbolicity in its flow. We introduce the concept of stochasticity of Sinai [65] and we refer some examples of continuous-time differential equations whose non-wandering solutions do not fill the whole manifold.

3.1 *The Classic Lorenz Attractor*

The phenomenon known (nowadays) as the *butterfly effect* was discovered by the meteorologist Edward Lorenz in 1961 while working on a simplified model of convection in the atmosphere. He published his findings in [40], but:

> it took some time before they *[the results]* were appreciated by meteorologists or known to mathematicians.
>
> Ian Stewart [68], *2011*

The system of differential equations under consideration is the following:

$$\begin{cases} \dot{x} = \sigma(y - x) \\ \dot{y} = rx - y - xz \,. \\ \dot{z} = xy - bz \end{cases} \tag{2}$$

Based on a computer assisted proof, Tucker [70] proved that for the Saltzman values $\sigma = 10$, $r = 28$ and $b = \frac{8}{3}$, there is an attractor, called *Lorenz butterfly*, containing (see Fig. 2a):

- an equilibrium point at the origin, where its linearization has eigenvalues λ_u, λ_s^1 and λ_s^2 satisfying:

$$\lambda_s^2 < \lambda_s^1 < 0 < \lambda_u \qquad and \qquad \lambda_u + \lambda_s^1 > 0; \tag{3}$$

- periodic solutions accumulating on the equilibrium;
- a dense orbit.

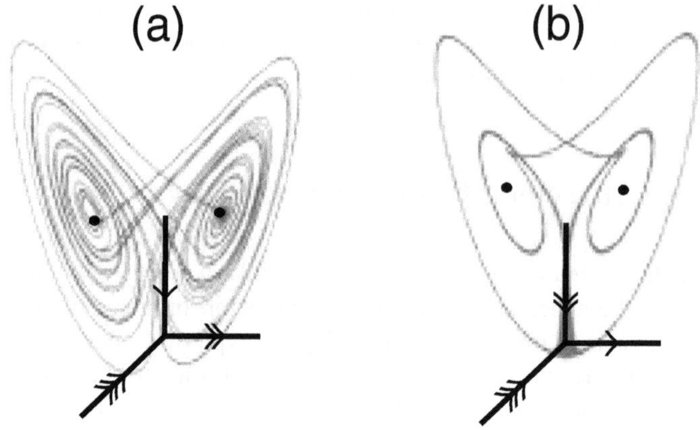

Fig. 2 Case (**a**)—classic Lorenz attractor: the sum of the leading eigenvalues of the linearization of (2) at the origin is positive. Case (**b**)—Lorenz-Rovella attractor: the sum of the leading eigenvalues of the linearization of (2) at the origin is negative

Attractors with such characteristics are called *Lorenz-like attractors*. Although the technical definition of "Lorenz-attractor" is more complicated, following [68] we give an heuristic definition:

Definition 2 An attractor is a region $\mathscr{A}$ of the phase space such that any initial condition starting near $\mathscr{A}$ converges towards a trajectory that lies on $\mathscr{A}$.

If $\mathscr{A}$ is neither an equilibrium point nor a periodic solution, this property means that although distinct trajectories on $\mathscr{A}$ may diverge, they remain on $\mathscr{A}$. So $\mathscr{A}$ is a Lyapunov-stable object. The behaviour on $\mathscr{A}$ is robust under perturbations in the following sense: if the system is subjected to a small modification, the trajectory can change dramatically; nevertheless, it still lies on $\mathscr{A}$, and in general densely fills $\mathscr{A}$ over infinite time. This is the notion of *robust transitivity* stated in [9, Section 3].

The classic theory of hyperbolic systems cannot be applied to compact flow-invariant sets containing equilibria accumulated by regular trajectories because the hyperbolic splitting $E_p^u \oplus E_p^c \oplus E_p^s$ cannot be extended *continuously* from regular trajectories to equilibria.

3.2 A New Theory Emerges

Although the Lorenz attractor is not uniformly hyperbolic, a peculiarity of Lorenz-type attractors is their "quasi"-similarity to the hyperbolic ones in terms of robustness. A weak form of hyperbolicity has been required and it became the source of the term *pseudo-hyperbolicity* used by Shilnikov and co-authors.

About ten years after Lorenz's work, a number of concrete Lorenz-like attractors were exhibited by Afraimovich, Bykov and Shilnikov [1] and Guckenheimer and Williams [30], for which the authors provided mathematical proofs that the models are sensitive to initial conditions, robust and non-hyperbolic. These are often called *geometric models*. Several papers describing the classic Lorenz attractor have been written—see [18, Ch. 9] and references therein.

Bautista [13] proved that if Λ is the geometric Lorenz attractor and $p \in \Lambda$ is a point lying in a hyperbolic periodic solution of Λ, then $\Lambda = \overline{W^u(p)}$. Moreover, Λ is a homoclinic class i.e., it can be seen as the topological closure of the transverse intersection of the invariant manifolds of a hyperbolic periodic solution in Λ. More formally:

$$\Lambda = \overline{W^u(p) \cap W^s(p)}.$$

Starting in the nineties, the systematic theory that explains the coexistence of robust flow-invariant sets containing equilibria and non-trivial sets of closed trajectories accumulating on them, has been developed. In three dimensions, Morales et al. [45] proved that robust sets containing equilibria are *singular hyperbolic sets*, i.e., they share the main properties of the geometric Lorenz attractor. The eigenvalues of the equilibria should be real and satisfy condition (3)—see [9, Lemma 3.22].

3.3 No Stable Solutions and Stochasticity

In the eighties, Sinai [65] introduced the following notion of attractor, very different from the usual one.

Definition 3 A *stochastic attractor* is an invariant closed set $\mathscr{A}$ in the phase space with the following properties:

1. There exists a neighbourhood U, $\mathscr{A} \subset U$, such that if $q \in U$, then

$$\lim_{t \to +\infty} dist(X^t(q), \mathscr{A}) = 0$$

2. For any initial probability distribution P_0 on $\mathscr{A}$, its shift as $t \to +\infty$ converges to an invariant distribution P on $\mathscr{A}$, independently of P_0.
3. The probability distribution P is *mixing*, i.e. the autocorrelation function tends to zero as $t \to +\infty$.

Both hyperbolic and Lorenz-type attractors are stochastic attractors and hence classical ergodic results may be used for their characterization. Small random perturbations essentially do not influence these attractors because the dynamic stochasticity tends to dominate the noise. The mixing condition 3. excludes the existence of attracting solutions. We refer the reader to [10, 41] for the ergodic theory of singular hyperbolic sets.

3.4 Examples

Now we give a brief discussion on examples of non-hyperbolic attractors that share some properties with the Lorenz-like sets. We refer the reader the source where a complete description of their dynamical properties is given. A good survey about these examples may be found in [35].

3.4.1 The Lorenz-Rovella Attractor

A new kind of attractor in three dimensions, the contracting *Lorenz attractor* or *Lorenz-Rovella attractor*, which is *probability persistent* but not *robust*, was obtained in [57] after two previous works by Arnéodo, Coullet and Tresser [11, 12]. It contains a hyperbolic equilibrium with real eigenvalues but now the sum of the leading eigenvalues is negative. It is persistent in terms of Lebesgue probability but not robust:

Persistence: there is a codimension two submanifold in the space of all vector fields, whose elements are full density points for the set of vector fields that exhibit an attractor of the same type in a generic family.

Non Robustness: for an open and dense set of perturbations, the attractors breaks into one or two stable periodic solutions, the equilibrium, a transitive and isolated set and heteroclinic connections between them.

This attractor combines a fold type behavior interacting with the dynamics associated to the presence of an equilibrium [18]—see the shape of its flow in Fig. 2b.

3.4.2 Cycles Involving at Least One Saddle-Focus

The theory concerning spiraling strange attractors containing saddle-foci is far from being completely understood; important examples have been described by Shilnikov [59, 60, 63], Tresser [69], Aguiar et al. [3], Rodrigues [52] and Rodrigues and Labouriau [55]. Although these spiralling sets are not robustly transitive, the three later examples may be persistent under symmetric perturbations.

The systematic study of the dynamics near a saddle-focus homoclinic cycle was pioneered by Shilnikov in the sixties. Hereafter, we assume that the eigenvalues of the linearization of the vector field at the equilibrium are given by: E and $-C \pm i\omega$, where $E, C, \omega \in \mathbf{R}^+\backslash\{0\}$—see Fig. 3. Under the eigenvalue condition $E > C$, infinitely many periodic solutions appear in any small neighborhood of the cycle [60, 63] . These periodic solutions are contained in suspended horseshoes and accumulate on the cycle.

Attractors with a spiral structure are expectable for perturbations of differential equations with two coexisting cycles. The periodic solutions near two symmetric saddle-focus homoclinic cycles are known to span every possible knot and link type.

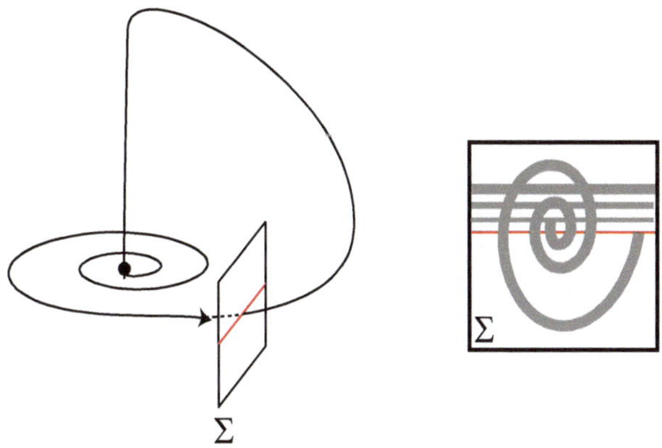

Fig. 3 Homoclinic cycle associated to a saddle-focus where the real expanding eigenvalue dominates the complex conjugate contracting eigenvalues. In any cross section to the cycle, we may find a Smale horseshoe leading to the occurrence of infinitely many periodic solutions of saddle-type. These periodic solutions accumulate on the cycle

The attractors associated to Shilnikov homoclinic cycles are characterized by the lack of uniform hyperbolicity, by the existence of a trajectory with positive Lyapunov exponent and by the existence of an open set in their basin of attraction. Under the condition $C < E < 2C$, Homburg [34] proved the existence of a dense set of homoclinic tangencies to hyperbolic periodic solutions and 2-periodic sinks nearby; see also [48].

Tresser [69, Section V] considered a heteroclinic cycle involving a saddle-focus and a saddle (non-focus) and obtained similar results, suggesting that the relevant part of the dynamics only depends on the presence of a saddle-focus—see Fig. 4. Motivated by the Lotka-Volterra systems, the author precised and generalized some of the Shilnikov conclusions for cycles involving not only saddle-foci.

After the classical homoclinic cycles associated to a single saddle-focus, *Bykov cycles* are the simplest heteroclinic cycles between two saddle-foci where one heteroclinic connection is structurally stable and the other is not. These cycles, also called by *T[erminal]-points*, are codimension two bifurcation cycles that involve two equilibria of different Morse indices. They have been first studied by Glendinning and Sparrow [25, 26] and later by Bykov [19]. Recently there has been a renewal of interest of this type of heteroclinic bifurcation in different scenarios—see for instance [23, 37, 53, 54] and references therein.

The authors of [3, 55] constructed explicit examples of vector fields containing a Bykov cycle on an attracting three-sphere. The construction has been amenable to the analytic proof of features that guarantee the existence of chaos. The explicit example consists of a vector field whose flow has a heteroclinic network with two saddle-foci, and a spiraling structure containing a hyperbolic transitive set. The cycle is structurally stable within the class of symmetric vector fields. It contains a two-dimensional connection that persists as a transverse intersection of invariant surfaces under symmetry-breaking perturbations.

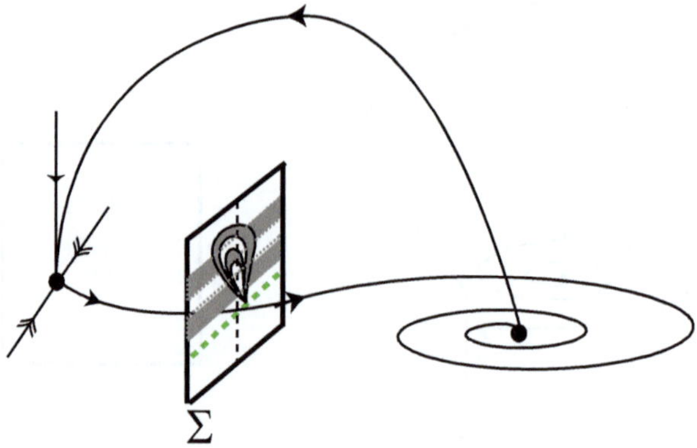

Fig. 4 Heteroclinic cycle studied in [69], involving a saddle-focus and a saddle (non-focus). Although Tresser did not present explicit examples, he made more precise some of Shilnikov's conclusions and generalized Shilnikov's results

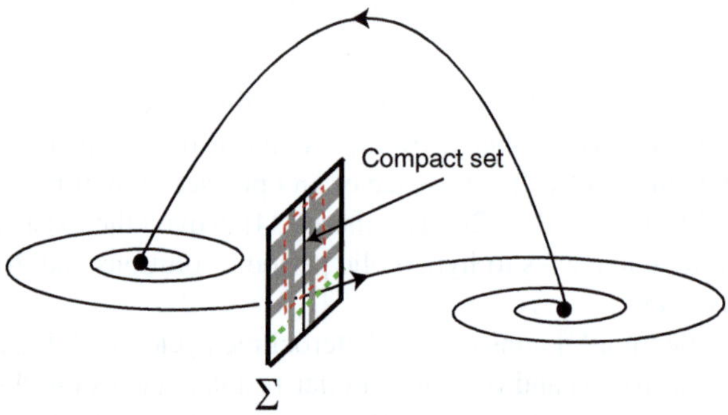

Fig. 5 Bykov cycle reported in [55]: heteroclinic cycle associated to two saddle-foci of different Morse indices, in which the one-dimensional invariant manifolds coincide and the two-dimensional invariant manifolds intersect transversely. There exists an increasing chain of suspended uniformly hyperbolic compact sets topologically conjugate to a full shift over a finite number of symbols, which accumulates on the cycle

Based on [37], by breaking the symmetry in a two-parameter family, the authors of [55] proved the existence of a wide range of dynamical behavior near the cycle: an attracting periodic trajectory; homoclinic orbits; heteroclinic connections that turn n times around the original cycle; suspended horseshoes and cascades of bifurcations of periodic trajectories near an unstable homoclinic cycle [38]. The coexistence of linked homoclinic orbits at the two saddle-foci has codimension 2 and takes place arbitrarily close to the cycle. Any invariant compact set in a section Σ transverse to the cycle, far from the invariant manifolds, is uniformly hyperbolic (for the first return map) [4]—see Fig. 5. This result is consistent with [9, Proposition 3.9].

Suggested by A. J. Homburg, Rodrigues [53] also proved that near the cycle, the shift dynamics does not trap most trajectories in the neighborhood of the network.

The example presented in [55] has an interesting relation with the works of [25, 26] about homoclinic cycles and T-points. The latter authors studied the existence of multiround heteroclinic cycles in a two-dimensional parameter diagram and found a logarithmic spiral of homoclinic cycles and more complicated Bykov cycles. In [26], the authors do not break the one-dimensional heteroclinic connection.

3.4.3 A Cycle Involving a Saddle-Focus and a Periodic Solution

Labarca and Pacífico [36] constructed a special case of robust non-hyperbolic flow whose first return map to a cross section resembles the Smale horseshoe map: the authors call it a *singular horseshoe*. It has been introduced as a model for stable non hyperbolic flows in the context of manifolds with boundary—see Fig. 6. In higher dimension than 3, the authors of [17] explored the dynamics near a heteroclinic network and showed that some cycles are preferred (under some conditions on the parameters). Some singular horseshoes could also coexist near the network.

We finish this section by noticing that the examples constructed in [3, 5, 55] are restrictions of symmetric polynomial vector fields in $\mathbf{R}^4$ and possess heteroclinic networks exhibiting heteroclinic switching [4, 5]. A simple polynomial form makes computations easier and allows the authors to prove the transverse intersection of two-dimensional invariant manifolds. All these heteroclinic networks are robust under any perturbation that does not break the symmetry of the system.

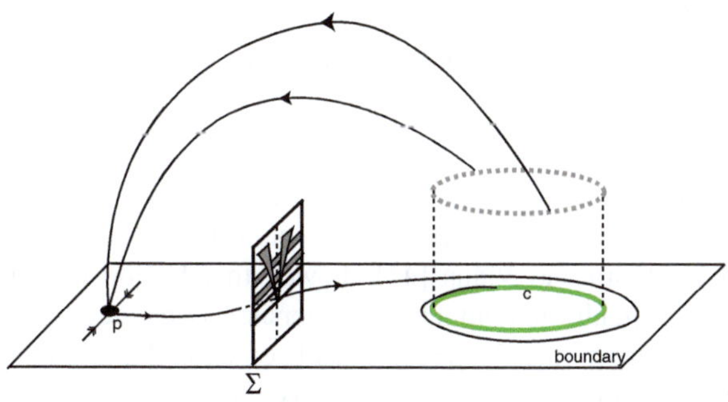

Fig. 6 A singular cycle and a singular horseshoe reported in [36]: the first return map for a singular horseshoe maps wedge-shaped regions to vertical strips, roughly contracting horizontal directions and expanding vertical directions (depicted here in a two-dimensional section Σ)

4 Wild Sets: Quasi-Stochastic Attractors

Like E. Lorenz, Hénon [32] provided computational arguments suggesting the existence of an attractor for a two-parameter family of quadratic maps of the plane—see details in [18]. Numerically, Hénon observed three important properties:

- the existence of folds,
- expansion along lines in the attractor and
- a fractal structure in a transversal direction.

After these numerical findings, the challenge was to provide a formal proof for the existence of a chaotic attractor with some "degree" of persistence.

4.1 Wild Attractors

First of all, let us define a wild attractor according to [62]. Suppose that the flow of a ODE possesses an attracting region embracing a hyperbolic (basic or not) set in which the stable and unstable manifolds are tangent. If it is so, such a hyperbolic set is called *wild* [46, 47].

By the last part of the seventies and going into the eighties, there were a series of very intriguing results that gave rise to new insights on how dynamics could develop in the future. Firstly, there was the work by Hénon [32], proposing a new kind of attractor. Following that, there was the work by Coullet and Tresser [20], concerning period doubling bifurcations for quadratic families of interval maps. The latter work forms part of a contribution that:

> ... sparkles a much more robust development than before.
>
> Jacob Palis [49], *2005*

The paper [32] has been a key to the remarkable work by Benedicks and Carleson [14] showing the existence of a probability persistent Hénon-like attractor, for some parameters. Mora and Viana [44] showed that Hénon-like attractors occur in the unfolding of quadratic homoclinic tangencies associated to dissipative fixed or periodic hyperbolic points. This result has been extended to higher dimensions, when the unstable manifold of the associated fixed or periodic point has dimension one (in the sectionally dissipative case). The suspension of this kind of sets is what Gonchenko [28, 29] calls *quasi-stochastic attractors*.

The term *quasi-stochastic attractor* denotes the limiting set enclosing periodic solutions of different Morse indices and structurally unstable homoclinic cycles, which may not be transitive. These attractors have not been sufficiently studied, particularly in the case where the dimension of the phase space is greater than 3.

4.2 Examples

Until a few years ago, Lorenz and hyperbolic attractors were the only ones that were classified as "genuine" attractors, which do not allow the appearance of periodic sinks under small perturbations. In [71], Turaev and Shilnikov provided a description of a wild hyperbolic spiral attractor that must be regarded as an attractor with irremovable sinks. There are not many explicit examples of this kind of sets. Recently, the authors of [39] found a mechanism to construct quasi-stochastic attractors—details in [51, Chapter 5].

In the context of Bykov cycles (see definition above), there are two different possibilities for the geometry of the flow around the cycle, depending on the direction trajectories turn around the heteroclinic connection of the one-dimensional invariant manifolds. In [3, 5, 37, 38, 53], the authors assumed (sometimes implicitly) that in the neighborhood of the two saddle-foci trajectories wind in the same direction around a heteroclinic connection.

In [39], the authors exhibit an example of a Bykov cycle where the different orientation of the flow around the one-dimensional manifolds has profound effects on the dynamics near the cycle; moreover, the authors of [39] proved that this phenomenon implies lack of uniform hyperbolicity near the cycle (for the first return map)—see Fig. 7. They found a condition defining an open subset in the space of parameters that determine the linear part of the vector field at the equilibria, inside which non-transverse intersections of the two-dimensional invariant manifolds of the two equilibria are dense. Such tangencies have been recognised as a mechanism

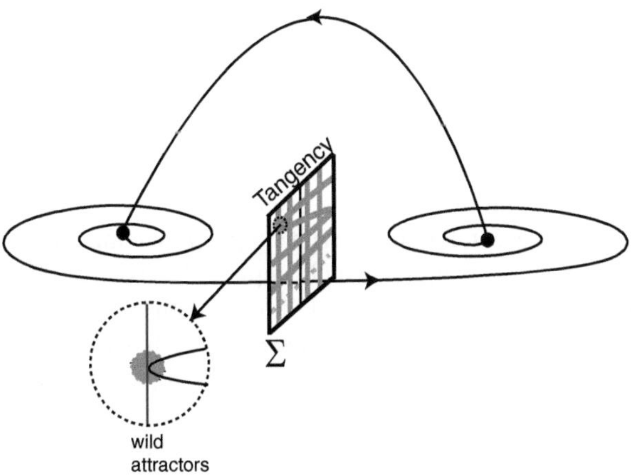

Fig. 7 Bykov cycle reported in [39]: heteroclinic cycle associated to two saddle-foci of different Morse indices, in which the one-dimensional invariant manifolds coincide and the two-dimensional invariant manifolds intersect transversely. For a full Lebesgue measure set of parameters that determine the linear part of the vector field at the equilibria, the authors found the coexistence of tangencies and transverse intersections of the two-dimensional invariant manifolds of the saddle-foci

for instability and lack of hyperbolicity in surface diffeomorphisms. Tangencies of invariant manifolds are associated to Newhouse phenomena: bifurcations leading to the birth of infinitely many asymptotically stable/unstable periodic solutions [46, 47].

The features of the invariant set constructed in [39, Section 6] fit in the properties of the quasi-stochastic attractors studied in [28]. The heteroclinic tangencies give rise to attracting periodic solutions which coexist with a basic set; the basins of attraction of some sinks lie in the gaps of the hyperbolic basic set. The transitive non-isolated set surrounds periodic solutions of different Morse indices, in sharp contrast to what is expected of attractors that are either uniformly hyperbolic or Lorenz-like. Hyperbolic sets and a countable set of stable solutions coexist in a set whose properties are far from being completely understood.

In this context, Shilnikov finishes the paper [62] with the opinion that:

> one should refrain from the fruitless ideology of complete description and turn to the study of some special but typical properties of the system.
>
> L.P. Shilnikov [62], *1997*

Finding "typical" properties will surely depend on the nature of the problem.

5 Final Remarks

In this survey, three types of chaotic sets have been identified: hyperbolic, Lorenz-type and quasi-attractors. Hyperbolic attractors are the limit sets of Smale's Axiom A systems and are structurally stable. Lorenz-type attractors are robustly transitive but are not structurally stable. Both types of attractors are stochastic in the sense that they have a mixing invariant measure. Quasi-stochastic attractors have periodic solutions of different Morse indices, where hyperbolic horseshoes and a countable set of Lyapunov-stable solutions may coexist.

In general, it is difficult to find explicit examples for which one can prove that the above attractors are present. Examples with simple polynomial forms of low degree are natural in symmetric contexts which implies the existence of flow-invariant submanifolds on which it is easier to find "fragile" homo/heteroclinic connections. Based on the vast catalog of exotic phenomena associated with heteroclinic cycles, we referred some examples of equivariant differential equations whose flows exhibit these structures and we characterized their non-wandering sets.

The theory of non-hyperbolic sets, even in dimension 3, is far from being understood. The study of quasi-stochastic attractors is almost untouched. We hope that this survey with the state of the art on the theory of hyperbolic and pseudo-hyperbolic attractors could be a starting point for a better understanding of the taxonomy of these sets.

Acknowledgements The author would like to express his gratitude to the referees for their helpull comments and also to Isabel Labouriau and Mário Bessa for suggestions and encouragement. CMUP is supported by the European Regional Development Fund through the programme COMPETE and by the Portuguese Government through the Fundação para a Ciência e a Tecnologia (FCT) under the project PEst-C/MAT/UI0144/2011. The author was supported by the grant with reference SFRH/BPD/84709/2012 of FCT.

References

1. Afraimovich, V.S., Bykov, V.V., Shilnikov, L.P.: On the appearence and structure of the Lorenz attractor. Dokl. Acad. Sci. USSR **234**, 336–339 (1977)
2. Aguiar, M.A.D., Castro, S.: Chaotic switching in a two-person game. Phys. D **239**, 1598–1609 (2010)
3. Aguiar, M.A.D., Castro, S.B., Labouriau, I.S.: Simple vector fields with complex behaviour. Int. J. Bifurcation Chaos **16**(2), 369–381 (2006)
4. Aguiar, M.A.D., Castro, S., Labouriau, I.: Dynamics near a heteroclinic network. Nonlinearity **18**, 391–414 (2005)
5. Aguiar, M.A.D., Labouriau, I.S., Rodrigues, A.A.P.: Swicthing near a heteroclinic network of rotating nodes. Dyn. Syst. **25**(1), 75–95 (2010)
6. Alekseev, V.: Quasirandom dynamical systems. I. Quasirandom diffeomorphisms. Math. Sbornik. Tom **76**(118), 1, 72–134 (1968)
7. Andronov, A., Pontryagin, L.: Systèmes grossiers. Dokl. Akad. Nauk USSR **14**, 247–251 (1937)
8. Anosov, D.V.: Geodesic flows on closed Riemannian manifolds of negative curvature. Proc. Steklov Math. Inst. **90**, 1–235 (1967)
9. Araújo, V., Pacífico, M.J.: Three-Dimensional Flows, Vol. 53 of Ergebnisse der Mathematik und ihrer Grenzgebiete. Springer, NewYork (2010)
10. Araújo, V. , Pacífico, M. J., Pujals, E., Viana, M.: Singular-hyperbolic attractors are chaotic. Trans. Am. Math. Soc. **361**(5), 2431–2485 (2009)
11. Arnéodo, A., Coullet, P., Tresser, C.: A possible new mechanism for the onset of turbulence. Phys. Lett. A **81**, 197–201 (1981)
12. Arnéodo, A., Coullet, P., Tresser, C.: Possible new strange attractors with spiral structure. Commun. Math. Phys. **79**, 573–579 (1981)
13. Bautista, S.: Sobre conjuntos hiperbólicos singulares, Ph.D. Thesis, IM.UFRJ, Rio de Janeiro (2005)
14. Benedicks, M., Carleson, L.:The dynamics of the Hénon map. Ann. Math. **133**, 73–169 (1991)
15. Birkhoff, G.D.: Dynamical systems. Am. Math. Soc. Colloq. Publ. **9**, 295 (1927)
16. Birkhoff, G.D.: Nouvelles recherches sur les systèmes dynamiques. Memorie Pont. Acad. Sci. Novo. Lyncaei **53**(1), 85–216 (1935)
17. Castro, S., Labouriau, I., Podvigina, O.: A heteroclinic network in mode interaction with symmetry. Dyn. Syst. Int. J. **25**(3), 359–396 (2010)
18. Bonatti, C., Díaz, L.J., Viana, M.: Dynamics Beyond Uniform Hyperbolicity. Springer, Berlin (2005)
19. Bykov, V.V.: Orbit structure in a neighbourhood of a separatrix cycle containing two saddle-foci. Am. Math. Soc. Transl. **200**, 87–97 (2000)
20. Coullet, P., Tresser, C.: Itérations d'endomorphims et groupe de renormalization. C. R. Acad. Sci. Paris Sér. I **287**, 577–580 (1978)
21. Devaney, R.: An Introduction to Chaotic Dynamical Systems, 2nd edn. Addison-Wesley, New York (1989)
22. Doering, C.: Persistently transtitive vector fields in three-dimensional manifolds. Proc. Dyn. Syst. Bifurcation Theory **160**, 59–89 (1987)

23. Fernández-Sánchez, F., Freire, E., Rodríguez-Luis, A.J.: T-points in a $\mathbf{Z}_2$-symmetric electronic oscillator. (I) analysis. Nonlinear Dyn. **28**, 53–69 (2002)
24. Franks, J.: Anosov diffeomorphisms, Global Analysis (Proc. Sympos. Pure Math., Vol. XIV, Berkeley, Calif.), vol. 1070, pp. 61–93. American Mathematical Society, Providence, RI (1968)
25. Glendinning, P., Sparrow, C.: Local and global behaviour near homoclinic orbits. J. Stat. Phys. **35**, 645–696 (1984)
26. Glendinning, P., Sparrow, C.: T-points: a codimension two heteroclinic bifurcation. J. Stat. Phys. **43**, 479–488 (1986)
27. Golubitsky, M.I., Stewart, I., Schaeffer, D.G.: Singularities and Groups in Bifurcation Theory, vol. II. Springer, New York (2000)
28. Gonchenko, S.V., Shilnikov, L.P., Turaev, D.V.: Dynamical phenomena in systems with structurally unstable Poincaré homoclinic orbit. Chaos **6**(1), 15–31 (1996)
29. Gonchenko, S.V. , Shilnikov, L.P., Stenkin, O.V., Turaev, D.V.: Bifurcations of systems with structurally unstable homoclinic orbits and moduli of Ω–equivalence. Comput. Math. Appl. **34**, 111–142 (1997)
30. Guckenheimer, J., Williams, R.F.: Structural stability of Lorenz attractors. Publ. Math. IHES **50**, 59–72 (1979)
31. Hayashi, S.: Connecting invariant manifolds and the solution of the C^1 stability and Ω-stability conjectures for flows. Ann. Math. **145**, 81–137 (1997)
32. Hénon, M.: A two dimensional mapping with a strange attractor. Comm. Math. Phys. **50**, 69–77 (1976)
33. Hirsch, M., Pugh, C., Shub, M.: Invariant Manifolds, Vol. 583 of Lecture Notes in Mathematics. Springer, New York (1977)
34. Homburg, A.J.: Periodic attractors, strange attractors and hyperbolic dynamics near homoclinic orbit to a saddle-focus equilibria. Nonlinearity **15**, 411–428 (2002)
35. Homburg, A.J., Sandstede, B.: Homoclinic and heteroclinic bifurcations in vector fields. In: Handbook of Dynamical Systems, vol. 3, pp. 379–524. North Holland, Amsterdam (2010)
36. Labarca, R., Pacífico, M.: Stability of singular horseshoes. Topology **25**, 337–352 (1986)
37. Labouriau, I.S., Rodrigues, A.A.P.: Global generic dynamics close to symmetry. J. Differ. Equ. **253**(8), 2527–2557 (2012)
38. Labouriau, I.S., Rodrigues, A.A.P.: Partial symmetry breaking and heteroclinic tangencies. In: Ibáñez, S., Pérez del Río, J.S., Pumariño, A., Rodríguez, J.A. (eds.) Progress and Challenges in Dynamical systems, Proceedings in Mathematics and Statistics, pp.281–299. Springer, NewYork (2013)
39. Labouriau, I.S., Rodrigues, A.A.P.: Dense heteroclinic tangencies near a Bykov cycle (2014). arXiv:1402.5455
40. Lorenz, E.: Deterministic nonperiodic flow. J. Atmos. Sci. **20**, 130–141 (1963)
41. Luzattto, S., Melbourne, I., Paccaut, F.: The Lorez attractor is mixing. Commun. Math. Phys. **260**, 393–401 (2005)
42. Mañé, R.: A proof of the C^1 stability conjecture. Publ. Math. IHES **66**, 161–210 (1988)
43. Melbourne, I.: Intermittency as a codimension-three Phenomenon. J. Dyn. Diff. Eqns. **1**(4), 347–367 (1989)
44. Mora, L., Viana, M.: Abundance of strange attractors. Acta Math. **171**, 1–71 (1993)
45. Morales, C.A., Pacífico, M.J., Pujals, E.R.: Robust transitive singular sets for 3-flows are partially hyperbolic attractors or repellers. Ann. Math. **160**(2), 375–432 (2004)
46. Newhouse, S.E.: Diffeomorphisms with infinitely many sinks. Topology **13**, 9–18 (1974)
47. Newhouse, S.E. : The abundance of wild hyperbolic sets and non-smooth stable sets for diffeomorphisms. Publ. Math. Inst. Hautes Études Sci. **50**, 101–151 (1979)
48. Ovsyannikov, I.M., Shilnikov, L.P.: On systems with saddle-focus homoclinic curve. Math. USSR Sbornik. **58**, 557–574 (1987)
49. Palis, J.: A global perspective for non-conservative dynamics. Ann. I. H. Poincaré **22**, 485–507 (2005)
50. Poincaré, H.: Sur le problème des trois corps et les équations de la dynamique. Acta Math. **13**, 1–270 (1890)

51. Rodrigues, A.A.P.: Heteroclinic Phenomena, PhD. Thesis, Department Matemática, Faculdade de Ciências da Universidade do Porto (2012)
52. Rodrigues, A.A.P.: Persistent Switching near a Heteroclinic Model for the Geodynamo Problem. Chaos Solitons Fractals **47**, 73–86 (2013)
53. Rodrigues, A.A.P.: Repelling dynamics near a Bykov cycle. J. Dyn. Diff. Equat. **25**(3), 605–625 (2013)
54. Rodrigues, A.A.P.: Moduli for heteroclinic connections involving saddle-foci and periodic solutions, Disc. Cont. Dyn. Syst. A **35**(7), 3155–3182 (2015)
55. Rodrigues, A.A.P., Labouriau, I.S.: Spiralling dynamics near a heteroclinic network. Phys. D **268**, 34–49 (2014)
56. Rodrigues, A.A.P., Labouriau, I.S., Aguiar, M.A.D.: Chaotic double cycling. Dyn. Syst. **26**(2), 199–233 (2011)
57. A. Rovella, A.: The dynamics of perturbations of contracting Lorenz maps. Bol. Soc. Brasil. Math. **24**, 233–259 (1993)
58. Shilnikov, L.P.: Strange attractors and dynamical models. J. Circuits Syst. Comput. **3**(1), 1–10 (1993)
59. Shilnikov, L.P.: Some cases of generation of periodic motion from singular trajectories. Math. USSR Sbornik **61**(103) 443–466 (1963)
60. Shilnikov, L.P.: A case of the existence of a denumerable set of periodic motions. Sov. Math. Dokl. **6**, 163–166 (1965)
61. Shilnikov, L.P.: A Poincaré-Birkhoff problem. Mat. Sb. **74**, 378–397 (1967)
62. Shilnikov, L.P.: Bifurcations and strange attractors. In: Proceedings of the International Congress of Mathematicians, vol. III, pp. 349–372. Higher Ed. Press, Beijing (2002)
63. Shilnikov, L.P.: The existence of a denumerable set of periodic motions in four dimensional space in an extended neighbourhood of a saddle-focus. Soviet Math. Dokl. **8**(1), 54–58 (1967)
64. Shub, M.: Global Stability of Dynamical Systems. Springer, NewYork (1987)
65. Sinai, Y.G.: Stochasticity of dynamical systems. In: Gaponov-Grekhov, A.V (ed.) Nonlinear Waves, pp. 192–212. Moskva Nauka, Moscow (1981)
66. Smale, S.: Diffeomorphisms with many periodic orbits. In: Cairus, S. (ed.) Differential Combinatorial Topology, pp. 63–86. Princeton University Press, Princeton (1960)
67. Smale, S.: Differentiable dynamical systems. Bull. Am. Math. Soc. **73**, 747–817 (1967)
68. Stewart, I.: Sources of uncertainty in deterministic dynamics: an informal overview. Phil. Trans. R. Soc. A **369**, 4705–4729 (2011)
69. Tresser, C.: About some theorems by L. P. Shilnikov. Ann. Inst. Henri Poincaré **40**, 441–461 (1984)
70. Tucker, W.: A rigorous ODE solver and Smale's 14th problem. Found. Comput. Math. **2**, 53–117 (2002)
71. Turaev, D., Shilnikov, L.P.: An example of a wild strange attractor. Mat. Sb. **189**(2), 137–160 (1998)

Dengue in Madeira Island

Helena Sofia Rodrigues, M. Teresa T. Monteiro, Delfim F.M. Torres, Ana Clara Silva, Carla Sousa, and Cláudia Conceição

Abstract Dengue is a vector-borne disease and 40 % of world population is at risk. Dengue transcends international borders and can be found in tropical and subtropical regions around the world, predominantly in urban and semi-urban areas. A model for dengue disease transmission, composed by mutually-exclusive compartments representing the human and vector dynamics, is presented in this study. The data is from Madeira, a Portuguese island, where an unprecedented outbreak was detected on October 2012. The aim of this work is to simulate the repercussions of the control measures in the fight of the disease.

1 Introduction

During the last decades, the global prevalence of dengue increased considerably. Madeira's dengue outbreak of 2012 is the first epidemics in Europe since the one recorded in Greece in 1928 [16]. Local transmission was also reported, for the first time, in France and Croatia in 2010 [15, 20] and the threat of possible

H.S. Rodrigues
Escola Superior de Ciências Empresariais, Instituto Politécnico de Viana do Castelo, Valença, Portugal
e-mail: sofiarodrigues@esce.ipvc.pt

M.T.T. Monteiro
ALGORITMI, Departamento de Produção e Sistemas, Universidade do Minho, Braga, Portugal
e-mail: tm@dps.uminho.pt

D.F.M. Torres (✉)
CIDMA, Departamento de Matemática, Universidade de Aveiro, Aveiro, Portugal
e-mail: delfim@ua.pt

A.C. Silva
Instituto de Administração da Saúde e Assuntos Sociais, IP-RAM, Funchal, Madeira, Portugal
e-mail: anaclarasilv@gmail.com

C. Sousa • C. Conceição
Instituto de Higiene e Medicina Tropical, Universidade Nova de Lisboa, Lisboa, Portugal
e-mail: CASousa@ihmt.unl.pt; claudiaconceicao@ihmt.unl.pt

J.-P. Bourguignon et al. (eds.), *Dynamics, Games and Science*, CIM Series in Mathematical Sciences 1, DOI 10.1007/978-3-319-16118-1_32

outbreaks of dengue fever in Europe is increasing. According to a recent study [3], 390 million dengue infections occur per year worldwide, of which 96 million with clinical symptoms. Methods considered by authorities for disease prevention include educational and vaccination campaigns, preventive drugs administration and surveillance programs.

Mathematical modeling plays a fundamental role in the study of the evolution of infectious diseases [1, 33, 36]. When formulating a model for a particular disease, a trade-off between simple and complex models is always present. The former, omit several details and are generally used for short-term and specific situations, but have the disadvantage of possibly being naive and unrealistic. The complex models have more details and are more realistic, but are generally more difficult to solve and analyze or may contain parameters whose estimates cannot be obtained [8]. Here we are interested in a dengue model defined by a system of ordinary differential equations, which enables the evaluation of the infectious disease transmission patterns.

The text is organized as follows. Section 2 presents some details about dengue, such as disease symptoms and vector transmission issues. The outbreak on Madeira island and measures to fight against the epidemics are described in Sect. 3. In Sect. 4 the mathematical model for the interaction between humans and mosquitoes is formulated, while numerical experiments using distinct levels of control are presented in Sect. 5. We end with Sect. 6 of conclusions and ideas for future work.

2 Dengue and the *Aedes* Mosquito

Dengue is a vector-borne disease transmitted from an infected human to an *Aedes* mosquito, commonly *Aedes aegypti* or *Aedes albopictus*, during a female blood-meal [4]. Then, the infectious mosquito, that needs regular meals of blood to mature their eggs, bites a potential healthy human and transmits the disease, thus completing the extrinsic cycle of the virus. Four dengue serotypes are known, designated as DEN-1, 2, 3 and 4, which cause a wide spectrum of human disease, from asymptomatic cases to classic dengue fever (DF) and more severe cases, known as dengue hemorrhagic fever (DHF). Symptoms include fever, headache, nausea, vomiting, rash, and pain in the eyes, joints, and muscles. Symptoms may appear up to two weeks after the bite of an infected mosquito and usually last for one week. In severe cases, symptoms may include intense stomach pain, repeated vomiting, and bleeding from the nose or gums and can lead to death. Recovery from infection by one virus provides lifelong immunity against that virus but only confers partial and transient protection against subsequent infection by the other three serotypes. There is good evidence that a sequential infection increases the risk of developing DHF [39].

Unfortunately, there is no specific treatment for dengue. Activities, such as triage and management, are critical in determining the clinical outcome of dengue. A rapid and efficient front-line response not only reduces the number of unnecessary hospital admissions but also saves lives. Although there is no effective and safe vaccine for dengue, a number of candidates are undergoing various phases of clinical trials [40]. With four closely related serotypes that can cause the disease, there is a need for an effective vaccine that would immunize against all four types; if not, a secondary infection could, theoretically, lead to a DHF case. Another difficulty in the vaccine production is that there is a limited understanding of how the disease typically behaves and how the virus interacts with the immune system. Research to develop a vaccine is ongoing and the incentives to study the mechanism of protective immunity are gaining more support, now that the number of outbreaks around the world is increasing [6]. Several mathematical models, including a few taking into account vaccination and optimal control, have been proposed in the literature: see [27–31] and references therein.

The life cycle of the mosquito has four distinct stages: egg, larva, pupa and adult. The first three stages take place in water, whilst air is the medium for the adult stage. *Aedes* females have a peculiar oviposition behavior: they do not lay all the eggs of an oviposition at once, in the same breeding site, but rather release them in different places, thus increasing the probability of successful births [23, 35]. In urban areas, *Aedes aegypti* breed on water collections in artificial containers such as cans, plastic cups, used tires, broken bottles and flower pots. With increasing urbanization and crowded cities, environmental conditions foster the spread of the disease that, even in the absence of fatal forms, breed significant economic and social costs (absenteeism, immobilization, debilitation and medication) [7].

It is very difficult to control or eliminate *Aedes* mosquitoes because they are highly resilient, quickly adapting to changes in the environment and they have the ability to rapidly bounce back to initial numbers after disturbances resulting from natural phenomena (e.g., droughts) or human interventions (e.g., control measures). We can safely expect that transmission thresholds will vary depending on a range of factors. Reduction of vector populations, both adult mosquitoes and in immature states, is currently the only way to prevent dengue.

3 Madeira's Dengue Outbreak

An outbreak of dengue fever, that lasted about 21 weeks between early October 2012 and late February 2013, occurred in Madeira, a Portuguese island, whose capital is Funchal. As March 12th, 2013, 2,168 probable cases of dengue fever have been reported, of which 1,084 were laboratory confirmed. All reported cases refer to the resident population of the island and no deaths or severe cases were reported. On the same day, according to the data available, the outbreak was considered finished by the Portuguese Health Authorities, since there was no autochthonous cases in the island [9]. The notified dengue fever cases in Madeira, by week, are in Fig. 1. Note that the number of confirmed dengue cases is lower than the notified ones.

In Fig. 2 it is possible to see the cumulative incidence of dengue cases along the island, by parish. Santa Luzia parish is the one that recorded the highest proportion

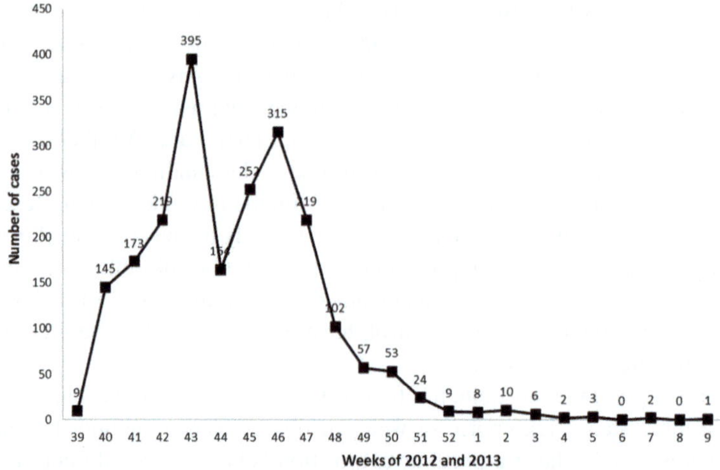

Fig. 1 Notified dengue fever cases in Madeira, by week, from October 2012 to February 2013 (*Source*: Instituto de Administração da Saúde e Assuntos Sociais, Região Autónoma da Madeira)

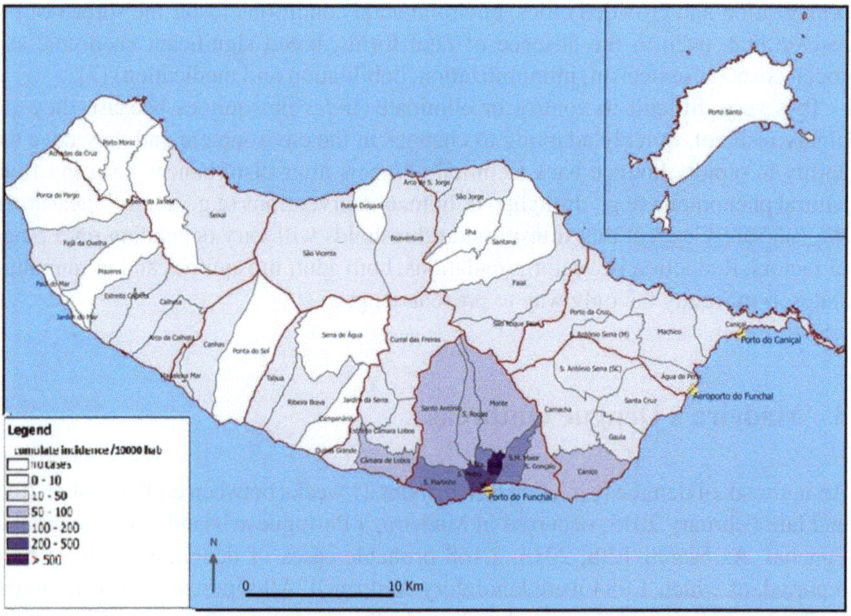

Fig. 2 Cumulative incidence of dengue in Madeira, by parish, from October 2012 to February 2013 (*Source*: Instituto de Administração da Saúde e Assuntos Sociais, Região Autónoma da Madeira)

of patients. As the mosquito lives mainly in urban areas with high human population density, and the vast majority of human cases were observed in this civil parish of the Funchal council, our study is constrained to this area.

The mosquito *Aedes aegypt* was detected in Madeira, for the first time, in 2005. The National Institute of Health Doutor Ricardo Jorge (INSA) performs reference laboratory diagnosis of dengue in Portugal. INSA conducted confirmatory laboratory diagnosis and identified the presence of DEN-1 virus in human samples [19]. Molecular analyses reported that the virus in Madeira island could have origin in Brazil or Venezuela, where the virus presents similar features and with whom there are intensive movements of trade and people [19].

After the acknowledgement of the presence of the dengue mosquito in Madeira, local Health authorities implemented several strategies to control this invasive specie of mosquitoes. However, the results showed small effects. *Aedes aegypti* in Madeira present a high resistance level to DDT, permethrin and deltamethrinm, the common tools allowed by the World Health Organization (WHO) [34]. Therefore, local measures changed to educational campaigns and entomological surveillance, to monitor the vector spread using traps, both for eggs and adult forms. Educational campaigns appealed the population to apply repellent and wear large clothes to avoid mosquito bites. Moreover, all recipients that could serve to breed the mosquito, like water collections in artificial containers (e.g., cans, plastic cups, used tires, broken bottles and flower pots), were asked to be removed or covered. Media-based tools were used to inform the population. These included newspapers, TV programs, TV spots, radio programs, radio spots, flyers, internet sites, announcements and specific talks in public places. A medical appointment dedicated to the dengue disease was also implemented in a health unit in Funchal, and a program for the monitoring of traps implemented. The number of eggs per trap, dispersedly placed along the island, with emphasis on the southern slope, were counted in order to understand their spatial distribution. Weekly entomological reports of *Aedes aegypti* in Madeira island were, and still are, broadcasted to sectorial partners. The application of insecticide was only applied in strategic places, such as the central hospital, the health unit dedicated to the attendance of dengue cases and a school identified as a transmission area [37].

4 The Mathematical Model

Taking into account the model presented in [10, 11] and the considerations of [25, 26], a temporal mathematical model to study Madeira's dengue outbreak is here proposed. It includes three epidemiological states for humans:

$S_h(t)$—susceptible (individuals who can contract the disease);
$I_h(t)$—infected (individuals who can transmit the disease);
$R_h(t)$—resistant (individuals who have been infected and have recovered).

These compartments are mutually-exclusive. There are three other state variables, related to the female mosquitoes (male mosquitos are not considered because they do not bite humans and consequently do not influence the dynamics of the disease):

$A_m(t)$—aquatic phase (includes egg, larva and pupa stages);
$S_m(t)$—susceptible (mosquitoes that can contract the disease);
$I_m(t)$—infected (mosquitoes that can transmit the disease).

In order to make a trade-off between simplicity and reality of the epidemiological model, some assumptions are considered:

- There is no vertical transmission, *i.e.*, an infected mosquito cannot transmit the disease to their eggs;
- Total human population N_h is constant: $S_h(t) + I_h(t) + R_h(t) = N_h$ at any time t;
- The population is homogeneous, which means that every individual of a compartment is homogeneously mixed with the other individuals;
- Immigration and emigration are not considered during the period under study;
- Homogeneity between host and vector populations, that is, each vector has an equal probability to bite any host;
- Humans and mosquitoes are assumed to be born susceptible.

To analyze the disease evolution, two control measures are considered in the model:

$c_m(t)$—proportion of insecticide (adulticide), $0 \leq c_m(t) \leq 1$;
$1 - \alpha(t)$—proportion of ecological control, $0 < \alpha(t) \leq 1$.

The application of adulticides is the most common control measure. However, its efficacy is often constrained by the difficulty in achieving sufficiently high coverage of resting surfaces and the insecticide resistance by the mosquito. Besides, the long term use of adulticide comports several risks: it can affect other species, it is linked to numerous adverse health effects, including the worsening of asthma and respiratory problems. The purpose of ecological control, that is, educational campaigns, is to reduce the number of larval habitat areas available to mosquitoes. The mosquitoes are most easily controlled by treating, cleaning and/or emptying containers that hold water, since the eggs of the specie are laid in water-holding containers. The ecological control must be done by both public health officials and residents in the affected areas. The participation of the entire population in removing still water from domestic recipients and eliminating possible breeding sites is essential [40].

Our dengue epidemic model makes use of the parameters described in Table 1 and consists of the system of differential equations

$$\begin{cases} \dfrac{dS_h(t)}{dt} = \mu_h N_h - \left(B\beta_{mh} \dfrac{I_m(t)}{N_h} + \mu_h \right) S_h(t) \\ \dfrac{dI_h(t)}{dt} = B\beta_{mh} \dfrac{I_m(t)}{N_h} S_h(t) - (\eta_h + \mu_h) I_h(t) \\ \dfrac{dR_h(t)}{dt} = \eta_h I_h(t) - \mu_h R_h(t) \end{cases} \tag{1}$$

Table 1 Parameters in the epidemiological model (1)–(2)

Parameter	Description	Range of values in literature	Value used	Source
N_h	Total population		112,000	[18]
B	Average daily biting (per day)		1/3	[12]
β_{mh}	Transmission probability from I_m (per bite)	[0.25, 0.33]	0.25	[12]
β_{hm}	Transmission probability from I_h (per bite)	[0.25, 0.33]	0.25	[12]
$1/\mu_h$	Average lifespan of humans (in days)		$1/79 \times 365$	[18]
$1/\eta_h$	Average viremic period (in days)	[1/15, 1/4]	1/7	[5]
$1/\mu_m$	Average lifespan of adult mosquitoes (in days)	[1/45, 1/8]	1/15	[14, 17, 22]
φ	Number of eggs at each deposit per capita (per day)		6	[32]
$1/\mu_A$	Natural mortality of larvae (per day)		0.2363	[2]
η_A	Maturation rate from larvae to adult (per day)	[1/11, 1/7]	1/9	[24]
k	Number of larvae per human		0.9	[13, 38]

coupled with the nonlinear control system

$$\begin{cases} \dfrac{dA_m(t)}{dt} = \varphi\left(1-\dfrac{A_m(t)}{\alpha(t)kN_h}\right)(S_m(t)+I_m(t)) - (\eta_A+\mu_A)\,A_m(t) \\ \dfrac{dS_m(t)}{dt} = \eta_A A_m(t) - \left(B\beta_{hm}\dfrac{I_h(t)}{N_h}+\mu_m+c_m(t)\right)S_m(t) \\ \dfrac{dI_m(t)}{dt} = B\beta_{hm}\dfrac{I_h(t)}{N_h}S_m(t) - (\mu_m+c_m(t))\,I_m(t) \end{cases} \tag{2}$$

subject to the initial conditions

$$\begin{array}{lll} S_h(0)=S_{h0}, & I_h(0)=I_{h0}, & R_h(0)=R_{h0}, \\ A_m(0)=A_{m0}, & S_m(0)=S_{m0}, & I_m(0)=I_{m0}. \end{array}$$

Note that the differential equation related to the aquatic phase does not involve the control variable c_m, because the adulticide does not produce effects in this stage of mosquito life. Figure 3 shows a scheme of the model.

In the next section we study the reality of Madeira's outbreak, using the most reliable information about the mosquito and the infected people. For a mathematical analysis of the model, in particular the analysis of equilibrium points and the basic reproduction number, we refer the reader to [28].

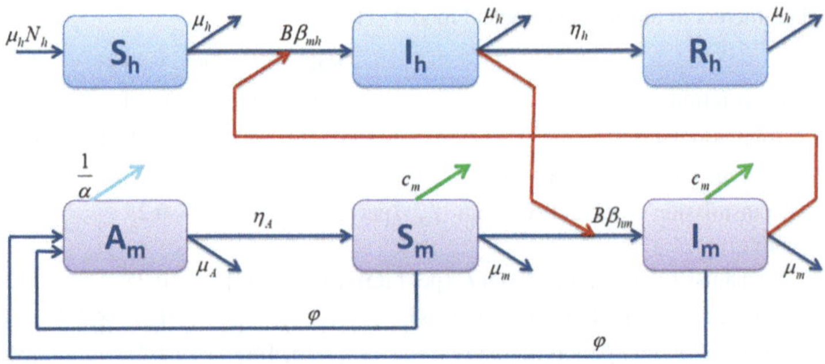

Fig. 3 Epidemiological model SIR (1) + ASI (2)

5 Numerical Experiments

In this section, numerical results are presented. Our aim is to show a simulation of the possible evolution of the dengue outbreak occurred on Madeira island, using parameterized and validated epidemiological and entomological data. The human data was adapted to the Madeira region through official data [9, 18]. Despite the research efforts, some information required for the model parametrization still lacks, especially entomological. This is due to the difficulty of obtaining it by laboratory assays. Even when experiments are possible, sometimes the mosquito behavior presents distinct features when in a controlled environment or in nature [21]. For this reason, a range of values for mosquito parameters were analyzed (see Table 1 for details). The initial values for the system of differential equations (1)–(2) are:

$$S_h(0) = 111991, \quad I_h(0) = 9, \quad R_h(0) = 0,$$
$$A_m(0) = 111900 \times 6, \quad S_m(0) = 111900 \times 3, \quad I_m(0) = 1000.$$

The software used in the simulations was `Matlab`, with the routine `ode45`. This function implements a Runge–Kutta method with a variable time step for efficient computation. Figure 4 shows different simulations for educational campaigns, without application of insecticide. This was the major control measure to fight the disease. It is possible to see that educational campaigns have an important role in the decrease of infected human and the best curve that fits the real data has an implementation of ecological control between 50 % and 75 %. Table 2 presents the total number of infected human individuals for the simulations done. Without any control measure, about 12 % of all population of Funchal would be infected. A common sense conclusion is that when we increase the proportion of control measures, the number of infected decreases considerably. In the same manner, the application of even small quantities of insecticide, seems to increase the effects of ecological control (compare simulations C and E). The explanation for this lies in the fact that, when an outbreak occurs, the application of an efficient adulticide

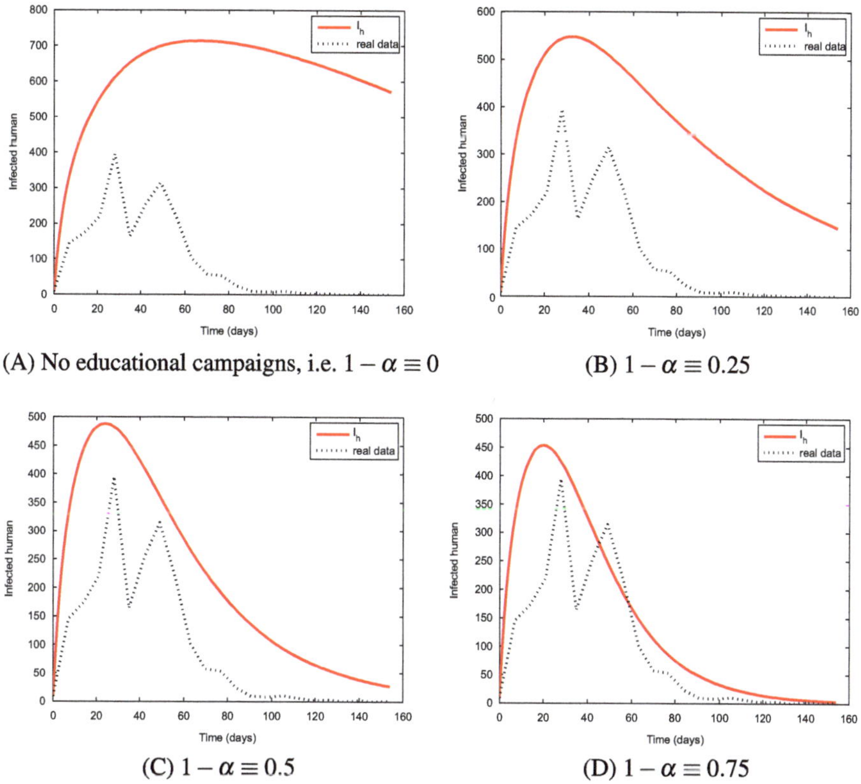

(A) No educational campaigns, i.e. $1-\alpha \equiv 0$

(B) $1-\alpha \equiv 0.25$

(C) $1-\alpha \equiv 0.5$

(D) $1-\alpha \equiv 0.75$

Fig. 4 Number of infected individuals without insecticide usage (i.e., $c_m = 0$) but with distinct levels of educational campaigns (*continuous line*) versus observed real data (*dotted line*)

Table 2 Total number of infected individuals

Simulations	Control values	Total number of infected
A	No control, i.e. $1-\alpha = c_m \equiv 0$	13,677
B	$1-\alpha \equiv 0.25$ and $c_m \equiv 0$	7,719
C	$1-\alpha \equiv 0.50$ and $c_m \equiv 0$	4,827
D	$1-\alpha \equiv 0.75$ and $c_m \equiv 0$	3,388
E	$1-\alpha \equiv 0.5$ and $c_m \equiv 0.01$	3,073
F	$1-\alpha \equiv 0.5$ and $c_m \equiv 0.02$	2,210
G	$1-\alpha \equiv 0.5$ and $c_m \equiv 0.05$	1,179
Real data		2,168

will immediately affect the transmission rate of the virus. Educational campaigns, even being a good strategy for the ecological control of the mosquito, imply time to promote the necessary motivation for the people to react to the disease. The graphs with education campaigns at 50 % and a variation of insecticide application are shown in Fig. 5. The curves that better illustrate the peak of the epidemics use

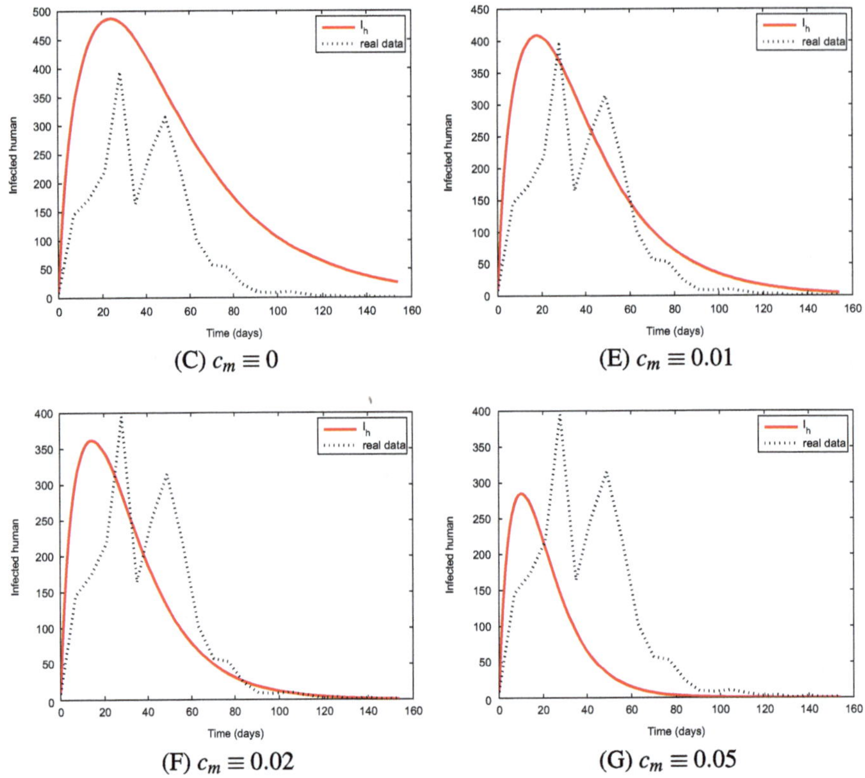

Fig. 5 Number of infected individuals for a constant educational campaign of $1-\alpha \equiv 0.5$ and distinct levels of insecticide (*continuous line*) versus observed real data (*dotted line*)

between 0 % and 1 % of insecticide. However, when compared to the table of total infected cases (Table 2), the nearest simulation is F ($\alpha = 50\,\%$ and $c_m = 2\,\%$). In fact, this difference can be explained by the fact that the simulation is made by comparing the total infected cases with the notified ones. It is well known by Health Authorities that, besides the asymptomatic cases, some patients do not go to the Health Centers: not only because they have light symptoms but also because their relatives or neighbors had already had dengue and they think they can handle the situation by themselves, at home.

Remark 1 A simple tuning of the control parameters α and c_m, by using some optimization technique like least square or curve fitting, does not seem appropriate here. Using the least square method to choose the best combination of the two controls, we obtained the proportion of adulticide $c_m = 0.0280$ and the value for educational campaigns $1-\alpha = 0$. This last value implies that the ecological control has no influence whatsoever in the system, which is not in agreement with the case under study.

6 Conclusions

One of the most important issues in epidemiology is to improve control strategies with the final goal to reduce or even eradicate a disease. In this paper a dengue model based on two populations, humans and mosquitoes, with educational and insecticide control measures, has been presented. Our study provides some important epidemiological insights about the impact of vector control measures into dengue in Madeira island. The work was done with collaboration of the *Instituto de Higiene e Medicina Tropical*, which provided us with valuable information about the disease characteristics and entomologic aspects, and *Instituto de Administração da Saúde e Assuntos Sociais* from Madeira, which gave us specific information about the outbreak, namely real numbers of the disease, affected areas and what kind of control was done in the island. Such cooperation and discussions with entomologists and doctors, was crucial to tune the parameter values of the mathematical model. Our results show how dengue burden can decrease with the help of vector control measures such as insecticide and ecological control. We concluded that small quantities of insecticide have a considerable impact in the short time intervention when an outbreak occurs. The application of educational campaigns decreases the disease burden and can act as a long time prevention. As future work, we intend to add an optimal control analysis to decide whether a given combination of control values is the best. Such analysis will be important for policy makers to know the optimal combination of the control strategies.

Acknowledgements This work was partially supported by The Portuguese Foundation for Science and Technology (FCT): Rodrigues and Torres through the Center for Research and Development in Mathematics and Applications (CIDMA) within project UID/MAT/04106/2013; Monteiro by the ALGORITMI Research Centre and project UID/CEC/00319/2013; Silva and Sousa by the project "Dengue in Madeira archipelago. Risk assessment for the emergence of Aedes aegypti mediated arboviroses and tools for vector control" with reference PTDC/SAU-EPI/115853/2009. The authors are very grateful to two anonymous referees, for valuable remarks and comments, which contributed to the quality of the paper.

References

1. Anderson, R.M., May, R.M.: Infectious Diseases of Humans: Dynamics and Control. Oxford University Press, Oxford (1991)
2. Barrios, J., Piétrius, A., Joya, G., Marrero, A., de Arazoza, H.: A differential inclusion approach for modeling and analysis of dynamical systems under uncertainty. Application to dengue disease transmission. Soft. Comput. **17**, 239–253 (2013)
3. Bhatt, S., et al.: The global distribution and burden of dengue. Nature **496**, 504–507 (2013)
4. Cattand, P., et al.: Tropical diseases lacking adequate control measures: dengue, leishmaniasis, and African trypanosomiasis. Disease Control Priorities in Developing Countries, pp. 451–466. DCPP Publications, Washington (DC) (2006)
5. Chan, M., Johansson, M.A.: The incubation periods of dengue viruses. PLoS ONE **7**(11), e50972 (2012)

6. Dengue Virus Net: http://denguevirusnet.com, Jan 2013
7. Derouich, M., Boutayeb, A.: Dengue fever: mathematical modelling and computer simulation. Appl. Math. Comput. **177**(2), 528–544 (2006)
8. Diekmann, O., Heesterbeek, J.A.P.: Mathematical Epidemiology of Infectious Diseases: Model Building, Analysis and Interpretation. Wiley, New York (2000)
9. DGS. Directorate-General of Health: Dengue Madeira. http://www.dgs.pt, Mar 2013
10. Dumont, Y., Chiroleu, F.: Vector control for the chikungunya disease. Math. Biosci. Eng. **7**(2), 313–345 (2010)
11. Dumont, Y., Chiroleu, F., Domerg, C.: On a temporal model for the chikungunya disease: modeling, theory and numerics. Math. Biosci. **213**(1), 80–91 (2008)
12. Focks, D.A., Brenner, R.J., Hayes, J., Daniels, E.: Transmission thresholds for dengue in terms of Aedes aegypti pupae per person with discussion of their utility in source reduction efforts. Am. J. Trop. Med. Hyg. **62**, 11–18 (2000)
13. Focks, D.A., Daniels, E., Haile, D.G., Keesling, J.E.: A simulation model of the epidemiology of urban dengue fever: literature analysis, model development, preliminary validation, and samples of simulation results. Am. J. Trop. Med. Hyg. **53**, 489–506 (1995)
14. Focks, D.A., Haile, D.G., Daniels, E., Mount, G.A.: Dynamic life table model for Aedes aegypti (Diptera: Culicidae): analysis of the literature and model development. J. Med. Entomol. **30**, 1003–1017 (1993)
15. Gjenero-Margan, I., et al.: Autochthonous dengue fever in Croatia, August–September 2010. Euro Surveill. **16**(9), 1–4 (2011)
16. Halstead, S.B., Papaevangelou, G.: Transmission of dengue 1 and 2 viruses in Greece in 1928. Am. J. Trop. Hyg. **29**(4), 635–637 (1980)
17. Harrington, L.C., et al.: Analysis of survival of young and old Aedes aegypti (Diptera: Culicidae) from Puerto Rico and Thailand. J. Med. Entomol. **38**, 537–547 (2001)
18. INE.: Statistics Portugal. http://censos.ine.pt
19. INSA. National Health Institute Doutor Ricardo Jorge.: Dengue Madeira. http://www.insa.pt/sites/INSA/Portugues/ComInf/Noticias/Paginas/DengueMadeiraDiagLab.aspx (2012)
20. La Ruche, G., et al.: First two autochthonous dengue virus infections in metropolitan France, September 2010. Euro Surveill. **15**(39), 1–5 (2010)
21. Luz, P.M., Codeço, C.T., Massad, E., Struchiner, C.J.: Uncertainties regarding dengue modeling in Rio de Janeiro, Brazil. Mem. Inst. Oswaldo Cruz **98**(7), 871–878 (2003)
22. Maciel-de-Freitas, R., Marques, W.A., Peres, R.C., Cunha, S.P., Lourenço-de-Oliveira, R.: Variation in Aedes aegypti (Diptera: Culicidae) container productivity in a slum and a suburban district of Rio de Janeiro during dry and wet seasons. Mem. Inst. Oswaldo Cruz **102**, 489–496 (2007)
23. Otero, M., Schweigmann, N., Solari, H.G.: A stochastic spatial dynamical model for aedes aegypti. Bull. Math. Biol. **70**(5), 1297–1325 (2008)
24. Padmanabha, H., et al.: Temperature induces trade-offs between development and starvation resistance in Aedes aegypti (L.) larvae. Med. Vet. Entomol. **25**(4), 445–453 (2011)
25. Rodrigues, H.S., Monteiro, M.T.T., Torres, D.F.M.: Optimization of dengue epidemics: a test case with different discretization schemes. AIP Conf. Proc. **1168**(1), 1385–1388 (2009)
26. Rodrigues, H.S., Monteiro, M.T.T., Torres, D.F.M.: Insecticide control in a dengue epidemics model. AIP Conf. Proc. **1281**(1), 979–982 (2010)
27. Rodrigues, H.S., Monteiro, M.T.T., Torres, D.F.M.: Dynamics of dengue epidemics when using optimal control. Math. Comput. Modell. **52**(9–10), 1667–1673 (2010)
28. Rodrigues, H.S., Monteiro, M.T.T., Torres, D.F.M.: Dengue in Cape Verde: vector control and vaccination. Math. Popul. Stud. **20**(4), 208–223 (2013)
29. Rodrigues, H.S., Monteiro, M.T.T., Torres, D.F.M.: Bioeconomic perspectives to an optimal control dengue model. Int. J. Comput. Math. **90**(10), 2126–2136 (2013)
30. Rodrigues, H.S., Monteiro, M.T.T., Torres, D.F.M.: Vaccination models and optimal control strategies to dengue. Math. Biosci. **247**(1), 1–12 (2014)

31. Rodrigues, H.S., Monteiro, M.T.T., Torres, D.F.M.: Seasonality effects on Dengue: basic reproduction number, sensitivity analysis and optimal control. Math. Methods Appl. Sci. Doi:10.1002/mma.3319
32. Rodrigues, H.S., Monteiro, M.T.T., Torres, D.F.M., Zinober, A.: Dengue disease, basic reproduction number and control. Int. J. Comput. Math. **89**(3), 334–346 (2012)
33. Rodrigues, P., Silva, C.J., Torres, D.F.M.: Cost-effectiveness analysis of optimal control measures for tuberculosis. Bull. Math. Biol. **76**(10), 2627–2645 (2014). Doi:10.1007/s11538-014-0028-6
34. Seixas, G.F.R.: Aedes (Stegomyia) aegypti (Diptera, Culicidae) da ilha da Madeira: origem geográfica e resistência aos insecticidas. Master Thesis. Universidade Nova de Lisboa, Lisboa (2012)
35. Shuman, E.K.: Global climate change and infectious diseases. Emerging infectious diseases. N. Engl. J. Med. **362**, 1061–1063 (2010)
36. Silva, C.J., Torres, D.F.M.: Optimal control for a tuberculosis model with reinfection and post-exposure interventions. Math. Biosci. **244**(2), 154–164 (2013)
37. Sousa, C.A., et al.: Ongoing outbreak of dengue type 1 in the autonomous region of Madeira, Portugal: preliminary report. Euro Surveill. **17**(49), 1–4 (2012)
38. Watson, T.M., Kay, B.H.: Vector competence of Aedes notoscriptus (Diptera: Culicidae) for Barmah Forest Virus and of this species and Aedes aegypti (Diptera: Culicidae) for dengue 1–4 viruses in Queensland, Australia. J. Med. Entomol. **36**, 508–514 (1999)
39. Wearing, H.J.: Ecological and immunological determinants of dengue epidemics. Proc. Natl. Acad. Sci. USA **103**(31), 11802–11807 (2006)
40. WHO: Dengue: Guidelines for Diagnosis, Treatment, Prevention and Control, 2nd edn. World Health Organization, Geneva (2009)

The Number of Saturated Numerical Semigroups with a Determinate Genus

J.C. Rosales, M.B. Branco, and D. Torrão

Abstract In this work we describe the saturated numerical semigroups, and characterize the SAT system of generators for them. We see how we can arrange them in a tree rooted in $\mathbb{N}$ and describe the sons of any vertex of this tree. Finally, we present an algorithm for computing the set of saturated numerical semigroups of a given genus

1 Introduction

Let $\mathbb{Z}$ and $\mathbb{N}$ be the set of integers and nonnegative integers, respectively. A **numerical semigroup** is a subset S of $\mathbb{N}$ that is closed under addition, $0 \in S$ and generates $\mathbb{Z}$ as a group. This last condition is equivalent to $\gcd(S) = 1$ (where gcd denotes the greatest common divisor). It is well known that if S is a numerical semigroup then $\mathbb{N}\backslash S$ has finitely many elements (see for instance [6]). The greatest integer not belonging to S is called **Frobenius number** of S, usually denoted $F(S)$, and the cardinality of $\mathbb{N}\backslash S$ is called the **genus** of S, denoted by **g(*S*)**. Moreover, S admits a unique minimal system of generators $\{n_1 < \cdots < n_p\}$ (see [1, 3, 6]). The integers n_1 and p are called **multiplicity** and **embedding dimension** of S, and denoted by $m(S)$ and $\mu(S)$, respectively. For $A \subseteq \mathbb{N}$, denote by $\langle A \rangle$ the submonoid of $\mathbb{N}$ generated by A, that is, $\langle A \rangle = \{\lambda_1 x_1 + \cdots + \lambda_n x_n \mid n \in \mathbb{N}\backslash\{0\}, x_1, \ldots, x_n \in A,$ and $\lambda_1, \ldots, \lambda_n \in \mathbb{N}\}$, which is a numerical semigroup if and only if gdc(A)=1.

J.C. Rosales
Departamento de Álgebra, Universidad de Granada, E-18071 Granada, Spain
e-mail: jrosales@ugr.es

M.B. Branco
Departamento de Matemática, Universidade de Évora, 7000 Évora, Portugal
e-mail: mbb@uevora.pt

D. Torrão (✉)
Universidade de Évora, Évora, Portugal
e-mail: denisetorrao@hotmail.com

J.-P. Bourguignon et al. (eds.), *Dynamics, Games and Science*, CIM Series in Mathematical Sciences 1, DOI 10.1007/978-3-319-16118-1_33

The goal of this work is to describe the saturated numerical semigroups and give an algorithmic method that computes the set of all saturated numerical semigroups with a given genus g.

2 Saturated Numerical Semigroups

In this section we start giving a characterization of the subsets of $\mathbb{N}$ that are saturated numerical semigroups.

A numerical semigroup S is **saturated** if the following condition holds: if s, $s_1,\ldots, s_r \in S$ are such that $s_i \leq s$ for all $i \in \{1,\ldots,r\}$ and $z_1,\ldots,z_r \in \mathbb{Z}$ are such that $z_1s_1 + \cdots + z_rs_r \geq 0$, then $s + z_1s_1 + \cdots + z_rs_r \in S$.
For $A \subseteq \mathbb{N}$ and $a \in A$, set

$$\mathrm{d}_A(a) = \gcd\{x \in A | x \leq a\}$$

The next results are known and can be founded in [5], as their proofs and they give some important properties of the saturated numerical semigroups.

Lemma 1 *Let S be a saturated numerical semigroup and let $s \in S$. Then $s + d_S(s) \in S$.*

Lemma 2 *Let A be a nonempty subset of $\mathbb{N}$ such that $gcd(A) = 1$ and $a + d_A(a) \in A$ for all $a \in A$. Then $a + kd_A(a) \in A$ for all $k \in \mathbb{N}$.*

Lemma 3 *Let A a nonempty subset of $\mathbb{N}$ such that $gcd(A) = 1$ and $a + d_A(a) \in A$ for all $a \in A$. Then $A \cup \{0\}$ is a numerical semigroup.*

From the previous results we can obtain the following theorem:

Theorem 1 *Let A be a nonempty subset of $\mathbb{N}$ such that $gcd(A) = 1$. The following conditions are equivalent:*

(1) A is a saturated numerical semigroup.
(2) $a + d_A(a) \in A$ for all $a \in A$.
(3) $a + kd_A(a) \in A$ for all $a \in A$ and $k \in \mathbb{N}$.

As we saw before, if S is a numerical semigroup, then $\mathbb{N}\backslash S$ has finitely many elements. This implies that the set of numerical semigroups containing S is also finite. Let X be a subset of $\mathbb{N}$ such that $gcd(X) = 1$, then every saturated numerical semigroup containing X also contains $\langle X \rangle$, and thus, there are finitely many of them. We call the intersection of all saturated numerical semigroups containing X the **saturated closure** of X, and denote it by $\mathrm{Sat}(X)$. Observe that $\mathrm{Sat}(X) = \mathrm{Sat}(\langle X \rangle)$, and, as consequence, we have that Sat(X) is the smallest saturated semigroup containing X. The following result gives us the guarantee that intersecting saturated numerical semigroups we obtain again a saturated numerical semigroup.

Proposition 1 *Let S_1 and S_2 be two saturated numerical semigroups. Then $S_1 \cap S_2$ is a saturated numerical semigroup.*

If S is a saturated numerical semigroup and X is a subset of $\mathbb{N}$ such that $gcd(X) = 1$ and Sat(X) $= S$, then we will say that X is a **SAT system of generators** of S. We say that X is a **minimal** SAT system of generators if in addition no proper subset of X is a SAT system of generators of S. We already know that every numerical semigroup is finitely generated (as a semigroup). Hence, for a given numerical semigroup S, there exists $\{n_1, \ldots, n_p\} \subset \mathbb{N}$ such that $S = \langle n_1, \ldots, n_p \rangle$. If S is a saturated numerical semigroup, then, it is obvious that Sat($n_1, \ldots, n_p$) $=$ Sat(S) $= S$, and thus every saturated numerical semigroup admits a finite SAT system of generators.

The arrow in $A = \{a, \cdots, b, \rightarrow\}$, with a and b integers, is used to express that the elements $b + k$ are also in A for all $k \in \mathbb{N}$.

Next, we give a formula that allows us to compute the elements of a saturated numerical semigroup.

Theorem 2 *Let $n_1 < n_2 < \cdots < n_p$ be positive integers such that $gcd(n_1, \ldots, n_p) = 1$. For every $i \in \{1, \ldots, p\}$, set $d_i = gcd(n_1, \ldots, n_i)$ and for all $j \in \{1, \ldots, p-1\}$ define $k_j = max\{k \in \mathbb{N} \mid n_j + kd_j < n_{j+1}\}$. Then $Sat(n_1, \ldots, n_p) = \{0, n_1, n_1 + d_1, \ldots, n_1 + k_1d_1, n_2, n_2 + d_2, \ldots n_2 + k_2d_2, \ldots, n_{p-1}, n_{p-1} + d_{p-1}, \ldots, n_{p-1} + k_{p-1}d_{p-1}, n_p, n_p + 1, \rightarrow\}$.*

Example 1 Let $\{n_1, n_2, n_3\} = \{4, 6, 13\}$. Then $d_1 = 4, d_2 = 2, d_3 = 1, k_1 = 0$, $k_2 = 3$. Hence

Sat (4,6,13) = {0,4,6,8,10,12,13,→}.

Moreover, it's easy to prove that every saturated numerical semigroup has a unique minimal SAT-system of generators (see again [5]).

Lemma 4 *Let S be a saturated numerical semigroup. Then*

$$\{s_1, \ldots, s_r\} = \{s \in S\backslash\{0\} | d_S(s) \neq d_S(s') \text{for all} s' < s, s' \in S\}$$

is the unique minimal SAT system of generators of S.

Example 2 Let S be the saturated numerical semigroup

$$\mathrm{S} = \{0, 4, 7, \rightarrow\}$$

It follows that $d_S(4) = 4$, $d_S(7) = 1 = d_S(7 + n)$, for all $n \in \mathbb{N}$. By the previous theorem the minimal SAT system of generators is {4,7}.

3 The Tree of Saturated Numerical Semigroups

A graph G is a pair (V, E), where V is a nonempty set whose elements are called vertices, and E is a subset of $\{(v, w) \in V \times V | v \neq w\}$. The elements of E are called the edges of G. A path of length n connecting the vertices x and y of G is a sequence of distinct edges of the form $(v_0, v_1), (v_1, v_2), \ldots, (v_{n-1}, v_n)$ with $v_0 = x$ and $v_n = y$. A graph G is a tree if there exists a vertex r (known as the root of G such that for every other x of G, there exists a unique path connecting x and r. If (x, y) is an edge of a tree, then we say that x is a son of y.

The next result also appears in [5, Proposition 17].

Lemma 5 *Let S be a saturated numerical semigroup different from $\mathbb{N}$. Then $S \cup \{F(S)\}$ is also a saturated numerical semigroup.*

Now, we are able to construct recursively the tree containing the set $\mathscr{L}$ of all saturated numerical semigroups. In fact, we define the graph $G(\mathscr{L})$ in the following way: the set of vertices of $G(\mathscr{L})$ is $\mathscr{L}$ and $(T, S) \in \mathscr{L} \times \mathscr{L}$ is an edge of $G(\mathscr{L})$ if and only if $T \cup \{F(T)\} = S$.

Lemma 6 *The graph $G(\mathscr{L})$ is a tree with a root equal to $\mathbb{N}$. Furthermore, the sons of a vertex $S \in \mathscr{L}$ are $S \backslash \{x_1\}, \ldots, S \backslash \{x_r\}$ where $x_1, \ldots, x_r$ are the elements of the minimal SAT-system of generators of S which are greater than $F(S)$.*

We can deduce from Lemma 5, the following result.

Lemma 7 *Let S be a numerical semigroup with minimal SAT-system of generators $A = \{n_1, \ldots, n_p\}$ and let $X = \{n_i \in A | n_i > F(S)\}$. Then $\{n_p\} \subseteq X \subseteq \{n_{p-1}, n_p\}$. Furthermore, $n_{p-1} \in X$ if and only if $n_{p-1} = n_p - 1$.*

Example 3

1) Let $S = Sat(\{3, 6, 11\}) = \{0, 3, 6, 8, 10, 11, \rightarrow\}$. Then 11 is the unique element in the minimal SAT-system of generators of S greater than F(S). Hence $S \in \mathscr{L}$ has a unique son, that is, $S \backslash \{11\}$.
2) Let $S = Sat(\{6, 18, 20, 21\}) = \{0, 6, 12, 18, 20, 21, \rightarrow\}$. Then 20 and 21 are the elements in the minimal SAT-system of generators of S greater than F(S). Hence the sons of $S \in \mathscr{L}$ are $S \backslash \{20\}$ and $S \backslash \{21\}$.

4 A Method for Computing the Set of All Saturated Numerical Semigroups of a Given Genus

Our aim in this section os to describe the minimal SAT-system of generators of the sons of a given saturated numerical semigroup from its minimal SAT-system.

Proposition 2 *Let S be a saturated numerical semigroup with minimal SAT-system of generators $\{n_1, \cdots, n_p\}$ and let $d_{p-1} = gcd\{n_1, \ldots, n_{p-1}\}$. Then the minimal*

SAT-system of generators of $S\backslash\{n_p\}$ *is equal to:*

1) $\{n_1,\ldots,n_{p-1},n_p+2\}$ *if* $d_{p-1}|n_p+1$;
2) $\{n_1,\ldots,n_{p-1},n_p+1\}$ *if* $gcd\{d_{p-1},n_p+1\}=1$;
3) $\{n_1,\ldots,n_{p-1},n_p+1,n_p+2\}$ *in other cases.*

Example 4
1) If $S=Sat(\{3,6,17\})$ then $S\backslash\{17\}=Sat(\{3,6,19\})$.
2) If $S=Sat(\{3,6,15\})$ then $S\backslash\{15\}=Sat(\{3,6,16\})$.
3) If $S=Sat(\{3,6,16\})$ then $S\backslash\{16\}=Sat(\{3,6,17,18\})$.

From Lemmas 6 and 7, we can deduce that, if S is a saturated numerical semigroup with minimal SAT-system of generators $\{n_1<\cdots<n_p\}$ then $S\backslash\{n_p\}$ is always a son of S. Moreover, S has another son, that is $S\backslash\{n_{p-1}\}$ if and only if $n_{p-1}=n_p-1$.

Proposition 3 *Let S be a saturated numerical semigroup with minimal SAT-system of generators* $\{n_1,\cdots,n_p\}$ *such that* $n_{p-1}=n_p-1$. *Then the minimal SAT-system of generators of* $S\backslash\{n_{p-1}\}$ *is equal to:*

a) $\{n_1+1,n_1+2\}$ *if* $p=2$;
b) *if* $p\geq 3$ *and* $d_{p-2}=gcd\{n_1,\ldots,n_{p-2}\}$ *then:*

  b.1) $\{n_1,\ldots,n_{p-2},n_p\}$ *if* $gcd\{d_{p-2},n_p\}=1$;
  b.2) $\{n_1,\ldots,n_{p-2},n_p,n_p+1\}$ *if other cases;*

Example 5
1) If $S=Sat(\{4,5\})$ then $S\backslash\{4\}=Sat(\{5,6\})$.
2) If $S=Sat(\{5,10,11\})$ then $S\backslash\{10\}=Sat(\{5,11\})$.
3) If $S=Sat(\{6,14,15\})$ then $S\backslash\{14\}=Sat(\{6,15,16\})$.

5 An Algorithm to Compute the Set of All Saturated Numerical Semigroups of a Given Genus

In this section we describe an algorithm to compute all the elements in $\mathscr{L}(g)$, that is the set of all saturated numerical semigroups with genus g. Clearly $\mathbb{N}=$ Sat($\{1\}$) has a unique son, which is Sat($\{2,3\}$) = $\{0,2,\rightarrow\}$. If we know $\mathscr{L}(g-1)$ then we can compute $\mathscr{L}(g)$, simply computing all the sons of $\mathscr{L}(g-1)$. For that, we need to know d_{p-1}, and, in some cases, d_{p-2} (see Propositions 2 and 3). To avoid repeating some calculus, and improve the efficiency of the algorithm we introduce the concept of α-representation of a saturated numerical semigroup. Let $S\neq\mathbb{N}$ be a saturated numerical semigroup, an α-representation of S is $[(n_1,n_2,\ldots,n_p),(x_1,x_2,\ldots,x_{p-1})]$ such that $\{n_1,n_2,\ldots,n_p\}$ is the minimal SAT-system of generators of S and $x_i=gcd\{n_1,\ldots,n_{p-i}\}$ for all $i\in\{1,\ldots,p-1\}$. Note that $x_1=gcd\{n_1,\ldots,n_{p-1}\}=d_{p-1}$ and $x_2=gcd\{n_1,\ldots,n_{p-2}\}=d_{p-2}$.

As an immediate consequence of Proposition 2, we have the next result.

Lemma 8 *Let* $[(n_1, \ldots, n_p), (x_1, \ldots, x_{p-1})]$ *be an* α*-representation of a saturated numerical semigroup* $S \neq \mathbb{N}$. *Then the* α*-representation of* $(S \backslash \{n_p\})$ *is equal to:*

1) $[(n_1, \ldots, n_{p-1}, n_p + 2), (x_1, \ldots, x_{p-1})]$ *if* $x_1 | n_p + 1$;
2) $[(n_1, \ldots, n_{p-1}, n_p + 1), (x_1, \ldots, x_{p-1})]$ *if* $gcd\{x_1, n_p + 1\} = 1$;
3) $[(n_1, \ldots, n_{p-1}, n_p + 1, n_p + 2), (gcd\{x_1, n_p + 1\}, x_1, \ldots, x_{p-1})]$ *if other cases.*

Example 6
1) If $S = Sat(\{3, 6, 17\})$ then the α-representation of S is $[(3, 6, 17), (2, 3)]$. Hence the α-representation of $S \backslash \{17\}$ is $[(3, 6, 19), (2, 3)]$.
2) If $S = Sat(\{3, 6, 15\})$ then the α-representation of S is $[(3, 6, 15), (2, 3)]$. Hence the α-representation of $S \backslash \{15\}$ is $[(3, 6, 16), (2, 3)]$.
3) If $S = Sat(\{3, 6, 16\})$ then the α-representation of S is $[(3, 6, 16), (2, 3)]$. Hence the α-representation of $S \backslash \{17\}$ is $[(3, 6, 17, 18), (2, 3)]$;

And from Proposition 3 we deduce the following result.

Lemma 9 *Let* $[(n_1, \ldots, n_p), (x_1, \ldots, x_{p-1})]$ *be an* α*-representation of a saturated numerical semigroup* $S \neq \mathbb{N}$ *such that* $n_{p-1} = n_p - 1$. *Then the* α*-representation of* $(S \backslash \{n_p - 1\})$ *is equal to:*

1) $[(n_1 + 1, n_1 + 2), (n_1 + 1)]$ *if* $p = 2$;
2) $[(n_1, \ldots, n_{p-2}, n_p), (x_2, \ldots, x_{p-1})]$ *if* $p \geq 3$ *and* $gcd\{x_2, n_p\} = 1$;
3) $[(n_1, \ldots, n_{p-2}, n_p, n_p + 1), (gcd\{x_2, n_p\}, x_2, \ldots, x_{p-1})]$ *if other cases.*

Example 7
1) If $S = Sat(\{4, 5\})$ then the α-representation of S is $[(4, 5), (4)]$. Hence the α-representation of $S \backslash \{4\}$ is $[(5, 6), (5)]$.
2) If $S = Sat(\{5, 10, 11\})$ then the α-representation of S is $[(5, 10, 11), (2, 5)]$. Hence the α-representation of $S \backslash \{10\}$ is $[(5, 11), (5)]$.
3) If $S = Sat(\{6, 14, 15\})$ then the α-representation of S is $[(6, 14, 15), (2, 6)]$. Hence the α-representation of $S \backslash \{14\}$ is $[(6, 15, 16), (3, 6)]$;

Finally, we are able to give the announced algorithm which shows us how to compute $\mathcal{L}(g)$.

Algorithm 1 *Input: g a positive integer.*
Output: $\mathcal{L}(g)$ *(the set of all saturated numerical semigroups with genus g)*

1) $A = \{[(2, 3), (2)]\}, i = 1, B = \emptyset$.
2) If $i = g$ *then return A.*
3) For each $[(n_1, \ldots, n_p), (x_1, \ldots, x_{p-1})] \in A$ *do*

3.1) If $x_1 | n_p + 1$ *then*
$B = B \cup \{[(n_1, \ldots, n_{p-1}, n_p + 2), (x_1, \ldots, x_{p-1})]\}$ *and go to Step 3.4).*
3.2) If $gcd\{x_1, n_p + 1\} = 1$ *then*
$B = B \cup \{[(n_1, \ldots, n_{p-1}, n_p + 1), (x_1, \ldots, x_{p-1})]\}$ *and go to Step 3.4).*
3.3) $B = B \cup \{[(n_1, \ldots, n_{p-1}, n_p + 1, n_p + 2), (gcd\{x_1, n_p + 1\}, x_1, \ldots, x_{p-1})]\}$.
3.4) If $n_{p-1} \neq n_p - 1$ *go to Step 4).*
3.5) If $p = 2$ *then* $B = B \cup \{[(n_1 + 1, n_1 + 2), (n_1 + 1)]\}$ *and go to Step 4).*

3.6) *If* $gcd\{x_2, n_p\} = 1$ *then*
$B = B \cup \{[(n_1, \ldots, n_{p-2}, n_p), (x_2, \ldots, x_{p-1})]\}$ *and go to Step 4).*

3.7) $B = B \cup \{[(n_1, \ldots, n_{p-2}, n_p, n_p + 1), (gcd\{x_2, n_p\}, x_2, \ldots, x_{p-1})]\}$ *and go to Step 3.4).*

4) $A = B, i = i + 1, B = \emptyset$ *and go to Step 2).*

Example 8 Let us compute all saturated numerical semigroups with genus 10.

First, and using the previous Algorithm, we compute the α-representation of all saturated numerical semigroup with genus less than 10 (denoted by A_i).

- $A_1 = \{[(2, 3), (2)]\}$;
- $A_2 = \{[(2, 5), (2)], [(3, 4), (3)]\}$;
- $A_3 = \{[(2, 7), (2)], [(3, 5), (3)], [(4, 5), (4)]\}$;
- $A_4 = \{[(2, 9), (2)], [(3, 7), (3)], [(4, 6, 7), (2, 4)], [(5, 6), (5)]\}$;
- $A_5 = \{[(2, 11), (2)], [(3, 8), (3)], [(4, 6, 9), (2, 4)], [(4, 7), (4)], [(5, 7), (5)], [(6, 7), (6)]\}$;
- $A_6 = \{[(2, 13), (2)], [(3, 10), (3)], [(4, 6, 11), (2, 4)], [(4, 9), (4)], [(5, 8), (5)], [(6, 8, 9), (2, 6)], [(7, 8), (7)]\}$;
- $A_7 = \{[(2, 15), (2)], [(3, 11), (3)], [(4, 6, 13), (2, 4)], [(4, 10, 11), (2, 4)], [(5, 9), (5)], [(6, 8, 11), (2, 6)], [(6, 9, 10), (3, 6)], [(7, 9), (7)], [(8, 9), (8)]\}$;
- $A_8 = \{[(2, 17), (2)], [(3, 13), (3)], [(4, 6, 15), (2, 4)], [(4, 10, 13), (2, 4)], [(4, 11), (4)], [(5, 11), (5)], [(6, 8, 13), (2, 6)], [(6, 9, 11), (3, 6)], [(6, 10, 11), (2, 6)], [(7, 10), (7)], [(8, 10, 11), (2, 8)], [(9, 10), (9)]\}$;
- $A_9 = \{[(2, 19), (2)], [(3, 14), (3)], [(4, 6, 17), (2, 4)], [(4, 10, 15), (2, 4)], [(4, 13), (4)], [(5, 12), (5)], [(6, 8, 15), (2, 6)], [(6, 9, 13), (3, 6)], [(6, 10, 13), (2, 6)], [(6, 11), (6)], [(7, 11), (7)], [(8, 10, 13), (2, 8)], [(8, 11), (8)], [(9, 11), (9)], [(10, 11), (10)]\}$.

And from the A_i we get the minimal SAT-system of generators of the set of saturated numerical semigroups with genus 10.

$$\{\{2, 21\}, \{3, 16\}, \{4, 6, 19\}, \{4, 10, 17\}, \{4, 14, 15\}, \{5, 13\}, \{6, 8, 17\},$$
$$\{6, 9, 14\}, \{6, 10, 15\}, \{6, 13\}, \{7, 12\}, \{8, 10, 15\}, \{8, 12, 13\}, \{9, 12, 13\},$$
$$\{10, 12, 13\}, \{11, 12\}\}$$

which are the sons of elements in A_9.

A procedure to compute the set of saturated numerical semigroups with a given genus can be done by calculating first all the numerical semigroups with this genus and then see which one of them are saturated. The problem is that even for small genus, this set can be very large. With the method presented in this paper the computation becomes much more efficient. The algorithm has been implemented in GAP (see [2, 4]). Next we give some timings.

For Genus 10,

```
gap> Length(SaturatedNumericalSemigroupsWithFixed
Genus(10));16
```

takes 0 ms, while computing the set of all saturated numerical semigroups with genus and then filtering those that are saturated takes 62 ms.

```
gap> Length(Filtered(NumericalSemigroupsWithGenus
(10), IsSaturatedNumericalSemigroup)); 16
```

As for 15, we get also 0 ms for

```
gap> Length(SaturatedNumericalSemigroupsWithFixed
Genus(15)); 40
```

while it takes 1,154 ms for

```
gap> Length(Filtered(NumericalSemigroupsWithGenus
(15), IsSaturatedNumericalSemigroup)); 40
```

For 25, we still get 0 ms with

```
gap> Length(SaturatedNumericalSemigroupsWithFixed
Genus(25)); 130
```

while it takes 289,803 ms with

```
gap> Length(Filtered(NumericalSemigroupsWithGenus
(25), IsSaturatedNumericalSemigroup)); 130
```

For genus 30 the time with this algorithm is 16 ms while with the filtering was not possible to calculate because it gets an error message: `"Error, exceeded the permitted memory"`.

In the following table there are the results obtained for genus up to 150. For each positive integer g we wrote the number of saturated numerical semigroups (n_g) of the given genus (g).

g	n_g	g	n_g	g	n_g	g	n_g	g	n_g	g	n_g	g	n_g	g	n_g	g	n_g	g	n_g
1	1	16	43	31	228	46	701	61	1,717	76	3,634	91	6,900	106	12,057	121	20,106	136	31,790
2	2	17	51	32	251	47	757	62	1,815	77	3,805	92	7,175	107	12,503	122	20,749	137	32,758
3	3	18	56	33	272	48	805	63	1,915	78	3,970	93	7,444	108	12,939	123	21,404	138	33,730
4	4	19	67	34	295	49	864	64	2,021	79	4,163	94	7,732	109	13,411	124	22,086	139	34,755
5	6	20	78	35	324	50	918	65	2,135	80	4,348	95	8,038	110	13,886	125	22,787	140	35,751
6	7	21	85	36	346	51	973	66	2,239	81	4,532	96	8,336	111	14,382	126	23,485	141	36,764
7	9	22	91	37	373	52	1,030	67	2,365	82	4,729	97	8,669	112	14,898	127	24,239	142	37,836
8	12	23	106	38	401	53	1,103	68	2,482	83	4,952	98	9,004	113	15,441	128	24,990	143	38,951
9	15	24	117	39	432	54	1,172	69	2,599	84	5,156	99	9,348	114	15,969	129	25,753	144	40,040
10	16	25	130	40	460	55	1,248	70	2,722	85	5,373	100	9,705	115	16,524	130	26,546	145	41,170
11	21	26	143	41	500	56	1,320	71	2,868	86	5,592	101	10,083	116	17,080	131	27,379	146	42,311
12	24	27	158	42	535	57	1,385	72	3,006	87	5,822	102	10,457	117	17,634	132	28,214	147	43,477
13	29	28	170	43	581	58	1,457	73	3,158	88	6,070	103	10,866	118	18,232	133	29,081	148	44,698
14	35	29	190	44	626	59	1,548	74	3,314	89	6,345	104	11,262	119	18,857	134	29,968	149	45,956
15	40	30	205	45	662	60	1,626	75	3,470	90	6,616	105	11,643	120	19,460	135	30,859	150	47,220

References

1. Barucci, V., Dobbs, D.E., Fontana, M.: Maximality properties in numerical semigroups and applications to the one-dimensional analytically irreducible local domais. Memoirs Am. Math. Sco. **598** (1997)
2. Delgado, M., García-Sánchez, P.A., Morais, J.: Numericalsgps: a GAP package on numerical semigroups. http://www.gap-system.org/Packages/numericalsgps.html (2008)
3. Fröberg, R., Gottlieb, G., Häggkvist, R.: On numerical semigroups. Semigroup Forum **35**, 63–83 (1987)
4. The GAP Group: GAP-Groups, Algorithms and Programming, Version 4.4. http://gap-system.org (2004)
5. Rosales, J.C., García-Sánchez, P.A., García-García, J.I., Branco, M.B.: Saturated numerical semigroups. Houston J. Math. **30**, 321–330 (2004)
6. Rosales, J.C., García-Sánchez, P.A.: Numerical Semigroups, Developments in Mathematics, vol. 20. Springer, New York (2009)

Modern Forecasting of NOEM Models

Manuel Sánchez Sánchez

Abstract In this paper we estimate a small structural model, in order to forecast the key macroeconomic variables of output growth and underlying inflation. In contrast to models with purely statistical foundations, the Bayesian Vector Autoregressive Dynamic Stochastic General Equilibrium (BVAR-DSGE) model, uses the theoretical information of a DSGE model to offset insample overfitting. We compare the forecast performance of BVAR-DSGE model with Minesota VAR and independently estimates DSGE model. The open economy DSGE model of Lubik and Schorfheide (2007) is implemented to provide prior information for the VAR.

1 Bayesian Vector Autoregression (BVAR)

The main difference with standard VAR models, lies in the fact that the model parameters are treated as random variables, and prior probabilities are assigned to them. We impose a prior distribution on a set of parameters that summarizes beliefs or knowledge about these parameters prior to observing the data. Priors reduce the sample variability in the parameter estimates by "shrinking" them toward a specific point in the parameter space—forecasting accuracy—In many BVARs the priors arise from statistics. The Minnesota prior shrinks the VAR parameters toward a unit root process.

2 DSGE Model Like a Prior

The DSGE model parameters describe the preferences of agents (tastes), the production function (technology), and other features of the economy. These parameters are called "deep parameters"—parameters that do not vary with policy—Lucas [18] critique implies that only models in which the parameters are deep—that is, models in which the parameters do not vary with policy—are suited to evaluate the impact

M.S. Sánchez (✉)
National University of Distance Education, Madrid, Spain
e-mail: mjsanchez@cee.uned.es

J.-P. Bourguignon et al. (eds.), *Dynamics, Games and Science*, CIM Series in Mathematical Sciences 1, DOI 10.1007/978-3-319-16118-1_34

of policy changes. Therefore BVAR-DSGE approach devises a framework which tries to imitate the forecasting accuracy of the BVAR(statistical) models, and simultaneously be immune to the Lucas critique [18]. Using general equilibrium models as priors—see, DeJong et al. [3]—means that the restrictions stemming from economic theory are imposed loosely instead of rigidly.[1] A key hyperparameter λ determines the weight attached to the theoretical DSGE model.

The approach to estimate the BVAR model has several steps:

2.1 Estimating the DSGE Model

The DSGE model is estimated using Bayesian methods. A fundamental result used in Bayesian analysis is that the posterior distribution is proportional to the likelihood function multiplied by the prior distribution—Bayes theorem:

$$P\left(\frac{\theta}{Y}\right) \propto P(Y/\theta)P(\theta) \tag{1}$$

Where: Y represents observed data, $\theta$are the unknown parameters, $P(\bullet)$are generic density functions.

2.2 A VAR Approximation to the DSGE Model

The log-linearized DSGE model—see, e.g., Lubik and Schorfheide [16]—can be written as a rational expectations (LRE) system of the form:

$$\Gamma_0(\theta) X_t = \Gamma_1(\theta) X_{t-1} + \Gamma_\epsilon(\theta)\epsilon_t + \Gamma_\eta(\theta)\eta_t \tag{2}$$

The solution can be expressed in state-space form as:

$$X_t = A(\theta)X_{t-1} + B(\theta)\epsilon_t$$
$$Y_t = C(\theta)X_t + D(\theta)\epsilon_t$$

Where: X_t: state vector, ϵ_t: vector of structural shocks, θ: vector of non-policy parameters. The matrices A,B,C and D, are non-linear functions of the structural parameters in the DSGE model.

[1]It allows incorporate subjective information about the parameters to be utilized in estimation Fukac and Pagan [8]. For details in Bayesian methods and state-space form see, Geweke [11] and Hamilton [12] respectively.

It is necessary to have the eigenvalues of $A - BD^{-1}C$ to be strictly less than one in modulus in order to have y_t with a infinite order VAR representation given by[2]:

$$y_t = \sum_{j=1}^{\infty} C(A - BD^{-1}C)^{j-1}BD^{-1}y_{t-j} + D\epsilon_t \tag{3}$$

If the largest eigenvalue is not close to unity, a low order VAR is likely to be a good approximation.[3]

2.3 Constructing a Prior for BVAR

We want to use a DSGE model to provide information about the parameters of the VAR.

One way of doing this is to simulate data from the DSGE and to combine it with the actual data when estimating VAR. However rather than literally simulating the artificial data, we can use the Theoretical Moments of the DSGE model instead of moments from simulated data, in order to avoid sampling variation. The prior distribution of the BVAR parameters:

$$P\left(\Phi, \sum_u /\theta\right) = c^{-1}(\theta)\left|\sum_u\right|^{\frac{-\lambda T+n+1}{2}} exp\left\{-\frac{1}{2}tr\left[\lambda T \sum_u^{-1}(\Omega)\right]\right\} \tag{4}$$

Where:
$\Omega = \left(\Gamma^*_{yy}(\theta) - \Phi'\Gamma^*_{xy}(\theta) - \Gamma^*_{yx}(\theta)\Phi + \Phi'\Gamma^*_{xx}(\theta)\Phi\right)$; $\Gamma^*_{yy}, \Gamma^*_{xy}$, $\Gamma^*_{yx}, \Gamma^*_{xx}$ be the theorist second-order moments of the variables in $Y$and X implied by the DSGE model.[4]

[2]This is the "poor man's invertibility condition" given in Fernandez-Villaverde et al. [7]. Previously, Wold [19] demonstrated that covariance-stationary processes have an infinite order moving average (MA) representation.

[3]The rate at which the autoregressive coefficients converge to zero is determined by the largest eigenvalue of $A - BD^{-1}C$. If this eigenvalue is close to unity, a low order VAR is likely to be a poor approximation to the infinite-order VAR implied by the DSGE model. If one or more of the eigenvalues of $A - BD^{-1}C$ are exactly equal to one in modulus, y_t does not have a VAR representation, i.e, the autoregressive coefficients do not converge to zero as the number of lags tend to infinity. Often, roots on the unit circle indicate that the observables have been overdifferenced.

[4]A VAR approximation of the DSGE model can be obtained from **restriction functions** that relate the **DSGE model parameters to the VAR parameters**: $\Phi^*(\theta) = \Gamma^{*-1}_{xx}(\theta)\Gamma^*_{xy}(\theta)$, $\sum^*_u(\theta) = \Gamma^*_{yy}(\theta) - \Gamma^*_{yx}(\theta)\Gamma^{*-1}_{xx}(\theta)\Gamma^*_{xy}(\theta)$.

The role of the hyperparameter λ is to determine the weight attached to the theoretical DSGE model. We can then formulate the prior for the BVAR parameters $P(\Phi, \sum_u / \theta)$, as conjugate, Inverted Wishart–Normal form: $\sum_u / \theta' IW$; $\Phi / \sum_u, \theta' N$

The joint prior density of both sets of parameters is then given by[5]:

$$P\left(\Phi, \sum_u, \theta\right) = P(\Phi, \sum_u / \theta) P(\theta) \tag{5}$$

2.4 Posterior Distribution

The posterior distribution of the BVAR parameters $\Phi$and $\sum_u$, $P(\Phi, \sum_u / Y, \theta)$—from which we will draw parameters when forecasting—Is obtained by combining the prior with information from the data, namely the likelihood function. We assume that the observable data vector y_t follows a vector autoregressive process of order p: $Y = X\Phi + U$

The likelihood function of the VAR model can be expressed as:

$$L(Y^{/} \Phi, \sum_u) \propto \left| \sum_u \right|^{-\frac{T}{2}} x \exp\left[-\frac{1}{2} tr \left\{ \sum_u^{-1} \left(Y'Y - \Phi' X' Y - Y' X \Phi + \Phi' X' X \Phi \right) \right\} \right] \tag{6}$$

(The likelihood function of the data is function of Φ, $\sum_u$)

Following Bayes Rule, the posterior is proportional to the likelihood times the Prior:

$$P\left(\Phi, \sum_u, / Y\right) \propto L(Y / \Phi, \sum_u) P(\Phi, \sum_u / \theta) P(\theta)$$

Since DSGE model prior and the likelihood function are conjugate, it is straight forward to show that the posterior distribution of $\Phi$and $\sum_u$ is also Inverted Wishart—Normal form.

[5]Our prior has hierarchical structure. We conduct a posterior predictive analysis in the spirit of Gelman et al. [10].

3 Optimal Mixture Model

The optimal mixture model, is the one associated with the value ofλ^6: that maximizes the marginal likelihood for the data, $\hat{\lambda}$:

$$P(Y/\lambda) = \int_{\Phi,\sum_u,\Theta} P(Y/\theta,\Phi,\sum_u)P(\theta,\Phi,\text{Âă}\sum_u/\lambda)d(\sum_u,\Phi,\theta) \tag{7}$$

The lowest value is 0, and in this case , the best representation for the data is the unrestricted VAR; The highest λ is ∞, i.e, the data are better fitted by the DSGE model. If $\hat{\lambda}$ is large, the theoretical model fits the data well, otherwise if $\hat{\lambda}$ tends to zero, the theoretical model does not describe the data.

4 A Small Open Economy Model

We use an open economy DSGE model with theoretical foundations closely related to the papers by Gali and Monacelli [9] and Lubik and Schorfheide [17] to provide prior information for the VAR.

DSGE models describe the general equilibrium allocations and prices of a model economy in which agents (households, firms, etc.) dynamically maximize their objectives (utility, profits, and so on) subject to their budget and resource constraints. The behaviour of actual and optimal policies in this kind of models has been a key focus of many papers, such as Benigno [2], Del Negro and Schorfheide [5].

4.1 General Modeling Features

The analysis is performed using a DSGE model for a small open economy integrated in a monetary union. Continuum of countries with a continuum of firms producing differentiated goods, in a monopolistically competitive environment. Firms set prices according to Calvo staggered pricing, production function is linear in labour, and Technology is assumed to follow a unit root process and is common to both the domestic and world economies. Consumers have constant intertemporal elasticity of substitution, and they aggregate consumption goods using Dixit-Stiglitz

[6]This λ represents the weight of the restrictions from the model imposed by the econometrician and it tells how much the economic model DSGE, is able to explain the real data.

aggregation. Monetary policy is specified by a flexible Taylor Rule. Financial markets are assumed to be perfect enabling risk-sharing between domestic and foreign consumers.

4.2 Household

A representative household maximizes utility given by $E_0 \sum_{t=0}^{\infty} \beta^t \left[\frac{(C_t/A_t)^{1-\sigma}-1}{1-\sigma} - \frac{N_t^{1+\varphi}}{1+\varphi} \right]$

Where, σ: household's risk aversion, φ: labour supply aversion, N_t: hours worked, A_t: world technology process, C_t: composite consumption index.

The composite good C is a Dixit-Stiglitz aggregator of goods produced at home and abroad and defined as: $C_t \equiv \left[(1-\alpha)^{\frac{1}{\eta}} (C_{H,t})^{\frac{\eta-1}{\eta}} + \alpha^{\frac{1}{\eta}} (C_{F,t})^{\frac{\eta-1}{\eta}} \right]^{\frac{\eta}{\eta-1}}$

Where, $0 \leq \alpha \leq 1$ is a share of imports in GDP (degree of openness), $\eta > 0$ is the substitutability between domestic and foreign goods from standpoint of domestic consumer., $C_{H,t}$: index of consumption of domestic goods given by the CES function, $C_{F,t}$: index of consumption of imported goods given by the CES function.

Under rational expectations, the household maximizes its utility subject to a borrowing constraint: $P_tC_t + D_t \leq R_tD_{t-1} + W_tN_t + T_t$. Where, P_t : consumer price index (CPI), R_t : return on investment D_{t-1} held at the end of period $t-1$(including shares in firms), W_t : nominal wage, T_t : lump-sum transfers.

4.3 Firms

A typical firm in the home economy produces a differentiated good with a linear technology represented by the production function: $Y_t = A_tN_t$. Where: $a_t = logA_t$is described by the AR(1) process[7]: $a_t = \rho_a a_{t-1} + \upsilon_t$. All firms face identical demand curves and take the aggregate price level and aggregate consumption index exogenously. Firms are price setting. However, each firm may change its price with probability $1-\theta$ every period, irrespective of the last time of adjustment. Therefore each period a fraction $1-\theta$ of firms reoptimizes its price, whereas the rest θ keep their prices unchanged. This price stickiness, θ is an important feature of the model because it allows monetary policy to affect real variables in the short run.

[7]A consequence of this is that some of the real variables (such as output) are normalized by technology before the log-linearisation.

4.4 Key Final Log-Linearised Equations

IS Equation[8]: $y_t = E_t y_{t+1} - \chi (R_t - E_t \pi_{t+1}) + \chi \rho_z z_t + \alpha \chi E_t \Delta q_{t+1} + \left(\frac{\chi}{\tau} - 1\right) E_t \Delta y^*_{t+1}$
New Keynesian Phillips curve[9]: $\pi_t = \beta E_t \pi_{t+1} + \alpha \beta E_t \Delta q_{t+1} - \alpha \Delta q_t + \frac{\kappa}{\lambda}(y_t - \bar{y}_t)$

4.5 Monetary Policy

Monetary policy are controlled by the ECB which sets the nominal interest rate according to the Taylor rule evaluated at the observed values of euro area variables[10]:
$R_t = \rho_R R_{t-1} + (1 - \rho_R)\left(\psi_1 \pi^{EA}{}_t + \psi_2 y^{EA}{}_t\right) + \epsilon_{R_t}$

4.6 Rest of the World

By assumption, the rest of the world corresponds to the rest of the monetary union, and therefore the nominal effective exchange rate is irrevocably set to unity, as all trade and financial flows are performed using the same currency.

Exogenous Processes[11]:

The exogenous processes are defined for the foreign output y^*_t, the change in terms of trade Δq_t, the worldwide technology shocks z_t,[12] and the foreign inflation π^*_t respectively as: $y^*_t = \rho_{y^*} y^*_{t-1} + \epsilon_{y^*_t}$; $\pi^*_t = \rho_{\pi^*} \pi^*_{t-1} + \epsilon_{\pi^*_t}$; $\Delta q_t = \rho_{\Delta q} \Delta q_{t-1} + \epsilon_{\Delta q_t}$; $z_t = \rho_z z_{t-1} + \epsilon_{z_t}$

[8]Implying that output depends on the expectations of future both home and abroad, the real interest rate, the expected changes in terms of trade and technology growth.

[9]Movements in the output gap $(y_t - \bar{y}_t)$, affect inflation as they are associated with changes in real marginal costs. Changes in the terms of trade enter the Phillip curve reflecting the fact that some consumer goods are imported.

[10]We assume Spain is too small to have a significant influence on the ECB's Taylor rule. Thus, changes in Spanish conditions do not affect R_t, which is determined by the Taylor rule above, evaluated at the observed values of euro area variables. Justiniano and Preston [14] include output growth as an additional argument in their policy rule.

[11]By this specification, we pin down the small open economy as a system affected by foreign data generating processes but which has no perceptible influence on the rest of the world.

[12]Technology is assumed to grow ate the rate z_t.

4.7 *Data, Priors and Estimation Results*

To estimate the structural parameters of the model we use Spanish and European quarterly (seasonally adjusted) data for real output growth, inflation , the nominal interest rate and terms of trade changes. All the time series are taken from the database developed for the REMS model (BDREMS). Sample period: we have decided to use only the period since the euro area was conceived, that is from 1997 onward.

The time series are made stationary by applying the Hodrick-Prescott Filter with smoothing parameter $\lambda = 1600$. By doing so, the analysis focuses on the business cycle frequencies, however filtering has important implications—see discussion in Del Negro and Schorfheide [4]. Our priors are selected in part by examining the results of recent DSGE modeling and by reference to economic theory. Additionally, we draw on past experience in modeling national economy by the Spanish Central Bank.

5 Forecasting Performance Comparison

In order to examine the forecasting gain from using priors from a DSGE model, we test—following Ingram and Whiteman [13]—whether the forecasts from the DSGE-VAR are competitive with forecasts from some benchmark models—unrestricted VAR, DSGE and Minnesota VAR—that historically have proven to be useful forecasting tools.

RMSE of BVAR-DSGE[13]			
2008:Q1-2012:Q4, VAR*(4)			
	One quarter ahead	Four quarters ahead	Eight quarters ahead
	Quarterly	Year-ended	Year-ended
Variable	Relative to unrestricted VAR		
Output growth	0.82	1.03	0.86
Underlying inflation	0.90	1.12	0.94
	Relative to DSGE		
Output growth	0.88	0.89	1.02
Underlying inflation	0.93	0.91	0.83
	Relative to Minnesota VAR		
Output growth	0.95	0.87	0.90
Underlying inflation	1.05	1.08	0.88
*We use Akaike information criterion to determine the optimal number of lags for the VAR.			

[13]To interpret this table, note that if the entry in a particular cell is less than one, then the BVAR-DSGE outperforms the corresponding benchmark model. Diebold and Mariano [6] provide a general framework for such tests.

We generate dynamic forecasting[14] for horizons of 1 up to 8 quarters—re-estimating the models each quarter over the out-of sample forecast horizon (2008–2012)—Forecasting accuracy is measured by univariate root mean squared forecast error (RMSE). To evaluate the forecasting performance of the models we construct out-of-sample forecasts and compute their RMSE.[15]

6 Conclusion

In this paper we evaluated the forecasting performance of a DSGE-VAR model estimated on Spanish data. We compare the performance of the DSGE-VAR to an unrestricted VAR, and a Bayesian VAR with Minnesota priors. The combination of a DSGE with a VAR model increases the number of free parameters, allowing for better fitting of the data, therefore we find[16] the DSGE-VAR model outperforms benchmark models.[17] DSGE priors are indeed useful as a means of improving the forecasting performance of the VAR. These results suggest that the theoretical information in the DSGE prior is a useful complement to the purely statistical Minnesota prior. Overall, the results show that the BVAR-DSGE is competitive at forecasting inflation and output. A natural extension to future work, would be to introduce richer DSGE models in order to improve the fit.

References

1. Adolfson, M., Lindé, J., Villani, M.: Forecasting performance of an open economy dynamic stochastic general equilibrium model. Sveriges Riksbank, Working Paper, 190 (2005)
2. Benigno, P.: Optimal monetary policy in a currency area. J. Int. Econ. **63**(2), 293–320 (2004)
3. DeJong, D.N., Ingram, B.F., Whiteman, C.H.: A Bayesian approach to dynamic macroeconomics. J. Econ. **98**(2), 203–223 (2000)
4. Del Negro, M., Schorfheide, F.: Take your model bowling: forecasting with general equilibrium models. Fed. Reserve Bank of Atlanta Econ. Rev. Q4, 35–50 (2003)
5. Del Negro, M., Schorfheide, F.: Priors from general equilibrium models for VARs. Int. Econ. Rev. **45**(2), 643–673 (2004)
6. Diebold, F.X., Mariano, R.S.: Comparing predictive accuracy. J. Bus. Econ. Stat. **13**(3), 253–263 (1995)

[14]The DSGE and DSGE-VAR forecasts are based on 100,000 Metropolis Hastings draws starting from the posterior mode. More detail about algorithm can be found in Koop [15].

[15]Notice that all the parameters in the DSGE model and the DSGE-VAR including the hyperparameter λ, that is re-estimated in each recursion.

[16]Based on univariate root mean squared forecast error (RMSE).

[17]This is the case for the one-quarter and eight quarters-ahead forecasts UVAR model. Compared to the DSGE and the Minnesota VAR models, the BVAR-DSGE forecasting outperforms inflation and output growth respectively at any horizon. Adolfson [1] examine out-of-sample forecast performance for DSGE models of the euro area.

7. Fernandez-Villaverde, J., Rubio-Ramirez, J., Sargent, T., Watson, M.: ABCs (and Ds) of understanding VARs. Am. Econ. Rev. **97**(3), 1021–1026 (2007)
8. Fukac, M., Pagan, A.: Issues in adopting DSGE models for use in the policy process. Australian National University, Centre for Applied Macroeconomic Analysis, CAMA Working Paper, 10/2006 (2006)
9. Gali, J., Monacelli, T.: Monetary policy and exchange rate volatility in a small open economy. Rev. Econ. Stud. **72**(3), 707–34 (2005)
10. Gelman, A., Carlin, J.B., Stern, H.S., Rubin, D.B.: Bayesian Data Analysis. Chapman and Hall/CRC, Boca Raton, Florida (2004)
11. Geweke, J.: Contemporary Bayesian Econometrics and Statistics. Wiley, Hoboken, New Jersey (2005)
12. Hamilton, J.D.: Time Series Analysis. Princeton University Press, Princeton, New Jersey (1994)
13. Ingram, B.F., Whiteman, C.H.: Supplanting the 'Minnesota' prior: Forecasting macroeconomic time series using real business cycle model priors. J. Monet. Econ. **34**(3), 497–510 (1994)
14. Justiniano, A., Preston, B.: Monetary policy and uncertainty in an empirical small open economy model. Mimeo, paper presented at Reserve Bank of New Zealand, Macroeconometrics and Model Uncertainty Conference (2006)
15. Koop, G.: Bayesian Econometrics, Wiley, Chichester, Sussex (1986); Litterman, R.B.: Forecasting with Bayesian vector autoregressions – five years of experience. J. Bus. Econ. Stat. **4**(1), 25–38 (2003)
16. Lubik, T., Schorfheide, F.: Testing for indeterminacy: An application to U.S. monetary policy. Am. Econ. Rev. **94**(1), 190–217 (2004)
17. Lubik, T., Schorfheide, F.: Do central banks respond to exchange rate movements? A structural investigation. J. Monet. Econ. **54**(4), 1069–1087 (2007)
18. Lucas, R.E., Jr.: Econometric policy evaluation: A critique. Carn.-Roch. Conf. Ser. Public Policy **1**, 19–46 (1976)
19. Wold, H.: A Study in the Analysis of Stationary Time Series. Almqvist and Wiksell, Uppsala (1938)

An Overview of Quantitative Continuous Compound Analysis

Rui Santos, João Paulo Martins, and Miguel Felgueiras

Abstract The application of compound tests in clinical analysis or acceptance sampling exults in resource savings. Furthermore, quantitative compound tests allow to infer whether the amount of some substance of any individual in the group is greater or lower than a prefixed threshold. However, the use of this type of tests must be done with caution to avoid having a high probability of misclassification. This work uses the weight of the tails of the underlying distribution as a measure of the adequacy of the application of continuous compounds tests.

1 Introduction

Compound, group or pooled analysis can be performed in order to reduce classification (to identify all the infected individuals) and estimation (to estimate the prevalence rate) costs in low infection prevalence rates in various application fields, cf. [3]. This work aims to provide an overview of a specific type of composite analysis, the quantitative continuous compound analysis. Thus, Sect. 2 outlines different possibilities to characterize infected and uninfected individuals. The implementation of continuous compound tests as well as the main measures to evaluate its accuracy are described in Sect. 3, which also contains an overview of the main applications of compound analysis. Two different methodologies to perform compound tests are described in Sect. 4, one to control the compound specificity and the other to control the compound sensitivity. Those methodologies are assessed by simulations (Sect. 5) in order to evaluate its performance under

R. Santos (✉) • J.P. Martins
School of Technology and Management, Polytechnic Institute of Leiria, CEAUL—Center of Statistics and Its Applications, Leiria, Portugal
e-mail: rui.santos@ipleiria.pt; jpmartins@ipleiria.pt

M. Felgueiras
School of Technology and Management, Polytechnic Institute of Leiria, CEAUL—Center of Statistics and Its Applications, CIIC—Computer Science and Communications Research Centre of Polytechnic Institute of Leiria, Leiria, Portugal
e-mail: mfelg@ipleiria.pt

J.-P. Bourguignon et al. (eds.), *Dynamics, Games and Science*, CIM Series in Mathematical Sciences 1, DOI 10.1007/978-3-319-16118-1_35

several continuous distributions with different right tail weights. Finally, the main conclusions are summarized in Sect. 6.

2 Infected and Uninfected Individuals in Quantitative Analysis

Let p be the prevalence rate of an infection which affects a population with N individuals, and the random variables (r.v.s) X_i denote the presence ($X_i = 1$) or absence ($X_i = 0$) of the infection in the i-th individual, hence $X_i \sim$ Ber(p), $i = 1, \ldots, N$. Qualitative analysis only intended to ascertain the presence or absence of some substance (e.g. antigenes, antibodies or some bacterial species) in the analyzed fluid (e.g. blood or urine) which allow to identify the presence of the infection. Thus, the presence of the substance in the fluid implies $X_i = 1$ and the absence entails $X_i = 0$. In a quantitative analysis the presence of the substance is not sufficient to classify each individual. Thus, let us suppose that the clinical trial for identification of the infected individuals is carried out by measuring the amount Y of a certain substance (e.g. the number of a particular species of bacteria) in a milliliter (*ml*) of blood. Moreover, in an infected individual ($X_i = 1$) the amount Y_i can be described by $Y_i = Y_i^+ \sim \mathbf{D}_1(\theta_1)$ where $\mathbf{D}_1$ is some distribution with support $\mathrm{S}_1 \subseteq \mathbb{R}$ and parameter vector θ_1. In an uninfected individual ($X_i = 0$) the amount Y_i can be characterized by $Y_i = Y_i^- \sim \mathbf{D}_0(\theta_0)$ where $\mathbf{D}_0$ denotes some distribution with support $\mathrm{S}_0 \subseteq \mathbb{R}$ and parameter vector θ_0. Whenever $\mathrm{S}_0 \cap \mathrm{S}_1 = \mathrm{S} = \emptyset$ the classification of each individual is straightforward once $Y_i \in \mathrm{S}_1 \Rightarrow X_i = 1$ and $Y_i \in \mathrm{S}_0 \Rightarrow X_i = 0$ (there is no problem of misclassification). Nevertheless, for most applications this is not true and $\mathrm{S}_0 \cap \mathrm{S}_1 = \mathrm{S} \neq \emptyset$, and, therefore, any classification methodology has a nonzero probability to return an erroneous classification for those individuals with $Y_i \in \mathrm{S}$, leading to the possibility of misclassification. Hereinafter it will be assumed that $\mathrm{S} \neq \emptyset$, although the opposite situation can be addressed analogously.

In addition, let Y_i be characterized by some distribution $\mathbf{D}(\theta)$ for the entire population, where θ denotes the parameter vector, i.e., $Y_i \sim \mathbf{D}(\theta)$, $i = 1, \ldots, N$. Thus, the distribution $\mathbf{D}(\theta)$ is a convex mixture of the two distributions $\mathbf{D}_0(\theta_0)$ and $\mathbf{D}_1(\theta_1)$ with weights $1 - p$ and p, respectively. Therefore, the distribution function F_Y of Y can be written as a convex combination of the distributions functions of F_{Y^-} and F_{Y^+}, i.e., through $F_Y(y) = (1 - p)F_{Y^-}(y) + pF_{Y^+}(y)$, $\forall y \in \mathbb{R}$.

In order to simplify, let us suppose that the presence of the infection leads to a high Y value, while the absence of the infection leads to the observation of a low value of Y (the opposite case is analogous). Hence, if Y_i exceeds a prefixed threshold $t \in \mathrm{S}$, then the individual is classified as infected (getting a positive test denoted by $X_i' = 1$), i.e., $Y_i > t \Rightarrow X_i' = 1$. Otherwise, it is considered uninfected (achieving a negative result $X_i' = 0$), thus $Y_i \leq t \Rightarrow X_i' = 0$.

The most commonly used measures to evaluate the single test accuracy are the sensitivity φ_s, i.e., the probability of getting a positive result ($X'_i = 1$) from an infected individual ($X_i = 1$), thus $\varphi_s = \mathrm{P}\left(X'_i = 1 | X_i = 1\right)$; and the specificity φ_e, i.e., the probability of getting a negative result from a not infected individual, thereby $\varphi_e = \mathrm{P}\left(X'_i = 0 | X_i = 0\right)$. Besides, φ_s assesses the test ability to identify an infected individual, crucial in epidemic cases, while φ_e evaluates the test ability to identify an uninfected individual.

3 Continuous Quantitative Compound Tests

Let n denote the chosen group size. To perform a compound analysis we begin by randomly choosing n individuals of the population. These individuals, denoted by the arbitrary indexes $j_1, \ldots, j_n$, will form the k-th group, i.e., $\mathrm{G}_k = \{j_1, \ldots, j_n\}$, for $k = 1, \ldots, \lceil \frac{N}{n} \rceil$ with $\lceil x \rceil$ denoting the smallest integer not less than x (the final group may contain fewer individuals but, for simplicity, we will assume that all groups have dimension n and for this specific last group the appropriate adaptations should be applied). Thus, the r.v.s X_i, for $i \in \mathrm{G}_k$, are mutually independent and the number $\mathrm{I}^{[n]}$ of infected members in the group is a binomial r.v., i.e., $\mathrm{I}^{[n]} \sim \mathbf{B}(n, p)$ with n trials and probability of success equals to p. Consequently, within each group the r.v.s Y_i are independent and identically distributed (i.i.d) to $\mathbf{D}(\theta)$. After that, one *ml* of blood from each one of the n elements is collected and then mixed together until uniformity is achieved. Hence, the amount of substance in the n *ml* of pooled blood is given by $B_n = \sum_{i \in \mathrm{G}_k} Y_i$. Then, one *ml* is randomly withdrawn from this homogeneous pooled blood in order to perform the compound test. Let the r.v. B_1 describes the amount of substance in this milliliter of pooled blood.

When $\mathbf{D}(\theta)$ is a count distribution, then B_1 can be modeled using hierarchical models and $B_1 \sim \mathbf{B}(B_n, \frac{1}{n})$. The distribution of B_1 for some of the most used count distributions may be found in [24].

However, hierarchical models cannot be applied when $\mathbf{D}(\theta)$ is absolutely continuous. In these cases we can consider a perfect mixed procedure, performed by some mechanical device, and consequently B_1 will be very close to the group mean, i.e., $B_1 \approx \overline{Y}_n = \frac{1}{n} B_n$, cf. [25], and then the $\overline{Y}_n$ distribution can be computed analytically or by simulation.

Finally, if the observed B_1 value exceeds the threshold $t^{[n]}$, which depends of the group size n, the group will be classified as infected (a positive compound test), otherwise the group is classified as uninfected (a negative compound result). A positive compound result ($X^{[n]} = 1$) means (if misclassification does not occur) that there is at least one infected individual in the group, i.e., identifies the groups in which $\sum_{i \in \mathrm{G}_k} X_i \geq 1$. Let us notice that these tests do not allow the identification of the infected individuals, but only the identification of the presence of at least one infected individual in the analyzed group. Insofar, a negative compound result ($X^{[n]} = 0$) means that no individual is infected within the group, i.e., aims to identify the groups in which $\sum_{i \in G_k} X_i = 0$. Note that to classify all individuals within the group as uninfected through the performance of individual tests, all Y_i, for $i = j_1, \ldots, j_n$, must fulfill $Y_i \leq t$, and thence $M_n = \max\left(Y_{j_1}, \ldots, Y_{j_n}\right) \leq t$. Therefore, quantitative compound tests use $B_1 \approx \overline{Y}_n$ in order to establish if the group maximum M_n exceeds the threshold t. The use of the mean to test the maximum of a group with n i.i.d. r.v.s applying several continuous distributions is investigated in [25].

Compound tests must be followed by individual tests whenever our main goal is the identification of all infected individuals in the population (the classification problem). The first and simplest classification methodology using pooled samples has been proposed by Dorfman in 1943 [5] for the identification of the syphilis infected soldiers. Whenever the compound analysis results are positive an individual test is performed to all the members of the group in this method. The main goal of compound analysis is to save resources and this is only possible when the large majority of the groups is not infected, and therefore are classified as uninfected through a single test (when the compound test is positive the number of required test are higher than if we only use individual tests). For this purpose, compound tests are only applied in low prevalence rates. In addition, the group size n is defined in order to ensure a high probability of getting negative compound tests.

More efficient methodologies in terms of the relative cost (expected number of tests for the classification *per* individual) have been developed ([13] provides an overview of its evolution). There are not only generalizations of Dorfman's methodology (the hierarchical algorithms in which positive groups are repeatedly divided into smaller non-overlapping subgroups until all members have been individually tested, cf. [6, 9, 10, 16, 18, 28, 29]), but also more complex sampling strategies using array-based group testing (which use overlapping pools, cf. [15, 22, 33]) or even multidimensional array algorithms (an extension to higher dimensional arrays, cf. [1, 23]).

Compound tests can also be applied without being required subsequent individual tests when the main goal is the estimation of the prevalence rate p (the estimation problem), cf. [27]. The estimators based in compound analysis can attain, under certain conditions, better performance than the estimators based on individual tests, allowing not only the reduction of the number of performed tests, but also the achievement of more accurate estimates with respect to the bias, efficiency as well as robustness, cf. [4, 7, 17, 20] among others. Moreover, some packages with applications of several compound testing estimators are available, such as *binGroup* for the R software [2].

Nevertheless, the main drawback of compound analysis is its higher probability of misclassification, mainly due to the dilution effect, cf. [11]. Therefore, in both cases (classification and estimation) the use of compound tests should only be performed if the problem of misclassification is monitored, enabling to balance the effects of cost and accuracy.

The misclassification in compound tests can be measured by the compound specificity $\varphi_e^{[n]} = \mathrm{P}(X^{[n]} = 0 | \sum_{i=1}^{n} X_i = 0) = \mathrm{P}(X^{[n]} = 0 | I^{[n]} = 0)$ which is usually higher than the single specificity φ_e as a consequence of the sample mean getting closer to the expected value as the group size n increases. On the other hand, the compound sensitivity $\varphi_s^{[n]}$ is defined as $\varphi_s^{[n]} = \mathrm{P}(X^{[n]} = 1|\sum_{i=1}^{n} X_i \geq 1) = \mathrm{P}(X^{[n]} = 1|I^{[n]} \geq 1)$. Moreover, $\varphi_s^{[n]}$ depends on the number of infected elements within the group due to the dilution factor—if we pool blood from one infected individual with the blood of many uninfected individuals, the substance will be diluted and the probability to detect whether there is an infected individual in the group can be quite low. Thus, the compound sensitivity $\varphi_s^{[j,n]}$ when there are j infected members in the group, using $\varphi_s^{[j,n]} = \mathrm{P}(X^{[n]} = 1|I^{[n]} = j)$ and $\varphi_s^{[n]} = \sum_{j=1}^{n} \varphi_s^{[j,n]}\, \mathrm{P}(I^{[n]} = j|I^{[n]} \geq 1)$, can be used to model the rarefaction of the substance, cf. [24]. Moreover, $\varphi_s^{[n]} \approx \mathrm{P}(X^{[n]} = 1|I^{[n]} = 1) = \varphi_s^{[1,n]}$ for low prevalence rates, cf. [8, 24], as a consequence of $\mathrm{P}(I^{[n]} = 1|I^{[n]} \geq 1) \approx 1$ for the usual applied group sizes. In addition, having just one infected individual in the group corresponds to the worst case scenario, i.e., $\varphi_s^{[1,n]} \leq \varphi_s^{[2,n]} \leq \cdots \leq \varphi_s^{[n,n]}$ and consequently $\varphi_s^{[1,n]} \leq \varphi_s^{[n]}$. Let us emphasize that in most compound analysis applications it is assumed that pooling does not affect misclassification (see, for instance, [19, 30, 31]), or do not take into account the number of infected members within the group, cf. [14]. In [32] and [34] hierarchical models are used to capture the dilution effect in HIV prevalence estimation, but the probability of misclassification has not been evaluated.

These misclassification measures can be generalized to the classification methodology itself, cf. [15, 24]. Hence, the $\mathcal{M}$ methodology specificity is the probability of an uninfected individual being classified as uninfected by the application of methodology $\mathcal{M}$ and, analogously, the $\mathcal{M}$ methodology sensitivity is the probability of an infected individual being classified as infected by the application of methodology $\mathcal{M}$. Therefore, the same definitions as in the individual tests are applied, but the probabilities are computed taking into consideration the application of the classification methodology $\mathcal{M}$ under investigation. For instance, the Dorfman's methodology specificity is given by (cf. [24])

$$\varphi_{e_n} = \mathrm{P}\left(X_i' = 0|\, X_i = 0, \mathcal{M}\right) = \sum_{j=0}^{n-1} \mathrm{P}\left(X_i' = 0|\, X_i = 0, \mathrm{I}^{[n-1]} = j\right) \mathrm{P}\left(\mathrm{I}^{[n-1]} = j\right),$$

and, analogously, the Dorfman's methodology sensitivity is given by

$$\varphi_{s_n} = \mathrm{P}\left(X_i' = 1|\, X_i = 1, \mathcal{M}\right) = \sum_{j=0}^{n-1} \mathrm{P}\left(X_i' = 1|\, X_i = 1, \mathrm{I}^{[n-1]} = j\right) \mathrm{P}\left(\mathrm{I}^{[n-1]} = j\right).$$

In the simulations (Sect. 5) we will restrict ourselves to the evaluation of the compound measures of misclassification $\varphi_e^{[n]}$ and $\varphi_s^{[n]}$ (without specifying any classification or estimation methodology) in the cases in which $\mathbf{D}(\theta)$ is a continuous distribution and applying two different methodologies to set up the cut off point $t^{[n]}$, which will be established in Sect. 4.

4 Methodologies to Set Up the Cut Off Point of Compound Tests

Two different methodologies to perform the compound tests are described in this section, following [26]. The underlying principle is to use each method in order to control the probability of a type of misclassification since it is impossible to improve both, such as in the usual statistical hypothesis tests.

The first methodology $\mathbf{M_1}$ aims to control the compound specificity and matches to the commonly used in compound tests applications, which can be formalized by the statistical hypotheses:

$$\mathbf{H_0}: \sum_{i=1}^{n} X_i = 0 \quad \text{versus} \quad \mathbf{H_1}: \sum_{i=1}^{n} X_i \geq 1, \qquad [\text{Methodology } \mathbf{M_1}]$$

or, analogously, $\mathbf{H_0}: \mathrm{I}^{[n]} = 0$ versus $\mathbf{H_1}: \mathrm{I}^{[n]} \geq 1$. In terms of comparison with the individual analysis, $\mathbf{H_0}$ implies $M_n \leq t$ and $\mathbf{H_1}$ entails $M_n > t$. Moreover, the test size is given by $\alpha = \mathrm{P}(X^{[n]} = 1 | \sum_{i=1}^{n} X_i = 0) = 1 - \varphi_e^{[n]}$. Hence, the compound specificity is hereby set at $1 - \alpha$ and the $\varphi_e^{[n]}$ is controlled by setting the value of the significance level α. Nevertheless, it neglects the compound sensitivity, and therefore the occurrence of false negatives is not monitored in this methodology. Consequently, the quality of the compound tests performed using this methodology shall be assessed by the compound sensitivity. In addition, under $\mathbf{H_0}$ the group G_k has no infected individuals, thus the r.v. Y_i, $i \in \mathrm{G}_k$, are i.i.d. to $\mathbf{D}_0(\theta_0)$ and the cut off point $t^{[n,\mathbf{M_1}]}$ is straightforward to compute.

The goal of the alternative methodology $\mathbf{M_2}$ is to control the compound sensitivity and therefore to prevent the occurrence of false negative results, which is crucial in epidemic situations. It was first proposed in [21] without any examination and was applied in discrete compound tests simulation in [26]. The $\mathbf{M_2}$ methodology is formalized by the following statistical hypotheses:

$$\mathbf{H_0}: \sum_{i=1}^{n} X_i \geq 1 \quad \text{versus} \quad \mathbf{H_1}: \sum_{i=1}^{n} X_i = 0, \qquad [\text{Methodology } \mathbf{M_2}]$$

which in terms of the r.v.s $\mathrm{I}^{[n]}$ corresponds to $\mathbf{H_0}: \mathrm{I}^{[n]} \geq 1$ versus $\mathbf{H_1}: \mathrm{I}^{[n]} = 0$. In this methodology the test size α is given by $\alpha = \mathrm{P}(X^{[n]} = 0 | \sum_{i=1}^{n} X_i \geq 1) = 1 - \varphi_s^{[n]}$, and therefore the compound sensitivity is fixed by setting the value of α. Nevertheless, there are different possible scenarios under $\mathbf{H_0}$ and consequently the computation of the cut off point $t^{[n,\mathbf{M_2}]}$ can be quite complex. In practice, a simplified methodology $\mathbf{M_2^\star}$ can be implemented in order to easily compute an approximate value of $t^{[n,\mathbf{M_2}]}$, performing the following hypothesis test:

$$\mathbf{H_0}: \sum_{i=1}^{n} X_i = 1 \quad \text{versus} \quad \mathbf{H_1}: \sum_{i=1}^{n} X_i = 0, \qquad [\text{Methodology } \mathbf{M_2^\star}]$$

i.e., $\mathbf{H_0} : \mathrm{I}^{[n]} = 1$ versus $\mathbf{H_1} : \mathrm{I}^{[n]} = 0$. The main goal is to use the threshold $t^{[n,\mathbf{M}_2^\star]}$, as it is a quite good approximation of the cut off point $t^{[n,\mathbf{M}_2]}$ as a consequence of $P(I^{[n]} = 1|I^{[n]} \geq 1) \approx 1$, cf. [8, 24], otherwise the use of compound analysis would not be advised. Thus, the results of applying this simplified $\mathbf{M}_2^\star$ are quite similar to $\mathbf{M_2}$ for low prevalence rates and the usual applied group sizes. Moreover, the significance level in $\mathbf{M}_2^\star$ is given by $\alpha = P(X^{[n]} = 0|\sum_{i=1}^{n} X_i = 1) = 1 - \varphi_s^{[1,n]}$, and therefore α will set $\varphi_s^{[1,n]}$. Thus, $\mathbf{M}_2^\star$ controls the compound sensitivity in the worst case scenario, and consequently controls indirectly the overall compound sensitivity $\varphi_s^{[n]}$. Hence, the applied significance level α will set up a lower limit for the compound sensitivity $\varphi_s^{[n]}$. In this alternative methodology the compound sensitivity is fixed and the quality of the test will be measured by the compound specificity.

5 Simulation

This section aims to evaluate the performance of continuous compound tests via simulations performed by the statistical software R, mainly its compound sensitivity $\varphi_s^{[n]}$ and compound specificity $\varphi_e^{[n]}$. Thus, the methodologies $\mathbf{M_1}$ and $\mathbf{M}_2^\star$ were applied and whereby the group mean $\overline{Y}_n$ was used to identify if the group maximum M_n exceeds the threshold $t = F_Y^{\leftarrow}(1-p)$ with $F_Y^{\leftarrow}$ being the generalized inverse function of F_Y, i.e., $F^{\leftarrow}(y) := \inf\{x : F(x) \geq y\}$.

In fact, the performance of those misclassification measures in continuous compound analysis depend decisively on the prevalence rate p, the group size n and mainly on the properties of the distribution $\mathbf{D}(\theta)$. Hence, in the simulations we used different prevalence rates $p \in \{0.05, 0.01, 0.001, 0.0001\}$. For each prevalence rate p it was applied the most efficient group size n in Dorfman's classification methodology, cf. [5, 24].

The continuous distributions used were the standard Gaussian distribution (denoted by $\mathscr{N}$), the Student's t distribution with m degrees of freedom ($t_{(m)}$), the chi-squared distribution with m degrees of freedom ($\chi^2_{(m)}$), the standard exponential distribution (**Exp**), the Pareto distribution with shape parameter α and density $f(x) = \alpha x^{-1-\alpha}$ for $x > 1$ ($\mathbf{P}_{(\alpha)}$), the standard log-Normal distribution ($\ln \mathscr{N}$), the Weibull distribution with shape parameter ζ ($\mathbf{W}_{(\zeta)}$), the slash normal distribution (**SN**), which is obtained by dividing a standard Gaussian r.v. by another independent r.v. with standard uniform distribution, and the standard Lévy distribution (**Lévy**). Moreover, the absolute value of some r.v.s was also used in order to avoid a bilateral heavy-tailed distribution because, in these cases, a large negative value can hide a large positive value in the compound analysis.

In [24] it is shown that the correlations $r(\overline{Y}_n, M_n)$ between the group mean $\overline{Y}_n$ and the group maximum M_n are indeed crucial for the quality of this type of tests. We also compute an approximate value of these correlations using simulations with 10^3 replicas of 10^4 groups. The mean and the standard deviation (within brackets) of the correlation coefficient for each case are shown in Tables 1 and 2. Another important factor highlighted in [24], although not measured, is the right tail weight of the distribution $\mathbf{D}(\theta)$. Thus, in this work we computed the right tail index τ_R defined by (cf. [12])

$$\tau_R = \left(\frac{F_{\mathbf{D}}^{-1}(0.99) - F_{\mathbf{D}}^{-1}(0.5)}{F_{\mathbf{D}}^{-1}(0.75) - F_{\mathbf{D}}^{-1}(0.5)}\right)\left(\frac{\Phi^{-1}(0.99) - \Phi^{-1}(0.5)}{\Phi^{-1}(0.75) - \Phi^{-1}(0.5)}\right)^{-1},$$

which compares the right tail weight of the distribution **D** with the right tail weight of the Gaussian distribution. Thus, for the Gaussian distribution $\tau_R = 1$ and heavier is the right tail of **D** higher is the τ_R index value of **D**.

The simulations results are displayed in Tables 1 and 2 in which the distributions have been ordered in ascending order of the right tail index τ_R. All the simulations presented, using 10^6 groups in each case, have been done considering the significance level $\alpha = 0.05$, however the results are representative of the general behaviour for other significance level values.

The results show that the correlations $r(\overline{Y}_n, M_n)$ steadily increase with the right tail index τ_R. The exceptions lie in the bilateral heavy-tailed distributions (such as **SN** and $t_{(1)}$) in which an individual with a quite negative value can hide other individual with a large positive value in the group mean computation. Nevertheless, when positive r.v.s are applied the association between $r(\overline{Y}_n, M_n)$ and τ_R is obvious. Thus, $r(\overline{Y}_n, M_n)$ are significant with low standard deviation for distributions with high right tail index τ_R (except in bilateral heavy-tailed distributions). It can also be inferred that correlations $r(\overline{Y}_n, M_n)$ decrease with the group size n, with a higher rate for low right tail distributions. For high right tail distributions the correlations $r(\overline{Y}_n, M_n)$ continues to attain very high values even for groups with quite higher dimensions, such as in the Lévy distribution with $n = 100$.

It is equally clear that methodology $\mathbf{M_1}$ can be actually applied to control the compound specificity and $\mathbf{M_2^\star}$ to control the compound sensitivity, with high accuracy. In all simulations $\varphi_e^{[n]} \approx 0.95$ in $\mathbf{M_1}$ and also $\varphi_s^{[n]} \approx 0.95$ in $\mathbf{M_2^\star}$, in agreement with our use of the significance level $\alpha = 0.05$.

The compound sensitivity $\varphi_s^{[n]}$ in $\mathbf{M_1}$ and the compound specificity $\varphi_e^{[n]}$ in $\mathbf{M_2^\star}$ have analogous performance. It performs poorly for distributions with low right tail index (as Table 1 clearly shows) and achieve high-quality accuracy in heavy right tail distributions (see Table 2). In addition, for low right tail distributions the performance gets worse very quickly when the group size n increases. For heavy right tail distributions the performance continues to attain quite good results even with a high group size n.

Table 3 shows the correlation between $r(\overline{Y}_n, M_n)$ and each of the non-controlled misclassification measure ($\varphi_s^{[n]}$ in the methodology $\mathbf{M_1}$ and $\varphi_e^{[n]}$ in $\mathbf{M_2^\star}$) including all the distributions analyzed in the previous tables, and in the second case excluding the bilateral heavy tailed distributions **SN** and $t_{(1)}$. The observed correlations are impressively high, mainly when the bilateral heavy tailed distributions are removed. Therefore, the correlations $r(\overline{Y}_n, M_n)$ (and consequently the right tail index τ_R) are

Table 1 Continuous compound tests simulations results with $\alpha = 0.05$

		$\mathcal{N}$	$t_{(5)}$	$\chi^2_{(5)}$	**Exp**	$\lvert t_{(5)} \rvert$	$\chi^2_{(1)}$	$\mathbf{P}_{(5)}$	$\ln \mathcal{N}$
τ_R		1	1.34	1.37	1.64	1.67	2.06	2.31	2.78
Simulations results for $p = 0.05$ and $n = 5$									
$r(\overline{Y}_n, M_n)$		.668 (.002)	.667 (.002)	.791 (.001)	.844 (.001)	.838 (.003)	.888 (.001)	.897 (.001)	.918 (.002)
$\mathbf{M_1}$	$\varphi_s^{[n]}$	34.1	42.8	49.4	58.7	58.0	68.4	96.5	74.8
$\mathbf{M_1}$	$\varphi_e^{[n]}$	94.9	95.0	95.0	94.9	95.0	95.0	95.0	95.1
$\mathbf{M_2^\star}$	$\varphi_s^{[n]}$	95.5	95.4	95.4	95.4	95.5	95.5	95.4	95.4
$\mathbf{M_2^\star}$	$\varphi_e^{[n]}$	31.7	32.2	55.4	67.8	63.9	78.4	95.6	80.9
Simulations results for $p = 0.01$ and $n = 10$									
$r(\overline{Y}_n, M_n)$		.539 (.002)	.559 (.002)	.677 (.002)	.744 (.001)	.741 (.005)	.804 (.001)	.825 (.002)	.858 (.004)
$\mathbf{M_1}$	$\varphi_s^{[n]}$	23.2	35.2	36.4	45.1	50.8	55.1	100	72.5
$\mathbf{M_1}$	$\varphi_e^{[n]}$	95.0	95.1	95.1	95.0	95.0	94.9	95.0	95.0
$\mathbf{M_2^\star}$	$\varphi_s^{[n]}$	95.2	95.3	95.1	95.3	95.1	95.2	95.1	95.1
$\mathbf{M_2^\star}$	$\varphi_e^{[n]}$	24.0	30.4	45.6	58.8	60.4	70.9	98.6	82.1

(continued)

Table 1 (continued)

		$\mathcal{N}$	$t_{(5)}$	$\chi^2_{(5)}$	**Exp**	$\lvert t_{(5)} \rvert$	$\chi^2_{(1)}$	$\mathbf{P}_{(5)}$	$\ln \mathcal{N}$
Simulations results for $p = 0.001$ and $n = 32$									
$r(\overline{Y}_n, M_n)$		.360 (.003)	.410 (.003)	.494 (.003)	.565 (.002)	.579 (.002)	.634 (.002)	.685 (.001)	.737 (.002)
$\mathbf{M_1}$	$\varphi_s^{[n]}$	14.9	28.3	24.0	30.8	44.8	38.5	100	70.9
$\mathbf{M_1}$	$\varphi_e^{[n]}$	95.0	95.0	95.1	95.1	95.0	95.0	95.0	95.0
$\mathbf{M_2^\star}$	$\varphi_s^{[n]}$	95.3	95.0	95.0	95.2	95.2	95.1	95.2	95.0
$\mathbf{M_2^\star}$	$\varphi_e^{[n]}$	15.0	26.6	29.3	40.4	52.2	52.2	99.8	80.9
Simulations results for $p = 0.0001$ and $n = 100$									
$r(\overline{Y}_n, M_n)$		.232 (.002)	.298 (.003)	.344 (.003)	.406 (.001)	.439 (.009)	.469 (.003)	.547 (.002)	.607 (.007)
$\mathbf{M_1}$	$\varphi_s^{[n]}$	10.5	25.1	16.1	21.8	42.2	27.0	100	71.9
$\mathbf{M_1}$	$\varphi_e^{[n]}$	94.9	94.9	95.1	95.0	94.9	95.1	95.0	95.0
$\mathbf{M_2^\star}$	$\varphi_s^{[n]}$	95.0	94.9	94.5	94.5	95.2	94.9	95.2	95.1
$\mathbf{M_2^\star}$	$\varphi_e^{[n]}$	10.9	23.2	19.4	25.9	45.9	34.5	100	78.5

Table 2 Continuous compound tests simulations results with $\alpha = 0.05$

		$\mathbf{W}_{(\frac{1}{2})}$	**SN**	$t_{(1)}$	$\lvert t_{(1)} \rvert$	**|SN|**	$\mathbf{P}_{(1)}$	$\mathbf{W}_{(\frac{1}{4})}$	**Lévy**
τ_R		4.17	7.87	9.23	12.9	13.2	14.2	37.6	241
Simulations results for $p = 0.05$ and $n = 5$									
$r(\overline{Y}_n, M_n)$		.949 (.002)	.622 (.309)	.663 (.309)	1.00 (.000)	1.00 (.000)	1.00 (.000)	.996 (.000)	1.00 (.000)
$\mathbf{M_1}$	$\varphi_s^{[n]}$	84.8	76.3	78.0	95.5	95.5	96.4	100	100
$\mathbf{M_1}$	$\varphi_e^{[n]}$	95.2	95.0	94.9	95.0	95.0	95.1	95.0	95.0
$\mathbf{M_2^\star}$	$\varphi_s^{[n]}$	95.5	95.3	95.4	95.4	95.6	95.4	95.5	95.5
$\mathbf{M_2^\star}$	$\varphi_e^{[n]}$	89.7	14.0	13.8	95.0	95.0	95.6	97.9	99.2
Simulations results for $p = 0.01$ and $n = 10$									
$r(\overline{Y}_n, M_n)$		.903 (.003)	.684 (.287)	.672 (.302)	1.00 (.000)	1.00 (.000)	1.00 (.000)	.991 (.000)	1.00 (.000)
$\mathbf{M_1}$	$\varphi_s^{[n]}$	80.6	91.3	90.9	100	100	100	100	100
$\mathbf{M_1}$	$\varphi_e^{[n]}$	95.1	94.9	95.0	95.0	95.0	95.1	95.1	95.0
$\mathbf{M_2^\star}$	$\varphi_s^{[n]}$	95.2	95.1	95.3	95.2	95.3	95.1	95.2	95.3
$\mathbf{M_2^\star}$	$\varphi_e^{[n]}$	88.9	69.5	59.7	98.6	98.7	98.6	98.4	99.3

(continued)

Table 2 (continued)

		$\mathbf{W}_{(\frac{1}{2})}$	**SN**	$t_{(1)}$	$\lvert t_{(1)} \rvert$	**|SN|**	$\mathbf{P}_{(1)}$	$\mathbf{W}_{(\frac{1}{4})}$	**Lévy**
Simulations results for $p = 0.001$ and $n = 32$									
$r(\overline{Y}_n, M_n)$		.789 (.003)	.633 (.312)	.630 (.312)	1.00 (.000)	1.00 (.000)	1.00 (.000)	.972 (.000)	1.00 (.000)
$\mathbf{M_1}$	$\varphi_s^{[n]}$	75.0	98.2	97.9	100	100	100	100	100
$\mathbf{M_1}$	$\varphi_e^{[n]}$	95.0	95.0	94.9	95.0	95.0	95.2	95.1	95.0
$\mathbf{M_2^\star}$	$\varphi_s^{[n]}$	95.3	95.1	95.0	95.2	95.3	95.1	95.2	95.2
$\mathbf{M_2^\star}$	$\varphi_e^{[n]}$	85.1	99.6	99.4	99.8	99.8	99.8	99.0	100
Simulations results for $p = 0.0001$ and $n = 100$									
$r(\overline{Y}_n, M_n)$		.651 (.006)	.619 (.307)	.607 (.317)	1.00 (.000)	1.00 (.000)	1.00 (.000)	.934 (.000)	1.00 (.000)
$\mathbf{M_1}$	$\varphi_s^{[n]}$	67.5	99.6	99.4	100	100	100	100	100
$\mathbf{M_1}$	$\varphi_e^{[n]}$	95.0	95.0	94.9	95.1	95.0	95.0	95.1	95.0
$\mathbf{M_2^\star}$	$\varphi_s^{[n]}$	94.9	95.1	95.3	94.9	95.3	94.8	95.2	95.2
$\mathbf{M_2^\star}$	$\varphi_e^{[n]}$	78.6	100	100	100	100	100	99.5	100

Table 3 Correlations between $r(\overline{Y}_n, M_n)$ and each of the non-controlled misclassification measure

	$p = 0.05$		$p = 10^{-2}$		$p = 10^{-3}$		$p = 10^{-4}$	
	$\mathbf{M_1}$	$\mathbf{M_2^\star}$	$\mathbf{M_1}$	$\mathbf{M_2^\star}$	$\mathbf{M_1}$	$\mathbf{M_2^\star}$	$\mathbf{M_1}$	$\mathbf{M_2^\star}$
All distributions	0.7273	0.9777	0.7813	0.9455	0.8176	0.8260	0.8432	0.8430
Without **SN** nor $t_{(1)}$	0.9435	0.9751	0.9366	0.9624	0.9267	0.9326	0.9073	0.9037

crucial to assess the suitability of applying continuous compound tests in order to control the probability of misclassification.

6 Final Remarks

Continuous compound analysis can be applied for classification and estimation purposes with high-quality accuracy in low prevalence rates whenever the distribution underlying the analyzed substance is an unilateral heavy-tailed distribution. Moreover, different methodologies to compute the cut off point of the compound analysis can be applied, in order to control the compound sensitivity or to control the compound specificity. The non-controlled misclassification measure will also return quite good results under the specified conditions. Furthermore, the right tail index can be applied to measure the suitability of continuous compound analysis. The performed simulations clearly reveal that a high τ_R value ensures simultaneously a high compound sensitivity and a high compound specificity in both proposed methodologies, and consequently a low probability of misclassification in the continuous compound analysis applications.

Acknowledgements Research partially sponsored by national funds through the Fundação Nacional para a Ciência e Tecnologia, Portugal—FCT under the project PEst-OE/MAT/UI0006/2014.

References

1. Berger, T., Mandell, J.W., Subrahmanya, P.: Maximally efficient two-stage screening. Biometrics **56**, 833–840 (2000)
2. Bilder, C.R., Zang, B., Schaarschmidt, F., Tebbs, J.M.: Bingroup: a package for group testing. R J. **2**, 56–60 (2010)
3. Boswell, M.T., Gore, S.D., Lovison, G., Patil, G.P.: Annotated bibliography of composite sampling, Part A: 1936–1992. Environ. Ecol. Stat. **3**, 1–50 (1996)
4. Chen, C., Swallow, W.: Sensitivity analysis of variable-sized group testing and its related continuous models. Biometrical J. **37**, 173–181 (1995)
5. Dorfman, R.: The detection of defective members in large populations. Ann. Math. Stat. **14**, 436–440 (1943)
6. Finucan, H.M.: The blood testing problem. Appl. Stat. J. Roy. St. C **13**, 43–50 (1964)

7. Garner, F.C., Stapanian, M.A., Yfantis, E.A., Williams, L.R.: Probability estimation with sample compositing techniques. J. Off. Stat. **5**, 365–374 (1989)
8. Gastwirth, J.L., Johnson, W.O.: Screening with cost-effective quality control: potential applications to HIV and drug testing. JASA **89**, 972–981 (1994)
9. Gastwirth, J.L.: The efficiency of pooling in the detection of rare mutations. Am. J. Hum. Genet. **67**, 1036–1039 (2000)
10. Gill, A., Gottlieb, D.: The identification of a set by successive intersections. In: Information and Control, 20–25. Ellis Horwood, Chichester (1974)
11. Hung, M., Swallow, W.: Robustness of group testing in the estimation of proportions. Biometrics **55**, 231–237 (1999)
12. Hoaglin, D.M., Mosteller, F., Tukey, J.W.: Understanding Robust and Exploratory Data Analysis. Wiley, New York (1983)
13. Hughes-Oliver, J.M.: Pooling experiments for blood screening and drug discovery. In: Dean, A., Lewis, S. (eds.) Screening: Methods for Experimentation in Industry, Drug Discovery, and Genetics, 48–68. Springer, Berlin (2006)
14. Hwang, F.K.: Group testing with a dilution effect. Biometrika **63**, 671–673 (1976)
15. Kim, H., Hudgens, M., Dreyfuss, J., Westreich, D., Pilcher, C.: Comparison of group testing algorithms for case identification in the presence of testing errors. Biometrics **63**, 1152–1163 (2007)
16. Johnson, N.L., Kotz, S., Wu, X.: Inspection Errors for Attributes in Quality Control. Chapman and Hall, New York (1991)
17. Lancaster, V.A., Keller-McNulty, S.: A review of composite sampling methods. J. Am. Stat. Assoc. **93**, 1216–1230 (1998)
18. Litvak, E., Tu, X.M., Pagano, M.: Screening for the presence of a disease by pooling sera samples. J. Am. Stat. Assoc. **89**, 424–434 (1994)
19. Liu, S.C., Chiang, K.S., Lin, C.H., Chung, W.C., Lin, S.H., Yang, T.C.: Cost analysis in choosing group size when group testing for potato virus Y in the presence of classification errors. Ann. Appl. Biol. **159**, 491–502 (2011)
20. Loyer, M.W.: Bad probability, good statistics, and group testing for binomial estimation. Am. Stat. **37**, 57–59 (1983)
21. Martins, J.P., Santos, R., Sousa, R.: Testing the maximum by the mean in quantitative group tests. In: Pacheco, A., et al. (eds.) New Advances in Statistical Modeling and Applications, Studies in Theoretical and Applied Statistics, Selected Papers of the Statistical Societies. Springer, 55–63 (2014)
22. Phatarfod, R.M., Sudbury, A.: The use of a square array scheme in blood testing. Stat. Med. **13**, 2337–2343 (1994)
23. Roederer, M., Koup, R.A.: Optimized determination of T cell epitope responses. J. Immunol. Methods **274**, 221–228 (2003)
24. Santos, R., Pestana, D., Martins, J.P.: Extensions of Dorfman's theory. In: Oliveira, P.E., et al. (eds.) Studies in Theoretical and Applied Statistics, Recent Developments in Modeling and Applications in Statistics, 179–189. Springer, New York (2013)
25. Santos, R., Felgueiras, M., Martins, J.P.: Known mean, unknown maxima? Testing the maximum knowing only the mean. Commun. Stat. Simul. Comput. (Published online: 23 Jan 2014)
26. Santos, R., Martins, J.P., Felgueiras, M.: Discrete compound tests and Dorfman's methodology in the presence of misclassification. In: Kitsos, C., et al. (eds.) Theory and Practice of Risk Assessment, Springer Proceedings in Mathematics & Statistics 136, Springer (2015)
27. Sobel, M., Elashoff, R.: Group testing with a new goal, estimation. Biometrika **62**, 181–193 (1975)
28. Sobel, M., Groll, P.A.: Group testing to eliminate efficiently all defectives in a binomial sample. Bell Syst. Tech. J. **38**, 1179–1252 (1959)
29. Sterret, A.: On the detection of defective members of large populations. Ann. Math. Stat. **28**, 1033–1036 (1957)

30. Tu, X.M., Litvak, E., Pagano, M.: Studies of AIDS and HIV surveillance, screening tests: can we get more by doing less? Stat. Med. **13**, 1905–1919 (1994)
31. Tu, X.M., Litvak, E., Pagano, M.: On the informativeness and accuracy of pooled testing in estimating prevalence of a rare disease: Application to HIV screening. Biometrika **82**(2), 287–297 (1995)
32. Wein, L.M., Zenios, S.A.: Pooled testing for HIV screening: capturing the dilution effect. Oper. Res. **44**, 543–569 (1996)
33. Woodbury, C.P., Fitzloff, J.F., Vincent, S.S.: Sample multiplexing for greater throughput in HPLC and related methods. Anal. Chem. **67**, 885–890 (1995)
34. Zenios, S., Wein, L.: Pooled testing for HIV prevalence estimation exploiting the dilution effect. Stat. Med. **17**, 1447–1467 (1998)

Varying the Money Supply of Commercial Banks

Martin Shubik and Eric Smith

Abstract We consider the problem of financing two productive sectors in an economy through bank loans, when the sectors may experience independent demands for money but when it is desirable for each to maintain an independently determined sequence of prices. An idealized central bank is compared with a collection of commercial banks that generate profits from interest rate spreads and flow those through to a collection of consumer/owners who are also one group of borrowers and lenders in the private economy. We model the private economy as one in which both production functions and consumption preferences for the two goods are independent, and in which one production process experiences a shock in the demand for money arising from an opportunity for risky innovation of its production function. An idealized, profitless central bank can decouple the sectors, but for-profit commercial banks inherently propagate shocks in money demand in one sector into price shocks with a tail of distorted prices in the other sector. The connection of profits with efficiency-reducing propagation of shocks is mechanical in character, in that it does not depend on the particular way profits are used strategically within the banking system. In application, the tension between profits and reserve requirements is essential to enabling but also controlling the distributed perception and evaluation services provided by commercial banks. We regard the inefficiency inherent in the profit system as a source of costs that are paid for distributed perception and control in economies.

M. Shubik (✉)
Cowles Foundation for Research in Economics, Yale University, New Haven, CT 06520-8281, USA

Santa Fe Institute, 1399 Hyde Park Road, Santa Fe, NM 87501, USA
e-mail: martin.shubik@yale.edu

E. Smith
Krasnow Institute for Advanced Study, George Mason University, Fairfax, VA 22030, USA

Santa Fe Institute, 1399 Hyde Park Road, Santa Fe, NM 87501, USA
e-mail: desmith@santafe.edu

J.-P. Bourguignon et al. (eds.), *Dynamics, Games and Science*, CIM Series in Mathematical Sciences 1, DOI 10.1007/978-3-319-16118-1_36

1 Preamble

1.1 The Problems of Decoupling Scale and Structure in Plumbing

Consider a problem faced by designers of plumbing for hotels. Trunk lines supply hot and cold water to many taps in many guest rooms. Sinks, showers, and toilets draw water from the trunks in uncoordinated and unpredictable ways. The water flow demanded from a trunk is a variable that aggregates across users who tap the trunk, the scale of which is subjected to ongoing shocks in the course of normal usage. Water also has pressure, and the relative pressure in hot and cold lines allows the guest taking a shower to set the desired temperature by adjusting two valves. Pressure might be called a "structural" feature of the plumbing system. In good circumstances—which even in crude plumbing systems may be approached under conditions of constant demand—the pressure across the trunk is stable over time and may even be constant across taps.[1] Stability both in space and in time are essential to the system's providing its key services.

However, under shocks to the scale of flow, in a plumbing system without well-designed reservoirs of pressure, scale shocks create pressure shocks. Showering in hotels a generation or two ago offered a well-known adventure: a guest somewhere would flush a toilet, and the patron in the shower would briefly scald and then freeze. This sequence might repeat dozens of times in the course of a single shower in a large hotel on a busy morning. Plumbing designed with inadequate technology fails to buffer scale shocks in demand from propagating into the structure variables of the water flow: the time-dependent sequence of pressure values at all the valves.

1.2 Scale and Structure Problems in Money Supply and Prices

A mathematically analogous problem arises in economies. Multiple sectors demand funds for ongoing business and to cope with cycles, take risks, and respond to their unknown outcomes and other surprises. Many of these are uncoordinated and unpredictable demand shocks to the money supply, which is a scale variable of the economy. The price system in an economy is a structural property dual to its money supply [3]. Prices across sectors, or across time within a sector, need not be constant, but in a well-functioning economy they should reflect a consistent response to agents' marginal utilities or other relevant measures of valuation. In particular, it may be a design objective for a banking system that prices in a sector not be subject to ongoing ripples or other disturbances that arise purely through financial frictions,

[1] This is true for taps at the same elevation; we leave aside corrections for gravity which are not central to the point of this illustration.

due to demand shocks for money in other sectors. Governments and bankers face problems of system design of a purely mechanistic nature—meaning that they apply in a wide range of strategic contexts—analogous to the problems faced by plumbing engineers who deliver a different quantity (water) also subject to inertia and friction and by its nature not compressible.

1.2.1 Innovation as a Source of Shocks that Require Economics Beyond General Equilibrium

A pair of articles by Shubik and Sudderth [6, 7] considers innovation as a process that creates demand shocks through the problem recognized by Schumpeter [4], of "breaking the circular flow of funds". The financial design problems that arise from contexts with innovation are inherently dynamical. They offer perhaps the most direct widespread class of economic phenomena that require a robust theory falling essentially outside the General Equilibrium paradigm.

In this paper we use the *cost innovation* model of Shubik and Sudderth as a testbed to study the banker's problem of decoupling sectors in an economy. Under idealized theoretical conditions, a model banking system can both function as a strategic dummy and also decouple production and consumption sectors if they are not otherwise coupled through substitution effects. However, actual banking systems do not operate under idealized theoretical conditions. They face uncertainty throughout the economy, and they require distributed and scalable services of perception and evaluation. The limited monolithic structure of a central bank that is easily modeled in theory, is replaced in operation with an ecology of one or more central banks and a collection of competitive commercial banks operating in the private sector and responsive to its fluctuating demands for service and its geographic and demographic distribution [1].

The policy tools used to grant commercial banks the independence they require to fulfill their functions, while still controlling the quality of risk in their portfolios, include reserve requirements (set reserve ratios and possibly also minimum reserve quantities) and profits which guide the banks' strategic actions and provide a layer of abstraction between the commercial banks operational decisions and their owners' preferences. Profits create an incentive to make bank money available, while reserve requirements control its scarcity. The design problem is to balance the forces of incentive and constraint to achieve policy objectives for the banking system as a whole.

Shubik and Sudderth consider the general problem of control in strategic reserve banking. Here we do not address that higher-level problem, but instead consider the pre-strategic (more purely mechanical) question of whether the existence of profits creates inherent limits in the extent to which banking systems can decouple demands for money from propagation of price shocks. We consider an explicitly time-dependent economy with many periods of production and consumption, in which stability of the price system within a sector is essential to planning an optimal program of output and distribution. We consider only 100 % reserve banking, so that profits of commercial banks arise only through the opening of interest rate spreads between the rates on loans and on deposits.

The paper compares the interface that a simple central bank could present to an economy if there were no need for perception and control, with the interface that a comparable for-profit commercial bank presents. In cases where the central bank can decouple sectors, we find that the introduction of interest rate spreads, which are needed to create a profit motive, inherently cross-couples sector prices.[2] In the comparison, all interest rates are treated as parameters (rather than strategic variables), so this is a purely mechanistic effect, holding independent of the strategic context to which profits might be put in more elaborate models that seek to capture larger-scale regulatory dynamics.

1.2.2 Stock and Flow Distinctions as a Further Measure of Cross-Sector Propagation of Disturbances

We use a continuous-time analogue of the discrete period innovation model of Shubik and Sudderth [6, 7] because scaling analysis on the approach to this limit makes precise the distinction between stocks and flows. Stocks include outstanding (revolving) loan levels and inputs to production, while flows include streams of interest payment, velocity of money, and consumption rates of goods. We show below that in an idealized economy where money demands and prices are buffered between systems, the inter-sector loan levels within the economy, and the overall money supply and its exchange with the banks, are also distinguished in their scaling behavior. Inter-sector loans scale as stocks that remain finite in the continuous-time limit, while total money supply and net private-sector credits or debts on which interest payments change the money supply, go to zero in the continuous-time limit as the velocity of money (a flow variable) becomes the regular property of that limit.

The introduction of profits that couples cross-sector production decisions also couples inter-sector and aggregate debt levels, breaking their independent scaling behavior in the idealized efficient economy. Thus profits that couple scale shocks to structure shocks do so in several dimensions.

2 Innovation, Chance, Growth, Cycle and the Money Supply

2.1 Efficiency, Arbitrage and Equilibrium

The no arbitrage and the efficiency conditions do not coincide with incomplete markets, but the property of no arbitrage can still be defined and reflects the

[2]The coupling is in linear proportion to the spread at sufficiently small spreads.

individualistic behavior property of the noncooperative equilibrium. Once we give up the comforting fiction of complete markets we still have the definition of Pareto Optimality as an ideal and a clean picture of efficiency; but we have no individualistic solution that guarantees its attainment. A welter of theoretical problems appear in the construction of indices to measure efficiency with incomplete markets. It is well known that one can construct comparative measures between two mechanisms and possibly decide that one is more efficient than the other over a given parametric range. There is also the important empirical problem of trying to measure just how inefficient is a market structure with incomplete markets when compared with the same structure with complete markets.

If we accept the position that any market mechanism requires resources to operate it, then even Pareto optimality is challenged.

2.2 *No Arbitrage and Varying the Money Supply*

If prices in a monetary economy are to be consistent with competitive markets[3] there are several scenarios that call for the variation of the money supply. They are exogenous uncertainty, strategic uncertainty, the presence of growth and cycles in the economy. All call for a flexible money supply if cash flow constraints are to be avoided. Possibly the most interesting scenario involves innovation where the financing of the risk involved in innovation calls for a flexible money supply. We use this as the context for much of the investigation below. We note that the ability to vary the money supply confers considerable economic power on the agent able to do so.

We address specifically cost innovation and the breaking of the circular flow of funds.

2.3 *A Closed Economy with Producers, Consumers, Commercial and Investment Banks and a Central Bank*

We preface our mathematical analysis with a verbal discussion of both the modeling problems, simplifications and basic questions.

The minimal number of agent types we need to illustrate a mechanism that varies the money supply is three. They are an aggregate set of consumers; producers and a central bank.

[3] An added condition is that prices are stationary when the real goods distribution is stationary. This raises further complications involving incentives and information conditions in an economy where all laws are not indexed against inflation or deflation. This problem is not considered further here.

The consumer/stockholder/passive saver is the one set of natural legal persons required. The others are corporate legal persons all owned by the "natural persons". They are the firms, and possibly a collection of commercial banks.
The central bank differs from the other legal persons as being part of government. We first describe the central bank.

2.3.1 The Central Bank

The central bank's powers may be modeled in many ways. The simplest is as a strategic dummy endowed with the ability to accept deposits or to make loans with unlimited issue of the only legal money in the system. A formal game can be defined if either the central bank sets interest rates at which it will lend or pay on deposits, or it sets a limit on the amount of money it offers in net supply.

- In virtually all of the existing national monetary systems, not only do central banks exist, but so do commercial banks. This raises the question: Why do commercial banks exist, if the central bank can vary the money supply by itself? As Bagehot noted [1] the commercial banks (and bill jobbers) perform as perceptors and evaluators of the state of business and the need for credit over the whole space of a nation. The Soviet Union did not utilize a commercial banking system internally. It utilized bureaucratically run branches. We do not consider their perception functions here; but observe that we may formulate the construction of a four agent model where there are consumers, producers, the central bank and commercial banks where the central bank has delegated much of the variation of the money supply to the commercial, for-profit banks. With this structure several questions must be answered:
- Can the commercial banking system be competitive?
- If so, in what dimensions do they compete?
- Can they be designed to transmit fully the policy of the central bank?
- Do reserve requirements play a role?
- What are the permitted strategies of the commercial banks?
- How are the banks' profits defined?

In our belief in the virtue of separating out problems we limit our analysis here to the influence of the commercial banks on shock transmission.

3 The Flexibility of Commercial Banks

In an enterprise economy the central bank may delegate the detailed adjustments of the money supply to a commercial banking system. The problems of economic coordination need to be resolved. The particular instruments and rules of this delegation, require a set of minimal models to demonstrate this systematically. For simplicity, to begin with, we consider the commercial banking system as a strategic dummy designed to provide a flexible money supply for an economy with variable monetary needs.

4 Preliminaries

> "The doorkeeper laughs and says: 'If you are so drawn to it, just try to go in despite my veto. But take note: I am powerful. And I am only the least of the door-keepers. From hall to hall there is one doorkeeper after another, each more powerful than the last. The third doorkeeper is already so terrible that even I cannot bear to look at him.' "
>
> – Franz Kafka, *Before the Law*

Above we have presented a brief verbal sketch of why one may need a flexible money supply. We now provide a formal model to achieve this goal.

What might appear to be relatively simple mechanisms require computation or simulation of specific examples in order to illustrate even behaviorally simplistic economics.

We offer a quote from Kafka that we deem apposite in dealing with economic models where the equations of motion can be tightly defined over the whole state space.

The task of abstracting the reason why a variable money supply is needed, and the construction of the minimal institutions that fill that need, is to acknowledge the diversity of instantiations both have taken historically. The rise of fractional reserve banking in London in the last half of the ninetieth century, and the real bills doctrine in a range of conceptions from Jean-Baptiste Say to Adam Smith, were formulations of parts of this problem. Contemporary discussions of the feasibility (and consequences) of control of the money supply through interest rates versus open-market operations, and of desirable reserve levels for banks, are different mechanistically but should be understood as addressing the same fundamental questions in an age where money and credit diversity are much larger than they were in the age of Smith.

5 Varying the Money Supply with Credit

5.1 Sources of Need for a Flexible Money Supply

The different needs for a flexible money supply can be captured in formal models in a variety of ways. Often one-period models suffice to illustrate limitations in the quantity or distribution of money. In these cases, the difference between efficient and inefficient function of the financial system may be defined in terms of the alternative between interior and boundary solutions.

5.2 Separating Scale from Structure

We abstract the need for a variable money supply as a need to *separate scale from structure* in production and exchange economies. All societies undergo variations in the desired volumes of trade. These may be cyclical as in harvest seasonalities, episodic driven by good or bad harvests, immigration and emigration, innovation, etc., or progressive driven by growth or decline of population or productivity. All these variations in the capacity for production and consumption, which drive variations in the desirable volume of trade, we regard as *scale* fluctuations. A well-functioning economy also must converge on a range of price systems, both inter-sectorial and inter-temporal, including interest rates for money loans. These we regard as properties of the *structure* characterizing equilibria or near-equilibria. The stability of these price systems and the extent to which they can approximate reservation prices of agents determine the efficiency of the economy in extracting surplus, and are essential to any program of rational planning.

In its most basic abstraction, the goal of monetary policy is to accommodate the needs for scale fluctuations in an economy, without causing scale shocks to propagate to cause disruptions of structure *where such propagation can be avoided by monetary design or regulation.* Obviously, many scale shocks inherently result in shocks to prices, production, or consumption, as when innovations in substitute goods change which consumption bundles are preferred. We regard as "avoidable" propagations those that result entirely from limits on the volume and distribution of money, across sectors in which production technologies or consumption preferences are not inherently coupled.[4] Informally, a money supply that is too "rigid" or "incompressible", such as a fixed stock of gold in circulation, will generically propagate shocks in the production or consumption volume in any sector into ripples of price change across all sectors and through time, until money can be redistributed to approximate a new equilibrium for the circular flow. Alleviating this rigidity is a goal of varying the money supply that can be recognized in a variety of monetary mechanisms across societies and in different eras.

An important and general hazard and technical challenge for institutions that provide a variable money supply is ensuring consistency in the quality of credit and the pricing of risk. These are essential to the stability particularly of intertemporal price systems. The problem of credit risk evaluation is not easily centralized, and is a primary driver to grant the status of legal tender to privately created bank credit. The Real Bills Doctrine of Adam Smith may be understood as an early mechanism to permit open-ended variability in bank credit while providing criteria for credit quality that could be used by banks evaluating a range of distinct contracts. Reserve levels in modern central banking and commercial banking systems are another mechanism that attempts to regulate credit quality implicitly through lending prices and leverage.

[4]This abstraction is easy to define in models. Validating the abstraction for actual economies may be more or less difficult depending on the sectors considered.

5.3 A Class of Minimal Models

As in previous work comparing the functionality of alternative money systems [8, 9], we construct a single underlying model of production, consumption, and trade, which creates a template for a family of strategic market games (differing in their financial system models) for which explicit non-cooperative equilibria can be computed. A formal specification of the models is given below; here we give a brief summary in order to explain the main purpose of the construction. Two kinds of storable goods define two production sectors. Production in each sector occurs by a simple input/output function, which converts an initial stock of the good into more of the same good at a rate that depends on the size of the working stock.[5] Working stocks are ideally storable, though they can be wasted (so that the constraint on the quantity of goods available is an inequality rather than an equality). Each good is also consumable. In solutions without waste, goods persist from the time they are produced until the time they are consumed.

Production within each sector is performed by competing firms which are jointly owned by individuals who are also consumers of the produced goods. Production, trade, and consumption all occur in a sequence of many simply-structured, equivalent periods, and the establishment of a circular flow is a feature of time-stationary non-cooperative equilibria that balance output rates by the firms against marginal utilities of consumption by the consumer/owners.

Innovation is modeled as the possibility for one group of firms to attempt to change the production function in a single (particular) period, at the cost of one-time consumption of a fraction of their working stock. The attempted change succeeds with a probability $\xi < 1$. Although the cost of production is reduced and the limiting output rate is raised for firms that successfully innovate, the initially-reduced working stock cuts their output rates until that stock can be rebuilt from the output, which may require many periods. Firms that attempt to innovate and fail suffer the stock reduction but retain the pre-innovation production function. The problem of whether innovation is desirable can be posed in either of the two goods-sectors independently,[6] and the general structure of solutions for the depletion and

---

[5]Our models resemble the von Neumann growth model, restricted to a single good. However, in our production function the rate of output is a non-linear rather than a linear function of the input stock.

[6]We do not digress to derive the solution for Robinson Crusoe here, because its important features are subsumed in the solutions we demonstrate. A more systematic introduction to this class of models, including a separate solution for Robinson Crusoe as a reference, will be given elsewhere.

There are essentially three levels of models that require consideration for a complete exposition of basic distinctions. They are

- Crusoe without money,
- the price-taking individual firm with money,
- the oligopolistic firm without money.

subsequent restoration of productive stocks serves as a reference for these sectors in a monetary economy.

The specific feature of this real-goods economy that allows us to measure efficiency of money and banking systems is that *only one good* undergoes the opportunity for innovation. The other good has time-stationary production and consumption parameters, which we choose to be separable. Therefore it has no intrinsic reason to be influenced by innovation in other sectors. We demonstrate, however, that buffering the two sectors in the economy becomes difficult if models are not permitted an unrealistic degree of fine-tuning, and this is a basis of the need for substructure within the banking sector.

5.3.1 Many-Period Models, and the Passage to Continuous-Time Limits

The Bellman equations for many-period strategic market game models, in which the non-cooperative equilibria are non-stationary, are generally difficult to solve if the periods cover non-infinitesimal quantities of goods produced, traded, and consumed (that is, if they correspond to non-infinitesimal intervals of real time).[7] Some of these difficulties diminish if we take model periods to correspond to infinitesimal time periods, and scale production, trade, and consumption to be infinitesimal accordingly. We will refer to this scaling limit as the *continuous time* (or "continuum") limit for a many-period strategic market game.

Singular events, such as the choice to innovate a firm's production function and the required consumption of stocks, remain events that occur within a single period, so in the continuum limit they become singular, but this creates no difficulties as long as the continuum is defined as a limit of discrete-period models.

5.3.2 Continuous Time Defined Through Equivalence Classes

Formally, we treat economic processes that occur in continuous time as processes that may be modeled with any of a sequence of discrete-time models, with time intervals Δt that go to zero along the sequence. One performs calculations in discrete time so that definitions of moves in the game are unambiguous, but then requires that all economically relevant structure in the solution does not depend on Δt. More formally: the continuous-time limit is defined if there is a *scaling* of the other quantities in the model with Δt for which observables evaluated at two different times t_1 and t_2, which are held fixed as Δt is varied, converge on finite

The first two should produce the same physical allocations but differ in the presence or absence of money.

[7]The source of the simplification is that difference equations and discrete series reduce to differential equations and integrals, though the structure and meaning of the Bellman equations remains unchanged.

limiting values as $\Delta t \to 0$. Therefore a continuous-time limit is associated with an *equivalence class* of discrete-time models.

The formalization of continuous time in terms of scaling and limits provides a systematic way to partition stocks from flows. Within any single discrete-time model, all quantities may be represented as stocks within periods or changes of stocks between periods. When an equivalence relation over Δt is introduced, those changes in stocks that are to be interpreted as flows are required to vanish in linear proportion to Δt. The constant of proportionality in this scaling relation– the *rate* of the flow—is held fixed and is one of the parameters that defines the equivalence class. By such scaling relations, in continuous-time models, stocks, flows, and shocks are distinguished *mathematically* as well as descriptively.

5.3.3 The Economic Meaning of Continuous-Time Limits: How Many Timescales Represent Economically Significant Commitments of a Model?

It is possible to take a more conceptual view of continuous-time limits than merely technical tricks that simplify Bellman equations. In conventional discrete-period models, the period length is a dynamically important time interval in the model. It interacts with other model features such as non-linear production functions or utilities, and this interaction is one source of complexity in Bellman equations. In a continuous-time limit, since stocks and flow converge on regular limits as $\Delta t \to 0$, the period length ceases to be a model property that influences economic dynamics. For problems such as shock and recovery in production, consumption, and the circular flow of funds, the natural timescales of economic dynamics are determined by production functions, utilities, and interest rates, and *only* by these model properties.

5.3.4 The Use of Continuum Limits to Separate Dimensions of Economic Dynamics

It is not necessary to use continuous-time limits only *at* the limit point $\Delta t \to 0$. The existence of a well-defined and regular limit ensures that solutions to discrete-period models at small but nonzero Δt also exist and that they are approximated (to various orders in Δt) by properties of the limiting solution. For many applications it is useful to approximate these short-period solutions in terms of the structural parameters at the limit point.

The most important pair of economic quantities in short-period models are the money supply and the money velocity. In the continuous-time limit, with production and consumption per period scaled in linear proportion to Δt, solutions with stable prices also have regular continuum limits for the velocity of money, and solutions for the money supply that scale as Δt times this velocity (by definition of the velocity of money).

When banks are introduced that can both inject or extract money in circulation, and also mediate loans between agents, the two quantities will generally scale differently. Changes in the money supply, in efficient or nearly-efficient solutions, scale as $\mathcal{O}(\Delta t)$, like the original money supply. Interest streams between agents in steady circular flows are rates, which thus approach regular limits as $\Delta t \to 0$. Hence any *inter-agent* balances at the bank likewise scale as $\mathcal{O}(1)$; that is: the debts accumulated between agents at the bank can become arbitrarily larger than the money in circulation, in efficient solutions.

A second way in which a money system can be inefficient is that it can couple inter-agent lending to changes in the whole-economy money supply. If such a coupling arises, it creates a severe instability. A drain of $\mathcal{O}(1)$ can deplete the money in circulation in a time of $\mathcal{O}(\Delta t)$. Conversely, an addition of money at $\mathcal{O}(1)$ can lead to prices that grow to $\mathcal{O}(1/\Delta t)$. The continuum limit therefore offers ways to test the monetary system's capacity to buffer quantities with different natural dependence on turnover time, as well as different sectors.

5.3.5 Relation of Consumer/Owners to Firms and Banks

The models provide a minimum level of distinction sufficient to define economic sectors for goods production, and centralized versus distributed banking activities. In order to make all strategic actors price-takers, each type is modeled on a continuum. In order to minimize strategic complexity in the relation of ownership to control, with respect to the risk of failed innovation, we distribute ownership through uniformly-distributed shares of firms or banks. We do, however, retain a distinction between owners of firms of the two types, so that the economy creates income consequences from innovation, which bear on the role the banks play.

The specific structure of firms, consumer/owners, and banks is:

Firms: The economy has two goods, and we index production or consumption associated with these with subscripts $i \in \{1, 2\}$. For each of the two goods, a continuum of firms exist, which we index with a coordinate in the continuous interval $[0, 1]$. (We will not denote this index explicitly to reduce notational clutter; any production function, consumption utility, working stock, etc., with subscript $i \in \{1, 2\}$ implicitly refers to a particular firm or individual.)

Consumer/owners: Each type of good is also associated with a group of consumer/owners, also indexed with a coordinate in the continuous interval $[0, 1]$. All owners of a given type own equal shares of all firms of that type, and no shares of firms of the other type. Share ownership determines how firms deliver profits to owners, and in the case of firms that can engage in risky innovation, uniform share distribution leads to the same decision (to innovate or not to innovate) for all firms of the same type, and distributes the profit risk over all owners of that type.

The central bank: In economies with a central bank, the central bank is an atomic actor and a strategic dummy. It is not owned by any agents in the economy, and does not define or deliver profits. Its function is both to define the rules of monetary function, and to control the injection or extraction of money (either directly, or through commercial banks).

Commercial banks: In economies with a commercial banking sector, a single kind of commercial bank exists. Commercial banks are in some cases modeled as strategic dummies acting according to fixed rules, but the purpose for which they are introduced ultimately requires that they be strategic profit-maximizers, so that the profit incentive in a context of regulatory constraint guides their function within the economy. Therefore from the start we introduce commercial banks (in the cases where they occur) as a continuum of competitive corporations, again indexed with a coordinate in the continuous interval [0, 1]. All consumer/owners (so, the owners of both types of firms) jointly own the commercial banks. Again each owner owns a uniform distribution of shares of all banks, so that each bank's profits are distributed uniformly among all owners.

5.3.6 The Different Models to be Considered

1. **Fixed money supply:** The minimal solution in the absence of banking assumes a fixed supply of money in circulation, without reserves. The money could be gold or government fiat. Its fixed supply causes the shock from innovation in one good to strongly impact prices and output levels in the other good. The non-cooperative equilibrium in this game is an inefficient outcome corresponding to the pre-institutional (with respect to banking) economy.
2. **An idealized central bank:** If the economy does not require distributed commercial banking, a benevolent central bank can vary the money supply and mediate borrowing and lending among agents internal to the economy without interest rate spreads or leveraging. We show that this solution, with finely tuned parameters, can perfectly decouple the two goods sectors, so that the shock from innovation in one sector does not affect output in the other. This outcome defines the efficient function of the monetary system, and shows that it is achievable in a constructive solution. The buffering of the two production sectors is possible despite the fact that agents of different types experience relative wealth variations, so their consumption of goods is altered by innovation.
3. **Commercial banks with interest rate spreads and 100 % reserves:** In a first step toward defining a profit-seeking commercial banking sector, we introduce commercial banks that borrow "fiat money" from the central bank, and are permitted to issue bank credit to consumers in 1:1 ratio[8] to their holdings of fiat. The commercial banks can still enable steady-state production outcomes with many of the scaling properties of the efficient solution, if interest rates are finely tuned. However, they necessarily transmit shocks from the innovated to the non-innovated good, in proportion to the size of the interest rate spread.

[8]We could introduce a k:1 gearing ratio here with a little extra work, but our illustration does not need it.

We formalize this model as though the central bank has set the spread parametrically thereby reducing the commercial banks to strategic dummies. We suggest that it is the unmodeled evaluation function followed by the decision to lend or not to lend where important competition enters into banking.

5.4 *Formal Definition: The Production and Consumption Problem*

This section defines the scaling relations for production and for utility of consumption, in which *rates* of production and consumption give the invariant functional relations.

All discrete-period games, from which the continuum limit is defined, consist of a long sequence of periods with a time index t. The index is incremented by Δt, and the maximal value taken by t is some number T, which is the last period of the game. (We will return below to the way this period is selected, in order to address problems of robustness and interpretation of terminal conditions.) The period in which innovation occurs is indexed $t = 0$. Dynamically equivalent games can be defined either by initiating the sequence of periods at some time $t = t_{\text{init}} \ll 0$, and allowing the economy to converge to a steady state by $t = 0$ (because the dynamics to be defined below does produce such convergence, as we will demonstrate), or we could take the starting period as $t = 0$. For simplicity we will use $t = 0$ as the initial period, and as initial conditions we will provide firms with working stocks of goods, and agents with quantities of money in hand, which equal the fixed-point values with pre-innovation production functions.

5.4.1 Production Functions in Continuous Time, and a Sell-Surplus Market for Goods

In order to associate a quantity of goods production with intervals of real time $[t_1, t_2]$, across a class of models which may have variable period length Δt, it is necessary to separate well-defined stock variables from well-defined flow variables. For any firm, we use a variable s_t (with further indices as needed to specify the firm's type, introduced in the next sub-section) to denote the firm's working stock at the beginning of the period indexed t. The firm's output is characterized fundamentally by a *rate of production*, which depends on the working stock, which we denote by $f(s_t)$ (with other indices as required to distinguish types). In discrete-period models with period length Δt, the *amount* of the good produced within a single period is therefore $f(s_t)\,\Delta t$. We take the incremental increase of goods through production to happen at the beginning of the period, following which firms may sell some of the goods, to be purchased and consumed by consumers within the same period.

Each firm chooses a quantity $q_t \Delta t$ of goods to offer at a buy-sell trading post [2], in which consumers bid money to purchase the good. This quantity is a strategic variable, and can be varied over the range $q_t \Delta t \in [0, s_t + f(s_t)\, \Delta t]$. However, we denote it with the factor Δt made explicit, because in non-cooperative equilibria, the quantity q_t will have a regular limit as an *offer rate* as $\Delta t \to 0$.

5.4.2 Two Production Sectors; One Can Innovate

Goods of types 1 and 2 are produced by firms having production functions denoted respectively f_1 and f_2. If we denote by $s_{i,t}$ the stock of a firm producing good i in period t, the forms we will assume for the production rates are[9]:

$$f_i(s_{i,t}) \equiv f_{i,\infty} - \rho_\pi e^{-2s_{i,t}}. \tag{1}$$

f_i has dimensions of a rate, so both $f_{i,\infty}$ and ρ_π are rates. We will choose ρ_π to equal the discount rate from the definition of firms' discounted profits (introduce below, after the market clearing rule has been defined), to simplify the forms of solutions in worked examples. Nothing apart from simplifying presentation depends on this choice. Well-defined models require that we choose $f_{i,\infty} \geqslant \rho_\pi$ so that production is non-negative for all $s_{i,t} \geqslant 0$. In order to use certain small-deviation approximations in examples below, we will set $f_{i,\infty}/\rho_\pi \gtrsim 1$, but nothing in the model depends on finely tuning the values of these parameters.

The production rate f_2 is assumed to be a fixed function at all periods in all models. The production function f_1 is eligible to change, in period $t = 0$, into a new production function

$$\tilde{f}_1(s_{1,t}) \equiv (1 + \theta) f_1(s_{1,t}), \tag{2}$$

for all periods $t \geqslant 0$, with $\theta > 0$ a fixed parameter. This change of form is the game's representation of successful innovation.[10]

If firms of type 1 try to innovate in period $t = 0$, they must consume a quantity $s^{(\text{cost})}$ from their stocks $s_{1,t}$ at $t = 0$. Innovation succeeds with probability $\xi < 1$.

---

[9]These forms are smoothed versions of a linear production function with a limiting output and corner solutions, developed by Shubik and Sudderth [6, 7]. Corner solutions provided a convenient way to truncate discrete-period models to a single period, but in the continuous-time setting, the smoothed production rate produces a simple decomposition of solutions.

[10]The form (2) is the smoothed counterpart to a combination of "cost innovation" and "capacity innovation" in the terminology introduced by Shubik and Sudderth [6, 7]. The rate of production for $s_{1,t} \lesssim 1/2$ is larger by the factor $(1 + \theta)$, generating the same output at less input cost. The saturation level $f_{1,\infty}$ likewise increases by the factor $(1 + \theta)$, so that maximum output capacity likewise increases. This combination is simpler, for the smoothed production function, than either cost innovation or capacity innovation alone.

Firms that attempt to innovate and fail still consume the stock $s^{(\mathrm{cost})}$, but are left with the previous production function f_1.

5.4.3 The Carry-Forward of Goods by Firms

The carry-forward equation for the working stock $s_{i,t}$ held by any firm of type i, at all values of i and t aside from the innovation event by firms of type-1, is

$$s_{i,t+\Delta t} \leqslant s_{i,t} + f_i(s_{i,t})\,\Delta t - q_{i,t}\Delta t. \tag{3}$$

The inequality indicates that the working stock could be wasted but cannot increase except by means of production.

The continuous-time limit is obtained from Eq. (3) by dividing by Δt, and replacing the difference $(s_{i,t+\Delta t} - s_{i,t})\,/\Delta t$ by the derivative ds_i/dt, to obtain a differential equation relating stocks to flows:

$$\frac{ds_i}{dt} \rightarrow f_i(s_i) - q_i. \tag{4}$$

We return to the definition of firms' profits after defining the consumption and trade problem for consumers.

5.4.4 Consumption Utilities in Continuous Time

We first introduce the functional form of utility. As a dummy index, let c_1 (without further subscripts) be the rate of consumption of good-1 by any consumer in any particular time period, and let c_2 be the rate of consumption of good-2, by that consumer.[11] Utility for the period's consumption must likewise be defined in terms of a rate in order to permit a well-defined continuous-time limit. The utility rate is a function of the two consumption rates. In this cascade of models we take the separable form

$$u(c_1, c_2) = -\rho\left(e^{-c_1/\gamma_1} + e^{-c_2/\gamma_2}\right). \tag{5}$$

ρ is a constant related to the natural rate of discount, which we define below, needed to provide the correct dimensions for u,[12] and γ_1 and γ_2 are two scale factors that determine the relative price elasticities of the two goods. Note that since c_1 and c_2

[11]Thus, in the discrete-period model, the *amounts* consumed in one period are $c_1\Delta t$ and $c_2\Delta t$.

[12]The absolute magnitude of this constant does not matter for the definition of $u(c_1, c_2)$; only the dimension of a rate is required. We use the rate ρ in the discount factor as this avoids introducing a further arbitrary parameter.

are rates, γ_1 and γ_2 must likewise have dimensions of rates, since the input to the exponential function must be a pure number.

From this base form, which is the same for all consumers, we can introduce an indexed notation for utilities of each of the two types of consumers, in terms of the goods produced by the firms they own and the goods produced by the firms they do not own.

For a consumer of type i, we denote by $c_{i,t}$ the rate of consumption of the good that his own firms produce (now indexing the good *relative to* the consumer's type), and $\tilde{c}_{i,t}$ the rate of consumption of the good produced by firms of the other type.[13] To define a notation that will allow us to refer to agents of either type, denote by u_i the utility rate for a consumer of type i. In terms of Eq. (5), $u_{1,2}$ are given by

$$\begin{aligned} u_1(c_1, \tilde{c}_1) &\equiv u(c_1, \tilde{c}_1) \\ u_2(c_2, \tilde{c}_2) &\equiv u(\tilde{c}_2, c_2) \, . \end{aligned} \tag{6}$$

The variables that define any consumer's state at the beginning of each period are a supply of money-in-hand $m_{i,t}$, and in cases where consumers may make deposits or take out loans with either a central bank or a commercial bank, a balance $a_{i,t}$ at the bank. The account balance $a_{i,t}$ may be of either sign as long as the conditions on money and credit permit.

The consumer's strategic variables within any period are quantities $b_{i,t}\Delta t$ of money to bid on goods made by the firms of his own type, and $\tilde{b}_{i,t}\Delta t$ to bid on goods of the other type, along with deposits $d_{i,t}\Delta t$ to make to the bank. (We refer to them as "deposits" to define the sign convention for the transfer of money between the consumer and the bank; if some $d_{i,t}$ is negative it is a withdrawal.) Therefore, like consumption levels, $b_{i,t}$, $\tilde{b}_{i,t}$, and $d_{i,t}$ are denominated as *rates*.

5.4.5 Market Clearing

The rate at which total bids are made on good i in the buy-sell trading post in any period t is related to the rates of bidding by the two agent types as

$$B_{i,t} = b_{i,t} + \tilde{b}_{\tilde{\imath},t}. \tag{7}$$

The price of good i in period t is denoted $p_{i,t}$. From the clearing rule for the Dubey-Shubik buy/sell model [2], it is given by

$$p_{i,t} = \frac{B_{i,t}\Delta t}{q_{i,t}\Delta t} = \frac{B_{i,t}}{q_{i,t}} = \frac{b_{i,t} + \tilde{b}_{\tilde{\imath},t}}{q_{i,t}}. \tag{8}$$

[13] To express this more didactically, ˜ is used to indicate exclusion, or opposition in binary sets: $\tilde{\imath}$ means whichever value in $\{1, 2\}$ that is not the value taken by index i. $\tilde{c}_i$ indicates the consumption rate of the good that is not the consumption rate c_i.

The price is defined either as a ratio of per-period bid and offer quantities, or as a ratio of their corresponding rates, since factors of Δt cancel in the ratio. Thus price level can converge to a regular continuous-time limit if the bid and offer rates do so.

The rates at which goods are delivered to consumers from trading posts are their consumption rates, which evaluate in the buy/sell game to

$$c_{i,t} = \frac{b_{i,t}}{p_{i,t}}$$

$$\tilde{c}_{i,t} = \frac{\tilde{b}_{i,t}}{p_{\bar{i},t}}. \tag{9}$$

5.4.6 Profit Rates for Firms and (When Applicable) Commercial Banks

Firms are defined in these games to carry forward goods between periods to use as working stocks, and thus they have no money expenses.[14] Their profits equal their proceeds from sale. The amount of profit made by a firm of type i in period t is denoted

$$\pi_{i,t}\Delta t = p_{i,t}q_{i,t}\Delta t = \left(b_{i,t} + \tilde{b}_{\bar{i},t}\right)\Delta t, \tag{10}$$

in which $\pi_{i,t}$ is the corresponding profit rate.

Each firm of type i distributes its profits uniformly among consumer/owners of type i as a source of income for those owners. Since both firms and owners are indexed on the same continuous interval [0, 1], the rate $\pi_{i,t}$ at which profit is delivered by a firm of type i is the same as the rate of income to the consumer of type i.

The firm's total discounted profit, which it seeks to maximize, is the sum

$$\Pi_i = \sum_{t=0}^{T} \beta_\pi^{t/\Delta t} \left[\pi_{i,t}\Delta t - \eta_{i,t}\left(s_{i,t+\Delta t} - s_{i,t} - \Delta t f_i(s_{i,t}) + q_{i,t}\Delta t\right)\right]. \tag{11}$$

The Lagrange multipliers $\eta_{i,t}$ enforce the inequality (3), and the profit discount factor β_π is given in terms of the profit rate of discount ρ_π by

$$\beta_\pi \equiv \frac{1}{1 + \rho_\pi \Delta t}. \tag{12}$$

This is the same ρ_π used to set a scale in the production rate functions (1), for reasons explained where these were introduced.

[14]This construction avoids most of the concerns with corporate financing.

Bank profits, when they are defined, will be particular to models, so at present we simply introduce a notation $\pi_{i,t}^{(B)}$ for the *rate of income* delivered from bank profits to owners of type i. In models without banking or without bank profits, this term is zero. (Recall that commercial banks, when introduced, will be indexed on a continuous interval [0, 1], but they will distribute profits to *two* types of consumers, each type also indexed on an interval [0, 1]. Therefore we will need to be careful with factors of 2 in relating banks' income to profits delivered to owners.)

From the foregoing definitions and the clearing rules (8), (9), the update equation for a consumer of type i's money-in-hand between the beginnings of two successive rounds is

$$m_{i,t+\Delta t} = m_{i,t} - d_{i,t}\Delta t - \left(b_{i,t} + \tilde{b}_{i,t}\right)\Delta t + \left(\pi_{i,t} + \pi_{i,t}^{(B)}\right)\Delta t. \tag{13}$$

5.4.7 The Consumer's Utility Maximization Problem

Trading posts and banks both transact in explicitly represented money (whether gold, fiat, or bank notes). Therefore bids on consumables, and bank deposits, are limited by a budget constraint, which takes the form for a consumer of type i in period t

$$d_{i,t}\Delta t + \left(b_{i,t} + \tilde{b}_{i,t}\right)\Delta t \leqslant m_{i,t}. \tag{14}$$

The consumer maximizes a discounted utility across all periods' consumption against the sequence of constraints (14) at each period t.

To define terminal conditions for the multi-period game, and to produce a salvage value for money, we introduce a "day of reckoning" at period $t = T + \Delta t$, in which any negative bank balance is penalized with a linear deduction $\Pi \min\left(a_{i,T+\Delta t}, 0\right)$ from the total utility. The linear default penalty is enforced by means of a Kuhn-Tucker multiplier on a finite interval $\Lambda_i \in [0, \Pi]$, as in [9]. We return in Sect. 5.4.9 to discuss information conditions, including when agents know the value of T.

The Lagrangian for the optimization problem of a consumer of type i contains a discounted sum of utilities from the rates defined in Eq. (6), constraint terms for the budget constraints, and constraint terms for final conditions. An appropriate form to produce a regular continuous-time limit is given by

$$\begin{aligned} U_i \equiv & \sum_{t=0}^{T} \beta^{t/\Delta t} \left\{u_i(c_{i,t}, \tilde{c}_{1,t})\,\Delta t + \lambda_{i,t}\left[m_{i,t} - d_{i,t}\Delta t - \left(b_{i,t} + \tilde{b}_{i,t}\right)\Delta t\right]\right\} \\ & + \beta^{(T+\Delta t)/\Delta t}\Lambda_i a_{i,T+\Delta t}. \end{aligned} \tag{15}$$

In models where banking does not exist, the terms $d_{i,t}$ and $a_{T+\Delta t}$ are omitted. Note that the factor $\Lambda_i a_{i,T+\Delta t}$ is discounted by $\beta^{(T+\Delta t)/\Delta t}$.

We also have not incorporated any terms constraining $a_{i,t}$ at intermediate times, such as might arise from limits on reserve requirements.

The per-period discount factor β in the utility function (15) is related to the period length Δt and the natural rate of discount ρ by

$$\beta \equiv \frac{1}{1 + \rho \Delta t}. \tag{16}$$

This convention leads to regular limits for the utility in continuous time. The increment Δt becomes a measure dt, and the sum over index t becomes an integral $\int dt$. The integrand will be a function only of rate-valued quantities, which in the continuous-time limit take piecewise-smooth trajectories. The ratio $m_{i,t}/\Delta t$ must likewise scale as a rate-valued quantity, which has the interpretation of the contribution from an agent of type i to the velocity of money, as the money supply scales linearly toward zero with Δt.

5.4.8 The Consumer's Bank-Balance Dynamics

The first two banking models demonstrated here permit unlimited revolving loans. Technically this means two things. The first is that the bank keeps an account, the balance of which is updated at a pre-specified interest rate within each period. The second is that the amount consumers deposit or withdraw is an unconstrained variable, apart from the penalty on unrepaid bank debts in the terminal conditions.

The bank's carry-forward equation for accounts it thus

$$a_{i,t+\Delta t} = \left(a_{i,t} + d_{i,t} \Delta t\right)\left(1 + \rho_{B,it} \Delta t\right). \tag{17}$$

Deposits or withdrawals are made at the beginning of the period, and interest accrues at a rate $\rho_{B,it}$.[15]

The bank may lend or accept deposits at different rates, in which case the interest rate for either type i is a function of time, evaluated to equal a lending or borrowing rate according to the rule

$$\rho_{B,it} = \begin{cases} \rho_{B,L} & \text{if } a_{i,t} < 0 \\ \rho_{B,D} & \text{if } a_{i,t} > 0. \end{cases} \tag{18}$$

To regularize the discontinuity at $a_t = 0$, we may adopt some convention such as $\rho_{B,t} = \left(\rho_{B,L} + \rho_{B,D}\right)/2$.[16] For a central bank acting as a public service, there is no

[15]Many alternative rules are well-defined: interest on deposits could accrue one period later than interest charged on loans, etc. Nothing depends on the intra-temporal order of interest charges and payments, in the continuous-time limit.

[16]Under conditions when the bank is actively used, $a_t = 0$ occurs only on time intervals of measure zero, so the results are not sensitive to the way the interest rate is regularized. Because, in this model, we assume initial conditions prior to the accumulation of bank balances, it is convenient to choose a regularization condition that will be consistent with the other simplifying assumptions made in the model.

need for interest rate spreads, but for a commercial bank a spread $\rho_{B,L} - \rho_{B,G} > 0$ will generally be required.

Using the money carry-forward relation (13) to express $d_{i,t}$ in terms of the bids, profits, and changes in agents' money-holdings, the account-balance carry-forward relation (17) may be written

$$\frac{a_{i,t+\Delta t}}{(1 + \rho_{B,it}\Delta t)} = a_{i,t} - (m_{i,t+\Delta t} - m_{i,t}) - \left(b_{i,t} + \tilde{b}_{i,t}\right) \Delta t + \left(\pi_{i,t} + \pi_{i,t}^{(B)}\right) \Delta t. \quad (19)$$

Using Eq. (10) for firms' profits, dividing Eq. (19) by Δt, and then taking $\Delta t \to 0$ produces the continuous-time expression for the bank balance in relation to the money-in-hand m_i of[17]

$$\left(\frac{d}{dt} - \rho_{B,i}\right) a_i \to \left(\tilde{b}_{\bar{i}} - \tilde{b}_i\right) - \frac{dm_i}{dt} + \pi_i^{(B)} + \mathcal{O}(\Delta t). \quad (20)$$

5.4.9 Terminal Conditions

The handling of terminal conditions in a class of extended-time games of this form, with lending at interest, a small number of events that can occur, and no stochasticity, is generally a somewhat artificial exercise as a model of decision making in real economies. On one hand, the attempt by consumers and firms to converge to a steady state that permits long-term regular behavior, and the degree to which monetary flexibility permits or impedes that attempt, is the aspect of decision making that the model probably captures robustly. On the other hand, the specification of terminal conditions is a requirement from the standpoint of experimental gaming, and this generally rules out a steady state. The artificial feature of a model that requires cancellation of all debts at a finite horizon, in an economy that has structurally changed in the interim in such a way that revolving debt permits it to accommodate the change, is that exponential growth of account balances can lead to sensitive and arbitrary coupling of terminal conditions to otherwise-negligible differences in interior solutions. A continuum of solutions to the first-order conditions exist with utilities and profits that differ by exponentially small factors in $\rho_\pi T$, but which involve very different response of the production decisions at the terminal conditions.

We resolve these ambiguities by making use of the following observation to single out the class of non-cooperative equilibria that robustly separate the responses to initial and terminal conditions in a non-arbitrary manner. These games possess non-cooperative equilibria that could be called "turnpike solutions". Consumers and

[17]The residual terms at $\mathcal{O}(\Delta t)$, which we denote explicitly despite the fact that they approach zero as $\Delta t \to 0$, come from time lags between the making of bids and the delivery of profits. As long as the rates are continuous (differentiable at order one) functions, these effects contribute terms $\sim (db_i/dt)\, \Delta t$ in Eq. (20).

firms, after a transient that occupies an interval in $\rho_\pi t$ much smaller than 1, can converge exponentially (in $\rho_\pi t$) toward stable production, trade, and consumption values that can be preserved indefinitely. In general, these solutions require non-zero bank balances, as some agents lend to others, with interest flows supporting asymmetries in their consumption that reflect the real structural changes in the production sector. These steady-state values are the turnpike values. The games also possess a class of unstable solutions, in which firms exponentially diverge from the turnpike values, depleting or hoarding stocks in response to exponentially diverging price levels created by consumer bidding, as consumers re-direct their money to return their bank balances to zero. The diverging solutions cannot be extended indefinitely because they become singular, so they never occur at intermediate times. They can be chosen, however, to accommodate a terminal condition that eliminates all debts to the banks. The turnpike solutions require a specific coordinated price-setting behavior by the two types of consumers, and of production decisions by the two types of firms (all of which can be computed non-cooperatively by each group of agents), in order that neither aggregate nor internal debt exist at the terminal time.

In addition to the turnpike solutions, a continuum of other solutions exist, in which very small uncanceled aggregate debts can grow exponentially, and require different behavior by the two types of consumers and the two types of firms, relative to turnpike solution, to cancel aggregate as well as internal debt. The final behavior of the agents in these solutions is sensitive to uncanceled aggregate debts that may be of order $e^{-\rho_\pi T}$ at the end of transient response to the initial conditions, and which constitute arbitrarily small deviations from the pure turnpike solution.

To isolate the turnpike solutions, the agents are not told the time T of the terminal round at the beginning of the game. Instead, they are told that, in each period Δt, a binary variable will be sampled. The first time t at which the variable equals 1, the terminal round will be announced to occur at a specified later time, such as $T = t + 5/\rho_\pi$ (so five times the discount horizon, out from the present). The probability to draw value 1 is made sufficiently small that the values of T will be Poisson distributed with a mean much longer than the discount horizon $1/\rho_\pi$. This look-ahead declaration provides sufficient time to implement the terminal behaviors starting from time t, with utility consequences of differing from the turnpike solution that are bounded above by $\mathcal{O}\left(e^{-(T-t)\rho_\pi}\right) \approx \mathcal{O}\left(e^{-5}\right)$. (We choose the look-ahead horizon $t + 5/\rho_\pi$ for convenience in examples below; this number may be chosen as large as desired to decouple the initial terminal intervals to any desired degree.) As long as the error $e^{-(T-t)\rho_\pi}$ is made $\ll \Delta t \rho_\pi$, it is a smaller correction than finite-period discretization effects that we are ignoring. Players who solve the initial transient to converge to the turnpike produce a solution that is within $\mathcal{O}\left(e^{-(T-t)\rho_\pi}\right)$ of any non-cooperative equilibrium solution for any large T. Any non-cooperative equilibrium not converging to the turnpike could be one of a range of exact solutions for a particular T, but which solution this would be would depend on aggregate debt levels of $\mathcal{O}\left(e^{-(T-t)\rho_\pi}\right)$, and the initial part of this trajectory would differ from any non-cooperative equilibrium, for any terminal time different from T by more than $\mathcal{O}(1/\rho_\pi)$, at more than $\mathcal{O}\left(e^{(T-t)\rho_\pi}\right)$.

We will not develop the full machinery of expected-utility maximization in this note, but will simply compute properties of the turnpike equilibria, with the understanding that all deviations from these by more than $\mathcal{O}\left(-e^{(T-t)\rho_\pi}\right)$ are incompatible with existence of any non-cooperative equilibrium over ranges of T where the terminal-condition sampling has large probability, and so will be ruled out by any generic expected-utility maximization.

We believe that this minimal use of a stochastic variable yields the kinds of solutions that would arise in an actual economy where money and banking are available to facilitate regular events of structural change in the production sector, and in which agents carry persistent debt and respond to new events of innovation by changing their debt structure as these arise, in ongoing sequences. The addition of *specific* finite-horizon debt could, of course, be introduced as a qualitative modification to these games, but it should then be justified by other criteria (lenders' limitations, etc.) besides the question whether a flexible money supply can alleviate constraints on the circular flow of funds, which is the topic addressed by the current class of games.

5.4.10 The Leading Contributions in $(\rho_B \Delta t)$ to the Time-Course of Monetized Private Credit and the Net Account Balance of Agents at the Bank

Now for the first time we may use small but nonzero Δt to distinguish the behavior of two components of the credit supply. One component comes from lending effectively by one type of consumers to the other, mediated by the bank. Promises to pay by consumers (enforced by the default penalty at the day of reckoning) are privately issued credit. Banks' promises to pay (whatever interest plus principle accrues) are met with bank credit. The part of loans and deposits that cancel among the consumers are effectively private credit from one group to another, monetized by the bank when it accepts private promises to pay and issues bankers' promises to pay. The part of loans or deposits that does not cancel when consumers are aggregated is the net injection or extraction of money in circulation. Injected money is also in the form of bank credit, while extraction may be whatever form of money was given to the consumers in the initial conditions. The two coordinates we use to represent intra-economy lending, and aggregate-economy lending, are respectively $(a_{1,t} - a_{2,t})$ and $(a_{1,t} + a_{2,t})$.

In a continuous-time model with regular prices, the supply of money in circulation scales as $\mathcal{O}(\Delta t)$. If banking is to leave prices regular, the change in money supply, driven by the sum of balances $(a_{1,t} + a_{2,t})$, must also scale linearly in $\mathcal{O}(\Delta t)$. In contrast, as we show now, the monetized private credit will normally scale as $\mathcal{O}(1)$ in economies operating at or near monetary efficiency. Thus some agents have outstanding, at any time, debts that are larger by $\mathcal{O}(1/\Delta t)$ than all money in circulation.

Personal Credit Monetized by Bank Accounting

From Eq. (20), the equation for $(a_{1,t} - a_{2,t})$ is

$$\left[\frac{d}{dt} - \frac{(\rho_{B,1} + \rho_{B,2})}{2}\right](a_1 - a_2) \to 2\left(\tilde{b}_2 - \tilde{b}_1\right) + \frac{(\rho_{B,1} - \rho_{B,2})}{2}(a_1 + a_2) - \frac{d\,(m_1 - m_2)}{dt} + \mathcal{O}(\Delta t)\,. \tag{21}$$

As long as both $(a_1 + a_2)$ and $d\,(m_1 - m_2)\,/dt$ are $\mathcal{O}(\Delta t)$ like the terms that have been dropped—a requirement if prices are not to diverge in the continuous-time limit—any $\mathcal{O}(1)$ contribution to $(a_1 - a_2)$ can only come from the term in $\left(\tilde{b}_{\bar{i}} - \tilde{b}_i\right)$.

Reducing Eq. (21) to quadrature gives the expression for the credit monetized by the banks within the economy,

$$\frac{a_{1,t} - a_{2,t}}{2} = \int_0^t dt'\, e^{\int_{t'}^{t} dt'' (\rho_{B,1t''} + \rho_{B,2t''})/2} \left(\tilde{b}_{2,t'} - \tilde{b}_{1,t'}\right) + \mathcal{O}(\Delta t)\,. \tag{22}$$

To determine the conditions under which these bank balances can approach a steady state turnpike solution that can extend indefinitely, we integrate Eq. (22) by parts to obtain the equivalent expression

$$\frac{a_{1,t} - a_{2,t}}{2} = e^{\int_0^t dt' (\rho_{B,1t'} + \rho_{B,2t'})/2} \left\{ \frac{\left(\tilde{b}_{2,0} - \tilde{b}_{1,0}\right)}{(\rho_{B,10} + \rho_{B,20})\,/2} + \int_0^t dt'\, e^{-\int_0^{t'} dt'' (\rho_{B,1t''} + \rho_{B,2t''})/2} \frac{d}{dt'} \frac{\left(\tilde{b}_{2,t'} - \tilde{b}_{1,t'}\right)}{(\rho_{B,1t'} + \rho_{B,2t'})\,/2} \right\} - \frac{\left(\tilde{b}_{2,t} - \tilde{b}_{1,t}\right)}{(\rho_{B,1t} + \rho_{B,2t})\,/2}. \tag{23}$$

Section "Steady Post-Innovation Output and Stable Money Supply Lead to Stable Bid Levels" in the Appendix shows that the intermediate-time bids $\left(\tilde{b}_{2,t'} - \tilde{b}_{1,t'}\right)$ converge on steady values as long as the money supply is asymptotically constant, which is the condition for a non-inflationary solution.[18] Hence the time derivative in the integral in Eq. (23) approaches zero for t' sufficiently large. We return in

[18]Without uncertainty it calls for the rate ρ defining the utilitarian rate of discount in Eq. (16) to equal the average of the two interest rates faced by the agents, as shown in Eq. (42) below. (In the worked example of the following sections, this will be the average of the borrowing and the lending rates.) With uncertainty there is a delicate correction depending on the variance.

Sect. 6.2.3 to the way this solution connects to a terminal transient that returns both of $(a_{1,t} \pm a_{2,t})$ to zero as $t \to T$.

The relation (23), which at large t is exponentially well-approximated by the vanishing of the term in curly braces, determines $\hat{\epsilon}$ from Eq. (31) and Eq. (30).[19] Because this equation is homogeneous of order one in the numéraire, it is not necessary to know the overall magnitude of the money supply to determine $\hat{\epsilon}$.

Aggregate Debt and Change in the Money Supply

The mechanism by which banks may change the money in circulation is lending to or accepting deposits from consumers at interest. For example, consumers may borrow an initial stock of money following the event in which innovation occurs, and over the course of restoring the principle to zero so that the money-in-circulation converges to a steady value, they pay some quantity of aggregate interest.

Summing Eq. (20) over both agent types gives the equation for $(a_{1,t} + a_{2,t})$:

$$\left[\frac{d}{dt} - \frac{(\rho_{B,1} + \rho_{B,2})}{2}\right](a_1 + a_2) \to -\frac{d}{dt}(m_1 + m_2) + \left(\pi_1^{(B)} + \pi_2^{(B)}\right) + \frac{(\rho_{B,1} - \rho_{B,2})}{2}(a_1 - a_2). \tag{24}$$

In models with interest rate spreads, we face the possibility that the term in $(\rho_{B,1} - \rho_{B,2})$ in the second line of Eq. (24) could destroy the stability of prices by coupling the quantity $(a_1 - a_2)$ which is $\mathcal{O}(1)$ to the change in the money supply which must scale as $\mathcal{O}(\Delta t)$ for prices to be stable. In appropriately defined models this potential instability will be avoided, because the total profits from commercial banks $\left(\pi_1^{(B)} + \pi_2^{(B)}\right)$ will be a revenue $-(\rho_{B,1}a_1 + \rho_{B,2}a_2)$, minus a stream paid to the central bank. As long as the stream to the central bank remains at $\mathcal{O}(\Delta t)$, the remaining revenue stream recirculates, canceling the term $(\rho_{B,1} - \rho_{B,2})(a_1 - a_2)/2$ to within $\mathcal{O}(\Delta t)$. Any component of $-(\rho_{B,1}a_1 + \rho_{B,2}a_2)$ that is $\mathcal{O}(1)$ is also assured to be positive, because it can only come from a difference $(a_1 - a_2)$ that is $\mathcal{O}(1)$,

---

[19] When the term in curly braces is exactly zero, the late-time steady-state relation becomes

$$\frac{(\rho_{B,1t} + \rho_{B,2t})}{2}\frac{(a_{1,t} - a_{2,t})}{2} = \left(\tilde{b}_{2,t} - \tilde{b}_{1,t}\right).$$

This expression is simply the interest paid to agents of type-1, plus their share of bank profits when profits are defined, which balances the deficit in the profits of type-1 firms relative to the bids made by type-1 agents (who will consume more). Thus a consistent circular flow is restored in the asymptotic steady state, in a context of asymmetric production, profits, depositing/borrowing, and consumption.

and the lending rate (on the negative account balance) will be higher than the rate on deposits (the positive balance).

The simplest case will be 100 % reserve banking, in which any aggregate loans a commercial bank makes to consumers cannot exceed supplies of "heavy money" the commercial bank borrows from the central bank and holds as reserves. In that case the total profit stream of the commercial bank takes the form

$$\pi^{(B)}_{1,t} + \pi^{(B)}_{2,t} = \rho_{C,t}\left(a_{1,t} + a_{2,t}\right) - \left(\rho_{B,1t}a_{1,t} + \rho_{B,2t}a_{2,t}\right), \tag{25}$$

where ρ_C is the interest rate charged by the central bank.

Substituting this into Eq. (24) gives

$$\left(\frac{d}{dt} - \rho_C\right)\left(a_1 + a_2\right) \to -\frac{d}{dt}\left(m_1 + m_2\right). \tag{26}$$

Note that, if there is no commercial bank, and the consumers borrow from or deposit into the central bank directly, Eq. (26) results directly from Eq. (24).

In the simplifying case where ρ_C is constant, Eq. (26) is integrated to give the result

$$\begin{aligned}(a_{1,t} + a_{2,t}) = e^{\rho_C t}&\left\{\left(a_{1,0} + a_{2,0} + m_{1,0} + m_{2,0}\right) - \int_0^t dt'\, \rho_C e^{-\rho_C t'}\left(m_{1,t'} + m_{2,t'}\right)\right\}\\ &- \left(m_{1,t} + m_{2,t}\right). \end{aligned}\tag{27}$$

Both in the model with only a central bank, and in the model with a commercial bank using 100 % reserves, we will set $a_{i,0} = 0$ as initial condition, and $(m_{1,0} + m_{2,0}) \equiv 2m_0$ to define the initial money supply. Agents may borrow an amount of money that scales as $\sim m_0$ from the bank in the period $t = 0$ when the innovation event occurs, changing both the initial money supply and the initial debt abruptly. Under any such borrowing, however, $\left(a_{1,t} + a_{2,t} + m_{1,t} + m_{2,t}\right)_{t\to 0+} = \left(a_{1,0} + a_{2,0} + m_{1,0} + m_{2,0}\right)$. Therefore both the initial value and the integral in Eq. (27) involve no singular terms even in the continuous-time limit.

The vanishing of the steady-state principle $-(a_{1,t} + a_{2,t})$ owed by the agents to the banks in Eq. (27) determines the initial borrowed amounts $(m_{1,t} + m_{2,t})_{t\to 0+} - 2m_0 = -(d_{1,0} + d_{2,0})\,\Delta t$, because these set the scale for the quantity $(m_{1,t'} + m_{2,t'})$ in the integral and the final term $(m_{1,t} + m_{2,t})$ relative to the initial term $(a_{1,0} + a_{2,0} + m_{1,0} + m_{2,0}) = 2m_0$, which is fixed. The vanishing of the term in curly braces in Eq. (27), taken as $t \to \infty$, given a value of $\hat{\epsilon}$ fixed by vanishing of the similar term in curly braces in Eq. (23), defines the turnpike response to the initial shock created by the innovation opportunity and the need to borrow.

5.5 *First-Order Conditions*

5.5.1 The Consumer's Goods-Consumption Problem

The first-order condition for consumption results from variation of $b_{i,t}$ and $\tilde{b}_{i,t}$ in Eq. (15), and takes the form

$$\frac{1}{p_{i,t}} \frac{\partial u_i}{\partial c_{i,t}} = \frac{1}{p_{\bar{i},t}} \frac{\partial u_i}{\partial \tilde{c}_{i,t}} = \sum_{t'=t}^{T} \beta^{(t'-t)/\Delta t} \lambda_{i,t'}. \tag{28}$$

Irrespective of how the Kuhn-Tucker multipliers for these constraints are set,[20] the ratios of first-order conditions (28) imply relations of relative consumption between the two types of agents, who purchase against a shared price system. In the remainder of this sub-section, we suppress the explicit time index, because the relations hold period-by-period at each t.

The two ratios of marginal utilities of consumption are both given in terms of prices by

$$\frac{\partial u_1 / \partial c_1}{\partial u_1 / \partial \tilde{c}_1} = \frac{p_1}{p_2} = \frac{\partial u_2 / \partial \tilde{c}_2}{\partial u_2 / \partial c_2}. \tag{29}$$

To solve for the consequences of this relation, we introduce a pair of coordinates to relate the consumption of the two types of agents to the offer levels $q_{i,t}$. Define

$$\begin{aligned} c_1 &\equiv \frac{q_1}{2} + \epsilon_1 & \tilde{c}_2 &\equiv \frac{q_1}{2} - \epsilon_1 \\ \tilde{c}_1 &\equiv \frac{q_2}{2} + \epsilon_2 & c_2 &\equiv \frac{q_2}{2} - \epsilon_2. \end{aligned} \tag{30}$$

The model choice of a separable exponential utility (5) leads to the result that the offsets $\epsilon_{1,2}$ from even division for the two goods are in a fixed proportion determined by the relative elasticities,

$$\frac{\epsilon_1}{\gamma_1} = \frac{\epsilon_2}{\gamma_2} \equiv \hat{\epsilon}. \tag{31}$$

The output rate q_i will always appear scaled by the factor γ_i in the utility, so we introduce a shorthand

$$\hat{q}_i \equiv \frac{q_i}{\gamma_i}. \tag{32}$$

[20] These multipliers are always nonzero, as the budget constraint is always tight.

Because prices are ratios of total bids to total outputs, Eq. (29) together with the condition (31) implies that

$$\frac{B_1}{B_2} = \frac{\hat{q}_1 e^{-\hat{q}_1/2}}{\hat{q}_2 e^{-\hat{q}_2/2}}. \tag{33}$$

The relation of the bid level for either good to the total money supply is then

$$\frac{B_i}{B_1 + B_2} = \frac{\hat{q}_i e^{-\hat{q}_i/2}}{\hat{q}_1 e^{-\hat{q}_1/2} + \hat{q}_2 e^{-\hat{q}_2/2}}. \tag{34}$$

Therefore prices are given in relation to the total money rate of circulation $B_1 + B_2$ by

$$p_i = \frac{1}{\gamma_i} \frac{e^{-\hat{q}_1/2}}{\hat{q}_1 e^{-\hat{q}_1/2} + \hat{q}_2 e^{-\hat{q}_2/2}} (B_1 + B_2). \tag{35}$$

5.5.2 Consumer's Banking Problem (When Applicable)

If the economy is one in which borrowing and lending are possible, a second condition for deposits or withdrawals results from variation of $d_{i,t}$. It there is no limit on consumers' account balances, the only two classes of Kuhn-Tucker multipliers come from the per-period budget constraint ($\lambda_{i,t}$) and the terminal conditions (Λ_i).[21] The first-order condition for deposits is then

$$\sum_{t'=t}^{T} \beta^{(t'-t)/\Delta t} \lambda_{i,t'} = \Lambda_i \prod_{t'=t}^{T} \left[\beta \left(1 + \rho_{B,it'} \Delta t\right)\right]. \tag{36}$$

Combining Eq. (28) with Eq. (36), and taking $\Delta t \to 0$, we arrive at the continuous-time relation among prices, output, interest rates, and a single Kuhn-Tucker multiplier for the terminal constraint:

$$\frac{1}{p_{i,t}} \frac{\partial u_i}{\partial c_{i,t}} = \frac{1}{p_{\tilde{i},t}} \frac{\partial u_i}{\partial \tilde{c}_{i,t}} \to e^{\int_t^T dt' \left(\rho_{B,it'} - \rho\right)} \Lambda_i. \tag{37}$$

Using the relations (29), (30), which hold at each time, we can evaluate the consumption first-order conditions (37) for the two types explicitly, to give

$$\frac{1}{p_{1,t}\gamma_1} e^{-\hat{q}_{1,t}/2} = \frac{1}{p_{2,t}\gamma_2} e^{-\hat{q}_{2,t}/2} \to e^{\hat{\epsilon}_t + \int_t^T dt' \left(\rho_{B,1t'} - \rho\right)} \frac{\Lambda_1}{\rho},$$

[21] If bounds were placed on the account balances, additional multipliers could arise within each period as shadow prices associated with these constraints.

$$\frac{1}{p_{1,t}\gamma_1} e^{-\hat{q}_{1,t}/2} = \frac{1}{p_{2,t}\gamma_2} e^{-\hat{q}_{2,t}/2} \to e^{-\hat{\epsilon}_t + \int_t^T dt' \left(\rho_{B,2t'} - \rho\right)} \frac{\Lambda_2}{\rho}. \tag{38}$$

The consumption asymmetry $\hat{\epsilon}$ must then satisfy

$$\hat{\epsilon}_t = \hat{\epsilon}_T - \frac{1}{2} \int_t^T dt' \left(\rho_{B,1t'} - \rho_{B,2t'}\right). \tag{39}$$

The Kuhn-Tucker multipliers for the two types of agents are related to the final-time value $\hat{\epsilon}_T$ as

$$e^{\hat{\epsilon}_T} \Lambda_1 = e^{-\hat{\epsilon}_T} \Lambda_2 \equiv \Lambda. \tag{40}$$

A Note on the Setting of the Default Penalty

We will show that, in general, $\hat{\epsilon}_T$ cannot equal zero, because consumers of different types have different incomes and consume at different levels. Therefore the shadow prices Λ_1 and Λ_2 cannot both be equal; hence, even in a game with artificially fine-tuned parameters, they could not both be set equal to the limiting value Π of the default penalty. Interior solutions can therefore only be obtained when at least one of $\Lambda_1 < \Pi$ or $\Lambda_2 < \Pi$ holds, and when both $a_{1,T} = 0$ and $a_{2,T} = 0$. This permits us to set Π "sufficiently large" that both $\Lambda_1 < \Pi$ and $\Lambda_2 < \Pi$, and to consider interior solutions without default and also with no savings at the day of reckoning. These two requirements define the terminal conditions for interior solutions with banking. We will illustrate their consequences for prices and production in Sect. 6.2.3.

The pair of first-order conditions (38) evaluate to a relation between the two prices and output levels to a single multiplier Λ (jointly determined by the agents' non-cooperative equilibria) and the (possibly time-dependent) interest rates of the two types:

$$\frac{1}{p_{1,t}\gamma_1} e^{-\hat{q}_{1,t}/2} = \frac{1}{p_{2,t}\gamma_2} e^{-\hat{q}_{2,t}/2} \to e^{\int_t^T dt' \left[\frac{1}{2}\left(\rho_{B,1t'} + \rho_{B,2t'}\right) - \rho\right]} \frac{\Lambda}{\rho}. \tag{41}$$

It was necessary that the relation between the output level of either good and its price in Eq. (41) be the same for the two goods, because by Eq. (35) either of these equals a relation between both output levels and the total money supply. Combining the two equations gives

$$\frac{\hat{q}_1 e^{-\hat{q}_1/2} + \hat{q}_2 e^{-\hat{q}_2/2}}{B_1 + B_2} \to e^{\int_t^T dt' \left[\frac{1}{2}\left(\rho_{B,1t'} + \rho_{B,2t'}\right) - \rho\right]} \frac{\Lambda}{\rho}. \tag{42}$$

Taking logarithms, and then differentiating with respect to t, then gives the relation between outputs, money supply, and interest rates

$$\frac{d}{dt}\log\left(\frac{\hat{q}_1 e^{-\hat{q}_1/2}+\hat{q}_2 e^{-\hat{q}_2/2}}{B_1+B_2}\right)=\left(\rho-\frac{\rho_{B,1}+\rho_{B,2}}{2}\right). \tag{43}$$

5.5.3 The Firms' Output Levels in Response to Prices

Firms attempt to maximize profits (11) in which the price sequences $p_{i,t}$ appear as parameters from Eq. (10).

Firms may respond to prices in either of two ways. Either

$$p_{i,t} \leq \eta_{i,t}, \tag{44}$$

and they set $q_{i,t} \to 0$, or else $q_{i,t} > 0$, and

$$\eta_{i,t-\Delta t} = \eta_{i,t}\beta_\pi\left(1+f_i'(s_{i,t})\right) \tag{45}$$

The former case can be realized by successfully-innovating firms in the early periods following innovation, in which they are better off to sit out of markets and rebuild their working stocks, while the type-1 firms that attempted to innovate and failed, provide the total supply in markets. Firms that failed in innovating can maintain market prices lower than the reservation prices of the successful firms, because their steady-state allocations at late times are not as high (we demonstrate this below), so that using their entire output to rebuild stocks is not as valuable to them as it is to the successful firms.

In the latter case, faced by all firms at sufficiently late times, by the type-1 firms that try and fail to innovate, and by all type-2 firms all the time, these firms optimize their output against the particular sequence of prices.

The recursive relation (45) among K-T multipliers becomes, in the continuous-time limit,

$$\eta_{i,t} = \eta_{i,T}e^{\int_t^T dt'\left[f_i'(s_{i,t'})-\rho_\pi\right]}. \tag{46}$$

When $p_{i,t} = \eta_{i,t}$, Eq. (46) dictates an intertemporal relation between prices and output which is the consequence of the profit-maximization criterion.

Setting $p_{i,t} = \eta_{i,t}$ in Eq. (46), combining this with the consumers' price/output/interest relations (41), taking logarithms, and differentiating with respect to t, produces a three-way relation among output levels, the stocks of all firms that are active offering in markets, the interest rates faced by consumers of both types, and the total money supply, in the form

$$\frac{d}{dt}\log\left(\frac{\hat{q}_1 e^{-\hat{q}_1/2} + \hat{q}_2 e^{-\hat{q}_2/2}}{B_1 + B_2}\right) = -\frac{1}{2}\frac{d}{dt}\hat{q}_i + \left[f_i'\left(s_i\right) - \rho_\pi\right]$$

$$= \left(\rho - \frac{\rho_{B,1} + \rho_{B,2}}{2}\right). \tag{47}$$

The equality in the second line applies only in the case that consumers set prices by varying the money in circulation through borrowing and lending.

5.5.4 Equation (47) Is the Main Relation

Equation (47) is the main relation that links output decisions by firms to the dynamics of the money supply. The right-hand side of the first equality is a second-order differential response function of the working stocks and aggregate output of firms of a given type, to a source term (the left-hand side of the first equality) which involves the output levels of both goods, and the total money supply $(B_{1,t} + B_{2,t})$. The $\hat{q}_i$ on the right-hand side represents a total output variable,[22] and originates in the separable exponential utility of consumption (5). In this respect the separability between the left and right-hand sides in an exact relation depends on the specific assumption of exponential utility, which we introduced in order to make the production/consumption model a sharp test case for monetary efficiency. If the second equality in Eq. (47) applies, it determines both the dynamics of the total money supply, and the source term for output decisions, in terms of the interest rates faced by the two kinds of consumers in relation to the natural rate of discount.

Working stock and output decisions for both firms are coupled to the same source term which is an aggregate property of the whole economy. Moreover, the production decisions in the two sectors are independent of one another *except* for this shared source term, and except for any initial and terminal conditions created, respectively, by the innovation-induced shock to the supplies of working stocks, and the requirement to nullify bank debts on the day of reckoning. The form imposed on these equations by a particular monetary system therefore determines whether that system can insulate production decisions in the two sectors from one another, and if it cannot, the manner and strength with which they are coupled.

[22]This term must be corrected with a measure term to relate it to individual firms' output levels if not all firms are active in markets, as we show below.

5.5.5 The Criterion of Monetary Efficiency

We may thus sharply define the criterion for efficiency of the monetary system. **If the banking system makes the supply of money in circulation sufficiently flexible** that the money supply $(B_{1,t} + B_{2,t})$ can exactly track the numerator term $\left(\hat{q}_1 e^{-\hat{q}_1/2} + \hat{q}_2 e^{-\hat{q}_2/2}\right)$, then the production decisions in the sectors for good-1 and good-2 are completely decoupled. Supply shocks in one sector do not affect production in the other. **Scale in the overall economy has been separated from the structure of production and consumption**, with the result that intertemporal coordination of production may be optimized for each good through a price system, delivering the same production profiles as if the two goods occupied two separate economies. Note that, in economies with banking, this is possible only if $\rho - (\rho_{B,1t} + \rho_{B,2t})/2$ is constant.

Although production decisions are decoupled in an efficient economy, the relative consumption levels of both types of agents, for both goods, may become responsive to the innovation shock because their relative incomes differ due to the dependence of profit rates $\pi_{i,t}$ on supply rates $q_{i,t}$, even in cases where the two price systems are decoupled.

The feature that production rates are coupled only through the total money supply and not through its instantaneous distribution depends on the exponential utility (5), through the cancellation (31) of $(\epsilon_1/\gamma_1 - \epsilon_2/\gamma_2)$. This kind of modeling choice is similar in spirit to the choice of strictly symmetric production technologies in the one-period models of [8]. It is a *minimal* form that permits the many functions of the price system as a separating hyperplane to be performed independently. The overall production sector is separated from the dynamics of consumption due to wealth effects by one variable (the total money supply), whether or not the production decisions by firms of different types are also separated from each other.

5.5.6 Expansions in Small Deviations About the Fixed-Point Production Rate

In order to produce simple approximate demonstrations of the behavior of models in this class, we consider innovation shocks that are small compared with background stock and production levels, and evaluate responses to leading order in small perturbations.

The steady-state condition for production stocks, with production function f_i and embedded in an economy with steady prices, is given by Eq. (46) as

$$f_i'(\bar{s}_i) \equiv \rho_\pi . \tag{48}$$

Whenever all firms of type i are offering in markets, the offer rate equals the total output rate for good i, so we can abuse notation and use $q_{i,t}$ for both quantities. If only a measure $(1-\xi)$ of firms are offering in markets, then the output level per firm equals $(1-\xi)$ times the offer rate of the active firms.

The *offer rate* is approximated at leading linear order in small departures $s_i - \bar{s}_i$ by

$$q_i \approx f_i(\bar{s}_i) + \left(f_i'(\bar{s}_i) - \frac{d}{dt}\right)(s_i - \bar{s}_i) = \bar{f}_i + \left(\rho_\pi - \frac{d}{dt}\right)(s_i - \bar{s}_i) \,. \tag{49}$$

The first-order expansion for the marginal productivity appearing in Eq. (46) is

$$f_i'(s_i) - \rho_\pi \approx f_i''(\bar{s}_i)\,(s_i - \bar{s}_i) \equiv \bar{f}_i''\,(s_i - \bar{s}_i) \,. \tag{50}$$

Using Eq. (49) to approximate q_i, and Eq. (50) to approximate the marginal productivity, in the first line of Eq. (47) gives

$$\frac{d}{dt}\log\left(\frac{\hat{q}_1 e^{-\hat{q}_1/2} + \hat{q}_2 e^{-\hat{q}_2/2}}{B_1 + B_2}\right) \approx \frac{\mu_i}{2\gamma_i}\frac{d}{dt}\left(\frac{d}{dt} - \rho_\pi\right)(s_i - \bar{s}_i) + \bar{f}_i''\,(s_i - \bar{s}_i) \tag{51}$$

Here we have introduced a measure term μ_i, which equals unity when all firms of type i are active in markets, and equals $(1-\xi)$ in the case when a measure ξ of type-1 firms that have successfully innovated are sitting out of markets.

The right-hand side of Eq. (51) is a *linear* second-order differential response function, which means that the responses to different source terms or within different time intervals can be constructed independently and added to produce the full solution for $(s_i - \bar{s}_i)$. Complex matching conditions only arise at points where the solution for the non-cooperative equilibrium changes structure in some way, as when a subset of firms first enters markets, or when a type of consumers switch from being borrowers to being lenders. These are economically meaningful changes that only occur at a few points in a continuous time interval corresponding formally to an infinite number of periods (each of infinitesimal duration), in contrast with period boundaries in discrete-period models, which create complex matching conditions in every period. This feature explains our statement that the continuous-time limit may be seen as one in which the model period length does not reflect an economically significant timescale, and therefore should not affect the structure of solutions.

6 Example Solutions

6.1 Exchange with Gold Money Only

In a gold economy without banking or any other reserve supply of gold, the circulation rate $(B_{1,t} + B_{2,t})$ is constant in every period. Consumers spend all money in their possession. Therefore the time derivative on the left-hand side of Eq. (47) cannot be zero if $q_{1,t}$ experiences the shock of the investment in innovation. The production decisions of the two goods that can appear on the right-hand side of Eq. (47) must be coupled. The failure of decoupling—which defines our criterion of optimal monetary performance—is the main result which shows that gold or any money with fixed supply provides poor support to an exchange economy in which the production functions and consumption utilities could otherwise be optimized separately. It is another realization of Schumpeter's general observation about difficulties in breaking the circular flow of funds.

6.2 *Innovation and Recovery in Utopia*

Having established in Sect. 6.1 that a fixed money supply couples the shock in good-1 to production decisions in good-2, we now consider the opposite case of banking that creates any required level of bank money and monetized private credit, to show that such a system can realize the ideal efficiency of decoupling the two production sectors by making the left-hand side of Eq. (51) equal zero. We call this economy "Utopia" because the constraint-functions of money and default penalties serve to coordinate the efficient allocation of goods, but money has no other explicit utility. Banking is likewise a public service, with the policy objective of maximizing monetary efficiency, no requirement for strategic action, and thus no need to produce profits.

6.2.1 Consumer Lending and Borrowing with a Central Bank that is a Strategic Dummy

A minimal bank for the Utopia model is an atomic central bank, which is a strategic dummy. It produces any desired quantity of central-bank credit (or effectively distributes government fiat), which is accepted in trading posts on par with gold, and it provides accounting services for both its own debt and private debt without cost. Its behavior is defined by two parameters, the central-bank interest rate ρ_C and the default penalty Π, which we take to be sufficiently severe to support whatever shadow price on consumption is required for solutions without strategic default.

The following two subsections show numerical solutions for 1) the initial transient that converges to the turnpike steady state with fixed bank balances and part of the circular flow conducted through interest payments, and 2) the terminal divergence from the turnpike that cancels bank balances.

The Utopia solution fully decouples the two production sectors only during the initial transient from the innovation shock to the long-term turnpike solution. The terminal transient, combined with a requirement (forced by our probabilistic announcement of the terminal time T) for agents to converge to the turnpike, breaks the decoupling of the two sectors. In order to cancel the intra-economy debts by one type of consumers to the others, without incurring a net debt of the consumers to the central bank, both types of consumers must bid in a way that induces both types of firms to alter their output levels, rather than just type-1 firms that experienced the innovation opportunity.[23] This is an economically appropriate solution property: the conditions of production are permanently changed in this economy; if the terminal conditions require the termination of bank loans under such changed production conditions, they cannot avoid distorting production because they create a condition of inflexible money supply and distribution. However, this distortion is limited to a finite horizon before the day of reckoning, and decouples from the main solution property of buffering the circular flow of funds.

6.2.2 Initial Transient: From the Innovation Shock to the Turnpike

The first-order conditions in Utopia begin with the general solutions derived in Sect. 5.5.

In order to permit a non-inflationary/non-deflationary price system, the central bank interest rate must be tuned relative to the utilitarian discount factor $\beta = 1/\left(1+\rho\Delta t\right)$ so that

$$\beta\left(1+\rho_C\Delta t\right)\to 1, \tag{52}$$

or $\rho_C = \rho$. Since the central bank is a public good, and the rate of discount is known, this is consistent with other assumptions of fine-tuning that define Utopia.

The parameter $\hat{\epsilon}$ determining the asymmetry in consumption by Eq. (30), (31) is constant in this model, and determined by the turnpike condition, which is vanishing of the two terms in curly braces in Eq. (23) and (27) for $t\to\infty$.

The shadow price on money from consumer purchases of goods is determined from Eq. (42), in the case where borrowing and lending rates are equal and both equal $\rho_C = \rho$. Since $(B_{1,t}+B_{2,t}) = (m_{1,t}+m_{2,t})$ in steady-state where outputs take their stationary production values (set by $f_i'(s_i) = \rho_\pi$), the shadow price is then given by

$$\frac{\hat{q}_1 e^{-\hat{q}_1/2}+\hat{q}_2 e^{-\hat{q}_2/2}}{B_1+B_2} = \frac{\Lambda}{\rho}. \tag{53}$$

---

[23] A continuum of solutions to the first-order conditions exists, in which the type-1 and type-2 firms deplete or hoard stocks in differing degrees so as to cancel the intra-economy debt $\left(a_{1,T}-a_{2,T}\right)$. This continuum includes a solution in which the type-2 firms continue to produce at the pre-innovation level, so they are buffered at all times. That solution, however, does not lead to a net *aggregate* balance $\left(a_{1,T}+a_{2,T}\right)=0$, if $\left(a_{1,t}+a_{2,t}\right)$ starts from a zero aggregate balance at $t\ll T$. Therefore the solution with $s_{2,t}=\bar{s}_2,\ \forall t$ can only be reached by leaving a finely tuned non-zero aggregate balance $\left(a_{1,t}+a_{2,t}\right)$ of $\mathcal{O}\left(e^{-(T-t)\rho_\pi}\right)$ at early times t following the transient. Such an initial condition would lead to a different terminal solution than $(s_{2,t}=\bar{s}_2,\ \forall t)$ at any slightly different value for T, and would be incompatible with any non-cooperative equilibrium solution at a value of T differing by more than $\mathcal{O}(1/\rho_\pi)$ from the value for T which $\left(a_{1,t}+a_{2,t}\right)$ was tuned.

The constancy of this ratio (equal to a constant shadow price), in Eq. (47), together with the initial condition $s_{2,t=0} = \bar{s}_2$ then gives $s_{2,t} = \bar{s}_2, \ \forall t$ as the unique turnpike solution, completing the proof that banking in Utopia buffers production of good-2 from the innovation shock in good-1.

A similar evaluation, starting from Eq. (41), produces the relation between prices and output levels in Utopia of

$$\frac{1}{p_{1,t}\gamma_1} e^{-\hat{q}_{1,t}/2} = \frac{1}{p_{2,t}\gamma_2} e^{-\hat{q}_{2,t}/2} \rightarrow \frac{\Lambda}{\rho}. \tag{54}$$

A Numerical Example

The following example evaluates the integrals (22), (27) and non-cooperative equilibrium conditions from in the preceding sections, to show how the characteristic recovery structure following innovation is realized and determines the monetary properties of the economy.

Input parameters are: asymptotic production rate $f_{1,\infty}/\rho_\pi = 2$; $\gamma_1/\rho_\pi = 1/2$; the probability of success for firms that try to innovate is $\xi = 1/5$; the innovation cost $j = 0.1$, and the innovation output multiplier $\theta = 1/5$. For convenience, to avoid introducing new parameters, we set $f_{2,\infty} = f_{1,\infty}$ and $\gamma_2 = \gamma_1$. The pre-innovation steady-state of supply is therefore $\hat{q}_1 = \hat{q}_2 = (f_{1,\infty} - \rho_\pi/2)/\gamma_1 = 3$. Section "Solutions for the Utopia Economy" in the Appendix computes details of the time constants and structure of the recovery trajectories.

The natural timescale in the model is set by the profit rate ρ_π, which determines the dynamics of production stocks and output levels by Eq. (47). Since revolving loans are permitted in any amount that agents demand, the bank interest rate ρ_C does not determine a dynamical timescale, though it does affect the quantities of borrowed money. For simplicity in the numerical example we also set $\rho_C = \rho_\pi$.

The asymmetry of consumption (30), (31) generated by the non-cooperative equilibria of this game as a consequence of innovation evaluates numerically to $\hat{\epsilon} \approx 0.0075751$. Relative to the similarly scaled pre-innovation rates of production $\hat{q}_1 = \hat{q}_2 = 3$, $\hat{\epsilon}$ provides a measure of the utilitarian asymmetry introduced by innovation in one good.

Properties of the solution are shown in the following series of figures.

The Two-Stage Recovery Involving Stocks of Successful and Failed Innovators

Figures 1 and 2 show that the type-1 firms undergo a two-stage recovery following the innovation event. Before period $t = 0$, all type-1 firms are equivalent, so when the average outcome of innovation leads to higher output, all firms attempt to innovate. The fraction $(1 - \xi)$ that fail continue to offer goods at market in all periods $t > 0$, and in an initial interval their offer rates exceed their production rates, so they deplete their working stocks $s_1^{(-)}$. The profit incentive for this strategy comes

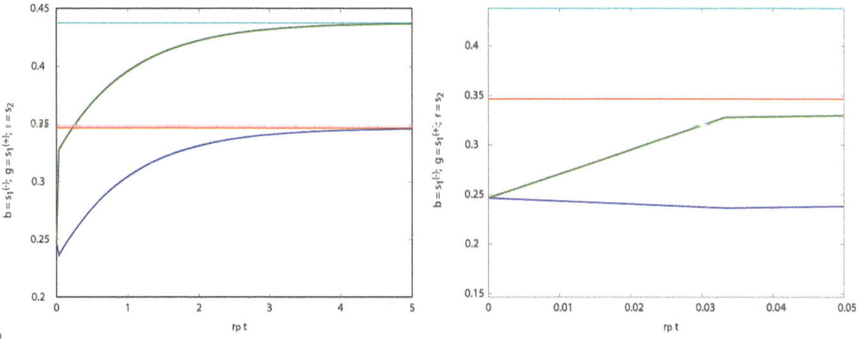

Fig. 1 The two-stage recovery associated with the cost and risk of failure in innovation. Firms of type-1 that try to innovate and fail follow recovery trajectories $s_1^{(-)}$ (*blue*) that initially deplete stocks while offering at an unsustainable rate in order to capture market share, by keeping prices below a level at which successful firms are willing to enter. When the stocks $s_1^{(+)}$ of successful firms (*green*) have grown and their shadow prices have decreased to equal market prices, both firms switch to offering at sustainable rates and converge with a fixed offset toward their late-time steady states (respectively *red and cyan*). Because of the choice (1) of functional form for f_i, the *red curve* is also the stock level $\bar{s}_2$, which is unaffected by innovation. *Left-hand panel* shows recovery over a long interval; *right-hand panel* gives a close-up of the interval following the innovation event

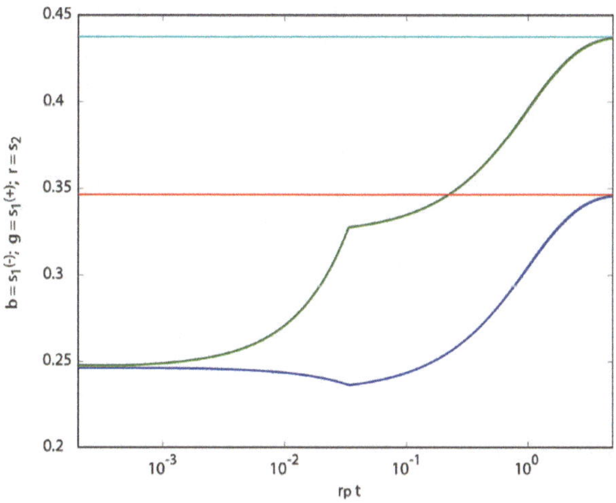

Fig. 2 Same timeseries as Fig. 1 with time $\rho_\pi t$ shown on log scale to make the initial phase more visible and to compress the subsequent recovery phase

from maintaining a price below the shadow price of the successfully-innovating firms, which will ultimately converge to a higher output level. The successful firms sit outside markets and accumulate stocks $s_1^{(+)}$, until their shadow prices fall to intersect the (rising) market prices maintained by the failed-innovation firms. After the two prices intersect, all firms offer in the markets, and the successful and failed type-1 firms both restore stocks to their (respective) steady-state production levels.

In the production functions (1), the steady-state stock is the same for both type-1 and type-2 firms, so the asymptotic level $\bar{s}_1$ to which failed-innovation type-1 firms recover is also the stock $\bar{s}_2$ maintained by type-2 firms throughout.

Rates at which Goods Are Delivered to Market for Consumption

Figure 3 shows the offer rates of the two groups of firms. Type-2 offer rates are constant. Type-1 offer rates are aggregated from the successful and failed-innovation firms. In the early interval, only a measure $(1 - \xi)$ of firms offer in markets, whereas in the later interval all firms offer. The discontinuous derivative in the stock $s_1^{(-)}$ visible in Fig. 1 exactly compensates for this jump in measure so that all of $s_1^{(-)}$, $s_1^{(+)}$, and q_1 are continuous through the transition.

Bid Levels and Money Supply in the Post-Innovation Interval

Figure 4 shows the bid levels on both types of goods by both groups of consumer/owners following the innovation shock. Total money supply in circulation $(B_{1,t} + B_{2,t})\,\Delta t$ is also shown (black curve) in the right-hand panel of the figure.

The amount of money in circulation per period is initially greater than $2m_0$ because agents of both types take out loans from the central bank. They borrow the maximum that they will be able to repay under the non-cooperative equilibrium trajectory. The money in circulation crosses (downward) through the pre-equilibrium value of $2m_0$ at $\rho_B t \approx 0.70050$ and continues to descend, as agents gradually pay down the principle.

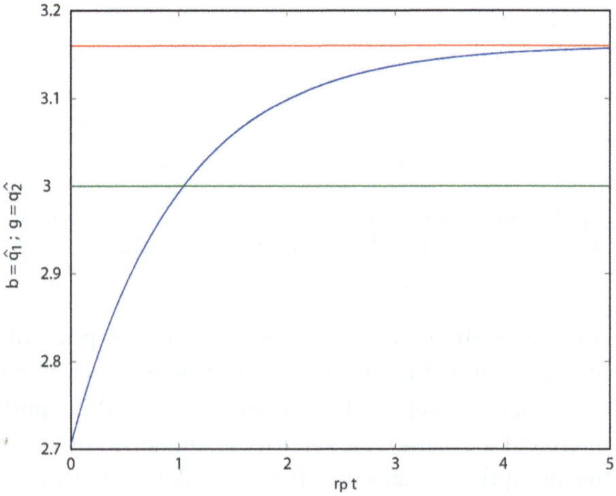

Fig. 3 Timeseries of the total rates $q_{1,t}/\gamma_1$ and $q_{2,t}/\gamma_2$ delivered to markets for consumption. $q_{2,t}$ (*green*) is constant at the pre-innovation solution over all time. $q_{1,t}$ (*blue*) begins in deficit relative to the pre-innovation solution, and ends in surplus relative to that solution

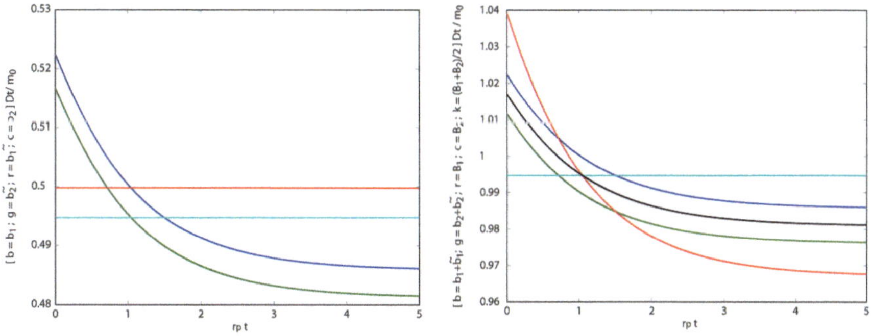

Fig. 4 Bid levels normalized by the pre-innovation money supply, $b_i \Delta t / m_0$, $\tilde{b}_i \Delta t / m_0$, aggregated in several ways. *Left panel*: by agents. *Blue* and *green* are bids on good-1 by consumers of types 1 and 2, respectively. *Red* and *cyan* are bids on good-2 by consumers of types 1 and 2 respectively. Type-1 consumers consume more of both goods, but in each period they pay out more than they receive in profits, a deficit that must be compensated by interest on bank savings. *Right panel*: by consumer-type or goods-type. Here *blue* and *green* are total expenditures by type-1 and type-2 consumers, respectively. *Red* and *cyan* are total bids offered on type-1 and type-2 goods, respectively. The *black curve* is $(B_1 + B_2) \Delta t / 2m_0$, which is the total money in circulation normalized by the pre-innovation value

Monetized Credit From a Persistent Internal Loan

Figure 5 shows the solution to Eq. (21) for $(a_{1,t} - a_{2,t})/2$ in relation to the excess of payment rates made by type-1 agents over payment rates by type-2 agents $(\tilde{b}_{1,t} - \tilde{b}_{2,t})$. The scale for the numéraire in this model is set by m_0, a quantity that scales $\sim \Delta t$, whereas the bid rates and inter-agent bank interest payment rates are regular quantities in the continuous-time limit. In order to normalize them to the numéraire, we compare the interest payments-per-period, which are $\rho_B (a_{1,t} - a_{2,t}) \Delta t / 2$, to m_0, and we likewise compare the excess bid amounts-per-period by type-1 over type-2 agents, which are $(\tilde{b}_{1,t} - \tilde{b}_{2,t}) \Delta t$, to m_0. These normalized curves are independent of Δt as $\Delta t \to 0$. The convergence of the two curves in Fig. 5 at late time verifies that the consumers converge to steady account balances at which interest payments via the bank provide part of the circular flow allowing type-1 agents to purchase and consume both goods at a constant excess rate ϵ over the rate of consumption by type-2 agents.

Aggregate Loan and Change in the Money Supply

Figure 6 shows the economy's aggregate balance with the banks $(a_1 + a_2)$ relative to the initial money supply of either agent type m_0. It also shows the excess money-in-circulation over the amount possible with the initial money supply, $(B_1 + B_2) \Delta t - 2m_0$, which is made possible by aggregate loans. Initially the two values are equal, but as the economy pays off the borrowed principle and also loses net money-in-

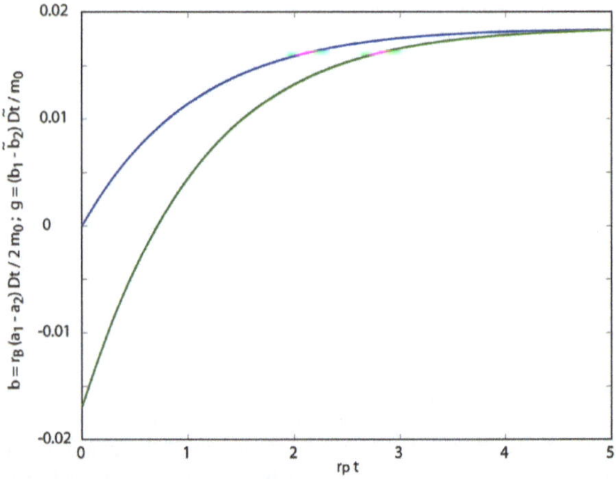

Fig. 5 Bank balance $\left(a_{1,t} - a_{2,t}\right)/2$ reflecting monetized private credit, normalized as $\rho_B \left(a_{1,t} - a_{2,t}\right) \Delta t / 2m_0$ (*blue*), which is the interest paid from type-2 agents to type-1 agents per period relative to the pre-innovation cash-per-agent, and excess bids per period of type-1 agents over type-2 agents similarly normalized, $\left(\tilde{b}_{1,t} - \tilde{b}_{2,t}\right) \Delta t / m_0$ (*green*), following the innovation event. The two converge to the same non-zero late-time steady state, as payment flows in the markets compensate interest flows through the bank. Note that the level of loans scales as $\left(a_{1,t} - a_{2,t}\right)/2 \sim \left(\tilde{b}_{1,t} - \tilde{b}_{2,t}\right)/\rho_B$, a quantity independent of Δt, which may be made arbitrarily larger than the money-in-circulation on the approach to the continuous-time limit

circulation to the payment of compounded interest, the money in circulation drops below $2m_0$. At late times, the principle is exactly repaid, and a new circular flow is established with asymptotically steady money supply $(B_{1,t} + B_{2,t}) \Delta t$ for $\rho_\pi t \gg 1$.

6.2.3 Terminal Conditions: Exiting the Turnpike to Cancel Bank Balances

A corresponding set of solutions for a terminal transient, which begins in the turnpike solutions for stocks, output, and prices, and terminates at a time T with zero bank balances, is shown in the next four figures. The overall behavior of the terminal transient is that type-1 consumers deplete their savings by increasing bids on goods, while type-2 consumers reduce their bids on goods to repay their outstanding account balances. These bids continue to respect all the non-cooperative equilibrium conditions, though now on an unstable diverging trajectory. In response to these changes in bidding behavior, the two types of firms either deplete or accumulate working stocks, altering their outputs to continue to maximize profits.

Working Stocks of the Firms

Figure 7 for the terminal transient may be compared with Fig. 1 for the behavior of stocks from the initial transient. Type-2 firms and failed-innovation type-1 firms

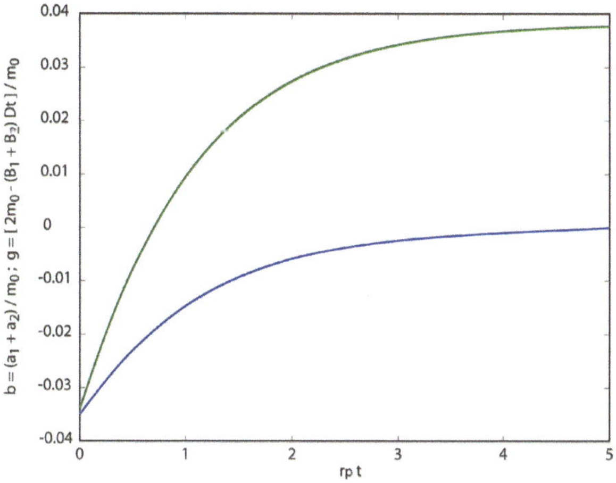

Fig. 6 Aggregate account balance the consumers hold at the bank, normalized as $(a_1 + a_2)/m_0$ (*blue*), and compared to total rate of money circulation in markets in excess of the circulation that would be possible with the initial gold-in-hand $2m_0$. The excess is normalized and plotted as $[2m_0 - (B_1 + B_2)\,\Delta t]/m_0$ (*green*). The amount borrowed in the period $t = 0$ when the innovation-cost is paid equals the excess of bids on goods over $2m_0$ (*green and blue curves* are equal at $t \to 0$ up to numerical imprecision). This loan amount is set using Eq. (27) so that the agents pay the principle to zero by time T. Note that $(B_{1,T} + B_{2,T})\,\Delta t < 2m_0$, so gold has left the private economy and is being held by the bank. Note also that the net loan $(a_1 + a_2)$ is a few percent of $m_0 \sim \Delta t$, whereas the difference of balances $(a_1 - a_2)$ in Fig. 5, which is credit from one agent type to the other monetized by the bank, is several percent of $m_0/(\rho_B \Delta t)$

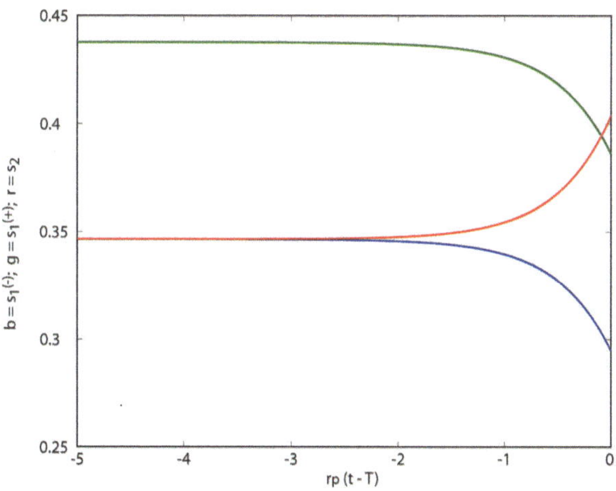

Fig. 7 Working stocks of the three types of firms in the terminal transient. Time is plotted as $\rho_\pi (t - T)$, which terminates at value 0. Trajectory $s_1^{(-)}$ (*blue*) and s_2 (*red*) begin at the same values but diverge in opposite directions. Trajectory $s_1^{(+)}$ of successful type-1 firms (*green*) moves in parallel to $s_1^{(-)}$ for unsuccessful type-1 firms. The initial working stocks of the terminal transient are the turnpike values to which the solutions in Fig. 1 converge at late times

both start with the same stocks $\bar{s}_2 = \bar{s}_1$, while successful innovation firms begin with stocks $\tilde{s}_1$. Because all type-1 firms optimize output against the same price system, both successful- and failed-innovation firms deplete stocks by the same amount, increasing output levels and lowering prices. Type-2 firms do the opposite, accumulating stocks and reducing outputs, and boosting prices.

Output Rates

Figure 8 shows the output rates produced by the stock trajectories from Fig. 7. Type-1 firms increase output rates, while type-2 firms reduce them. Recall that prices are given by Eq. (54).

Elimination of Intra-Economy Lending

Figure 9 shows the intra-economy debt, due to type-2 consumer borrowing from the central bank and type-1 consumer lending to the bank. The quantity $(a_{1,t} - a_{2,t})$ is plotted. Recall that this quantity is $\mathcal{O}(1)$ and thus generally much larger than the money in circulation. Hence, within $\mathcal{O}(\Delta t)$, $(a_{1,t} - a_{2,t})/2 \approx a_{1,t} \approx -a_{2,t}$.

In the initial steady state, the interest stream to/from the bank, $\rho_B(a_{1,t} - a_{2,t})\Delta t/2$, equals the excess bids by type-1 agents per period relative to bids from type-2 agents, $(\tilde{b}_{1,t} - \tilde{b}_{2,t})\Delta t$. The interest stream from bank accounts exactly provides the excess bids by type-1 agents to support their higher consumption levels. As the terminal transient develops, the bid excess by type-1 agents increases to deplete the

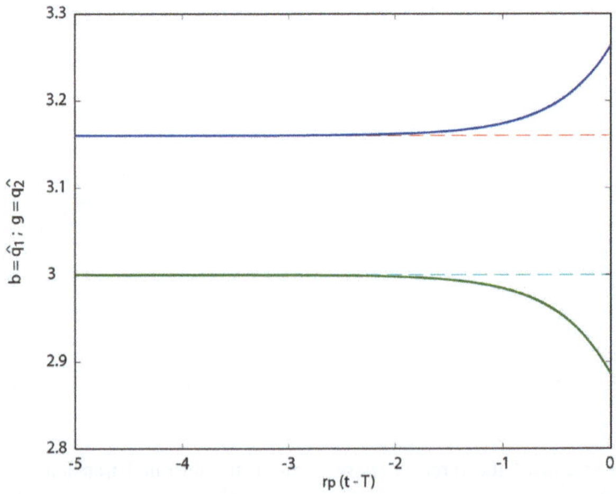

Fig. 8 Output rates $q_{1,t}/\gamma_1$ (*blue*) and $q_{2,t}/\gamma_2$ (*green*) delivered to markets for consumption in the terminal transient. The initial output levels for the terminal transient are the turnpike values to which the solutions in Fig. 3 converge, shown as *dashed lines*

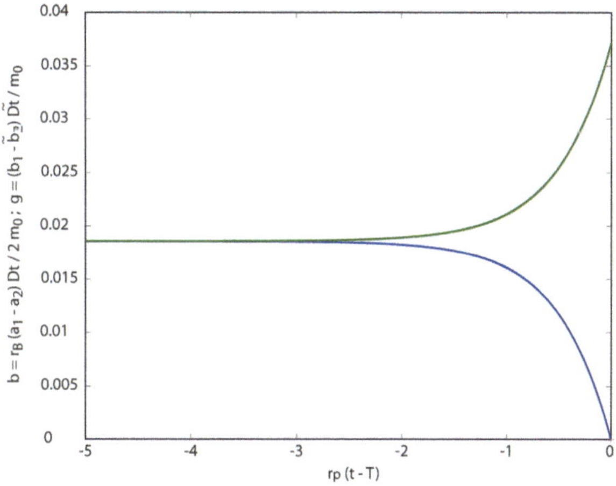

Fig. 9 Difference of bank balances scaled as $\rho_B\left(a_{1,t}-a_{2,t}\right)\Delta t/2m_0$ (*blue*), and excess bids per period of type-1 agents over type-2 agents similarly normalized, $\left(\tilde{b}_{1,t}-\tilde{b}_{2,t}\right)\Delta t/m_0$ (*green*). The turnpike value of steady intra-economy loans $\left(a_{1,t}-a_{2,t}\right)$ to which the solutions in Fig. 5 converge is returned to zero in the terminal transient

principle in their account, at the same time as type-2 agents repay principle. These differences continue to respect the consumption relations (30) at fixed $\hat{\epsilon}$, because changes in the output levels by firms have adjusted the price levels consistently with the changes in bids.

Non-Accrual of Aggregate Debt by Either the Economy or the Bank

Finally, Fig. 10 shows the aggregate account balance $(a_{1,t}+a_{2,t})$ through the terminal transient. Because innovation has made the collection of type-1 firms distinct from the type-2 firms, it is not possible for them to maintain an exactly fixed money supply through the entire terminal transient. Therefore, the bids by the two types of agents, and the output levels by the two types of firms, must be adjusted so that any non-zero aggregate balance acquired early in the transient is repaid by time T, leading in general to a change in the money-in-circulation from the turnpike value. Because the innovation shock we have assumed in this example is small, the two types of firms remain broadly similar. In order for their net contribution to debt to cancel, their output levels must be roughly mirror images, and this is the reason for the opposite behavior of the stock transients in Fig. 7 and the output transients in Fig. 8. The changes in money supply throughout the transient therefore remain small relative to money-in-circulation.

Further properties of the economy in the terminal transient can be computed, along the same lines as those presented for the initial transient.

6.2.4 Summary of Banking in Utopia

The preceding model has used a context in which a rigid money supply leads to a failure of output efficiency, to illustrate how a simple banking scheme can

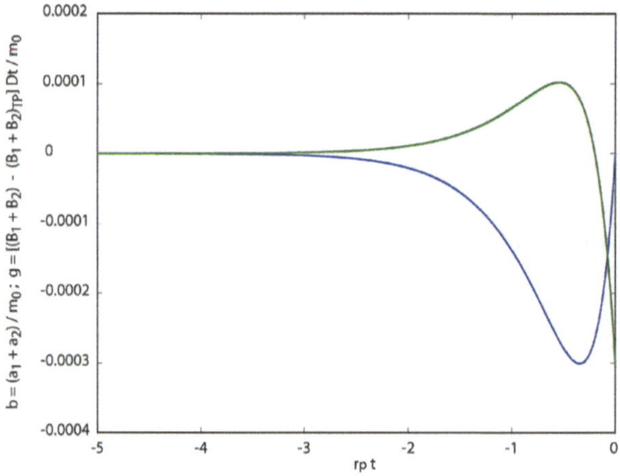

Fig. 10 Aggregate account balance the consumers hold at the bank, normalized as $(a_1 + a_2)/m_0$ (*blue*), and compared to total rate of money circulation in markets, now in excess of the turnpike money supply from the late-time asymptote in Fig. 6, which we denote $(B_1 + B_2)_{\text{TP}}\Delta t$. The money supply is plotted as $[(B_1 + B_2) - (B_1 + B_2)_{\text{TP}}]\,\Delta t/m_0$ (*green*). The total balance $\left(a_{1,T} + a_{2,T}\right) = 0$ as a property of the non-cooperative equilibrium solution

restore this efficiency. The main features of the Utopia model are that a single bank can change the money-in-circulation both transiently and persistently when this is required to stabilize the price system against which producers optimize, and can also monetize personal credit within the society to support emergent differences in purchasing power. The outcome of a many-period game is economically realistic: the owners of a technology that undergoes an innovative improvement in output capacity can become net holders of the debt of other members of the society, and the interest on this debt can support an indefinite increase in their relative purchasing power. It is an important feature of the banking model that members of the society can arrive at non-cooperative equilibria in which new steady states of money supply and the circular flow of funds are established, in which the bank withdraws from participation in the economy except as a keeper of its internal accounts.

6.3 Commercial Banking, Profit, and the Consequences of Interest Rate Spreads

In economies with distributed banking sectors, a criterion governing strategic action by the banks is profit maximization. Profits may come either from interest rate spreads or from permitting the banks to issue credit that receives the protection of law but is backed by only a fraction of its value in reserves of some form of "heavy money", which could be gold, government fiat, or central-bank credit. We consider first the introduction of interest rate spreads as a sole modification to the Utopia model, the resulting problems in the definition of profits, and the consequences of spreads for efficiency, which may be expressed in terms of the spread values independently of how (or whether) they are used strategically by the banks.

A non-zero spread exists whenever $\rho_{B,L} > \rho_{B,D}$ in Eq. (18), for the banks that serve consumers. In this section we consider the spread a fixed parameter and do not yet consider strategic action by banks.

The main features of (both transient and persistent) change in the money supply and monetization of private credit can be retained by profitable banks (if they are owned by the consumers and distribute their profits to consumers), but the efficiency of Utopia is lost in proportion to sizes of the spreads. We will show that the introduction of interest rate spreads inherently couples the innovation shock in good-1 to production and output decisions in good-2, with a strength proportional to the spread. Profit increases with increasing spreads, but so does cross-coupling among sectors and the consequent inefficiency. Therefore any regulatory system that requires spreads as a control mechanism carries an inherent efficiency cost.

6.3.1 The Continued Need for a Central Bank Even at 100 % Reserves

Even in the absence of fractional-reserve lending, purely mechanistic problems of defining profits from interest, using bank credit to vary the money supply, while also permitting asymptotically steady states in the absence of innovation, will require that we regard the banks with which consumers interact directly as *commercial banks*, and that we retain a *central bank* as a distinct entity.

In Utopia, it was important that the central bank *not* be merely a publicly-owned pass-through entity. One of its main functions was the injection or withdrawal of net quantities of money from the supply-in-circulation. This was achieved by accumulation of interest payments on (fully-repaid) net initial deposits or loans by the consumers. A feature of the Utopia model that makes a non-pass-through bank into a problem, however, when interest rate spreads are introduced, is that consumers in a two-good economy asymptotically make revolving loans from one type to the other, mediated by the bank, as shown in Fig. 5. If the bank collects a steady stream of payments proportional to $(\rho_{B,L} - \rho_{B,D})$ from these loans, and those are not passed back into the economy, no steady-state money supply and price system are possible. Yet the game must not pass all interest payments back to consumers, or else the banks lose the capability to vary the money supply.

To preserve both essential functions of the Utopian central bank when interest rate spreads are introduced, we must define one component of the net interest stream paid by consumers to the commercial bank as profit which is returned to the consumers, and a remainder that is not profit (because it is passed through to the central bank), with this remainder used to vary the money supply in circulation. The net interest paid by consumers to commercial banks will be $-(\rho_{B,1t}a_{1,t} + \rho_{B,2t}a_{2,t})$. The net interest paid by the commercial banks to the central bank, on money it must borrow to change the total money-in-circulation, is $-\rho_{C,t}(a_{1,t} + a_{2,t})$. The profit rate, which is a sum of two equal streams $\pi_{1,t}^{(B)} + \pi_{2,t}^{(B)}$ paid to the two types of agents, is then given in Eq. (25). As long as $\rho_{B,L} \geqslant \rho_C \geqslant \rho_{B,D}$, profits are never negative. In this section, we take the central bank rate ρ_C to be constant, as in Utopia.

For convenience of exposition here, since $\rho_{B,L}$ and $\rho_{B,D}$ are parameters, we take their average to equal the central bank rate, $(\rho_{B,L} + \rho_{B,D})/2 = \rho_C$. The central bank continues to be a public service, so we will set $\rho_C = \rho$ (the utilitarian rate of discount) to enable non-inflationary/non-deflationary turnpike solutions. More general solutions with steady-state production rates, but inflating or deflating prices, are also well-defined through Eq. (47), but are more complicated. The single new parameter for the commercial banks is then $(\rho_{B,L} - \rho_{B,D})/2$.

6.3.2 Interest Rate Spreads and Efficiency

For the single, simple event of innovation used in this class of games, the interest rates make a single transition at a time we may denote t_{split}. In terms of this transition time, instead of setting the left-hand side of Eq. (51) equal to zero as it is in Utopia, the equation satisfied by $s_{2,t}$ becomes

$$\left[\frac{d}{dt}\left(\frac{d}{dt} - \rho_\pi\right) + 2\gamma_2 \bar{f}_2''\right](s_2 - \bar{s}_2) = \pm\Theta\left(t_{\text{split}} - t\right)\gamma_2\left(\rho_{B,L} - \rho_{B,D}\right). \tag{55}$$

(Θ denotes the Heaviside function, which takes value one for $t < t_{\text{split}}$ and zero otherwise.) The boundary conditions for this second-order equation are that $s_2 = \bar{s}_2$ at $t = 0$ and again at $t \to \infty$. For the production function f_2 from Eq. (1), $\bar{f}_2'' = -2\rho_\pi$.[24] The value of t_{split} must be determined self-consistently with the signs of the bank balances in the solution that it yields. For small spreads, it is well approximated from the Utopia solutions shown in Fig. 5 and Fig. 6. We return to the determination of t_{split} in Sect. 6.3.3.

The solution to Eq. (55) is a sum of growing and decaying exponentials on the interval $0 \leqslant t \leqslant t_{\text{split}}$, and a decaying exponential for $t > t_{\text{split}}$. The magnitude of

---

[24] Firms of type-1, in the period when both are offering in the markets, have equations identical in form to Eq. (55), for the *deviations* of their stocks from the Utopia solutions. For the firms that attempt to innovate and fail, we denote these deviations $\delta\left(s_1^{(-)} - \bar{s}_1\right)$, and for the firms that attempt to innovate and succeed, the corresponding quantity is $\delta\left(s_1^{(+)} - \tilde{s}_1\right)$. In the initial period, when firms that successfully innovated are sitting outside the markets, their inventory growth is governed only by internal production and they do not optimize against prices. The type-1 firms that failed to innovate satisfy a slightly modified equation given by

$$\left[(1-\xi)\frac{d}{dt}\left(\frac{d}{dt} - \rho_\pi\right) + 2\gamma_1 \bar{f}_1''\right](s_1 - \bar{s}_1) = \pm\Theta\left(t_{\text{split}} - t\right)\gamma_1\left(\rho_{B,L} - \rho_{B,D}\right),$$

because their measure is $(1-\xi)$ and the level of output than can contribute scales by the same factor.

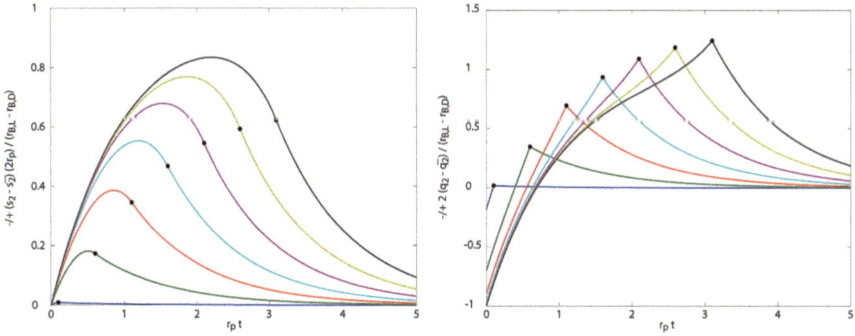

Fig. 11 Response of stocks and offers of good-2 to a discontinuity in interest rate by an amount $\pm\Theta\left(t_{\rm split} - t\right)\left(\rho_{B,L} - \rho_{B,D}\right)/2$. *Left panel*: the response descaled by the strength of the spread, given by $\mp\left(s_2 - \bar{s}_2\right) \times \left(2\rho_\pi\right) / \left(\rho_{B,L} - \rho_{B,D}\right)$, with a range of times $\rho_\pi t_{\rm split}$ (markers) from 0.1 to 3.1 in increments of 0.5. *Right panel*: the response of offers descaled by the spread, given by $\mp\left(q_2 - \bar{q}_2\right) \times 2/\left(\rho_{B,L} - \rho_{B,D}\right)$, for the same cases

the excursion can be determined by the condition that both the stocks and offer level be continuous through the transition. The matching conditions can always be met because the growing solution has a shorter time constant than the decaying solution.

Figure 11 shows the excursion in stock levels and offer rates by the type-2 firms in response to such a shock, at a sequence of increasing values of $t_{\rm split}$. The quantities plotted in the figure are $\mp\left(s_2 - \bar{s}_2\right) \times \left(2\rho_\pi\right) / \left(\rho_{B,L} - \rho_{B,D}\right)$, and $\mp\left(q_2 - \bar{q}_2\right) \times 2/\left(\rho_{B,L} - \rho_{B,D}\right)$. The $\mp$ sign corresponds to the $\pm$ sign in Eq. (55), and thus determines the direction of the excursion in stocks and offers.

The Sign of the Excursion

If, in the immediate aftermath of the innovation, both types borrow from the bank, then $\rho_{B,1} = \rho_{B,2} = \rho_{B,L}$, and the sign in Eq. (55) is negative. The effect is that good-2 firms try to optimize production against a larger discount rate than ρ_π, which means increasing the target s_2. This is done transiently by reducing offers and accumulating. Later, when the bank rates split, and one group lends while the other borrows, the target stock level returns to $\bar{s}_2$, and offer rates are increased to return toward it.

6.3.3 Approximating the Effect on Output Using a Small-Parameter Expansion

We will not pursue a full self-consistent solution to the production/trade model with interest rate spreads. The major qualitative features that result from the introduction of spreads may be illustrated with an approximate solution. The approximation is valid if the spread $(\rho_{B,L} - \rho_{B,R})/\rho \ll 1$. In this limit, output, prices, and consumption allocation are dominated by the properties of the Utopia solution. If we choose a small but nonzero period length $\rho_\pi \Delta t \ll 1$, then the money in circulation $(B_{1,t} + B_{2,t})\Delta t$ (along with all changes in that money supply) scale as $\sim \rho_\pi \Delta t$ relative to the long-term indebtedness within the economy $(a_{1,t} - a_{2,t})$, for $\rho_\pi t \gg 1$.

Solution Part I: Relating Money Supply to Outstanding Private Debt and Determining t_{split}

In this solution, $\rho_\pi \Delta t$ is used to relate the quantity $\rho_\pi \Delta t\,(a_{1,t} - a_{2,t})/2m_0$ from Fig. 5, to $(a_{1,t} + a_{2,t})/m_0$ from Fig. 6. From these two, values a_1/m_0 and a_2/m_0 are obtained. The leading order approximation for the crossing time t_{split} is then its value in the Utopia solution. This approximation is then used in Eq. (55) and its counterparts for successful and failed innovating firms of type-1, to obtain the linear-order corrections to the production stocks and output rates. In an iterative solution, these profiles could then be fed back into equations for prices and allocations to update t_{split}, and the process could be repeated, but for this example we will stop with the leading-order approximation.

A Numerical Example

To provide a numerical example, we take a very coarse discretization $\rho_\pi \Delta t = 0.1$ to scale the money supply relative to the acquired internal debt. This number is of course much too large to be well-approximated with the continuous-time recovery trajectory in the Utopia example, and we use it only to produce effects in the plots that are large enough to see. The same methods we illustrate here continue to apply as $\rho_\pi \Delta t$ is made arbitrarily smaller, and the response sizes scale in proportion.

With this large value of $\rho_\pi \Delta t$, the crossing time when a_1 passes through zero (consumers of type-1 change from being net borrowers to net lenders) is given by $\rho_\pi t_{\mathrm{split}} \approx 0.099$. The corresponding values of a_1/m_0 and a_2/m_0, and the perturbations in the goods-stocks, are shown in Fig. 12.

The solution combines three distinct output programs. Type-2 firms and type-1 firms that try to innovate and fail follow nearly the same trajectories of accumulation of goods $(s_2 - \bar{s}_2)$ and $\delta\left(s_1^{(-)} - \bar{s}_1\right)$ (see Footnote 24). Type-1 firms that successfully innovate do not optimize against the price system initially, so their accumulation of stocks is unaffected. After they enter markets, they follow a similar but less-extensive period of accumulation for $\delta\left(s_1^{(+)} - \tilde{s}_1\right)$, shown in the figure in cyan.

Correcting Stocks and Outputs, and Checking for Consistency

The modified stock trajectories for the three types of firms are shown in Fig. 13. For simplicity we take $f_{2,\infty} = f_{1,\infty}$ in Eq. (1), so that the two goods are completely

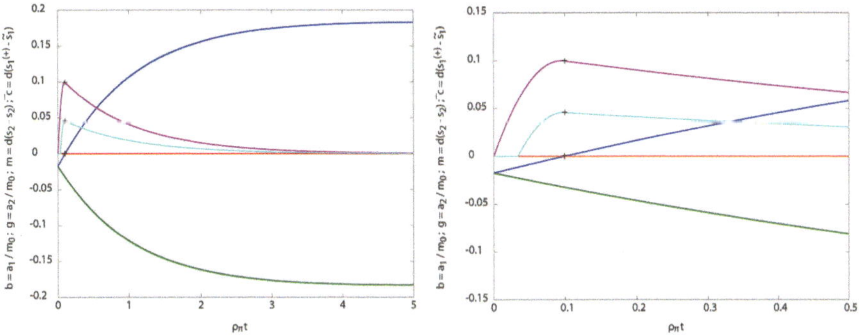

Fig. 12 Bank balances a_1/m_0 (*blue*) and a_2/m_0 (*green*) from Figs. 5 and 6 taking $\rho\Delta t = 0.1$. *Magenta curve* shows the change in response $\delta\left(s - \bar{s}\right)$ due to the interest rate discontinuity, which applies both to s_2 and to $s_1^{(-)}$, since both types of firms optimize output against the price system at all times. *Cyan curve* shows the response $\delta\left(s_1^{(+)} - \tilde{s}\right)$, which is zero in the interval when the successfully innovating firms are not optimizing their output against the price system, and nonzero when these firms enter the market. The two curves are shown to scale, and normalized so that the maximum of $\delta\left(s_2 - \bar{s}\right)$ is set to 0.1 for viewing purposes. The time when a_1 crosses through zero, and the interest rate $\rho_{B,1}$ changes from $\rho_{B,L}$ to $\rho_{B,D}$ is the time used for the matching conditions of both stock s_i and output q_i, marked with a *black cross*. *Left panel* is an extended recovery interval; *right panel* is a close-up of the initial interval following the innovation shock, during which balances are accumulated

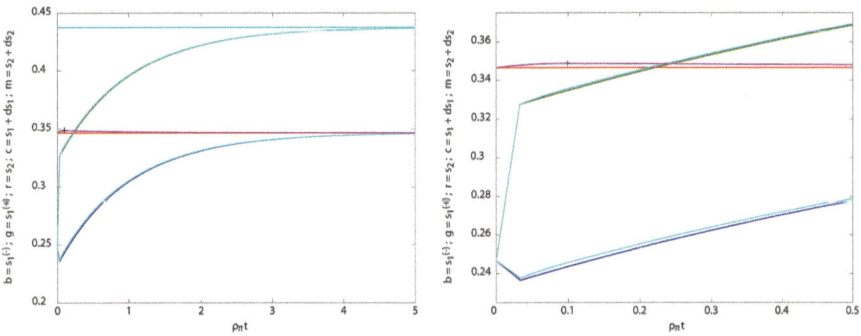

Fig. 13 Time-course of goods-stocks for the three kinds of firms. The Utopia solution of Fig. 1 is the leading order approximation for $s_1^{(-)}$ (*blue*), $s_1^{(+)}$ (*green*), and (in the simple case where $f_{1,\infty} = f_{2,\infty}$) s_2 (*red*). The perturbed stock levels taking $\left(\rho_{B,L} - \rho_{B,D}\right)/2\rho = 0.25$ are shown in *cyan* for both of $s_1^{(\pm)}$, and in *magenta* for s_2. Profile $s_1^{(+)}$ shows no change while successfully-innovating firms sit out of the market, and then undergoes a smooth deviation in output between the time it enters and the time the interest rates shift to their asymptotic late-time values. (Although the interest rate spread is set very large in order to produce a visible effect on output, the corrections to s_2 remain small, justifying the small-parameter approximations used)

equivalent in their production characteristics before the innovation event. The production-stock trajectories are obtained by adding the corrections from Eq. (55) (and its counterparts for type-1 firms) to the Utopia solution.

In the figure, we have again chosen a very coarse perturbation, $(\rho_{B,L} - \rho_{B,D}) / 2\rho = 0.25$, so the interest rate spread is fully *one half* of the average rate charged by the central bank. Again we do this to obtain results that are large enough to see easily in plots; for more realistic spreads the corrections scale proportionally. Even so, the figure shows that the perturbations to the histories of maintained stocks are small. A plot of the same recovery solutions with time on a logarithmic scale is shown in Fig. 14.

The offer levels, which depend on the time derivatives of the stocks, show coarser perturbations, in keeping with this large interest rate spread, as shown in Fig. 15. The most important feature is the initial drop in output (and therefore consumption) of good-2 (shown in magenta), which in the Utopia solution was unaffected by the innovation event in good-1. The output of good-1 also falls (shown in cyan in the figure) relative to its Utopia trajectory. Here we see the first feature of the small-parameter approximation indicating its incompleteness as a solution. In an initial post-shock interval, only failed-innovation type-1 firms offer, and they have measure $(1 - \xi)$. When prices have risen suitably, the successfully-innovating type-1 firms enter (as in Utopia and in the previous chapters), so that all type-1 firms are offering. Simply adding these two corrections to the Utopia solution produces a discontinuity that is an approximation error. In a full solution, adjustment of the matching conditions would absorb this correction (which

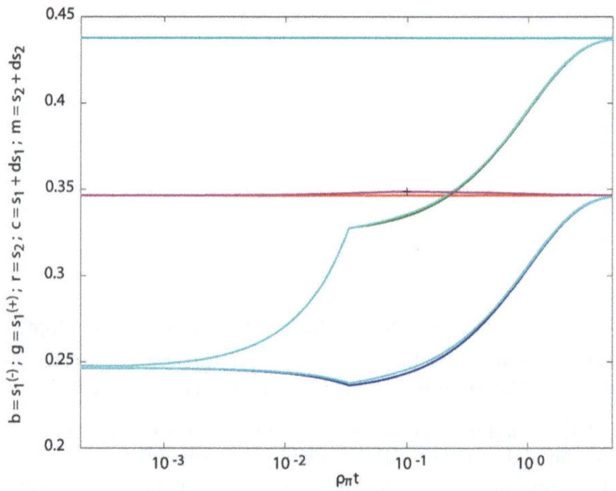

Fig. 14 Same recovery trajectories as in Fig. 13, with $\rho_\pi t$ plotted on logarithmic scale as in Fig. 2

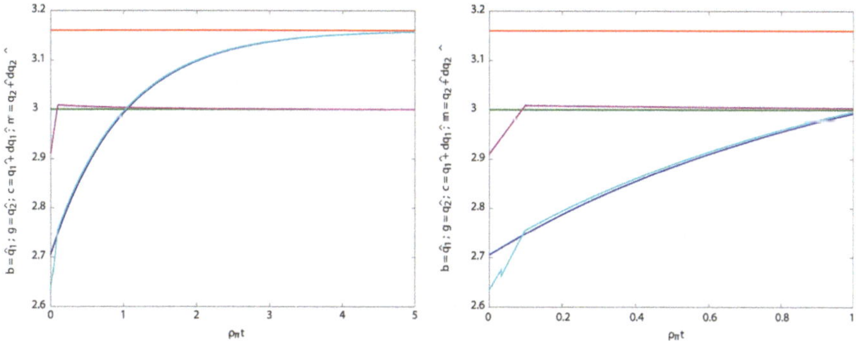

Fig. 15 Offer levels for the two goods in the Utopia solution and with nonzero interest rate spread under the parameters of Fig. 13. Utopia offer rates from Fig. 3 are $\hat{q}_1$ (*blue*) and $\hat{q}_2$ (*green*). Perturbed output due to the interest-rate discontinuity for $\hat{q}_1$ is *cyan* and for $\hat{q}_2$ is magenta. *Left panel* is the full relaxation trajectory (post-innovation asymptote for $\hat{q}_1$ shown in *red*); *right panel* expands the interval following the innovation shock. Unlike the stocks from Fig. 13, which show only a small perturbation, the consumption rates $\hat{q}$ show the larger effect that might have been expected for the large interest rate spread $(\rho_{B,L} - \rho_{B,D})/2\rho = 0.25$ chosen to make the effects visible. The small discontinuity in $\hat{q}_1$ visible in the *right-hand panel* comes from the fact that the perturbations in output levels were not fed back—in this leading-order approximation—to the leading-order optimization problem; doing so would have lead to a correction in the matching conditions for the offer levels of the type-1 firms by a small fraction of its Utopia solution value, to absorb this discontinuity

is only $\approx$ 0.3 % even for a wide spread) and restore continuity to the offer rates.[25]

6.3.4 Further Properties

Solutions for bank balances, bid levels, and other properties, can be carried through, and are qualitatively like those in the Utopia model. The equations for both intra-economy and aggregate account balances have already been presented in Sect. 5.4.10 in a form compatible with this model. The change in total money supply is responsive only to the central bank rate ρ_C, and initial loans can be fully repaid to converge to a turnpike solution, as in Utopia. The structure of intra-economy lending differs from that in the Utopia solution because borrowers and lenders pay at different rates, while bank profits are distributed to both types of consumers in equal measure. These differences change the quantitative properties of account

[25]In a true small-parameter expansion with both $\rho_\pi \Delta t \ll 1$ and $(\rho_{B,L} - \rho_{B,D})/2\rho \ll 1$, the value t_{split} would be shorter than the natural recovery time for stocks $s_1^{(\pm)}$, so that the output of the successfully-innovating firms would never even respond to the interest-rate spread. The resulting solution would be simpler in structure than the one presented here, as well as smaller in magnitude.

dynamics and their steady state values, but not their qualitative character. The terminal transient differs minimally from the Utopia solution, because the account balances do not change sign, so one type of consumers remains a lender with a fixed rate throughout the transient, while the other remains a borrower also with a fixed rate throughout the transient.

6.3.5 Concluding Comments Regarding Interest Rate Spreads

At small spreads, the introduction of bank profits creates small quantitative change but no qualitative change to the Utopia solution. This result demonstrates that the independent scaling between money supply and private debt (with respect to powers of $\rho_B \Delta t$) is not a fragile or fine-tuned property of Utopia, and can be retained in more institutionally complex models. The functions of varying the money supply and monetizing private debt are likewise robust. However, any active response by banks to consumer demand, which either limits the money supply or makes consumers' discounting of money time-dependent, in the sense of Eq. (47), propagates shocks from innovation across sectors and impairs optimal planning of production schedules.

6.3.6 A Note on Fractional-Reserve Lending

Although we do not present a formal model of banking with fractional reserves we conjecture that many of the basic qualitative aspects the models deliver the same message. In particular an official currency, a central bank and commercial banks are all artifacts to deal with evaluation, perception and substitutes for trust needed to promote and protect trade. The lack of natural physical laws for creating and destroying money call for the apparatus of sociopolitical laws to replace the laws for the creation and consumption of physical goods. The specifics of reserve ratio banking, reserves and excess reserves are discussed elsewhere [5].

Although we have presented an analysis on varying the money supply when there is, in essence no basic uncertainty in our models beyond one innovation decision, this is only the tip of an iceberg. We did not deal with the presence of a stream of random events that more closely characterizes ongoing innovation. The modeling considerations indicate that there is a welter of worthwhile case distinctions that depend on factors other than the mechanism of varying the money supply. In particular it is our belief that a key factor in the existence of a two tiered mechanism involving both a central bank and commercial banks is (as Bagehot observed) the importance of the banking system as a distributed perception device. Our efforts were devoted to variation of the money supply. In doing so we were able to illustrate how the failure to so adequately can cause considerable fluctuation that might be otherwise avoided.

Acknowledgements MS and ES are both external faculty of the Santa Fe Institute. The current work grows out of collaborations at SFI on the theory of money and its relations to problems of organization in physical and biological sciences. ES gratefully acknowledges support from William Melton and from Insight Venture Partners.

Appendix: Supporting Algebra for Non-Cooperative Equilibria of Game Models

Steady Post-Innovation Output and Stable Money Supply Lead to Stable Bid Levels

This section shows from the first-order conditions for consumption that, if output levels converge to steady late-time values, and if the money supply converges to a steady value, then bid levels by both type-1 and type-2 agents also converge to steady values. This condition is not an accounting identity, but part of the optimization problem that agents must solve. It requires only one strategic degree of freedom to be met, which is the overall consumption asymmetry $\hat{\epsilon}$ that governs agents' bid levels throughout the post-innovation consumption schedule.

From the notation of Eq. (30) in the main text, for the consumption asymmetries ϵ_1 and ϵ_2, and the fact that consumption rates c_i and $\tilde{c}_{\bar{\imath}}$ are related to bid rates b_i and $\tilde{b}_{\bar{\imath}}$ through the same prices p_i, the ratios of consumption levels of the same good by the two types of agents may be written in terms of the $\hat{q}_i$ and $\hat{\epsilon}$ as

$$\frac{c_1}{\tilde{c}_2} = \frac{b_1}{\tilde{b}_2} = \frac{1 + 2\hat{\epsilon}/\hat{q}_1}{1 - 2\hat{\epsilon}/\hat{q}_1}$$
$$\frac{\tilde{c}_1}{c_2} = \frac{\tilde{b}_1}{b_2} = \frac{1 + 2\hat{\epsilon}/\hat{q}_2}{1 - 2\hat{\epsilon}/\hat{q}_2}. \tag{56}$$

Introducing two further notational abbreviations

$$x_1 \equiv \frac{2\hat{\epsilon}}{\hat{q}_1} \qquad\qquad x_2 \equiv \frac{2\hat{\epsilon}}{\hat{q}_2}, \tag{57}$$

the bid rates by either agent type are written in terms of the total bid rates B_1 and B_2 as

$$b_1 = \frac{1 + x_1}{2} B_1 \qquad\qquad \tilde{b}_2 = \frac{1 - x_1}{2} B_1$$
$$\tilde{b}_1 = \frac{1 + x_2}{2} B_2 \qquad\qquad b_2 = \frac{1 - x_2}{2} B_2 \tag{58}$$

The bid rates B_1 and B_2 are then related to the total money supply by Eq. (34) in the main text.

As long as the values of the late-time interest rates are well-defined, $\hat{\epsilon}_t$ at late t has a fixed value, by Eq. (40). Then, as long as production levels q_i converge to steady values, the ratios of both B_i to the total money supply converge to steady values by Eq. (34). Finally, under these two conditions, the relations of all b_i and $\tilde{b}_i$ to the total money supply also converge, by Eq. (58).

This completes the result, and shows that steady credit and debt balances for the two agents $a_{1,T}$ and $a_{2,T}$ can be attained with a suitably chosen $\hat{\epsilon}$ by Eq. (23).

Solutions for the Utopia Economy

This section provides solutions for the non-cooperative equilibria of the Utopia model of Sect. 6.2. We begin with the output equations for good-2, which does not undergo an innovation shock.

The Unshocked Good Remains at Steady State Unperturbed

The main Eq. (51) for the response of output decisions to prices, under the condition (40) on shadow prices, becomes

$$\left[\frac{d}{dt}\left(\frac{d}{dt} - \rho_\pi\right) + 2\gamma_2 \bar{f}_2''\right](s_2 - \bar{s}_2) = 0. \tag{59}$$

Since the initial condition from the pre-shock equilibrium was $s_{2,0} = \bar{s}_2$, the unique bounded solution is $s_{2,t} = \bar{s}_2$ for all t.

Recovery of the Shocked Good

$s_{1,t}^{(-)}$ denotes the stock of the type-1 firms that tried to innovate and failed, and $s_{1,t}^{(+)}$ denotes the stock of the type-1 firms that succeeded. The initial conditions for both stocks in the periods immediately following the innovation are $s_{1,0+}^{(\pm)} \rightarrow \bar{s}_1 - j$. The steady-state stock for failed-innovation firms is $\bar{s}_1 \equiv (1/2)\log 2$, and the steady-state stock for successfully-innovating firms is $\tilde{s}_1 = \bar{s}_1 + (1/2)\log(1+\theta)$.

In an initial interval following the shock, only a measure $(1 - \xi)$ of firms offer in markets. The recovery equation (51) for these firms becomes

$$\frac{(1-\xi)}{2\gamma_1}\frac{d}{dt}\left(\frac{d}{dt} - \rho_\pi\right)\left(s_1^{(-)} - \bar{s}_1\right) \approx -\bar{f}_1''\left(s_1^{(-)} - \bar{s}_1\right). \tag{60}$$

This solution will govern offers $q_{1,t}$ until the shadow prices of successfully-innovating firms fall to intersect market prices. Thereafter the successfully-innovating firms also begin to offer.

Once both firms have entered, both relax to the new steady states with the converging solution to the equation

$$\frac{1}{2\gamma_1}\frac{d}{dt}\left(\frac{d}{dt}-\rho_\pi\right)\left(s_1^{(-)}-\bar{s}_1\right)\approx -\bar{f}_1''\left(s_1^{(-)}-\bar{s}_1\right). \tag{61}$$

These are both second-order linear equations, which possess growing and decaying solutions. We first introduce notations for characteristic rates in the two regimes:

$$\begin{aligned}
\omega_+^2 &\equiv -2\gamma_1\bar{f}_1'' \quad \text{evaluates on Eq. (1) to} \quad 4\gamma_1\rho_\pi \\
\omega_-^2 &\equiv -\frac{2\gamma_1}{(1-\xi)}\bar{f}_1'' = \frac{\rho_+^2}{(1-\xi)}
\end{aligned} \tag{62}$$

In terms of these, the solutions for the relaxation time constants are

$$\begin{aligned}
\frac{1}{\tau} &= \pm\sqrt{\omega_-^2+\frac{\rho_\pi^2}{4}}-\frac{\rho_\pi}{2} && t\leqslant t_1 \\
\frac{1}{\tau} &= \pm\sqrt{\omega_+^2+\frac{\rho_\pi^2}{4}}-\frac{\rho_\pi}{2} && t> t_1.
\end{aligned} \tag{63}$$

Both the positive and negative roots are needed in the initial transient for $t \leqslant t_1$. Only the positive root is required for relaxation toward the turnpike solution in the initial transient for $t > t_1$. The negative root in the second line of Eq. (63) will become important again, however, for the growing solution in the terminal transient.

Equivalent expressions exist for production by type-2 firms. In the numerical example, where the production and consumption parameters are set to equal values for the two types, the type-2 dynamics will depend on the same time constants as the dynamics for type-1 firms in the interval $t > t_1$.

Relaxation and Matching Conditions

The timescale for relaxation shared among models is the discount rate in the profit criterion ρ_π. Therefore introduce a dimensionless coordinate

$$z \equiv \rho_\pi t. \tag{64}$$

Two scale factors that define local timescales relative to z are given shorthand notations $\sqrt{\pm}$, which denote

$$\sqrt{+} \equiv \sqrt{1+\frac{4\rho_+^2}{\rho_\pi^2}} = \sqrt{1+\frac{8\gamma_1\bar{f}_1''}{\rho_\pi^2}} = \sqrt{1+\frac{16\gamma_1}{\rho_\pi}}$$

$$\sqrt{-} \equiv \sqrt{1 + \frac{4\rho_-^2}{\rho_\pi^2}} = \sqrt{1 + \frac{16\gamma_1}{(1-\xi)\,\rho_\pi}}. \tag{65}$$

The two trajectories in the initial interval after the innovation event are

$$\begin{aligned} s_{1,z}^{(+)} - \tilde{s}_1 &= -j - \frac{1}{2}\log(1+\theta) + \frac{f_{1,\infty}(1+\theta)}{\rho_\pi}\left(e^z - 1\right) \\ s_{1,z}^{(-)} - \bar{s}_1 &= e^{z/2}\left[-j\,\mathrm{ch}\left(\frac{z}{2}\sqrt{-}\right) + \sigma\,\mathrm{sh}\left(\frac{z}{2}\sqrt{-}\right)\right]. \end{aligned} \tag{66}$$

The trajectory for $s_{1,z}^{(+)}$ is fully determined by the production function because these firms are not responsive to markets. The trajectory for $s_{1,z}^{(-)}$ is determined by its initial conditions up to a single parameter σ which will be determined by matching conditions when successful innovators enter the markets.

The market prices and the shadow prices of successful type-1 firms become equal at some time z_1, which we will identify numerically. (The existence of a unique intersection is assured because the shadow prices of successful firms are falling while the market prices that can be maintained by the unsuccessful firms are rising, during the initial post-innovation interval.)

When the successful type-1 firms have entered the markets, their stocks relax with a fixed offset equal to the difference of late-time steady-state stocks, according to the functions

$$s_{1,z}^{(+)} - \tilde{s}_1 = s_{1,z}^{(-)} - \bar{s}_1 = \left(s_{1,z_1}^{(-)} - \bar{s}_1\right) e^{(z-z_1)\left(\sqrt{+}-1\right)/2} \tag{67}$$

The undetermined parameter σ in Eq. (66) is set by the requirement that the total offering $q_{1,t}$ be continuous through the transition at $z = z_1$, because continuity of q_1 is required for continuity of the price against which firms perform their discounting.

In the numerical solutions of Sect. 6.2.2, the radicals determining the relaxation time constants (65) evaluate to $\sqrt{+} = 3$ and $\sqrt{-} = \sqrt{11} \approx 3.3166$. The resulting time constants (63) are given by $1/\rho_\pi\tau = \left(\pm\sqrt{11} - 1\right)/2$ for $t \leqslant t_1$; $1/\rho_\pi\tau = 1$ for $t > t_1$. The matching parameter that makes both prices and quantities continuous is $\sigma \approx -0.14536$. The remaining features of these solutions are presented as plots in the main text.

Terminal Transient

A terminal transient is solved in terms of the divergences of the three working stocks from their steady-state turnpike values. The functional forms (using properties of non-cooperative equilibria previously derived for stocks when all firms optimize against a shared price system) are given by

$$s_{1,t}^{(+)} - \tilde{s}_1 = s_{1,t}^{(-)} - \bar{s}_1 = \left(s_{1,T}^{(-)} - \bar{s}_1\right) e^{(t-T)/\tau}$$

$$s_{2,t} - \bar{s}_2 = \left(s_{2,T} - \bar{s}_2\right) e^{(t-T)/\tau}. \tag{68}$$

When (as is the case in the numerical example) $\gamma_1 = \gamma_2 = \rho_\pi/2$, the time constant in both divergences is given by the negative root in the second line of Eq. (63), which evaluates to

$$\frac{1}{\tau} = -\frac{\left(\sqrt{+} + 1\right)}{2}\rho_\pi = -2\rho_\pi. \tag{69}$$

The two parameters in the solution (68), $s^{(-)}_{1,T}$ and $s_{2,T}$, are determined by the requirements that $(a_{1,T} - a_{2,T}) = 0$ and $(a_{1,T} + a_{2,T}) = 0$. Initial conditions are $(a_{1,t} + a_{2,t}) = 0$ as $t \to -\infty$, and $\rho_C\,(a_{1,t} - a_{2,t}) = \tilde{b}_1 - \tilde{b}_2$ of the turnpike solution for $t \to -\infty$. Results of numerical solution are shown in the figures of Sect. 6.2.3.

References

1. Bagehot, W.: Lombard Street, 3rd edn. Henry S. King, London (1873)
2. Dubey, P., Shubik, M.: The noncooperative equilibria of a closed trading economy with market supply and bidding strategies. J. Econ. Theory **17**, 1–20 (1978)
3. Samuelson, P.A.: Structure of a minimum equilibrium system. In: Stiglitz, J.E. (ed.) The Collected Scientific Papers of Paul A. Samuelson, pp. 651–686. MIT Press, Cambridge, MA (1966)
4. Schumpeter, J.A.: The Theory of Economic Development. Harvard University Press, Cambridge (1955)
5. Shubik, M., Smith, E.: The control of an enterprise economy (2014)
6. Shubik, M., Sudderth, W.: Cost innovation: Schumpeter and equilibrium. part 1: Robinson crusoe, cFDP # 1786, pp. 1–32 (2012)
7. Shubik, M., Sudderth, W.: Cost innovation: Schumpeter and equilibrium. part 2: Innovation and the money supply, cFDP # 1881, pp. 1–41 (2012)
8. Smith, E., Shubik, M.: Strategic freedom, constraint, and symmetry in one-period markets with cash and credit payment. Econ. Theory **25**, 513–551 (2005). sFI preprint # 03-05-036
9. Smith, E., Shubik, M.: Endogenizing the provision of money: costs of commodity and fiat monies in relation to the valuation of trade. J. Math. Econ. **47**, 508–530 (2011)

[illegible] (58)

[illegible] the maximal example [illegible] The [illegible] given by the maximal [illegible] which reads as

[illegible] (60)

The two parameters in the solution (58) [illegible] and [illegible] are determined by the [illegible] for [illegible] = 0 and [illegible] = 0. Initial conditions are [illegible] and [illegible] of the [illegible]. Results of a numerical solution are shown in the figures of Sect. [illegible]

References

1. [illegible]
2. [illegible] The [illegible] and bidding strategies. [illegible] Theory 17, [illegible] (1978)
3. [illegible] equilibrium system. In: Stiglitz, J.E. (ed.) The Collected Scientific Papers of Paul A. Samuelson, pp. [illegible] MIT Press, Cambridge, MA ([illegible])
4. [illegible] The Theory of Economic Development. Harvard University Press, Cambridge ([illegible])
5. [illegible] M., Sparta [illegible] (2013)
6. [illegible] M., [illegible] W. [illegible] and [illegible] (2013)
7. [illegible]
8. [illegible] 75 [illegible] (1985)
9. [illegible] M. [illegible] the Validation of [illegible] J. Math. [illegible] 47, [illegible] (2011)

Optimal Control of Tuberculosis: A Review

Cristiana J. Silva and Delfim F. M. Torres

Abstract We review the optimal control of systems modeling the dynamics of tuberculosis. Time dependent control functions are introduced in the mathematical models, representing strategies for the improvement of the treatment and cure of active infectious and/or latent individuals. Optimal control theory allows then to find the optimal way to implement the strategies, minimizing the number of infectious and/or latent individuals and keeping the cost of implementation as low as possible. An optimal control problem is proposed and solved, illustrating the procedure. Simulations show an effective reduction in the number of infectious individuals.

1 Introduction

Mycobacterium tuberculosis is the cause of most occurrences of tuberculosis (TB) and is usually acquired via airborne infection from someone who has active TB. It typically affects the lungs (pulmonary TB) but can affect other sites as well (extra-pulmonary TB). Only approximately 10 % of people infected with *M. tuberculosis* develop active TB disease. Therefore, approximately 90 % of people infected remain latent. Latent infected TB people are asymptomatic and do not transmit TB, but may progress to active TB through either endogenous reactivation or exogenous reinfection [52, 53]. Following the World Health Organization (WHO), between 1995 and 2011, 51 million people were successfully treated for TB in countries that adopted the WHO strategy, saving 20 million lives [60]. However, the global burden of TB remains enormous. In 2011, there were an estimated 8.7 million new cases of TB (13 % co-infected with HIV) and 1.4 million people died from TB [60]. The increase of new cases has been attributed to the spread of HIV, the collapse of public health programs, the emergence of drug-resistant strains of *M. tuberculosis* [19, 37, 38] and exogenous re-infection, where a latently-infected individual acquires a new infection from another infectious (see [6, 12, 17] and references cited therein). In the absence of an effective vaccine, current control

C.J. Silva • D.F.M. Torres (✉)
Department of Mathematics, Center for Research and Development in Mathematics and Applications (CIDMA), University of Aveiro, 3810–193 Aveiro, Portugal
e-mail: cjoaosilva@ua.pt; delfim@ua.pt

J.-P. Bourguignon et al. (eds.), *Dynamics, Games and Science*, CIM Series in Mathematical Sciences 1, DOI 10.1007/978-3-319-16118-1_37

programs for TB have focused on chemotherapy. Lack of compliance with drug treatments not only may lead to a relapse but to the development of antibiotic resistant TB, called multidrug-resistant TB (MDR-TB), which is one of the most serious public health problems facing society today [27]. The progress in responding to multidrug-resistant TB remains slow. There are critical funding gaps for TB care and control, which is critical to sustain recent gains, make further progress and support research and development of new drugs and vaccines [60].

Mathematical models are an important tool in analyzing the spread and control of infectious diseases [26]. Understanding the transmission characteristics of the infectious diseases in communities, regions and countries, can lead to better approaches to decrease the transmission of these diseases [25, 43, 49]. There are many mathematical dynamic models for TB, see, e.g., [4, 8, 13, 14, 21, 48, 58]. Most models consider that there are two different ways to progress to active disease after infection: "fast progressors" and "slow progressors". It is also considered that only 5–10 % of the infected individuals are fast progressors. The remaining are able to contain the infection (latent infected individuals) and have a much lower probability to develop active disease by endogenous reactivation. More recent models also consider the possibility of latent and treated individuals being reinfected, since it was already recognized that infection and/or disease do not confer full protection [57]. Models show that reinfection can be an important component of TB transmission and can have impact on the efficacy of interventions [13, 21, 40, 58]. Here we focus on TB models that consider: development of drug resistant TB [7]; exogenous reinfection [5, 6, 16, 17, 22, 35]; fast and slow progression to infection [5, 6, 16, 22]; post-exposure interventions [22]; immigration of infectious individuals [35]; and time-dependent parameters [59]. These models can be particularly useful in comparing the effects of various prevention, therapy and control programs [25, 32]. Since a variety of these programs are available, it is a natural objective to design optimal programs in terms of some pre-assumed criterion. This calls for the application of optimal control tools [33].

Optimal control has a long history of being applied to problems in biomedicine, particularly, to models for cancer chemotherapy [15, 29–32, 34, 54–56]. But until recently, little attention has been given to models in epidemiology [3, 20, 32, 41, 42, 44, 46]. In this paper we review the application of optimal control to TB mathematical models. The first paper appeared in 2002 [27], and considers a mathematical model for TB based on [7] with two classes of infected and latent individuals (infected with typical and resistant strain TB) where the aim is to reduce the number of infected and latent individuals with resistant TB. Two control strategies are proposed to achieve the objective: a *case finding* control measure, referring to the identification of individuals latently infected with typical TB and who are at high risk of developing the disease and who may benefit from prevention therapy (reducing the number of latent individuals that develop the disease) [27, 39]; and a *case holding* control, representing the effort that prevents the failure of the treatment in the typical TB infectious individuals and referring to activities and techniques used to ensure regularity of drug intake for a duration adequate to achieve a cure (reducing the incidence of acquired drug-resistant TB) [11, 27].

In [24] the authors consider the problem of minimizing the number of infectious individuals with a control intervention representing the effort on the prevention of the exogenous reinfection. The authors of [35] propose the implementation of a *case finding* control, representing the fraction of active infectious individuals that are identified and will be isolated in a facility, for an effective treatment and prevention of contact with susceptible and latent individuals, and a control measure based on the medical testing/screening of new immigrants before they are allowed into the population. In [59] three control interventions are studied with the aim of reducing the number of latent and active infectious individuals: *distancing* control, representing the effort of reducing susceptible individuals that become infected, such as, isolation of infectious individuals or educational campaigns; *case finding* control applied to latent individuals; and *case holding* control for infectious individuals. In [5, 6] *case finding* and *case holding* control measures are proposed for the minimization of the number of active infected individuals. In [16] the authors propose optimal control strategies for reducing the number of individuals in the class of *the lost to follow up individuals*. In [47, 50], optimal strategies for the minimization of the number of active TB infectious and persistent latent individuals are proposed.

The study of optimal control strategies produce valuable theoretical results, which can be used to suggest or design epidemic control programs. Depending on a chosen goal (or goals), various objective criteria may be adopted [5]. Although the implementation of the control policies, suggested by the mathematical analysis, can be difficult, they can be a support for the public health authorities and simulation of optimal control problems applied to mathematical models may become a powerful tool in their hands (see [5] and references cited therein).

The manuscript is organized as follows. In Sect. 2 mathematical models for TB dynamics are reviewed. They form, after introduction of the control functions, the control system of the optimal control problems on TB epidemics under consideration. The models with controls are presented in Sect. 3. A general optimal control problem is formulated in Sect. 4, where we explain how to obtain the analytic expression for the optimal controls, using the Pontryagin minimum principle [36]. In Sect. 5 we recall the numerical methods used to compute the optimal controls and associated dynamics. The main conclusions, derived from the numerical simulations, are resumed. Finally, in Sect. 6, an example is given, illustrating the effectiveness of the implementation of the control strategies on a TB control disease. We end with Sect. 7 of conclusions and future research.

2 Uncontrolled TB Models

Mathematical models have become important tools in analyzing the spread and control of infectious diseases [25]. In this section we present different mathematical TB models which are, after some modifications, the control system of optimal control problems on TB epidemics (see Sect. 3).

In an infectious disease model, the total population is divided into epidemiological subclasses. Some of the standard classes are: susceptible individuals (S), latently infected individuals (infected but not infectious) (E), infectious (I), and the recovered and cured individuals (R). Eight possible compartmental models, described by their flow patterns, are: *SI*, *SIS*, *SEI*, *SEIS*, *SIR*, *SIRS*, *SEIR* and *SEIRS*. For example, in a *SEIRS* model, susceptible become exposed in the latent period, then infectious, then recovered with temporary immunity and then susceptible again when the immunity wears off [25]. Here, we choose to denote the class of latently infected individuals by L and the class of recovered and cured individuals by T.

In [7] the authors present a *SEIRS* model for TB. The latently infected and infectious individuals with typical TB are denoted by L_1 and I_1, respectively. The model is given by

$$\begin{cases} \dot{S}(t) = \Lambda - \beta c S(t) \frac{I_1(t)}{N(t)} - \mu S(t), \\ \dot{L}_1(t) = \beta c S(t) \frac{I_1(t)}{N(t)} - (\mu + k_1 + r_1) L_1(t) + \sigma \beta c T(t) \frac{I_1(t)}{N(t)}, \\ \dot{I}_1(t) = k_1 L_1(t) - (\mu + r_2 + d_1) I_1(t), \\ \dot{T}(t) = r_1 L_1(t) + r_2 I_1(t) - \sigma \beta c T(t) \frac{I_1(t)}{N(t)} - \mu T(t), \end{cases} \tag{1}$$

where N denotes the total population, $N(t) = S(t) + L_1(t) + I_1(t) + T(t)$, Λ is the recruitment rate, β and $\sigma\beta$ are the probabilities that susceptible and treated individuals become infected by one infectious individual I_1 per contact per unit of time, respectively, c is the per-capita contact rate, μ is the per-capita natural death rate, k_1 is the rate at which an individual leaves the latent class L_1 by becoming infectious, d_1 is the per-capita TB induced death rate, and r_1 and r_2 are per-capita treatment rates for latent and infectious individuals, respectively. It is assumed that an individual can be infected only through contacts with infectious individuals.

In the same paper [7], a two-strain model is presented which considers resistant TB strain. Two subclasses of the total population are added: L_2 (latent) and I_2 (infectious), representing the developmental stages of resistant strains. It is assumed that I_2 individuals can infect S, L_1 and T individuals. The model is given by the following system:

$$\begin{cases} \dot{S}(t) = \Lambda - \beta c S(t) \frac{I_1(t)}{N(t)} - \mu S(t) - \beta^* c S(t) \frac{I_2(t)}{N(t)}, \\ \dot{L}_1(t) = \beta c S(t) \frac{I_1(t)}{N(t)} - (\mu + k_1 + r_1) L_1(t) + \sigma \beta c T(t) \frac{I_1(t)}{N(t)} + p r_2 I_1(t) \\ \qquad\quad - \beta^* c L_1(t) \frac{I_2(t)}{N(t)}, \\ \dot{I}_1(t) = k_1 L_1(t) - (\mu + r_2 + d_1) I_1(t), \\ \dot{L}_2(t) = q r_2 I_1(t) - (\mu + k_2) L_2(t) + \beta^* c \left(S(t) + L_1(t) + T(t)\right) \frac{I_2(t)}{N(t)}, \\ \dot{I}_2(t) = k_2 L_2(t) - (\mu + d_2) I_2(t), \\ \dot{T}(t) = r_1 L_1(t) + (1 - p - q) r_2 I_1(t) - \sigma \beta c T(t) \frac{I_1(t)}{N(t)} - \mu T(t) - \beta^* c T(t) \frac{I_2(t)}{N}, \end{cases} \tag{2}$$

with $N(t) = S(t) + L_1(t) + I_1(t) + L_2(t) + I_2(t) + T(t)$ and where β^* is the probability that treated individuals become infected by one resistant-TB infectious individual I_2 per contact per unit of time, d_2 and k_2 have similar meanings as d_1 and k_1 for resistant-TB, and $p + q$ is the proportion of those treated infectious individuals who did not complete their treatment. The proportion p modifies the rate that departs from the latent class, and $qr_2I_1(t)$ gives the rate at which individuals develop resistant-TB due to an incomplete treatment of active TB. Therefore, $p \geq 0$, $q \geq 0$ and $p + q \leq 1$.

The results of [17] suggest that exogenous reinfection has a drastic effect on the qualitative dynamics of TB. If we introduce into model (1) the term $\rho\beta cL_1I_1/N$, which represents exogenous reinfection, we obtain the exogenous reinfection tuberculosis model developed in [17]. The parameter ρ represents the level of reinfection. A value of $\rho \in (0, 1)$ implies that reinfection is less likely than a new infection. In fact, a value of $\rho \in (0, 1)$ implies that a primary infection provides some degree of cross immunity to exogenous reinfections. A value of $\rho \in (1, \infty)$ implies that TB infection increases the likelihood of active TB. The authors take the conservative view that $0 < p < 1$ (see (6) in Sect. 3 for the model with controls).

In [35] a mathematical model is presented, which takes into account immigration of infectious individuals as well as isolation of the infectious individuals for treatment. The model without controls is an extension of that of [17]: one subclass of the total population, the class of isolated infectious individuals with typical TB, is added. The corresponding controlled model is given in Sect. 3, by (7).

In [5, 6, 16] fast and slow progression to the infectious class are considered and both models consider exogenous reinfection, chemoprophylaxis of latently infected individuals and treatment of active infected individuals. In [6] a *SEI* model is proposed, where the infective class is divided into two subclasses: diagnosed infectious (those who have an active TB confirmed after an examination in a hospital) and undiagnosed infectious (i.e., those who have an active TB but not confirmed by an examination in a hospital), denoted by I_1 and I_2, respectively. The model in [6] is given by the following system of ordinary differential equations:

$$\begin{cases} \dot{S} = \Lambda - \beta\frac{I_1}{N}S - \mu S\,, \\ \dot{L}_1 = (1-g)\beta\frac{I_1}{N}S + r_2I_1 + r_3I_2 - (1-r_1)\sigma\lambda L_1 - [\mu + k_1(1-r_1)]L_1\,, \\ \dot{I}_1 = gf\beta\frac{I_1}{N}S + h(1-r_1)(k_1 + \sigma\beta\frac{I_1}{N})L_1 - (\mu + d_1 + r_2)I_1\,, \\ \dot{I}_2 = g(1-f)\lambda S + (1-h)(1-r_1)(k + \sigma\lambda)E - (\mu + d_3 + r_3)J\,, \end{cases} \tag{3}$$

where the fraction g of newly infected individuals are assumed to undergo a fast progression directly to TB, while the remainder is latently infected and enter the latent class L_1. Among the newly infected individuals that undergo a fast progression to TB, a fraction f of them is detected, and will enter the diagnosed infectious class I_1, while the remaining $1 - f$ is undetected and will be transferred into the undiagnosed infectious class I_2. In this model r_2 is the rate of effective per capita therapy of diagnosed infectious individuals I_1. It is assumed that undiagnosed infectious individuals can naturally recover and will be transferred into the latent class L_1 at a constant rate $r_3 < r_2$. Here σ is the factor reducing the risk of infection as a result of acquiring immunity for latently infected individuals L_1. Among latently infected individuals who become infectious, the fraction h of them is diagnosed and treated, while the remaining $1 - h$ is not diagnosed and enters the undiagnosed infectious class I_2. The parameter d_3 is the per capita TB induced death rate for undiagnosed infectious individuals. If we consider $f = 1$, $h = 1$, $r_3 = 0$ and $d_3 = 0$, then we obtain the model proposed in [5].

In [22] the authors present a model for TB that considers exogenous reinfection and post-exposure interventions. The class L_3 denotes the fraction of early latent individuals, that is, individuals that were recently infected (less than 2 years) and are not yet infectious; while L_4 denotes the class of persistent latent individuals who where infected and remain latent. The other classes are S, I_1 and T, with the same meaning has in the previous models. The model of [22] is given by the following system:

$$\begin{cases} \dot{S}(t) = \mu N - \frac{\beta}{N} I_1(t) S(t) - \mu S(t), \\ \dot{L_3}(t) = \frac{\beta}{N} I_1(t) \left(S(t) + \sigma L_4(t) + \sigma_R T(t)\right) - (\delta + \tau_1 + \mu) L_3(t), \\ \dot{I}_1(t) = k_1 \delta L_3(t) + \omega L_4(t) + \omega_R T(t) - (\tau_0 + \mu) I_1(t), \\ \dot{L_4}(t) = (1 - k_1)\delta L_3(t) - \sigma \frac{\beta}{N} I_1(t) L_4(t) - (\omega + \tau_2 + \mu) L_4(t), \\ \dot{T}(t) = \tau_0 I_1(t) + \tau_1 L_3(t) + \tau_2 L_4(t) - \sigma_R \frac{\beta}{N} I_1(t) T(t) - (\omega_R + \mu)\, T(t)\,. \end{cases} \tag{4}$$

Here σ has the same meaning has in the model (3) but applies to persistent latent individuals, L_4, and σ_R represents the same parameter factor but for treated patients; δ denotes the rate at which individuals leave the L_3 compartment; ω is the rate of endogenous reactivation for persistent latent infections (untreated latent infections); ω_R is the rate of endogenous reactivation for treated individuals (for those who have undergone a therapeutic intervention); τ_0 is the rate of recovery under treatment of active TB (assuming an average duration of infectiousness of 6 months); τ_1 and τ_2 apply to latent individuals L_3 and L_4, respectively, and are the rates at which chemotherapy or a post-exposure vaccine is applied. In this model it is assumed that the total population is constant, i.e., the rate of birth and death, μ, are equal and there are no disease-related deaths.

3 Controlled TB Models

The model (2) is the basis of the work developed in [27], where two control functions, u_1 and u_2, are introduced, representing control strategies for the two-strain TB model. The control system is given by

$$
\begin{cases}
\dot{S}(t) = \Lambda - \beta c S(t)\frac{I_1(t)}{N(t)} - \mu S(t) - \beta^* c S(t)\frac{I_2(t)}{N(t)}, \\
\dot{L}_1(t) = \beta c S(t)\frac{I_1(t)}{N(t)} - (\mu + k_1 + u_1(t) r_1) L_1(t) + \sigma\beta c T(t)\frac{I_1(t)}{N(t)} \\
\qquad + (1 - u_2(t)) p r_2 I_1(t) - \beta^* c L_1(t)\frac{I_2(t)}{N(t)}, \\
\dot{I}_1(t) = k_1 L_1(t) - (\mu + r_2 + d_1) I_1(t), \\
\dot{L}_2(t) = (1 - u_2(t)) q r_2 I_1(t) - (\mu + k_2) L_2(t) + \beta^* c \left(S(t) + L_1(t) + T(t)\right) \frac{I_2(t)}{N(t)}, \\
\dot{I}_2(t) = k_2 L_2(t) - (\mu + d_2) I_2(t), \\
\dot{T}(t) = u_1(t) r_1 L_1(t) + \left(1 - ((1 - u_2(t)))(p + q)\right) r_2 I_1(t) - \sigma\beta c T(t)\frac{I_1(t)}{N(t)} \\
\qquad - \mu T(t) - \beta^* c T(t)\frac{I_2(t)}{N(t)}.
\end{cases}
\tag{5}
$$

The control u_1 represents the fraction of typical TB latent individuals, L_1, that is identified and put under treatment (to reduce the number of individuals that may be infectious). The coefficient $1 - u_2(t)$ represents the effort that prevents the failure of the treatment in the typical TB infectious individuals (to reduce the number of individuals developing resistant TB). When the control u_2 is near 1, there is low treatment failure and high implementation costs.

In [24] the authors consider the exogenous reinfection TB model presented in [17] and introduce a control which simulates the effect of exogenous reinfection, that is, they consider a fixed value for ρ, $\rho = 0.4$, and multiply the term $\rho\beta c L_1 I_1/N$ by $1 - u$. The coefficient $1 - u$ represents the effort that prevents the exogenous reinfection in order to reduce the contact between the infectious and exposed individuals, thus decreasing the number of infectious individuals. The exogenous reinfection TB model with control, proposed in [24], is given by

$$
\begin{cases}
\dot{S}(t) = \Lambda - \beta c S(t)\frac{I_1(t)}{N(t)} - \mu S(t), \\
\dot{L}_1(t) = \beta c S(t)\frac{I_1(t)}{N(t)} - p\beta c(1 - u(t)) L_1(t)\frac{I_1(t)}{N(t)} - (\mu + k_1) L_1(t) + \sigma\beta c T(t)\frac{I_1(t)}{N(t)}, \\
\dot{I}_1(t) = p\beta c(1 - u(t)) L_1(t)\frac{I_1(t)}{N(t)} + k_1 L_1(t) - (\mu + r_2 + d_1) I_1(t), \\
\dot{T}(t) = r_2 I_1(t) - \sigma\beta c T(t)\frac{I_1(t)}{N(t)} - \mu T(t),
\end{cases}
\tag{6}
$$

with $N(t) = S(t) + L_1(t) + I_1(t) + T(t)$.

In [35] the model takes into account immigration of infectious individuals as well as isolation of the infectious for treatment. Two control functions are considered: u_1 and u_2. The control u_1 accounts for medical testing/screening of new immigrants, before they are allowed into the population, while the coefficient $1 - u_1$ is the effort that sustains such a testing policy. The control u_2 is a *case finding* control that represents the fraction of active individuals that are identified and will be isolated in a special facility, like a hospital, for effective treatment and prevention of contacts with susceptible and latent individuals. Hence, the term $1 + u_2$ represents the effort that sustains the isolation policy. The model with controls is given by

$$\begin{cases} \dot{S} = \Lambda^* + (1 - (1 - u_1(t))(p^* + q^*))A - \beta c S \frac{I_1 + lJ}{N} - \mu S, \\ \dot{L}_1 = (1 - u_1(t))p^* A + (1 - m)\beta c S \frac{I_1 + lJ}{N} - p\beta c L_1 \frac{I_1 + lJ}{N} + \sigma \beta c T \frac{I_1 + \sigma J}{N} \\ \qquad -(k_1 + \mu)L_1, \\ \dot{I}_1 = (1 - u_1(t))q^* A + m\beta c S \frac{I_1 + lJ}{N} + p\beta c L_1 \frac{I_1 + lJ}{N} + k_1 L_1 - (\mu + d_3 + r_2)I_1 \\ \qquad -(1 + u_2(t))\xi I_1, \\ \dot{J} = (1 + u_2(t))\xi I_1 - (r_3 + \mu + d_4)J, \\ \dot{T} = r_2 I_1 + r_3 J - \sigma \beta c T \frac{I_1 + \sigma J}{N} - \mu T. \end{cases} \tag{7}$$

The constant A represents the number of new members arriving into the population, per unit of time; p^* is the fraction of A arriving infected with latent TB; and q^* is the fraction of A arriving infected with active TB, so that $0 \leq p^* + q^* \leq 1$. It is assumed that $1 - (p^* - q^*)A$ individuals are free from the disease. The parameter Λ^* is the recruitment rate. Here the population is replenished from births and immigration; d_3 and d_4 are the typical TB-induced mortality rates for active TB individuals, that were not isolated from the population, and for isolated TB cases, respectively; r_3 is the treatment rate for isolated infectious individuals. The parameter l is the isolation level and lies in the range $0 \leq l \leq 1$, where $l = 0$ indicates absolute isolation for active infectious TB cases and $l = 1$ indicates no effective isolation. The parameter $0 \leq \sigma^* \leq 1$ determines the level of contact that treated individuals have with isolated individuals. The authors assume that $\sigma^* < l$ and that the treated individuals have a reduced contact with the isolated infectious group, as some of the treated individuals are from the J class. By m, $0 < m < 1$, it is denoted the fraction of persons with new infections who develop to TB fast, per unit of time, while ξ is the rate of isolation. The parameters μ, β, c, $\sigma\beta$, k_1, p, σ and r_2, have the same meaning as in the previous models (see Table 1).

In [59] the authors modified a model from [2] in order to study the transmission dynamics for TB in South Korea in the 40 years period from 1970 to 2009. The total population, N, is divided into susceptible individuals (S), high-risk latent (L_1) that are recently infected but not infectious, active-TB infectious (I) and permanently latent (L_5) with low risk. The main difference from the other TB models is the incorporation of time-dependent parameters. The birth and mortality rates are assumed as the time-dependent functions $b(t)$ and $\mu(t)$, respectively. The time-dependent function $k(t)$ is the per-capita rate of progression to active-TB from the recently latent class L_1. Individuals who do not progress from the class L_1 to the class I and those who are treated in the class L_1, are moved to the class L_5 at the per-capita rate α and $r(t)$, respectively. The time-dependent function $s(t)$ is the proportion of treated infectious individuals who did not complete their treatment; $1 - s(t)$ is the treatment success rate for active tuberculosis. As previously, the parameter β is the number of new infections with active-TB per unit of time. The

Table 1 Parameters that are used in the mathematical models for TB transmission (with and without controls)

Symbol	Description
Λ	Recruitment rate
μ	Per-capita natural death rate
b	Effective birth rate
d_1	Per-capita typical TB induced death rate
d_2	Per-capita resistant TB induced death rate
β	Rate at which susceptible individuals become infected by an infectious individual with typical TB
β^*	Rate at which susceptible individuals become infected by one resistant-TB infectious individual
$\sigma\beta$	Rate at which treated individuals become infected by an infectious individual with typical TB
c	Per-capita contact rate
k_1	Rate of progression to active TB
k_2	Rate of progression to active resistant TB
r	Per-capita treatment rate
r_1	Treatment rate of individuals with latent typical TB
r_2	Treatment rate of individuals with infectious typical TB
r_3	Treatment rate of undiagnosed infectious individuals
$1-s$	Treatment success rate
p	Level of exogenous reinfection
$u+v$	Proportion of treated infectious individuals who did not complete their treatment
g	Fraction of newly infected individuals that undergo a fast progression to the infectious class
f	Fraction of newly infected individuals that undergo a fast progression to TB
h	Fraction of infectious individuals that are diagnosed and treated
σ	Factor reducing the risk of infection as a result of acquiring immunity for latently infected individuals
σ_R	Factor reducing the risk of infection as a result of acquiring immunity for treated individuals
δ	Rate at which individuals leave L_3 compartment
α	Non-progress rate from L_1 to I
ω	Rate of endogenous reactivation for persistent latent infections
ω_R	Rate of endogenous reactivation for treated infections
τ_0	Rate of recovery under treatment of active TB
τ_1	Rate of recovery under treatment of latent individuals L_3
τ_2	Rate of recovery under treatment of latent individuals L_4
N	Total population

authors propose optimal control treatment strategies of TB in South Korea, for the period from 2010 to 2030, for various possible scenarios. Since it is not feasible to have the mortality data or the total population data for the future, the authors used the averaged constant values from the year 2001–2009 instead of using $b(t)$, $\mu(t)$, $s(t)$ and $r(t)$. The estimated time-dependent $k(t)$ from the year 1970–2009 is, however, used to find the optimal treatment strategy for the future. Three time-dependent controls are introduced into the TB system. The control $u_1(t)$ is the *distancing control* and the coefficient $1 - u_1(t)$ represents the effort of reducing susceptible individuals that become infected by infectious individuals, such as isolation of infectious people or educational programs/campaigns for healthy control. The *case finding* control, $u_2(t)$, represents the effort of decreasing the number of individuals that may be infectious, such as identification through screening of latent individuals who are in high risk of developing TB and who may benefit from prevention intervention. The *case holding* control, $1 - u_3(t)$, represents the effort of reducing the reinfection individuals, such as taking care of patients until they complete their treatment. The control system is given by

$$\begin{cases} \dot{S}(t) = bN(t) - \mu S(t) - (1 - u_1(t))\beta \frac{S(t)}{N(t)} I(t) \,, \\ \dot{L}_1(t) = (1 - u_1(t))\beta \frac{S(t)}{N(t)} I(t) - (k(t) + u_2(t)\alpha + \mu) L_1(t) + (1 - u_3(t)) srI(t), \\ \dot{I}(t) = k(t) L_1(t) - (r + \mu) I(t) \,, \\ \dot{L}_5(t) = (1 - (1 - u_3(t))s) rI(t) + u_2(t)\alpha L_1(t) - \mu I(t) \,, \end{cases}$$

where $N(t) = S(t) + L_1(t) + I(t) + L_5(t)$.

In [5] the author formulates an optimal control problem where one control, u, is introduced in the TB model. The control represents the effort on the chemoprophylaxis parameter (r_1) of latently infected individuals to reduce the number of individuals that may develop active TB. The model with control is given by (3) with $1 - u_1 r_1$ instead of $1 - r_1$ and considering $f = h = 1$ and $d_2 = r_3 = 0$. In [6], additionally to the control u_1, a second control u_2 is included in the model (3), which represents the effort on detection (h) of infectious, to increase the treatment rate of infectious and, consequently, to reduce the number of infectious and the source of infection. The model with controls is given by (3) with $1 - u_1 r_1$ instead of $1 - r_1$ and $u_2 h$ instead of h.

In [16] the authors propose a model adapted to Africa, in particular to Cameroon. A new class of individuals, called *the lost to follow up individuals*, is introduced. The individuals in this class are active infectious individuals who didn't take the treatment until the end, due to a brief relief of a long time treatment. Some of the lost to follow up individuals can transmit the disease without presenting any symptom. The authors present control measures for the reduction of the number of individuals that progress to the class of the lost to follow up individuals, L.

In [47, 50] two control functions, u_1 and u_2, and two real positive constants, ϵ_1 and ϵ_2, were introduced in the model (4). The control u_1 represents the effort in preventing the failure of treatment in active TB infectious individuals I_1 (*case holding*), and the control u_2 governs the fraction of persistent latent individuals L_4

that is put under treatment (*case finding*). The parameters $\epsilon_i \in (0, 1)$, $i = 1, 2$, measure the effectiveness of the controls u_i, $i = 1, 2$, respectively, i.e., these parameters measure the efficacy of treatment interventions for active and persistent latent TB individuals, respectively. In [47] the model is applied to Angola.

In [27, 47, 50] it is assumed that the total population N is constant, that is, the recruitment rate is equal to μN, $\Lambda = \mu N$, and the TB induced death rates are equal to zero. In [5, 6, 16, 24, 35, 59] the total population is not considered to be constant.

4 Optimal Control Problems

The control strategies for the reduction of infectious and/or latent individuals imply a cost of implementation. This implementation cost depends on many factors, for example, costs for activities to facilitate *case holding*. Those activities can be challenging because of the fact that chemotherapy must be maintained for several months to ensure a lasting cure, but patients usually recover their sense of well-being after only a few weeks of treatment and may often stop taking medications [27, 39]. For *case finding*, the control policies consider actions for the prevention of disease development with preventive therapy of latently infected individuals, which can be done in different ways, for example, identifying TB cases where the first initiative patient/provider contact is taken by health providers (*active case finding*) or by the patient (*passive case finding*), and screening activities among population groups at high risk of TB (for example, immigrants from high prevalence countries) [27, 35]. The implementation cost is taken into account in the formulation of an optimal control problem and is mathematically traduced by a functional.

Let L and I denote the latent infected and infectious individuals, respectively, without any specific characteristic, and $u = (u_1, \ldots, u_n)$, with $n \in \{1, 2, 3\}$ for the models described in Sect. 3, be the bounded Lebesgue measurable control function. Different cost functionals have been considered on the previously cited works on optimal control applied to TB models:

$$C_1(u) = \int_0^{t_f} \left[A_1 I(t) + A_2 L(t) + \sum_{i=1}^{n} \frac{B_i}{2} u_i^2(t) \right] dt\,,$$

$$C_2(u) = \int_0^{t_f} \left[A_1 I(t) + \sum_{i=1}^{n} \frac{B_i}{2} u_i^2(t) \right] dt\,,$$

and

$$C_3(u) = \int_0^{t_f} \left[A_2 L(t) + \sum_{i=1}^{n} \frac{B_i}{2} u_i^2(t) \right] dt\,.$$

It is assumed that the cost of the treatments are nonlinear and take a quadratic form. The coefficients, $A_j, j \in \{1,2\}$, and B_i, $i \in \{1,2,3\}$, are balancing cost factors due to the size and importance of the three parts of the objective functional.

For the cost functional C_2 and C_3, the aim is to minimize the infectious and latent individuals, respectively, while keeping the cost low. For the cost functional C_1, both infectious and latent individuals are wished to be minimized, keeping the cost of control interventions low.

A cost functional of type C_1 is adopted by Jung et al. [27], Silva and Torres [47, 50], and Whang et al. [59], C_2 is chosen in [5, 6, 24, 35] and C_3 is the objective functional in [16].

Let $(\mathscr{S})$ denote a mathematical model for TB with controls (see Sect. 3) given by a finite number, m, of differential equations. Assume that the control system $(\mathscr{S})$ is given by $\dot{X} = f(X,u)$, where f is a Lipschitz continuous function with respect to the state variable X, $X \in \mathbb{R}^m$, on the time interval $[0,t_f]$ and $X(0) = X_0$ be the initial condition. Moreover, let $g(X,u)$ denote the integrand of the cost functional C under consideration and assume that g is convex with respect to the control u. The optimal control problem consists in finding a control u^* such that the associated state trajectory X^* is solution of the control system $(\mathscr{S})$, in the time interval $[0,t_f]$ with initial conditions $X^*(0)$, and minimizes the cost functional C,

$$C(u^*) = \min_{\Omega} C(u)\,, \tag{8}$$

where Ω is the set of admissible controls (bounded and Lebesgue integrable functions) given by

$$\Omega = \left\{u \in L^1(0,t_f)\,|\,0 \le u_i \le 1\,,\ i = 1,\ldots,n\right\}.$$

According to the Pontryagin minimum principle [36], if $u^*(\cdot) \in \Omega$ is optimal for the optimization problem (8) subject to the control system $(\mathscr{S})$ with fixed initial conditions X_0 and fixed final time t_f, then there exists a nontrivial absolutely continuous mapping $\lambda : [0,t_f] \to \mathbb{R}^m$, called the *adjoint vector*, such that

$$\dot{X} = \frac{\partial H}{\partial \lambda} \tag{9}$$

and

$$\dot{\lambda} = -\frac{\partial H}{\partial X}\,, \tag{10}$$

where the function H defined by

$$H = H(X,\lambda,u) = g\,(X,u) + \langle \lambda, f(X,u)\rangle$$

is called the *Hamiltonian*, and the minimality condition

$$H(X^*(t), \lambda^*(t), u^*(t)) = \min_{0 \leq u \leq 1} H(X^*(t), \lambda^*(t), u) \tag{11}$$

holds almost everywhere on $[0, t_f]$. Moreover, the transversality conditions

$$\lambda_i(t_f) = 0, \quad i = 1, \ldots, m, \tag{12}$$

hold. This approach was considered in [6, 16, 24, 27, 35, 47, 59] for obtaining an analytic expression of the optimal control u^*. In [5] the analytical expression of the optimal control u^* is derived, using an algebraic approach, by solving a Riccati equation.

5 Numerical Methods and Simulations

In [27] the optimal treatment strategy is obtained by solving the optimality system, consisting of 12 ODEs from (5) and adjoint equations (10). An iterative method is used for solving the optimality system. The authors start to solve the state equations with a guess for the controls over the simulated time using a forward fourth order Runge–Kutta scheme. Because of the transversality conditions (12), the adjoint equations are solved by a backward fourth order Runge–Kutta scheme using the current iteration solution of the state equations. Then, the controls are updated by using a convex combination of the previous controls and the value from the characterizations derived by (11). This process is repeated and iteration is stopped if the values of unknowns at the previous iteration are close enough to the ones at the present iteration. The same numerical procedure is applied in [16, 35, 59]. In [47, 50] the authors use also the software IPOPT (https://projects.coin-or.org/Ipopt), the Matlab Optimal Control Software PROPT (http://tomdyn.com) and the algebraic modeling language AMPL (http://www.ampl.com). See, for example, [1] for details on numerical simulations of optimal control applied to life sciences using Matlab. In [24] the authors apply a semi-implicit finite difference method developed by Gumel et al. [23] and presented in [28]. For a gentle overview see [45].

In [27] different optimal control strategies are presented, which depend on the population size, cost of implementing treatment controls and the control parameters. The authors conclude that programs that follow the proposed control strategies can effectively reduce the number of latent and infectious resistant-strain TB cases. In [24] the numerical results show the effectiveness to introduce the control that prevents the exogenous reinfection, which reactivates the bacterium tuberculosis at the latent individuals. Analogously, in [5, 6] the results emphasize the importance of controlling exogenous reinfection using chemoprophylaxis and detection methods in reducing the number of actively infected individuals with tuberculosis. The numerical simulations in [35] show that the proposed control interventions can effectively reduce the number of latent and infectious TB cases. More precisely, the optimal control results show that a cost effective combination of screening/medical testing of immigrants, as well as isolation of infectious persons for treatment, may depend on cost of implementation of the controls and the parameters of the model, specially, the rate of isolation ξ, isolation level l, fraction of immigrants with latent TB p, and fraction of immigrants with active TB q.

6 Example: Optimal Control for the TB SEIRS Model

In this section we introduce a *case finding* control function u to the SEIRS mathematical model for TB (1) from [7]. The coefficient $1 - u(t)$ represents the effort that sustains the success of the treatment of latent individuals L_1. We assume that the total population N is constant, that is, $d_1 = 0$. This assumption is appropriate when the time period is short or when the natural deaths or the immigration balances the emigration (see [25]).

The controlled model is given by (see Table 1 for the meaning of the parameters)

$$\begin{cases} \dot{S}(t) = \Lambda - \frac{\beta}{N} c S(t) I_1(t) - \mu S(t), \\ \dot{L}_1(t) = \frac{\beta}{N} c S(t) I_1(t) - (\mu + r_1) L_1(t) - (1 - u(t)) k_1 L_1(t) + \sigma \frac{\beta}{N} c T(t) I_1(t), \\ \dot{I}_1(t) = (1 - u(t)) k_1 L_1(t) - (\mu + r_2 + d_1) I_1(t), \\ \dot{T}(t) = r_1 L_1(t) + r_2 I_1(t) - \sigma \frac{\beta}{N} c T(t) I_1(t) - \mu T(t). \end{cases} \tag{13}$$

Our aim is to minimize the number of infectious individuals I_1, while keeping the cost of control strategies implementation low, that is, (we choose a cost functional of type C_2 of Sect. 4)

$$C(u) = \int_0^{t_f} \left[A I_1(t) + \frac{B}{2} u^2(t) \right] dt\,. \tag{14}$$

In this example we propose to solve the optimal control problem that consists in finding a control u^* such that the associated state trajectory (S^*, L_1^*, I_1^*, T^*) is solution of the control system (13) in the time interval $[0, t_f]$ with initial conditions $(S(0), L_1(0), I_1(0), T(0))$ and minimize the cost functional C,

$$C(u^*) = \min_{\Omega} C(u)\,, \tag{15}$$

where Ω is the set of admissible controls given by

$$\Omega = \{u \in L^1(0, t_f) \,|\, 0 \leq u \leq 1\}.$$

Theorem 1 *The optimal control problem* (13), (15) *with fixed initial conditions* $S(0)$, $I_1(0)$, $L_1(0)$ *and* $T(0)$ *and fixed final time* t_f, *admits an unique solution* $(S^*(\cdot), I_1^*(\cdot), L_1^*(\cdot), T^*(\cdot))$ *associated to an optimal control* $u^*(\cdot)$ *on* $[0, t_f]$. *Moreover, there exists adjoint functions* $\lambda_1^*(\cdot)$, $\lambda_2^*(\cdot)$, $\lambda_3^*(\cdot)$ *and* $\lambda_4^*(\cdot)$ *such that*

$$\begin{cases} \dot{\lambda}_1^*(t) = \lambda_1^*(t)\left(\frac{\beta}{N}cI_1^*(t) + \mu\right) - \lambda_2^*(t)\frac{\beta}{N}cI_1^*(t), \\ \dot{\lambda}_2^*(t) = \lambda_2^*(t)((\mu + r_1) + (1 - u^*(t))k_1) - \lambda_3^*(t)(1 - u^*(t))k_1 - \lambda_4^*(t)r_1, \\ \dot{\lambda}_3^*(t) = -A + \lambda_1^*(t)\frac{\beta}{N}cS^*(t) - \lambda_2^*(t)\left(\frac{\beta}{N}cS^*(t) + \sigma\frac{\beta}{N}cT^*(t)\right) \\ \qquad +\lambda_3^*(t)(\mu + r_2 + d_1) - \lambda_4^*(t)\left(r_2 - \sigma\frac{\beta}{N}cT^*(t))\right), \\ \dot{\lambda}_4^*(t) = -\lambda_2^*(t)\sigma\frac{\beta}{N}cI_1^*(t) + \lambda_4^*(t)\left(\sigma\frac{\beta}{N}cI_1^*(t) - \mu\right), \end{cases} \tag{16}$$

with transversality conditions

$$\lambda_i^*(t_f) = 0, \quad i = 1, \ldots, 4.$$

Furthermore,

$$u^*(t) = \min\left\{\max\left\{0, \frac{k_1}{B}L_1^*(t)\left(\lambda_3^*(t) - \lambda_2^*(t)\right)\right\}, 1\right\}. \tag{17}$$

Proof Existence of an optimal solution (S^*, L_1^*, I_1^*, T^*), associated to an optimal control u^*, comes from the convexity of the integrand of the cost functional (14) with respect to the control u and the Lipschitz property of the state system with respect to state variables (S, L_1, I_1, T) (see, e.g., [10, 18]). System (16) is derived from the Pontryagin minimum principle (see (10), [36]) and the optimal control (17) comes from the minimality condition (11). The optimal control given by (17) is unique along all time interval due to the boundedness of the state and adjoint functions, the Lipschitz property of systems (13) and (16) and the fact that the problem is autonomous.

We end by presenting some numerical simulations with the following parameter values: $\mu = 0.0143$, $c = 1$, $\beta = 13$, $\sigma = 1$, $r_1 = 2$, and $r_2 = 1$ (see [7]). The initial conditions are: $S(0) = (76/120)N$, $L_1(0) = (38/120)N$, $I_1(0) = (5/120)N$, and $T(0) = (1/120)N$ (see [27]). We start showing that the implementation of the control has a positive impact on the reduction of infectious individuals. In Fig. 1

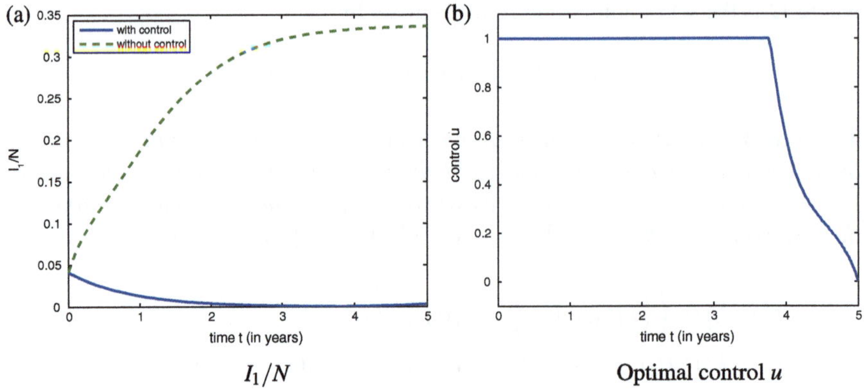

Fig. 1 Fraction of infectious individuals, with and without control, and optimal control (for $k_1 = 1, A = 1, B = 100$ and $N = 10{,}000$). (**a**) I_1/N; (**b**) Optimal control u

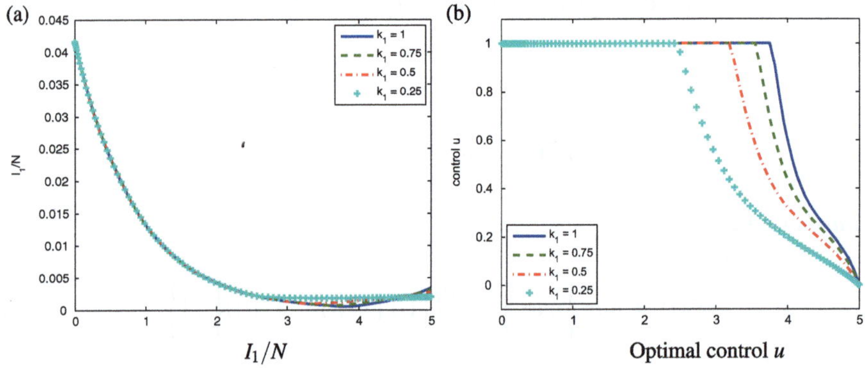

Fig. 2 Fraction of infectious individuals and optimal control for $k_1 \in \{0.25, 0.5, 0.75, 1\}$ (with $B = 100, A = 1$ and $N = 10{,}000$). (**a**) I_1/N; (**b**) Optimal control u

we observe that the fraction of infectious individuals decreases significatively when control strategies are implemented. If our aim is to reduce the number of infectious individuals giving special attention to keep the cost of implementation of the control measures low, then the weight constant B should take bigger values than A. Take, without loss of generality, $A = 1$ and $B \geq 50$. In this case, we observe that the fraction of infectious individuals I_1/N and the optimal control u depend on the rate of progression to active TB (see Fig. 2) and the size N of total population (see Fig. 3). The period of time that the optimal control attains its maximum value decreases with B (see Fig. 4). However, contrary to what is desired, the fraction of

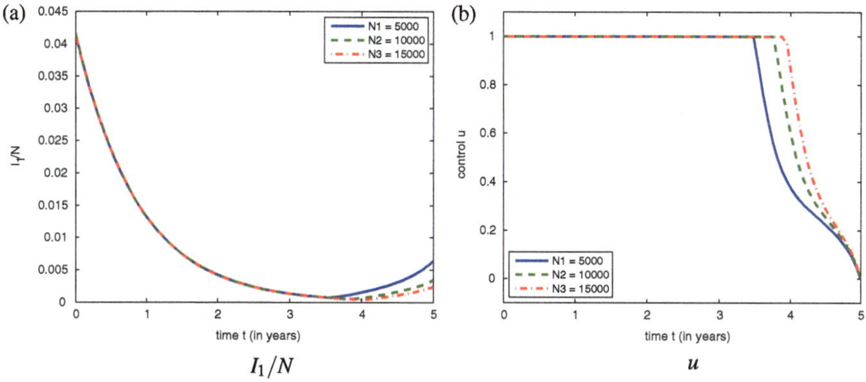

Fig. 3 Fraction of infectious individuals and optimal control for $N \in \{5000, 10{,}000, 15{,}000\}$ (with $B = 100$, $A = 1$ and $k_1 = 1$). (**a**) I_1/N; (**b**) u

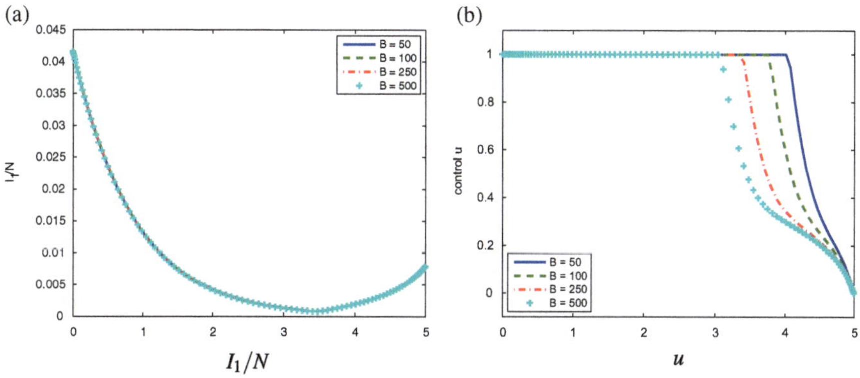

Fig. 4 Fraction of infectious individuals and optimal control for $B \in \{50, 100, 250, 500\}$ (with $A = 1$, $N = 10{,}000$ and $k_1 = 1$). (**a**) I_1/N; (**b**) u

infectious individuals starts increasing after some specific period of time. This can be avoided if the rate k of progression to active TB is low (see Fig. 2), or if we give more importance to the decrease of the number of infectious individuals than to the cost of implementation of the control policies, that is, if we increase the value of the weight constant A. In fact, for $A \geq B$ the fraction of infectious individuals never increases in all treatment period, regardless the size of the population N or the value of k (see Fig. 5). On the other hand, the optimal control attains the maximal value almost all the treatment period, which implies a higher cost implementation of control measures (see Fig. 6).

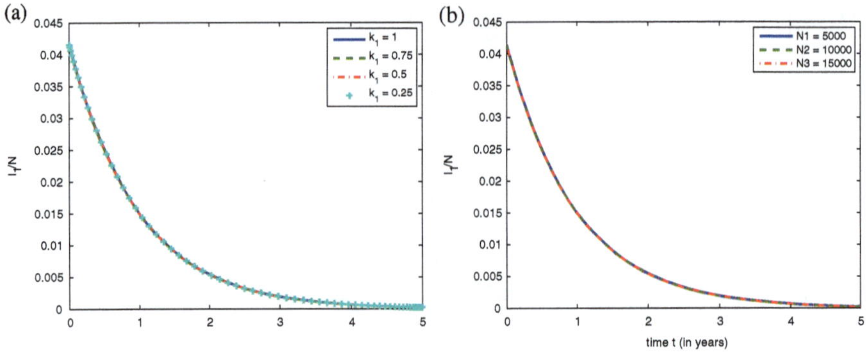

I_1/N I_1/N

Fig. 5 Fraction of infectious individuals for $A = B = 100$ (with $k_1 \in \{0.25, 0.5, 0.75, 1\}$ and $N \in \{5000, 10{,}000, 15{,}000\}$). **(a)** I_1/N; **(b)** I_1/N

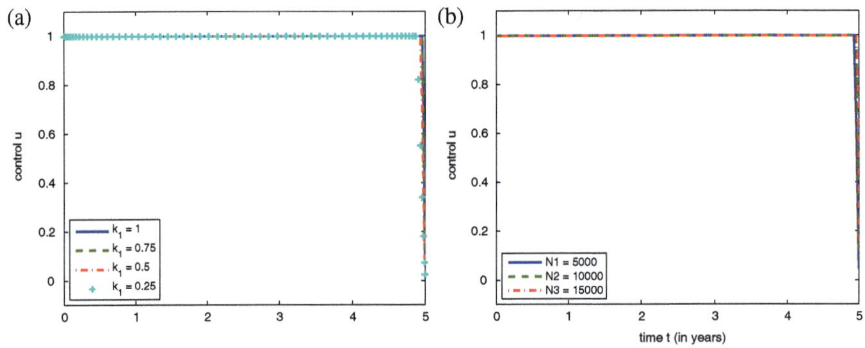

u u

Fig. 6 Optimal control for $A = B = 100$ (with $k_1 \in \{0.25, 0.5, 0.75, 1\}$ and $N \in \{5000, 10{,}000, 15{,}000\}$). **(a)** u; **(b)** u

7 Conclusion

A state of the art of uncontrolled and controlled mathematical models for tuberculosis (TB) has been presented. In particular, the paper reviews the works on optimal control of various models for the disease transmission dynamics of TB. Several results related to the dynamics and optimal control of TB have been reviewed and summarized. Two control strategies, "case finding" and "case holding", are used to demonstrate the optimal control analysis.

The topics covered do not provide an exhaustive survey but rather an illustrative overview. For instance, a TB vaccine called BCG (Bacillus of Calmette and Guérin) has been used especially for children for several decades, and in some papers a dynamical system with vaccination has been formulated and analyzed (see, e.g., [9]), but the subject has not been covered here. The example provided (see Sect. 6) is also very simple: only a single-strain TB dynamics with SEIRS model is presented. The reader interested in a model to study the optimal control of a two-strain (drug-sensitive and drug-resistant) TB dynamics is referred to [27].

Current research includes the development of co-infection mathematical models for TB and human immunodeficiency virus (HIV) transmission dynamics [51]. The novelty of [51], with respect to available results in the literature, is considering both TB and acquired immune deficiency syndrome (AIDS) treatment for individuals with both infectious diseases. Results show that TB treatment for individuals with only TB infection reduces the number of individuals that become co-infected with TB and HIV/AIDS, and reduces the diseases (TB and AIDS) induced deaths. They also show that the treatment of individuals with only AIDS also reduces the number of co-infected individuals. Further, TB-treatment for co-infected individuals in the active and latent stage of TB disease, implies a decrease of the number of individuals that passes from HIV-positive to AIDS. Application of optimal control to such combined TB-HIV/AIDS co-infection models poses a number of numerical challenges and is under investigation. This will be addressed in a forthcoming paper.

Acknowledgements This work was partially presented at the Thematic session *Control of diseases and epidemics*, MECC 2013—International Conference Planet Earth, Mathematics of Energy and Climate Change, 25–27 March 2013, Calouste Gulbenkian Foundation (FCL), Lisbon, Portugal. It was supported by Portuguese funds through the *Center for Research and Development in Mathematics and Applications* (CIDMA), and *The Portuguese Foundation for Science and Technology* (FCT), within project UID/MAT/04106/2013. Silva was also supported by FCT through the post-doc fellowship SFRH/BPD/72061/2010, Torres by project PTDC/EEI-AUT/1450/2012. The authors are very grateful to two anonymous referees, for valuable remarks and comments, which significantly contributed to the quality of the paper.

References

1. Anita, S., Arnautu, V., Capasso, V.: An Introduction to Optimal Control Problems in Life Sciences and Economics: From Mathematical Models to Numerical Simulation with MATLAB. Modeling and Simulation in Science, Engineering and Technology, XII. Birkhäuser, Basel (2011)
2. Aparitio, J.P., Capurro, A.F., Castillo-Chavez, C.: Markers of disease evolution: the case of tuberculosis. J. Theor. Biol. **212**(2), 227–237 (2002)
3. Behncke, H.: Optimal control of deterministic epidemics. Optim. Control Appl. Methods **21**, 269–285 (2000)
4. Blower, S., Small, P., Hopewell, P.: Control strategies for tuberculosis epidemics: new models for old problems. Science **273**, 497–500 (1996)
5. Bowong, S.: Optimal control of the transmission dynamics of tuberculosis. Nonlinear Dyn. **61**(4), 729–748 (2010)
6. Bowong, S., Alaoui, A.M.A.: Optimal interventions strategies for tuberculosis. Commun. Nonlinear Sci. Numer. Simul. **18**, 1441–1453 (2013)
7. Castillo-Chavez, C., Feng, Z.: To treat or not to treat: the case of tuberculosis. J. Math. Biol. **35**(6), 629–656 (1997)

8. Castillo-Chavez, C., Feng, Z.: Mathematical models for the disease dynamics of tuberculosis. In: Horn, M.A., Simonett, G., Webb, G. (eds.) Advances in Mathematical Population Dynamics-Molecules, Cells and Man, pp. 117–128. Vanderbilt University Press, Nashville (1998)
9. Castillo-Chavez, C., Feng, Z.: Global stability of an age-structure model for TB and its applications to optimal vaccination strategies. Math. Biosci. **151**(2), 135–154 (1998)
10. Cesari, L.: Optimization — Theory and Applications. Problems with Ordinary Differential Equations. Applications of Mathematics, vol. 17. Springer, New York (1983)
11. Chaulet, P.: Treatment of Tuberculosis: Case Holding Until Cure. WHO/TB/83, 141. World Health Organization, Geneva (1983)
12. Chiang, C.Y., Riley, L.W.: Exogenous reinfection in tuberculosis. Lancet Infect. Dis. **5**, 629–636 (2005)
13. Cohen, T., Murray, M.: Modeling epidemics of multidrug-resistant *M. tuberculosis* of heterogeneous fitness. Nat. Med. **10**(10), 1117–1121 (2004)
14. Dye, C., Garnett, G.P., Sleeman, K., Williams, B.G.: Prospects for worldwide tuberculosis control under the who dots strategy. Directly observed short-course therapy. Lancet **352**(9144), 1886–1891 (1998)
15. Eisen, M.: Mathematical Models in Cell Biology and Cancer Chemotherapy. Lectures Notes in Biomathematics, vol. 30. Springer, Berlin (1979)
16. Emvudu, Y., Demasse, R., Djeudeu, D.: Optimal control of the lost to follow up in a tuberculosis model. Comput. Math. Methods Med. 2011, 12 (2011). Art. ID 398476
17. Feng, Z., Castillo-Chavez, C., Capurro, A.F.: A model for tuberculosis with exogenous reinfection. Theor. Popul. Biol. **57**(3), 235–247 (2000)
18. Fleming, W.H., Rishel, R.W.: Deterministic and Stochastic Optimal Control. Springer, New York (1975)
19. Frieden, T., Driver, R.C.: Tuberculosis control: pas 10 years and future progress. Tuberculosis **83**, 82–85 (2003)
20. Gaff, H., Schaefer, E.: Optimal control applied to vaccination and treatment strategies for various epidemiologic models. Math. Biosci. Eng. **6**, 469–492 (2009)
21. Gomes, M., Franco, A., Gomes, M., Medley, G.: The reinfection threshold promotes variability in tuberculosis epidemiology and vaccine efficacy. Proc. R. Soc. B **271**(1539), 617–623 (2004)
22. Gomes, M.G.M., Rodrigues, P., Hilker, F.M., Mantilla-Beniers, N.B., Muehlen, M., Paulo, A.C., Medley, G.F.: Implications of partial immunity on the prospects for tuberculosis control by post-exposure interventions. J. Theor. Biol. **248**(4), 608–617 (2007)
23. Gumel, A.B., Shivakumar, P.N., Sahai, B.M.: A mathematical model for the dynamics of HIV-1 during the typical course of infection. In: Proceedings of the Third World Congress of Nonlinear Analysts, vol. 47, pp. 2073–2083 (2001)
24. Hattaf, K., Rachik, M., Saadi, S., Tabit, Y., Yousfi, N.: Optimal control of tuberculosis with exogenous reinfection. Appl. Math. Sci. (Ruse) **3**(5–8), 231–240 (2009)
25. Hethcote, H.: A thousand and one epidemic models. In: Levin, S.A. (ed.) Frontiers in Theoretical Biology, pp. 504–515. Springer, Berlin (1994)
26. Hethcote, H.: The mathematics of infectious diseases. SIAM Rev. **42**, 599–653 (2000)
27. Jung, E., Lenhart, S., Feng, Z.: Optimal control of treatments in a two-strain tuberculosis model. Discrete Contin. Dyn. Syst. Ser. B **2**(4), 473–482 (2002)
28. Karrakchou, J., Rachik, M., Gourari, S.: Optimal control and infectiology: application to an HIV/AIDS model. Appl. Math. Comput. **177**, 807–818 (2006)
29. Ledzewicz, U., Schättler, H.: Optimal bang-bang controls for a 2-compartment model in cancer chemotherapy. J. Optim. Theory Appl. **114**, 609–637 (2002)
30. Ledzewicz, U., Schättler, H.: Anti-Angiogenic therapy in cancer treatment as an optimal control problem. SIAM J. Control. Optim. **46**, 1052–1079 (2007)
31. Ledzewicz, U., Schättler, H.: Optimal and suboptimal protocols for a class of mathematical models of tumor anti-angiogenesis. J. Theor. Biol. **252**, 295–312 (2008)

32. Ledzewicz, U., Schättler, H.: On optimal singular controls for a general SIR-model with vaccination and treatment. Discrete Contin. Dyn. Syst. Supplement, 981–990 (2011)
33. Lenhart, S., Workman, J.T.: Optimal Control Applied to Biological Models. Chapman & Hall/CRC, Boca Raton (2007)
34. Martin, R., Teo, K.L.: Optimal Control of Drug Administration in Cancer Chemotherapy. World Scientific, Singapore (1994)
35. Okuonghae, D., Aihie, V.U.: Optimal control measures for tuberculosis mathematical models including immigration and isolation of infective. J. Bio. Syst. **18**(1), 17–54 (2010)
36. Pontryagin, L., Boltyanskii, V., Gramkrelidze, R., Mischenko, E.: The Mathematical Theory of Optimal Processes. Wiley Interscience, New York (1962)
37. Raviglione, M.C.: Evolution of WHO, 1948–2001 policies for tuberculosis control. Lancet **359**, 775–780 (2002)
38. Raviglione, M.C., Dye, C., Schmizt, S., Kochi, A.: For the global surveillance and monitoring project: assessment of worldwide tuberculosis control. Lancet **350**, 624–629 (1997)
39. Reichman, L.B., Hershfield, E.S.: Tuberculosis: A Comprehensive International Approach. Dekker, New York (2000)
40. Rodrigues, P., Rebelo, C., Gomes, M.G.M.: Drug resistance in tuberculosis: a reinfection model. Theor. Popul. Biol. **71**, 196–212 (2007)
41. Rodrigues, H.S., Monteiro, M.T.T., Torres, D.F.M.: Dynamics of dengue epidemics when using optimal control. Math. Comput. Model. **52**(9–10), 1667–1673 (2010)
42. Rodrigues, H.S., Monteiro, M.T.T., Torres, D.F.M.: Bioeconomic perspectives to an optimal control dengue model. Int. J. Comput. Math. **90**(10), 2126–2136 (2013)
43. Rodrigues, H.S., Monteiro, M.T.T., Torres, D.F.M.: Dengue in Cape Verde: vector control and vaccination. Math. Popul. Stud. **20**(4), 208–223 (2013)
44. Rodrigues, P., Silva, C.J., Torres, D.F.M.: Optimal control strategies for reducing the number of active infected individuals with tuberculosis. In: Proceedings of the SIAM Conference on Control and Its Applications (CT13) 8–10 July, pp. 44–50. SIAM, San Diego (2013)
45. Rodrigues, H.S., Monteiro, M.T.T., Torres, D.F.M.: Optimal control and numerical software: an overview. In: Systems Theory: Perspectives, Applications and Developments, pp. 93–110. Nova Science Publishers, New York (2014)
46. Rodrigues, H.S., Monteiro, M.T.T., Torres, D.F.M.: Vaccination models and optimal control strategies to dengue. Math. Biosci. **247**(1), 1–12 (2014)
47. Silva, C.J., Torres, D.F.M.: Optimal control strategies for tuberculosis treatment: a case study in Angola. Numer. Algebra Control Optim. **2**(3), 601–617 (2012)
48. Silva, C.J., Torres, D.F.M.: Optimal control applied to tuberculosis models. The IEA-EEF European Congress of Epidemiology 2012: epidemiology for a fair and healthy society. Eur. J. Epidemiol. **27**, S140–S141 (2012)
49. Silva, C.J., Torres, D.F.M.: An optimal control approach to malaria prevention via insecticide-treated nets. In: Conference Papers in Mathematics, 8 pp. (2013). Art. ID 658468
50. Silva, C.J., Torres, D.F.M.: Optimal control for a tuberculosis model with reinfection and post-exposure interventions. Math. Biosci. **244**(2), 154–164 (2013)
51. Silva, C.J., Torres, D.F.M.: Modeling TB-HIV syndemic and treatment. J. Appl. Math. (2014). Art. ID 248407. http://dx.doi.org/10.1155/2014/248407
52. Small, P.M., Fujiwara, P.I.: Management of tuberculosis in the United States. N. Engl. J. Med. **345**(3), 189–200 (2001)
53. Styblo, K.: State of art: epidemiology of tuberculosis. Bull. Int. Union Tuberc. **53**, 141–152 (1978)
54. Swan, G.W.: Role of optimal control in cancer chemotherapy. Math. Biosci. **101**, 237–284 (1990)
55. Swierniak, A.: Optimal treatment protocols in leukemia – modelling the proliferation cycle. In: Proceedings of the 12th IMACS World Congress, vol. 4, pp. 170–172. Baltzer, Basel, Paris (1988)
56. Swierniak, A.: Cell cycle as an object of control. J. Biol. Syst. **3**, 41–54 (1995)

57. Verver, S., Warren, R.M., Beyers, N., Richardson, M., van der Spuy, G.D., Borgdorff, M.W., Enarson, D.A., Behr, M.A., van Helden, P.D.: Rate of reinfection tuberculosis after successful treatment is higher than rate of new tuberculosis. Am. J. Respir. Crit. Care Med. **171**, 1430–1435 (2005)
58. Vynnycky, E., Fine, P.E.: The natural history of tuberculosis: the implications of age-dependent risks of disease and the role of reinfection. Epidemiol. Infect. **119**(2), 183–201 (1997)
59. Whang, S., Choi, S., Jung, E.: A dynamic model for tuberculosis transmission and optimal treatment strategies in South Korea. J. Theor. Biol. **279**, 120–131 (2011)
60. WHO.: Global Tuberculosis Control. WHO Report, Geneva (2012)

A Bayesian Modelling of Wildfires in Portugal

Giovani L. Silva, Paulo Soares, Susete Marques, M. Inês Dias, M. Manuela Oliveira, and José G. Borges

Abstract In the last decade wildfires became a serious problem in Portugal due to socieconomic and climate change trends. In order to analyse wildfire data, we employ beta regression for modelling the proportion of burned wild area, under a Bayesian perspective. Our main goal is to find out fire risk factors that influence the proportion of area burned and what may make a wild area susceptible or resistant to fire. Then, we analyse wildfire data in Portugal during 1990–1994 through Bayesian normal and beta regression models that use Markov chain Monte Carlo methods for estimating quantities of interest.

1 Introduction

In Portugal, wildfires (related to natural forests and other plant areas) have been increasing in the last years. Fire is indeed an important issue in Mediterranean region affecting namely the ecological and economic aspects of forest areas and causing loss of human life. Many factors have contributed to the increasing number of wildfires, e.g., climate change [7]. Some studies have identified changes in the number of fires, burned area and fire size distribution depending on topographical variables and vegetation type, e.g., in the Spanish region Catalonia [10] and Portugal [13].

Gomes [9] pointed out many causes and consequences of forest fires in Portugal, e.g., currently, rural and forest areas in Portugal are considerably deserted due to

G.L. Silva (✉) • P. Soares
CEAUL and Department Mathematics, Instituto Superior Técnico, Universidade de Lisboa, Avenida Rovisco Pais, 1, 1049-001 Lisbon, Portugal
e-mail: giovani.silva@tecnico.ulisboa.pt; paulo.soares@tecnico.ulisboa.pt

M.I. Dias • M.M. Oliveira
CIMA and Department Mathematics, Universidade de Évora, Évora, Portugal
e-mail: misd@uevora.pt; mmo@uevora.pt

S. Marques • J.G. Borges
FRC and Department Natural Resources, Environment and Territory, Instituto Superior de Agronomia, Universidade de Lisboa, Lisboa, Portugal
e-mail: smarques@isa.ulisboa.pt; joseborges@isa.ulisboa.pt

J.-P. Bourguignon et al. (eds.), *Dynamics, Games and Science*, CIM Series in Mathematical Sciences 1, DOI 10.1007/978-3-319-16118-1_38

population migrations from these areas to the main cities, which began in the 1950s. Fernandes et al. [5] proposed a fuel modelling and fire hazard assessment, used to evaluate and compare the fire hazard potential between forest types defined by their composition and structure. They found that potential fire behaviour is primarily driven by stand structure, rather than by cover type.

Marques et al. [13] presented an approach of the characterisation fire occurrence in Portugal, combining the use of geographic information systems (GIS) and generalised linear models (GLM). They emphasised the relationship between ecological and socioeconomic features on the proportion of area burned, recording also the number of fires and fire size for three 5-year periods, including the period 1990–1994. Descriptive statistics indicated variations in the distribution of fires over recent decades, with a significant increase in number of very large fires. Regression models underlined the impact of the forest cover type and the proximity to roads on the proportion of area burned.

For modelling wildfires, GLM [1, 14] have been often adopted, even as that is based on the Gaussian distribution by transforming the response [13]. Ferrari and Cribari-Neto [6] proposed a regression model where the response is beta distributed using a parameterisation of the beta law that is indexed by mean and dispersion parameters. Beta regression can be used for modelling the proportion of area burned that is restricted to the interval (0, 1). The regression parameters of the beta regression model are interpretable in terms of the mean of the response and, when the *logit* link is used, of an odds ratio, unlike the parameters of a linear regression that employs a transformed response [6].

This work proposes to model the proportion of burned area due to wildfires in Portugal, based on beta regression and under a Bayesian perspective (see e.g. [8, 17] for some Bayesian GLMs). The rest of the article is organised as follows. Section 2 succinctly describes the motivation of this work and the different modelling of wildfires. In Sect. 3 we present Bayesian beta regression for modelling the proportion of area burned, taking the use of Markov chain Monte Carlo (MCMC) methods for estimating quantities of interest. Some results of Bayesian beta regression related to the wildfire data analysis in the entire Portuguese mainland between 1990 and 1994, and concluding remarks are done respectively in Sects. 4 and 5, including the identification of the fire risk factors.

2 Motivation and Methods

In Portugal, burned area mapping, obtained by semi-automated classification of high-resolution remote sensing data from Instituto Superior de Agronomia (ISA)—Universidade de Lisboa, identified 35,198 fire perimeters with burned areas equal to or greater than 5 ha in the period 1975–2007 and the corresponding area burned is about 3.8×10^6 ha that is equivalent to nearly 40 % of the country area [13].

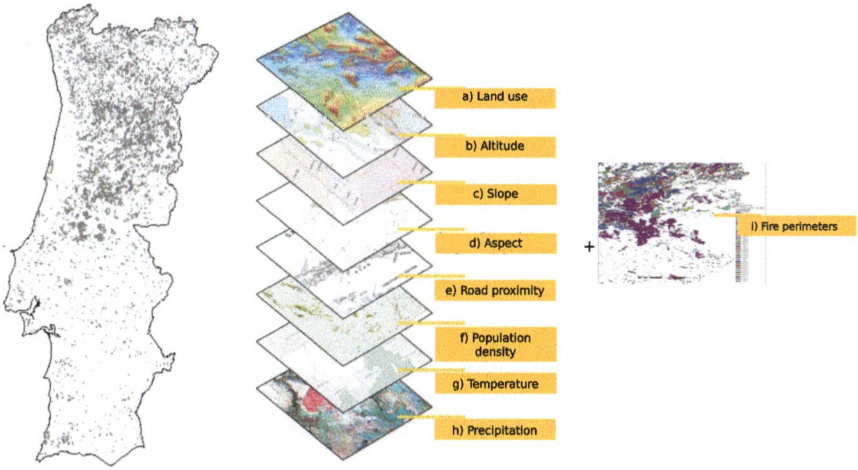

Fig. 1 Fire perimeters between 1990 and 1994 in Portugal (*left*), a zoom over a burned area is shown in the *right*, based on the classes of the covariates: (**a**) land use, (**b**) altitude, (**c**) slope, (**d**) slope orientation (aspect), (**e**) road proximity, (**f**) population density, (**g**) temperature, (**h**) precipitation, (**i**) layer indicating the fire perimeters

In the period 1990–1994, Marques et al. [13] pointed out that: (1) 5706 Portuguese wildfires were recorded and the total burned area extended over 442,745 ha, burning about 4.97 % of the country area, (2) the average area burned per wildfire was 77 ha, (3) 149 wildfires extended over 500 ha, accounting for 44 % of the burned area, (4) none extended over 10,000 ha. Figure 1 (left side) exemplifies the distribution (frequency) of these fires identifying high and critical fire zones that are specially located in the northern and central interior of Portugal.

In order to analyse variations in Portuguese wildfires in 1990–1994, the areas burned were included as map layers in the GIS database according to eight fire features (covariates), which were initially categorised, based on extensive preliminary data analysis and referred in the paper [13], into several classes: altitude (m), slope (%), slope orientation, population (hab/km^2), roads proximity (m), number of days with precipitation greater than 1 mm in the fire season (from May to October), number of days with maximum temperature higher than 25 °C in the fire season, and land cover (Table 1), including the observed proportion of the each land use classes in parentheses. Theses classes were also chosen based on some studies using the same data such as Moreira et al. [15] and Pereira et al. [19].

Figure 1 illustrates the fire perimeters used for constructing burned area data from related map layers. Notice that land cover map used in this study further included a map at the scale 1/25,000 (*Carta de Ocupação do Solo—COS'90*) produced by *Instituto Geográfico Português* using cartographic information from aerial photography mostly dated from 1990 [13], as well as that road proximity

Table 1 Description of the classes of the eight fire features used in the wildfire data

Roads proximity (m)	Population (hab/m^2)	Slope (%)	Altitude (m)	Slope orientation	Precipitation (number of days ≥1 mm)[a]	Temperature (number of days >25 °C)[a]
≥1000	<25	0–10	< 200	Flat	0–6	0–3
<1000	25–100	10–20	200–400	North	7–13	4–48
	>100	20–30	400–700	East	14–18	49–71
		>30	>700	South	19–22	72–92
				West	23–26	93–112
					≥27	≥113

Land cover: annual crop (5.3 %); eucalyptus (10.9 %); hardwoods (7 %); hardwoods and softwoods mixed with eucalyptus (HSME) (8.9 %); agro-forestry (5.8 %); permanent crop (3.5 %); shrubs (27.4 %); resinous or softwoods (RS) (18.8 %); softwoods mixed with eucalyptus (SME) (8.6 %); others (e.g. social areas) (3.8 %)
[a]Number of days in the fire season (from May to October)

included trails and was defined (1000 m distance) based on previous work e.g. Catry et al. [3]. Although continuous covariates as temperature and precipitation could be better explored in their natural form, we chose to categorize them because of a matter of simplicity and interpretation for the data collection and the model parameters, respectively.

For the modelling of wildfires, we record the observed proportion of burned area, denoted by r_i that is the burned area out of total area for the ith combination of levels for the covariates in study, $i = 1, \ldots, k$. We propose to model the proportion of burned area from these eight underlying covariates by assuming beta distribution for r_i, i.e.,

- Beta model: $r_i \sim Beta(\mu_i\phi, (1-\mu_i)\phi)$, with mean $E(r_i) = \mu_i$ and variance $Var(r_i) = \frac{\mu_i(1-\mu_i)}{\phi+1}$.

Alternative GLM can model the proportion r_i, for instance:

- Gaussian model: $logit(r_i) \equiv \log(r_i/(1-r_i)) \sim Normal(\mu_i, \sigma^2)$;
- Gamma model: $-\log(r_i) \sim Gamma(\nu, \nu/\mu_i)$, with $E(r_i) = \mu_i$ and $Var(r_i) = \frac{\mu_i^2}{\nu}$.

These two models and other GLM based on transformations of r_i, such as $\arcsin(\sqrt{r_i})$ and Box-Cox transformation, are discussed and developed in [1, 14]. Figure 2 displays histograms of the observed proportion of area burned without and with *logit* transformation in Portugal during the period 1990–1994, indicating that transforming response may not be the best way of wildfire modelling, what happens in the proportions close to one in Fig. 2 and notice that the beta model does not transform the response.

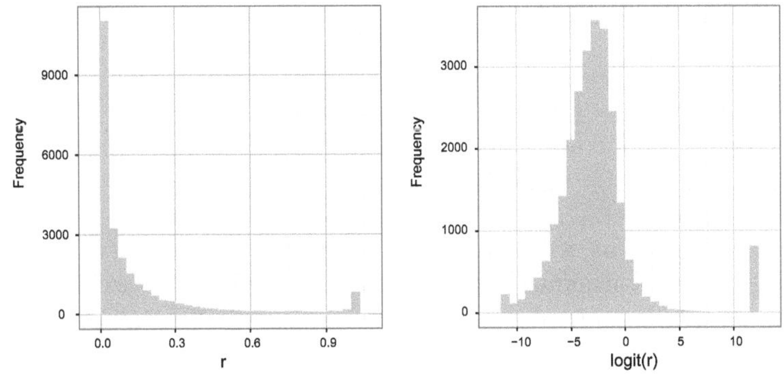

Fig. 2 Histograms of the observed proportion of area burned without (*right*) and with (*left*) *logit* transformation in Portugal during the period 1990–1994

3 Bayesian Beta Regression

Let $r_1, \ldots, r_k$ be random variables, where r_i follows a beta distribution with mean μ_i and unknown precision ϕ, whose probability density function is

$$f(r_i|\mu_i, \phi) = \frac{\Gamma(\phi)}{\Gamma(\mu_i\phi)\Gamma((1-\mu_i)\phi)} r_i^{\mu_i\phi-1}(1-r_i)^{(1-\mu_i)\phi-1}, \quad 0 < r_i < 1, \tag{1}$$

where $\Gamma(\cdot)$ is the gamma function, $0 < \mu_i < 1$ and $\phi > 0$, $i = 1, \ldots, k$. Notice that the parameterisation of the beta distribution (1) was suggested by Ferrari and Cribari-Neto [6] in order to model response variable that is continuous and restricted to the interval (0, 1) and is related to other variables through a regression structure.

The beta regression model is obtained from Eq. (1) by assuming that the mean μ_i can be written as

$$g(\mu_i) = \mathbf{z}_i^T \boldsymbol{\beta} \equiv \eta_i, \tag{2}$$

where $\boldsymbol{\beta} = (\beta_1, \ldots, \beta_p)^T$ is the regression parameter vector associated with the covariate vector $\mathbf{z}_i = (z_{i1}, \ldots, z_{ip})^T$ for the ith observation, $i = 1, \ldots, k$, and $g(\cdot)$ is a *logit* link function $g(\mu) = \log[\mu/(1-\mu)]$ (for other link functions, see [1, 6, 14]).

3.1 Posterior Distribution

For the likelihood, we can assume different sampling distributions for the proportion r_i, e.g., beta distribution defined in Eq. (1) or normal distribution for the transformed proportion, as referred in Sect. 2. Based on the former distribution with *logit* link function in Eq. (2), the likelihood function is given by

$$L(\boldsymbol{\beta},\phi|\mathbf{x}) = \prod_{i=1}^{k} \frac{\Gamma(\phi)}{\Gamma(\mu_i\phi)\,\Gamma((1-\mu_i)\phi)} r_i^{\mu_i\phi-1}(1-r_i)^{(1-\mu_i)\phi-1}, \tag{3}$$

where $\mathbf{x} = \{r_i; \mathbf{z}_i, i=1,\ldots,k\}$ is the data, and $\mu_i = e^{\mathbf{z}_i^T\boldsymbol{\beta}}/(1+e^{\mathbf{z}_i^T\boldsymbol{\beta}})$, $i=1,\ldots,k$.

In Bayesian analysis, we also consider information a priori that here consists of assuming independent normal distributions with zero mean and variances v_j^2 for the regression coefficients, $j = 1,\ldots,p$, and inverse gamma distribution with shape a and scale b parameters for the precision parameter ϕ (or gamma distribution with shape a and scale b parameters for the variance σ^2 related to normal regression). In fact, we assigned non-informative prior distribution, i.e., highly dispersed, but proper normal and inverse gamma prior distributions for the model parameters $\boldsymbol{\beta}$ and ϕ (or $1/\sigma^2$), respectively. In that case, one expects that inferential results on the model parameters are not too different from those ones under a frequentist approach.

Assuming a priori independence amongst the model parameters, we can construct the joint posterior density related to the beta regression model (2), which is denoted by

$$\pi(\boldsymbol{\beta},\phi|\mathbf{x}) \equiv \frac{L(\boldsymbol{\beta},\phi|\mathbf{x})\pi_1(\boldsymbol{\beta})\pi_2(\phi)}{\int\int L(\boldsymbol{\beta},\phi|\mathbf{x})\pi_1(\boldsymbol{\beta})\pi_2(\phi)\, d\boldsymbol{\beta} d\phi}, \tag{4}$$

where $\pi_1(\boldsymbol{\beta})$ and $\pi_2(\phi)$ are the normal and inverse gamma prior distributions of $\boldsymbol{\beta}$ and ϕ, respectively, being the distribution (4) proportional to

$$\prod_{i=1}^{k} \frac{\Gamma(\phi)\, r_i^{\mu_i\phi-1}(1-r_i)^{(1-\mu_i)\phi-1}}{\Gamma(\mu_i\phi)\,\Gamma((1-\mu_i)\phi)}\; e^{-\frac{1}{2}\sum_{j=1}^{p}(\beta_j^2/v_j^2)}\; \phi^{-(a+1)} e^{-b/\phi}. \tag{5}$$

Notice that the mean μ_i is a function of the linear predictor $\eta_i = \mathbf{z}^T\boldsymbol{\beta}$, $i=1,\ldots,k$.

The joint posterior distribution (5) is awkward to work with, since the marginal posterior distributions of some parameters are not easy to obtain explicitly. These posteriors can be evaluated using MCMC methods (see e.g. [8, 11, 17]). In particular Gibbs sampling that works by iteratively drawing samples for each parameter from the corresponding full conditional distribution, which is friendly implemented in software WinBUGS [12]. Other MCMC method, proposed by Hoffman and Gelman [11], is the No-U-Turn Sampler (NUTS) that is a variant of the Hamiltonian Monte Carlo (HMC), also known as hybrid Monte Carlo. Neal [16] presented HMC method in order to avoid a long time to converge to the posterior distribution as e.g. in Gibbs sampling by using a clever auxiliary variable scheme that transforms the problem of sampling from a posterior distribution into the problem of simulating Hamiltonian dynamics.

3.2 Evaluating and Comparing Models

An important issue in Bayesian data analysis is to choose among postulated sub-models of a statistical model, e.g. the beta regression model (2). Some summary measures of model comparison, such as the posterior mean of Deviance $D(\boldsymbol{\theta})$, where $\boldsymbol{\theta}$ is the model parameter vector, are easily evaluated with MCMC methods. Other two measures of predictive accuracy are Deviance Information Criterion (DIC) and Watanabe-Akaike Information Criterion (WAIC) (see [8, 21]). DIC is here defined as

$$DIC = D(\overline{\boldsymbol{\theta}}) + Var(D(\boldsymbol{\theta})), \tag{6}$$

where $\overline{\boldsymbol{\theta}}$ and $Var(D(\boldsymbol{\theta}))$ denote the posterior mean of model parameter $\boldsymbol{\theta}$ and the posterior variance of the deviance, respectively, whereas WAIC is defined by

$$WAIC = D(\overline{\boldsymbol{\theta}}) + 2\sum_{i=1}^{k} Var(D_i(\boldsymbol{\theta})), \tag{7}$$

where $Var(D_i(\boldsymbol{\theta}))$ denotes the posterior variance of the ith term of the deviance. DIC and WAIC handle Bayesian models of any degree of complexity, and models with smaller (6) and (7) should be preferred to models with larger ones.

4 Wildfire Data Analysis

For the wildfire data described in Sect. 2, we fitted several regression models based on the response, proportion of the burned area in Portugal during the period 1990–1994, as in Marques et al. [13], but now focusing on the beta regression instead of normal regression, and under a Bayesian perspective. One of the eight covariates presented in Table 1, i.e. slope orientation, was removed from the analysis by not showing any difference among its categories.

Let M_1 and M_3 denote regression beta model (1) with eight covariates showed in Table 1, apart from the covariate slope orientation, whereas M_2 and M_4 represent the corresponding normal models. Table 2 lists these sub-models of the beta and normal model with only main covariate effects, M_1 and M_2, and also with interactions between two covariates. Based on the comparison model measures DIC (6) and WAIC (6), fitted models with interactions had better evaluation than models with only main effects for both beta and normal models. These evaluating values were calculated taking into account the same response $r_i/(1-r_i)$ (so-called *odd*), which generates a sampling log-normal and second-kind beta distributions for normal and beta distributions, respectively. So, normal regression had better fitting than beta

Table 2 Model comparison measures of four fitted regression models for wildfire data

Regression models	WAIC	DIC
M_1: beta regression with only main effects	−10,161.37	−10,165.68
M_2: normal regression with only main effects	−14,593.76	−14,605.28
M_3: beta regression with interactions between two covariates	−10,723.74	−10,728.94
M_4: normal regression with interactions between two covariates	−15,263.72	−15,285.41

regression, and that can namely be associated with the large number of observations ($k = 25{,}388$). However, we chose to select model M_3 in order to illustrate the beta regression model that has not been employed in the analysis of wildfire burned areas, even as it can be considered the natural choice.

For all models showed in Table 2, we assumed prior normal distribution with mean zero and variance 10^4 for the regression parameters and prior inverse gamma and gamma distributions with shape parameter 1 and scale parameter 0.01 for the precision parameter ϕ (beta regression) and the variance σ^2 (normal regression), respectively. That is, highly dispersed, but proper prior distributions. MCMC samples of size 5000 were obtained for all models, after 2500 iterations of burn-in, implemented in software Stan [20]. A study of convergence of the samples was carried out with no worrying features.

For selected beta model M_3, Table 3 displays the model parameter estimates: posterior mean, standard deviation (SD) and 95 % highest posterior density (HPD) credible intervals (CI) for the model parameters. Note that related to the proportion of area burned in Portugal during the period 1990–1994:

1. There is no significant effect of annual and permanent crops in contrast to the other categories of land cover;
2. The land covers with larger likelihood to have wildfires are (in increasing order) agro-forestry, hardwoods, hardwoods and softwoods mixed with eucalyptus (HSME), resinous or softwoods (RS), eucalyptus, softwoods mixed with eucalyptus (SME), and shrubs (the most likelihood).
3. The proportion increases for larger categories of slope and altitude, whereas population and roads proximity display a decreasing effect in the proportion.
4. Because temperature and precipitation had an unexpected negative effect in the proportion, we decided to look for a potential interaction effect between the two covariates. We found significant interaction between temperature and precipitation in model M_3, even as that is not clear for the smaller categories of both covariates.
5. As large the categories of temperature and precipitation as large is the odd of burned area (interaction effect). Notice that the largest category did not have observation enough for confirming that.
6. The estimates in Table 3 also indicates that there is some dispersion in the proportion of area burned (see 95 % HPD credible interval of ϕ).

Table 3 Estimates of the regression parameters and dispersion parameter (ϕ) for model M_3

			95 % CI					95 % CI	
Parameter	Mean	SD	Lower	Upper	Parameter	Mean	SD	Lower	Upper
Roads proximity (β_{17})					Temperature ($\beta_{25}, \ldots, \beta_{29}$)				
<1000	−0.05	0.01	−0.08	−0.02	4–48	−2.63	0.12	−2.87	−2.39
Population (β_{18}, β_{19})					49–71	−2.63	0.13	−2.87	−2.38
25–100	−0.09	0.02	−0.12	−0.05	72–92	−2.67	0.13	−2.92	−2.43
≥100	−0.15	0.02	−0.19	−0.11	93–112	−2.94	0.16	−3.24	−2.63
Slope ($\beta_{14}, \beta_{15}, \beta_{16}$)					≥113	−2.78	0.20	−3.18	−2.41
10–20	0.21	0.02	0.18	0.24	Land cover ($\beta_2, \ldots, \beta_{10}$)				
20–30	0.66	0.02	0.61	0.70	Annual crop	−0.04	0.04	−0.11	0.03
≥30	2.16	0.06	2.05	2.27	Eucalyptus	0.49	0.04	0.42	0.56
Altitude ($\beta_{11}, \beta_{12}, \beta_{13}$)					Hardwoods	0.18	0.04	0.11	0.25
200–400	0.12	0.02	0.08	0.16	HSME	0.32	0.04	0.24	0.39
400–700	0.15	0.02	0.11	0.19	Agro-forestry	0.08	0.04	0.00	0.15
≥700	0.40	0.02	0.35	0.44	Permanent crop	−0.05	0.04	−0.13	0.02
Precipitation ($\beta_{20}, \ldots, \beta_{24}$)					Shrubs	0.61	0.03	0.54	0.68
7–13	−0.58	0.13	−0.83	−0.30	RS	0.48	0.04	0.42	0.55
14–18	0.08	0.18	−0.29	0.43	SME	0.59	0.04	0.51	0.66
19–22	0.43	0.19	0.06	0.80					
23–26	−0.62	0.17	−0.96	−0.29	Intercept (β_1)	0.81	0.13	0.57	1.05
≥27	−2.72	0.96	−4.62	−0.84	ϕ	1.36	0.01	1.34	1.39
(β_{30}) Temperature (4–48) × Precipitation (7–13)						0.70	0.14	0.40	0.95
(β_{31}) Temperature (4–48) × Precipitation (14–18)						0.03	0.19	−0.33	0.40
(β_{32}) Temperature (4–48) × Precipitation (19–22)						−0.11	0.19	−0.46	0.29
(β_{33}) Temperature (4–48) × Precipitation (23–26)						0.62	0.18	0.28	0.98
(β_{34}) Temperature (4–48) × Precipitation (≥27)						2.70	0.96	0.79	4.57
(β_{35}) Temperature (49–71) × Precipitation (7–13)						0.54	0.14	0.27	0.82
(β_{36}) Temperature (49–71) × Precipitation (14–18)						−0.15	0.19	−0.53	0.21
(β_{37}) Temperature (49–71) × Precipitation (19–22)						−0.35	0.19	−0.72	0.04
(β_{38}) Temperature (49–71) × Precipitation (23–26)						0.73	0.18	0.39	1.09
(β_{39}) Temperature (49–71) × Precipitation (≥27)						2.96	0.96	1.09	4.88
(β_{40}) Temperature (72–92) × Precipitation (7–13)						0.47	0.15	0.19	0.75
(β_{41}) Temperature (72–92) × Precipitation (14–18)						−0.04	0.19	−0.42	0.34
(β_{42}) Temperature (72–92) × Precipitation (19–22)						−0.35	0.19	−0.72	0.04
(β_{43}) Temperature (72–92) × Precipitation (23–26)						0.85	0.18	0.49	1.21
(β_{44}) Temperature (72–92) × Precipitation (≥27)						3.41	0.96	1.50	5.30
(β_{45}) Temperature (93–112) × Precipitation (7–13)						0.60	0.18	0.25	0.96
(β_{46}) Temperature (93–112) × Precipitation (14–18)						0.70	0.23	0.27	1.17
(β_{47}) Temperature (93–112) × Precipitation (19–22)						1.38	0.26	0.88	1.89
(β_{48}) Temperature (93–112) × Precipitation (23–26)						3.26	0.23	2.81	3.73
(β_{49}) Temperature (≥113) × Precipitation (7–13)						0.57	0.22	0.10	0.98

5 Concluding Remarks

This analysis of wildfire data in Portugal allow us to figure out the influence of the observed combinations of eight fire risk features on the proportion of burned area. Our results of beta regression are essentially consistent with those ones of normal regression, presented in Marques et al. [13], whose analysis did not include the explanatory variables: slope orientation, precipitation, temperature and the interaction between the last two ones. In fact, our model and conclusions bring improvements on the results reported by them based on a similar data set. So, we also identified changes in the proportion of burned area depending on topographical variables and vegetation type. Pereira et al. [18] pointed out that some variability of the burned area in Portugal is partly due both to the amount of precipitation in the fire season and in the preceding late spring season and to the occurrence of atmospheric circulation patterns that induce extremely hot and dry spells.

In addition, our intuition about interaction between precipitation and temperature was corrected, and we also believe that some latent variables can explain some unobserved heterogeneity in these wildfire data, e.g. spatial extra-variation across fire regions. For instance, Amaral-Turkman et al. [2] proposed a spatio-temporal model to analyse jointly the probability of ignition and fire sizes in Australia and New Zealand. Further research is being developed for capturing the spatio-temporal effects on the proportion of burned area, more proper sampling distributions and link functions. Notice that 4 % of observed burned areas were 0 or 1 being replaced by 10^{-10} and 1×10^{-10}, respectively, for simplicity, We intend to include that issue in future work, as well as to do a full sensitivity analysis of our prior options (see e.g. [8]) and some simulation to clarify the impact of a big data as our wildfires in the results. For the our choice of beta regression instead of normal regression, we also believe that a comprehensive simulation study must be done in order to verify the second choice, as well as the residual analysis for understanding that unexplained situation of the observed proportions close to one (see e.g. Espinheira et al. [4]).

Acknowledgements This paper was partially supported by the project PEst-OE/MAT/UI0006/2014 of the Fundação para a Ciência e a Tecnologia (FCT). We also thank FCT for funding the Post-Doctoral fellowship of Susete Marques "SFRH/BPD/96806/2013". In addition the authors would like to thank the two referees for the valuable and comprehensive comments that have improved the final version of the paper.

References

1. Amaral-Turkman, M.A., Silva, G.L.: Modelos Lineares Generalizados - da teoria à prática. SPE Edition, Lisbon (2000)
2. Amaral-Turkman, M.A., Turkman, K.F., Le Page, Y., Pereira, J.M.: Hierarchical space-time models for fire ignition and percentage of land burned by wildfires. Environ. Ecol. Stat. **18**, 601–617 (2011)

3. Catry, F., Rego, F., Bação, F., Moreira, F.: Modelling and mapping wildfire ignition risk in Portugal. Int. J. Wildland Fire **18**, 921–931 (2009)
4. Espinheira, P.L., Ferrari, S.L.P., Cribari-Neto, F.: On beta regression residuals. J. Appl. Stat. **35**, 407–419 (2008)
5. Fernandes, P., Luz, A., Loureiro, C., Ferreira-Godinho, P., Botelho, H.: Fuel modelling and fire hazard assessment based on data from Portuguese National Forest Inventory. For. Ecol. Manage. **234S**, S229 (2006)
6. Ferrari, S.L.P., Cribari-Neto, F.: Beta regression for modeling rates and proportions. J. Appl. Stat. **31**, 799–815 (2004)
7. Flannigan, M.D., Amiro, B.D., Logan, K.A., Stocks, B.J., Wotton, B.M.: Forest fires and climate change in the 21st century. Mitig. Adapt. Strat. Glob. Chang. **11**, 847–859 (2005)
8. Gelman, A., Carlin, J.B., Stern, H.S., Dunson, D.B., Vehtari, A., Rubin, D.B.: Bayesian Data Analysis, 3rd edn. CRC Press, London (2014)
9. Gomes, J.F.P.: Forest fires in Portugal: how they happen and why they happen. Int. J. Environ. Stud. **63**, 109–119 (2006)
10. González, J.R., Pukkala, T.: Characterization of forest fires in Catalonia (Northeast Spain). Eur. J. For. Res. **126**, 421–429 (2007)
11. Hoffman, M.D., Gelman, A.: The No-U-Turn sampler: adaptively setting path lengths in Hamiltonian Monte Carlo. arXiv **1111**, 4246 (2011). http://arxiv.org/abs/1111.4246
12. Lunn, D.J., Thomas, A., Best, N., Spiegelhalter, D.: WinBUGS – a Bayesian modelling framework: concepts, structure, and extensibility. Stat. Comput. **10**, 325–337 (2000)
13. Marques, S., Borges, J.G., Garcia-Gonzalo, J., Moreira, F., Carreiras, J.M.B., Oliveira, M.M., Cantarinha, A., Botequim, B., Pereira, J.M.C.: Characterization of wildfires in Portugal. Eur. J. For. Res. **130**, 775–784 (2011)
14. McCullagh, P., Nelder, J.A.: Generalized Linear Models, 2nd edn. CRC Press, Boca Raton (1989)
15. Moreira, F., Rego, F.C., Godinho-Ferreira, P.: Temporal (1958–1995) pattern of change in a cultural landscape of northwestern Portugal: implications for fire occurrence. Landsc. Ecol. **16** 557–567 (2001)
16. Neal, R.: Handbook of Markov Chain Monte Carlo, Chap. 5: MCMC Using Hamiltonian Dynamics. CRC Press, Chichester (2011)
17. Paulino, C.D., Amaral-Turkman, M.A., Murteira, B.: Estatística Bayesiana. Fundação Calouste Gulbenkian, Lisboa (2003)
18. Pereira, M.G., Trigo, R.M., da Camara, C.C., Pereira, J.M.C., Leite, S.M.: Synoptic patterns associated with large summer forest fires in Portugal. Agr. Forest. Meteorol. **129**, 11–25 (2005)
19. Pereira, J.M.C., Carreiras, J.M.B., Silva, J.M.N., Vasconcelos, M.J.: Alguns conceitos básicos sobre fogos rurais em Portugal. In: Pereira, J.S., Pereira, J.M.C., Rego, F.C., Silva, J.M.N., Silva, T.P. (eds.) Incêndios Florestais em Portugal, pp. 133–161. ISA Press, Lisboa (2006)
20. Stan Development Team.: Stan: A C++ Library for probability and sampling, Version 2.2. (2014). http://mc-stan.org/
21. Watanabe, S.: Asymptotic equivalence of Bayes cross validation and widely applicable information criterion in singular learning theory. J. Mach. Learn. Res. **11**, 3571–3591 (2010)

Minimum H-Decompositions of Graphs and Its Ramsey Version: A Survey

Teresa Sousa

Abstract The subject of H-decompositions of graphs was first introduced by Erdős, Goodman and Pósa in 1966. Given graphs G and H, an H-*decomposition* of G is a partition of the edge set of G, such that, each part is either a single edge or forms a graph isomorphic to H. Let $\phi(n, H)$ be the smallest number ϕ, such that, any graph G with n vertices admits an H-decomposition with at most ϕ parts. The exact computation of $\phi(n, H)$ for an arbitrary H is still an open problem. In this paper we will survey recent results about H-decompositions of graphs and we will also introduce its Ramsey or coloured version together with recent results on this problem.

1 Introduction

All graphs in this paper are finite, undirected and simple. For standard notation and terminology the reader is referred to [3].

Given two graphs G and H, an H-*decomposition* of G is a partition of the edge set of G such that each part is either a single edge or forms an H-subgraph, i.e., a graph isomorphic to H. We allow partitions only, that is, every edge of G appears in precisely one part. Let $\phi(G, H)$ be the smallest possible number of parts in an H-decomposition of G.

An H-*packing* of a graph G is a set of pairwise edge disjoint H-subgraphs of G. The H-*packing number* of G, denoted by $p_H(G)$, is the maximum cardinality of an H-packing of G. It is easy to see that, for non-empty H, we have

$$\phi(G, H) = e(G) - p_H(G)(e(H) - 1),$$

where $e(G)$ denotes the number of edges in G. Dor and Tarsi [4] showed that if H has a component with at least three edges then the problem of checking whether an input graph G is perfectly decomposable into H-subgraphs is NP-complete. Thus, it

T. Sousa (✉)
Departamento de Matemática and Centro de Matemática e Aplicações, Faculdade de Ciências e Tecnologia, Universidade Nova de Lisboa, Campus de Caparica, 2829-516 Caparica, Portugal
e-mail: tmjs@fct.unl.pt

J.-P. Bourguignon et al. (eds.), *Dynamics, Games and Science*, CIM Series in Mathematical Sciences 1, DOI 10.1007/978-3-319-16118-1_39

is NP-hard to compute the function $\phi(G,H)$ for such H. Nonetheless, many results were proved about the extremal function

$$\phi(n,H) = \max\{\phi(G,H) \mid v(G) = n\},$$

which is the smallest number such that any graph G of order n admits an H-decomposition with at most $\phi(n,H)$ parts. Here $v(G)$ denotes the number of vertices in the graph G.

This function was first studied, in 1966, by Erdős, Goodman and Pósa [5], who were motivated by the problem of representing graphs by set intersections. They proved that $\phi(n,K_3) = t_2(n)$, where K_s denotes the complete graph (clique) of order s, and $t_{r-1}(n)$ denotes the number of edges in the Turán graph of order n, $T_{r-1}(n)$, which is the unique complete $(r-1)$-partite graph on n vertices where every partition class has either $\lfloor \frac{n}{r-1} \rfloor$ or $\lceil \frac{n}{r-1} \rceil$ vertices. A decade later, Bollobás [2] proved that $\phi(n,K_r) = t_{r-1}(n)$, for all $n \geqslant r \geqslant 3$.

General graphs were only considered recently by Pikhurko and Sousa [18], who studied $\phi(n,H)$ for arbitrary graphs H. They have determined the asymptotic value of $\phi(n,H)$ for any fixed graph H as n tends to infinity (see Theorem 3). In the special case when H is a bipartite graph, they were able to determine $\phi(n,H)$ with a constant additive error term. Furthermore, their proof provides an algorithm returning the exact value of $\phi(n,H)$ with running time polynomial in $\log n$ (see Theorem 1). Since then, a few papers have been published about this problem. In Sect. 2 we will present recent results about H-decompositions of graphs. In Sect. 3 we will introduce the Ramsey or coloured version of this problem and state some new results.

2 H-Decompositions of Graphs

Let K_s denote the complete graph (or clique) on s vertices and let $T_{r-1}(n)$ denote the Turán graph of order n. $T_{r-1}(n)$ is the unique complete $(r-1)$-partite graph on n vertices where every partition class has either $\lfloor \frac{n}{r-1} \rfloor$ or $\lceil \frac{n}{r-1} \rceil$ vertices. *Turán's Theorem* [27] states that $T_{r-1}(n)$ is the unique graph on n vertices that has the maximum number of edges and contains no complete subgraph of order r.

Let $\mathrm{ex}(n,H)$ denote the maximum number of edges in a graph on n vertices not containing H as a subgraph, that is,

$$\mathrm{ex}(n,H) = \max\{e(G) \mid v(G) = n, H \not\subseteq G\}.$$

The function $\mathrm{ex}(n,H)$ is usually called the *Turán function* for H. Recall that for a complete graph on r vertices we have $\mathrm{ex}(n,K_r) = t_{r-1}(n)$.

In 1966, Erdős, Goodman and Pósa [5] proved that $\phi(n,K_3) = t_2(n)$. A decade later, Bollobás [2] proved that $\phi(n,K_r) = t_{r-1}(n)$, for all $n \geqslant r \geqslant 3$. General graphs have been considered recently by Pikhurko and Sousa and since then a few papers have been published about this problem. In this section we will bring together all results about H-decompositions of graphs. We will start with the case when H is a bipartite graph and then consider the case when H is non-bipartite.

Let $K_{m,n}$ denote the complete bipartite graph with parts of size m and n. For a bipartite graph H it is easy to determine the asymptotic value of the function $\phi(n, H)$ (see Sousa [22]):

Lemma 1 ([22]) *For any non-empty graph H with m edges and any integer n, we have*

$$\phi(n,H) \leqslant \frac{1}{m}\binom{n}{2} + \frac{m-1}{m}\,\mathrm{ex}(n,H). \tag{1}$$

In particular, if H is a fixed bipartite graph with m edges and $n \to \infty$, then

$$\phi(n,H) = \left(\frac{1}{m} + o(1)\right)\binom{n}{2}. \tag{2}$$

Proof To prove (1) remove greedily one by one the edge-sets of H-subgraphs of a given graph G and then remove the remaining edges. The bound (1) follows as at most $\mathrm{ex}(n, H)$ parts are single edges.

The upper bound in (2) follows from (1) and the equality

$$\mathrm{ex}(n, K_{t,t}) = O(n^{2-1/t}), \tag{3}$$

of Kővari, Sös and Turán [12]. The lower bound in (2) follows from $\phi(n,H) \geqslant \phi(K_n, H) \geqslant \frac{1}{m}\binom{n}{2}$. □

Pikhurko and Sousa [18] managed to determine $\phi(n,H)$ for any fixed bipartite graph H with an $O(1)$ additive error (see Theorem 1 below). Furthermore, their proof gives a procedure for computing the exact value of $\phi(n,H)$ for all large n, that runs in polylogarithmic time.

For a non-empty graph H, let $\gcd(H)$ denote the greatest common divisor of the degrees of H. For example, $\gcd(K_{6,4}) = 2$ while for any tree T with at least 2 vertices we have $\gcd(T) = 1$. We have the following result.

Theorem 1 ([18]) *Let H be a bipartite graph with m edges and let $d = \gcd(H)$. Then there is $n_0 = n_0(H)$ such that for all $n \geqslant n_0$ the following statements hold.*

If $d = 1$, then if $\binom{n}{2} \equiv m - 1 \pmod{m}$,

$$\phi(n,H) = \phi(K_n,H) = \left\lfloor \frac{n(n-1)}{2m} \right\rfloor + m - 1, \tag{4}$$

otherwise,

$$\phi(n,H) = \phi(K_n^*,H) = \left\lfloor \frac{n(n-1)}{2m} \right\rfloor + m - 2 \tag{5}$$

where K_n^* denotes any graph obtained from K_n after deleting at most $m-1$ edges in order to have $e(K_n^*) \equiv m-1 \pmod m$. Furthermore, the only graph attaining $\phi(n,H)$ is either K_n or K_n^*.

If $d \geqslant 2$, then

$$\phi(n,H) = \frac{nd}{2m}\left(\left\lfloor \frac{n}{d} \right\rfloor - 1\right) + \frac{1}{2}n(d-1) + O(1). \tag{6}$$

Moreover, there is a procedure with running time polynomial in $\log n$ which determines $\phi(n,H)$ and describes a family $\mathcal{D}$ of n-sequences such that a graph G of order n satisfies $\phi(G,H) = \phi(n,H)$ if and only if the degree sequence of G belongs to $\mathcal{D}$. (It will be the case that $|\mathcal{D}| = O(1)$ and each sequence in $\mathcal{D}$ has $n - O(1)$ equal entries, so $\mathcal{D}$ can be described using $O(\log n)$ bits.)

Later, Sousa [24] determined the exact value of $\phi(n, C_4)$ for n sufficiently large, where C_4 denotes the cycle on 4 vertices.

Theorem 2 ([24]) *There is $n_0 = n_0(C_4)$ such that for all $n \geqslant n_0$ the following statements hold.*

(i) If n is even then $\phi(n, C_4) = \frac{n^2}{8} + \frac{n}{4} + 1$.
(ii) If $n \equiv 1 \pmod 8$ then $\phi(n, C_4) = \frac{n^2}{8} + \frac{n}{8} + \frac{14}{8}$.
(iii) If $n \equiv 3 \pmod 8$ then $\phi(n, C_4) = \frac{n^2}{8} + \frac{n}{8} + \frac{3}{2}$.
(iv) If $n \equiv 5 \pmod 8$ then $\phi(n, C_4) = \frac{n^2}{8} + \frac{n}{8} + \frac{10}{8}$.
(v) If $n \equiv 7 \pmod 8$ then $\phi(n, C_4) = \frac{n^2}{8} + \frac{n}{8} + 2$.

We will now consider the case when H is not a bipartite graph. Recall that the *chromatic number* of a graph G, denoted by $\chi(G)$, is the minimum number of colours needed to colour the vertices of G, such that, no edge joins two vertices with the same colour. Observe that $\chi(G) = 2$ if and only if G is a bipartite graph and $\chi(G) \geq 3$ otherwise. Pikhurko and Sousa [18] proved the following result about H-decompositions of graphs for a fixed non-bipartite graph H.

Theorem 3 ([18]) *Let H be any fixed graph of chromatic number $r \geqslant 3$. Then,*

$$\phi(n,H) = t_{r-1}(n) + o(n^2).$$

The lower bound follows from the trivial inequalities $\phi(n,H) \geq \text{ex}(n,H) \geq t_{r-1}(n)$. To prove the upper bound one needs more sophisticated tools. In outline, Pikhurko and Sousa proof is the following. First, they apply Szemerédi's Regularity Lemma [26] to the graph G that they want to decompose. The regularity partition of G gives a weighted graph K with large but bounded number k of vertices. By generalizing the method of Bollobás [2] they were able to decompose K into weighted copies of K_r and K_2 with aggregate weight at most $t_{r-1}(k) + o(k^2)$. Then, the graph G is splitted into subgraphs that correspond to the cliques from the above decomposition of K. Finally, each of the obtained r-partite subgraphs of G can be almost perfectly decomposed into copies of H by using a theorem of Pippenger and Spencer [19].

Pikhurko and Sousa [18] also made the following conjecture.

Conjecture 1 ([18]) For any graph H of chromatic number $r \geqslant 3$, there exists $n_0 = n_0(H)$ such that $\phi(n,H) = \mathrm{ex}(n,H)$ for all $n \geqslant n_0$.

The exact value of the function $\phi(n,H)$ is far from being known, however, this conjecture has been verified for some special graphs. Sousa [22, 23, 25] verified the conjecture for the cycles of length 5 and 7 and for clique-extensions. Her results are the following.

Theorem 4 ([22, 25]) *Let C_t denote the cycle on t vertices. Then,*

(i) $\phi(n, C_5) = t_2(n) = \lfloor n^2/4 \rfloor$, *for all* $n \geq 6$;
(ii) $\phi(n, C_7) = t_2(n) = \lfloor n^2/4 \rfloor$, *for all* $n \geq 10$.

For $r \geq 3$, *a clique-extension of order* $r+1$ is a connected graph that consists of a K_r plus another vertex, say x, adjacent to at most $r-1$ vertices of K_r. For $i = 1, \cdots, r-1$, the $H_{r,i}$ be the clique-extension of order $r+1$ that has $\deg x = i$.

Theorem 5 ([23]) *For all $n \geq 4$ and $i = 1, 2$ we have*

$$\phi(n, H_{3,i}) = t_2(n) = \lfloor n^2/4 \rfloor.$$

Theorem 6 ([23]) *Let $r \geq 4$ and let H be any clique-extension of order $r+1$. For all $n \geq r+1$ we have*

$$\phi(n, H) = t_{r-1}(n).$$

A graph H is *edge-critical* if there exists an edge $e \in E(H)$ such that $\chi(H) > \chi(H-e)$. Özkahya and Person [17] verified Conjecture 1 for all edge-critical graphs with chromatic number $r \geq 3$, extending the results obtained previously by Sousa for the cycles of length 5 and 7 and for clique-extensions. Their result is the following.

Theorem 7 ([17]) *For any edge-critical graph H with chromatic number $r \geq 3$, there exists $n_0 = n_0(H)$ such that $\phi(n,H) = \mathrm{ex}(n,H)$, for all $n \geq n_0$. Moreover, the only graph attaining $\mathrm{ex}(n,H)$ is the Turán graph $T_{r-1}(n)$.*

Recently, Allen, Böttcher, and Person [1] improved the error term obtained by Pikhurko and Sousa in Theorem 3, this result is also an extension of the result of Özkahya and Person in Theorem 7. Before stating the result we need the following definition.

Given a graph H with $\chi(H) = r$, the *decomposition family $\mathscr{F}_H$ of H* is the set of bipartite graphs which are obtained from H by deleting $r-2$ colour classes in some r-colouring of H. Observe that $\mathscr{F}_H$ may contain graphs which are disconnected, or even have isolated vertices. Let $\mathscr{F}_H^*$ be a minimal subfamily of $\mathscr{F}_H$, such that, for any $F \in \mathscr{F}_H$, there exists $F' \in \mathscr{F}_H^*$ with $F' \subseteq F$. We define

$$\mathrm{biex}(n, H) := \mathrm{ex}(n, \mathscr{F}_H) = \mathrm{ex}(n, \mathscr{F}_H^*).$$

Allen, Böttcher, and Person main result states that the $o(n^2)$ error term in Theorem 3 can be replaced by $O(\mathrm{biex}(n, H))$, which is $O(n^{2-\gamma})$ for some $\gamma > 0$ by the result of Kövari, Sós and Turán [12]. Furthermore, they also proved that this error term has the correct order of magnitude.

Theorem 8 ([1]) *For every integer $r \geq 3$ and every graph H with $\chi(H) = r$ there are constants $c = c(H) > 0$ and $C = C(H)$ and an integer n_0 such that for all $n \geq n_0$ we have*

$$\mathrm{ex}(n, K_r) + c \cdot \mathrm{biex}(n, H) \leq \phi(n, H) \leq \mathrm{ex}(n, K_r) + C \cdot \mathrm{biex}(n, H).$$

Observe that for every edge-critical graph H and every n we have $\mathrm{biex}(n, H) = 0$, therefore, Allen, Böttcher, and Person result is indeed an extension of the result of Özkahya and Person.

Finally, Liu and Sousa [14] verified Conjecture 1 for the k-fan graph. The *k-fan graph*, denoted by F_k, is the graph on $2k + 1$ vertices consisting of k triangles which intersect in exactly one common vertex, called the *centre* of F_k. Observe that $\chi(F_k) = 3$ and for $k \geq 2$ the graph F_k is not edge-critical.

In 1995, Erdős, Füredi, Gould, and Gunderson [6] have determined the value of the function $\mathrm{ex}(n, F_k)$ as well as the F_k-extremal graphs for every fixed k and n sufficiently large. They have proved the following result.

Theorem 9 ([6]) *Let $\mathscr{F}_{n,k}$ be the following family of graphs.*

- *If k is odd and $n \geqslant 4k - 1$, then a member of $\mathscr{F}_{n,k}$ is a Turán graph $T_2(n)$ with two vertex-disjoint copies of K_k added into one class.*
- *If k is even and $n \geqslant 4k - 3$, then a member of $\mathscr{F}_{n,k}$ is a $T_2(n)$ with a graph having $2k - 1$ vertices, $k^2 - \frac{3}{2}k$ edges and maximum degree $k - 1$ added into one class.*

For $k \geqslant 1$ and $n \geqslant 50k^2$, we have

$$\mathrm{ex}(n, F_k) = \left\lfloor \frac{n^2}{4} \right\rfloor + g(k) = \begin{cases} \left\lfloor \frac{n^2}{4} \right\rfloor + k^2 - k & \text{if } k \text{ is odd}, \\ \left\lfloor \frac{n^2}{4} \right\rfloor + k^2 - \frac{3}{2}k & \text{if } k \text{ is even}. \end{cases} \tag{7}$$

Moreover, the only F_k-free graphs with $\mathrm{ex}(n, F_k)$ edges are the members of $\mathscr{F}_{n,k}$.

Liu and Sousa [14] proved the following result.

Theorem 10 ([14]) *For $k \geqslant 1$, there exists $n_0 = n_0(k)$, such that, for all $n \geqslant n_0$ we have*

$$\phi(n, F_k) = \mathrm{ex}(n, F_k).$$

Moreover, the only graphs attaining $\mathrm{ex}(n, F_k)$ are the members of $\mathscr{F}_{n,k}$.

3 Monochromatic Decompositions of Graphs

Motivated by the recent work published about H-decompositions of graphs, a natural problem to consider is the Ramsey or coloured version of this problem. More precisely, let G be a graph on n vertices whose edges are coloured with k colours, for some $k \geq 2$ and let $\mathscr{H} = (H_1, \ldots, H_k)$ be a k-tuple of fixed graphs, where repetition is allowed. A *monochromatic $\mathscr{H}$-decomposition* of G is a partition of its edge set, such that, each part is either a single edge or forms a monochromatic copy of H_i in colour i, for some $1 \leqslant i \leqslant k$. Let $\phi_k(G, \mathscr{H})$ be the smallest number, such that, for any k-edge-colouring of G there is a monochromatic $\mathscr{H}$-decomposition of G with at most $\phi_k(G, \mathscr{H})$ elements. The aim is to study the function

$$\phi_k(n, \mathscr{H}) = \max\{\phi_k(G, \mathscr{H}) \mid v(G) = n\}, \tag{8}$$

which is the smallest number such that, any k-edge-coloured graph of order n admits a monochromatic $\mathscr{H}$-decomposition with at most $\phi_k(n, \mathscr{H})$ elements. In the case when $H_i \cong H$ for every $1 \leqslant i \leqslant k$, we simply write $\phi_k(G, H) = \phi_k(G, \mathscr{H})$ and $\phi_k(n, H) = \phi_k(n, \mathscr{H})$.

The first open instance of this problem is the case when we want to decompose our graph G into monochromatic copies of a fixed K_r, with $r \geq 3$. The function $\phi_k(n, K_r)$, for $k \geqslant 2$ and $r \geqslant 3$, has been studied by Liu and Sousa [15], who obtained results involving the Ramsey numbers and the Turán numbers.

Recall that for $r \geq 3$ and $k \geqslant 2$, the *Ramsey number for K_r*, denoted by $R_k(r)$, is the smallest value of s for which every k-edge-colouring of K_s contains a monochromatic K_r. The Ramsey numbers are notoriously difficult to calculate, even though, it is known that their values are finite for all $r \geq 3$ and $k \geqslant 2$ [21]. In fact, for the Ramsey numbers $R_k(r)$, only three of them are currently known. In 1955, Greenwood and Gleason [7] were the first to determine $R_2(3) = 6$, $R_3(3) = 17$ and $R_2(4) = 18$.

We will also consider 'blow-up' versions of k-edge-colourings. A more precise definition of a 'blow-up' is as follows. For $s \geqslant 2$, let G be an s-partite graph with partition classes $V_1, \ldots, V_s$, let f be a k-edge-colouring of G, and let f' be a k-edge-colouring of K_s. We say that f, or G, is a *blow-up* of f' if the vertices of K_s can be labelled $v_1, \ldots, v_s$ such that, for all $x \in V_i$ and $y \in V_j$ with $1 \leqslant i \neq j \leqslant s$, we have $f(xy) = f'(v_i v_j)$. We can easily prove a lower bound on the value of $\phi_k(n, K_r)$ for all $r \geq 3$ and $k \geq 2$.

Lemma 2 ([15]) *Let $r \geq 3$, $k \geq 2$ and $n \geqslant R_k(r)$. Then,*

$$\phi_k(n, K_r) \geq t_{R_k(r)-1}(n). \tag{9}$$

Proof By the definition of $R_k(r)$, there exists a k-edge-colouring f' of the complete graph $K_{R_k(r)-1}$ with no monochromatic K_r. Now, consider the Turán graph $T_{R_k(r)-1}(n)$ with a k-edge-colouring f which is a blow-up of f'. The graph $T_{R_k(r)-1}(n)$

with the k-edge-colouring f has no monochromatic K_r and thus we have

$$\phi_k(n, K_r) \geqslant \phi_k(T_{R_k(r)-1}(n), K_r) = t_{R_k(r)-1}(n).$$

□

Liu and Sousa [15] proved that the lower bound of $t_{R_k(r)-1}(n)$ is asymptotically correct for $k \geq 4$ and $r = 3$ (see Theorem 11 below), and exact for $k = 2, 3$ and $r = 3$ (see Theorem 13) and for $k \geq 2$ and $r \geq 4$ (see Theorem 14), with n sufficiently large in both cases.

Theorem 11 ([15]) *For all $k \geqslant 2$, we have*

$$\phi_k(n, K_3) = t_{R_k(3)-1}(n) + o(n^2). \tag{10}$$

In particular, it is known that $R_2(3) = 6$ and $R_3(3) = 17$. Indeed, for two colours, it is easy to see that the only 2-edge-colouring of K_5 not containing a monochromatic K_3 is the one where each colour class induces a cycle of length 5, as shown in Fig. 1. Let f_2 denote this 2-edge-colouring of K_5. For three colours, the Ramsey number $R_3(3) = 17$ was first determined, in 1955, by Greenwood and Gleason [7]. Later, in 1968, Kalbfleisch and Stanton [11] considered the structures of all possible 3-edge-colourings of K_{16} not containing a monochromatic K_3. Their result is stated in terms of the *Clebsch graph*, which is a well-known 5-regular, Hamiltonian, K_3-free graph on 16 vertices and 40 edges.

Theorem 12 ([11]) *There exist exactly two different 3-edge-colourings of K_{16} with no monochromatic K_3. In each case, each colour class induces the Clebsch graph.*

Let f_3 and f_3' be the two 3-edge-colourings of K_{16} as stated in Theorem 12. Liu and Sousa [15] improved the upper bound in (10) for the cases $k = 2, 3$, as follows.

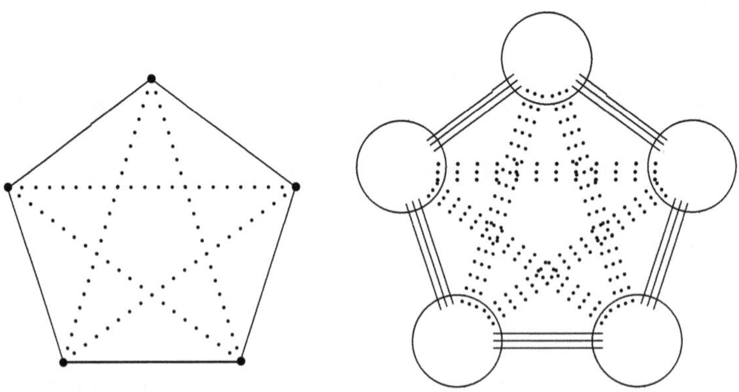

Fig. 1 The 2-edge-colouring of K_5, and its blow-up

Theorem 13 ([15]) *There is an n_0 such that, for all $n \geq n_0$, we have the following*

$$\phi_2(n, K_3) = t_5(n),$$
$$\phi_3(n, K_3) = t_{16}(n).$$

Moreover, the only graph attaining $\phi_k(n, K_3)$ is $T_5(n)$, with a blow-up of the 2-edge-colouring f_2 for $k = 2$, or $T_{16}(n)$ with a blow-up of the 3-edge-colourings f_3 or f_3' for $k = 3$.

The authors also make the following conjecture for monochromatic K_3-decompositions.

Conjecture 2 ([15]) Let $k \geq 4$. Then $\phi_k(n, K_3) = t_{R_k(3)-1}(n)$ for $n \geqslant R_k(3)$.

For larger cliques Liu and Sousa [15] have also determined the exact value of the function $\phi_k(n, K_r)$ for all $k \geq 2$ and $r \geq 4$, provided that n is sufficiently large.

Theorem 14 ([15]) *Let $r \geq 4$, $k \geqslant 2$. There is an $n_0 = n_0(r, k)$ such that, for all $n \geq n_0$, we have*

$$\phi_k(n, K_r) = t_{R_k(r)-1}(n). \tag{11}$$

In particular, $\phi_2(n, K_4) = t_{17}(n)$.

Moreover, the only graph attaining $\phi_k(n, K_r)$ is the Turán graph $T_{R_k(r)-1}(n)$ with a k-edge-colouring that does not contain a monochromatic copy of K_r.

The proof of Theorem 14 is simple and requires few results, therefore it will be included. Let us first introduce the necessary tools.

For $r \geq 3$, a *K_r-cover* in a graph is a set of edges meeting all K_r's, that is, the removal of a K_r-cover results in a K_r-free graph. A *K_r-packing* in a graph is a set of pairwise edge-disjoint K_r's. The *K_r-covering number* of a graph G, denoted by $\tau_r(G)$, is the minimum size of a K_r-cover of G and the *K_r-packing number* of G, denoted by $\nu_r(G)$, is the maximum size of a K_r-packing of G.

One can easily observe that

$$\nu_3(G) \leq \tau_3(G) \leq 3\nu_3(G). \tag{12}$$

In 1981, Tuza [28] conjectured that the second inequality of (12) is not optimal.

Conjecture 3 ([28]) For every graph G, we have $\tau_3(G) \leq 2\nu_3(G)$.

Conjecture 3 remains open, although many partial results have been proved. By using the earlier results of Krivelevich [13], and Haxell and Rödl [10], Yuster [29] proved the following theorem, which states that, asymptotically, Tuza's conjecture holds, and he also extended the result to larger cliques.

Theorem 15 ([10, 13, 29]) *Let G be a graph on n vertices. Then,*

(i) $\tau_3(G) \leqslant 2\nu_3(G) + o(n^2)$;
(ii) $\tau_r(G) \leqslant \lfloor \frac{r^2}{4} \rfloor \nu_r(G) + o(n^2)$, *for* $r \geq 4$.

Next, we recall the following result of Győri [8, 9] about the existence of edge-disjoint copies of K_r in graphs on n vertices with more than $t_{r-1}(n)$ edges.

Theorem 16 ([8, 9]) *Let $r \geqslant 3$ and G be a graph on n vertices with $e(G) = t_{r-1}(n) + m$, where $m = o(n^2)$. Then G contains $m - O(\frac{m^2}{n^2}) = (1 - o(1))m$ edge-disjoint copies of K_r.*

We are now able to present the proof of Theorem 14.

Proof (Proof of Theorem 14) The lower bound was proved in Lemma 2. Let us now prove the upper bound. Let G be any k-edge-coloured graph on n vertices and for the sake of simplicity let $R = R_k(r)$. We will show that $\phi_k(G, K_r) \leq t_{R-1}(n)$ with equality if and only if $G = T_{R-1}(n)$.

Let $e(G) = t_{R-1}(n) + m$, where m is an integer. If $m < 0$, we can decompose G into single edges and there is nothing to prove. If $m = 0$ and G contains a monochromatic copy of K_r then G admits an edge-monochromatic K_r-decomposition with at most

$$t_{R-1}(n) - \binom{r}{2} + 1 < t_{R-1}(n)$$

parts and we are done. If G does not contain a monochromatic K_r, then the definition of the Ramsey number implies that G does not contain a copy of K_R. Therefore, $G = T_{R-1}(n)$, by Turán's Theorem. Now, let $m > 0$ and let ℓ be the maximum number of edge-disjoint monochromatic K_r's in G. If $\ell > \frac{m}{\binom{r}{2}-1}$, then

$$\phi_k(G, K_r) \leq \ell + e(G) - \binom{r}{2}\ell < t_{R-1}(n). \tag{13}$$

Therefore, it suffices to show that $\ell > \frac{m}{\binom{r}{2}-1}$.

Consider first the case $m = o(n^2)$. By Theorem 16 the graph G contains $(1 - o(1))m$ edge-disjoint copies of K_R. Since each K_R contains a monochromatic copy of K_r, this implies that $\ell > \frac{m}{\binom{r}{2}-1}$ and we are done.

Finally, assume that $m \geqslant Cn^2$, for some constant $C > 0$. In order to get a contradiction, suppose that $\ell \leq \frac{m}{\binom{r}{2}-1}$. For $1 \leq i \leq k$ let G_i be the subgraph of G on n vertices that contains all edges coloured with colour i. By Theorem 15, our assumption implies that

$$\begin{aligned}\sum_{i=1}^{k}\tau_r(G_i) &\le \sum_{i=1}^{k}\left\lfloor\frac{r^2}{4}\right\rfloor\nu_r(G_i)+o(n^2)\\ &\le \left\lfloor\frac{r^2}{4}\right\rfloor\ell+o(n^2)\\ &\le \left\lfloor\frac{r^2}{4}\right\rfloor\frac{m}{\binom{r}{2}-1}+o(n^2)\\ &\le \frac{4}{5}m+o(n^2),\ \text{since } r\ge 4.\end{aligned} \tag{14}$$

That is, by deleting at most $\frac{4}{5}m + o(n^2)$ edges from G, we obtain a subgraph G' that does not contain a monochromatic copy of K_r. But then we have

$$e(G') \ge e(G) - \frac{4}{5}m - o(n^2) \ge t_{R-1}(n) + \frac{1}{5}m - o(n^2) > t_{R-1}(n).$$

Therefore, Turán's Theorem implies that G' must contain a copy of K_R which contains a monochromatic copy of K_r. This is a contradiction and the proof is complete. □

As an extension of the monochromatic K_r-decomposition problem Liu, Pikhurko and Sousa considered the problem when the clique K_r is replaced by a fixed k-tuple of cliques $\mathscr{C} = (K_{r_1}, \ldots, K_{r_k})$. Their results involve the Turán numbers and the (generalized) Ramsey numbers. Let us recall the latter.

For $k \geqslant 2$ and integers $r_1, \ldots, r_k \geqslant 3$, the *Ramsey number for* $K_{r_1}, \ldots, K_{r_k}$, denoted by $R(r_1, \ldots, r_k)$, is the smallest value of s, such that, whenever K_s is given a k-edge-colouring, there exists a monochromatic K_{r_i} in colour i, for some $1 \leqslant i \leqslant k$. For the case when $r_1 = \cdots = r_k = r$, for some $r \geqslant 3$, we simply write $R_k(r) = R(r_1, \ldots, r_k)$. Since $R(r_1, \ldots, r_k)$ does not change under any permutation of $r_1, \ldots, r_k$, without loss of generality, we may assume throughout that $3 \leqslant r_1 \leqslant \cdots \leqslant r_k$. The Ramsey numbers are notoriously difficult to calculate, even though, it is known that their values are finite for all $k \geqslant 2$ and $3 \leqslant r_1 \leqslant \cdots \leqslant r_k$ [21]. To this date, the values of $R(3, r_2)$ have been determined exactly only for $3 \leqslant r_2 \leqslant 9$, and these are shown in the following table [20].

r_2	3	4	5	6	7	8	9
$R(3, r_2)$	6	9	14	18	23	28	36

The remaining Ramsey numbers that are known exactly are $R(4, 4) = 18$, $R(4, 5) = 25$, and $R(3, 3, 3) = 17$ [20]. The gap between the lower bound and the upper bound for the Ramsey numbers is still quite large.

Liu, Pikhurko and Sousa [16] proved the following theorem, which extends the results obtained previously by Liu and Sousa [15]. Furthermore, it also verifies Conjecture 2 provided that n is sufficiently large.

Theorem 17 ([16]) *Let $k \geq 2$, $3 \leqslant r_1 \leqslant \cdots \leqslant r_k$, and $R = R(r_1, \ldots, r_k)$. Let $\mathscr{C} = (K_{r_1}, \ldots, K_{r_k})$. Then, there is an $n_0 = n_0(r_1, \ldots, r_k)$ such that, for all $n \geq n_0$, we have*

$$\phi_k(n, \mathscr{C}) = t_{R-1}(n). \tag{15}$$

Moreover, the only order-n graph attaining $\phi_k(n, \mathscr{C})$ is the Turán graph $T_{R-1}(n)$ (with a k-edge-colouring that does not contain a colour-i copy of K_{r_i} for every $1 \leqslant i \leqslant k$).

In particular, when all the cliques in $\mathscr{C}$ are equal to K_3, Theorem 17 completes the results obtained previously by Liu and Sousa in Theorem 11. In fact, one can easily extract the following corollary from Theorem 17.

Corollary 1 ([16]) *Let $k \geq 2$, $r \geqslant 3$ and n be sufficiently large. Then,*

$$\phi_k(n, K_r) = t_{R_k(r)-1}(n).$$

Moreover, the only order-n graph attaining $\phi_k(n, K_r)$ is the Turán graph $T_{R_k(r)-1}(n)$ (with a k-edge-colouring that does not contain a monochromatic copy of K_r).

Acknowledgements This work was partially supported by Fundação para a Ciência e a Tecnologia (Portuguese Foundation for Science and Technology) through the project UID/MAT/00297/2013 (Centro de Matemática e Aplicações).

References

1. Allen, P., Böttcher, J., Person, Y.: An improved error term for minimum H-decompositions of graphs. J. Combin.Theory Ser. B **109**, 120–133 (2014)
2. Bollobás, B.: On complete subgraphs of different orders. Math. Proc. Camb. Philos. Soc. **79**(1), 19–24 (1976)
3. Bollobás, B.: Modern Graph Theory. Graduate Texts in Mathematics, vol. 184. Springer, New York (1998). doi:10.1007/978-1-4612-0619-4. http://dx.doi.org/10.1007/978-1-4612-0619-4
4. Dor, D., Tarsi, M.: Graph decomposition is NP-complete: a complete proof of Holyer's conjecture. SIAM J. Comput. **26**(4), 1166–1187 (1997). doi:10.1137/S0097539792229507. http://dx.doi.org/10.1137/S0097539792229507
5. Erdős, P., Goodman, A.W., Pósa, L.: The representation of a graph by set intersections. Can. J. Math. **18**, 106–112 (1966)
6. Erdős, P., Füredi, Z., Gould, R.J., Gunderson, D.S.: Extremal graphs for intersecting triangles. J. Combin. Theory Ser. B **64**(1), 89–100 (1995). doi:10.1006/jctb.1995.1026. http://dx.doi.org/10.1006/jctb.1995.1026

7. Greenwood, R.E., Gleason, A.M.: Combinatorial relations and chromatic graphs. Can. J. Math. **7**, 1–7 (1955)
8. Győri, E.: On the number of edge-disjoint triangles in graphs of given size. In: Combinatorics (Eger, 1987), Colloque Mathematical Society János Bolyai, vol. 52, pp. 267–276. North-Holland, Amsterdam (1988)
9. Győri, E.: On the number of edge disjoint cliques in graphs of given size. Combinatorica **11**(3), 231–243 (1991). doi:10.1007/BF01205075. http://dx.doi.org/10.1007/BF01205075
10. Haxell, P.E., Rödl, V.: Integer and fractional packings in dense graphs. Combinatorica **21**(1), 13–38 (2001). doi:10.1007/s004930170003. http://dx.doi.org/10.1007/s004930170003
11. Kalbfleisch, J.G., Stanton, R.G.: On the maximal triangle-free edge-chromatic graphs in three colors. J. Combin. Theory **5**, 9–20 (1968)
12. Kövari, T., Sós, V.T., Turán, P.: On a problem of K. Zarankiewicz. Colloq. Math. **3**, 50–57 (1954)
13. Krivelevich, M.: On a conjecture of Tuza about packing and covering of triangles. Discrete Math. **142**(1–3), 281–286 (1995). doi:10.1016/0012-365X(93)00228-W. http://dx.doi.org/10.1016/0012-365X(93)00228-W
14. Liu, H., Sousa, T.: Fan decompositions of graphs (Submitted)
15. Liu, H., Sousa, T.: Monochromatic K_r-decompositions of graphs. J. Graph Theory **76**, 92–101 (2014)
16. Liu, H., Pikhurko, O., Sousa, T.: Monochromatic clique decompositions of graphs. J. Graph Theory (to appear)
17. Özkahya, L., Person, Y.: Minimum H-decompositions of graphs: edge-critical case. J. Combin. Theory Ser. B **102**(3), 715–725 (2012). doi:10.1016/j.jctb.2011.10.004. http://dx.doi.org/10.1016/j.jctb.2011.10.004
18. Pikhurko, O., Sousa, T.: Minimum H-decompositions of graphs. J. Combin. Theory Ser. B **97**(6), 1041–1055 (2007). doi:10.1016/j.jctb.2007.03.002. http://dx.doi.org/10.1016/j.jctb.2007.03.002
19. Pippenger, N., Spencer, J.: Asymptotic behavior of the chromatic index for hypergraphs. J. Combin. Theory Ser. A **51**(1), 24–42 (1989)
20. Radziszowski, S.P.: Small Ramsey numbers. Electron. J. Combin. **DS01**, Research Paper ds1, 27 (1996)
21. Ramsey, F.P.: On a problem of formal logic. Proc. Lond. Math. Soc. **30**, 264–286 (1930)
22. Sousa, T.: Decompositions of graphs into 5-cycles and other small graphs. Electron. J. Combin. **12**, Research Paper 49, 7 pp. (electronic) (2005). http://www.combinatorics.org/Volume_12/Abstracts/v12i1r49.html
23. Sousa, T.: Decompositions of graphs into a given clique-extension. Ars Combin. **100**, 465–472 (2011)
24. Sousa, T.: 4-Cycle decompositions of graphs. Open J. Discret. Math. **2**, 125–130 (2012)
25. Sousa, T.: Decompositions of graphs into cycles of length seven and single edges. Ars Combin. (to appear)
26. Szemerédi, E.: Regular partitions of graphs. In: Problèmes combinatoires et théorie des graphes (Colloq. International CNRS, Univ. Orsay, Orsay, 1976), Colloq. Internat. CNRS, vol. 260, pp. 399–401. CNRS, Paris (1978)
27. Turán, P.: On an extremal problem in graph theory. Mat. Fiz. Lapok **48**, 436–452 (1941)
28. Tuza, Zs.: Finite and infinite sets. In: Colloquia Mathematica Societatis János Bolyai, vol. 37, p. 888. North-Holland Publishing Co., Amsterdam (1984)
29. Yuster, R.: Dense graphs with a large triangle cover have a large triangle packing. Combin. Probab. Comput. **21**, 952–962 (2012)

7. [illegible]wood, R. [illegible], [illegible] A.M.: [illegible] relations and [illegible] graphs. [illegible] (1995)
8. [illegible]: On the [illegible] decompositions of graphs. In: [illegible] Mathematics [illegible] (1968)
9. [illegible]: On the number of [illegible] decompositions [illegible]. http://dx.doi.org/10.1007/BF01 [illegible]
10. [illegible], E., [illegible] and [illegible] graphs. [illegible] (2011), [illegible] http://dx.doi.org/10.1007/s[illegible]
11. Kahnberg, [illegible]: On the [illegible] graphs. [illegible] (1958)
12. [illegible], T., [illegible], V., [illegible], P., [illegible]
13. [illegible] http://dx.doi.org/10.1016/S0012-365X(00)00[illegible]
14. [illegible], H., [illegible], T.: [illegible] decompositions of graphs [illegible]
15. [illegible], H., [illegible] graphs. [illegible]
16. [illegible]
17. [illegible], C., [illegible], L.: [illegible] decompositions of graphs [illegible] 2010 [illegible]
18. [illegible], S., [illegible]: [illegible] graphs [illegible] (1988)
19. [illegible]
20. [illegible]
21. [illegible], T.: [illegible] graphs [illegible] (1955)
22. [illegible] (1976)
23. [illegible]
24. [illegible] of graphs into cycles of length [illegible] (to appear)
25. [illegible], F.: Regular partitions of graphs. In: Problèmes combinatoires et théorie des graphes (Colloq. Internat. CNRS, Univ. Orsay, Orsay, 1976), Colloq. Internat. CNRS, vol. 260, pp. [illegible] CNRS, Paris (1978)
26. [illegible], P.: On an extremal problem in graph theory. Mat. Fiz. Lapok 48, 436–452 (1941)
27. [illegible], Zs.: [illegible]. In: Colloquia Mathematica Societatis János Bolyai, vol. [illegible], pp. [illegible]. North-Holland Publishing Co., Amsterdam (1980)
28. Yuster, R.: [illegible] packings. Comb. Probab. Comput. [illegible] (2012)

Appendix A: CIM International Planet Earth Events DGS II, 2013

In 2013 the CIM organized the International Conference on the Mathematics of Planet Earth: DGS II, 2013—International Conference Planet Earth, Dynamics, Games and Science, 2–4 September 2013. Furthermore, the CIM organized the Advanced School Planet Earth directly before and after the International Conference: School DGS II, 2013—Advanced School Planet Earth, Dynamics, Games and Science, 26–31 August and 5–7 September 2013.

The CIM Mathematics of Planet Earth events stemmed from the CIM's role as a partner institution of the International Program Mathematics of Planet Earth 2013 (MPE 2013). We were pleased that the CIM-MPE events were announced, for example, in the ICIAM newsletter for January 2013 and the EMS newsletter for March 2013.

These events were enthusiastically supported by many Portuguese institutions, including: the SPM; SPE; APDIO; CEMAPRE; CEAUL; CMA-UNL; CMAF-UL; CMUP; INESCTEC; ISR; IT; UECE FCUL; ISEG; Calouste Gulbenkian Foundation (FCG) and Ciência Viva (CV).

The International Conference DGS II, 2013 was hosted by the Calouste Gulbenkian Foundation.

The Advanced School Planet Earth, Dynamics, Games and Science was hosted by the Escola Superior de Economia e Gestão, Universidade Técnica de Lisboa (ISEG-UTL).

In addition, the CIM would especially like to thank Irene Fonseca for her scientific guidance, João Paulo Almeida for his guidance and coordination of the events, Antónia Turkman for her assistance in coordinating with the Calouste Gulbenkian Foundation, Telmo Parreira for organizing and compiling the proceedings, and Paulo Mateus, Pedro Baltazar and Telmo Parreira for developing and maintaining the conference website. The CIM would like to thank the CGF staff and members of the local organizing committee as well as the Calouste Gulbenkian Foundation for their incredible hospitality throughout the event and for providing to speakers and participants the opportunity to experience the beautiful city of Lisbon in a friendly ambiance.

J.-P. Bourguignon et al. (eds.), *Dynamics, Games and Science*, CIM Series in Mathematical Sciences 1, DOI 10.1007/978-3-319-16118-1

The CIM would like to thank the following 18 keynote speakers of DGS II, 2013 for their insightful lectures:

- Elvio Accinelli, UASLP, Mexico
- Michel Benaïm, Université de Neuchâtel, Switzerland
- Fabio Chalub, Universidade Nova de Lisboa, Portugal
- Jim Cushing, University of Arizona, USA
- João Lopes Dias, Universidade Técnica de Lisboa, Portugal
- Pedro Duarte, Universidade de Lisboa, Portugal
- Marta Faias, Universidade Nova de Lisboa
- Lorens Imhof, University of Bonn, Germany
- Yunping Jiang, City University of New York, USA
- José Martins, I.P. Leiria, Portugal
- Bruno Oliveira, Universidade do Porto, Portugal
- Jorge Pacheco, Universidade do Minho, Portugal
- Joana Pais, ISEG/Technical University of Lisbon, Portugal
- Alberto A. Pinto Universidade do Porto
- Martin Shubik, Yale University, USA
- Renato Soeiro, Universidade do Porto, Portugal
- Satoru Takahashi, National University of Singapore
- Jorge Zubelli, IMPA, Brasil.

The CIM would also like to thank the 120 invited speakers for their valued presentations, as well as the 29 session organizers, who contributed their hard work and dedication to make the event a success:

- Elvio Accinelli, Facultad de Economia de la UASLP
- João Paulo Almeida, Instituto Politécnico de Bragança
- Mário Bessa, Universidade de Beira interior
- Domingos Cardoso, Universidade de Aveiro
- Marta Faias, FCT-UNL
- Sara Fernandes, Universidade de Évora
- Jorge Freitas, Universidade de Porto
- José Pedro Gaivão, ISEG, UTL
- Orlando Costa Gomes, ISCAL/IPL
- Clara Grácio, Universidade de Évora
- Fátima Leite and Antonio Pascoal, UC and UTL
- Alberto A. Álvarez López, UNED
- José Martins, Polytechnic Institute of Leiria
- Célia Moreira, Universidade de Porto
- Cláudia Nunes, IST/CEMAT
- Bruno Oliveira, FCNA, Universidade de Porto
- Isabel Pereira, Universidade de Aveiro
- Edgard Pimentel, Universidade Técnica de Lisboa
- Alberto Pinto, Universidade do Porto
- Ana Margarida Ribeiro and Rita Ferreira, FCT-UNL and IST-UTL

- José Leonel Rocha, Instituto Superior de Engenharia de Lisboa
- Alexandre Rodrigues, Universidade de Porto
- Ricardo Serrão Santos, Universidade de Açores
- Luís Silva, Universidade de Açores
- Luís Silva, ISEL
- Nico Stollenwerk, Universidade de Lisboa
- Ricardo Teixeira, Universidade de Açores
- Paulo B. Vasconcelos, Universidade de Porto
- Juha Videman and Gonçalo Dias, CAMGSD/IST.

The CIM would like to thank the members of the local organizing committee of DGS II, 2013 for their outstanding support: Michel Benaïm (Université de Neuchâtel), Henrique Silveira (IST), Renato Soeiro (FCUP), Filipe Martins (FCUP), João Passos Coelho (FCUP), Joana Becker (FCUP), João Paulo Almeida (IPB), Carla Azevedo (FCUP), Ricardo Cruz (FCUP), José Martins (IPL), Renato Fernandes (FCUP), Isabel Figueiredo (FCUP), Telmo Parreira (UM) and Joel Teixeira (FCUP).

The book of abstracts of DGS II, 2013 can be found in the link:
http://www.alunos.dcc.fc.up.pt/~up200405927/cim/bookDGS.pdf

President of CIM Alberto Adrego Pinto

Appendix B: Interviews MPE: DGS II

CIM thanks the participants Elvio Accinelli (UASLP, Mexico), Alberto Álvarez-López (UNED, Spain), Michel Benaïm (Université de Neuchâtel, Switzerland), Mário Bessa (Universidade da Beira interior), Fabio Chalub (Universidade Nova de Lisboa, Portugal), Ana Dias (Universidade do Porto, Portugal), Orlando Gomes (ISCAL/IPL, Portugal), Clara Grácio (Universidade de Évora, Portugal), Filipe Martins (LIAAD INESC TEC, Portugal), Bruno Oliveira (Universidade do Porto, Portugal), Joana Pais (Universidade de Lisboa, Portugal), Alexandre Rodrigues (Universidade do Porto, Portugal), Luís Filipe Silva (CIBIO Universidade dos Açores, Portugal), Luís Silva (ISEL Lisboa, Portugal), and Paulo Vasconcelos (Universidade do Porto, Portugal) of the International Conference and Advanced School Planet Earth, Dynamics, Games and Science II (DGS II), Portugal, 28 August to 6 September 2013, for sharing their ideas and points of view with us in this interview.

The questions presented here are based on several interviews; in particular, the interviews published in previous CIM bulletins. CIM thanks Renato Araujo and Alberto Pinto for organizing this interview (see also CIM Bulletin 35).

On the meeting

What is your general impression of the meeting?

Elvio Accinelli: These kinds of meetings are of great interest for making progress in different areas of applied mathematics, and they create networks on research topics of common interest.

Alberto Álvarez-López: I can talk about the DGS meetings II and III, held in Lisbon and in Porto, respectively. I found them very interesting. I met people who work in areas similar to mine, and I could hear some colleagues' opinions about my own work. In addition, I enjoyed them very much for their social aspects.

J.-P. Bourguignon et al. (eds.), *Dynamics, Games and Science*, CIM Series in Mathematical Sciences 1, DOI 10.1007/978-3-319-16118-1

Michel Benaïm: Very good. It was very friendly and gave me the opportunity to meet and discuss with researchers having different backgrounds and mathematical cultures.

Mário Bessa: It was a good opportunity to meet several mathematicians working in related areas and develop some connections. I think that the Portuguese mathematical community should be more involved in this event.

Ana Dias: I found the meeting very interesting.

Orlando Gomes: The International Conference on Dynamics, Games and Science is, in my view, an extremely useful forum to discuss ideas and progress in research in a variety of fields concerning applied mathematics. In the events in which I have been present I have learned a lot about subjects on multiple areas ranging from evolutionary games to chaotic dynamics, stochastic optimization, and network analysis, just to cite a few.

Clara Grácio: I think this congress was an enjoyable opportunity to fulfill the objectives that I described in other questions as important for students and researchers who attended this event. I participated in the meeting held in September at the Calouste Gulbenkian Foundation. This meeting allowed us to talk to other colleagues, presenting our works in progress and discussing possibilities for continuing and improving that work, as well as future projects.

Filipe Martins: I think the Dynamics, Games and Science II conference was an amazing meeting featuring a wide range of topics and keynote speakers. My general opinion is that it was very well organized and featured many brilliant presentations. I think these kinds of conferences are very important. For me, as a student, it was a huge boost in terms of encouragement to pursue a Ph.D., as I was finishing my Master's thesis at the time.

Bruno Oliveira: It was an excellent meeting where I had the opportunity to exchange ideas with many colleagues and learn from them.

Joana Pais: Very well organized. Amazing capacity of the organizers to put together an extremely interesting program, with a substantial group of well-known researchers. Very interesting talks, even though, in my opinion, they covered topics that were probably too diverse. Filipe Silva: The importance of these types of meetings is the possibility of joining researchers who use mathematical tools in very different contexts, contributing to the transferability of knowledge between the different fields.

Luís Silva: The meeting was very interesting, bringing together an outstanding group of researchers, both domestic and foreign, in a fantastic and inspiring place.

Paulo Vasconcelos: The overall quality of the papers presented was great. The location was attractive, and the group lunches were full of life.

Something you would like to highlight?

Elvio Accinelli: These kinds of meetings are of great interest for making progress in different areas of applied mathematics and creating networks on research topics of common interest.

Alberto Álvarez-López: People from different "countries" of the world of mathematical applications could meet there, from pure mathematicians to biologists, economists, and engineers: a very interesting mixture. In addition, I would like to highlight the format of the sessions: short talks related to each other, which is perfect for cultivating interplay among senior and junior scholars. In fact, these events were a good opportunity for young researchers: they could show their own work and also listen to very relevant opinions from senior colleagues.

Ana Dias: The quality of the talks, the variety of the themes addressed at the talks, and the event location, the Calouste Gulbenkian Foundation, all made these meetings very pleasant to experience.

Orlando Gomes: I believe that the strong feature of the Dynamic, Games and Science meetings is their interdisciplinary nature. With the use of mathematical methods as a unifying force, conferences offer a large variety of studies in a large variety of fields. Applications to economics and finance coexist with studies in themes relating to biology, ecology, or physics.

Luís Silva: I would like to highlight the quality of the plenary talks.

Paulo Vasconcelos: The intensive preparation by the organizing committee resulted in a smooth learning experience for the participants in a very pleasant setting.

How important do you think that events like this are for students and researchers?

Elvio Accinelli: Students in the process of completing their theses can find places to develop their research and to finish their work.

Mário Bessa: These types of meetings are quite important both for students and researchers because we have the chance of contact with related fields of expertise, thus gaining a deeper perspective on the application of our theoretical models in several different contexts.

Ana Dias: Very important not only for learning about new mathematical studies, but also for interchanging ideas, sometimes between mathematicians with different backgrounds.

Orlando Gomes: These events are a very good opportunity to share ideas, to learn, and to create networks among researchers. They are, of course, particularly important for young researchers who are starting a career by allowing them to present their work and establish the contacts they need to progress in their research effort. Graduate and undergraduate students have the opportunity at these events to have their first contact with the world of science.

Clara Grácio: In my opinion the scientific meetings are an excellent opportunity for researchers to present their work to their colleagues in order to receive feedback at an early stage of their research and are therefore an integral part of the process of science. These presentations also serve as informal reviews by peers, which may help researchers to develop, clarify, and improve their work and will no doubt help in the final phase of writing and submission to final publication. Also, and very important, the meetings allow researchers to hear about what others are studying, to develop relations with related disciplines by talking to colleagues from different institutions around the world, and to learn about new tools and research techniques that can be relevant to their work, other programs, and projects in common. These are truly scientific meetings arising in an academic environment where the questions and answers are natural, objective, honest, and fearless, and where the only goals are help, cooperation, and the development and dissemination of knowledge.

Bruno Oliveira: Events like this give researchers an opportunity to report their results to the scientific community. More importantly, in my opinion, they also open channels of communication between researchers, which enhances the work we develop. Regarding students, I think that these events give them a wonderful way of obtaining state-of-the-art knowledge from experts in these subjects.

Joana Pais: Very important. Research dissemination and networking are essential.

Filipe Silva: This might broaden their views, and make them see their daily research with different eyes.

Luís Silva: These kinds of events are extremely important both for students and for researchers. For the students they provide an excellent opportunity to make contact with senior researchers, learn about the most current issues, and even help them to decide about their future topics of research. These events allow the researchers to publicize their work and to exchange ideas with their colleagues.

Paulo Vasconcelos: The advanced school is an important meeting point for students with high level researchers, which can be a rare opportunity. Researchers enjoy the outstanding opportunity to publish proceedings within a prestigious and exigent editorial brand as well as participate in a book series devoted to applied mathematics.

How do you see the impact of this meeting on your field and outside of your field?

Michel Benaïm: This type of meeting allows people with different backgrounds (game theory, dynamical systems, probability) but common interests (in the present case "dynamics in games") to meet and is a good opportunity for cross-fertilization of ideas.

Fabio Chalub: Most of the meetings in the field of mathematics are "technique-based"; i.e., a number of professionals who have mastered the same techniques get together and discuss problems where these techniques were applied. This was a different kind of meeting in the sense that we had a large number of problems but no predefined mathematical technique. All areas of mathematics were represented in the conference, and the researchers could see where their expertise and abilities were required. This can forge a new generation of students who are more "problem-oriented" and who necessarily will learn more subjects, as opposed to the precocious specialization we see today.

Ana Dias: A good impact due, also, to the fact that some of the works will be published in a Springer book, which is a very good way of reaching readers from other fields.

Orlando Gomes: There are not many international quality scientific conferences or series of conferences in Portugal. This is a good example of a well-organized series of conferences that, I believe, has a good impact in promoting applied mathematics. As I see it, it is an interdisciplinary meeting with repercussions that go beyond mathematics; for instance, it is also an important event in my own research field, i.e., economics.

Joana Pais: I believe that, even though the impact on the field may be substantial, the outside impact is limited. This is not an exclusive feature of this particular event, but it is common to most (if not all) of the events of this nature. Clearly, it is a difficult exercise to translate the language of science into a language that the general public can understand. In fact, while there is no ambiguity in mathematics, when we translate mathematical language into words, our messages are probably not perceived the way we meant it. Still, disseminating scientific knowledge in the public sphere, particularly in the domain of social sciences, is important. It makes us think about why we believe that our research is necessary and useful.

Luís Silva: I think that this meeting may have a significant impact in the field, especially because this subject is relatively recent, and a meeting with this dimension of topics is not very common. In the particular case of Portugal, I think it presented a lot of subjects to several people.

What would you say is, generally, the impact of these events on specific areas, as they relate to and on the interplay between different areas or fields of knowledge?

Elvio Accinelli: These events are of great importance for creating networks between groups of different countries; consequently, they have a great impact on the work area as they allow one to learn about progress elsewhere.

Ana Dias: Good impact.

Orlando Gomes: This type of meeting is, as stated in previous answers, a way to promote the cross-fertilization of knowledge in various fields where game theory and dynamic processes matter. It is an extremely helpful event for all those who want to develop competence and explore new territories in applied science. New research projects, of an interdisciplinary nature, will certainly arise from the contacts researchers make in these conferences.

Bruno Oliveira: Of benefit to both students and researchers was the fact that this meeting covered a broad area of subjects in mathematics, in particular dynamical systems and game theory, and an even broader area of applications in the sciences, presenting research in several distinct topics of, for instance, economics and biology. This diversity can build bridges between different problems, allowing the attendees to further improve their work.

On your research

Did you always want to be a mathematician?

Alberto Álvarez-López: Well, when I was a child, besides math I also liked language (I mean grammar and so on). But to tell you the truth, I always wanted to be a mathematician. Anyway, upon finishing my Bachelor's degree in Mathematics, I landed a position in a faculty of economics. Through the years I have discovered a wonderful field in which to apply mathematics that is very rich and interesting by itself!

Fabio Chalub: In fact, I graduated and received my Master's degree in Physics. During this time, I followed as many disciplines in mathematics as I could, and I got the impression that the most fundamental results in physics could only be entirely appreciated with a deep understanding of the mathematics behind them. In the end, I decided to do my Ph.D. in Mathematics involving the work on the border between math and physics.

Ana Dias: Looking back, my answer is yes.

Orlando Gomes: I am an economist, with research interests related to the mathematical modeling of economic phenomena. Economic processes have always fascinated me, and I believe that mathematics is necessarily the language through which economic events can be rigorously addressed and explained. My interest in modeling socio-economic events goes back to my undergraduate studies in economics (more than 20 years ago).

Clara Grácio: As we know, mathematics can be, sometimes, frustrating indeed, but it is in this struggle where the challenge itself lies. You experience a sense of accomplishment, even contentment, when you discover the missing piece of the puzzle, and mentally exclaim: That was it! Also, when you can establish unsuspected relationships between different areas of mathematics and/or other sciences, the coherence, connection, and immensity of mathematics emerge. I always liked the interconnection between the various areas of knowledge, from language or history to physics or biology, the wealth that allows us to move forward. And in order to advance in this way, mathematics is essential and indispensable. To the question of whether I always wanted to study mathematics, the answer is that mathematics has always been the first choice as long as the studies allow the monitoring of other areas.

Filipe Martins: I only thought of being a mathematician very recently. I decided to study mathematics as an undergraduate just two months before the start of the academic year. It was a pretty quick decision. I was trying to choose between mathematics and physics. The decision was taken completely by impulse, in 5 minutes.

Bruno Oliveira: It's a yes and no answer. Ever since I was young I had a fondness for mathematics. Later I gained an interest in physics, informatics and astronomy (from watching the TV series Cosmos by Carl Sagan). So, mathematics was always there, but linked to other sciences.

Alexandre Rodrigues: No, I did not always want to be a mathematician. In fact, I do not consider that I am a mathematician. I prefer to say that I am a researcher in mathematics. After completing my undergraduate degree I was convinced that I would like to be a high school teacher, but my desired career direction became clear while I was pursuing my M.Sc. degree. Even during that stage I considered exploring a different subject and switching to physics.

Filipe Silva: No, I always wanted to be a biologist, considering that life is probably the most complex and evolved form of matter/energy in the universe. However, during my research and teaching activities, I became progressively aware of the importance of using mathematical and statistical tools in biology, and in science in general.

Luís Silva: No. During most of my time in secondary school I was convinced that I wanted to be a psychologist.

Paulo Vasconcelos: Not always ... but almost always!

How did you start working in this area? What was the motivation? Could you tell us about your mathematical beginnings and subsequent career development?

Elvio Accinelli: Motivated by social problems, I felt a vocation for economics. In the last year of primary school my teacher made me see that mathematics could be an excellent tool for thought. Later, when I was in prison as a political prisoner, I met José Luis Massera, who greatly influenced my thinking. Some years later, in the IMPA I had the opportunity to learn mathematical economics. Since then I feel real pleasure working in this area.

Michel Benaïm: I have always worked at the interface of probability and dynamics. My interest in game theory started in Nefeli Cafe, a coffee shop located in Berkeley, near the math department, 20 years ago. At this time I was working with Moe Hirsch on some applications of topological dynamics for investigating the long-term behavior of certain stochastic processes called "stochastic approximations." A friend of mine, Paolo Ghirardato, at this time a Ph.D. student in economics, suggested that I look at a preprint by Drew Fudenberg and David Kreps on "stochastic fictitious play." It turned out that the techniques I was developing with Moe Hirsch proved to be very useful for analyzing stochastic fictitious play and more generally leaning processes in game theory.

Mário Bessa: After I finished my Bachelor's in Mathematics at the University of Porto, a colleague of mine asked me if I would like to go to some informal conversations about mathematics, taking place once a week, with Professor Jorge Rocha at the University of Porto. Since Jorge Rocha is a dynamicist, I started learning about this area, and immediately I began to enjoy dynamics. Then, I finished my Master's thesis in dynamical systems with Jorge Rocha and I went to IMPA for a Ph.D. program with a thesis also in dynamical systems, supervised by Marcelo Viana. Finally, I returned to Portugal where I completed six years of a post-doc program and taught at the Polytechnic Institute of Coimbra. Now, I am an associate professor at the University of Beira Interior.

Fabio Chalub: During my Ph.D. study, I followed a course in the mathematical models used in ecology and, immediately after that, some colleagues and I started a discussion group in math-biology. I became fascinated with the topic and decided to work on it during my post-doc, in Vienna. I studied models for cell motility and had some relevant results during that time. I also enjoyed the fact that the math-biology group in Vienna is very well established, and I could learn new topics. At a meeting in Vienna, I met Jose Francisco Rodrigues, from the Lisbon University, and he told me about a post-doc position in Lisbon and his particular interest in starting a group in mathematical biology. I went to Lisbon intending to stay 2 years, but after a few months my wife and I were seeking opportunities for a longer stay. This was 12 years ago! In 2005, I got a position at Universidade Nova de Lisboa, and since then I have been there, first as an assistant professor, then as an associate professor, and now as an "investigador FCT" researcher.

Ana Dias: Professor Isabel Labouriau introduced me to the area of dynamical systems for my Master's thesis. I would say that the contact with Professor Isabel Labouriau in the Applied Mathematics Department and the job I got at the University of Porto were the main starting points in my becoming a mathematician. Any trip to any place for work has a story, and when we return we bring memories. For sure my period at Warwick University during my Ph.D. study was the most important period of my research, because during that time I found out what I really liked to work on, and my supervisor, Professor Ian Stewart, had a fundamental role in that discovery.

Orlando Gomes: My work in theoretical economic research started with my Master's course (1995–1996). The possibility of approaching economic processes through the use of mathematical tools, namely dynamic systems (linear and non-linear, deterministic and stochastic, in discrete and in continuous time), fascinated me, and I have pursued studies in this area ever since. The first models that I approached related economic growth processes. Economic growth was the theme of my Master's thesis and of my Ph.D. thesis (which I completed in 2002). Later, I diversified my studies to areas that involve business cycles, monetary policy, international trade, individual decision-making, social interaction, and others. The common denominator of all this research is related to the use of tools of dynamic analysis and dynamic optimization.

Filipe Martins: After my undergraduate studies I had no real idea about the nature of research in mathematics, but after three years as an undergraduate, I decided to undertake a Master's degree in Mathematics, specializing in statistics and probability. Really, I only became more aware of research in mathematics when I was working on my Master's thesis. I liked it very much and noticed that to continue research in mathematics could be a good idea, and then I started thinking about taking a Ph.D. in the subject, and, happily, I got a Ph.D. scholarship. I would describe my areas of interest concisely as applied mathematics, which is what I like. What I studied for my Master's thesis was financial mathematics. Now I'm continuing on that topic, but I am also working on applications of dynamical systems to biology and economics. Again, the best way to designate it is applied mathematics. There is a wide range of topics for future work in this area. The rate at which work possibilities arise in applied mathematics is far greater than the rate at which you solve them. For each one you work on, a lot more appear as possible continuations. My favorite theorem in mathematics is possibly Banach's fixed point theorem.

Bruno Oliveira: After my degree in Astronomy, I completed a Master's degree in Applied Mathematics and, later, a Ph.D. in Applied Mathematics. My motivation has been a desire to understand how things work: from the universe to quantum mechanics, passing through humans in diverse subjects such as immune responses by T cells, price formation in random markets, firms competing with investment in R&D, children's growth, dietary patterns, or obesity treatment. And the tools that I have been using are rooted in mathematics, in particular, dynamical systems, game theory, and statistics, with links to computer modeling, and also requiring my knowledge of physics when studying interaction phenomena. In my career, I have taught subjects in astronomy, physics, and biostatistics. In particular, in these latter years I have been teaching biostatistics applied to nutrition and food sciences, which led me to do a Habilitation in Basic Sciences of Clinical Nutrition.

Alexandre Rodrigues: I really started my work in this area during my Ph.D. study, as after my M.Sc. it became clear to me that I really wanted to do research in dynamical systems. At the beginning, the motivation was the challenge of completing a Ph.D. in Mathematics. I remember quite well that I had two main concerns. (i) Could I discover something new in mathematics? (ii) Could I develop some important step towards an open problem? In fact, I do not know if I have achieved these goals. The main motivation was to complete a Ph.D. in Mathematics in a subject that I tried to pursue during my M.Sc. At the time, it seemed unattainable.

Filipe Silva: Working mostly in quantitative ecology, I became more and more interested in the use of statistical models to describe ecological phenomena. I became aware that statistical thinking evolved in close connection with biology and other sciences, and that its historical evolution had a parallel in the development of the other sciences. I also became involved in teaching biostatistics and quantitative methods to different student at levels, which further developed my interest in the area.

Luis Silva: My main motivation came from J. Sousa Ramos. He taught Introduction to Computation in my first year, and he had an uncommon point of view about that (and any other) discipline. He strongly believed that the students should be challenged from the beginning, so in the first classes he presented us with some of the most important math problems of that time: Fermat's theorem, Collatz, P/NP, and Poincaré's conjecture, etc., then he taught us Pascal and stimulated us to start exploring Julia sets, Mandelbrot sets, the Lorenz attractor, and so on. I think that he was mainly responsible for my decision of trying to be a mathematician instead of a high school teacher. Then I finished my undergraduate work and immediately got a job as assistant professor at FMH-UTL. At the same time I started a Master's study at IST-UTL and made my thesis with Sousa Ramos, then changed to the University of Évora, then finished my Ph.D. with Sousa Ramos again, and after ten years came back to Lisbon, to ISEL, where I am now.

How would you describe the essence of your own research to a young student?

Elvio Accinelli: Mathematical economics is both an intellectual challenge and an important tool for understanding the economy, for better social development.

Mário Bessa: Well, first I would describe how dynamical systems is not exactly an area but a confluence of several areas and so offers a good opportunity to study different aspects of mathematics. Then, I will emphasize that dynamical systems problems are often easy to formulate and to understand, although they are usually hard to solve. I would also like to say that working in dynamics is quite amusing, because our objects are continuously changing when time evolves and we should be aware that intuition frequently tricks us. Finally, I really like to work with my co-authors, because we then enjoy enormous creativity. Indeed, when we try to explain to each other the questions, problems, solutions, and arguments that we are interested in, again and again we say to each other, 'Imagine that....

Fabio Chalub: I work in applied mathematics; therefore, I decide the problems I want to solve, but I do not decide the mathematical techniques necessary to solve them. My general interest is in population dynamics, and currently I work on two fronts: population genetics and epidemiology. In the former case, I am interested in exploring the mathematical richness of widely used models. In the latter case, we study the interaction between deterministic models and human behavior, in particular, how the course of an outbreak is affected by changes of behavior in the society. Sometimes, we find predictions in models that were not known; other times we find that some consequences do not follow from the models, contrary to the general belief; finally, we provide solid grounds for the models that appear in the literature and explicitly show their limitations. Our main goal is related to the conceptual understanding of the area, not to providing better models for specific problems.

Ana Dias: When we have interactions between units that are evolving with time, there are consequences for the dynamics that come just from the fact that there are interactions.

Orlando Gomes: I would say that economics is the field of knowledge where one can most successfully apply mathematical rigor to human decision and human action and that this is a fascinating combination independently of the type of phenomena under examination, this being of a micro or of a macroeconomic nature. Furthermore, I would say as well that my studies address dynamic processes in economics, because time is the most fundamental variable in this science; all economic issues necessarily involve a temporal dimension.

Alexandre Rodrigues: I work with dynamical systems. Roughly speaking, a dynamical system is a concept in mathematics where a rule describes how a point evolves (in time) in a geometrical space. The evolution rule may be given by the solution of a differential equation. Finding the explicit solutions of these equations is, in general, impossible. Sometimes these equations have some additional structures: algebraic symmetries which might help us to understand the qualitative behavior of the system. Heteroclinic cycles are a common feature of symmetric differential equations and persist under perturbations that preserve the symmetry. The dynamics near a heteroclinic cycle are well known and it is characterized by intermittency: a solution remaining near the cycle spends long periods of time close to a particular kind of sets and makes fast transitions among them. The rigorous analysis of the intermittent dynamics associated to the structure of the sets close to heteroclinic cycles is an exciting and challenging field of research. The characterization of the dynamics near these kinds of cycles is what I have been studying.

Filipe Silva: The fascinating idea of being able to see parts of the complexity of biological entities reflected in much more simple models, resulting from the systematic but creative activity of human mind.

Luís Silva: I work in symbolic dynamics; basically, I study the combinatorial aspects of dynamical systems.

Which would you say are the most interesting/challenging open (or recently solved) problems in your area, and what do you think the future holds in your area and in your line of research?

Elvio Accinelli: I think that understanding how the markets work could be helpful to obtain a sustainable development of mankind. The mathematical economy is a path toward that goal.

Alberto Álvarez-López: Roughly speaking, I work in elaborating mathematical models to describe some aspects of economic behavior, especially in the presence of uncertainty. Of course, there are many problems under this umbrella to be studied. I point out a very general one: we agree that the economic agent (a consumer, a firm, etc.) is not rational; well, I think a new non-rational theory describing his/her behavior is necessary—I mean a completely new theory, with a very different approach.

Mário Bessa: My preference goes to the well-known "closing lemma" problem. This is a question that dates back to seminal works of Poincaré on celestial mechanics. Like I told before, this is a good example of a problem that is easy to formulate as you will see: if an orbit returns near to a place where it was before, is it possible to perturb the system in order to close the orbit? Of course, several aspects should be clarified; for example, what do we mean by "perturb"? Indeed, closing orbits requiring coarse approximations are well established; however, the problem is very hard when we demand finer approximations. If the requirements on the approximation increase too much, then it is known that the closing lemma has no solution!

Ana Dias: In my line of research on dynamics of coupled cell networks, I would say that it is important to have a theory for coupled cell networks like there is one for symmetric dynamical systems based on representation group theory.

Orlando Gomes: Since I study economic problems in general, I believe that although this is a very active science that has produced many meaningful results and advances in the last few decades, there are still many open questions. In macroeconomics, for instance, the permanent conflict between neoclassical and Keynesian economics and the difficulty in handling concrete aggregate problems (such as high rates of unemployment and deep recessions) reveal that much work still has to be developed in order to reach a unifying macroeconomic theory. At the micro level, a well-established theory of decision and individual behavior based on revealed preferences is now being challenged by advances in neuroscience, which indicate that one must go beyond the effective choices of economic agents and focus on the processes inside the human brain that trigger the decisions.

Alexandre Rodrigues: We do not know persistent classes of dynamical systems for which there is a set of positive measure which consists of initial points of orbits with historic behavior. For special dynamical systems, i.e., with boundary or with symmetry, historic behavior may persist. The main problem, however, remains open for dynamical systems without such constraints. In this context, R. Bowen described a system of differential equations on the plane whose flow has a heteroclinic cycle consisting of a pair of saddle equilibria connected by two trajectories. The eigenvalues of the derivative of the vector field at the two saddles are such that the cycle attracts solutions that start inside it. In this case each solution in the domain has historic behavior. Breaking the cycle, the flow loses this feature. This type of behavior may become persistent for dynamical systems in manifolds with boundary or in the presence of symmetry.

Filipe Silva: There is considerable excitement about the growing use of Bayesian statistics in different fields of biology. But, the future might bring new conceptual developments that will link or eventually merge frequentist and Bayesian statistics.

How do you see your area in terms of its importance in mathematics and in other fields of knowledge, the impact on and from other areas, and how do you expect this interplay to develop further?

Elvio Accinelli: I think that economic theory is in actuality a source of challenges for mathematics, whose resolution can achieve progress of both sciences. I would venture to say that economic theory, at present, can be as important for mathematics as it was physical in the nineteenth and early twentieth centuries.

Mário Bessa: Since the area of dynamical systems is a junction of several areas, there is intrinsically a large connection between mathematical subjects that are sometimes apparently unrelated. Moreover, its relation with other sciences (life, exact, social, computer, etc.) greatly enlarge these types of interactions. I believe that nowadays the classical nomenclature of dynamical systems is also used in other areas and turns out to be part of the language of these fields.

Fabio Chalub: The importance is growing a lot, in the world in general and in some particular countries like the USA, UK, France, the Netherlands, Spain, Germany, Sweden, and others. Fortunately, Portugal is no exception. It is still difficult to go from the theory to real applications, as this cannot be done by the same person or even the same groups. We have to talk to people with completely different backgrounds, and this is not easy. Generally, it is not difficult to get funding from government agencies, but for young Ph.D. graduates it is still difficult to find positions, as most of the mathematicians do not see "mathematical biology" as an area differently from "mathematical physics." It is seen as a topic of research, but not as a division of mathematics, like algebra, geometry, or analysis.

Ana Dias: As most real-world applications are governed by dynamics that can be interpreted as units interacting, any theory for coupled cell networks that develops model-independent kinds of results is important and of interest for science in general.

Orlando Gomes: The nineteenth century philosopher Stanley Jevons once stated that if economics is to acquire the status of science, it needs to be a mathematical science. In fact, since then, the studies that contributed to the undeniable self-affirmation of economics as an autonomous scientific field have essentially adapted tools, concepts, and techniques from mathematics. Game theory, differential calculus, linear algebra, recursive analysis, optimal control, and other powerful instruments provided by mathematics have contributed to build what economic science is today. Moreover, some mathematical concepts were created and developed as specific tools of the economic theory and then served other fields of knowledge as well. The interplay between mathematics and economics is a fruitful one, and it will certainly be explored in more depth in the future.

Paulo Vasconcelos: Computational mathematics is crucial for applied mathematics. Bringing mathematics to solve problems is the ultimate purpose of our research. Other fields of knowledge depend on the knowledge transfer, and there is nothing like computers to help simulate natural, physical, chemical, or even human processes.

Do you have a favorite result, your own and/or from others?

Elvio Accinelli: Yes, my favorite result is the explanation of the economic crisis as the result of small perturbations on the fundamentals of so-called singular economies.

Ana Dias: My favorite result is on ODE-equivalent networks and concerns the idea of different graphs leading to the same kinds of dynamics—they just have be linearly equivalent: a nonlinear result that has a linear question. The part that I like more in my work is the fact that every time we have a problem, we have a challenge in hand that we try to address. When we have success, it is a good feeling: the feeling of contributing to science with something, even if it is a small contribution.

Orlando Gomes: There are many powerful and appealing results in economics. Personally, I am a fan of the so-called Ramsey growth model: a simple and elegant optimal control problem that indicates how a representative agent chooses, in an intertemporal perspective, how to optimally allocate resources between consumption and savings, in order to maximize expected utility.

Is it difficult to get funding for research in your area?

Ana Dias: Until now not so difficult. The amounts asked are not so much compared with other research areas, so that might help.

Orlando Gomes: In the last few years in Portugal it became, in my view, difficult to get funding for doing research in any scientific area.

Clara Grácio: Research and higher education have been maltreated in recent years, for decades, with an unacceptable government underfunding which translates into immense difficulties for both higher education institutions and research centers and institutes of state laboratories. Even with the dedication of Portuguese researchers, integrated or not, this policy did not allow the scientific development that would have been possible, resulting in wasted potential and resources. Combined with a real reduction in funds invested in vacancies for teachers or researchers in institutions of higher education, laboratories, and others, there has been a lack of coordination and a lack of transparency and programming in resources invested. Mathematics is no exception, and in this sense is not easy to get funds for the development of scientific work.

Filipe Silva: Yes, it's easier to get funding for applied research, such as the study of forest resources, than for more fundamental research, for instance that devoted to new methods. We thus try to mix both.

Luís Silva: Yes, but unfortunately that problem is not restricted to my specific area. On the contrary, in Portugal it is generalized to the majority of scientific activities.

Paulo Vasconcelos: Since part of my research depends on new computer architectures, yes, it is very difficult, especially in Portugal, where we do not have state laboratories or research centers with high-end machinery.

On research, more generally

What would you say are the most important things to keep a research group going?

Elvio Accinelli: A common interest in the research topics and the possibility for all team members to develop their lines of work.

Alberto Álvarez-López: Keeping in contact (personal if possible) for discussion, holding brain-storming sessions, a good coordination among members, deadlines to have the work done... I do not know if they are the most important things, but they are useful.

Fabio Chalub: All members should be engaged in the research, so it is crucial to find a topic of general interest that involves everybody in the production of results. We cannot think of our colleagues, even if we are leading the group, as a bunch of employees. Everybody should have autonomy to produce their own results, and be judged by the quality of the output produced. This is the case in mathematics and other more theoretical subjects; however, I am perfectly aware that we cannot apply this policy to run a lab.

Ana Dias: Not to stop and to have people that really like what they are doing. Another thing is that people have to respect each other's work.

Orlando Gomes: A common goal, the capacity to work with others and to accept their criticisms, and gaining the notion that one is contributing to the advancement of science.

Clara Grácio: Respect for the successes obtained by each of the elements of this group but fundamental support at certain times, less good, that each of the elements can benefit by. Transparency, quality, and consistency are important in defining the group's strategic line, making it a key element. When these features come together, the group is a team, it is a school. I had the privilege of belonging to one of those rare schools, coordinated by the very bright (in all these respects) Professor Sousa Ramos.

Bruno Oliveira: Motivation. People should like what they are doing and feel that their work is recognized within the group.

Filipe Silva: Leadership, commitment, cohesion.

Luís Silva: In the first place, people must trust and respect each other; then I think it is very important to define a leader for each task.

Paulo Vasconcelos: Focus, dynamics, and a good working environment.

How do you see the relation between traveling and research?

Elvio Accinelli: It is very important to travel and see the results that other people have obtained. Travel expenses are one of the better investments in research, even if the results are not displayed immediately.

Alberto Álvarez-López: Well, if you do not have an assistant to arrange the details... organizing travel consumes energy. Anyway, I find traveling a very good way to meet colleagues and interchange ideas. In some workshops social aspects are very important: scholars are persons in the end, and they need to talk and share opinions and ideas with other persons.

Michel Benaïm: Traveling is a good way to meet people and develop new research. It's often much easier to talk with someone in front of a blackboard rather than to read a math paper. However, with emails, skype, and other communication technologies things are changing rapidly, and traveling is not as important as it used to be.

Ana Dias: It is important, although now there ways to interact without having to travel that are also good, not expensive, and save time.

Orlando Gomes: Research is many times an individual and solitary effort that we make in our offices or homes, but no meaningful research contributions gain life without a discussion with others. Our colleagues can help us improve our original ideas and assist us in transforming them into relevant scientific results. The meeting between researchers in the same or in related fields is a fundamental stage of any scientific endeavor. Therefore, it is my opinion that the participation in conferences in seminars and conferences around the world is a key step for the progress of science.

Bruno Oliveira: Traveling to meet other researchers and present our results is the best way to get feedback from our research. I have made big steps in my work after speaking with others about what I have found and after hearing from others what they have found. The positive input can come from new results by others, different methodologies to apply to our work, or a simple change of perspective that will allow new insight into a problem.

Joana Pais: Even though technology for communicating with other researchers is available nowadays, so that communicating is extremely easy and virtually costless, I believe that traveling, whether to attend conferences or to visit other researchers, is essential. Luís Silva: Particularly for young researchers, the contact with different research teams can be particularly beneficial, particularly when different skills can be developed in this way.

Filipe Silva: It is extremely important. Nowadays we have easy access to a huge quantity of information, but there are lots of things that are much easier to learn in a good conversation than by reading books or papers.

Paulo Vasconcelos: It is good in a very natural way. Research is widespread. A researcher needs to communicate with others, not only to share his research and to broadcast, but also to gather expertise from other colleagues in the field.

On teaching

What do you think about the relation between teaching and researching?

Alberto Álvarez-López: I think there are three main aspects to our task as scholars: research, teaching, and simply studying (knowledge in itself). Every one of us shares these three aspects in some proportion. The system should allow someone with a strong proportion in one of them (any of them, with no prevalence) to feel comfortable. However, this is not always true. On the other hand, we sometimes have a fourth task: the administrative labor—and this is often the first task. Anyway, I do find that my courses are richer if there is a research related to them.

Ana Dias: Good.

Orlando Gomes: They are, undoubtedly, complements. The creation and the diffusion of knowledge are two sides of the same coin: without research, no knowledge would be available to pass to students; without any one to teach, research would be simply useless.

Joana Pais: I used to believe that research helped to improve the quality of teaching. While I still believe this can be true when we talk about teaching at the advanced/graduate level, it is certainly not the case at the undergraduate level, where we have very good teachers that do not do research. The positive effects of teaching on research are even more difficult to grasp.

Filipe Silva: It's crucial; it really is a dialectic relationship, with many ideas and skills developed in one activity, easily transferable to the other.

Luís Silva: I think that the majority of the fundamental research should be done in the universities and that all university teachers must do research and that the majority of the researchers also should teach. On the other hand, I think that the university career should be more flexible in the sense of permitting large periods for doing just one of these two things. Nowadays we feel permanently pressed to do both things simultaneously, and I don't think that this is good for either of the two activities.

Paulo Vasconcelos: A teacher without research cannot convey a message of future, of challenge.

Any thoughts on what's crucial for a university teacher and or student?

Alberto Álvarez-López: You have to find pleasure in studying. And you have to learn to say "wait a moment, and let me analyze that," instead of giving a quick "yes" or "no."

Ana Dias: A good and enthusiastic teacher and a good and enthusiastic student.

Orlando Gomes: For both, the curiosity, the will to learn, and not being afraid to make mistakes.

Filipe Silva: A never-ending curiosity and the will to continue learning.

Luis Silva: Planning.

Alexandre Rodrigues: In a few words, I would say that a teacher should view a classroom as a pool of potential researchers and honor students. Students bring enthusiasm and a fresh perspective to our research. There is always the possibility that questions that come up in class will inspire new directions for our research. I find that stimulating interaction, encouraging independent thought, and accepting criticism are crucial in a classroom. And one should have a sense of humor—students love it. Technically, I believe that a teacher should give to the student a sense of the field, its past, present, and future directions, and the origins of its ideas and concepts. He/she should present facts and concepts from related fields. Theoretically, these are achievable goals; nevertheless, I realize that combining all these points might be difficult.

Paulo Vasconcelos: The duality research/teacher is difficult to keep equilibrated. In reality, usually teaching hours may be counterproductive for academic progression.

What are your thoughts on the relation between high school and university in terms of education?

Alberto Álvarez-López: I do not find them, at least in my country, as close as they should be. In mathematics, for instance, there is a gap between the level in high school and the requirements in university, especially in some grades. This causes a delay in the correct evolution of students. The high school teacher is not necessarily the guilty party: from the university we must better connect with him/her. Anyway, a deep change in the educational scheme should be considered.

Ana Dias: So and so. There is not a smooth transition between the two.

Orlando Gomes: The university should, more than any other school level, be capable of showing to students that what they learn, how they learn, and what use they make of this learning are essentially in their own hands and dependent on their own will.

Bruno Oliveira: In Portugal, university admission is based on high school grades, and the method of evaluation places too much emphasis on memorization to the detriment of problem-solving skills. I think that the system should aim to guide the students to the degree that is more fitted to their skills.

Filipe Silva: In Portugal I presently feel a considerable gap between those two levels. It's probably not a matter of the amount of knowledge that students have, but it is the way that they face their studies. It takes them all of the first year at the university to adapt to their new habits and to eventually develop a new, more independent way of studying.

Luís Silva: Particularly in mathematics, the relation was too bad for too long. Over a long time, the high school programs changed, and the university programs for the first years took too long to adapt. At the same time students arrived at the university poorly prepared, and people from the two school systems have had great difficulty in getting together to talk about what to do.

Paulo Vasconcelos: Completely wrong. Schools tend to prepare their students to take exams so that they can enroll in good universities. But critical thinking and creativity are neither exploited nor encouraged.

Do you have any advice for students starting their research?

Elvio Accinelli: Courses must be completed within the scheduled time, and then one should begin and continue working on the thesis without interruption until it is finalized. In general, those who leave their thesis for a time will fail to finish.

Alberto Álvarez-López: Yes: prepare a question (or a list of questions) to be answered. The question should be interesting. The answer should be relevant as well as technically correct.

Mário Bessa: Be persistent, resilient, curious, patient, and especially be able to scribble through large amounts of paper with flaws, mistakes, and wrong computations. Never believe that your supervisor has a magic wand to answer your questions and solve all your problems. It is you who should make the magic wand!

Ana Dias: They should try to work in what they like.

Orlando Gomes: Enjoy it. If you plan to go into research thinking only about career or monetary rewards, do not do it. You will need to have a passion for knowledge, or else you will feel frustrated.

Bruno Oliveira: Having a degree or a Ph.D. in an area does not mean that you will do the same thing for the rest of your life. You can use the expertise you have obtained in one area and apply it to a different one. The interface you create can be extremely rich in content and very motivating to explore.

Alexandre Rodrigues: The four years of Ph.D. work can be very frustrating—you need real determination to stick to a handful of projects and get the job done. You should be completely sure that you love doing research in that specific field. You will enjoy it sometimes, but other times it will be very frustrating. It is, in general, solitary work; you speak to a few people including your advisor, but it is still solitary. The results will be unconvincing many times; basically, you will end up with a thesis for which only a few individuals in the world can assess the exact value. If you have started your Ph.D., do not give up. Make an effort to make the difference; be really good. Even when the proof of a result is already given, try to do it by yourself.

Filipe Silva: I consider it a privilege to be able to devote our lives, or at least a part of them, to research, that is to try to better understand our world. Also, research activities have the potential to develop scientific reasoning and many other skills (e.g., persistence, creativity, statistical reasoning) that can be useful in other fields of activity.

And for the ones who are hesitating between pursuing a Ph.D. and looking for a different job?

Alberto Álvarez-López: Well, If you like to study, if you really like to work hard studying, go ahead with your Ph.D. The job of a scholar is one of the best you can choose, in the sense that almost everything you do is a direct "investment" in yourself. There are, of course, several contras: the wages are usually low, labor promotion is sometimes difficult, you work a lot of hours, bureaucratic tasks often feed you....

Ana Dias: They should try to do what they prefer.

Orlando Gomes: It is a matter of vocation. There are many appealing and well-paid jobs that do not require a Ph.D. It is all a matter of making the choices that we are most comfortable with.

Filipe Silva: I don't like to push students into academic activities, since the path to the Ph.D. is as important as the final result, so they have to be fully committed to endure (and enjoy) their own research voyage.

Have all of your research students chosen academic careers?

Alberto Álvarez-López: Most of them. I have to say that a few students were part-time students; they were also working out of the university.

Mário Bessa: Since academic jobs, in the area of mathematics, experienced a large decrease in supply in the last decade, Ph.D. students, after finishing their Ph.D. program, try to find business and finance jobs. Fortunately, my former students (Master and Ph.D.) are working as risk analysts in a bank. I point out that their employers are very satisfied with their skills and competence.

Ana Dias: I just have two and both are academics, although not yet with stable jobs.

Filipe Silva: No, several students have professions as teachers or in areas related to the environment. I think that the society, namely the private sector, should interact much more closely with researchers and they should eventually think how their skills can be put to work for the common interest, even if they are not directed to pure research. But it seems that we are still at a considerable distance from a complete integration of researchers in the society as a whole.

Luís Silva: Three out of four.

Paulo Vasconcelos: No, mainly lately they are finding jobs outside of academia.

On other issues

Do you have hobbies?

Alberto Álvarez-López: I very much enjoy good literature, and reading and writing in general.

Ana Dias: Right now, maybe just cooking, due to the lack of time.

Orlando Gomes: I like to take long walks and to enjoy the company of my family.

Filipe Martins: I am a proud Portuguese, and enjoy my country very much. My main hobbies are reading, music, and playing the piano and watching Boavista F.C. play. I am an avid reader.

Filipe Silva: Jogging and swimming in the ocean. Fortunately, I can do it all the year round in the Azores.

Luís Silva: I am a big fan of enduro mountain biking.

Do you have a connection to Portugal? How do you see its development?

Elvio Accinelli: I have an excellent relationship with Portugal, especially with the group of applied mathematics from Porto. With my work group in México we could make many joint projects with the group led by Alberto Pinto. The development of joint work with this group is of particular interest to us.

Alberto Álvarez-López: I feel as if I had a brother in Portugal. My visits to this brother are not very frequent, but when I am with him, I always feel exactly as if I were at home.

Ana Dias: I am working at the University of Porto. I see that Portugal is progressing with many people working hard, and I hope they will not lose their enthusiasm.

Orlando Gomes: I am Portuguese. I think Portugal is a victim of a drifting European Union and of the poor quality of its own economic policies. Visible setbacks in the areas of culture and science are, for me, the most painful.

Filipe Silva: Living in the Azores islands, I am aware of the consequences that can arise from unplanned development. Development without knowledge will hardly be development at all, and surely not sustainable. That's why universities and other innovation/research institutions play a crucial role in training the new generations and in contributing to a development that will not compromise Earth's resources and future generations.